Second Edition

VARIABLE SPEED GENERATORS

Second Edition

VARIABLE SPEED GENERATORS

Ion Boldea

IEEE Life Fellow
University Politehnica Timisoara
Timisoara, Romania

CRC Press is an imprint of the
Taylor & Francis Group, an informa business

CRC Press
Taylor & Francis Group
6000 Broken Sound Parkway NW, Suite 300
Boca Raton, FL 33487-2742

Printed by CPI on sustainably sourced paper
Version Date: 20150817

International Standard Book Number-13: 978-1-4987-2357-2 (Hardback)

Visit the Taylor & Francis Web site at
http://www.taylorandfrancis.com

and the CRC Press Web site at
http://www.crcpress.com

Contents

Preface to the Second Edition

The first edition of this single-author, two-book set was published in 2006. Since then, electric energy, "produced" mostly via electric generators, has become one of the foremost activities in our global economy world. The subject of electric generators (*Synchronous Generators* and *Variable Speed Generators* as two books) attracted special attention worldwide both from industry and academia in the last decade. Electric generators' design and control may constitute a new graduate course in universities with electric power programs.

Also, in the design and control of electric generators for applications ranging from energy conversion to electric vehicles (transportation) and auxiliary power sources, new knowledge and developments have been published in the last ten years. In the last ten years, in wind generators alone, the installed power has increased from some 40,000 MW to 300,000 MW (in 2014).

In view of these developments, we decided to come up with a new edition that

- Keeps the structure of the first edition to avoid confusion for users
- Keeps the style with many numerical worked-out examples of practical interest, together with more complete case studies
- Includes text and number corrections
- Adds quite a few new paragraphs in both books, totaling around 100 pages, to illustrate synthetically the progress in the field in the last decade

The new additions in the second edition are

Synchronous Generators

- Chapter 2 (Section 2.9): High Power Wind Generators, with less or no PM—an overview
- Chapter 4 (Section 4.15): PM-Assisted DC-Excited Salient Pole Synchronous Generators
  (Section 4.16): Multiphase Synchronous Machine Inductances via Winding Function Method
- Chapter 6 (Section 6.17): Note on Autonomous Synchronous Generators' Control
- Chapter 7 (Section 7.21): Optimization Design Issues
  (Section 7.21.1): Optimal Design of a Large Wind Generator by Hooke–Jeeves Method
  (Section 7.21.2): Magnetic Equivalent Circuit Population-Based Optimal Design of Synchronous Generators
- Chapter 8 (Section 8.10): Online Identification of SG Parameters
  (Section 8.10.1): Small-Signal Injection online Technique
  (Section 8.10.2): Line Switching (On or Off) Parameter Identification for Isolated Grids
  (Section 8.10.3): Synthetic Back-to-Back Load Testing with Inverter Supply

Variable Speed Generators

- Chapter 2 (Section 2.14): Ride-Through Control of DFIG under Unbalanced Voltage Sags
  (Section 2.15): Stand-Alone DFIG Control under Unbalanced Nonlinear Loads
- Chapter 5 (Section 5.8): Stand-Alone SCIG with AC Output and Low Rating PWM Converter
  (Section 5.10): Twin Stator Winding SCIG with 50% Rating Inverter and Diode Rectifier
  (Section 5.11): Dual Stator Winding IG with Nested Cage Rotor
- Chapter 6 (Section 6.8): IPM Claw-Pole Alternator System for More Vehicle Braking Energy Recuperation: A Case Study
- Chapter 8 (Section 8.12): 50/100 kW, 1350–7000 rpm (600 N m Peak Torque, 40 kg) PM-Assisted Reluctance Synchronous Motor/Generator for HEV: A Case Study
- Chapter 9 (Section 9.11): Double Stator SRG with Segmented Rotor
- Chapter 10 (Section 10.16): Grid to Stand-Alone Transition Motion-Sensorless Dual-Inverter Control of PMSG with Asymmetrical Grid Voltage Sags and Harmonics Filtering: A Case Study
- Chapter 11 (Section 11.5): High Power Factor Vernier PM Generators

We hope that the second edition will be of good use to graduate students, to faculty, and, especially, to R&D engineers in industry that deal with electric generators, design control, fabrication, testing, commissioning, and maintenance. We look forward to the readers' comments for their confirmation and validation and for further improvement of the second edition of these two books: *Synchronous Generators* and *Variable Speed Generators.*

Professor Ion Boldea
IEEE Life Fellow
Romanian Academy
University Politehnica Timisoara
Timisoara, Romania

Preface to the First Edition

Electric energy is a key factor for civilization. Natural (fossil) fuels such as coal, natural gas, and nuclear fuel are fired to produce heat in a combustor and then the thermal energy is converted into mechanical energy in a turbine (prime mover). The turbine drives an electric generator to produce electric energy. Water potential and kinetic and wind energy are also converted to mechanical energy in prime movers (turbine) to drive an electric generator.

All primary energy resources are limited, and they have a thermal and chemical (pollutant) effect on the environment.

Currently, much of electric energy is produced in constant-speed-regulated synchronous generators that deliver electric energy with constant AC voltage and frequency into regional and national electric power systems, which further transport and distribute it to consumers.

In an effort to reduce environment effects, electric energy markets have been recently made more open, and more flexible distributed electric power systems have emerged. The introduction of distributed power systems is leading to an increased diversity and growth of a wider range power/unit electric energy suppliers. Stability, quick and efficient delivery, and control of electric power in such distributed systems require some degree of power electronics control to allow lower speed for lower power in electric generators to tap the primary fuel energy.

This is how *variable-speed* electric generators have come into play recently [up to 400 (300) MVA/unit], as for example, pump storage wound-rotor induction generators/motors have been in used since 1996 in Japan and since 2004 in Germany.

This book deals in depth with both constant- and variable-speed generator systems that operate in stand-alone and power grid modes.

Chapters have been devoted to topologies, steady-state modeling and performance characteristics, transients modeling, control, design, and testing, and the most representative and recently proposed standard electric generator systems.

The book contains most parameter expressions and models required for full modeling, design, and control, with numerous case studies and results from the literature to enforce the understanding of the art of electric generators by senior undergraduate and graduate students, faculty, and, especially, industrial engineers who investigate, design, control, test, and exploit electric generators for higher energy conversion ratios and better control. This 20-chapter book represents the author's unitary view of the multifacets of electric generators with recent developments included.

Chapter 1 introduces energy resources and fundamental solutions for electric energy conversion problems and their merits and demerits in terms of efficiency and environmental effects. In Chapter 2, a broad classification and principles of various electric generator topologies with their power ratings and main applications are presented. Constant-speed-synchronous generators (SGs) and variable-speed wound-rotor induction generators (WRIGs); cage rotor induction generators (CRIGs); claw pole rotors; induction; PM-assisted synchronous, switched reluctance generators (SRGs) for vehicular and other applications; PM synchronous generators (PMSGs); transverse flux (TF); and flux reversal (FR) PMSGs, and, finally, linear motion PM alternators are all included.

Chapter 3 treats the main prime movers for electric generators from topologies to basic performance equations and practical dynamic models and transfer functions.

Steam, gas, hydraulic, and wind turbines and internal combustion (standard, Stirling, and diesel) engines are dealt with. Their transfer functions are used in subsequent chapters for speed control in corroboration with electric generator power flow control.

Chapters 4 through 8 deal with SGs steady state, transients, control, design, and testing with plenty of numerical examples and sample results that cover the subject comprehensively.

This part of the book is dedicated to electric machines and power systems professionals and industries.

Chapters 9 through 11 deal with WRIGs that have a bidirectional rotor connected AC–AC partial rating PWM converter for variable-speed operation in stand-alone and power grid modes. Steady-state transients (Chapter 9), vector and direct power control (Chapter 10), and design and testing (Chapter 11) are treated in detail with plenty of applications and digital simulation and test results to facilitate in-depth assessment of WRIG systems currently built from 1 MVA to 400 MVA per unit.

Chapters 12 and 13 discuss cage rotor induction generators (CRIG) in self-excited modes used as power grid and stand-alone applications with small speed regulation by a prime mover (Chapter 12) or with full-rating PWM converters connected to a stator and wide-variable speed (Chapter 13) with ± 100% active and reactive power control and constant (or controlled) output frequency and voltage in both power grid and stand-alone operations.

Chapters 9 through 13 are targeted to wind, hydro, and, in general, to distributed renewable power system professionals and industries.

Chapters 14 through 17 deal with representative electric generator systems proposed recently for integrated starter alternators (ISAs) on automobiles and aircraft, all operating at variable speed with full power ratings electronics control. Standard (and recently improved) claw pole rotor alternators (Chapter 14), induction (Chapter 15), PM-assisted synchronous (Chapter 16), and switched reluctance (Chapter 17) ISAs are discussed thoroughly. Again, with numerous applications and results, from topologies, steady state, and transients performance, from modeling to control design and testing for the very challenging speed range constant power requirements (up to 12 to 1) typical to ISA. ISAs have reached the markets, used on a mass-produced (since 2004) hybrid electric vehicles (HEVs) for notably higher mileage and less pollution, especially for urban transport.

This part of the book (Chapters 14 through 17) is targeted at automotive and aircraft professionals and industries.

Chapter 18 deals extensively with radial and axial air gaps, surfaces, and interior PM rotor permanent magnet synchronous generators that work at variable speed and make use of full power rating electronics control. This chapter includes basic topologies, thorough field and circuit modeling, losses, performance characteristics, dynamic models, and bidirectional AC–AC PWM power electronics control in power grid and in stand-alone applications with constant DC output voltage at variable speed. Design and testing issues are included, and case studies are treated using numerical examples and transient performance illustrations.

This chapter is directed at professionals interested in wind and hydraulic energy conversion, generator set (stand-alone) with power/unit up to 3–5 MW (from 10 rpm to 15 krpm) and 150 kW at 80 krpm (or more).

Chapter 19 investigates with numerous case study designs two high-torque density PM synchronous generators (transverse flux [TFG] and flux reversal [FRG]), introduced in the last two decades that take advantage of non-overlapping multipole stator coils. They are characterized by lower copper losses/N m and kg/N m and find applications in very-low-speed (down to 10 rpm or so) wind and hydraulic turbine direct and transmission drives, and medium-speed automotive starter-alternators.

Chapter 20 investigates linear reciprocating and linear progressive motion alternators. Linear reciprocating PMSGs (driven by Stirling free-piston engines) have been introduced (up to 350 W) and are currently used for NASA's deep-mission generators that require fail-proof operation for 50,000 h. Linear reciprocating PMSGs are also pursued aggressively as electric generators for series (full electric propulsion) vehicles for power up to 50 kW or more; finally, they are being proposed for combined electric (1 kW or more) and thermal energy production in residences with gas as the only prime energy provider.

The author thanks the following:

- Illustrious people that have done research, wrote papers, books, patents, and built and tested electric generators and their control over the last decades for providing the author with "the air beneath his wings"
- The author's very able PhD students for electronic editing of the book
- The highly professional, friendly, and patient editors of CRC Press

Professor Ion Boldea
IEEE Life Fellow
University Politehnica Timisoara
Timisoara, Romania

Author

Ion Boldea:

- MS (1967), PhD (1973) in electrical engineering; IEEE member (1977), fellow (1996), and life fellow (2007)
- Visiting scholar in the United States (15 visits, 5 years in all, over 37 years), the United Kingdom, Denmark, South Korea
- Over 40 years of work and extensive publications (most in IEEE trans. and conferences and with IET (former IEE), London, in linear and rotary electric motor/generator modeling, design, their power electronics robust control, and MAGLEVs; 20 national and 5 international patents
- Eighteen books in the field, published in the United States and the United Kingdom
- Technical consultant for important companies in the United States, Europe, South Korea, and Brazil for 30 years
- Repeated intensive courses for graduate students and industries in the United States, Europe, South Korea, and Brazil
- Four IEEE paper prize awards
- Cofounding associate (now consulting) editor from 1977 for *EPCS Journal*
- Founding (2000) and current chief editor of the Internet-only international *Electronic Engineering Journal*: www.jee.ro
- General chair in 10 biannual consecutive events of the International Conference OPTIM (now IEEE-tech-sponsored, ISI and on IEEExplore)
- Member of the European Academy of Arts and Sciences located in Salzburg, Austria
- Member of the Romanian Academy of Technical Sciences (since 1997)
- Correspondent member of Romania Academy (2011)
- IEEE-IAS Distinguished Lecturer (2008–2009) with continued presence ever since
- IEEE "Nikola Tesla" Award 2015

1

Wound-Rotor Induction Generators: Steady State

1.1 Introduction

Wound-rotor induction generators (WRIGs) are provided with three-phase windings on the rotor and on the stator. They may be supplied with energy at both rotor and stator terminals. For this reason, such generators are called "doubly fed induction generators" (DFIGs), or double-output induction generators (DOIGs). WRIGs can operate in both monitoring and generating modes, provided the power electronics converter that supplies the rotor circuits, in general, via slip-rings and brushes is capable of handling power in both directions.

WRIGs provide constant or controlled voltage V_s and frequency f_1 power through the stator, while the rotor is supplied through a static power converter at variable voltage V_r and frequency f_2, the rotor circuit may absorb or deliver electric power. As the number of poles of both stator and rotor windings is the same at steady state. Therefore, according to the frequency theorem, the speed ω_m is as follows:

$$\omega_m = \omega_1 \pm \omega_2; \quad \omega_m = \Omega_R \cdot p_1 \tag{1.1}$$

where

p_1 is the number of pole pairs

Ω_R is the mechanical rotor speed

The positive (+) sign in Equation 1.1 indicates that the phase sequence in the rotor is the same as in the stator and $\omega_m < \omega_1$, that is, subsynchronous operation. The negative (–) sign in Equation 1.1 corresponds to an inverse phase sequence in the rotor when $\omega_m > \omega_1$, that is, supersynchronous operation.

For constant frequency output, the rotor frequency ω_2 has to be modified "in step" with the speed variation. This way, variable speed at constant frequency (and voltage) can be maintained by controlling the voltage, frequency, and phase sequence in the rotor circuit.

It may be argued that WRIGs work as synchronous generators (SGs) with three-phase alternating current (AC) excitation at slip (rotor) frequency $\omega_2 = \omega_1 - \omega_m$. However, as $\omega_1 \neq \omega_m$, the stator induces voltages in the rotor circuits even at steady state, which is not the case in conventional SGs. As a result, additional power components occur.

The main operational modes of the WRIG are depicted in Figure 1.1a through d (See Figure 1.1a for basic configuration). The first two modes (Figure 1.1b and c) refer to subsynchronous and supersynchronous generations defined earlier. For motoring, the reverse is true for the rotor circuit. In addition, the stator absorbs active power for motoring. The slip S is defined as follows:

$$S = \frac{\omega_2}{\omega_1} \begin{array}{ll} >0; & \text{subsynchronous operation} \\ <0; & \text{supersynchronous operation} \end{array} \tag{1.2}$$

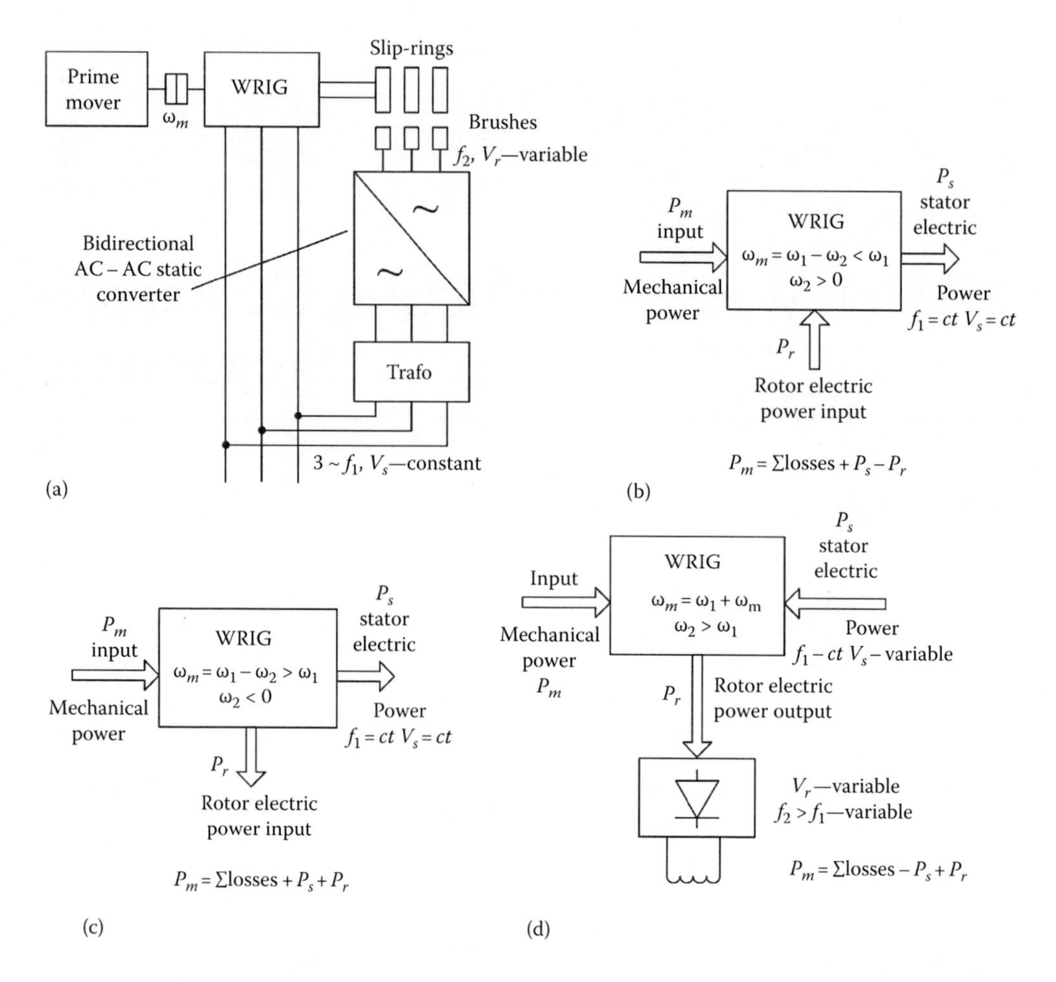

FIGURE 1.1 Main operating modes are (a) basic configuration, (b) subsynchronous generating ($\omega_r < \omega_1$), (c) supersynchronous generating ($\omega_r > \omega_1$), and (d) rotor output WRIG (brushless exciter).

As WRIGs work, in general, for $\omega_2 \neq 0$ ($S \neq 0$), they retain the characteristics of an induction machine. The main output active power is delivered through the stator; however, in supersynchronous operation, a good part, about slip stator powers (SPs), is delivered through the rotor circuit. With limited speed-variation range, that is from S_{max} to $-S_{max}$, the rotor-side static converter rating—for zero-reactive power capability on the rotor side—would be $P_{conv} \approx |S_{max}|P_s$. With S_{max} typically equal to ±0.2 to 0.25, the static power converter ratings and costs would correspond to 20%–25% of the stator-delivered output power.

At maximum speed, the WRIG will deliver increased electric power, P_{max}:

$$P_{max} = P_s + P_{rmax} = P_s + |S_{max}|P_s \tag{1.3}$$

with the WRIG designed at P_s for $\omega_m = \omega_1$ speed. The increased power is delivered at higher than rated speed:

$$\omega_{mmax} = \omega_1\left(1 + |S_{max}|\right) \tag{1.4}$$

Consequently, the WRIG is designed electrically for P_s at $\omega_m = \omega_1$, but mechanically at ω_{mmax} and P_{max}.

The capability of WRIGs to deliver power at variable speed but at constant voltage and frequency represents an asset in providing more flexibility in power conversion and also better stability in frequency and voltage control in the power systems to which such generators are connected.

The reactive power delivery by WRIGs depends heavily on the capacity of the rotor-side converter to provide it. When the converter works at unity power factor delivered on the source side, the reactive power in the machine has to come from the rotor-side converter. However, such a capability is paid for by the increased ratings of the rotor-side converter. As this means increased converter costs, in general, WRIGs are adequate for working at unity power factor at full load on the stator side.

Large reactive power releases to the power system are still to be provided by existing SGs or from WRIGs working at synchronism ($S = 0$, $\omega_2 = 0$) with the back-to-back pulse-width-modulated (PWM) voltage converters connected to the rotor controlled adequately for the scope.

Wind and small hydroenergy conversions in units of 1 megawatt (MW) and more per unit require variable speed to tap the maximum of energy reserves and to improve efficiency and stability limits. High-power units in pump-storage hydropower (400 MW [1]) and thermopower plants with WRIGs provide for extra flexibility for the ever-more stressed distributed power systems of the near future. Even existing (old) SGs may be retrofitted into WRIGs by changing the rotor and its static power converter control.

The WRIG can also be used to generate power solely on the rotor side for rectifier loads (Figure 1.1d). To control the direct voltage (or direct current [DC]) in the load, the stator voltage is controlled, at constant frequency ω_1, by a low-cost AC three-phase voltage changer. As the speed increases, the stator voltage has to be reduced such that the current in the DC load connected to the rotor ($\omega_2 = \omega_1 + \omega_m$) is kept constant. If the machine has a large number of poles ($2p_1 = 6, 8, 12$), then the stator AC excitation input power becomes rather low, as most of the output electric power comes from the shaft (through motion).

Such a configuration is adequate for brushless exciters needed for SMs or for SGs, where field current is needed from zero speed, that is, when full-power converters are used in the stator of the respective SMs or SGs.

With $2p_1 = 8$, $n = 1500$ rpm, and $f_1 = 50$ Hz, the frequency of the rotor output $f_2 = f_1 + np_1 = 50 + (1500/60) \times 4 = 150$ Hz. Such a frequency is practical with standard iron core laminations and reduces the contents in harmonics of the output-rectified load current.

The following subjects related to WRIG steady state are detailed in this chapter:

- Construction elements
- Basic principles
- Inductances
- Steady-state model (equations, phasor diagram, and equivalent circuits)
- Steady-state characteristics at power grid
- Steady-state characteristics for isolated loads
- Losses and efficiency

1.2 Construction Elements

The WRIG topology contains the following main parts:

- Stator laminated core with N_s uniformly distributed slots
- Rotor laminated core with N_r uniformly distributed slots
- Stator three-phase winding placed in insulated slots
- Rotor shaft
- Stator frame with bearings
- Rotor copper slip-rings and stator (placed) brushes to transfer power to (from) rotor windings
- Rotor three-phase AC winding in insulated rotor slots
- Cooling system

1.2.1 Magnetic Cores

The stator and rotor cores are made of thin (typically 0.5 mm) nonoriented grain silicon steel lamination provided with uniform slots through stamping (Figure 1.2a). To keep the airgap reasonably small, without incurring large core surface harmonics eddy current losses, only the slots on one side should be open. On the other side of the airgap, they should be half closed or half open (Figure 1.2b).

Though, in general, the use of radial–axial ventilation systems leads to the presence of radial channels between 60 and 100 mm long elementary stacks, at least for powers up to 2–3 MW, axial ventilation with single lamination stacks is feasible (Figure 1.3a and b). As the airgap is slightly increased in comparison with standard induction motors, the axial airflow through the airgap is further facilitated. The axial channels (Figure 1.3a) in the stator and rotor yokes (behind the slot region) play a key role in cooling the stator and the rotor, as do the radial channels (Figure 1.3b) for the radial–axial ventilation.

The radial channels, however, are less efficient, as they are "traveled" by the windings. Thus, additional phase resistance and leakage inductance are added by the winding zones in the radial channel contributions. In very large, or long, stack machines, radial–axial cooling may be inevitable; but as explained before, below 3 MW, the axial cooling in unistack cores, already in industrial use for induction motors, seems to be the way of the future.

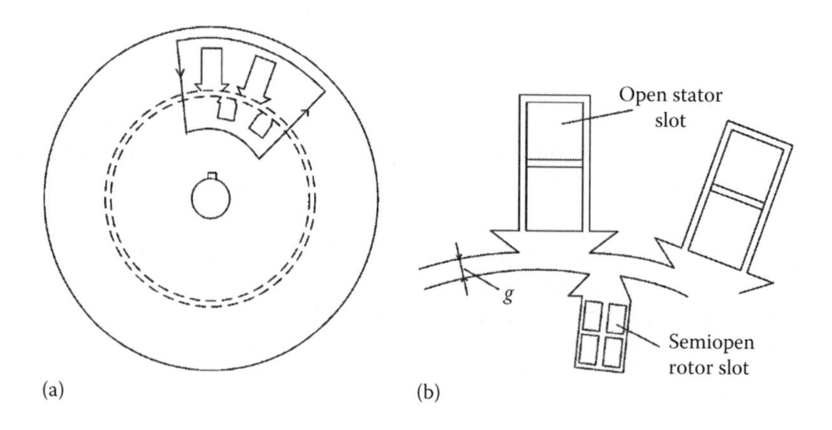

FIGURE 1.2 (a) Stator and (b) rotor slotted lamination.

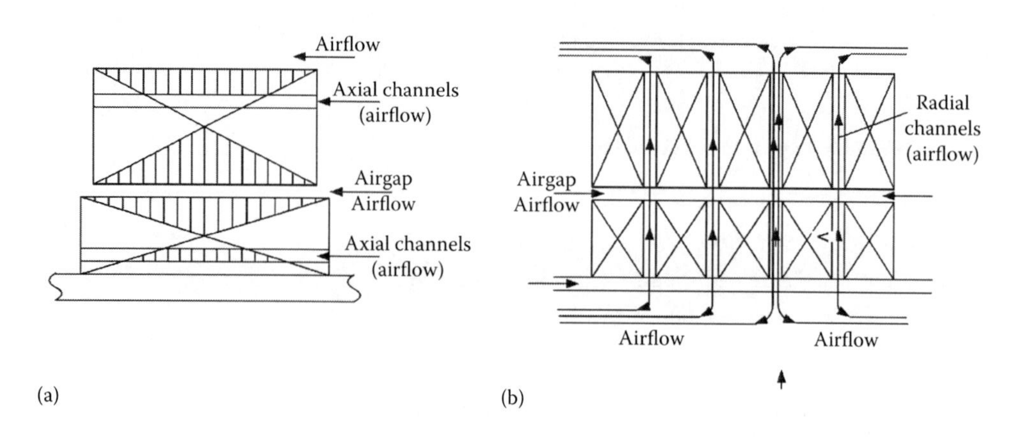

FIGURE 1.3 Stator and rotor stacks (a) for axial cooling and (b) for radial–axial cooling.

1.2.2 Windings and Their Magnetomotive Forces

The stator and rotor three-phase windings are similar in principle. In Chapter 4 in *Synchronous Generators*, their design is described in some detail. Here, only the basic issues are presented. The three-phase windings are built to provide for traveling magnetomotive forces (mmfs) capable of producing a traveling magnetic field in the uniform airgap (slot openings are neglected or considered through the Carter coefficient $K_C = 1.02$–1.5):

$$B_g(x,t) = \frac{\mu_0 F_{s,r}(x,t)}{g K_C (1 + K_s)} \tag{1.5}$$

where

$F_{s,r}(x,t)$ is equal to the mmfs per pole produced by either stator or rotor windings
g is the airgap
K_C is the Carter coefficient to account for airgap increase due to slot openings
K_s is the iron core contribution to equivalent magnetic reluctance of the main flux path (see Figure 1.2a)

To produce a traveling airgap field, the stator and the rotor mmfs, seen from the stator and the rotor, respectively, need to be as follows:

$$F_s(\theta_s, t) = F_{1s}\cos(p_1\theta_s - \omega_1 t) \tag{1.6}$$

$$F_r(\theta_r, t) = F_{1r}\cos(p_1\theta_r \pm \omega_2 t) \tag{1.7}$$

where p_1 is the number of electrical periods of the magnetic field wave in the airgap or of pole pairs. The rotor mmf is produced by currents of frequency ω_2.

At constant speed, the rotor and stator geometrical angles are related by

$$p_1\theta_r = p_1\theta_s - \omega_r t + \gamma; \quad \omega_r = \Omega_r \cdot p_1; \quad p_1\theta_s = \omega_1 t \tag{1.8}$$

where ω_r is the rotor speed in electrical radians per second (rad/s). Consequently, $F_r(\theta_s, t)$ becomes

$$F_r(\theta_s, t) = F_1\cos\left[p_1\theta_s - (\omega_r \pm \omega_2)t + \gamma\right] \tag{1.9}$$

The average electromagnetic torque and power per electric period is nonzero only if the two mmfs are stationary with respect to each other. That is,

$$\omega_1 = \omega_r \pm \omega_2; \quad S = \frac{\omega_2}{\omega_1} \tag{1.10}$$

The positive sign (+) is used when $\omega_r < \omega_1$, and thus the rotor and stator mmf waves rotate in the positive direction. The negative sign (−), used when $\omega_r > \omega_1$, refers to the case when the rotor mmf wave moves in the direction opposite to that of the stator. In addition, the torque is nonzero when the angle $\gamma \neq 0$, that is, when the two mmfs are phase shifted.

To produce a traveling mmf, three phases, space lagged by 120° (electrical), have to be supplied by AC currents with 120° (electrical) time-lag angles between them (see Chapter 4 in *Synchronous Generators*, on the SG).

So, all three-phase windings for, say, maximum value of current, should independently produce a sinusoidal spatial mmf:

$$(F_{sA,B,C}(\theta_s,t))_{t=0} = \frac{2}{3}F_{1s}\cos\left(p_1\theta_s - (i-1)\frac{2\pi}{3}\right) \quad (1.11)$$

Each phase mmf has to produce $2p_1$ semiperiods along a mechanical period. With only one coil per pole per phase, there would be $2p_1$ coils per phase and $2p_1$ slots per phase if each coil occupies half of the slot (Figure 1.4a).

From the rectangular distribution of phase mmf (Figures 1.4a and b and 1.3b), a fundamental is extracted:

$$F_{sA}(p_1,\theta_s) = \frac{4}{\pi}n_c I\sqrt{2}\cos p_1\theta_s; \quad n_s\text{—turns/coil} \quad (1.12)$$

The harmonics content of the phase mmf in Figure 1.4b is hardly acceptable, but more steps in its distribution (more slots) and chorded coil would drastically reduce these space harmonics (Figure 1.5).

For the 2-pole 24-slot winding with chorded coils (coil span/pole pitch = 10/12), the number of steps in the phase mmf is larger, and thus, the harmonics are reduced (Figure 1.5). For the fundamental component (based on Figure 1.5b), the expression of the mmf per pole and phase is obtained:

$$F_{sA1} = \frac{2W_1 k_{W1} K_{y1}\cdot I\sqrt{2}}{\pi p_1}; \quad W_1\text{—turns/phase} \quad (1.13)$$

For the space harmonic ν, in a similar way,

$$F_{sA\nu} = \frac{2W_1 K_{d\nu} K_{y\nu}\cdot I\sqrt{2}}{\nu p_1} \quad (1.14)$$

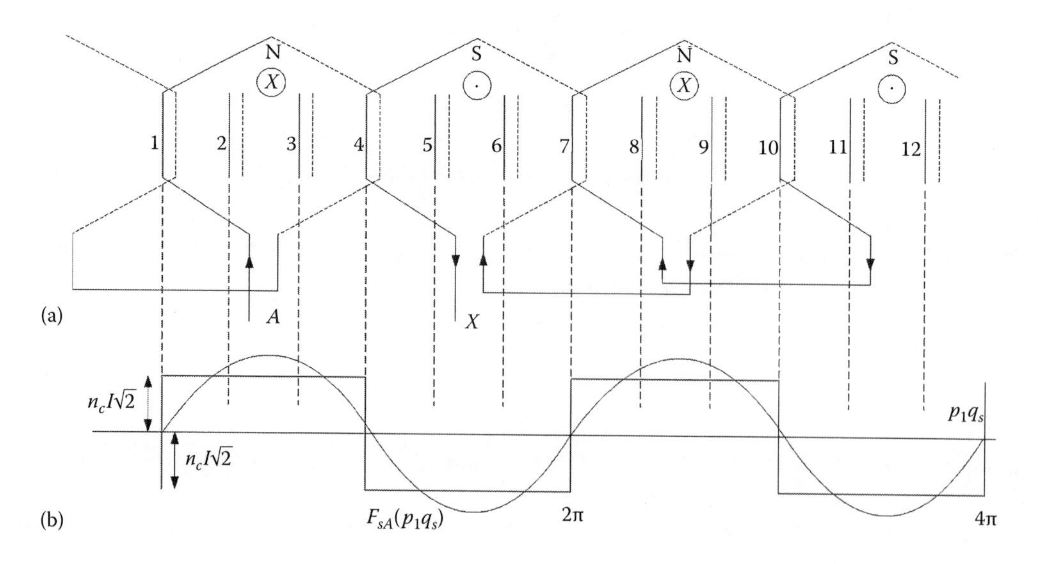

FIGURE 1.4 Elementary three-phase winding with $2p_1 = 4$ poles and $N_s = 12$ slots: (a) coils of phase A in series and (b) phase A mmf for maximum phase current.

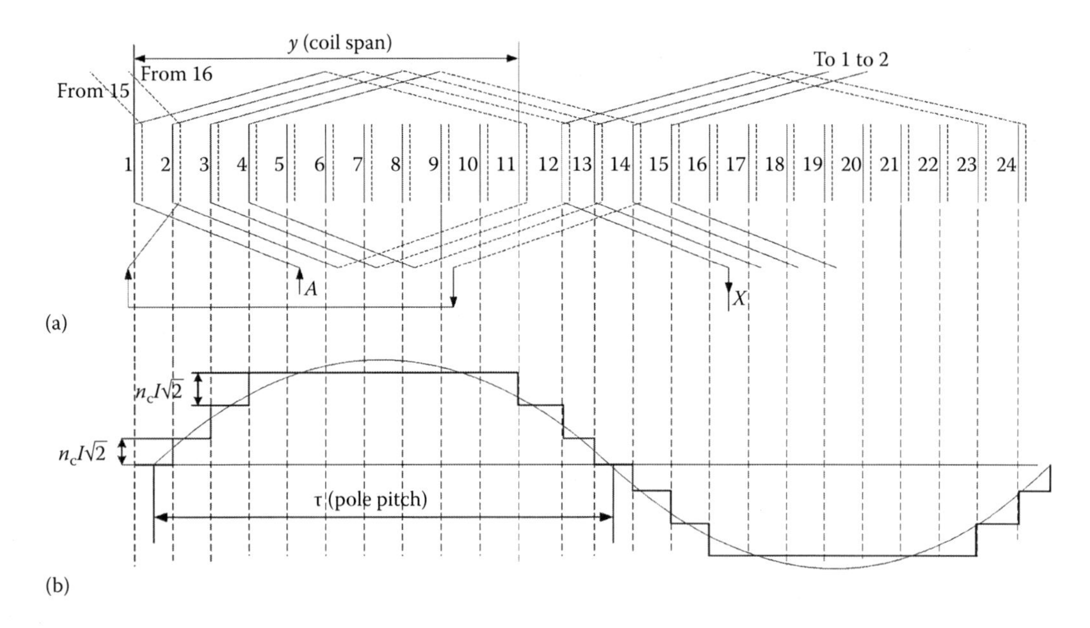

FIGURE 1.5 Two-pole ($2p_1 = 2$), $N_s = 24$ slots three-phase winding, with two layers in slot, coil span $y/\tau = 10/12$: (a) slot-to-phase allocation for layer 1 and coils of phase A and (b) phase A mmf for maximum current.

with K_{dv} and K_{yv} known as distribution and chording factors:

$$K_{dv} = \frac{\sin v\pi/6}{q\sin(v\pi/6q)} K_{yv} = \sin\frac{vy}{\tau}\frac{\pi}{2} \tag{1.15}$$

where q is the number of slots per pole per phase:

$$q_{s,r} = \frac{N_{s,r}}{2p_1m_1} = \frac{N_{s,r}}{6p_1} \tag{1.16}$$

Only the odd harmonics are present, in general, as the positive and negative mmf poles are identical, while the multiples of three harmonics are zero for symmetric currents (equal amplitude, 120° phase shift): $\nu = 1, 5, 7, 11, 13, 17, 19, \ldots$ It was proven (Chapter 4, in *Synchronous Generators*) that harmonics 7, 13 and 19 are positive, and 5, 11, 17, ... are negative in terms of sequence. By adding the contributions of the three phases, we obtain the mmf amplitude per pole $F_{s\nu}$:

$$F_{sv} = \frac{3}{2}F_{sAv} = \frac{3W_1K_{dv}K_{yv}I\sqrt{2}}{vp_1} \tag{1.17}$$

Similar expressions may be derived for the rotor. To avoid parasitic synchronous torques, the number of slots of the stator and the rotor has to differ:

$$N_s - N_r; \quad q_s - q_r \tag{1.18}$$

Harmonics have to be treated carefully, as the radial magnetic pull due to rotor eccentricity tends to be larger in WRIG than in cage-rotor induction generators (IGs) [2].

In general, WRIGs tend to be built with integer q both in the stator and in the rotor. Additionally, current paths in parallel may be used to reduce elementary conductor cross-sections.

Frequency (skin) effects have to be reduced, specifically in large WRIGs, with bar-made windings where transposition may be necessary (Roebel bar, see Chapter 7, in *Synchronous Generators*).

Finally, the rotor winding end connections have to be protected against centrifugal forces through adequate bandages, as for cylindrical rotor SGs.

Whenever possible, the rated (design) voltage of the rotor winding has to be equal to that in the stator as required in the control of the rotor-side static power converter at maximum slip. This way, a voltage-matching transformer can be avoided on the supply side of the static converter. Consequently, the rotor-to-stator turns ratio a_{rs} is as follows:

$$a_{rs} \cong \frac{W_r K_{q1}^r K_{d1}^r}{W_s K_{q1}^s K_{d1}^s} \cong \frac{1}{|S_{max}|} \tag{1.19}$$

Care must be exercised in such designs to avoid connecting the stator at the full-voltage power grid at zero speed ($S = 1$), as the voltage induced in the rotor windings will be a_{rs} times larger than the rated one, jeopardizing the rotor winding insulation and the rotor-side static power converter.

If starting as a motor is required (for pump storage, etc.), it is done from the rotor, with the stator short-circuited, using the rotor-side bidirectional power flow capabilities. Then, at certain speed $\omega_{rmin} > \omega_{rn}(1 - |S_{max}|)$, the stator circuit is opened. The machine is cruising while the control prepares the synchronization conditions by using the inverter on the rotor to produce adequate voltages in the stator. After synchronization, motoring (for pump storage) can be performed safely.

In WRIGs, a considerable amount of power (up to $|S_{max}| \cdot P_{sN}$) is transferred in and out of the rotor electrically through slip-rings and brushes. With $|S_{max}| = 0.20$, it is about 20% of the rated power of the machine. Remember that in SGs, the excitation power transfer to rotor by slip-rings and brushes is about 5–10 times less.

The question is if those multimegawatts could be transferred through slip-rings and brushes to the rotor in large-power WRIGs. The answer seems to be "yes," as 200 and 400 MW units have been in operation for over 5 years at up to 30 MW power transfer to the rotor.

In contrast to SGs, WRIGs have to use higher voltage for the power transfer to the rotor to reduce the slip-ring current. Multilevel voltage source bidirectional PWM MOSFET-controlled thyristor (MCT) converters are adequate for the scope of our discussion here. If the rotor voltage is increased in the kilo-volt (and above) range, the insulation provisions for the rotor slip-rings and on the brush framing side are much more demanding.

Note that SG brushless exciters based on the WRIG principle with rotor rectified output do not need slip-rings and brushes. In WRIGs with large stator voltage (V_n = 18 kV, 400 MW), it may be more practical to use lower rated (maximum) voltage in the rotor, that is up to 4.5 kV, and then use a step-up voltage adapting transformer to match the rotor connected static power converter voltage (4.5 kV) to the local (stator) voltage (i.e., 18 kV). Such a reduction in voltage may reduce the eventual costs of the static power converter so much as to overcompensate the costs of the added transformer.

1.2.3 Slip-Rings and Brushes

A typical slip-ring rotor is shown in Figure 1.6. It is obvious that three copper rings serve each phase as the rotor currents are large.

FIGURE 1.6 Slip-ring wound rotor.

1.3 Steady-State Equations

The electromagnetic force (emf) self-induced by the stator winding, with the rotor winding open, E_1, is as follows:

$$E_1 = \pi\sqrt{2} f_1 W_1 K_{W1} \phi_{10} \quad \text{(RMS)} \tag{1.20}$$

$$K_{W1} = K_{d1} \cdot K_{y1} \tag{1.21}$$

The flux per pole ϕ_{10} is

$$\phi_{10} = \frac{2}{\pi} B_{g10} \tau l_i \tag{1.22}$$

where
- l_i is the stack length
- τ is the pole pitch
- B_{g10} is the airgap fundamental flux density peak value:

$$B_{g10} = \frac{\mu_0 F_{s10}}{K_C g(1 + K_s)} \tag{1.23}$$

where F_{s10} is the amplitude of stator mmf fundamental per pole

From Equation 1.17, with $\nu = 1$,

$$F_{s10} = \frac{3 W_1 K_{W1} I_o \sqrt{2}}{\pi p_1} \tag{1.24}$$

But the same emf E_1 may be expressed as follows:

$$E_1 = \omega_1 L_{1m} \cdot I_{10} \tag{1.25}$$

Therefore, the main flux, magnetization (cyclic) inductance of the stator—with all three phases active and symmetric—L_{1m} is as follows (from Equations 1.20 through 1.25):

$$L_{1m} = \frac{6\mu_0 (W_1 K_{W1})^2 \tau l_i}{\pi^2 p_1 K_C g(1+K_s)} \tag{1.26}$$

The Carter coefficient $K_C > 1$ accounts for both stator and rotor slot openings ($K_C \approx K_{C1}K_{C2}$). The saturation factor K_s, which accounts for the iron core magnetic reluctance, varies with stator mmf (or current for a given machine), and so does magnetic inductance L_{1m} (Figure 1.7).

Besides L_{1m}, the stator is characterized by the phase resistance R_s and leakage inductance L_{sl} [2]. The same stator current induces an emf E_{2s} in the rotor open-circuit windings. With the rotor at speed ω_r—slip $S = (\omega_1 - \omega_r)/\omega_1$—$E_{2s}$ has the frequency $f_2 = Sf_1$:

$$\begin{aligned} E_{2s}(t) &= E_{2s}\sqrt{2}\cos\omega_2 t \\ E_{2s} &= \pi\sqrt{2}Sf_1 W_2 K_{W2}\phi_{10} \end{aligned} \tag{1.27}$$

Consequently,

$$\frac{E_{2s}}{E_1} = S\frac{W_2 K_{W2}}{W_1 K_{W1}} = S \cdot K_{rs} \tag{1.28}$$

This rotor emf at frequency Sf_1 in the rotor circuit is characterized by phase resistance R_r^r and leakage inductance L_{rl}^r. Also, the rotor is supplied by a system of phase voltages at the same frequency ω_2 and at a prescribed phase.

The stator and rotor equations for steady state/phase may be written in complex numbers at frequency ω_1 in the stator and ω_2 in the rotor:

$$(R_s + j\omega_1 L_{sl})\underline{I}_s - \underline{V}_s = \underline{E}_1 \quad \text{at } \omega_1 \tag{1.29}$$

$$\left(R_r^r + jS\omega_1 L_{rl}\right)\underline{I}_r^r - \underline{V}_r^r = \underline{E}_{2s} \quad \text{at } \omega_2 \tag{1.30}$$

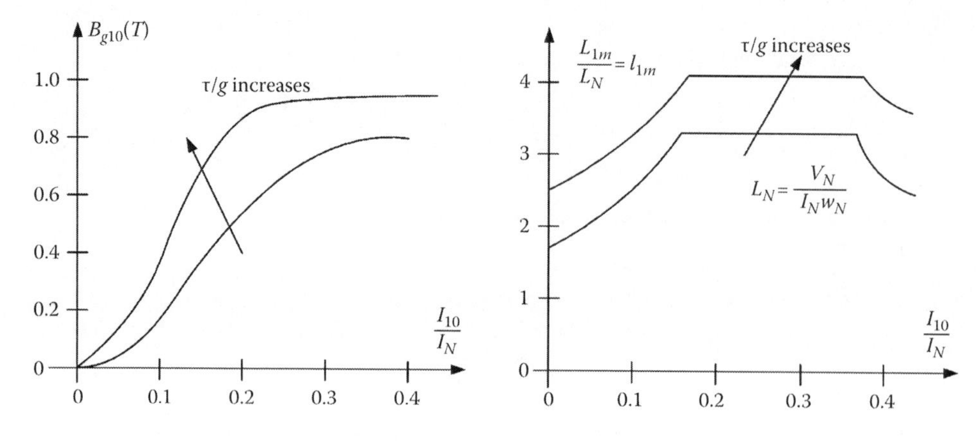

FIGURE 1.7 Typical airgap flux density (B_{g10}) and magnetization inductance (in per unit [P.U.]) vs. P.U. stator current.

According to Equation 1.28, we may multiply Equation 1.30 by $1/K_{sr}$ to reduce the rotor to the stator:

$$(R_r + jS\omega_1 L_{rl})\underline{I}_r - \underline{V}_r = \frac{E_{2s}}{K_{sr}}; \quad E_{2s} = SE_1 K_{sr}$$

$$R_r = \frac{R_r^r}{K_{rs}^2 L_{rl}} = \frac{L_{rl}^r}{K_{rs}^2} \tag{1.31}$$

$$\underline{V}_r = \frac{\underline{V}_R^r}{K_{rs}\underline{I}_r} = \underline{I}_r^r \cdot K_{rs}$$

The division of Equation 1.31 by slip S yields the following:

$$\left(\frac{R_r}{S} + j\omega_1 L_{rl}\right)\underline{I}_r - \frac{\underline{V}_r}{S} = \frac{S\underline{E}_1}{S} \tag{1.32}$$

But, Equation 1.31 may also be interpreted as being "converted" to frequency ω_1, as E_1 is at ω_1 ($E_{2s}/S = E_1$):

$$\left(\frac{R_r}{S} + j\omega_1 L_{rl}\right)\underline{I}_r - \frac{\underline{V}_r}{S} = \underline{E}_1; \quad \text{at } \omega_1 \tag{1.33}$$

In Equation 1.33, the rotor voltage V_r and current I_r vary with the frequency ω_1 and, thus, are written (in fact) in stator coordinates. A "rotation transformation" has been operated this way. In addition, all variables are reduced to the stator. Physically, this would mean that Equation 1.33 refers to a rotor at standstill, which may produce or absorb active power to cover the losses and delivers in motoring the mechanical power of the actual machine it represents.

Finally, the emf E_1 may now be conceived to be produced by both I_s and I_r (at the same frequency ω_1), both acting upon the magnetization inductance L_{1m} as the rotor circuit is reduced to the stator:

$$\underline{E}_1 = -j\omega_1 L_{1m}(\underline{I}_s + \underline{I}_r) = -j\omega_1 L_{1m}\underline{I}_m \tag{1.34}$$

1.4 Equivalent Circuit

The equivalent circuit corresponding to Equations 1.29, 1.33, and 1.34 is illustrated in Figure 1.8. Two remarks about Figure 1.8 are in order:

- The losses in the machine occur as stator- and rotor-winding losses $p_{cos} + p_{cor}$, core losses p_{Fe}, and mechanical losses p_{mec}:

$$p_{cos} = 3R_s I_s^2; \quad p_{cor} = 3R_r I_r^2; \quad p_{Fe} = 3R_{1m}(S\omega_1)I_{so}^2 \tag{1.35}$$

- The resistance R_{1m} that represents the core losses depends slightly on slip frequency $\omega_2 = S\omega_1$, as non-negligible core losses also occur in the rotor core for $Sf_1 > 5$ Hz.
- The active power balance equations are straightforward, from Figure 1.8, as the difference between input electrical powers P_s and P_r and the losses represents the mechanical power P_m:

$$P_m = \left[3\frac{R_r I_r^2}{S} - 3\frac{\text{Re}\left(\underline{I}_r^* \underline{V}_r\right)}{S}\right](1-S) = T_e \frac{\omega_1}{p_1}(1-S) = P_{elm}(1-S)$$

$$\sum P = p_{cos} + p_{cor} + p_{mec} + p_{Fe} \tag{1.36}$$

P_{elm} is the electromagnetic (through airgap) power.

$$P_s + P_r' = 3\text{Re}\left(\underline{V}_s \underline{I}_s^*\right) + 3\text{Re}\left(\underline{V}_r \underline{I}_r^*\right) = P_m + \sum p \tag{1.37}$$

T_e is the electromagnetic torque. The sign of mechanical power for given motion direction is used to discriminate between motoring and generating. The positive sign (+) of P_m is considered here for motoring (see the association of directions for $\underline{V}_s, \underline{I}_s$ in Figure 1.8).

The motor/generator operation mode is determined (Equation 1.36) by two factors: the sign of slip S and the sign and relative value of the active power input (or extracted) electrically from the rotor P_r (Table 1.1). Therefore, the WRIG may operate as a generator or a motor both subsynchronously ($\omega_r < \omega_1$) and supersynchronously ($\omega_r > \omega_1$). The power signs in Table 1.1 may be portrayed as in Figure 1.9.

If all the losses are neglected, from Equations 1.36 and 1.37:

$$P_m \approx -P_r \frac{(1-S)}{S} \approx P_s + P_r \tag{1.38}$$

Consequently,

$$P_r = -SP_s \tag{1.39}$$

The higher the slip, the larger the electric power absorption or delivery through the rotor. In addition, it should be noted that in supersynchronous operation, both stator and rotor electric powers add up to convert the mechanical power. This way, up to a point, oversizing, in terms of torque capability, is not required when operation at $S = -S_{\max}$ occurs with the machine delivering $P_s(1 + |S_{\max}|)$ total electric power.

Reactive power flow is similar. From the equivalent circuit,

$$Q_s + Q_r = 3\text{Imag}\left(\underline{V}_s \underline{I}_s^*\right) + 3\text{Imag}\left(\frac{\underline{V}_r \underline{I}_r^*}{S}\right) = 3\omega_1\left(L_{sl} I_s^2 + L_{rl} I_r^2 + L_m I_m^2\right) \tag{1.40}$$

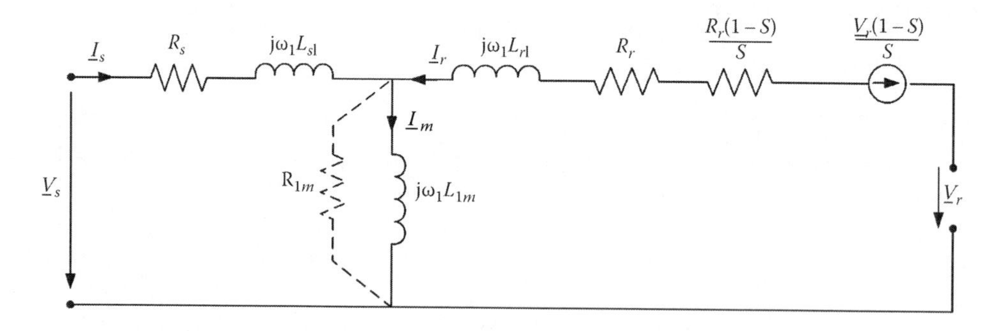

FIGURE 1.8 Wound-rotor induction generator (WRIG) equivalent circuit for steady state.

TABLE 1.1 Operation Modes

S	$0 < S < 1$ Subsynchronous ($\omega_r < \omega_1$)		$S < 0$ Supersynchronous ($\omega_r > \omega_1$)	
Operation Mode	Motoring	Generating	Motoring	Generating
P_m	>0	<0	>0	<0
P_s	>0	<0	>0	<0
P_r	<0	>0	>0	<0

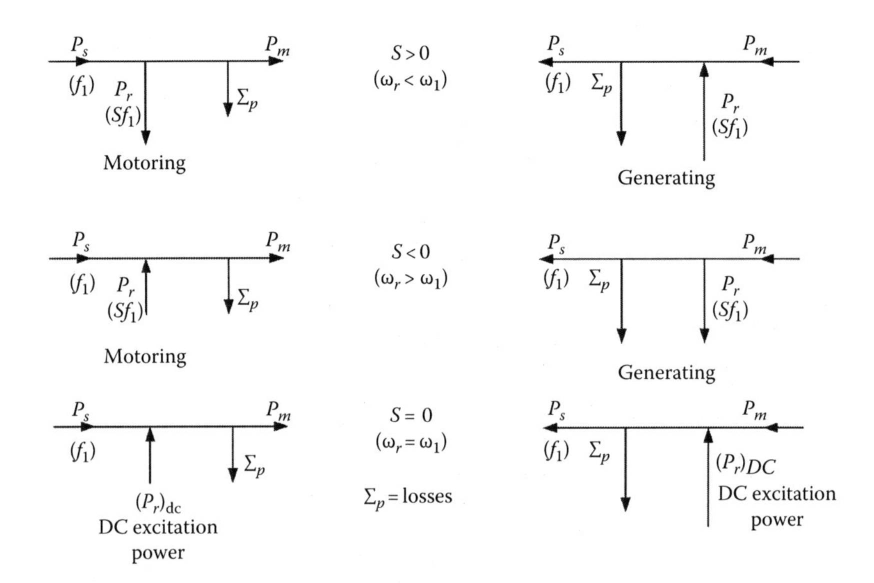

FIGURE 1.9 Operation modes of wound-rotor induction generator (WRIG) at $S > 0$, $S < 0$, and $S = 0$.

Therefore, the reactive power required to magnetize the machine may be delivered by the rotor or by the stator or by both. The presence of S in Equation 1.40 is justified by the fact that machine magnetization is perceived in the stator at stator frequency ω_1.

As the static power converter rating depends on its rated apparent power rather than active power, it seems to be practical to magnetize the machine from the stator. In this case, however, the WRIG absorbs reactive power through the stator from the power grids or from a capacitive-resistive load. In stand-alone operation mode, however, the WRIG has to provide for the reactive power required by the load up to the rated lagging power factor conditions. If the stator operates at unity power factor, the rotor-side static power converter has to deliver reactive power extracted either from inside itself (from the capacitor in the DC link) or from the power grid that supplies it.

As magnetization is achieved with lowest kVAR in DC, when active power is not needed, the machine may be operated at synchronism ($\omega_r = \omega_1$) to fully contribute to the voltage stability and control in the power system. To further understand the active and reactive power flows in the WRIG, phasor diagrams are used.

1.5 Phasor Diagrams

To make better use of the phasor diagram, we will expose in the steady-state equations (Equations 1.29, 1.33, and 1.34) the phase flux linkages in the stator $\underline{\Psi}_s$, in the airgap $\underline{\Psi}_m$, and in the rotor $\underline{\Psi}_r$:

$$\begin{aligned}
&\underline{\Psi}_m = L_{1m}\underline{I}_m; \quad \underline{I}_m = \underline{I}_s + \underline{I}_r \\
&\underline{\Psi}_s = \underline{\Psi}_m + L_{sl}\underline{I}_s; \quad \underline{\Psi}_s = L_s\underline{I}_s + L_{1m}\underline{I}_r; \quad L_s = L_{sl} + L_{1m} \\
&\underline{\Psi}_r = \underline{\Psi}_m + L_{rl}\underline{I}_r; \quad \underline{\Psi}_r = L_r\underline{I}_r + L_{1m}\underline{I}_s; \quad L_r = L_{rl} + L_{1m}
\end{aligned} \tag{1.41}$$

All quantities in Equation 1.41 are reduced to the stator and "in-stator coordinates"—same frequency f_1. With these new symbols, Equations 1.29, 1.33, and 1.34 become

$$\underline{I}_s R_s - \underline{V}_s = -j\omega_1\underline{\Psi}_s; \quad \underline{I}_r R_r - \underline{V}_r = -j\omega_1 S\underline{\Psi}_r = +\underline{E}_r' \tag{1.42}$$

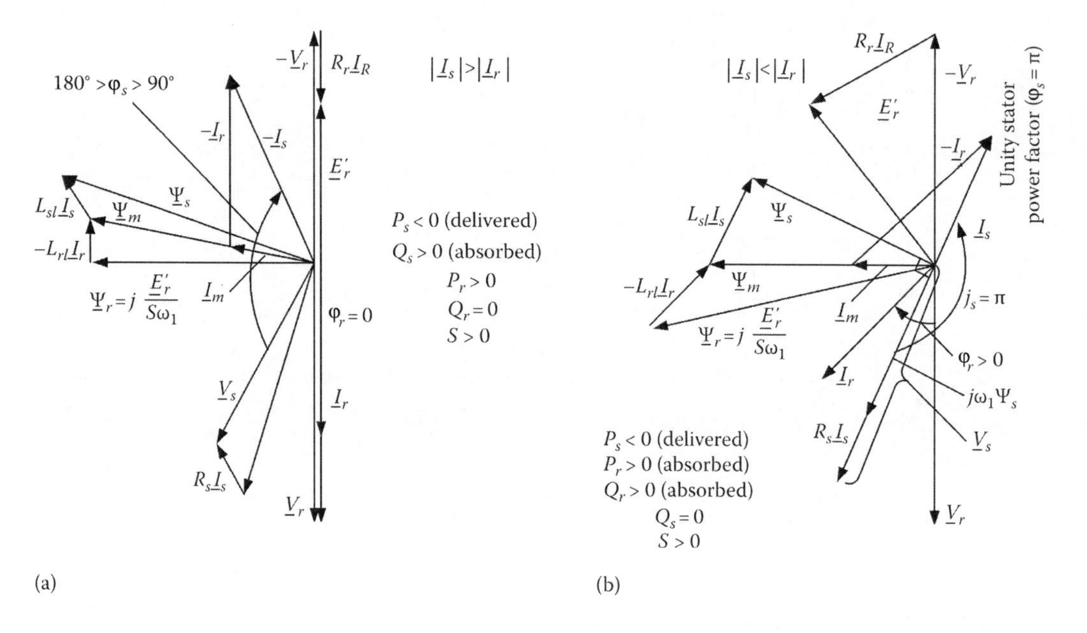

FIGURE 1.10 Phasor diagrams for wound-rotor induction generator (WRIG) in generator mode, $S > 0$ ($\omega_r < \omega_1$): (a) for rotor unity power factor and (b) stator unity power factor.

To build the phasor diagrams, the value and sign of S and the phase shift φ_r between $\underline{V}_r$ and $\underline{I}_r$ in the rotor have to be known, together with machine parameters and the amplitude $|\underline{V}_r|$ of V_r. Let us explore two cases: underexcitation and overexcitation, that is, respectively, with stator magnetization and rotor magnetization of the machine (cos φ_s—leading and, respectively, lagging). For underexcitation conditions, we may assume unity power factor in the rotor ($\varphi_r = 0$), as the magnetization is provided by the stator (Figure 1.10a), and start by drawing the V_r and I_r pair of phasors and then continue by using Equations 1.41 and 1.42, alternatively, until V_s is obtained.

The phasor diagrams indicate that when the machine is underexcited, $\Psi_r < \Psi_s (I_r < I_s)$. When it is overexcited, $\Psi_r > \Psi_s (I_r > I_s)$. The operation of WRIG may also be approached from the viewpoint of a synchronous machine.

From Equation 1.42,

$$\underline{I}_s(R_s + j\omega_1 L_s) - \underline{V}_s = \underline{E}_p = -j\omega_1 L_{1m} \underline{I}_r \tag{1.43}$$

Now, the problem is that the apparent synchronous reactance of the machine is L_s, the no-load inductance, while the emf E_p is produced only by the rotor current at stator frequency f_1. As the slip $S \neq 0$, there is also interference between stator and rotor currents, therefore such an interpretation does not hold much promise in terms of practicality. However, the rotor flux Ψ_r in Equation 1.42 seems to be determined solely by the rotor voltage and current for given slip. To make use of this apparent decoupling, express $\underline{\Psi}_s$ as a function of $\underline{\Psi}_r$ and $\underline{I}_s$ from Equation 1.41:

$$\underline{\Psi}_s = \underline{\Psi}_r \cdot \frac{L_r}{L_{rm}} - L_{sc}\underline{I}_s; \quad L_r = L_{rl} + L_{1m}; \quad L_{sc} = L_s - \frac{L_{1m}^2}{L_r} \approx L_{sl} + L_{rl} \tag{1.44}$$

Introducing Equation 1.44 into Equation 1.42 yields the following:

$$\underline{I}_s(R_s + j\omega_1 L_{sc}) - \underline{V}_s = -j\omega_1 \frac{L_r}{L_m} \underline{\Psi}_r = \underline{E}_{\Psi r}^s \tag{1.45}$$

Keeping the rotor flux constant, the machine behaves like a synchronous machine with a synchronous reactance that is the short-circuit reactance X_{sc}. As $X_{sc} \ll X_s$, this "new machine" behaves much better in terms of stability and voltage regulation.

Controlling the WRIG to keep the rotor flux constant is practical and, in fact, it was extensively used in vector-controlled AC drives [3].

We may now totally eliminate $\underline{I}_r$ from Equation 1.42, with the following:

$$\underline{I}_r = \frac{\underline{\Psi}_r - L_{1m}\underline{I}_s}{L_r} \tag{1.46}$$

$$-R_r \frac{L_m}{L_r}\underline{I}_s + \left(\frac{R_r}{L_r} + jS\omega_1\right)\underline{\Psi}_r = \underline{V}_r \tag{1.47}$$

$$(R_s + j\omega_1 L_{sc})\underline{I}_s + j\omega_1 \frac{L_r}{L_m}\underline{\Psi}_r = \underline{V}_s \tag{1.48}$$

This set of equations is easy to solve, provided the stator voltage V_s, power P_s, and SP factor angle φ_s are given:

$$I_s = \frac{P_s}{3V_s \cos\varphi_s} \tag{1.49}$$

With V_s in the horizontal axis, the stator current phasor I_s is obtained:

$$\begin{aligned} \underline{V}_s &= V_s \\ \underline{I}_s &= I_s(\cos\varphi_s - j\sin\varphi_s) \end{aligned} \tag{1.50}$$

From Equation 1.48, $\underline{\Psi}_r$ is determined as amplitude and phase with respect to stator voltage. Then, rotor current I_r—in stator phase coordinates—can be computed from Equation 1.46, both in amplitude and phase. Finally, if the speed ω_r is known, the slip S is known ($S = 1 - \omega_r/\omega_1$) and, thus, from Equation 1.47, the required rotor voltage phasor V_r (in stator coordinates) is computed (V_r, δ_{Vr}).

Example 1.1

Consider a WRIG with the following data: $P_{SN} = 12.5$ MW, cos $\varphi_N = 1$, $V_{SN1} = 6$ kV/(star connection) at $S_{max} = -0.25$, the turn ratio $K_{rs} = 1/S_{max} = 4.0$, $r_s = r_r = 0.0062$ (P.U.), $r_m = \infty$, $l_{sl} = l_{rl} = 0.0625$ (P.U.), $l_{1m} = 5.00$ (P.U.), $f_{1N} = 50$ Hz, $2p_1 = 4$ poles. Calculate the following:

- The parameters R_s, R_r, X_{sl}, X_{rl}, X_{1m} in Ω
- For $S = -S_{max}$ and maximum power P_{max} at cos $\varphi_s = 1$, calculate the rotor current, rotor voltage, and its angle δ_{Vr} with respect to the stator voltage, rotor active and reactive power P_r, Q_r^r, and total electric generator power $P_g = P_s + P_r$.

Solution

- The stator current at P_{SN} and cos $\varphi_s = 1$ is

$$(I_s)_{S=-Smax} = \frac{P_{SN}}{\sqrt{3}V_{SN}\cos\varphi_N} = \frac{12.5 \cdot 10^6}{\sqrt{3} \cdot 6000 \cdot 1} = 1.204 \cdot 10^3 \text{ A}$$

Based on the definition of base reactance X_N, the latter is

$$X_N = \frac{V_{SNl}/\sqrt{3}}{I_{SN}} = \frac{6000}{\sqrt{3}\cdot 1204} = 2.88\ \Omega$$

$$R_s = R_r = r_s \cdot X_N = 0.00625 \cdot 2.88 = 0.018\ \Omega$$
$$X_{sl} = X_{rl} = l_{sl} \cdot X_N = 0.0625 \cdot 2.88 = 0.18\ \Omega$$
$$X_{1m} = l_{rm} \cdot X_N = 5 \cdot 2.88 = 14.4\ \Omega$$

- The maximum current $\underline{I}_s$ is in phase opposition with the stator voltage as $\varphi_s = -180°$ in the generator mode, and as in Equations 1.47 and 1.48, absorbed powers are positive. The phasor diagram for this case is shown in Figure 1.11. From Equation 1.48, the rotor flux phasor $\underline{\Psi}_r$ is obtained:

$$\underline{\Psi}_r = -j\left(\frac{6000}{\sqrt{3}} - 1204 \cdot (0.018 + j \cdot 2 \cdot 0.18)\right) \cdot \frac{14.4}{2\pi \cdot 50} = 1.363 - j10.9756$$

The rotor current $\underline{I}_r$ is as follows (Equation 1.46):

$$\underline{I}_r = \frac{(1.363 - j10.9765)\cdot 314}{(0.18 + 14.4)} - \frac{14.4 \cdot (-1204)}{0.18 + 14.4} = 1218.49 - j236.3$$

From Equation 1.47, we can now compute the rotor voltage phasor $\underline{V}_r$ for $S_{max} = -0.25$:

$$\underline{V}_r = -0.018 \cdot \left(\frac{14.4}{14.4 + 0.18}\right)(-1204) + \left[\frac{0.018 \cdot 314}{14.4 + 0.18} + j(-0.25)\cdot 314\right](1.363 - j10.9756)$$
$$= -840 - j111.24$$

The reactive power through rotor Q_r, perceived at stator frequency, is (Figure 1.11)

$$Q_r = 3\mathrm{Imag}\left(\frac{V_r I_r^*}{S}\right) = 3\mathrm{Imag}\frac{(-840 - j111.14)(1218 + j2363)}{-0.25} = 4.004\ \mathrm{MVAR}$$

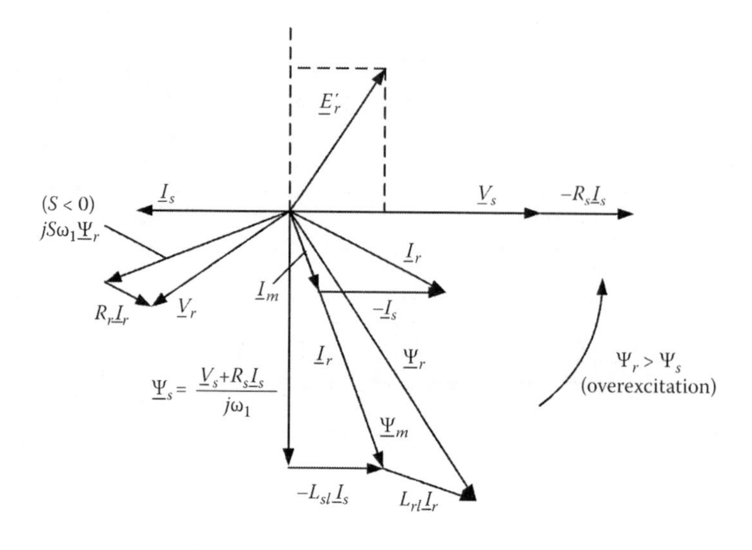

FIGURE 1.11 Phasor diagram for generating and unity stator power factor for $S < 0$.

In our case, $Q_s = 0$, so Q_r has to completely cover the reactive in the WRIG at stator frequency:

$$Q_r = 3X_{ls}I_s^2 + 3X_{ls}I_r^2 + 3X_{lm}\,|\,\underline{I}_s + \underline{I}_r\,|^2$$

$$= 3 \cdot 0.018(1204^2 + 1218^2 + 236.3^2) + 3 \cdot 14.4 \cdot |-1204 + 1218 - j236.3|^2 = 4.04\ \text{MVAR}$$

As expected, the two values of Q_r are very close to each other. Positive Q_r means absorbed reactive power, as it should, to fully magnetize the machine from the rotor ($Q_s = 0$). Q_r should not be confused with the reactive power Q_r^r that is measured at the slip-rings, at frequency $S\omega_1$:

$$Q_r^r = |S|Q_r = |0.25| \cdot 4.04\ \text{MVA} = 1.01\ \text{MVA}$$

The absolute value of slip is used to account for both subsynchronous and supersynchronous situations correctly, preserving the sign of the rotor-side reactive power.

The winding losses in the machine are the only losses considered in our example:

$$\sum_p = 3R_sI_s^2 + 3R_rI_r^2 = 3 \cdot 0.018 \cdot (1204^2 + 1218^2 + 236.3^2)$$

$$= 161.404 \cdot 10^3\ \text{W} = 161.404\ \text{kW}$$

The active power P_r^r through the rotor slip-rings is as follows:

$$P_r^r = 3\text{Re}\left(V_rI_r^*\right) = 3 \cdot (-840 - j111.24)(1218 + j236.3) = -2.9905 \cdot 10^6\ \text{W} = -2.9905\ \text{MW}$$

The mechanical power (Equation 1.36) is as follows:

$$P_m = \left[3R_rI_r^2 - 3Re\left(V_rI_r^*\right)\right]\left(\frac{1-S}{S}\right)$$

$$= \left[\frac{3 \cdot 0.018(1218^2 + 236^2)}{-0.25} + \frac{2.995 \cdot 10^6}{-0.25}\right]\left(1 + 0.25\right) = -15.39 \cdot 10^6\ \text{W} = -15.39\ \text{MW}$$

Checking the power balance Equation 1.37 shows the errors in our calculations (the losses in the machine are rather small at 1%):

$$P_s + P_r^r = -12.5 - 2.9905 = -15.4905\ \text{MW}$$

The mechanical power P_m absolute value should have been larger than $|P_s + P_r^r|$ by the losses in the machine. This is not the case, and care must be exercised when doing complex number calculations in order to be precise, especially for very high-efficiency machines. The computation of reactive power showed very good results because it has been rather large. Now, the megavoltampere (MVA) rating of the rotor-side converter considered for $S_{\max} = -0.25$ and unity power factor in the stator is as follows:

$$P_{ap}^r = \sqrt{P_r^{r2} + Q_r^{r2}} = \sqrt{2.9905^2 + 1.01^2} = 3.156\ \text{MVA} \tag{1.1.1}$$

The oversizing of the converter is not notable for unity power factor in the stator. With a turn ratio $a_{rs} = 4/1$, at $S_{\max} = -0.25$, the rotor circuit will be fed at about the rated voltage of the stator and at rotor current reduced by a_{rs} time with respect to that calculated:

$$V_r^{\text{real}} = |\underline{V}_r|a_{rs} = \sqrt{840^2 + 111.24^2} \cdot 4 = 3389\ \text{V} \tag{1.1.2}$$

$$I_r^{\text{real}} = \frac{|\underline{I}_r|}{a_{rs}} = \frac{\sqrt{1218^2 + 236.3^2}}{4} = 310 \text{ A} \tag{1.1.3}$$

It should also be noted that for overexcitation, when $\Psi_r > \Psi_s$ and $I_r = 1240.7$ A $> I_s = 1204$ A and when the WRIG is used in a configuration with a large number of poles, the magnetization reactance decreases (in P.U.) notably, and thus, the reactive power requirement from the rotor is larger. Consequently, the static power converter connected to the rotor should provide for it, directly, if the latter also works at the unity power factor at the source side. The back-to-back (bidirectional) PWM voltage source converter seems to be fully capable of providing for such requirements through the right sizing of the DC link capacitor bank.

1.6 Operation at the Power Grid

The connection of a WRIG to the power grid is similar to the case of an SG [4–6,8]. There is, however, an exceptional difference: the rotor-side static converter provides the conditions of synchronization at any speed in the interval $\omega_r(1 \pm |S_{max}|)$ and electronically brings the stator open-circuit voltages at the same frequency and phase with the power grid. In fact, the control system (Chapter 2) has a sequence for synchronization. Always successful, synchronization is feasible in a short time, in contrast to SGs, for which frequency and phase may be adjusted only through refined speed control by the turbine governor that tends to be slow due to high mechanical inertia. Furthermore, the WRIG may be started as a motor with the stator short-circuited, and then, above $\omega_r(1 - |S_{max}|)$, the stator circuit is opened. Subsequently, the synchronization control may be triggered, and, after synchronization, the machine is loaded either as a motor ($P_s > 0$) or as a generator ($P_s < 0$) through adequate closed-loop fast control (Figure 1.12).

Once connected to the power grid, it is important to describe its active and reactive power capabilities at constant voltage and frequency ω_1, but at variable speed ω_r (and $\omega_2 = \omega_1 - \omega_r$).

To describe the operation at the power grid, the powers P_s and P_r vs. power angle, for given speed (slip) and rotor voltage are considered to be representative. To simplify the characteristics $P_s(\delta_{Vr})$ and $P_r(\delta_{Vr})$, the stator resistance is neglected. The power angle is taken as the angle between $\underline{V}_s$ and $\underline{V}_r$ (in stator coordinates) (Figure 1.13).

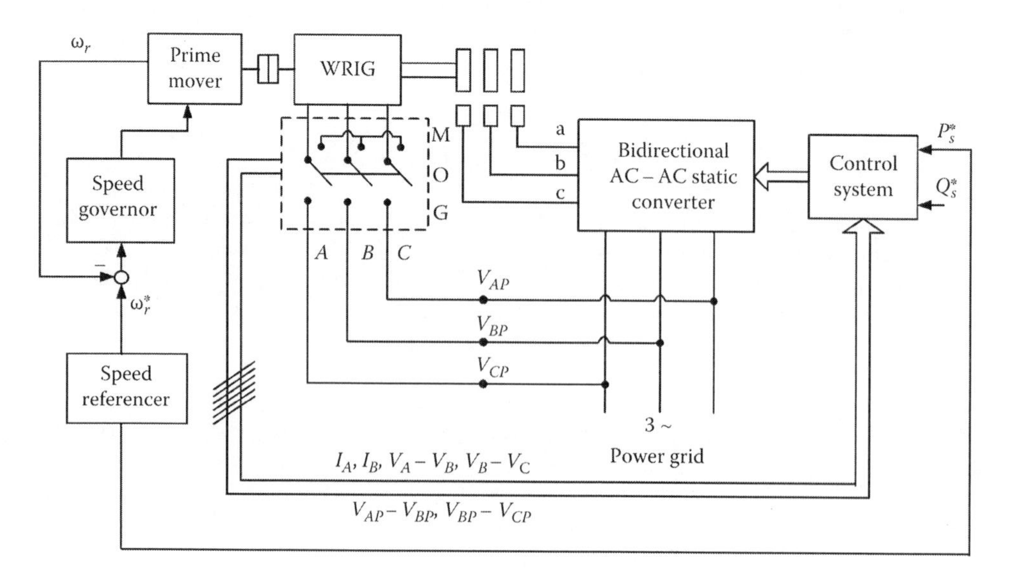

FIGURE 1.12 Synchronization arrangement for wound-rotor induction generator (WRIG): M—motor starting, O—synchronization preparation mode, and G—generator at power grid.

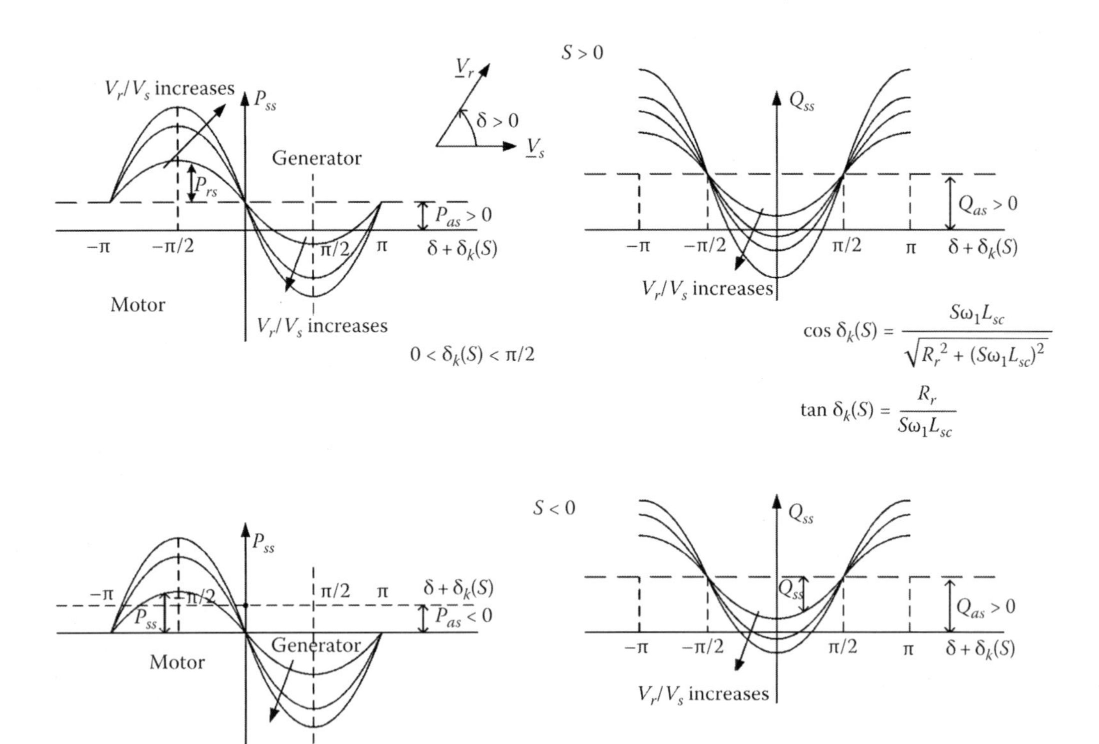

FIGURE 1.13 Powers P_s, Q_s vs. power angle $\delta + \delta_k(S)$.

1.6.1 Stator Power versus Power Angle

The machine steady-state Equations 1.41 and 1.42 with currents $\underline{I}_s$ and $\underline{I}_r$ for $R_s = 0$ are as follows:

$$\underline{V}_s = j\omega_1(L_s\underline{I}_s + L_{1m}\underline{I}_r) \tag{1.51}$$

$$\underline{V}_r = R_r\underline{I}_r + jS\omega_1(L_r\underline{I}_r + L_{1m}\underline{I}_s) = V_r(\cos\delta + j\sin\delta) \tag{1.52}$$

Eliminating $\underline{I}_s$ from Equation 1.51 yields the following:

$$R_r + \left(1 + jS\omega_1\left(L_s - \frac{L_{1m}^2}{L_s}\right)\right)\underline{I}_r = V_r(\cos\delta + j\sin\delta) - SV_s\frac{L_{1m}}{L_s} \tag{1.53}$$

With $L_r - L_{1m}^2/L_s \approx L_{sc}$,

$$\underline{I}_r = \frac{\left(V_r\cos\delta - SV_s\frac{L_{1m}}{L_s} + jV_r\sin\delta\right)(R_r - jS\omega_1L_{sc})}{R_r + (S\omega_1L_{sc})^2} \tag{1.54}$$

The stator active and reactive powers P_s, Q_s from Equation 1.51 are

$$P_s + jQ_s = 3V_s\underline{I}_s^* = 3L_{1m}\frac{V_s}{L_s}\left(\frac{jV_s}{\omega_1 L_{1m}} - \underline{I}_r^*\right)$$

$$= 3\frac{jV_s^2}{\omega_1 L_s} - \frac{(V_r\cos\delta - SV_s(L_{1m}/L_s) - jV_r\sin\delta)(R_r + jS\omega_1 L_{sc})}{R_r^2 + S^2\omega_1^2 L_{sc}^2}\cdot 3\frac{L_{1m}}{L_s}V_s \tag{1.55}$$

Equation 1.57 becomes

$$P_s = \underbrace{-3V_sV_r\frac{L_{1m}}{L_s}\frac{\sin(\delta+\delta_k(S))}{\sqrt{R_r^2 + (S\omega_1 L_{sc})^2}}}_{\text{synchronous active power }(P_{ss})} + \underbrace{3V_s^2\left(\frac{L_{1m}}{L_s}\right)^2\frac{R_rS}{R_r^2 + (S\omega_1 L_{sc})^2}}_{\text{asynchronous active power with short-circuited rotor }(P_{as})} \tag{1.56}$$

$$Q_s = \underbrace{\frac{3V_s^2}{\omega_1 L_s}\left[1 + \frac{(S\omega_1 L_{1m})^2 L_{sc}}{\left[R_r^2 + (S\omega_1 L_{sc})^2\right]\cdot L_s}\right]}_{\text{absorbed reactive power}} \underbrace{- 3V_sV_r\frac{L_{1m}}{L_s}\frac{\cos(\delta+\delta_k(S))}{\sqrt{R_r^2 + (s\omega_1 L_{sc})^2}}}_{\text{synchronous reactive power }(Q_{ss})\text{ with short-circuited rotor }(Q_{as})} \tag{1.57}$$

The resemblance to the nonsalient-pole SG is evident. However, the second term in P_s is produced asynchronously and is positive (motoring) for positive slip and negative (generating) for negative slip. The first term in Q_s represents the reactive power absorbed by the machine reactances. The angle δ_k depends heavily on slip S and R_r:

$$\delta_k = 0 \quad \text{for } |S\omega_1 L_{sc}| \gg R_r \tag{1.58}$$

$$\begin{aligned} \delta_k &= \frac{\pi}{2} && \text{for } S = 0 \\ 0 < \delta_k &< \frac{\pi}{2} && \text{for } S > 0 \\ \frac{\pi}{2} < \delta_k &< \pi && \text{for } S < 0 \end{aligned} \tag{1.59}$$

To bring more generality to the P_s and Q_s dependences on δ, we represent P_s, Q_s as a function of (δ + $\delta_k(S)$) (Figure 1.13). We may separate the two components in P_s and Q_s:

$$P_s = P_{ss} + P_{as} \tag{1.60}$$

$$Q_s = Q_{ss} + Q_{as} \tag{1.61}$$

P_{ss} and Q_{ss} are dependent on ($\delta + \delta_k(S)$), while P_{as} and Q_{as} are slip-dependent only.

The variable $\delta + \delta_k(S)$ greatly simplifies the graphs, but care must be exercised when the actual power angle operation zone is computed. It is evident that for a voltage-fed rotor circuit—(V_r, δ) given—there is a certain difference between motor and generator operation zones, because the asynchronous power is positive (motoring) for $S > 0$ and negative (generating) for $S < 0$.

The sign of S does not influence reactive power Q_s ($\delta + \delta_k(S)$), but again, $\delta_k(S)$ depends on slip. To "produce" zero reactive power stator conditions, the rotor voltage ratio V_r/V_s has to be increased.

The peak active power is larger in motoring for $S > 0$ (subsynchronous) operation and, respectively, in generating for $S < 0$ (supersynchronous). Notice that WRIG peak stator active power is determined by the short-circuit ($\omega_1 L_{sc}$) rather than no-load ($\omega_1 L_s$) reactance. However, as $V_r/V_s \ll 1$, the peak active power is not very large, though larger than in SGs in general. The electromagnetic power ($R_s = 0$, $P_{Fe} = 0$) is as follows:

$$P_{elm} = P_s = T_e \frac{\omega_1}{p_1} \tag{1.62}$$

So, the electromagnetic torque is strictly proportional to stator active power P_s (for zero stator losses).

1.6.2 Rotor Power versus Power Angle

The rotor electric active and reactive powers P_r^r, Q_r^r are as follows:

$$P_r^r + jQ_r^r = 3\underline{V}_r \underline{I}_r^* \tag{1.63}$$

The rotor "produced" equivalent reactive power Q_r seen from the stator (at stator frequency) is

$$Q_r = \text{Imag}\left(\frac{3\underline{V}_r \underline{I}_r^*}{S}\right) \tag{1.64}$$

From Equations 1.52 and 1.54,

$$P_r^r = \underbrace{\frac{3V_r^2 R_r}{R_r^2 + (S\omega_1 L_{sc})^2}}_{\text{rotor copper losses}} + \underbrace{3V_r V_s \frac{L_{1m}}{L_s} \frac{S \cdot \sin(\delta - \delta_k)}{\sqrt{R_r^2 + (S\omega_1 L_{sc})^2}}}_{\text{synchronous rotor power with shorted stator}} \tag{1.65}$$

$$Q_r^r = \underbrace{\frac{3V_r^2 S\omega_1 L_{sc}}{R_r^2 + S\omega_1 L_{sc}^2}}_{\text{reactive power absorbed}} - \underbrace{3V_r V_s \frac{L_{1m}}{L_s} \frac{S \cdot \cos(\delta - \delta_k)}{\sqrt{R_r^2 + (S\omega_1 L_{sc})^2}}}_{\text{synchronous reactive with shorted stator rotor power}} \tag{1.66}$$

Similar graphs $P_r^r(\delta - \delta_k)$ and $Q_r^r(\delta - \delta_k)$ may be drawn by using these expressions, but they are of a minimal practical use than P_s and Q_s. They are, however, important for designing the rotor-side static power converter and for determining the total rotor electric power delivery, or absorption, during subsynchronous or supersynchronous operation.

1.6.3 Operation at Zero Slip ($S = 0$)

At zero slip, from Equation 1.59, it follows first that $\delta_k = \pi/2$. Finally, from Equations 1.56 and 1.57,

$$P_s = -3V_s V_r \frac{L_m}{R_r L_s} \sin\left(\delta + \frac{\pi}{2}\right) \tag{1.67}$$

$$Q_s = \frac{3V_s^2}{\omega_1 L_s} - 3\frac{V_s V_r L_m}{R_r L_s} \cos\left(\delta + \frac{\pi}{2}\right) \tag{1.68}$$

$$I_r = \frac{V_r}{R_r} \tag{1.69}$$

Note again that the rotor voltage is considered in stator coordinates. The power angle ($\delta + (\pi/2)$) is typical for SGs, where it is denoted by δ_u (the phase shift between rotor-induced emf and the phase voltage).

For operation at zero slip ($S = 0$), when the rotor circuit is DC fed, all the characteristics of SGs hold true [7]. In fact, it seems adequate to run the WRIG at $S = 0$ when massive reactive power delivery (or absorption) is required. Though active and reactive power capability circles may be defined for WRIGs, it seems to us that, due to decoupled fast active and reactive power control through the rotor-connected bidirectional power converters (Chapter 8, in *Synchronous Generators*), such graphs may become somewhat superfluous.

1.7 Autonomous Operation of WRIGs

Insularization of WRIGs, in case of need, from the power grids, caused by excess power in the system or stability problems, leads to autonomous operation. Autonomous operation is characterized by the fact that voltage has to be controlled, together with stator frequency (at various rotor speeds in the interval $[1 \pm |S_{max}|]$), in order to remain constant under various active and reactive power loads. Whatever reactive power is needed by the consumers, it has to be provided from the rotor-side converter after covering the reactive power required to magnetize the machine. When large reactive power loads are handled, it seems that running at constant speed and zero slip ($S = 0$) would be adequate for taking full advantage of the rotor-side static converter limited ratings and for limiting rotor windings and converter losses. On the other hand, for large active loads, supersynchronous operation is suitable, as the WRIG may be controlled to operate around unity power factor while keeping the stator voltage within limits. Subsynchronous operation should be used when part loads are handled in order to provide for better efficiency of the prime mover for partial loads. The equivalent circuit (Figure 1.8) may easily be adapted to handle autonomous loads under steady state (Figure 1.14).

For autonomous operation, the stator voltage V_s is replaced by the following:

$$\underline{V}_s = -(R_{load} + jX_{load})\underline{I}_s \tag{1.70}$$

In these conditions, retaining the power angle δ as a variable does not seem to be so important. The rotor voltage "sets the tone" and may be considered in the real axis: $\underline{V}_r = V_r$. Neglecting the stator resistance R_s does not bring any simplification, as it is seen in series with the load (Equation 1.70):

$$\begin{aligned}[R_s + R_{load} + j(X_{load} + X_{sl})]\underline{I}_s &= -jX_{1m}(\underline{I}_s + \underline{I}_r) = \underline{E}_m(\underline{I}_m)\\ (R_r + jSX_{rl})\underline{I}_r - \underline{V}_r &= -jSX_{1m}(\underline{I}_s + \underline{I}_r) = \underline{E}_m \cdot S; \quad \underline{I}_m = \underline{I}_s + \underline{I}_r\end{aligned} \tag{1.71}$$

Both equations are written in stator coordinates (at frequency ω_1 for all reactances). We may consider now the WRIG as being supplied only from the rotor, with the stator connected to an external impedance. In other words, the WRIG becomes a typical induction generator fed through the rotor, having

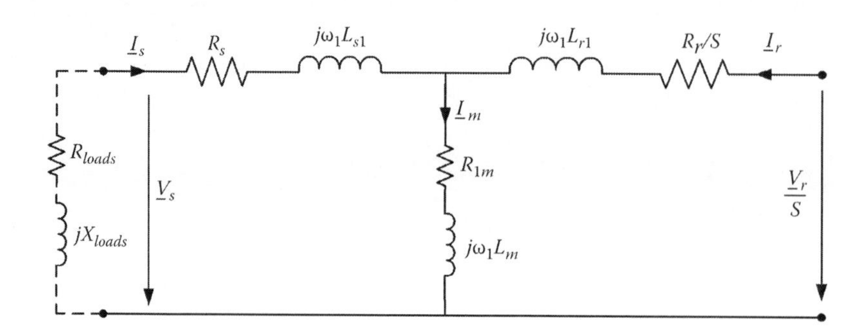

FIGURE 1.14 Equivalent circuit of wound-rotor induction generator (WRIG) for autonomous operation.

stator load impedance. It is expected that such a machine would be a motor for positive slip ($S > 0$, $\omega_r < \omega_1$) and a generator for negative slip ($S < 0$, $\omega_r > \omega_1$).

This is a drastic change of behavior with respect to the WRIG connected at a fixed frequency and voltage (strong) power grid, where motoring and generating are practical both subsynchronously and supersynchronously.

By properly adjusting the rotor frequency ω_2 with speed ω_r to keep ω_1 constant and controlling the amplitude and phase sequence of rotor voltage V_r, the stator voltage may be kept constant until a certain stator current limit, for given load power factor.

To obtain the active and reactive powers of the stator and the rotor P_s, Q_s, P_r^r, Q_r^r, solving first for the stator and rotor currents in Equation 1.71 is necessary. Neglecting the core loss resistance R_{1m} ($R_{1m} = 0$) yields the following:

$$\begin{aligned}
&\underline{I}_s = \frac{V_r}{R_{se}(S) + jX_{se}(S)}\\
&R_{se}(S) = \frac{-R_r X_{s+l} - SX_r R_{s+l}}{X_{1m}}\\
&X_{se}(S) = \frac{-SX_r X_{s+l} + R_{s+l} R_r}{X_{1m}} + SX_{1m}\\
&R_{s+l} = R_s + R_{loads}; \quad X_{s+l} = X_s + X_{loads}\\
&X_s = X_{sl} + X_{1m}; \quad X_r = X_{rl} + X_{1m}
\end{aligned} \tag{1.72}$$

The active and reactive powers of stator and rotor are straightforward:

$$\begin{aligned}
&P_s = 3I_s^2 R_{loads} > 0\\
&Q_s = 3I_s^2 X_{loads} <> 0
\end{aligned} \tag{1.73}$$

Also, $\underline{I}_r$ from Equation 1.71 is

$$\underline{I}_r = j\frac{(R_{s+l} + jX_{s+l})\underline{I}_s}{X_{1m}} \tag{1.74}$$

$$P_r^r + jQ_r^r = 3\left(V_r \cdot I_r^*\right) \tag{1.75}$$

The mechanical power P_m is simply

$$P_m = \frac{1-S}{S}\left[3R_r I_r^2 - P_r^r\right] \tag{1.76}$$

$$Q_r^r - 3\left(X_{sl} I_s^2 + X_r I_r^2 + X_{1m} I_m^2\right) = Q_s \tag{1.77}$$

As the machine works as an induction machine fed to the rotor, with passive impedance in the stator, all characteristics of it may be used to describe its performance. The power balance for motoring and generating is described in Figure 1.15a and b.

Note that subsynchronous operation as a motor is very useful when self-starting is required. The stator is short-circuited ($R_{load} = X_{load} = 0$), and the machine accelerates slowly (to observe the rotor-side converter P_{low} rating) until it reaches the synchronization zone $\omega_r(1 \pm |S_{\max}|)$. Then the stator circuit is opened, but the induced voltage in the stator has a small frequency. Consequently, the phase sequence in the rotor voltages has to be reversed to obtain $\omega_1 > \omega_r$ for the same direction of rotation. This is the beginning of the resynchronization control mode when the machine is free-wheeling. Finally, within a

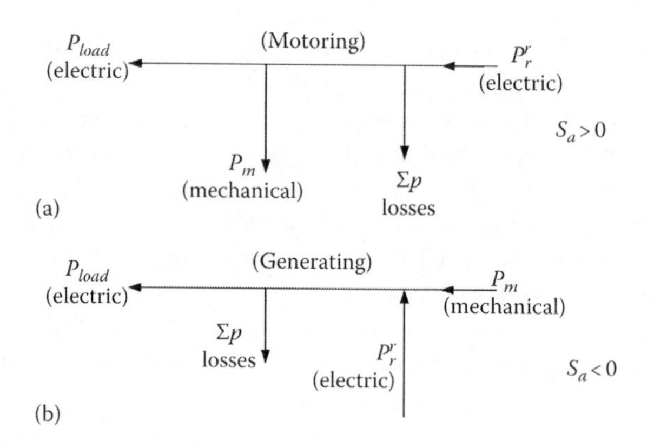

FIGURE 1.15 Power balance: (a) $S_a > 0$ and (b) $S_a < 0$.

few milliseconds, the stator voltage and frequency conditions are met, and the machine stator is reconnected to the load.

Induction motoring with a short-circuited stator is also useful for limited motion during bearing inspections or repairs.

Autonomous generating (on now-called ballast load) may be used as such and when, after load rejection, fast braking of the mover is required, to avoid dangerous overspeeding until the speed governor takes over.

The stator voltage regulation in generating may be performed through changing the rotor voltage amplitude while the frequency ω_1 is controlled to stay dynamically constant by modifying frequency ω_2 in the rotor-side converter.

Example 1.2

For the WRIG in Example 1.1 at $S = -0.25, f_1 = 50$ Hz, $I_s = I_{sN}/2 = 602$ A, $V_r = V_{r\max} = V_s$, $\cos\varphi_s = 1$, compute the following:

- The load resistance R_{loads} per phase in the stator
- The load (stator voltage) V_s and load active power P_s
- The rotor current and active and reactive power in the rotor P_r, Q_r
- The no-load stator voltage for this case and the phasor diagram

After the computations are made, discuss the results.

Solution

- We have to go straight to Equation 1.72, with, $R_s = R_r = 0.018\ \Omega$, $X_{sl} = X_{rl} = 0.018\ \Omega$
- $X_{lm} = 14.4\,\Omega, X_s = X_r = 14.58\,\Omega$ (calculated in stator terms), $V_{rphase} = (6000/\sqrt{3}V)$ actual value,

  $S = -0.25, I_s = 602$ A, $X_{load} = 0 (\cos\varphi_s = 1), W_r/W_s = 4/1$, where the only unknown is R_{loads}:

$$R_{se}(S) = -0.01366 + 0.2531 \cdot R_{loads}$$

$$X_{se}(S) = 1.25 \cdot 10^{-3} \cdot R_{loads} + 0.09$$

$$\underline{I}_s = 602 = \frac{-V_r \cdot (W_s/W_r)}{R_{se}(S) + jX_{se}(S)}$$

Consequently, $R_{loads} \approx 5.678\ \Omega$.

- The stator voltage per phase V_s is simply ($\cos\varphi_s = 1$).

$$(V_s)_{phase} = R_{load} I_s = 5.678 \cdot 602 = 3418\ \text{V}.$$

$$P_s = -3R_{loads} I_s^2 = -3 \cdot 5.678 \cdot 602 = -6.173\ \text{MW}$$

- The rotor current (Equation 1.74) is

$$\underline{I}_r = j\frac{(R_s + R_{loads} + jX_{s+l})\underline{I}_s}{X_{1m}}$$

with

$$\underline{I}_s = 602 \cdot (0.9884 - j0.0674)$$

Therefore,

$$\underline{I}_r = -586.5 + j287.084$$

The active and reactive powers in the rotor are as follows:

$$P_r^r + jQ_r^r = 3V_r I_r^* \cdot \frac{W_s}{W_r} = 3 \cdot \frac{6000}{4\sqrt{3}} \cdot (-586.5 + j287.084) = -1.5255\ \text{MW} + j0.7467\ \text{MVAR}$$

Therefore, the rotor circuit absorbs reactive power to magnetize the machine, but it delivers active power, together with the stator. The mechanical power covers for all losses in the machine and produces both P_s and P_r:

$$P_m - \sum_p = |P_s + P_r| = -1.5255 - 6.173 = -7.6985\ \text{MW}$$

The losses considered in our example are only the winding losses:

$$\sum_p = 3R_s I_s^2 + 3R_r I_r^2 = 3 \cdot 0.018 \cdot (602^2 + 658.95^2) = 42.592\ \text{kW}$$

Thereby, the mechanical power is as follows:

$$P_m = 7.6985 + 0.04259 = 7.741\ \text{MW}$$

- The no-load voltage in the stator for the above conditions is simply

$$E_m = X_{1m} I_m; \quad \underline{I}_m = \underline{I}_s + \underline{I}_r$$

$$\underline{I}_m = 9 + j246$$

This is the magnetization current for the airgap flux:

$$E_m = X_{1m} I_m = 14.4 \cdot 246 = 3542.4\ \text{V}$$

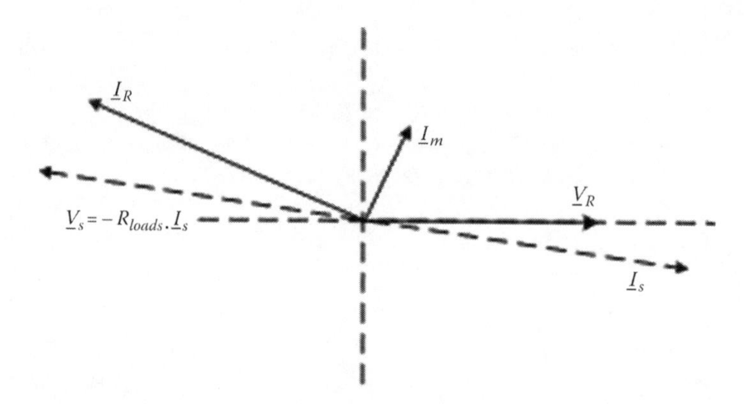

FIGURE 1.16 Phasors for autonomous wound-rotor induction generator (WRIG) operation at $S = -0.25, f_1 = 50$ Hz.

The voltage regulation is very small:

$$\Delta V = \frac{E_m - V_s}{V_s} = \frac{3542 - 3418}{3418} = 3.64\%$$

The current and voltage phasors are shown in Figure 1.16.

1.8 Operation of WRIGs in the Brushless Exciter Mode

With a brushless exciter, the power is delivered through the rotor, after rectification, to the excitation circuit of an SG (Figure 1.17). The commutation in the diode rectifier causes small harmonics in the rotor current, but for its fundamental, the power factor may be considered as unity. The diode rectifier commutation causes some voltage reduction as already shown in Chapter 6 in *Synchronous Generators* (the paragraph on excitation systems).

The rotor rotates opposite to the stator mmf, and thus

$$\omega_2 = \omega_1 + \omega_r > \omega_1 \tag{1.78}$$

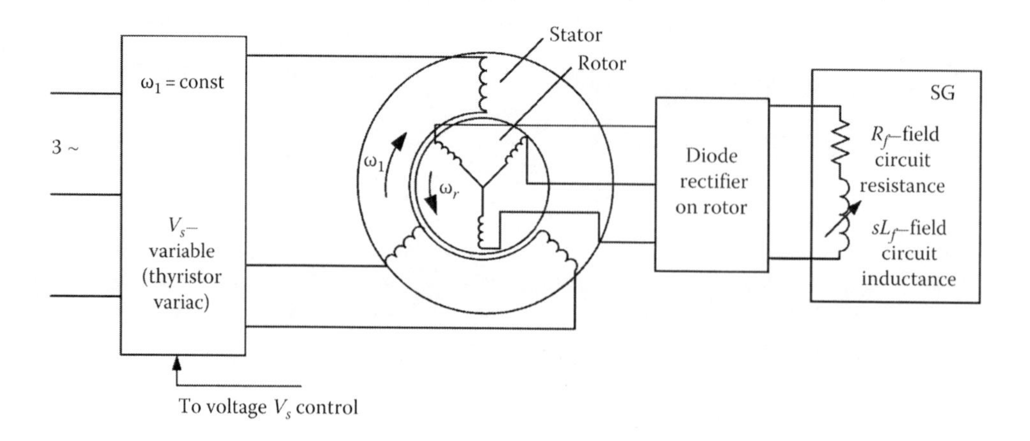

FIGURE 1.17 Wound-rotor induction generator (WRIG) as brushless exciter.

The frequency in the rotor is at its minimum at zero speed and then increases with speed. If the WRIG is provided with a number of poles that is notably larger than that of the SG, then the frequency ω_2 would be notably larger than ω_1:

$$\omega_2 = \omega_1 + 2\pi n_1 p_1; \quad n_1 = \frac{f_1}{p_g} \tag{1.79}$$

with p_g-pole pairs in the SG with excitation that is fed from the WRIG exciter:

$$\omega_2 = \omega_1\left(1 + \frac{p_1}{p_g}\right) \tag{1.80}$$

The larger p_1/p_g, the higher the rotor (slip) frequency; with $p_1/p_g = 3{,}4$, good results may be obtained. The WRIG-exciter is supplied through a static variance at constant frequency ω_1, so the converter's cost is low.

The machine equations (Equation 1.42) remain valid, but we will use ω_2 instead of $S\omega_1$:

$$\begin{aligned} \underline{I}_s R_s - \underline{V}_s &= -j\omega_1 \underline{\Psi}_s = -j\omega_1(L_s \underline{I}_s + L_{1m}\underline{I}_r) \\ \underline{I}_r R_r + \underline{V}_r &= -j\omega_2 \underline{\Psi}_r = -j\omega_2(L_r \underline{I}_r + L_{1m}\underline{I}_s) \end{aligned} \tag{1.81}$$

The speed ω_r is now negative ($\omega_r < 0$), that is, $\omega_1 > 0$ and $\omega_2 > 0$. The slip $S = \omega_2/\omega_1 > 1$.

We already used the positive sign (+) on the left side of rotor equation to have positive power for generating. Also, for simplicity, a resistance load will be considered:

$$\underline{V}_r = \underline{I}_r \cdot R_{load\,r} \tag{1.82}$$

Equation 1.81 with Equation 1.82 may be solved simply for stator and rotor currents:

$$\underline{I}_s = \frac{j\underline{I}_r(R_r + j\omega_2 L_r + R_{loads})}{\omega_2 L_{1m}} \tag{1.83}$$

$$\underline{I}_r = \frac{\underline{V}_s}{j((R_r + R_{load} + j\omega_2 L_r)(R_s + j\omega_1 L_s)/(\omega_2 L_{1m})) + j\omega_1 L_m} \tag{1.84}$$

The electromagnetic torque T_e is

$$T_e = 3p_1 \text{Real}\left(j\underline{\Psi}_s \underline{I}_s^*\right) = 3p_1 L_{1m} \text{Real}\left(j\underline{I}_s I_r^*\right) \tag{1.85}$$

At zero speed, the WRIG-exciter works as a transformer, and all the active and reactive power is delivered by the stator. When the speed increases—with resistive load in the rotor circuit—the stator "delivers" the reactive power to magnetize the machine and the active power to cover the losses and some part of the load active power.

The bulk of the active power to the load comes, however, from the mechanical power P_m. The higher the ratio ω_2/ω_1, the higher is the P_m contribution to P_r (rotor-delivered active power).

Example 1.3: WRIG as Brushless Exciter

Consider a WRIG with the main data: $R_s = R_r = 0.015$ P.U., $L_{sl} = L_{rl} = 0.14$ P.U., $L_{1m} = 3$ P.U., $V_{SNl} =$ 440 V (star), $I_{SN} = 1000$ A, the frequency $f_1 = 60$ Hz, and the rotor speed $n_N = 1800$ rpm. The number of pole pairs is $p_1 = 6$. The rotor-to-stator turns ratio is $a_{rs} = 1$. Determine the following:

- The rotor frequency $f_2(\omega_2)$ and the ideal maximum no-load rotor voltage
- The rotor-side load resistance voltage, current, power P_{r0} at zero speed, and $I_r = 1000$ A in the rotor
- The required stator voltage, current, and input active and reactive powers P_s, Q_s, for the same load resistance R_{load} and current load $I_r = 1000$ A, but at $n_N = 1800$ rpm

Solution

- The rotor-side frequency $f_2(\omega_2)$ is simply as follows (Equation 1.79):

$$\omega_2 = \omega_1 + \frac{2\pi n_1 p_1}{\omega_1} = \left(1 + \frac{2\pi \cdot (1800/60) \cdot 6}{2\pi 60}\right) = 4\omega_1$$

So, $f_2 = 4f_1 = 240$ Hz.

The ideal no-load rotor voltage V_r^r (unreduced to the stator, for full stator voltage at speed n_N) is as follows:

$$V_{r0}^r = a_{rs} \cdot V_s \cdot \frac{\omega_2}{\omega_1} = 1 \cdot 440 \cdot \frac{4}{1} = 1760 \text{ V (line voltage, RMS)}$$

The rotor circuit might be designed to comply with this voltage during an excitation 4/1 forcing.

At zero speed ($\omega_2 = \omega_1$), the ideal rotor voltage would be

$$\left(V_{r0}^r\right)_{stall} = a_{rs} \cdot V_s \cdot \frac{\omega_1}{\omega_1} = 1 \cdot 440 \cdot \frac{1}{1} = 440 \text{ V}$$

- The machine parameters in Ω (all reduced to the stator) are as follows:

$$X_n = \frac{V_{SNl}/\sqrt{3}}{I_{SN}} = \frac{(440/\sqrt{3})}{1000} = 0.2543\,\Omega$$

Therefore,

$$R_s = R_r = (R_s)_{\text{P.U.}} \cdot X_n = 0.015 \cdot 0.2543 = 3.8145 \cdot 10^{-3}\,\Omega$$

$$L_{sl} = L_{rl} = (X_{sl})_{\text{P.U.}} \cdot \frac{X_n}{\omega_1} = 0.14 \cdot \frac{0.2543}{2\pi 60} = 9.45 \cdot 10^{-5} \text{ H}$$

$$L_{1m} = (X_{1m})_{\text{P.U.}} \cdot \frac{X_n}{\omega_1} = 3 \cdot \frac{0.2543}{2\pi 60} = 2.0247 \cdot 10^{-3} \text{ H}$$

At zero speed, $\omega_2 = \omega_1$, the rotor current I_r may be calculated from the following (Equation 1.84): Finally,

$$R_{load} = 0.226\,\Omega$$

Therefore, the rotor voltage V_r (reduced to the stator) is

$$V_r = R_{load} \cdot I_r = 0.226 \cdot 1000 = 226\,\text{V}$$

For voltage regulation,

$$\Delta V = \frac{V_s - V_r}{V_s} = \frac{254 - 226}{254} = 0.1102 = 11.02\%$$

The large leakage reactances of the stator and the rotor are responsible for this notable voltage drop (notable for a transformer or an induction machine, but small for an SG of any type).

The rotor-delivered power P_r is as follows:

$$P_r = 3V_r \cdot I_r = 3 \cdot 226 \cdot 1000 = 678\,\text{kW}$$

- Now, we make use of Equation 1.84 to calculate the stator voltage required for $I_r = 1000$ A, with

$$V_s = 96.8\,\text{V (RMS per phase)}$$

The stator $\underline{I}_s$ from Equation 1.83 is

$$\underline{I}_s = j\underline{I}_r \frac{(R_r + j\omega_2 L_r + R_{load})}{\omega_2 L_{1m}} = j\frac{1000 \cdot (0.02298 + j3.1938)}{3.0516} = -1046.6 + j75.3;$$

$$I_s = 1049.3\,\text{A} > I_r$$

The stator active and reactive powers are as follows:

$$P_s + jQ_s = 3\underline{V}_s \underline{I}_s^* = 3 \cdot (-64.06 - j72.1) \cdot (-1046.6 - j75.3)$$

$$P_s = 184.752\,\text{kW}, \quad Q_s = 240.751\,\text{kVAR}$$

The delivered electric power through the rotor P_r is still 678 kW, as the load resistance and current were kept the same, but much of the power now comes from the shaft as $P_s \ll P_r$.

A few remarks are in order:

- As the machine is rotated, less active power is delivered through the stator, with much of it "extracted" from the shaft (mechanically). This is a special advantage of this configuration.
- With the machine in motion ($\omega_r = 3\omega_1$), the required stator voltage decreases notably. A static variac may be used to handle such a 1/5 voltage reduction easily.
- The machine magnetization is provided by the stator, and because $\omega_2 = 4\omega_1$, the power factor in the stator is poor.
- The magnetization by the stator is also illustrated by $I_s > I_r$.
- The stator voltage reserve at full speed may be used for forcing the excitation (load) current in the supplied synchronous machine excitation, but, in that case, the rotor voltage would increase above the rated value (440 V root-mean-squared [RMS]/line). The rotor winding insulation and the flying diode rectifier have to be sized for such events.
- The capability of the WRIG to serve as an exciter from zero speed—demonstrated in this example—makes it a good solution when the excitation power is required from zero speed, as is the case in variable-speed large SMs or SGs.

- The internal reactance of the WRIG is important to know in order to assess voltage regulation and to model the machine with rectified output.
- To emphasize the "synchronous" reactance of WRIG as an exciter, the stator current is eliminated from the stator equation by introducing the stator flux $\underline{\Psi}_s$:

$$\underline{\Psi}_r = \frac{L_m}{L_s}\underline{\Psi}_s + L_{sc}\underline{I}_r \tag{1.3.1}$$

The rotor equation (Equation 1.81) may now be written as follows:

$$\underline{V}_r = -j\omega_2\frac{L_m}{L_s}\underline{\Psi}_s - (R_r + j\omega_2 L_{sc})\underline{I}_r = \underline{E}_r - \underline{Z}_{ex}\underline{I}_r \tag{1.3.2}$$

$$\underline{Z}_{ex} = R_r + j\omega_2 L_{sc} \tag{1.3.3}$$

The term "$\underline{Z}_{ex}$" represents the internal (synchronous) impedance of WRIG as an exciter source.

The first term in Equation 1.3.2 is the emf $\underline{E}_r^*$:

$$\underline{E}_r = -j\omega_2\frac{L_m}{L_s}\underline{\Psi}_s \tag{1.3.4}$$

The stator flux may be considered variable, with stator voltage as follows ($R_s = 0$):

$$-j\omega_1\underline{\Psi}_s \approx \underline{V}_s \tag{1.3.5}$$

Consequently,

$$E_r = \frac{V_s L_m \omega_2}{L_s \omega_1} \tag{1.3.6}$$

An equivalent circuit based on Equations 1.3.2 and 1.3.6 may be built (Figure 1.18).

Basically, the emf $\underline{E}_r'$ varies with ω_2 (i.e., with speed for constant ω_1) and with the stator voltage V_s.

The "synchronous" reactance of the machine is, in fact, the short-circuit reactance.

Therefore, the voltage regulation is reasonably small, and the transient response is expected to be swift; a definite asset for excitation control.

As the frequency in the rotor is large ($\omega_2 > \omega_1$), the core losses in the machine have to be considered.

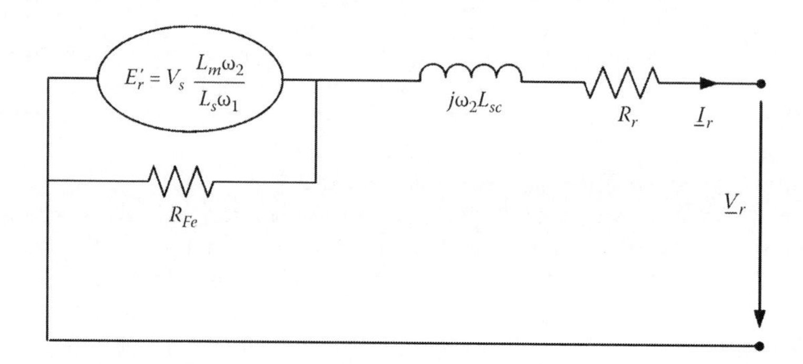

FIGURE 1.18 Equivalent circuit (phase) for wound-rotor induction generator (WRIG) as an exciter source.

One way of doing this is to "hang" a core resistance R_{Fe} in parallel with the emf E_r', R_{Fe} may be taken as a constant, to be determined either from measured or calculated core losses P_{Fe}:

$$p_{Fe} = 3\frac{E_r'^2}{R_{Fe}} \tag{1.3.7}$$

1.9 Losses and Efficiency of WRIGs

The loss components in WRIGs may be classified as follows:

- Stator-winding losses
- Stator core losses
- Rotor-winding losses
- Mechanical losses

The stator-winding losses are due to alternative currents flowing into the stator windings. With constant frequency ($f_1 = 50(60)$ Hz), only in medium and large power machines is the skin effect important. Roebel bars may be used in large power WRIGs to keep the influence of the skin effect coefficient below 0.33 (i.e, 33% additional losses):

$$p_{cos} = 3I_s^2(R_s)_{DC}\cdot\left(1 + K_{skin}^s\right) \tag{1.86}$$

In the rotor, the frequency $f_2 = Sf_1$, and, with WRIGs, $|f_2| < 0.3f_1$. The rotor-to-stator turn ratio a_{rs} is chosen to be larger than 1 ($a_{rs} = 1/|S_{\max}|$) for low stator voltage WRIGs (up to 2–3 MW). In this case, the skin effect in the rotor is negligible.

However, in large machines, as the rotor voltage will probably not go over 4–6 kV (line voltage), even in the presence of specially built slip-rings, the rotor currents are large, in the range of thousands of amperes, again, transposed conductors are needed for the rotor windings. There will be some skin effect, but, as the rotor frequency $|f_1| < 1/3f_1$, in general, its influence will be less important than in the stator ($K_{skin}^r < K_{skin}^s < 0.3$):

$$p_{cor} = 3I_{r1}^2(R_r)_{DC}\cdot\left(1 + K_{skin}^r\right) \tag{1.87}$$

For details on skin effect, see Chapter 7 in *Synchronous Generators*.

The fundamental stator core and rotor core losses may be approximated by an aggregated core-loss resistance R_{Fe}:

$$R_{Fe} = R_{Fes}(\omega_1) + R_{Fer}(\omega_2) \tag{1.88}$$

This is exposed to the airgap emf E_m:

$$\underline{E}_m = -jX_m I_m;\quad \underline{I}_m = \underline{I}_s + \underline{I}_r \tag{1.89}$$

Therefore,

$$p_{Fes} = \frac{3(X_m I_m)^2}{R_{Fe}(\omega_1)} \tag{1.90}$$

$$p_{Fer} = \frac{3(X_m I_m)^2}{R_{Fer}(\omega_2)} \tag{1.91}$$

The values of stator and rotor core loss resistances R_{Fe} and R_{Fer} may be obtained through experiments or from the design process.

When $|\omega_2| < \omega_1$, the rotor core losses are definitely smaller than those in the stator. This is not so when the WRIG is used as an exciter ($|\omega_2| \gg \omega_1$), and thus, even though the core volume is larger in the stator, the rotor core losses are larger.

Additional losses occur in the stator and rotor windings in relation to the circuit time harmonics due (mainly) to the static power converter connected to the rotor. They are strongly dependent on the PWM strategy and on the switching frequency.

Additional core losses occur due to space and time harmonics in the mmf of stator and rotor windings, in the presence of double slotting. Current time harmonics bring additional core losses.

The additional space harmonics core losses occur on the rotor and stator surface toward the airgap. Generally, only the first slot harmonics $\nu_{ss} = (N_s/p_1) \pm 1$, $\nu_{rs} = (N_r/p_1) \pm 1$, as influenced by the corresponding first-order airgap magnetic conductance harmonics, are considered to produce surface core losses that deserve attention [2]. Current time harmonics, on the other hand, produce additional core losses mainly along a thin layer along the slot walls.

Mechanical losses include ventilator (if any) losses, bearing-friction losses, brush-friction losses, and windage losses (P_{mec}):

$$\eta = \frac{P_s + P_r}{P_m} = \frac{P_s + P_r}{P_s + P_r + \sum p} \tag{1.92}$$

$$\sum p = p_{cos} + p_{cor} + p_{Fe} + p_s + p_{mec} + p_{sr}$$

For generating, P_s, in Equation 1.92, is always considered positive (delivered), while P_r is positive (delivered) for supersynchronous operation, and $P_r < 0$ (absorbed) for subsynchronous operation. P_m is the mechanical (input) power. The slip-ring losses are denoted by p_{sr}, and the strayload losses are denoted by p_s.

For details on efficiency (through iso-efficiency curves), see Reference 9.

1.10 Summary

- WRIGs are provided with three-phase AC windings on the rotor and on the stator. WRIGs are also referred to as DFIGs or DOIGs [10–12].
- WRIGs are capable of producing constant frequency (f_1) and voltage stator output power at variable speed if the rotor windings are controlled at variable frequency (f_2) and variable voltage. The rotor frequency f_2 is determined solely by speed n (rps) and $f_1 : f_2 = f_1 - np_1$; p_1 equals the number of pole pairs; and p_1 is the same in the stator and in the rotor windings.
- WRIGs may operate as both motors and generators subsynchronously ($n < f_1/p_1$) and supersynchronously ($n > f_1/p_1$), provided the static power converter that supplies the rotor winding is capable of bidirectional power flow.
- The slip is defined as $S = \omega_2/\omega_1 = f_2/f_1$ and is positive for subsynchronous operation and negative for supersynchronous operation, and so is f_2. Negative f_2 means the opposite sequence of phases in the rotor is followed.
- WRIGs are adequate in applications with limited speed control range ($|S_{max}| < 0.2$–0.3), as the rating of the rotor-side static converter is around $P_{SN}|S_{max}|$, where P_{SN} is the rated SP. The electric power P_r in the rotor is delivered for generating in supersynchronous operation and is absorbed in subsynchronous operation: $P_r \approx P_{SN}^* S$. The total maximum power P_t delivered supersynchronously is thus,

$$P_t = P_s + P_r = P_{SN}\left(1 + |S_{max}|\right)$$

- Consequently, in supersynchronous operation, WRIGs can produce significantly more total electric power than the rated power at synchronous speed ($S = 0$).
- WRIGs may also operate at synchronism, as a standard synchronous machine, provided the rotor-side static power converter is able to handle DC power. Back-to-back voltage source PWM converters are adequate for the scope. It may be argued that, in this case ($S = 0$), a WRIG acts like a damperless SG. True, but this apparent disadvantage is compensated for by the presence of fast close-loop control of active and reactive power, which produces the necessary damping any time the machine deviates from synchronism. WRIG is also adequate to work as a synchronous condensator and contribute massively, when needed, to voltage control and stability in the power grid.
- WRIGs have laminated iron cores with uniform slots to host the AC windings. Integer q (slots/pole/phase) windings are used. Open slots may be used only on one side of the airgap. Axial cooling unistack cores are now in use up to 2–3 MW, while axial–radial cooling multistack cores are necessary above 3 MW.
- To avoid parasitic synchronous torque, it suffices to have different numbers of slots in the rotor and the stator. With large q and chorded coils, the main mmf harmonics are reduced and also reduced are their asynchronous parasitic torques.
- The rotor-to-stator turn ratio a_{rs} may be chosen as unity, but in this case, a voltage matching transformer is needed between the static converter in the rotor and the local power grid. Alternatively, $a_{rs} = 1/|S_{max}| > 1$ when the transformer is eliminated.
- A WRIG may be magnetized either from the stator or from the rotor, so the magnetization curve may be calculated (or measured) from both sides.
- When the reactive power is delivered through the rotor (overexcitation), the stator may operate at the unity power factor. The lagging power factor in the stator seems to be a moderate to large burden on the rotor-side static converter kilovoltampere rating.
- A minimum kilovoltampere rating of the rotor-side static power converter is obtained when the SP factor is leading (underexcitation $\Psi_r < \Psi_s$). Ψ_r and Ψ_s are, respectively, the rotor and the stator flux linkage amplitudes per phase.
- For operation at the power grid, synchronization is required. However, synchronization is much faster and easier than with SGs, because it may be performed at any speed $\omega_r > \omega_1\left(1-|S_{max}|\right)$ by controlling the rotor-side converter in the synchronization mode to make the power grid and stator voltages of the WRIG equal to each other and in phase. The whole synchronization process is short, as the rotor voltage and frequency (phase) are controlled quickly by the static power converter without any special intervention by the prime mover's governor.
- The values of active power and reactive powers P_s, Q_s, P_r, Q_r, vs. the rotor voltage (power) angle δ are somewhat similar to those in the case of cylindrical rotor SGs, but additional asynchronous power terms are present, and the stable operation zones depend heavily on the value and sign of slip (Figure 1.12). However, the decoupled active and reactive power control (see Chapter 2) eliminates such inconveniences to a great extent.
- The peak value of synchronous power components in P_s, P_r, for constant rotor flux, depend on the short-circuit reactance (impedance) of the machine. Voltage regulation is moderate, for the same reason.
- The reactive power Q_r^r, absorbed from the rotor-side converter at $f_2 = S_1 f_1$ frequency, is "magnified" in the machine to the frequency f_1, $Q_r = Q_r^r / |S|$, as it has to produce the magnetic energy stored in the short-circuit and magnetization inductances. Operation at unity power factor in the stator at full power leads, thus, to a moderate increase in rotor-side static converter kilovoltampere rating for $|S_{max}| < 0.25$.
- The WRIG may also operate as a standalone generator. It was demonstrated that such an operation is preferred for low reactive power requirements at low negative slips. Constant frequency, constant voltage output in the stator with autonomous load does not seem to be advantageous

when the speed varies by more than ±5%. Ballast loads may be handled at any speed effectively, at smaller slip.

- With the stator short-circuited, the WRIG may be run as a motor to start the prime mover, say, for pumping in a pump-storage plant.
- After acceleration to $\omega_r > \omega_1 \cdot \left(1-\left|S_{max}\right|\right)$, the stator circuit is opened, the sequence of rotor voltages is changed, and their frequency f_2 and amplitude are reduced to produce the conditions necessary for quick stator synchronization. After that, motoring or generating operation is commanded subsynchronously or supersynchronously.
- The WRIG may operate in the brushless exciter mode to produce DC power on the rotor side with a diode rectifier and thus feed the excitation of a synchronous machine from zero speed up to the desired speed.
- The stator is supplied through a static voltage changer (soft starter type) at constant frequency ω_1, while the rotor moves such that the rotor frequency $\omega_2 = \omega_1 + |\omega_r| > \omega_1$. With $\omega_2/\omega_1 = 3, 4$, good performance is obtained. In all situations, the magnetization (reactive power) is delivered through the stator, but most of the load active power comes from the shaft mechanical power, and only a small part comes from the stator. At zero speed, however, all the excitation power is delivered by the stator, electrically.
- When the speed increases, for constant rotor voltage, the stator voltage of the WRIG exciter is reduced considerably. So, there is room for excitation forcing needs in the SG, provided the WRIG exciter insulation can handle the voltage. The internal impedance of WRIG for brushless exciter mode is, again, the short-circuit impedance. Thus, the commutation of diode reduction of the DC output voltage should be moderate.
- Besides fundamental winding and core losses, additional losses occur in the windings and magnetic cores of WRIGs due to space and time harmonics.
- The WRIG was proven to be reliable for delivering power at variable speed with very fast decoupled active and reactive control in industry up to 400 MW/unit. It is yet to be seen if the WRIG will get a large share in the electric power generation of the future, at low, medium, and high powers per unit.
- Swapping rotor with stator roles by using a rotary transformer, at full power and grid/load frequency, to eliminate the brushes may prove practical if the speed is low or the output/load frequency is large (400 Hz).
- DFIG has been recently introduced in industry (at 100 MVA/unit) to connect two electric energy systems of slightly different frequencies.

References

1. T. Kuwabara, A. Shibuya, H. Feruta, E. Kita, and K. Mitsuhashi, Design and dynamic response characteristics of 400 MW adjustable speed pump storage unit OHKAWACHI station, *IEEE Trans.*, EC-11(2), 1996, 376–384.
2. I. Boldea and S.A. Nasar, *Induction Machine Handbook*, CRC Press, Boca Raton, FL, 2001, p. 327.
3. I. Boldea and S.A. Nasar, *Electric Drives*, CRC Press, Boca Raton, FL, 1998, Chapter 14.
4. J. Tscherdanze, Theory of double-fed induction machine, *Archiv fur electrotechnik*, 15, 1925, 257–263 (in German).
5. A. Leonhard, Asynchronous and synchronous running of the general doubly fed three phase machine, *Archiv fur Electrotechnik*, 30, 1936, 483–502 (in German).
6. F.J. Bradly, A mathematical model for the doubly-fed wound rotor generator, *IEEE Trans.*, PAS-103(4), 1998, 798–802.
7. M.S. Vicatos and J.A. Tegopoulos, Steady state analysis of a doubly-fed induction generator under synchronous operation, *IEEE Trans.*, EC-4(3), 1989, 495–501.

8. I. Cadirci and M. Ermi, Double output induction generator operating at subsynchronous and supersynchronous speeds: Steady state performance optimization and wind energy recovery, *Proc. IEE*, 139B(5), 1992, 429–442.
9. A. Masmoudi, A. Toumi, M.B.A. Kamoun, and M. Poloujadoff, Power flow analysis and efficiency optimization of a doubly fed synchronous machine, *EMPS J.*, 21(4), 1993, 473–491.
10. D.G. Dorrel, Experimental behavior of unbalanced magnetic pull in three-phase induction motors with excentric rotors and the relationship to teeth saturation, *IEEE Trans.*, EC-14(3), 1999, 304–309.
11. G. Abad, J. López, M. Rodríguez, L. Marroyo, and G. Iwanski, *Doubly Fed Induction Machine: Modeling and Control for Wind Energy Generation Applications*, Wiley, IEEE, 2011.
12. A. Merkhouf, P. Doyon, and S. Upadhyay, Variable frequency transformer-concept and electromagnetic design evaluation, *IEEE Trans.*, EC-23(4), 2008, 989–996.

2

Wound-Rotor Induction Generators: Transients and Control

2.1 Introduction

Wound-rotor induction generators (WRIGs) are used as variable-speed generators. These generators are connected to a strong or a weak power grid or as motors with the same specifications. Moreover, WRIGs may operate as stand-alone generators at variable speed.

In all these operating modes, WRIGs undergo transients. Transients may be caused by the following:

- Prime mover torque variations for generator mode
- Load machine torque variations for motor mode
- Power grid faults for generator mode
- Electric load variations in stand-alone generator mode

During transients, in general, speed, voltage and current amplitudes, power, torque, and frequency vary in time, until they eventually stabilize to a new steady state.

Dynamic models for typical prime movers (Chapter 3), such as hydraulic, wind, or steam (gas) turbines or internal combustion engines, are needed to investigate the complete transients of WRIGs.

An adequate WRIG model for transients is imperative, along with closed-loop control systems to provide stability in speed, voltage, and frequency response when the active and reactive power demands are varied.

Typical static power converters capable of up to four-quadrant operation (super- and undersynchronous speed) also need to be investigated as a means for WRIG control for constant stator voltage and frequency, for limited variable speed range.

Vector or direct power control methods with and without motion sensors are described, and sample transient response results are obtained. Behavior during power grid faults is also explored, as, in some applications, WRIGs are not to be disconnected during faults, in order to contribute quickly to power balance in the power grid right after fault clearing. Let us now proceed to tackle these issues one by one.

2.2 WRIG Phase Coordinate Model

The WRIG is provided with laminated stator and rotor cores with uniform slots in which three-phase windings are placed (Figure 2.1). Usually, the rotor winding is connected to copper slip-rings. Brushes on the stator collect (or transmit) the rotor currents from (to) the rotor-side static power converter. For the time being, the slip-ring brush system resistances are lumped into rotor phase resistances, and the converter is replaced by an ideal voltage source.

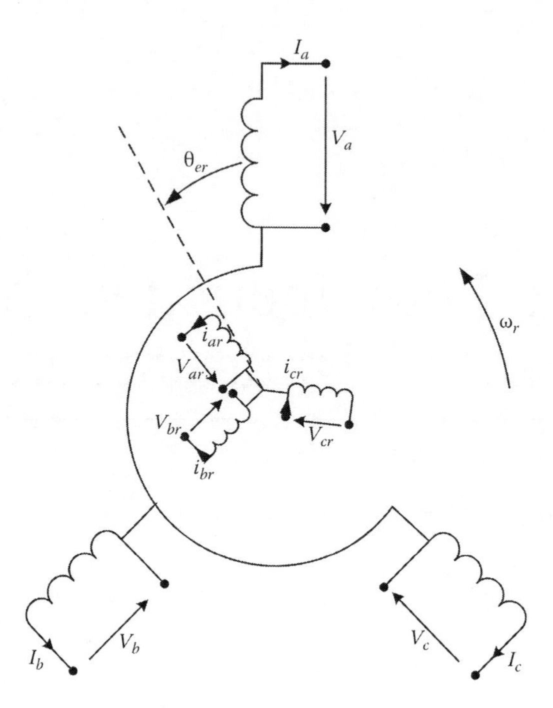

FIGURE 2.1 Wound-rotor induction generator (WRIG) phase circuits.

As already pointed out in Chapter 1, the windings in slots produce a quasi-sinusoidal flux density distribution in the rather uniform air gap (slot openings are neglected). Consequently, the main flux self-inductances of various stator and, rotor phases, respectively, are independent of rotor position. The stator–rotor phase main flux mutual inductances, however, vary sinusoidally with rotor position θ_{er}. The mutual inductances between stator phases are also independent of rotor position, as the air gap is basically uniform. The same is valid for mutual inductances between rotor phases. When mentioning stator and rotor phases and leakage inductances, all phase circuit parameters are included, with the exception of parameters to account for core losses (fundamental and stray load core losses). Winding stray load losses are basically caused by frequency effects in the windings and may be accounted for in the phase resistance formula. Therefore, the phase coordinate model of WRIG is straightforward:

$$
\begin{aligned}
I_aR_s + V_a &= -\frac{d\Psi_a}{dt} \quad & I_{ar}R_r + V_{ar} &= -\frac{d\Psi_{ar}}{dt}\\
I_bR_s + V_b &= -\frac{d\Psi_b}{dt} \quad & I_{br}R_r + V_{br} &= -\frac{d\Psi_{br}}{dt}\\
I_cR_s + V_c &= -\frac{d\Psi_c}{dt} \quad & I_{cr}R_r + V_{cr} &= -\frac{d\Psi_{cr}}{dt}
\end{aligned}
\tag{2.1}
$$

The stator equations are written in stator coordinates and the rotor equations are written in rotor coordinates, and this explains the absence of motion-induced voltages. Generator mode association of voltage signs for both the stator and the rotor is evident. Therefore, delivered electric powers are positive.

We may convert Equation 2.1 into matrix form:

$$
\left|i_{abc,a_rb_rc_r}\right|\left|R_{abca_rb_rc_r}\right| + \left|V_{abca_rb_rc_r}\right| = -\frac{d\left|\Psi_{abca_rb_rc_r}\right|}{dt}
\tag{2.2}
$$

$$
\begin{aligned}
\left|R_{abca_rb_rc_r}\right| &= Diag\left|R_s, R_s, R_s, R_r^r, R_r^r, R_r^r\right| \\
\left|V_{abca_rb_rc_r}\right| &= Diag\left|V_a, V_b, V_c, V_{ar}^r, V_{br}^r, V_{cr}^r\right|^T \\
\left|I_{abca_rb_rc_r}\right| &= Diag\left|I_a, I_b, I_c, I_{ar}^r, I_{br}^r, I_{cr}^r\right|^T \\
\left|\Psi_{abca_rb_rc_r}\right| &= Diag\left|\Psi_a, \Psi_b, \Psi_c, \Psi_{ar}^r, \Psi_{br}^r, \Psi_{cr}^r\right|^T
\end{aligned}
\tag{2.3}
$$

The relationship between the flux linkages and currents is expressed as follows:

$$
\left|\Psi_{abca,b_r,c_r}\right| = \left|L_{abca_rb_rc_r}(\theta_{er})\right|\left|i_{abca_rb_rc_r}\right| \tag{2.4}
$$

$$
L_{abcqb_rc_r}(\theta_{er}) = \begin{bmatrix}
L_{sl}+L_{os} & \frac{L_{os}}{2} & \frac{L_{os}}{2} & M\cos\theta_{er} & M\cos\left(\theta_{er}+\frac{2\pi}{3}\right) & M\cos\left(\theta_{er}-\frac{2\pi}{3}\right) \\
-\frac{L_{os}}{2} & L_{sl}+L_{os} & \frac{L_{os}}{2} & M\cos\left(\theta_{er}-\frac{2\pi}{3}\right) & M\cos\theta_{er} & M\cos\left(\theta_{er}+\frac{2\pi}{3}\right) \\
\frac{L_{os}}{2} & \frac{L_{os}}{2} & L_{sl}+L_{os} & M\cos\left(\theta_{er}+\frac{2\pi}{3}\right) & M\cos\left(\theta_{er}-\frac{2\pi}{3}\right) & M\cos\theta_{er} \\
M\cos\theta_{er} & M\cos\left(\theta_{er}-\frac{2\pi}{3}\right) & M\cos\left(\theta_{er}+\frac{2\pi}{3}\right) & L_{rl}+L_{or} & -\frac{L_{or}}{2} & -\frac{L_{or}}{2} \\
M\cos\left(\theta_{er}+\frac{2\pi}{3}\right) & M\cos\theta_{er} & M\cos\left(\theta_{er}-\frac{2\pi}{3}\right) & -\frac{L_{or}}{2} & L_{rl}+L_{or} & -\frac{L_{or}}{2} \\
M\cos\left(\theta_{er}-\frac{2\pi}{3}\right) & M\cos\left(\theta_{er}+\frac{2\pi}{3}\right) & M\cos\theta_{er} & -\frac{L_{or}}{2} & -\frac{L_{or}}{2} & L_{rl}+L_{or}
\end{bmatrix} \tag{2.5}
$$

The constant mutual inductances on the stator and the rotor are $-L_{os}/2$ and $-L_{or}/2$, because they are derived from $L_{os}\cos(2\pi/3)$ and $L_{or}\cos(2\pi/3)$, respectively, on account of assumed sinusoidal winding (inductance) distributions. The electromagnetic torque may be derived from Equation 2.2 after multiplication by $(i_{abca_rb_rc_r})^T$ using the principle of power balance:

$$
\begin{aligned}
\left|V_{abcarbrcr}\right|\cdot\left|I_{abcarbrcr}\right| = &-R_{abcarbrcr}\cdot\left|I_{abcarbrcr}\right|^2 - \frac{d}{dt}\left(\frac{1}{2}\left|I_{abcarbrcr}\right|^T\left|L_{abcarbrcr}(\theta_{er})\right|\left|I_{abcarbrcr}\right|\right) \\
&-\frac{1}{2}\left|I_{abcarbrcr}\right|T\frac{d}{d\theta_{er}}\left|L_{abcarbrcr}\right|\left|I_{abcarbrcr}\right|\frac{d\theta_{er}}{dt}
\end{aligned} \tag{2.6}
$$

The "substantial" (total) derivative d_s/dt masks the second term of Equation 2.6, which represents the stored magnetic energy variation in time, while the third term is the electromagnetic (electric) power P_{elm}, which crosses the air gap from the rotor to the stator or vice versa.

The first term in Equation 2.6 is the winding losses. P_{elm} should be positive for generating:

$$
P_{elm} = -\frac{1}{2}\left|I_{abca_rb_rc_r}\right|^T\frac{d}{d\theta_{er}}\left|L_{abca_rb_rc_r}(\theta_{er})\right|\left|I_{abca_rb_rc_r}\right|\omega_r = -T_e\frac{\omega_r}{p_1};\quad \omega_r = \frac{d\theta_{er}}{dt} \tag{2.7}
$$

For generating, with $P_{elm} > 0$, the electromagnetic torque T_e has to be negative (for braking the rotor):

$$
T_e = +\frac{p_1}{2}\left|I_{abca_rb_rc_r}\right|^T\left|\frac{dL_{abca_rb_rc_r}(\theta_{er})}{d\theta_{er}}\right|\left|I_{abca_rb_rc_r}\right| \tag{2.8}
$$

The motion equations are as follows:

$$J\frac{d\omega_r}{dt} = T_{mech} + T_e \frac{d\theta_{er}}{dt} = \omega_r \tag{2.9}$$

with $T_e < 0$ and $T_{mech} > 0$ for generating and $T_e > 0$ and $T_{mech} < 0$ for motoring. An eighth-order system of first-order differential equations was obtained. Some of its coefficients are dependent on rotor position θ_{er}, that is, of time.

Such a complex model, where, in addition, magnetic saturation (implicit in L_{os}, L_{or}, and M) is not easy to account for, is to be used mainly for asymmetrical (unbalanced) conditions in the power supply, in the static power converter, or in the parameters (short-circuited coils in one phase or between phases).

2.3 Space-Phasor Model of WRIG

For stability computation or control equation system design, the phase coordinate (variable) model has to be replaced with the recent widely accepted space-phasor (vector, or d–q) model obtained through the modified Park complex transformation [1]:

$$\begin{aligned}\overline{I}_s^{\,b} &= I_d^b + jI_q^b = \frac{2}{3}\left(i_a + i_b e^{j(2\pi/3)} + i_c e^{-j(2\pi/3)}\right)e^{-j\theta_b}\\ \overline{I}_r^{\,b} &= I_{dr}^b + jI_{qr}^b = \frac{2}{3}\left(i_{ar} + i_{br} e^{j(2\pi/3)} + i_{cr} e^{-j(2\pi/3)}\right)e^{-j(\theta_b - \theta_{er})}\end{aligned} \tag{2.10}$$

The same transformation, in general, orthogonal coordinates, rotating at the general electric speed $\omega_b = (d\theta_b/dt)$, is valid for voltages and flux linkages $\overline{V}_s^{\,b}, \overline{V}_r^{\,b}, \overline{\Psi}_s^{\,b}, \overline{\Psi}_r^{\,b}$. The space phasors represent the three-phase induction machine (IM) completely, only if one more variable component in the stator and in the rotor are introduced. This is the so-called zero-sequence (homopolar) component:

$$\begin{aligned}I_{os} &= \frac{1}{3}(i_a + i_b + i_c)\\ I_{or} &= \frac{1}{3}(i_{ar} + i_{br} + i_{cr})\end{aligned} \tag{2.11}$$

The zero-sequence component, which is inherent to the $dq0$ model (see Chapter 6), is independent from the others and does not, in general, participate in the electromagnetic power production.

The inverse transform is as follows:

$$i_a(t) = \text{Real}\left|\overline{I}_s^{\,b} e^{j\theta_b}\right| + I_{os} \tag{2.12}$$

$$i_{ar}(t) = \text{Real}\left|\overline{I}_r^{\,b} e^{j(\theta_b - \theta_{er})}\right| + I_{or} \tag{2.13}$$

To proceed from the phase-coordinate (variable) to the space-phasor model, let us first reduce the rotor to stator variables:

$$\begin{aligned}&\frac{M}{L_{os}} = \frac{W_2 k_{w2}}{W_1 k_{w1}} = K_{rs}; \quad \frac{L_{or}}{L_{os}} = K_{rs}^2; \quad I_{ar} = I_{ar}^r \cdot K_{rs}\\ &L_{rl} = \frac{L_{rl}^r}{K_{rs}^2}; \quad R_r = \frac{R_r}{K_{rs}^2}; \quad V_{ar} = \frac{V_{ar}^r}{K_{rs}}\end{aligned} \tag{2.14}$$

The transformations in Equation 2.14 "replace" the actual rotor winding with an equivalent one with the same number of turns and slots as that of the stator, while conserving the losses in the windings, the rotor electric power input, and the magnetic energy stored in the leakage inductances.

By applying the space-phasor transformations, after the reduction to stator variables, we simply obtain the voltage currents and flux/current relations for the space-phasor model:

$$\begin{aligned}
&\bar{I}_s R_s + \bar{V}_s = -\frac{d\bar{\Psi}_s}{dt} - j\omega_b \bar{\Psi}_s \\
&\bar{I}_r R_s + \bar{V}_r = -\frac{d\bar{\Psi}_r}{dt} - j(\omega_b - \omega_r)\bar{\Psi}_s \\
&\bar{\Psi}_s = L_s \bar{I}_s + L_m \bar{I}_r; \quad L_s = L_{sl} + L_m; \quad L_m = \frac{3}{2} L_{os} \\
&\bar{\Psi}_r = L_r \bar{I}_r + L_m \bar{I}_s; \quad L_r = L_{rl} + L_m
\end{aligned} \tag{2.15}$$

The torque may be derived through the power balance principle, either from the stator or from the rotor equation:

$$\frac{3}{2}\text{Real}(\bar{I}_s^* \bar{V}_s^*) = -\frac{3}{2} R_s I_s^2 - \text{Real}\left(\frac{3}{2} \bar{I}_s^* \frac{d\bar{\Psi}_s}{dt}\right) - \text{Real}\left(\frac{3}{2} j\omega_b \bar{I}_s \bar{\Psi}_s\right) \tag{2.16}$$

The electromagnetic power is represented by the last term:

$$P_{elm} = -T_e \frac{\omega_b}{p_1} = -\frac{3}{2}\omega_b \text{Imag}[\bar{\Psi}_s \bar{I}_s^*] \tag{2.17}$$

Therefore, the electromagnetic torque is

$$T_e = \frac{3}{2} p_1 \text{Imag}[\bar{\Psi}_s \bar{I}_s^*] = \frac{3}{2} p_1 (\Psi_d i_q - \Psi_q i_d) = -\frac{3}{2} p_1 (\Psi_{dr} i_{qr} - \Psi_{qr} i_{dr}) \tag{2.18}$$

Again, $Te < 0$ for generating. The factor 3/2 stems from the complete power balance between the three-phase machine and its space-phasor model. The superscript b has been dropped for simplicity in writing.

The motion equations are the same as in Equation 2.9. The space-phasor model is to be completed with the zero-sequence equations that also result from these transformations:

$$\begin{aligned}
&I_{so} R_s + V_{so} \approx -L_{sl} \frac{di_{so}}{dt} \\
&I_{ro} R_r + V_{ro} \approx -L_{rl} \frac{di_{ro}}{dt}
\end{aligned} \tag{2.19}$$

The zero sequence is irrelevant for the power transfer by the magnetomotive force (mmf) fundamental, but it produces additional stator (rotor) losses. For star connection or for symmetric transients or steady-state modes, they are, however, zero, as the sum of the phase currents is zero. The instantaneous active and reactive powers P_s, Q_s, P_r', Q_r', from the stator and the rotor are as follows:

$$\begin{aligned}
&P_s = \frac{3}{2}\text{Real}(\bar{V}_s \bar{I}_s^* + 2\bar{V}_{so}\bar{I}_{so}^*) = \frac{3}{2}[(V_d i_d + V_q i_q) + 2\,\text{Real}(\bar{V}_{so}\bar{I}_{so}^*)] \\
&Q_s = \frac{3}{2}\text{Imag}(\bar{V}_s \bar{I}_s^* + 2\bar{V}_{so}\bar{I}_{so}^*) = \frac{3}{2}[(V_d i_q - V_q i_d) + 2\,\text{Imag}(\bar{V}_{so}\bar{I}_{so}^*)] \\
&P_r^r = \frac{3}{2}\text{Real}(\bar{V}_r \bar{I}_r^* + 2\bar{V}_{ro}\bar{I}_{ro}^*) = \frac{3}{2}[(V_{dr} i_{dr} - V_{qr} i_{qr}) + 2\,\text{Real}(\bar{V}_{ro}\bar{I}_{ro}^*)] \\
&Q_r^r = \frac{3}{2}\text{Real}(\bar{V}_r \bar{I}_r^* + 2\bar{V}_{ro}\bar{I}_{ro}^*) = \frac{3}{2}[(V_{dr} i_{qr} - V_{qr} i_{dr}) + 2\,\text{Imag}(\bar{V}_{ro}\bar{I}_{ro}^*)]
\end{aligned} \tag{2.20}$$

2.4 Space-Phasor Equivalent Circuits and Diagrams

The space-phasor equations (Equations 2.15 and 2.19) may be represented in equivalent circuits that use any combination of two variables from $\overline{\Psi}_s, \overline{I}_s, \overline{\Psi}_r, \overline{I}_r$, with the other two eliminated, based on flux/current relationships (Equation 2.15). Current variables are typical:

$$\overline{\Psi}_s = \overline{\Psi}_m + L_{sl}\overline{I}_s; \quad \Psi_m = L_m(\overline{I}_s + \overline{I}_r)$$
$$\overline{\Psi}_r = \overline{\Psi}_m + L_{rl}\overline{I}_r; \quad \overline{I}_s + \overline{I}_r = \overline{I}_m \tag{2.21}$$

Consequently, Equation 2.15 becomes

$$(R_s + (s + j\omega_b)L_{sl})\overline{I}_s + \overline{V}_s = -L_{mt} \cdot s(\overline{I}_s + \overline{I}_r) - j\omega_b L_m(\overline{I}_s + \overline{I}_r) \tag{2.22}$$

$$(R_r + (s + j(\omega_b - \omega_r))L_{rl})\overline{I}_r + \overline{V}_r = -L_{mt} \cdot s(\overline{I}_s + \overline{I}_r) - j(\omega_b - \omega_r)L_m(\overline{I}_s + \overline{I}_r) \tag{2.23}$$

L_{mt} is the transient magnetization inductance of the WRIG. The equivalent circuit is shown in Figure 2.2.

Magnetic saturation of the main flux path is accounted for in the space-phasor model simply by the functions $L_{mt}(i_m)$ and $L_m(i_m)$, which may be determined experimentally or online. The motion-induced voltages are also visible in the coordinates system rotating at electrical speed ω_b. The coordinates system speed ω_b may be arbitrary:

- Stator coordinates: $\omega_b = 0$
- Rotor coordinates: $\omega_b = \omega_r$
- Synchronous coordinates: $\omega_b = \omega_1$ (stator voltage or current frequency) are preferred—for steady-state and symmetrical stator voltages:

$$V_{abc}(t) = V_s\sqrt{2}\cos\left(\omega_1 t - (i-1)\frac{2\pi}{3}\right) \tag{2.24}$$

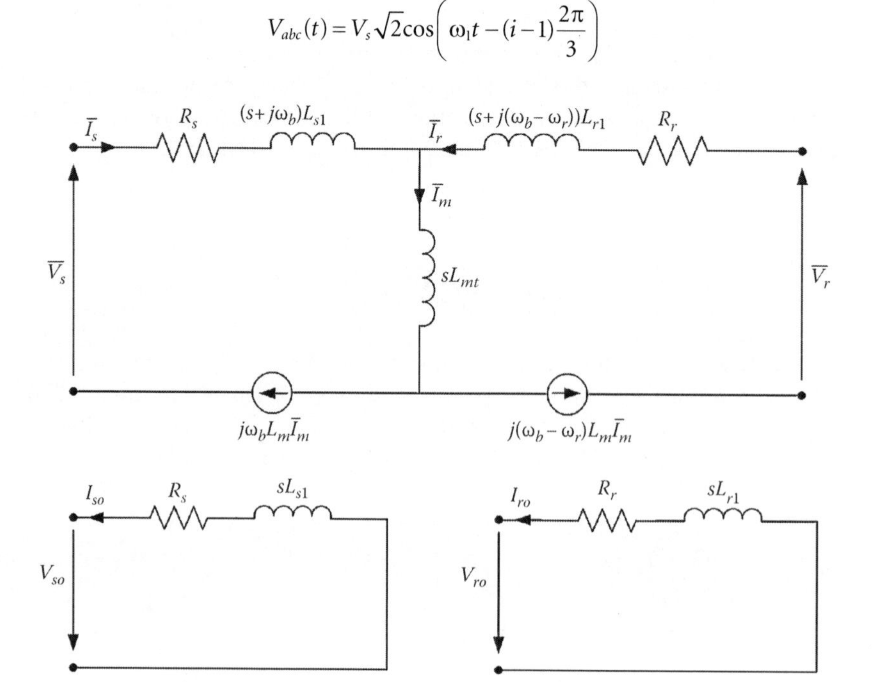

FIGURE 2.2 Space-phasor equivalent circuit of the WRIG.

With Equation 2.10,

$$\overline{V}_s = V_s\sqrt{2}\left[\cos(\omega_1 - \omega_b)t - j\sin(\omega_1 - \omega_b)t\right] \tag{2.25}$$

Also, with

$$V_{a_r b_r c_r}(t) = V_r\sqrt{2}\cos\left((\omega_1 - \omega_r)t + \gamma - (i-1)\frac{2\pi}{3}\right) \tag{2.26}$$

$$\overline{V}_r = V_r\sqrt{2}[\cos((\omega_1 - \omega_b)t + \gamma) - j\sin((\omega_1 - \omega_b)t + \gamma)] \tag{2.27}$$

The actual frequency of the rotor voltages for steady-state ω_2 is known to be $\omega_2 = \omega_1 - \omega_r$ (see also Chapter 1).

The stator and rotor voltage space phasors have the same frequency under steady state: $\omega_1 - \omega_b$.

Therefore, for steady state, $s =$:

- j_{ω_1} in stator coordinates ($\omega_b = 0$)
- j_{ω_2} ($\omega_2 = \omega_1 - \omega_r$) in rotor coordinates ($\omega_b = \omega_r$)
- 0 in synchronous coordinates ($\omega_b = \omega_1$)

At steady state, in synchronous coordinates, the WRIG voltages, currents, and flux leakages are direct current (DC) quantities. Synchronous coordinates are, thus, frequently used for WRIG control. Other equivalent circuits, with $\overline{\Psi}_s$, $\overline{I}_r$ and $\overline{\Psi}_r$, $\overline{I}_s$ pairs as variables, may also be developed but with little gain. For steady state, the space-phasor circuit has the same form irrespective of ω_b as $s = j(\omega_1 - \omega_b)$. And, if and only if magnetic saturation is ignored, $L_{mt} = L_m$ (Figure 2.3).

What differs in steady state when the reference system speed ω_b varies is the frequency of space phasors, which is $\omega_1 - \omega_b$. The equivalent circuit in Figure 2.3 is similar to the per phase equivalent (phasor) circuit in Chapter 1, but it has a distinct meaning. The homopolar (zero-sequence) part still depends on the reference system speed ω_b. The space-phasor model may also be illustrated through the space-phasor voltage diagram (Figure 2.4).

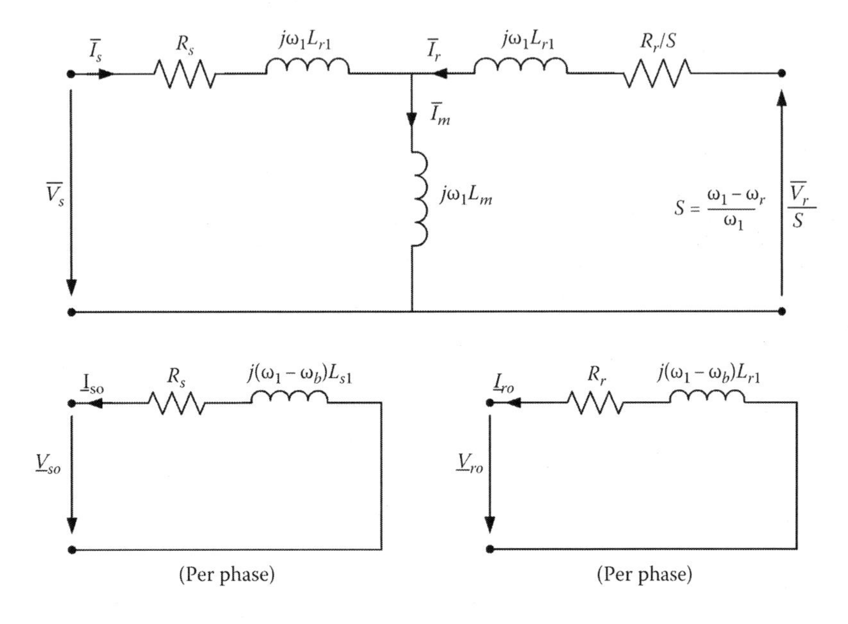

FIGURE 2.3 Steady-state space-phasor circuit model of unsaturated WRIG.

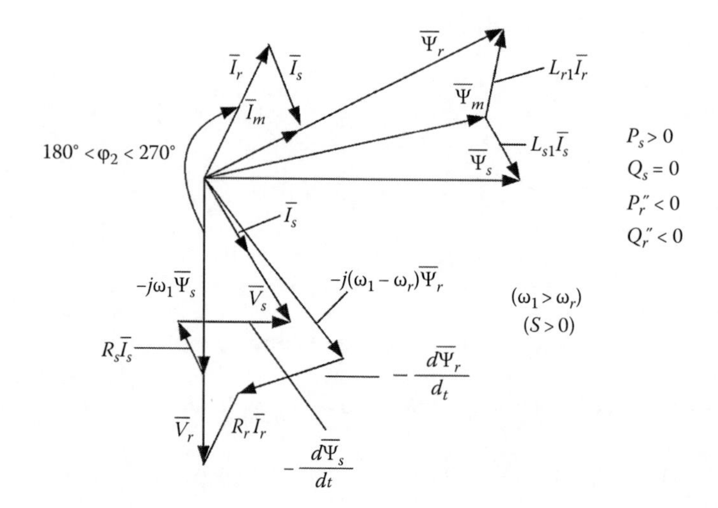

FIGURE 2.4 Space-phasor diagram of WRIG for subsynchronous generator operation at unity stator power factor.

Consider cos $\varphi = 1$ in the stator, generating operation under synchronous speed. (Active and reactive powers are absorbed through the rotor and delivered through the stator.) Consequently, the phase angle between rotor voltage and current φ_2 is $180° < \varphi_2 < 270°$. It is zero between the stator voltage and current, as cos $\varphi_1 = 1$. I, as the magnetization is produced through the rotor, $I_r > I_s$. For negative torque, $\overline{\Psi}_s$ is ahead of $\overline{I}_s$, and $\overline{\Psi}_r$ is behind $\overline{I}_r$ for the same situation. This makes the drawing of the space-phasor diagrams during transients fairly easy, especially if we fix the coordinate system space ω_b, for example, $\omega_b = \omega_1$.

The machine is overexcited, as $\Psi_r > \Psi_s$, to produce unity power factor in the stator, at steady state. During steady state, for synchronous coordinates ($\omega_b = \omega_1$), d/dt terms, along the stator and the rotor, flux linkage derivatives are zero.

Another way of representing the WRIG transients is by the structural (block) diagram. To derive it, we have to choose the pair of variables. The stator and rotor flux linkage space phasors $\overline{\Psi}_s$ and $\overline{\Psi}_r$ seem to be most appropriate, as they lead to a rather simplified structural diagram. First, the stator and the rotor currents are eliminated from Equation 2.15:

$$\overline{I}_s = \frac{\left(\frac{\overline{\Psi}_s}{L_s} - \overline{\Psi}_r\frac{L_m}{L_sL_r}\right)}{\sigma}, \quad \sigma = 1 - \frac{L_m^2}{L_sL_r}$$
$$\overline{I}_r = \frac{\left(\frac{\overline{\Psi}_r}{L_r} - \overline{\Psi}_s\frac{L_m}{L_sL_r}\right)}{\sigma} \tag{2.28}$$

and then,

$$\tau_s'\frac{d\overline{\Psi}_s}{dt} + (1 + j\omega_b\tau_s')\overline{\Psi}_s = -\tau_s'\overline{V}_s + K_r\overline{\Psi}_r$$
$$\tau_s'\frac{d\overline{\Psi}_r}{dt} + (1 + j(\omega_b - \omega_r)\tau_r')\overline{\Psi}_r = -\tau_r'\overline{V}_r + K_s\overline{\Psi}_s \tag{2.29}$$

$$K_s = \frac{L_m}{L_s} \approx 0.9 - 0.97 K_r = \frac{L_m}{L_r} \approx 0.91 - 0.97$$
$$\tau_s' = \frac{L_s}{R_s}\cdot\sigma\tau_r' = \frac{L_r}{R_r}\cdot\sigma T_e = \frac{3}{2}p_1\,\text{Imag}(\overline{\Psi}_s\overline{I}_s^{\,*}) \tag{2.30}$$

As the equations of motion are not included, Equation 2.29 represents the equation for electromagnetic transients (constant speed). Also, in general, $\omega_b = \omega_1 = ct.$ for power grid operation of a WRIG.

There is, as expected, some coupling of the stator and the rotor equations through flux linkages, but the time constants involved may be called the stator and the rotor transient time constants τ_s' and τ_r', both in the order of milliseconds to a few tens of milliseconds for the entire power range of WRIGs.

As the flux linkages can vary quickly, so can the stator and the rotor currents because there is a linear relationship between them (if saturation is neglected).

Equations 2.29 and 2.30 lead to the structural diagram shown in Figure 2.5. The presence of the current calculator in the structural diagram is justified, because, generally, either flux linkage or current (or torque) control is performed.

The space-phasor (vector) model may be easily broken into dq variables based on the following definitions (Equation 2.10):

$$
\begin{aligned}
&\frac{d\Psi_d}{dt} = -V_d - R_s i_d + \omega_b \Psi_q \\
&\frac{d\Psi_q}{dt} = -V_q - R_s i_q - \omega_b \Psi_d \\
&\Psi_{d,q} = L_s I_{d,q} + L_m I_{dr,qr} \\
&\frac{d\Psi_{dr}}{dt} = -V_{dr} - R_r i_d + (\omega_b - \omega_r)\Psi_{qr} \\
&\frac{d\Psi_{qr}}{dt} = -V_{qr} - R_r i_q - (\omega_b - \omega_r)\Psi_{dr} \\
&\Psi_{dr,qr} = L_r I_{dr,qr} + L_m I_{d,q} \\
&T_e = \frac{3}{2} p_1 (\Psi_d I_q - \Psi_q I_d)
\end{aligned}
\tag{2.31}
$$

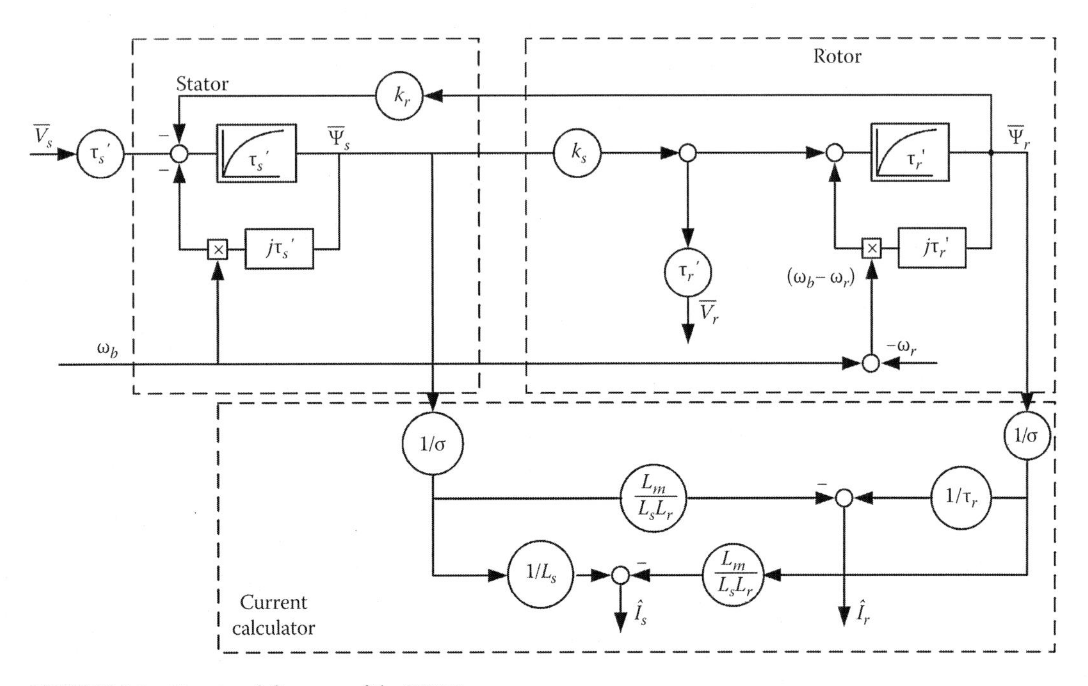

FIGURE 2.5 Structural diagram of the WRIG.

$$\begin{vmatrix} V_d \\ V_q \\ V_{os} \end{vmatrix} = |P(\theta_b)| \begin{vmatrix} V_a \\ V_b \\ V_c \end{vmatrix} \quad \begin{vmatrix} I_d \\ I_q \\ I_{os} \end{vmatrix} = |P(\theta_b)| \begin{vmatrix} I_a \\ I_b \\ I_c \end{vmatrix}$$

$$\begin{vmatrix} V_{dr} \\ V_{qr} \\ V_{or} \end{vmatrix} = \| P(\theta_b - \theta_{er})| \begin{vmatrix} V_{ar} \\ V_{br} \\ V_c \end{vmatrix} \quad \begin{vmatrix} I_{dr} \\ I_{qr} \\ I_{or} \end{vmatrix} = | P(\theta_b - \theta_{er})| \begin{vmatrix} I_{ar} \\ I_{br} \\ I_{cr} \end{vmatrix}$$

$$P(\theta) = \frac{2}{3} \begin{vmatrix} \cos(-\theta) & \cos\left(-\theta + \frac{2\pi}{3}\right) & \cos\left(-\theta - \frac{2\pi}{3}\right) \\ \sin(-\theta) & \sin\left(-\theta + \frac{2\pi}{3}\right) & \sin\left(-\theta + \frac{2\pi}{3}\right) \\ \frac{1}{2} & \frac{1}{2} & \frac{1}{2} \end{vmatrix} \tag{2.32}$$

$$\| P(\theta)|^{-1} = \frac{3}{2} \| P(\theta)|^T$$

$$\frac{d\Psi_{os}}{dt} = -V_{os} - R_s I_{os}, \quad \frac{d\Psi_{or}}{dt} = -V_{or} - R_r I_{or}; \quad \Psi_{os} \approx L_{sl} I_{os}, \quad \Psi_{or} \approx L_{sl} I_{or} \tag{2.33}$$

$$\frac{J}{p_1}\frac{d\omega_r}{dt} = T_e + T_{mech}; \quad T_e < 0 \text{ for generating}$$
$$\frac{d\theta_{er}}{dt} = \omega_r; \quad \omega_r = p_1 \Omega_1 = p_1 \cdot 2\pi n \tag{2.34}$$

The generalized matrices for stator and rotor quantities were included for completeness.

2.5 Approaches to WRIG Transients

First, we must discriminate between the closed-loop controlled and open-loop operation at constant rotor voltage or current, with slip frequency ω_2 adjusted to speed, to preserve constant stator frequency $\omega_1 = \omega_2 + \omega_r$. When connected to a stiff power grid, the stator voltage is fixed in terms of amplitude and phase:

$$V_{abc} = V_s \sqrt{2} \cos\left(\omega_1 t - (i-1)\frac{2\pi}{3}\right) \tag{2.35}$$

with

$$V_{a_r b_r c_r} = V_r \sqrt{2} \cos\left((\omega_1 - \omega_r)t - (i-1)\frac{2\pi}{3} + \gamma_v\right) \tag{2.36}$$

for voltage "open-loop" control, and

$$I_{a_r b_r c_r} = I_r \sqrt{2} \cos\left((\omega_1 - \omega_r)t - (i-1)\frac{2\pi}{3} + \gamma_i\right) \tag{2.37}$$

for rotor "open-loop" control.

After using the Park transformation, in synchronous coordinates $\theta_b = \omega_1 t$, for example, for voltage "open-loop" control,

$$\begin{aligned} V_d &= V_s\sqrt{2} \\ V_q &= 0 \\ V_{dr} &= V_r\sqrt{2}\cos\gamma_V \\ V_{qr} &= -V_r\sqrt{2}\sin\gamma_V \end{aligned} \tag{2.38}$$

For constant stator voltage V_r, the influences of the phase advance of the latter with respect to stator voltage, γ_V, and the speed ω_r (or slip $S = (\omega_1 - \omega_r)/\omega_1$) on transients are paramount. In general, during transients, both electromagnetic and mechanical variables vary in time. However, in very fast transients, the speed may be considered constant; thus, the motion equations (Equation 2.34) may be ignored in the *dq* model.

Laplace transform and linear control techniques may be used in such cases, but they are limited to fast rotor voltage amplitude (V_r) or phase angle (γ_r) variations. Also, a short circuit on the stator side at WRIG terminals or somewhere along a transmission line may be approached in this way [2]. As such cases are straightforward, we leave them out here, while treating them more realistically later, in the presence of speed variation.

The linearization of the *dq* model equations is used to study small deviation transients when the two motion equations are included. Subsequently, the eigenvalue method may be used to investigate small signal stability performance. Alternatively, the simplified synchronizing and damping torque coefficient method may be used for the same scope. For voltage and current (scalar) control, the eigenvalue method was successfully applied to WRIG [3,4] with conclusions such as the following:

- The current-controlled WRIG steady-state stability increases considerably with load, while under voltage control, only a small increase occurs.
- The power angle limits are independent of load.
- Under voltage control, for low values of rotor voltage V_r, only motor operation is stable for subsynchronous speeds ($S > 0$), and only generator operation is stable for supersynchronous speeds ($S < 0$). For current control, motor and generator operation are stable for both positive and negative slip ($S \neq 0$).

To reduce the eigenvalue computation effort, the stator resistance may be neglected. As the presence of stator resistance is known to increase damping, it follows that the results for $R_s = 0$ are conservative (on the safe side) [5].

The dynamic behavior of WRIG may also be approached by solving the *dq* model equations (Equations 2.31 through 2.34) through numerical methods for various perturbations [6].

The introduction of vector control, or, very recently, of direct power control of WRIGs changed the perspectives on WRIG stability and transients, because the new controls tend to linearize the system. Very fast, almost speed-independent, active and reactive power control was obtained this way. The quality of transient response is still paramount, as there are always limitations on the mechanical side or electrical side of the WRIG system.

As the control of WRIG is strongly dependent on the static converter used on the rotor side, we will first dwell on this issue for a while.

2.6 Static Power Converters for WRIGs

The static power converter for WRIGs is connected, in general, to the rotor's three-phase windings through brushes and slip-rings. It is rated approximately at $P_{rN} = |S_{max}|P_{SN}$, where S_{max} is the maximum slip value and P_{SN} is the stator rated electric power. In general, $|S_{max}| < 0.2–0.3$ and decreases with the power rating per unit, reaching less than 0.05–0.1 in the hundreds of megawatt power machines, to limit the static power converter rating. There are a few static power converter configurations suitable for WRIGs:

- The uncontrolled (or controlled rectifier) + current source inverter, with low-cost thyristors (the DC current link AC–AC converter)
- The cycloconverter: a voltage source commutated direct AC–AC converter with limited output/input frequency ratio $\omega_2/\omega_1 < 0.33$, with low-cost thyristors
- The matrix converter: a voltage source direct AC–AC converter with bidirectional insulated gate bipolar transistors (IGBTs) or integrated gate commutated thyristor (IGCTs) (or MOSFET-controlled thysistor (MCTs)) with free output/input frequency ratio
- The forced commutated rectifier-voltage source pulse-width modulated (PWM) inverter with IGBTs or IGCTs (the DC voltage link AC–AC converter)
- The three active switch (Vienna) rectifier—inverter system

All these static converters may be built in two- or multiple-level topologies for low-, respectively, medium-voltage applications.

These configurations differ in terms of costs, two- or four-quadrant operation under subsynchronous and supersynchronous operation, current harmonics content, and response quickness in active and reactive power control of the stator in the WRIG.

Traditionally, the first static converter introduced for WRIG (motor also) comprised a machine-side diode rectifier with a DC choke and a source-side commutated current–source thyristor inverter (Figure 2.6).

When a diode rectifier is used on the machine side, the power flow is unidirectional—from the machine to the power grid via the step-up transformer. It means that the WRIG may operate as a motor undersynchronously ($S > 0$) and as a generator supersynchronously ($S < 0$). That is, two-quadrant operation. In addition, it is not possible to go through or operate at synchronism ($S = 0$).

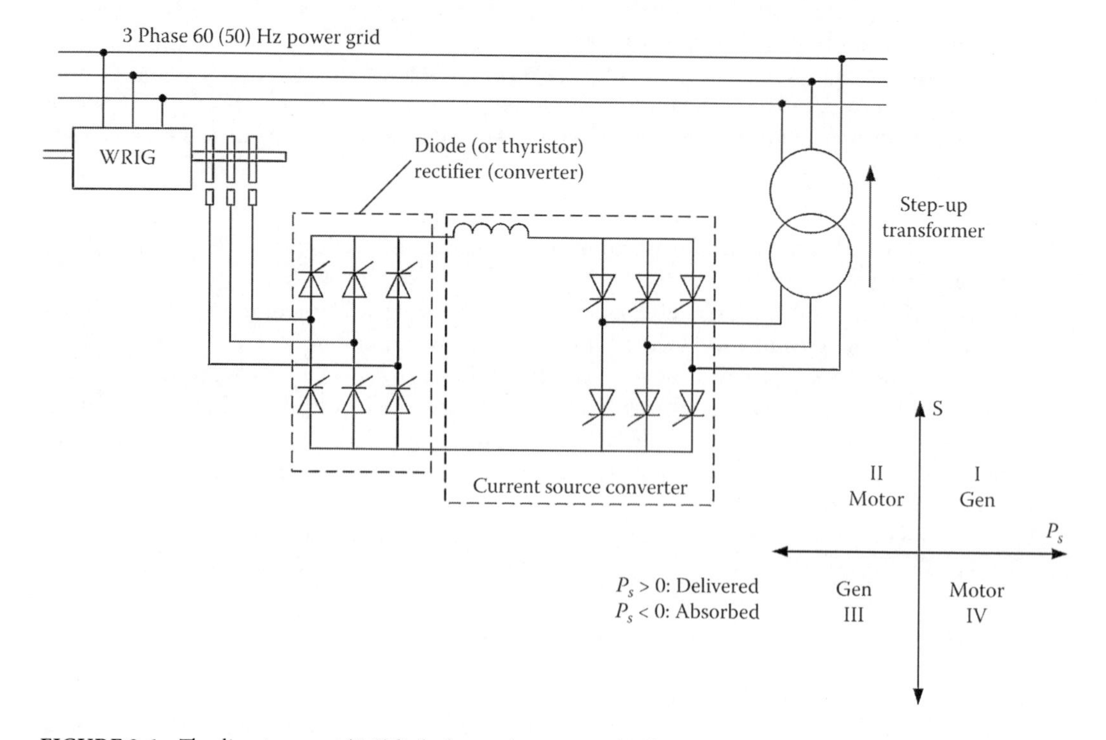

FIGURE 2.6 The direct current (DC) link alternating current (AC)–AC converter.

Finally, the current harmonics content in the rotor and in the stator is rich, and the power factor on the source side is rather modest.

For more flexibility, the current machine-side converter may be made with thyristors to allow bidirectional power flow and thus provide for four-quadrant operation: motoring and generating for both $S > 0$ and $S < 0$. The severity of harmonics content remains. Sustained operation at synchronism ($S = 0$) is not feasible, but going through synchronous speed is feasible with careful control. See References 7,8 for a thorough analysis of the current link AC–AC converter WRIG for four-quadrant operation.

Based on Reference 7, typical experimental waveforms or thyristor voltage V_{thr}, DC link voltage V_{DC}, rotor current, rotor voltage, and line current for supersynchronous and subsynchronous operation are shown in Figure 2.7.

The rotor and stator waveforms and their harmonious spectrum are shown in Figure 2.8 for $S = -0.27$.

The magnitude of supply distortion currents is less than 1% for harmonics higher than fifth and seventh, but the latter may reach 10%. At low slip values, fifth and seventh rotor current harmonics may result in stator subharmonics currents. The harmonics content of supply currents increases with slip.

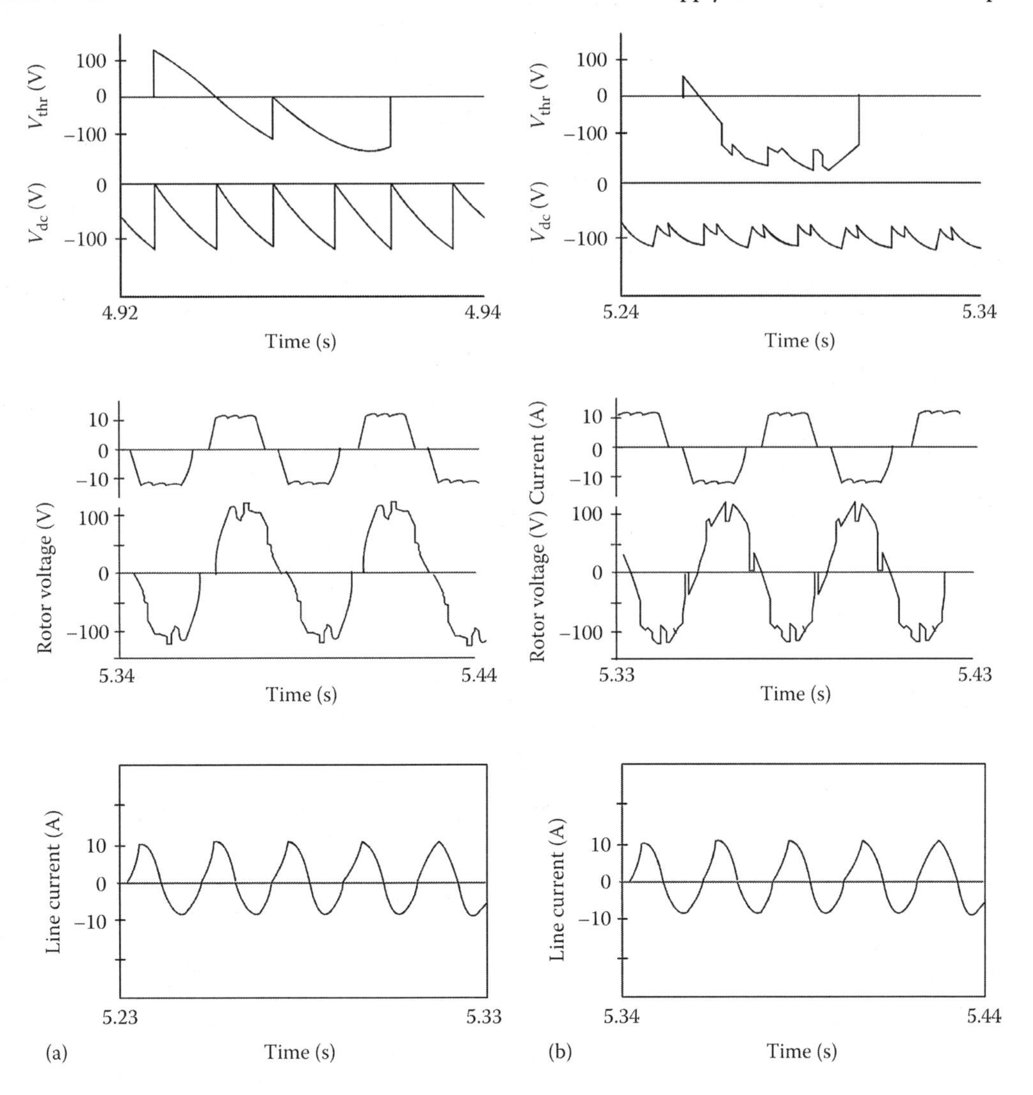

FIGURE 2.7 Direct current (DC) link alternating current AC–AC converter WRIG typical waveforms: (a) $S < 0$ and (b) $S > 0$.

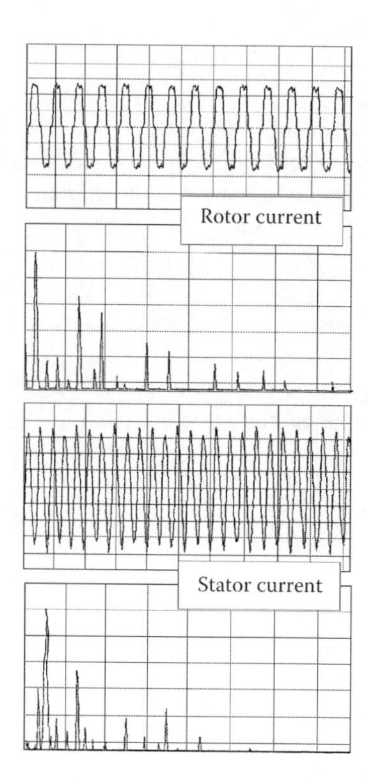

FIGURE 2.8 WRIG with direct current (DC) link alternating current AC–AC converter: rotor and stator current harmonics spectrum at $S = -0.27$.

Overlapping of thyristor commutation and harmonics sets severe limitations on WRIG operation range. Commutation failure bars operation close to (or at) synchronous speed. The indirect AC–AC converter shown in Figure 2.6 has both component converters as source (naturally) commutated. Forced commutation may improve the situation both in terms of faster and safer commutation and in power factor, but the costs become large, and the presence of the large DC choke remains a serious drawback.

2.6.1 Direct AC–AC Converters

Direct AC–AC converters rely, again, on source commutation and are thus rather simple and inexpensive. The number of pulses may be 6, 12, or 18 to reduce current harmonics by using step-down transformers with dual or multiple delta–star (D–Y) connected secondary (Figure 2.9).

Each phase is fed from a double, controlled, rectifier—one for each current polarity—to produce operation in four quadrants (positive and negative output voltage and current). The output voltage (at rotor terminals) is "assembled" from sections of neighboring phase waveforms of the same polarity pertaining to the input (source) voltage at frequency ω_1. The output frequency ω_2 is thus a fraction of ω_1, in practice: $|\omega_2| < \omega_1/3$. This is enough for WRIG operation, however, in case starting and synchronization as a motor are required, this is not enough, as the synchronization should take place at much higher speeds than obtainable with rotor supply and short-circuited stator motoring. Reactive power for the commutation of thyristors is drawn from the power source, and, if magnetization through rotor (unity stator power factor) is desired, overrating of the cycloconverter and transformer is required. In special configurations, ω_2 could be increased further, and this means to add capacitors for reactive power production. But in this case, the cycloconverter's essential advantages of low cost and reliability

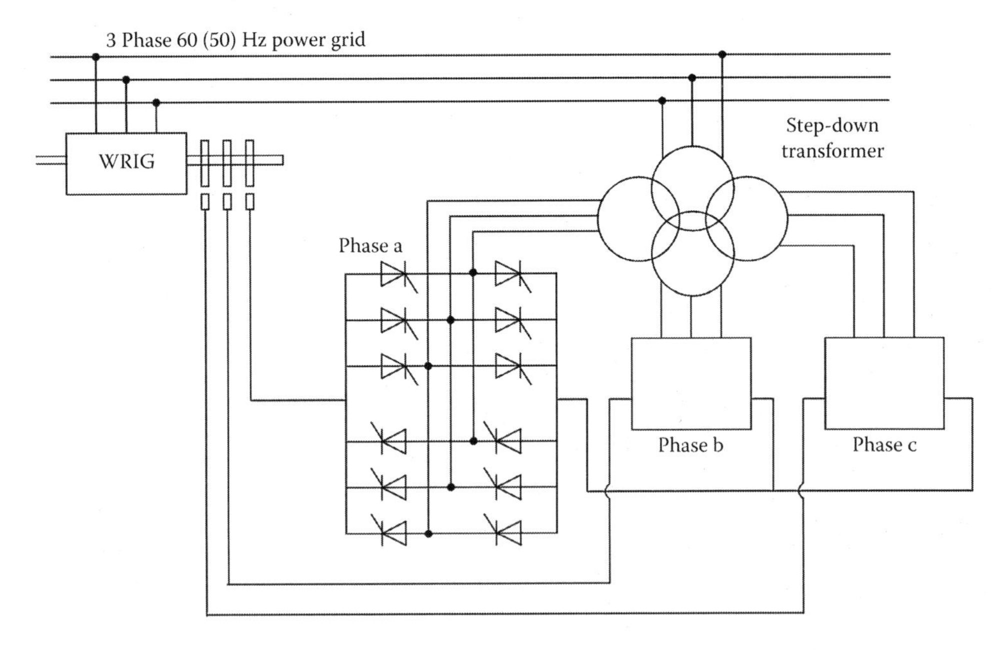

FIGURE 2.9 High-power cycloconverter for WRIG.

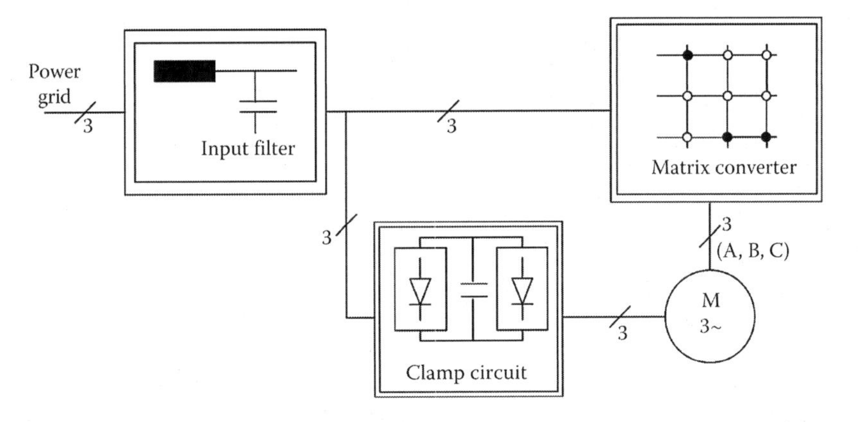

FIGURE 2.10 Typical matrix converter (with IGBTs).

are partially lost. Harmonics with cycloconverter WRIGs are pertinently described in Reference 9. Large power WRIGs (above 100 MW) are still provided with cycloconverters in the rotor circuit. A superior version of a direct AC–AC converter is the so-called matrix converter. The matrix converter (Figure 2.10) uses faster, bidirectional power switches and is, basically, a voltage source. The output voltage is made of sections of input voltages, intelligently "extracted" to form a PWM AC output voltage. There are some storage elements for voltage clamping (mainly) on the input side; but still, the power balance should be provided inside each cycle in the absence of strong DC link energy storage elements. The matrix converter is claimed to show high power/volume and ultrafast power control with a controlled input power factor and low harmonics, and all this with four-quadrant natural operation and output frequency independent of ω_1, which is essential for motors starting with WRIGs from the rotor side with a short-circuited stator.

2.6.2 DC Voltage Link AC–AC Converters

Two voltage source PWM six-leg inverters may be connected back-to-back to form a DC voltage link AC–AC indirect converter (Figure 2.11).

Bidirectional power flow is inherent, while output (rotor-side) frequency ω_2 is limited only by the switching frequency of power switches that may be gate turn-offs (GTOs), IGBTs, or IGCTs (for high power).

The two-level converter is generally used today up to 2–3 MW with IGBTs and line voltage outputs of 690 V. The presence of the large capacitor in the DC link provides for generous reactive power handling capacity. The rather large switching frequency (above 1 kHz) provides for lower current harmonics, both in the rotor and on the supply side.

Furthermore, fast commutation of power switches provides for very fast active and reactive power response, drawing temporarily on the mechanically stored energy in the prime mover generator rotors. Consequently, intervention in the power system for power balance or voltage control is very fast, notably faster than with synchronous generators (SGs). Moreover, operation and running through synchronism ($S = 0$) comes naturally. This makes the DC voltage link AC–AC converter "ideal" for WRIGs.

But, as the power unit goes up, the maximum rotor voltage has to increase in order to keep the current through slip-rings and brushes at reasonably large levels. Voltages in the kilovolt range are typical for rotor powers up to 10 MW and above 10 kV for rotor powers above 10 MW. Multilevel DC voltage link AC–AC converters are suitable for such voltage levels. A typical three-level GTO converter is shown in Figure 2.12a and b. Other "cellular-type" multilevel converters are now commercially available in the kilovolt and megawatt range.

To further reduce the harmonics content, it is also feasible to use two six-pulse (leg) converters in parallel on the same DC bus to work together as a 12-pulse converter when connected to a Δ,Y dual secondary step-up transformer on the source side [10].

Multiple cells in series are used to handle higher voltages, but the multilevel voltage principle holds good. Such systems are also called high-voltage direct current (HVDC)-light.

After presenting the various AC–AC converters for WRIGs, it seems clear that the trend is in favor of DC voltage link AC–AC converters with IGBTs and IGCTs, up to rotor powers of megawatts and, respectively, tens of megawatts, which, in fact, cover the whole range of WRIG power per unit, up to 400 MW.

Consequently, control systems of WRIGs will be investigated in what follows only in association with DC voltage link AC–AC converters.

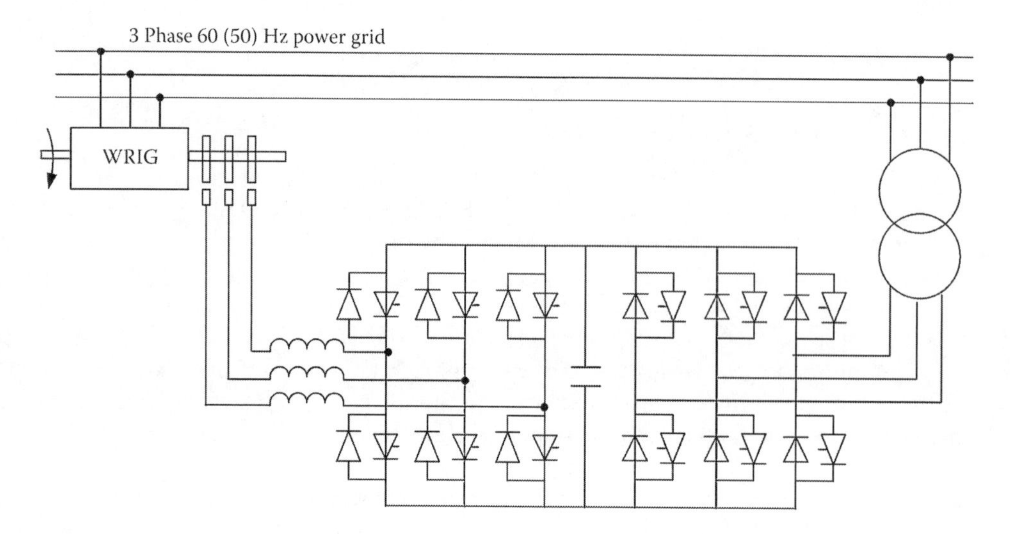

FIGURE 2.11 Two-level voltage link AC–AC converter.

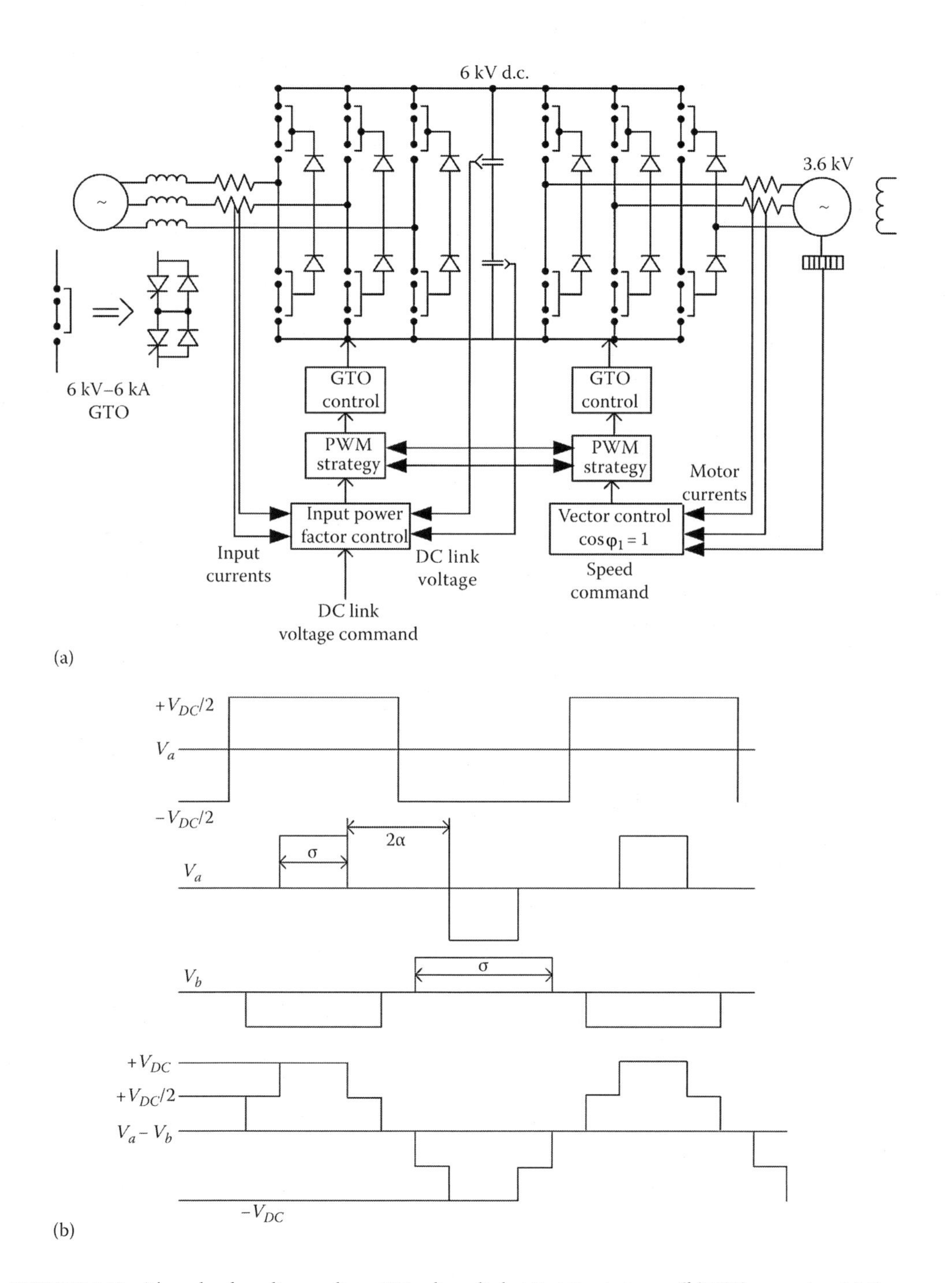

FIGURE 2.12 Three-level medium-voltage DC voltage link AC–AC gate turn-off (GTO) converter: (a) the configuration and (b) typical voltage waveforms.

2.7 Vector Control of WRIG at Power Grid

Vector control stems from decoupled flux-current and torque-current control in AC drives. It resembles the principle of decoupled control of excitation and armature current in DC brush machines. Vector control means, in fact, decoupled flux and torque control.

Intuitively, adding two more outer loops—one stator voltage loop to produce the reference flux current and one frequency outer loop to control generator speed—for autonomous operation is realized. When the WRIG is connected to the power grid, active and reactive powers are close-loop controlled, and they produce the reference flux and torque currents in vector control.

As motor and generator operation alternates, the control system has to be designated to handle both without hardware modifications. This makes the control system design more complicated.

2.7.1 Principles of Vector Control of Machine (Rotor)-Side Converter

Let us restate here the WRIG space-phasor model in synchronous coordinates:

$$\begin{aligned} \overline{I}_s R_s + \overline{V}_s &= -\frac{d\overline{\Psi}_s}{dt} - j\omega_1 \overline{\Psi}_s; \quad \overline{\Psi}_s = L_s \overline{I}_s + L_m \overline{I}_r \\ \overline{I}_r R_r + \overline{V}_r &= -\frac{d\overline{\Psi}_r}{dt} - j(\omega_1 - \omega_r)\overline{\Psi}_r; \quad \overline{\Psi}_r = L_r \overline{I}_r + L_m \overline{I}_s \end{aligned} \tag{2.39}$$

Aligning the system of coordinates to stator flux seems most useful, as, at least for power grid operation, $\overline{\Psi}_s$ is almost constant, because the stator voltages are, in general, constant in amplitude, frequency, and phase:

$$\overline{\Psi}_s = \Psi_s = \Psi_d; \quad \Psi_q = 0; \quad \frac{d\Psi_q}{dt} = 0 \tag{2.40}$$

Therefore, the stator equation, split into dq axes, becomes as follows:

$$\begin{aligned} I_d R_s + V_d &= -\frac{d\Psi_d}{dt} \\ I_q R_s + V_q &= -\omega_1 \Psi_q \end{aligned} \tag{2.41}$$

As the stator flux ψ_d does not vary much, $(d\Psi_d/dt) \approx 0$, and, with $R_s = 0$,

$$\begin{aligned} V_d &= 0 \\ V_q &= -\omega_1 \Psi_q \end{aligned} \tag{2.42}$$

Now the active and reactive stator powers P_s, Q_s are as follows:

$$\begin{aligned} P_s &\approx \frac{3}{2}(V_d I_d + V_q I_q) = \frac{3}{2} V_q I_q = \frac{3}{2}\omega_1 \Psi_d \cdot \frac{L_m I_{qr}}{L_s}; \quad \omega_1 = \omega_r \\ Q_s &\approx \frac{3}{2}(V_d I_q - V_q I_d) = \frac{3}{2}\omega_1 \Psi_d I_d = \frac{3}{2}\omega_1 \frac{\Psi_d}{L_s}(\Psi_d - L_m I_{dr}) \end{aligned} \tag{2.43}$$

Equation 2.43 clearly shows that under stator flux orientation (vector) control, the active power delivered (or absorbed) by the stator, P_s, may be controlled through the rotor current I_{qr}, while the reactive power (at least for constant $\Psi_s = \Psi_d$) may be controlled through the rotor current I_{dr}.

Both powers depend heavily on stator flux ψ_s and frequency ω_1 (i.e., on stator voltage). This constitutes the basis for vector control of P_s and Q_s by controlling the rotor currents I_{dr} and I_{qr} in synchronous coordinates.

As pulse-width modification on the machine-side converter is generally performed on rotor voltages, voltage decoupling in the rotor is required, again in synchronous coordinates.

At steady state, from Equation 2.39,

$$\bar{\Psi}_r = \frac{L_m \Psi_d}{L_s} + L_{sc}(I_{dr} + jI_{qr}); \quad \Psi_s = \Psi_d, \quad \Psi_q = 0, \quad L_{sc} = L_r - \frac{L_m^2}{L_s} \tag{2.44}$$

$$V_{dr} + jV_{qr} = -R_r(I_{dr} + jI_{qr}) - jS\omega_1\left(\frac{L_m}{L_s}\Psi_d + L_{sc}(I_{dr} + jI_{qr})\right) \tag{2.45}$$

From Equation 2.45,

$$\begin{aligned} V_{dr} &= -R_r I_{dr} + L_{sc} S\omega_1 I_{qr} : S = 1 - \frac{\omega_r}{\omega_1} \\ V_{qr} &= -R_r I_{qr} - S\omega_1\left(\frac{L_m}{L_s}\Psi_d + L_{sc} I_{dr}\right) \end{aligned} \tag{2.46}$$

Equation 2.46 constitutes the rotor voltage decoupling conditions. The resistance terms may be dropped as *dq* rotor current closed loops are added anyway.

It also remains to estimate the stator flux linkage space-phasor ψ_s in amplitude and instantaneous position. The inductances L_m, L_s, and L_{sc} also have to be known. The machine-side converter has to produce the correspondents of V_{dr} and V_{qr} in rotor coordinates:

$$\begin{aligned} V_{ar}^* &= V_{dr}^*\cos(\theta_s - \theta_{er}) - V_{qr}^*\sin(\theta_s - \theta_{er}) \\ V_{br}^* &= V_{dr}^*\cos\left(\theta_s - \theta_{er} - \frac{2\pi}{3}\right) - V_{qr}^*\sin\left(\theta_s - \theta_{er} - \frac{2\pi}{3}\right) \\ V_{cr}^* &= -V_{ar}^* - V_{br}^* \end{aligned} \tag{2.47}$$

The zero voltage sequence reference voltage $V_{ar}^* = 0$. θ_s represents the stator flux angle with respect to stator phase *a* axis and is

$$\theta_s = \int \omega_1 dt + \theta_{so} \tag{2.48}$$

The rotor electrical position θ_{er} is

$$\theta_{er} = p_1\theta_r; \quad p_1\text{—polepairs} \tag{2.49}$$

with θ_r the rotor phase a_r axis position with respect to stator phase *a* axis. Under steady state, V_{dr} and V_{qr} are DC, $\theta_s = \omega_1 t$, and $\theta_{er} = \omega_r t + \theta_{ero}$ and, as expected, the rotor voltages will have the slip frequency $\omega_2 = \omega_1 - \omega_2 = S\omega_1$. These principles of vector control are illustrated in Figure 2.13.

As evident in Figure 2.13, the rotor currents have to be measured. Also, the Ψ_d and θ_s have to be estimated. The rotor position is needed to be adjusted. Rotor speed ω_r is also needed, both in the voltage decoupler and for prime-mover speed control. The mechanical power P_m versus speed ω_r is obtained through an optimization criterion:

$$P_m = P_s + P_r^r + \sum \text{losses} \tag{2.50}$$

In general, $P_m(\omega_r)$ is linear with speed (Figure 2.14) and depends on prime-mover and WRIG characteristics.

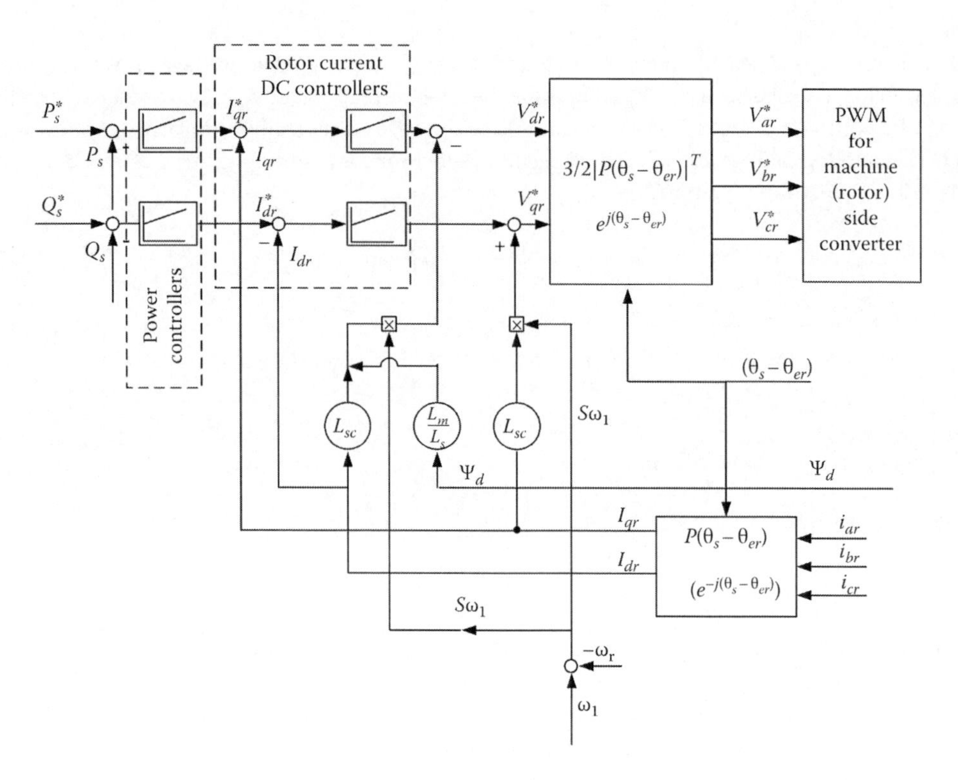

FIGURE 2.13 P_s, Q_s, vector control structural diagram for a WRIG.

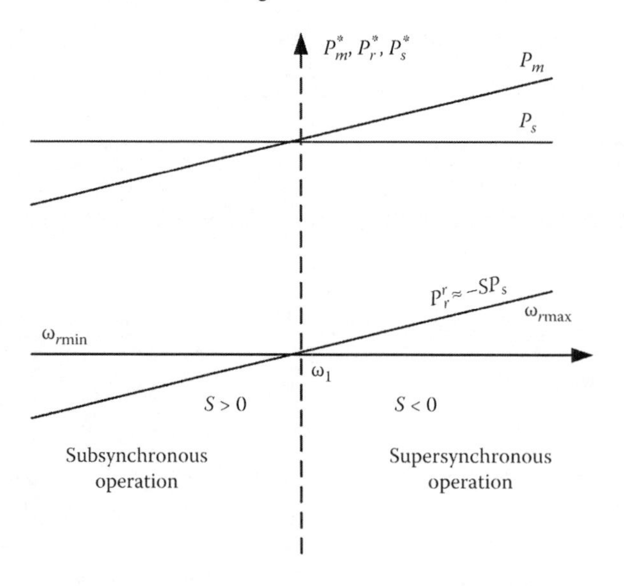

FIGURE 2.14 Power reference (envelope) division vs. speed in a WRIG, in generator mode.

The stator delivered power envelope increases slowly with speed, but the mechanical power varies notably as the rotor electric power changes sign at synchronous speed (ω_1).

A given total electric power requirement ($P_s + P_r^r \approx P_m$) asks for an optimum speed to deliver it, and the prime-mover speed generator (Chapter 3) has to be able to produce it through closed-loop control. This is the second reason why speed feedback is required. Also, reference power P_s versus speed is obtained from curves as those shown in Figure 2.14.

It is clear that either rotor electrical position or speed ω_r are measured or estimated. A resolver or a rugged (magnetic) encoder would be enough hardware to produce both θ_r and ω_r, in practice. As such a device tends to be costly, not so reliable, and affected by noise in terms of precision, motion estimators are preferred, leading to the so-called sensorless control. Note that as sensorless control may also be used for direct power control at the power grid or for stand-alone operation of WRIG, the estimators for sensorless control and performance with them will be treated after such control alternatives are exposed in forthcoming paragraphs.

2.7.2 Vector Control of Source-Side Converter

The source-side converter is connected to the power grid eventually via a step-up transformer in some embodiments. However, the transformer is eliminated by designing the rotor windings with a higher than unity turn ratio K_{rs}:

$$K_{rs} \approx \frac{1}{|S_{\max}|} > 1 \tag{2.51}$$

At maximum slip, the rotor voltage equals the stator voltage. In general, the source-side voltage converter uses a power filter to reduce current harmonics flow into the power source (Figure 2.15). The presence of the source-side power filter is imperative for stand-alone operation mode.

The voltage equations across the inductors (L and R) are as follows:

$$\begin{vmatrix} V_a \\ V_b \\ V_c \end{vmatrix} = R \begin{vmatrix} I_{as} \\ I_{bs} \\ I_{cs} \end{vmatrix} + L\frac{d}{dt}\begin{vmatrix} I_{as} \\ I_{bs} \\ I_{cs} \end{vmatrix} + \begin{vmatrix} V_{as} \\ V_{bs} \\ V_{cs} \end{vmatrix} \tag{2.52}$$

These equations may be translated into dq synchronous coordinates that may be aligned to axis d voltage ($V_q = 0$, $V_d = V_s$):

$$\begin{aligned} V_d &= Ri_{ds} + L\frac{di_{ds}}{dt} - \omega_1 L i_{qs} + V_{ds} \\ V_q &= Ri_{qs} + L\frac{di_{qs}}{dt} + \omega_1 L i_{ds} + V_{qs} \end{aligned} \tag{2.53}$$

where ω_1 is the speed of the reference system or the supply frequency.

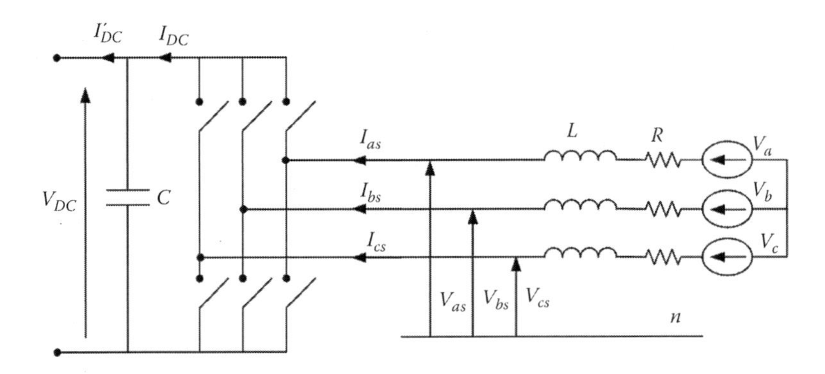

FIGURE 2.15 Source-side voltage converter.

Neglecting the harmonics due to switching in the converter and the machine losses and converter losses, the active power balance equation is as follows:

$$V_{DC}I_{DC} = \frac{3}{2}V_d I_d = P_r; \quad V_q = 0 \tag{2.54}$$

But, with the PWM depth, m_1, as known,

$$V_d = \frac{m_1}{2\sqrt{2}}V_{DC} \tag{2.55}$$

Then, from Equation 2.54,

$$I_{DC} = \frac{3m_1 I_d}{4\sqrt{2}} \tag{2.56}$$

The DC link voltage link equation is

$$C\frac{dV_{DC}}{dt} = I_{DC} - I'_{DC} = \frac{3m_1 I_d}{4\sqrt{2}} - I'_{DC} \tag{2.57}$$

It is evident from Equation 2.57 that DC link voltage V_{DC} may be controlled through I_d control (Equation 2.56). The reactive power from (to) the source Q_r is

$$Q_r = \frac{3}{2}(V_d I_q - V_q I_d) = \frac{3}{2}V_d I_q; \quad V_q = 0 \tag{2.58}$$

Consequently, the reactive power from the power source to (from) the source-side converter may be controlled through I_q.

But, the voltage decoupler (from Equation 2.53) is required:

$$\begin{aligned} V'_{ds} &= \omega_1 L I_q + V_d \\ V'_{qs} &= -\omega_1 L I_d \end{aligned} \tag{2.59}$$

In general, the reactive power from power source through the source-side converter is set to zero ($I_q = 0$), but $Q_r \neq 0$, positive or negative, is feasible, according to Equation 2.58.

These vector control principles are illustrated in the generic scheme shown in Figure 2.16. The DC link voltage is, in general, kept constant to take advantage of full voltage for capacitor energy storage in the DC link. There is a DC voltage controller outside the DC current (I_d, I_q) controllers. The voltage decoupler (Equation 2.59) is also included.

The Park transformation is operated in two stages, *abc*-αβ (3/2 stator coordinates) and αβ-*dq* in synchronous coordinates, aligned to supply voltage space phasor ($d\theta_e/dt = \omega_1$). Some filtering is needed for ω_1 calculation from θ_e, though for $\omega_1 \approx$ ct. If ω_r varies, the angle θ_e^r of the voltage time integral (a fictitious flux linkage) should be calculated and then advanced by 90° to get $\theta_e = \theta_e^r\, \pi/2$.

Further on, the speed of ideal stator flux linkage may be calculated and filtered properly for less noise. As the stator flux $\Psi_s(\Psi_d)$ is estimated for the vector control of the machine-side converter, this latter method may prove to be more practical. The reactive power exchange θ_r^r with the source-side converter is included in Figure 2.16. A complete design methodology with digital simulations and laboratory-scale results for the vector control of WRIG at power grid is given in Reference 11. Instead of reference reactive power θ_r^*, a given small power factor angle (positive or negative) φ^* may be defined as follows:

$$I_q^* = I_d^* \tan\varphi^* \tag{2.60}$$

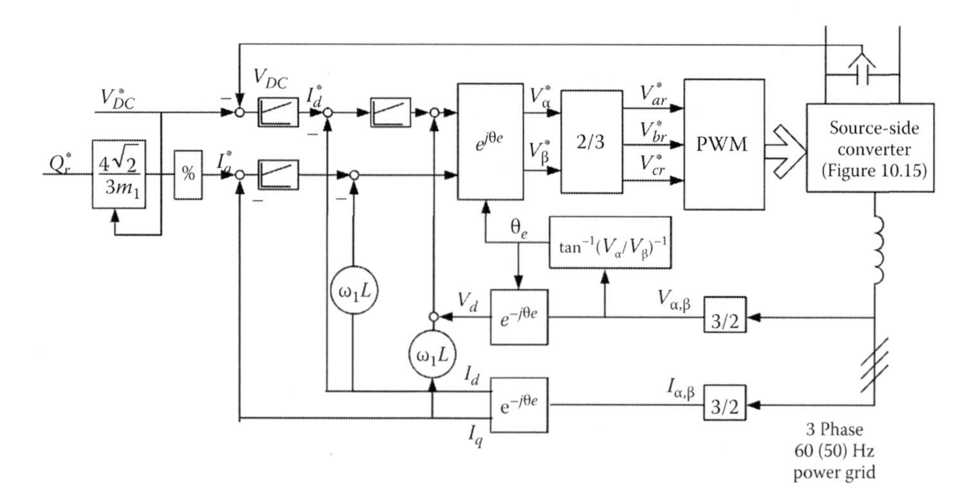

FIGURE 2.16 Vector control principle of the source-side converter.

This solution may simplify the implementation, as only multiplication by a constant is performed. We will now present through digital simulations a description of a case study for a 2 MW wind-generator application [12]. A hydraulic turbine prime mover would impose slightly different optimum mechanical power P_m versus speed and mechanical time constants in the speed governor. It might also need motor starting for pump storage. These aspects will be treated later on in this chapter.

2.7.3 Wind Power WRIG Vector Control at the Power Grid

A schematic diagram of the overall system is shown in Figure 2.17.

Two back-to-back voltage-fed PWM converters are inserted in the rotor circuit, with the supply-side PWM converter connected to the grid by a resistance inductance (RL) filter, which limits the high-frequency ripple due to switching harmonics. The complete simulation model of the system was implemented in MATLAB® and Simulink®, and its parts are explained in what follows.

2.7.3.1 Wind Turbine Model

The wind turbine is modeled in terms of optimal power tracking, to provide maximum energy capture from the wind. The implemented characteristics are presented in Figure 2.18a and b.

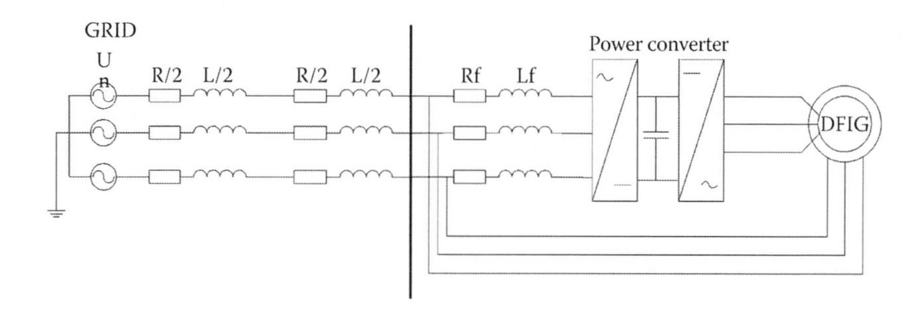

FIGURE 2.17 WRIG connected to the power grid.

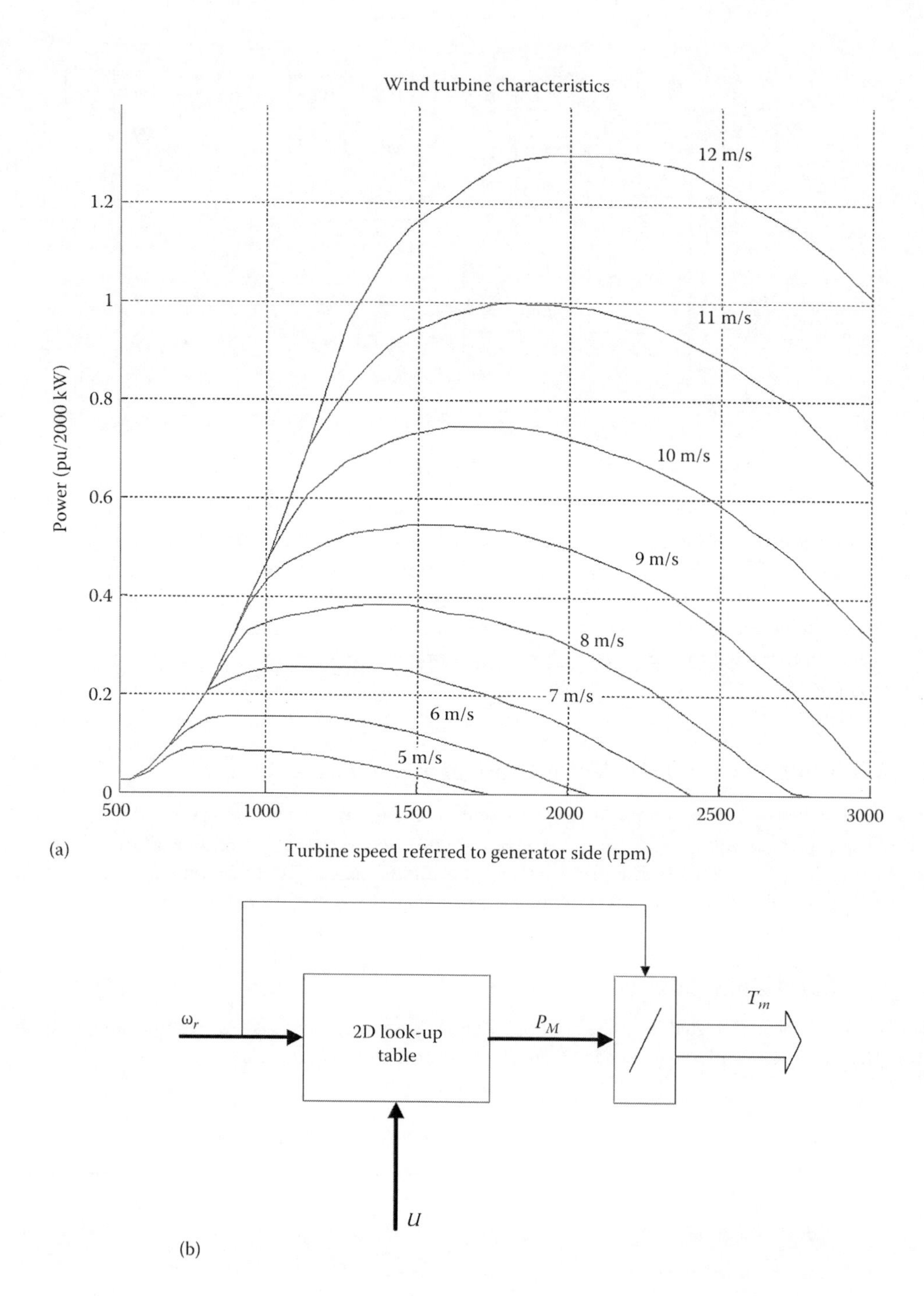

FIGURE 2.18 Implemented wind turbine characteristics: (a) aerodynamic characteristics and (b) the turbine torque model.

These characteristics are based on the following:

$$P_M = \frac{1}{2}\rho_{air} \cdot C_p(\lambda,\beta) \cdot \pi \cdot R^2 \cdot U^3 \tag{2.61}$$

where
C_p is the power efficiency coefficient
U is the wind velocity
β is the pitch angle
R is the blade radius
P_M is the mechanical power produced by the wind turbine
ρ_{air} is the air density (usually 1.225 kg/m^3)

The tip speed ratio λ is

$$\lambda = \frac{\omega_r R}{U} \tag{2.62}$$

Equation 2.61 becomes

$$P_M^{opt} = \frac{1}{2}\rho_{air} \cdot \pi \cdot R^5 \cdot \frac{C_p^{opt}}{\lambda_{opt}^3} \cdot \omega_r^3 = K_W \cdot \omega_r^3 \tag{2.63}$$

where K_w is a wind-turbine-dependent coefficient.

Equation 2.63 holds the key to the optimization of the variable-speed wind turbine by tracking the optimal turbine speed at a given wind speed. The result of the optimization is the optimal power efficiency C_p at the given speed. In the simulation, the optimum power is returned by the model through a two-dimensional lookup table with interpolation of the turbine characteristics. The action of the governor is considered instantaneous.

Stator and rotor voltages expressed in terms of d and q axes quantities rotating in an arbitrary reference frame with the speed ω_b are represented by Equations 2.31 through 2.34.

The WRIG parameters are as follows: P_N = 2.0 MW, (V_{SN})line = 690 V root mean squared (RMS), $2p_1 = 4$, f_N = 50 Hz, l_m = 3.658 per unit (P.U.), l_{sl} = 0.0634 (P.U.), l_{rl} = 0.08466 (P.U.), $r_s = 4.694 \times 10^{-3}$ (P.U.), $r_r = 4.86 \times 10^{-3}$ (P.U.), H = 3.611 (s).

Usually, a speed range of ±25% is chosen around the synchronous speed. To have a lower current at the rotor side of the machine, which is equivalent to a lower converter sizing, and the same voltage at the rotor side and stator side at the limits of the speed range (the voltage adapting transformer between the converter and the power system is avoided), a transformer ratio K_{rs} of the generator windings of four is chosen:

$$V_{r\max} = K_{rs}\left|S_{\max}\right| \cdot (V_s)_{rated} \tag{2.64}$$

2.7.3.2 Supply-Side Converter Model

As inferred in a previous discussion, the supply-side converter is used to control the DC link voltage, regardless of the level and the direction of the rotor power (Figure 2.19). A vector-control strategy is used, with the reference frame oriented along the stator voltage. The converter is current regulated, with the direct axis current used to control the DC link voltage; meanwhile, the transverse axis current is used to regulate the displacement between the voltage and the current (and, thus, the power factor).

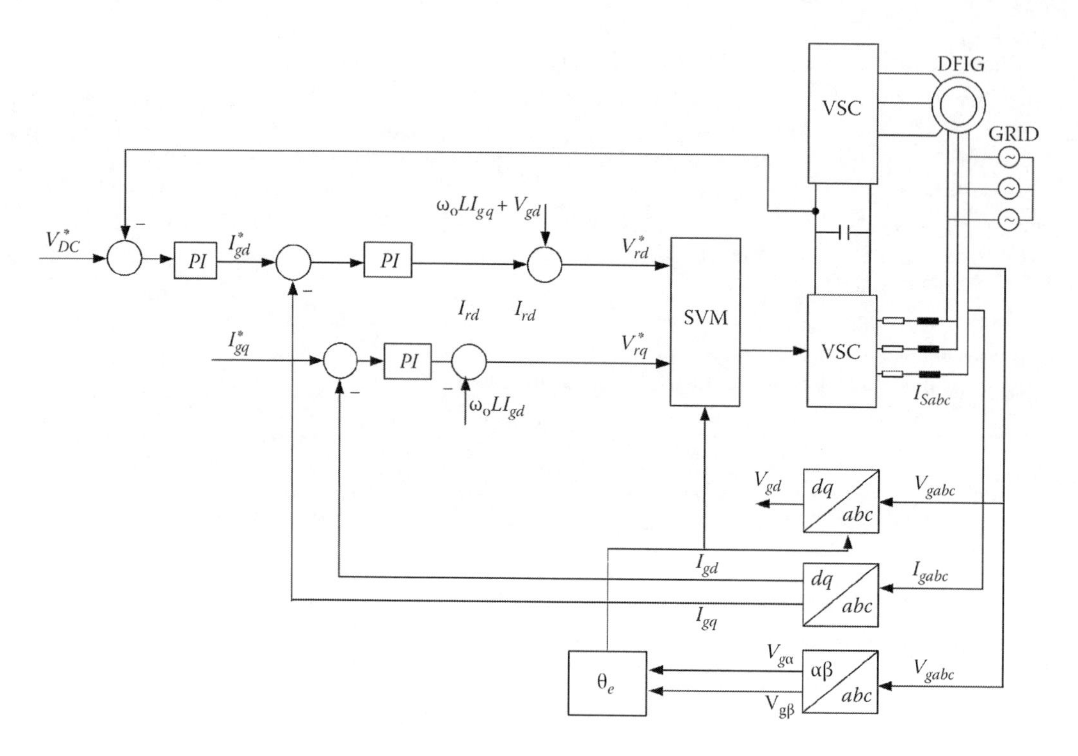

FIGURE 2.19 The block diagram of the supply-side converter control.

The angular position of the supply voltage θ_e is calculated:

$$\theta_e = \tan^{-1}\left(\frac{V_{g\beta}}{V_{g\alpha}}\right) \tag{2.65}$$

where $V_{g\alpha}$ and $V_{g\beta}$ are the stator voltages expressed in terms of α and β axes (stationary).

The design of the current controllers follows from the transfer function of the plant in the *z*-domain, the sampling time being 0.45 ms. For a nominal closed-loop natural frequency of 125 Hz and a damping factor of 0.8, the transfer function of the proportional integral (PI) controller is 18.69 $(z - 1335.4)/(z - 1)$. The current loop is designed to be much faster than the DC voltage loop and is thus considered ideal. The transfer function of the DC voltage controller results in $17.95(z - 2640)/(z - 1)$.

2.7.3.3 Generator-Side Converter Model

The generator is controlled in a synchronously rotating *dq* frame, with the *d* axis aligned with the stator flux vector position, which ensures a decoupled control between the electromagnetic torque and the rotor excitation contribution (Figure 2.20). In fact, decoupled active and reactive power control of the generator is obtained (Figure 2.21), as explained earlier.

Again, knowing the transfer function of the plant and imposing the natural closed-loop frequency and the damping factor, the transfer function of the current controllers results in $12(z - 0.995)/(z - 1)$. The power loops are designed to be much slower than the current loops, and the transfer function for the power controllers are thus obtained: $0.00009(z - 0.9)/(z - 1)$.

The reference voltages generated by the current control loops with electromagnetic force (emf) compensation are transformed back to the rotor referenced frame and are space vector modulated to generate the voltage pulses for the inverter.

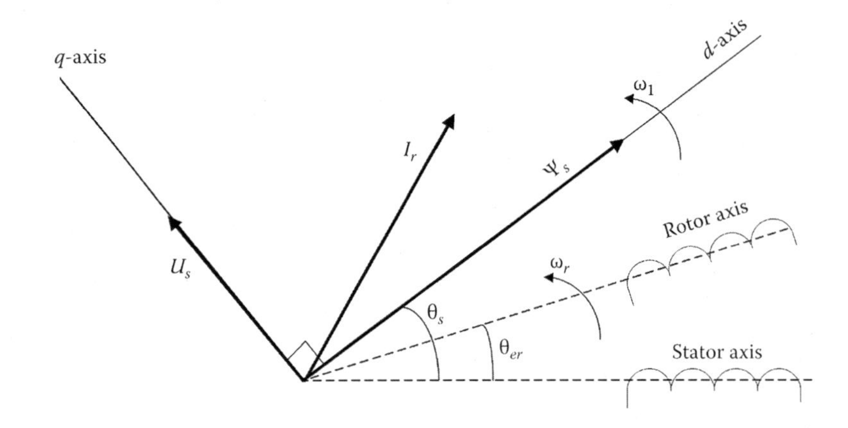

FIGURE 2.20 Location of different vectors in stationary coordinates.

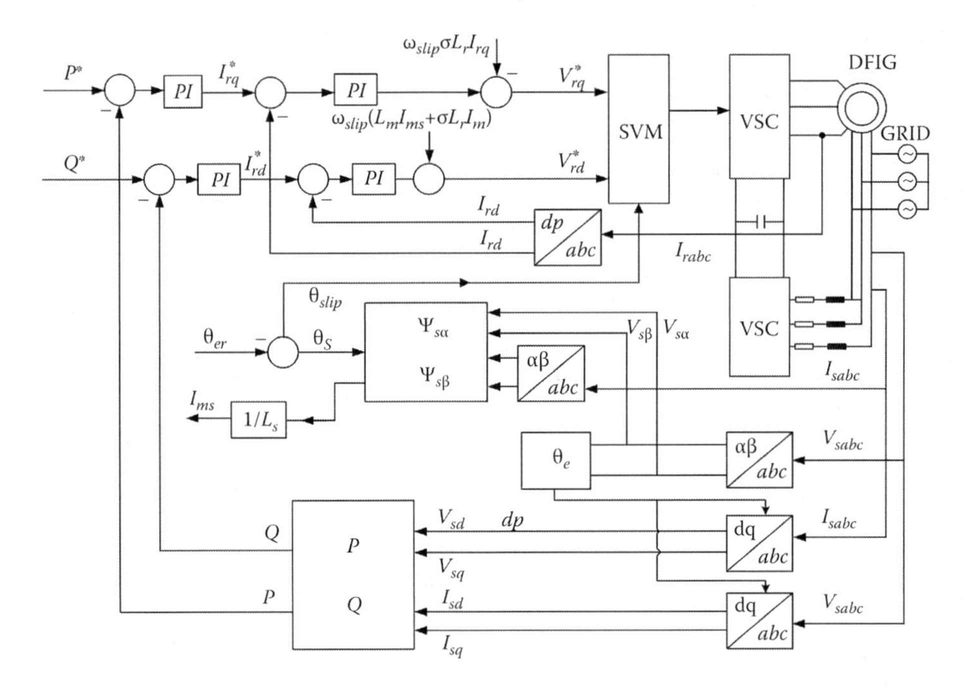

FIGURE 2.21 The block diagram of the machine-side converter control in a doubly fed wind turbine.

2.7.3.4 Simulation Results

Extended simulations were performed on the model thus implemented. At 1.5 s, the wind speed is increased from 7 to 11 m/s, then reduced to 6 m/s; the reference of active power is increased from zero to 1.2 MW, while the reference reactive power is maintained at 300 kVAR. The various stator and rotor variable waveforms are shown in Figure 2.22a through f. The smooth transition through synchronous speed is noteworthy.

The stator active and reactive powers are shown in Figure 2.23a and b, respectively.

Decoupled control is evident as also demonstrated in Reference 13. Safe passage through synchronous speed ($n_1 = f_1/p_1$) is visible.

The system response is stable and rather quick in terms of active power control.

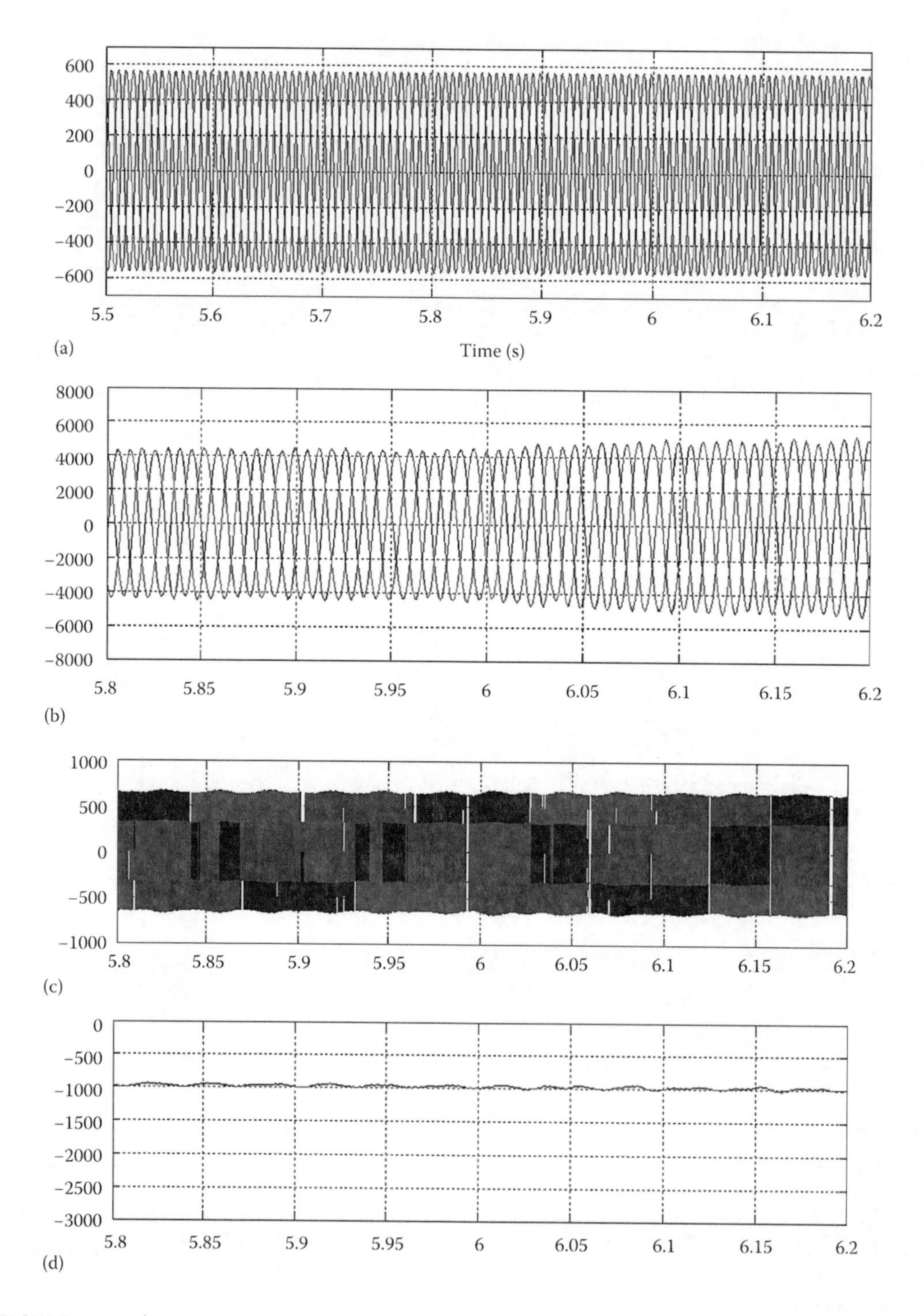

FIGURE 2.22 The generator waveforms: (a) stator voltages, (b) stator currents, (c) rotor voltages, and (d) DC link voltage.

(Continued)

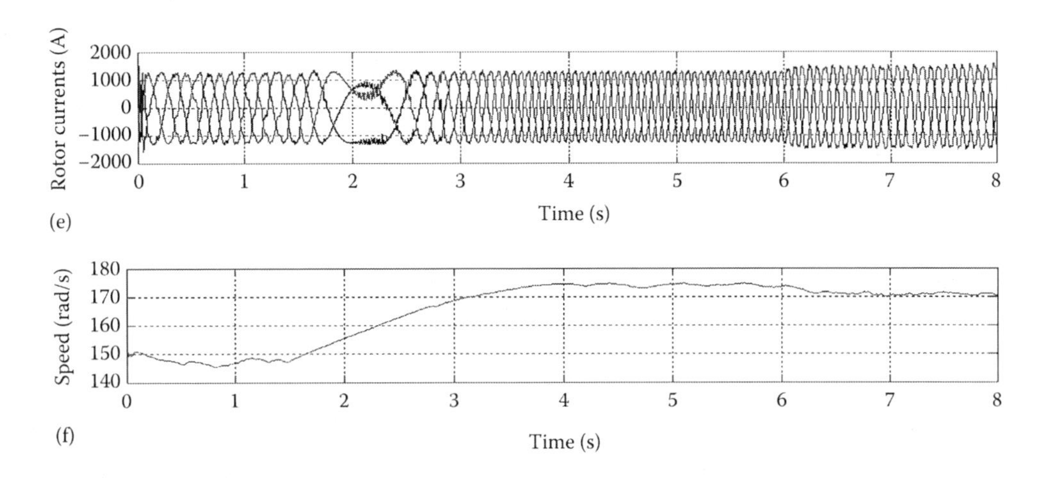

FIGURE 2.22 (*Continued*) The generator waveforms: (e) rotor currents and (f) speed.

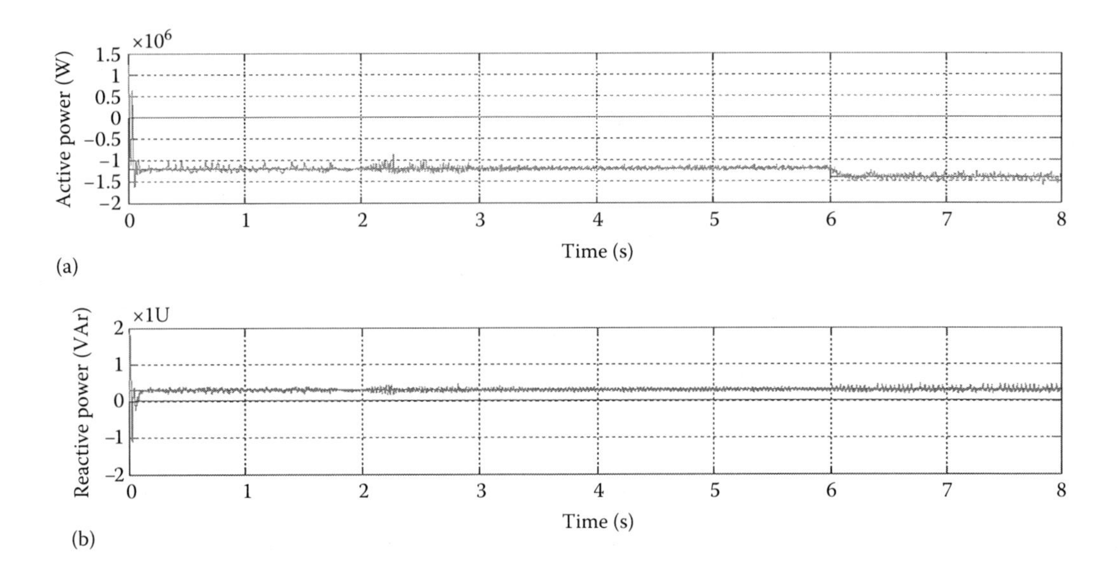

FIGURE 2.23 The (a) active and (b) reactive stator power control.

2.7.3.5 Three-Phase Short Circuit on the Power Grid

To study the behavior of the whole system under a three-phase short circuit on the power grid, without taking any protective measures, the grid is modeled as in Figure 2.17 and divided into two parts. The short circuit is applied at the middle of the line. Sample results are shown in Figure 2.24a through h.

It can be noted that there are two dominant current peaks, one at the initiation of the short circuit, and the other, even more severe, when the short circuit is cleared. Two severe transients in torque also occur, and a transient in speed, which means that during the fault the electromagnetic torque is very small, and the generator accelerates. The torque of the turbine naturally decreases during the fault, as the generator speed increases and the wind speed remains constant (see the characteristics in Figure 2.18).

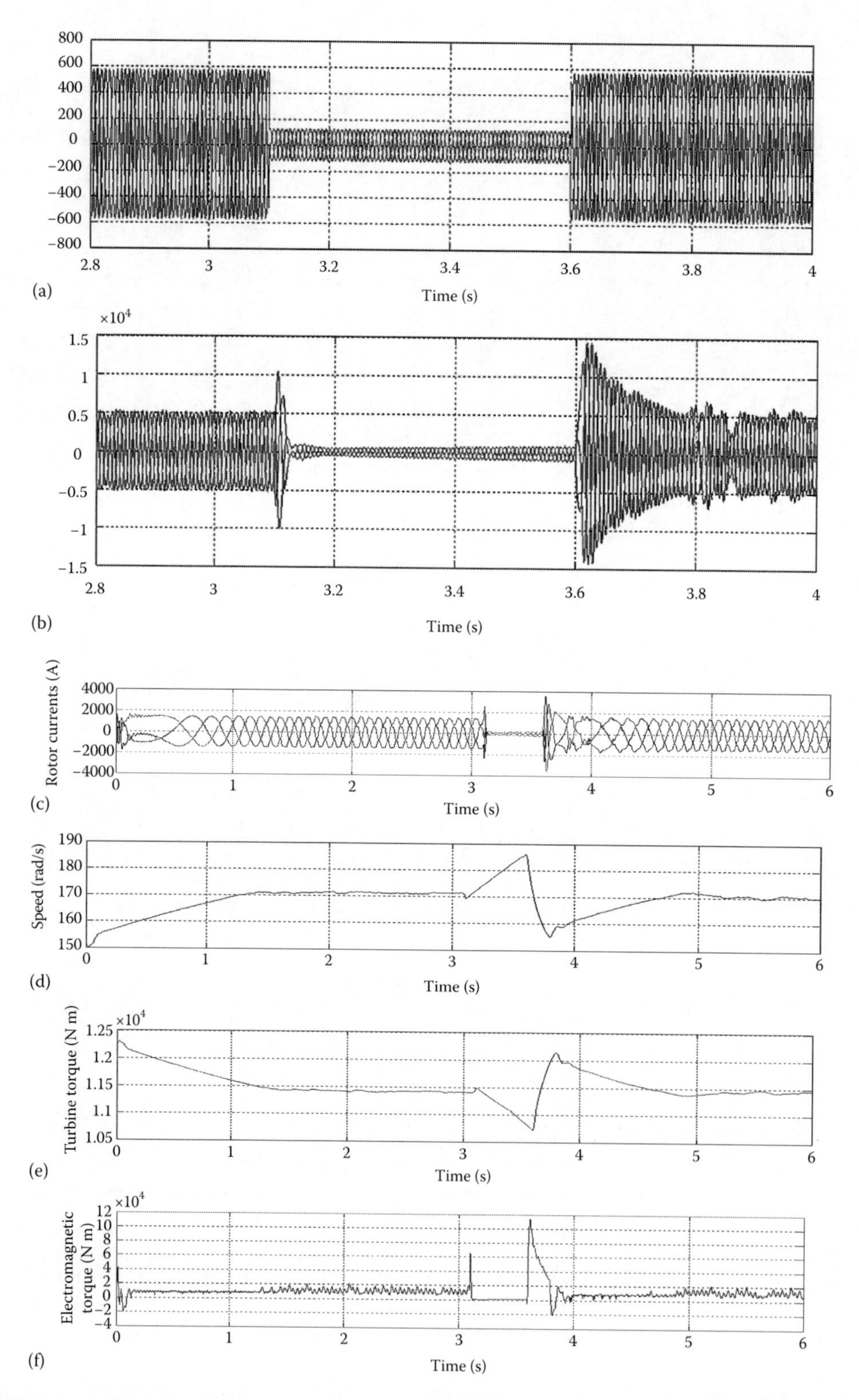

FIGURE 2.24 Three-phase short circuit on the power grid: (a) stator voltage, (b) stator currents, (c) rotor currents, (d) speed, (e) turbine torque, and (f) electromagnetic torque.

(Continued)

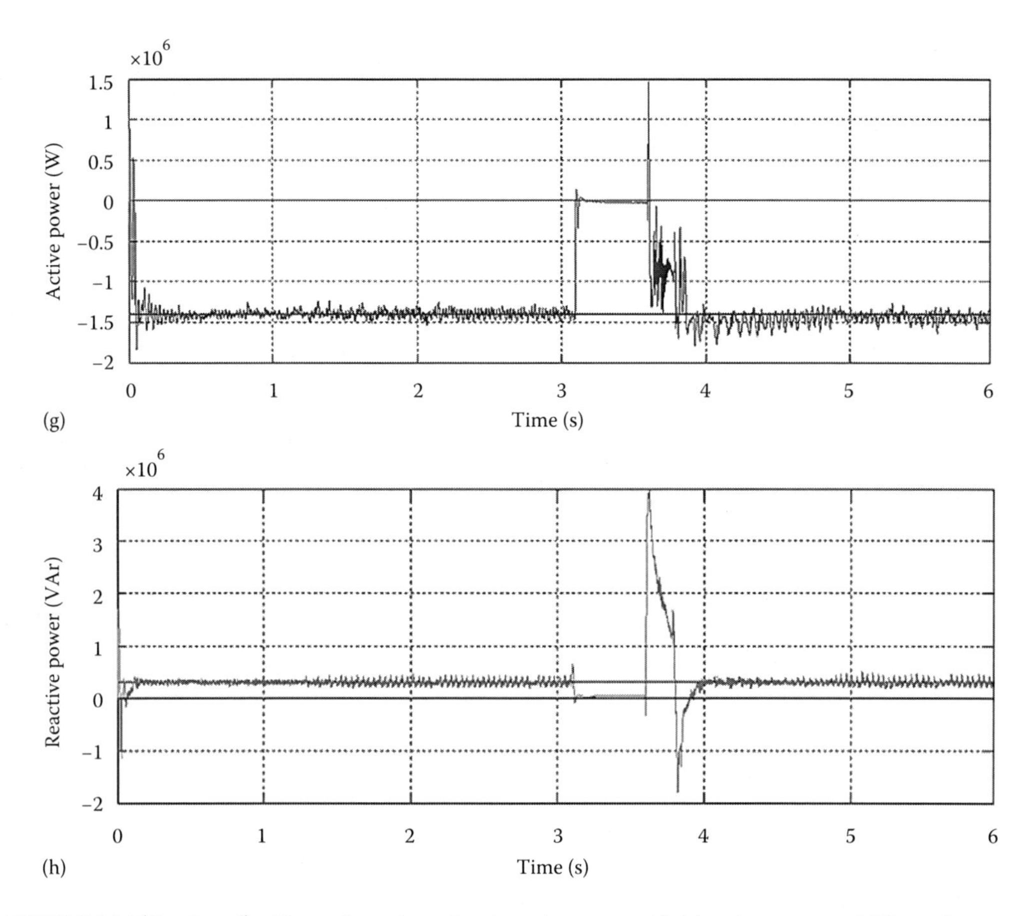

FIGURE 2.24 (*Continued*) Three-phase short circuit on the power grid: (g) active power, and (h) reactive power.

2.7.3.6 Mechanism to Improve Performance during Fault

To reduce the transients due to short circuit, a few methods were investigated. The first was to reduce the active and reactive power references immediately after the fault occurs. But it did not produce any improvement.

The most efficient method, according to this study, was to limit the rotor currents. This was done by limiting the reference rotor currents, in fact, the output of the active and reactive power PI controllers at the ±1.5 I_{rN} on the q axis and ±0.5 I_{rN} on the d axis. The results are presented in what follows and in Figure 2.25a through e [12].

Improvements can easily be noted. First and the most important is the reduction of the second and highest rotor current peak, when short circuit is cleared, by 50%. The stator current second peak is also reduced by the same amount. The acceleration of the generator is not avoided, but when the system is restoring, the deceleration is less pronounced, and the generator speed stabilizes earlier. Practically, the system is recovering 1 s more quickly than in the case with no limitations in the rotor currents.

Note that for large power wind farms, continuous operation during short circuit faults is required. This mode becomes more important, as the wind turbine should be able to contribute with power just after a short circuit. The good dynamic performance and accuracy of the active and reactive power control were demonstrated. Also, the waveforms of the generator in a three-phase short circuit are presented. For better continuous operation during short-circuit faults, different methods were investigated.

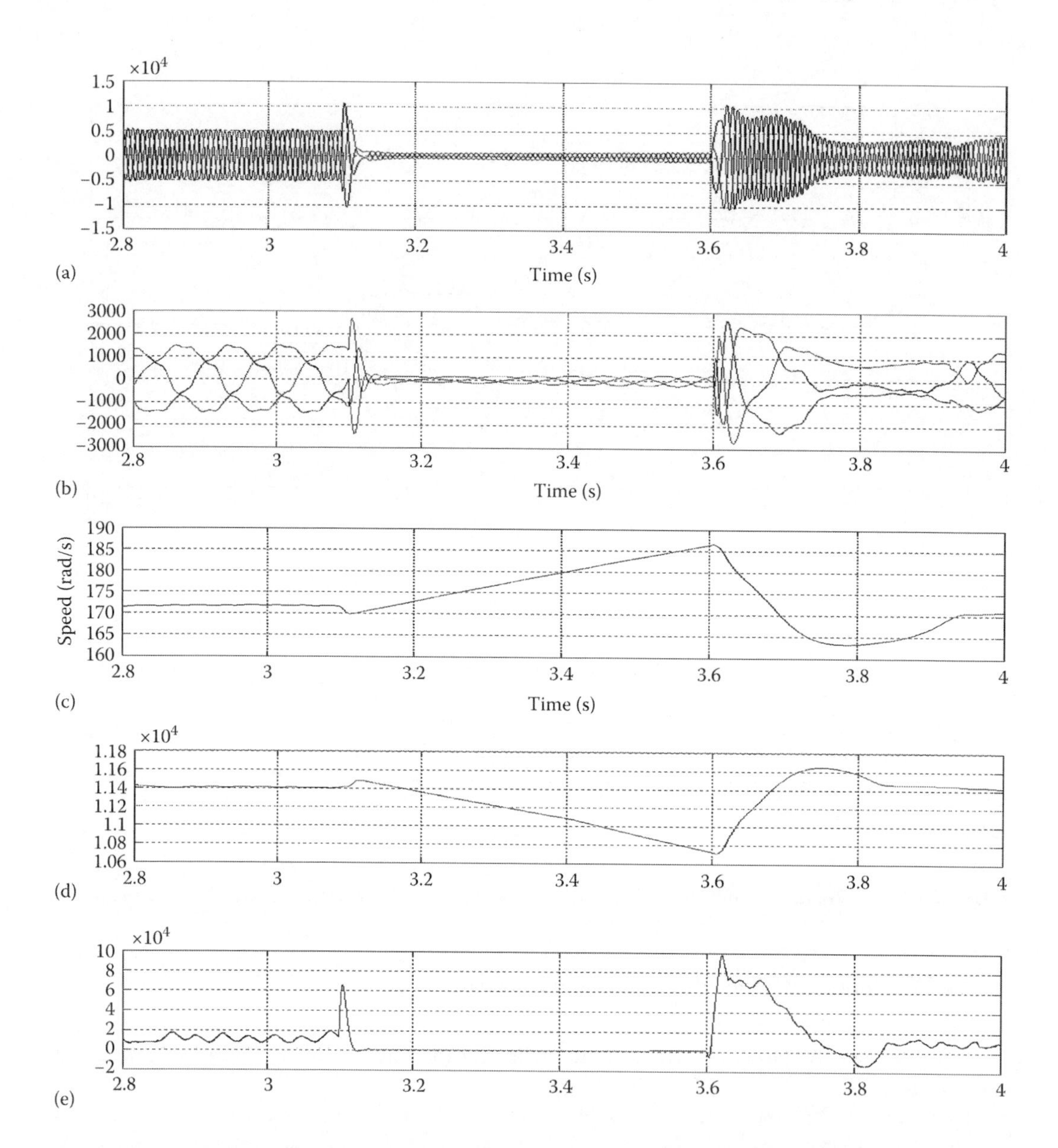

FIGURE 2.25 Three-phase short circuit on the power grid, with limited rotor currents in the generator: (a) stator currents, (b) rotor currents, (c) speed, (d) turbine torque (N m), and (e) electromagnetic torque.

The reduction of the power references during the fault in order to prevent the rotor overcurrents and generator overspeeding was performed, but it did not produce expected results. The best situation was obtained with limitation of the rotor current references. Other methods might produce even better results.

2.8 Direct Power Control of WRIG at Power Grid

Direct active and reactive power control of WRIGs stems from direct torque and flux control (DTFC) in AC electric drives [14]. In this perspective, vector control of active and reactive power of WRIG as presented in previous paragraphs should be considered as indirect control.

The active and reactive stator powers are calculated from measured voltages and currents and are closed-loop controlled by hysteresis or more advanced regulators. These regulators directly trigger a sequence of voltage vectors in the machine-side converter, based on power errors and also on the position of the stator flux in one of the six 60°-wide sectors. Consequently, coordinate transformation is apparently eliminated, and is therefore the necessity to acquire (or estimate) the rotor position.

The speed estimation (or measurement) is still needed for prime-mover control as optimum P_m (and P_s) depends on speed, and, in fact, it increases almost linearly with speed.

2.8.1 Concept of DPC

As the stator voltage V_s is considered rather constant, in general, the stator flux Ψ_s (for constant frequency ω_1) is

$$\Psi_s \approx \frac{V_s}{\omega_1} \tag{2.66}$$

with $R_s \approx 0$. At steady state, as demonstrated for vector control, in stator flux coordinates (Equation 2.43), the active stator power P_s is dependent on I_{qr}, and reactive power Q_s is dependent on I_{dr}. The rotor currents were written in stator flux coordinates. Here, we will use an alternative approach. The DPC implies directly triggering the rotor voltage vectors in rotor coordinates. Consequently, the rotor space-phasor equation is written in rotor coordinates as follows:

$$\bar{I}_r^r R_r + \bar{V}_r^r = -\frac{d\bar{\Psi}_r^r}{dt} \tag{2.67}$$

Though neglecting the resistive voltage is producing a significant error in Equation 2.67 at small values of slip S when V_r (its fundamental) is small, for the time being, we neglect it to see the greater picture of rotor flux variation that takes place along the applied rotor voltage vector:

$$\bar{\Psi}_{rf} - \bar{\Psi}_{ri} = -\int \bar{V}_r^r dt \tag{2.68}$$

The rotor flux change (increment) falls opposite to the applied voltage vector's direction, as the generator association of signs was adopted. But, the rotor flux $\bar{\Psi}_r$ is as follows:

$$\bar{\Psi}_r = \frac{L_m}{L_s}\bar{\Psi}_s + L_{sc}\bar{I}_r \tag{2.69}$$

The active and reactive stator powers P_s, Q_s (with zero stator losses) are

$$\begin{aligned} P_s &\approx -\frac{3}{2}p_1\omega_1 \mathrm{Imag}(\bar{\Psi}_r \bar{I}_r^*) \\ Q_s &\approx -\frac{3}{2}p_1\omega_1 \mathrm{Real}(\bar{\Psi}_r \bar{I}_r^*) \end{aligned} \tag{2.70}$$

with Equation 2.69, and introducing a flux power angle δ_Ψ between $\bar{\Psi}_s$ and $\bar{\Psi}_r$, P_s and Q_s become

$$\begin{aligned} P_s &\approx \frac{3}{2}p_1\omega_1 \frac{L_m \Psi_s \Psi_r}{L_s L_{sc}} \sin\delta_\Psi \\ Q_s &\approx \frac{3}{2}p_1\omega_1 \frac{L_m}{L_s L_{sc}} \Psi_r\left(\Psi_r - \Psi_s \cos\delta_\Psi\right) \end{aligned} \tag{2.71}$$

The flux angle δ_Ψ is positive for $\bar{\Psi}_r$ ahead of $\bar{\Psi}_s$ (Figure 2.4). Positive power means delivered power. Equations 2.71 lead us to make the following remarks:

- To increase active stator and reactive power delivery, for constant stator flux (constant V_s/ω_1), the rotor flux $\bar{\Psi}_r$ should increase in amplitude.
- To increase only active power P_s delivery, the rotor flux (power) angle δ_Ψ has to be increased. For a given flux power angle δ_Ψ, Equations 2.71 suggest that the reactive power generated by the machine is increased by increasing the rotor flux amplitude $\bar{\Psi}_r$.

In other words, the rotor flux space phasor has to be decreased or increased in amplitude and accelerated or decelerated. This rationale presupposes the same coordinates for rotor and stator flux (Figure 2.26).

The only problem is that the applied rotor voltage space phasor (vector) by the machine-side converter is in rotor coordinates. Under synchronous operation, the rotor voltage vector travels in the direction of motion (in rotor coordinates). Therefore, accelerating the rotor flux really means accelerating it in the direction of motion under synchronous speed. The synchronous speed acceleration of rotor flux has to be performed opposite to motion direction. This rationale is illustrated in Figure 2.27a and b.

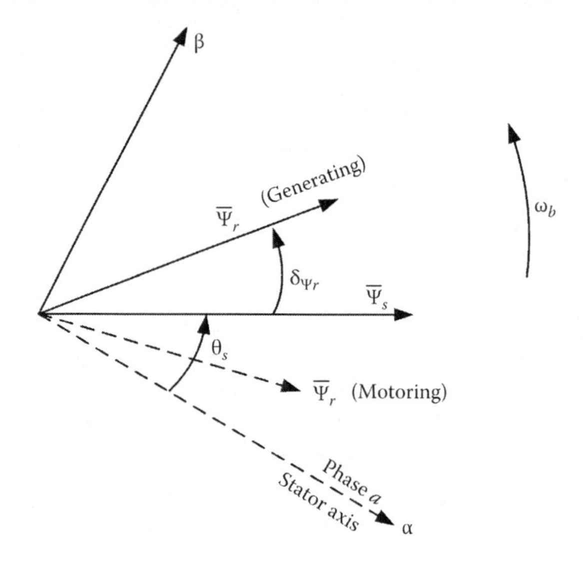

FIGURE 2.26 Stator/rotor flux angle.

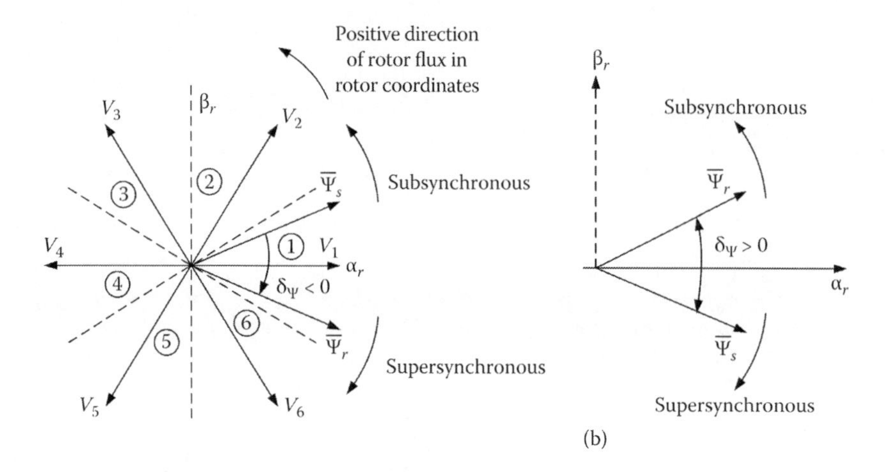

FIGURE 2.27 Stator/rotor flux vector ω_2 rotor coordinates: (a) for motoring and (b) for generating.

The rotor flux $\overline{\Psi}_r$ is estimated directly in stator coordinates from the current model:

$$\overline{\Psi}_r^s = \frac{L_r}{L_m}\overline{\Psi}_s^s - L_{sc}\overline{I}_s^s \tag{2.72}$$

with

$$\overline{\Psi}_s = -\int(\overline{V}_s - R_s\overline{I}_s)dt$$

That is, with measured stator quantities and known parameters—L_{sc} (short-circuit inductance) and coupling ratio $L_r/L_m = 1.02$ to 1.10—the rotor flux may be estimated. Based on the sign of active and reactive power errors,

$$\begin{aligned} \varepsilon_{PS} &= P_s^* - P_s \\ \varepsilon_{QS} &= Q_s^* - Q_s \end{aligned} \tag{2.73}$$

with rotor flux vector in sector K, the application of voltage vectors $V_{(K+1)}$ and $V_{(K+2)}$ would increase the delivered stator active power, while the vectors $V_{(K-1)}$ and $V_{(K-2)}$ would reduce it. Moreover, the application of $V_{(k)}$, $V_{(K-1)}$, and $V_{(K-2)}$ would decrease the delivered reactive power, while $V_{(k)}$, $V_{(K+1)}$, $V_{(K+2)}$, and $V_{(K+3)}$ would increase it. The reactive power control is the same in motor and generator operation modes.

Note that the order of voltage vectors is opposite in subsynchronous operation with respect to supersynchronous operation (Figure 2.27). The zero voltage vectors of the converter $V_{(0)}$, $V_{(7)}$ stall the rotor flux vector without affecting its magnitude. Their effect on active power is thus opposite in sub- and supersynchronous operation. Seen from the rotor side, in subsynchronous operation, V_0 increases δ_Ψ as $\overline{\Psi}_s$ travels at slip speed. Hence, the active power increases in this case.

Also, there is some influence of the zero voltage vector on the reactive power through the flux power angle δ_Ψ (Equation 2.71) increment. When δ_Ψ increases, cos δ_Ψ decreases, and thus, Q_s increases. The reverse is true when δ_Ψ decreases. Therefore, the zero voltage vector effects are different in the four modes of operation (quadrants) (Table 2.1).

It becomes rather simple to adopt the adequate rotor flux sectors.

The vector flux sector may be found by the direction of change in Q_s when a certain voltage vector is applied in sub- or supersynchronous modes, avoiding the use of the observer (Equations 2.71 and 2.72) [15]. Table 2.2 illustrates the selection of voltage vector by ε_{P_s} and ε_{Q_s} (Equation 2.73) and the sector. Zero voltage vectors are illustrated in Table 2.3.

TABLE 2.1 Zero Voltage Vector Effects on P_s, Q_r

Speed	Motoring	Generating
Subsynchronous	$\delta\Psi_r\uparrow \rightarrow Q_s\downarrow$	$\delta\Psi_r\downarrow \rightarrow Q_s\uparrow$
Supersynchronous	$\delta\Psi_r\downarrow \rightarrow Q_s\uparrow$	$\delta\Psi_r\uparrow \rightarrow Q_s\downarrow$

TABLE 2.2 Active Voltage Vectors

P_s	Q_s	Sector 1	Sector 2	Sector 3	Sector 4	Sector 5	Sector 6
$\varepsilon_{P_s} \le 0$	$\varepsilon_{Q_s} < 0$	V_3	V_4	V_5	V_6	V_1	V_2
$\varepsilon_{P_s} \le 0$	$\varepsilon_{Q_s} \ge 0$	V_2	V_3	V_4	V_5	V_6	V_1
$\varepsilon_{P_s} > 0$	$\varepsilon_{Q_s} < 0$	V_5	V_6	V_1	V_2	V_3	V_4
$\varepsilon_{P_s} > 0$	$\varepsilon_{Q_s} \ge 0$	V_6	V_1	V_2	V_3	V_4	V_5

TABLE 2.3 Zero Voltage Vectors

Speed	Motor	Generator
Subsynchronous	$\varepsilon_{Q_s} \leq 0$ and $\varepsilon_{P_s} \leq 0$	$\varepsilon_{Q_s} > 0$ and $\varepsilon_{P_s} \leq 0$
Supersynchronous	$\varepsilon_{Q_s} > 0$ and $\varepsilon_{P_s} < 0$	$\varepsilon_{Q_s} \leq 0$ and $\varepsilon_{P_s} < 0$

The influence of a voltage vector on a Q_s sign change, for a given sector, is summarized in Table 2.4.

In practice, when applying a given (known) voltage vector, the measured reactive power change sign is obtained. If the actual sign of change is opposite to that in Table 2.4, the decision is to change the sector according to Table 2.5. The source-side converter may be vector controlled, as in the previous paragraphs, for vector control. Typical experimental results are shown in Figure 2.28a and b [15].

Figure 2.29a and b show the vector information (the step-like function), P_s and Q_s for steady state, the passing-through synchronous speed (the rotor current), and Q_s and sector information during starting on the fly [15].

During starting on the fly, at the beginning, the sector information is not correct (Figure 2.29c), but it comes "to its senses" within 1 ms. A safe run through synchronous speed is visible.

More advanced P_s and Q_s controllers may be used, with a space vector modulator, to improve rotor current waveforms and make the DPC strategy even better, while still keeping it simple enough and robust.

Note that apart from vector control and DPC, feedback linearized control was recently proposed for the machine-side converter with promising results [16].

TABLE 2.4 Q_s Changes Sign (Expected)

Sector	V_0	V_1	V_2	V_3	V_4	V_5	V_6	V_7
1	0	+	+	−	−	−	+	0
2	0	+	+	+	−	−	−	0
3	0	−	+	+	+	−	−	0
4	0	−	−	+	+	+	−	0
5	0	−	−	−	+	+	+	0
6	0	+	−	−	−	+	+	0

Source: Adapted from Datta, R. and Ranganathan, V.T., *IEEE Trans.*, PE-16(3), 390, 2001.

TABLE 2.5 Direction of Sector Change (Advance)

Sector	V_0	V_1	V_2	V_3	V_4	V_5	V_6	V_7
1	0	0	−1	+1	0	−1	+1	0
2	0	+1	0	−1	+1	0	−1	0
3	0	−1	+1	0	−1	+1	0	0
4	0	0	−1	+1	0	−1	+1	0
5	0	+1	0	−1	+1	0	−1	0
6	0	−1	+1	0	−1	+1	0	0

Source: Adapted from Datta, R. and Ranganathan, V.T., *IEEE Trans.*, PE-16(3), 390, 2001.

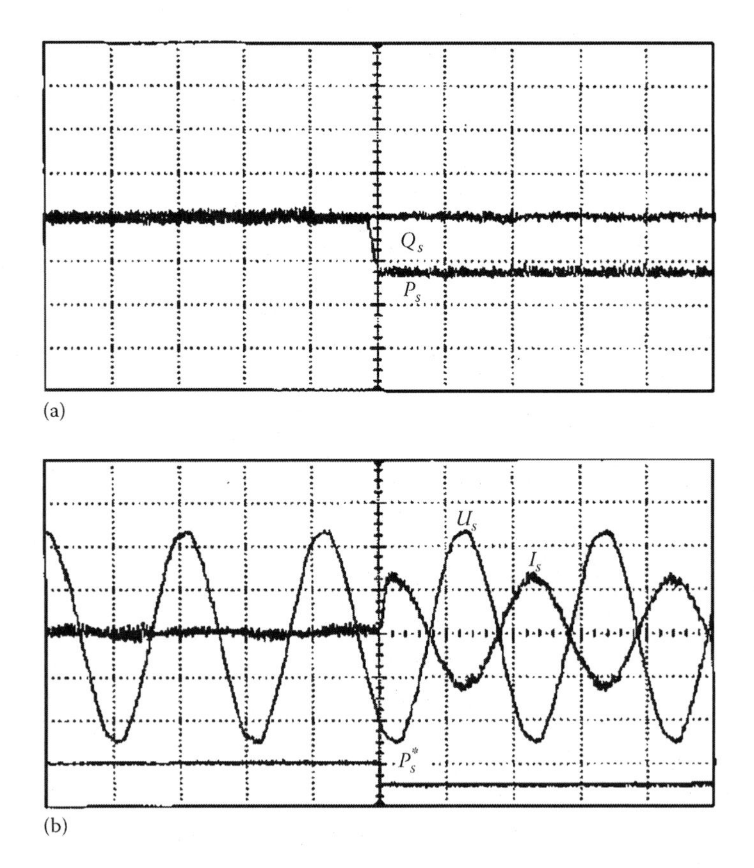

FIGURE 2.28 Direct power control (DPC) of the WRIG: (a) P_s step variation from 0 to −0.5 P.U., $Q_s = 0$ and (b) V_s, I_s transients.

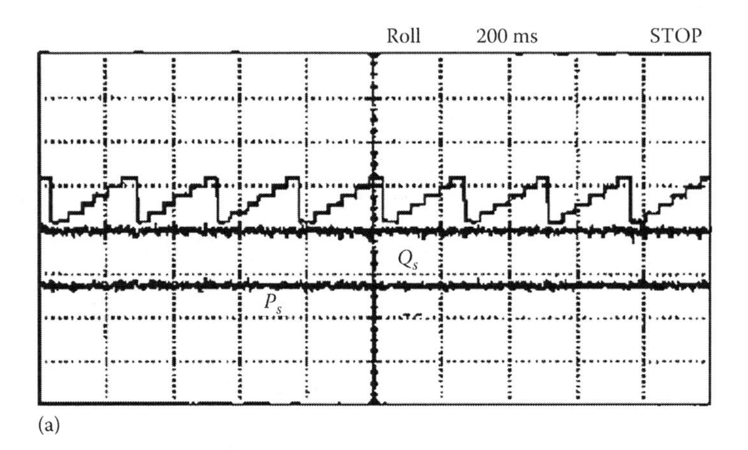

FIGURE 2.29 (a) Test results P_s, Q_s and sector step change. *(Continued)*

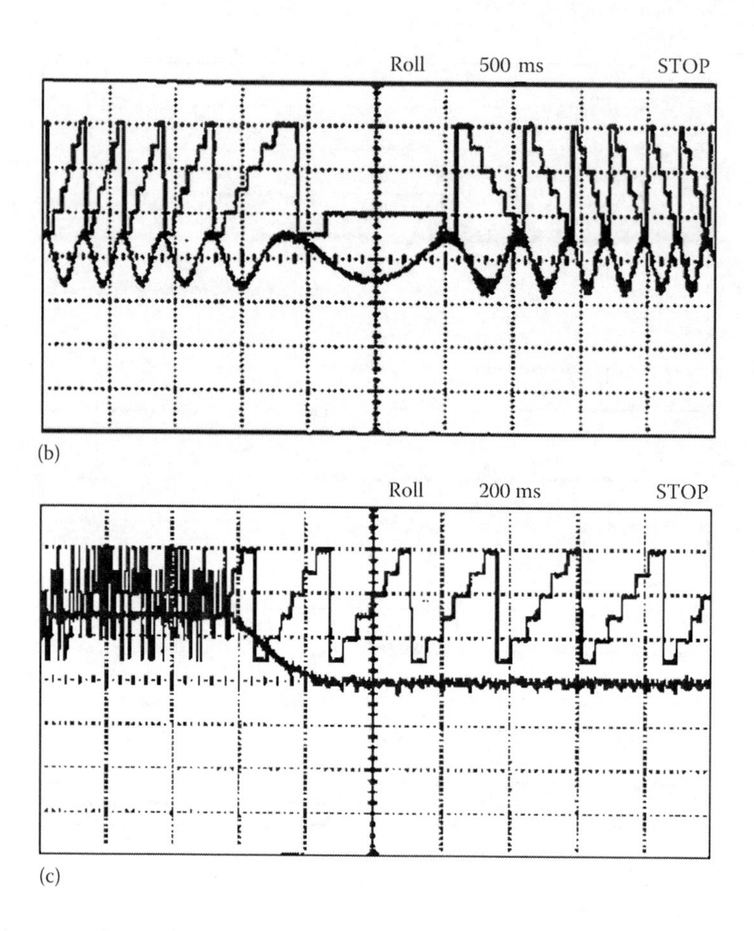

FIGURE 2.29 (*Continued*) (b) rotor current I_r during transition through synchronous speed and (c) Q_s slow ramping during start-up.

2.9 Independent Vector Control of Positive and Negative Sequence Currents

Unbalanced voltage conditions in the power network may occur, and they greatly influence the standard vector control of DPC schemes. It is, however, possible to superimpose vector control of direct and inverse sequence stator current, in order to gain some immunity to grid disturbances. Alternatively, separate phase stator current control is feasible with one stator current as zero. A typical such scheme is presented in Figure 2.30 [17].

The system response in conventional (positive sequence) and, respectively, positive plus negative sequence vector control with a 10% negative-sequence voltage is shown in Figure 2.31a and b, respectively, for a 200 kV A prototype [17] with l_m = 2.9 P.U., $l_s = l_r$ = 3.0 P.U., and $r_s = r_r$ = 0.02 P.U.

The reduction in stator phase current imbalance is notable. It is feasible even to drive the current in phase a to zero and work with only two active phases. As expected, the stator and the rotor instantaneous active powers will pulsate, but the system is capable of feeding two-phase loads with zero current summation [17] (Figure 2.32).

These enhanced possibilities of vector control add to the power quality delivered by WRIGs.

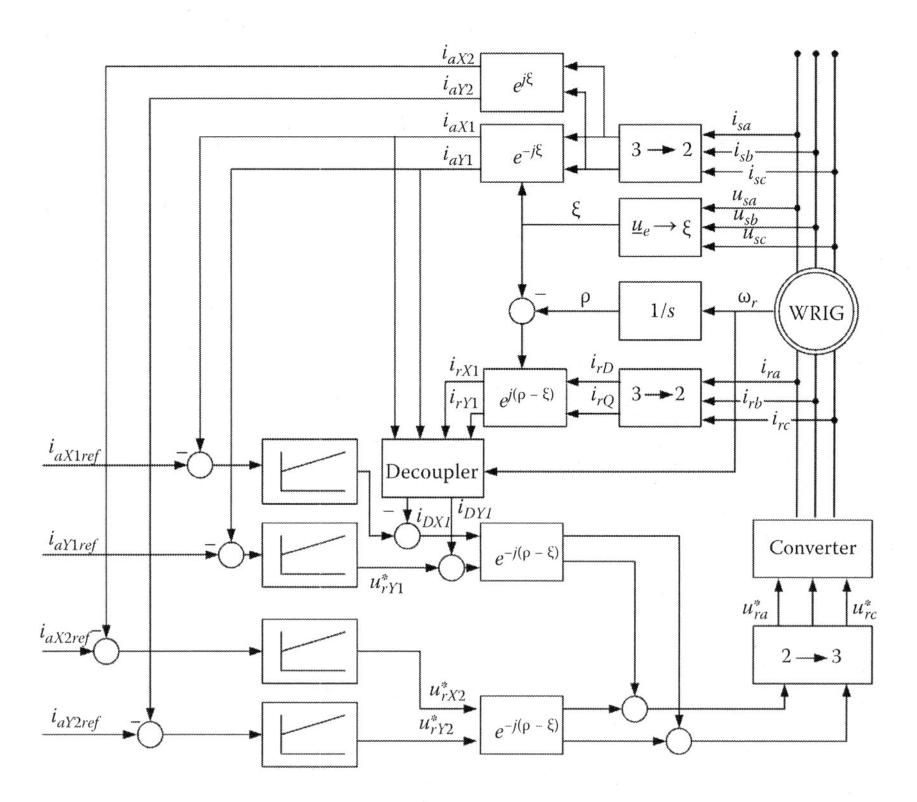

FIGURE 2.30 Positive and negative sequence current vector control of WRIG.

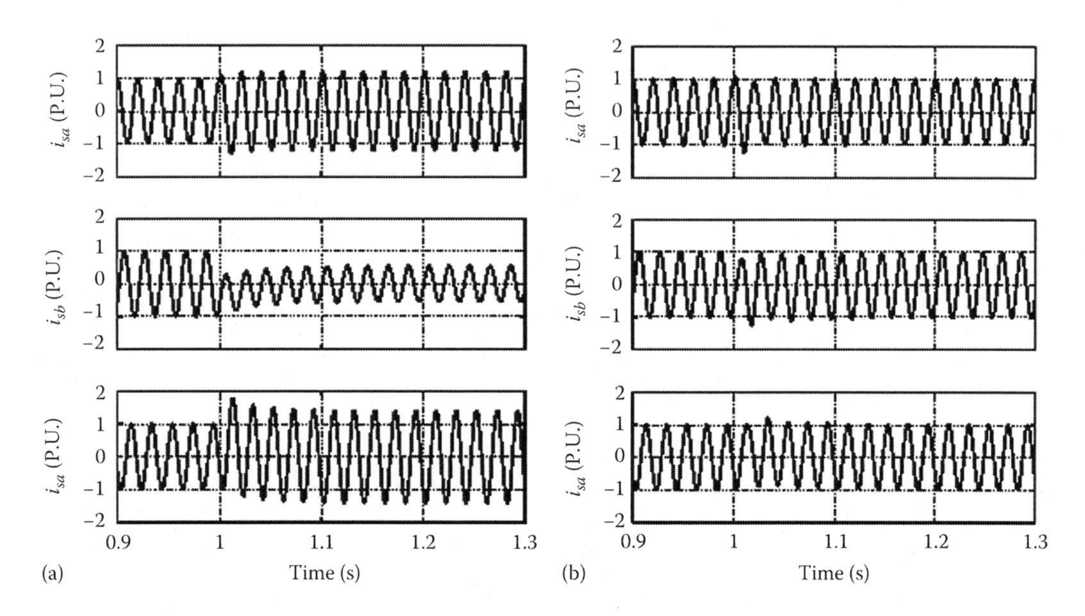

FIGURE 2.31 Response of stator phase currents with 10% negative sequence voltage: (a) positive sequence vector control and (b) positive and negative sequence control.

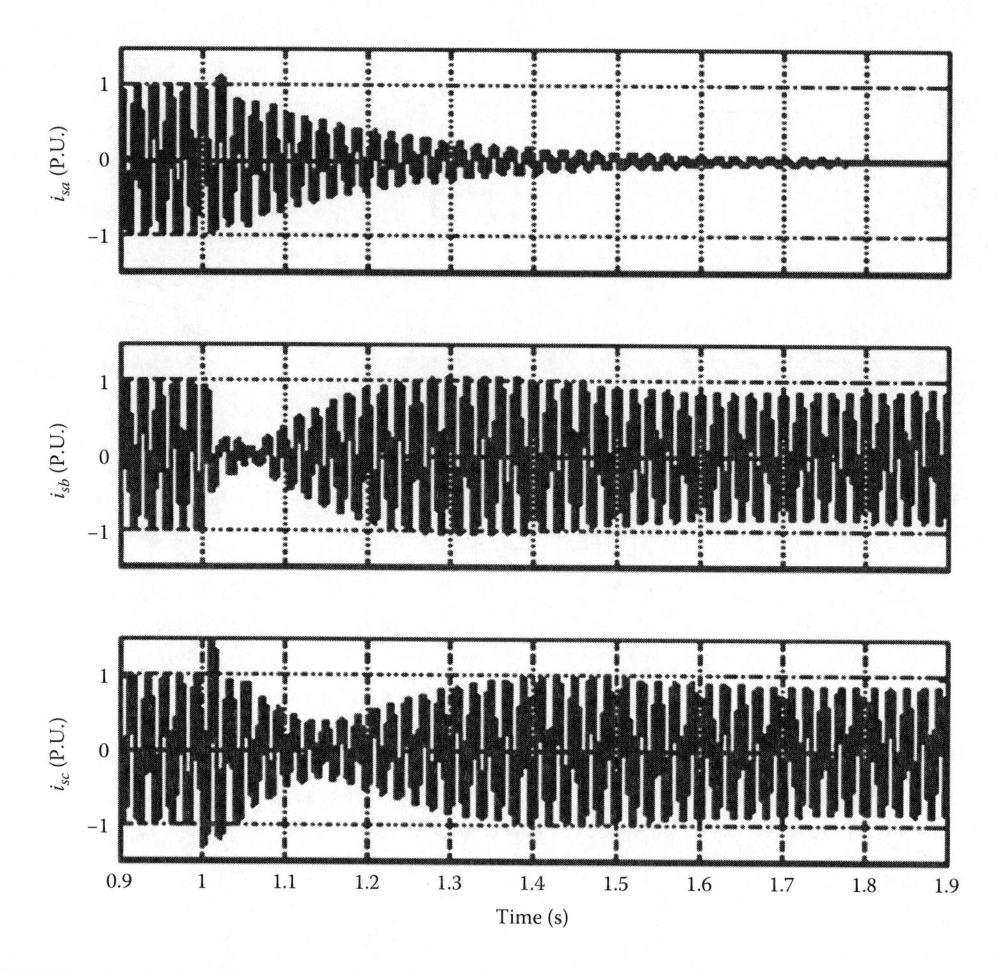

FIGURE 2.32 Stator currents in two-phase operation.

2.10 Motion-Sensorless Control

While apparently direct power control works without a rotor position sensor, vector control requires one for Park transformations. Both schemes need speed feedback to control the prime mover. Therefore, estimators or observers for rotor electrical position θ_{er} and speed ω_r are required for sensorless control.

The availability of stator and rotor currents through measurements is a great asset of a WRIG, for rotor position estimation. Also, remember that vector control of the machine (rotor)-side converter is performed in stator flux coordinates with

$$\bar{\Psi}_s = \Psi_s = L_s I_{ms}; \quad L_s = L_{sl} + L_m \tag{2.74}$$

where I_{ms} is the stator flux magnetization current.

In stator coordinates,

$$\bar{\Psi}_s = L_s \bar{I}_{ms} = L_s \bar{I}_s + L_m \bar{I}_r^s \tag{2.75}$$

The location of rotor current vector with respect to stator flux and phase a and rotor phase axis is shown in Figure 2.33.

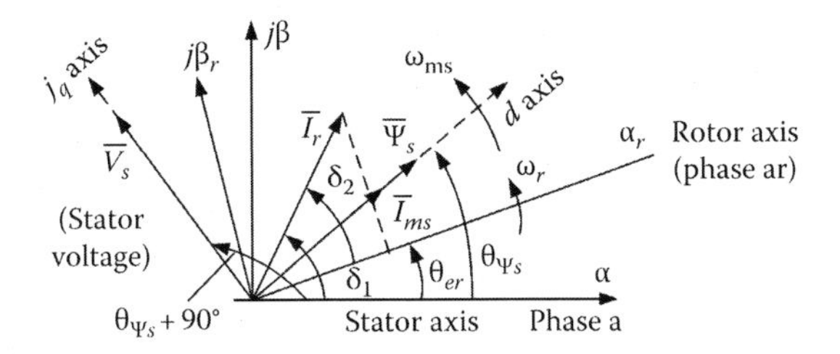

FIGURE 2.33 Location of rotor current vector with respect to stator, rotor, and stator flux axes.

The stator flux magnetization current vector $\overline{I}_{ms}$ in stator (α,β) coordinates is simply as follows (from Figure 2.33):

$$I_{ms\alpha} = I_{ms}\cos\theta_{\Psi_s}$$
$$I_{ms\beta} = I_{ms}\sin\theta_{\Psi_s} \tag{2.76}$$

Consequently, from Equation 2.76—in stator coordinates—the rotor current components are as follows:

$$I_{r\alpha} = (I_{ms\alpha} - I_{s\alpha})\frac{L_s}{L_m}; \quad I_{s\alpha} = I_a \tag{2.77}$$

$$I_{r\beta} = (I_{ms\beta} - I_{s\beta})\frac{L_s}{L_m}; \quad I_{s\beta} = \frac{1}{\sqrt{3}}(2I_b + I_a) \tag{2.78}$$

$$I_r = \sqrt{I_{r\alpha}{}^2 + I_{r\beta}{}^2} \tag{2.79}$$

But, the rotor currents are directly measured in rotor coordinates:

$$\cos\delta_2 = \frac{I_{r\alpha r}}{I_r}$$
$$\sin\delta_2 = \frac{I_{r\beta r}}{I_r} \tag{2.80}$$

with $I_{r\alpha r}$, $I_{r\beta r}$ calculated from measured i_{ar}, i_{br}, and i_{cr}:

$$I_{r\alpha r} = I_{ar}, \quad I_{r\beta r} = (2I_{br} + I_{ar})/\sqrt{3} \tag{2.81}$$

The unknowns here are the $\sin\theta_{er}$ and $\cos\theta_{er}$, that is, the sine and cosine of the rotor electrical angle θ_{er}:

$$\theta_{er} = \delta_1 - \delta_2 \tag{2.82}$$

Therefore,

$$\sin\theta_{er} = \sin(\delta_1 - \delta_2) = \sin\delta_1\cos\delta_2 - \sin\delta_2\cos\delta_1 = \frac{(I_{r\beta}\cdot I_{r\alpha r} - I_{r\alpha}\cdot I_{r\beta r})}{I_r^2}$$
$$\cos\theta_{er} = \cos(\delta_1 - \delta_2) = \cos\delta_1\cos\delta_2 + \sin\delta_2\sin\delta_1 = \frac{(I_{r\alpha}\cdot I_{r\alpha r} - I_{r\beta}\cdot I_{r\beta r})}{I_r^2} \tag{2.83}$$

In fact, provided the stator flux magnetization current $\bar{I}_{ms}$ is known, sin θ_{er} and cos θ_{er} are obtained in the next time sampling step without delay. The fastest response in rotor position information is thus obtained. The rotor speed ω_r may be calculated from sin θ_{er} and cos θ_{er} as follows:

$$\frac{d\theta_{er}}{dt} = \hat{\omega}_r = -\sin\theta_{er}\frac{d(\cos\theta_{er})}{dt} + \cos\theta_{er}\frac{d(\sin\theta_{er})}{dt} \tag{2.84}$$

A digital filter has to be used to reduce noise in Equation 2.84 and thus obtain a usable speed signal.

Though at constant stator voltage and frequency, Ψ_s is constant, and so is I_{ms}, under faults or voltage disturbances in the power grid, Ψ_s (and thus I_{ms}) varies. It is, however, practical to start with the rated value of I_{ms} and calculate the first pair of rotor current components in stator coordinates $I'_{r\alpha r}$, $I'_{r\beta r}$:

$$\begin{aligned}
I'_{r\alpha}(k) &= I_{r\alpha r}(k)\cos\theta_{er}(k-1) - I_{r\beta r}(k)\sin\theta_{er}(k-1)\\
I'_{r\beta}(k) &= I_{r\beta r}(k)\cos\theta_{er}(k-1) + I_{r\alpha r}(k)\sin\theta_{er}(k-1)\\
I'_{ms\alpha}(k) &= I_{s\alpha}(k) + \frac{L_m}{L_s}I'_{r\alpha}(k)\\
I'_{ms\beta}(k) &= I_{s\beta}(k) + \frac{L_m}{L_s}I'_{r\beta}(k)
\end{aligned} \tag{2.85}$$

These new values of $\left(I'_{ms\alpha}, I'_{ms\beta}\right)$ will then be used in the next time step computation cycle. Therefore, I_{ms} is always one time step behind; but as the sampling time is small, it produces very small errors, even if a 1 ms delay low-pass filter on $I'_{ms}(k)$ is used. The magnetic saturation curve is required to account for saturation with rotor position estimation to yield good results when the stator flux varies notably, as in grid faults. Typical rotor position estimation results are shown in Figure 2.34 [18].

The speed estimation before and after filtering is shown in Figure 2.35a and b [18].

As $L_m/L_s = 1 - L_{sl}/L_s$ is close to unity, even a ±50% error in L_{sl}/L_s leads to negligible magnetization current estimation errors. The present θ_{er} and ω_r estimation method, with some small changes, also works when the stator currents are zero. Such a situation occurs when the synchronization conditions to connect the WRIG to the power grid are prepared. Other θ_{er} and ω_r estimators were proposed and shown to give satisfactory results [19,20].

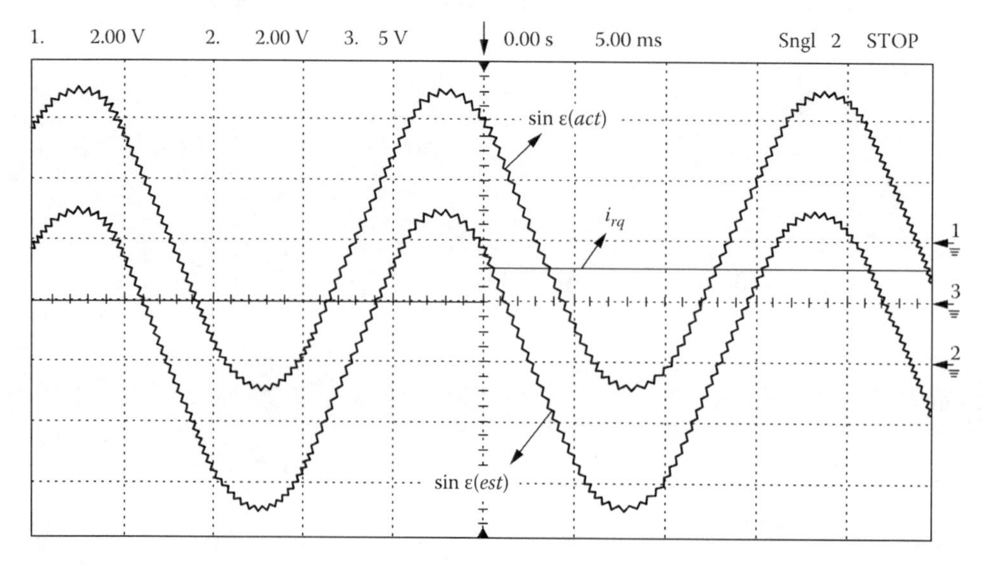

FIGURE 2.34 Estimated and actual rotor position (sin θ_{er}) for step increase V^*_{rq}.

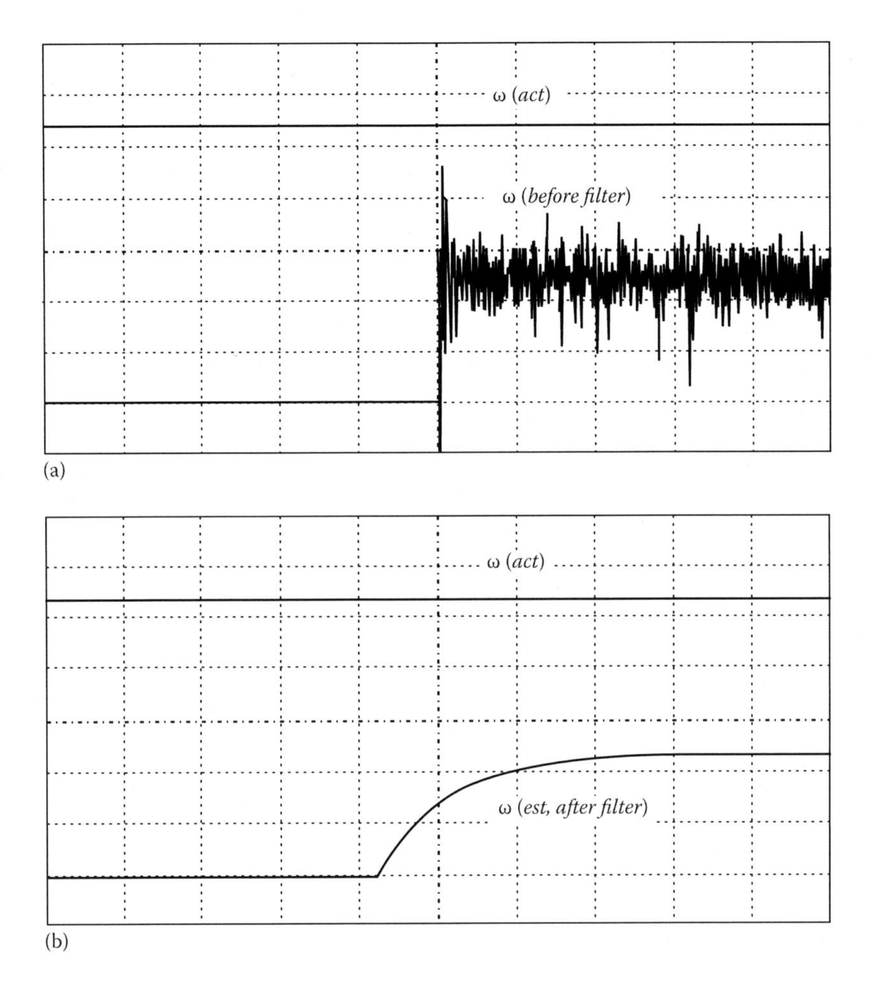

FIGURE 2.35 Estimated speed transients: (a) before filtering and (b) after filtering.

2.11 Vector Control in Stand-Alone Operation

The stand-alone operation mode may be accidental, when the power grid is not a good recipient of power, or intentional. Ballast loads are supplied in the first case. In principle, it is possible to maintain P_s, Q_s vector control as described earlier. But, the stator flux Ψ_s and frequency ω_1 are imposed, rather than estimated.

When the WRIG is isolated, self-excitation is required first, on no-load. But, after that, the vector control of the rotor source-side converter in stator voltage (or flux) orientation has a problem, as the stator voltage has harmonics even if a filter is used. Consequently, to "clean up" the noise, the stator voltage vector position θ_{V_s} is calculated as follows:

$$\theta_{V_s} = \int \omega_1^* dt + \frac{\pi}{2} = \theta_{\Psi_s} + \frac{\pi}{2} \tag{2.86}$$

where ω_1^* is the reference stator frequency. A controlled ballast load is needed to compensate for the difference between mechanical power and electrical load. The control of the source-side converter remains as that for grid operation, with the q axis (reactive power) current reference as zero, in general.

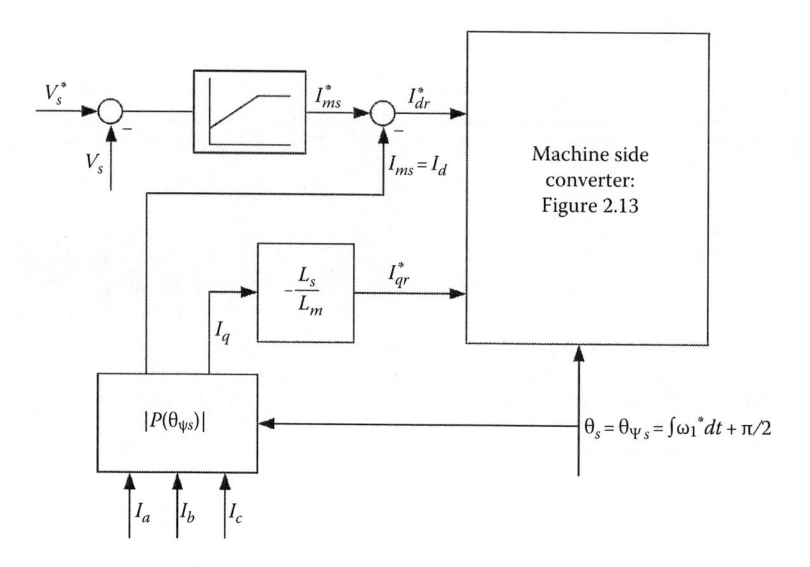

FIGURE 2.36 Machine-side converter vector control for stand-alone operation.

The machine-side converter control will have the active and reactive power regulators replaced by magnetization current regulators (to keep stator voltage constant), while along axis q,

$$I_{qr}^* = -\frac{L_s I_q}{L_m} \tag{2.87}$$

To provide for alignment along the stator flux axis,

$$\Psi_{ds} = \Psi_{\alpha s}\cos\theta_{\Psi_s} + \Psi_{\beta s}\sin\theta_{\Psi_s} \tag{2.88}$$

$$\Psi_{\alpha s,\beta s} = -\int (V_{s\alpha,\beta} - R_s I_{s\alpha,\beta})dt \tag{2.89}$$

$$I_{ms} = \frac{\Psi_s}{L_s} \tag{2.90}$$

It would be adequate to add one more regulator for magnetization current.

The corroboration of V_{DC}^* in the DC link with V_s^* is a matter of optimization (Figure 2.36). The turbine is commanded in the torque mode, and the torque regulator produces the control variable for ballast load to match the mechanical and electrical power when the actual load varies. More on stand-alone operation is available in the literature [21,22].

2.12 Self-Starting, Synchronization, and Loading at the Power Grid

There are situations, such as with pump-storage hydropower plants, when the WRIG needs to start as an induction motor with short-circuited stator until slowly the speed reaches $\omega_o > \omega_{r\min} = \omega_1(1 - S_{\max})$. Then, the stator terminals are opened, and the machine floats with the speed slightly decreasing. During a short period (40–50 ms), the synchronization conditions are prepared by adequate control. In essence, the stator voltage amplitude and phase angle differences (errors) are driven to zero by close-loop control of I_{dr} and I_{qr}—the rotor current components. In the process, the rotor phase sequence used for starting to ω_o has to be reversed to prepare for synchronization after the stator windings are opened.

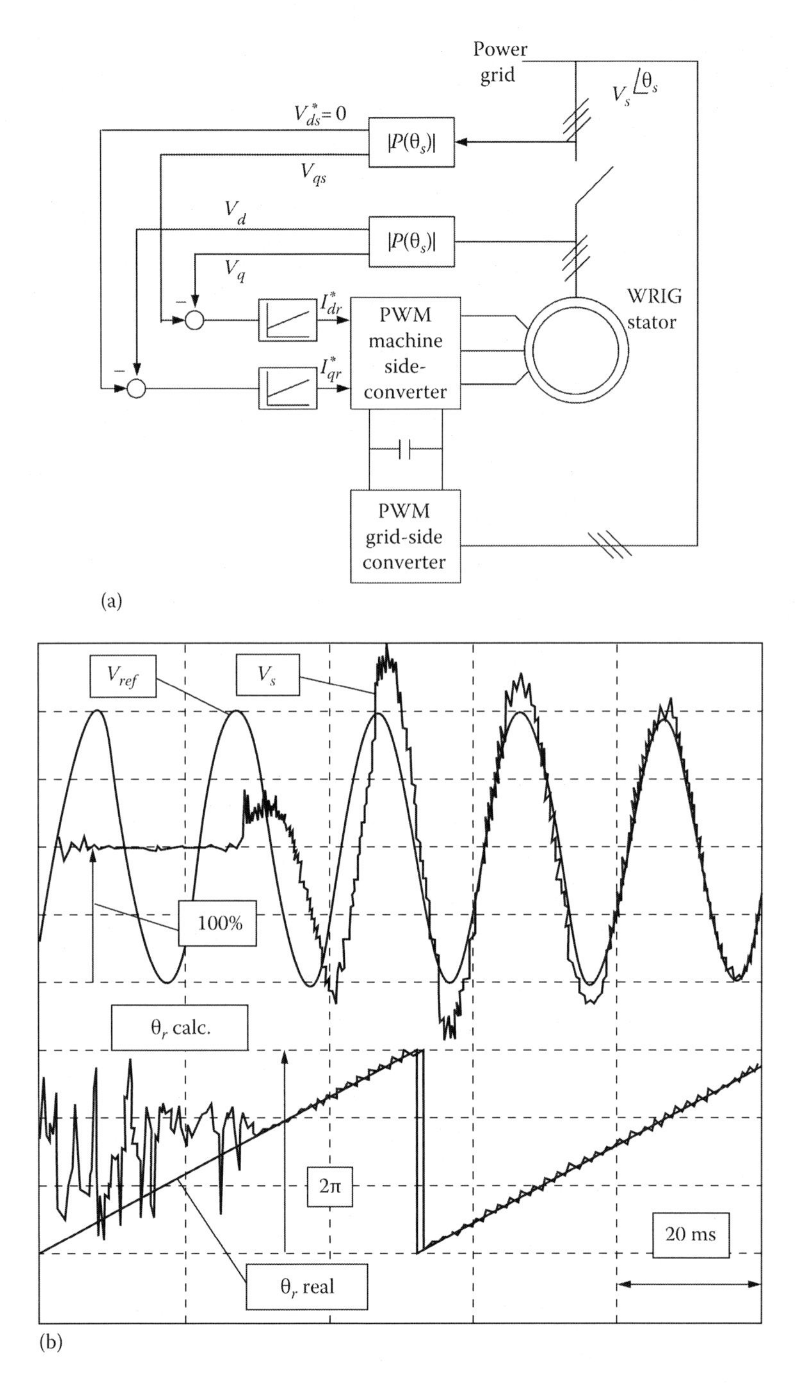

FIGURE 2.37 Mode II (synchronization): (a) control scheme and (b) results.

Therefore, instead of active and reactive power regulators (Figure 2.13), the voltage amplitude and phase angle error regulators are introduced (Figure 2.37).

Fast synchronization can be seen in Figure 2.37. Starting as a motor with the short-circuited stator may be done with or without an autotransformer. The autotransformer is disconnected at a certain speed ω_{ins} above a certain synchronization speed ω_o, in order to take advantage of higher voltage and, thus, allow for a lower rating static power converter to be connected to the rotor (Figure 2.38a and b) [19].

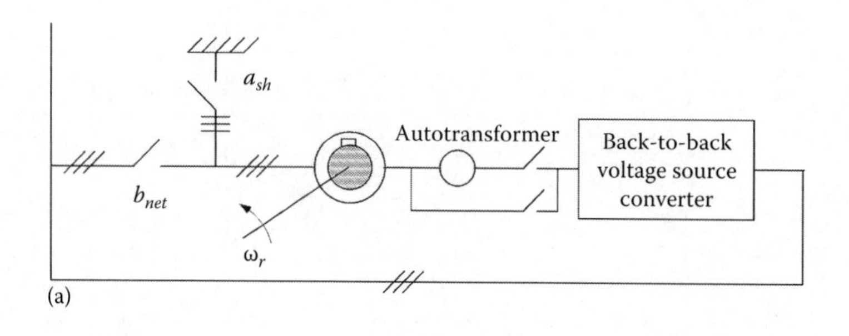

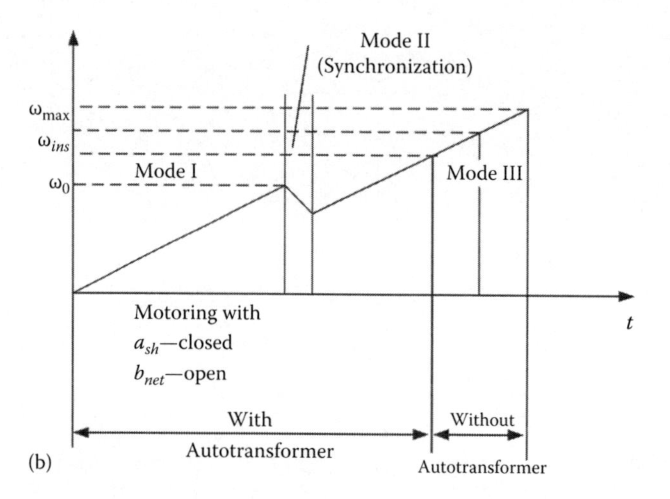

FIGURE 2.38 The three-operation modes (a) and the speed vs. time (b).

Control during motor starting with a short-circuited stator may be done, and also with the vector control scheme of Figure 2.21, it must only be slightly modified. For example, the reactive power regulator will be bypassed. The reactive rotor reference current I_{dr}^* is kept constant and positive as the machine magnetization is produced through the rotor. The current I_{dr}^* should correspond to the rated no-load current of the WRIG. The active power regulator in Figure 2.13 should be replaced by a speed regulator (Figure 2.39).

To illustrate the transient response of WRIG for generating and pumping operation at the power grid, results for a 400 MW unit are shown in Figures 2.40 and 2.41 [22]. This large WRIG was also started with a shorted stator for the pumping operation mode.

The response is smooth and fast, proving the vector control capability to handle decoupled active and reactive power control.

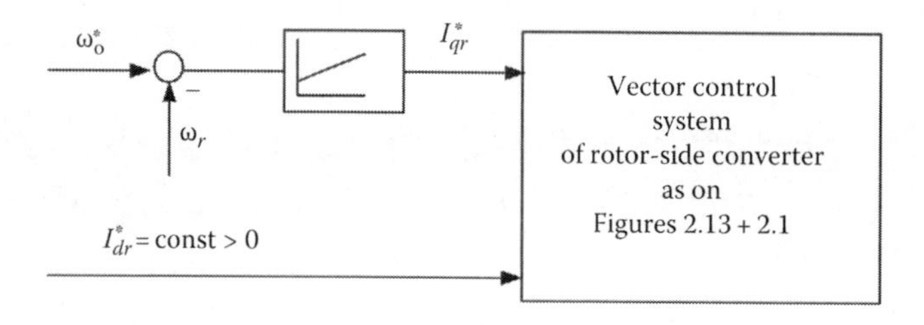

FIGURE 2.39 Rotor reference currents during shorted-stator motor starting to speed $\omega_o > \omega_{rmin}$.

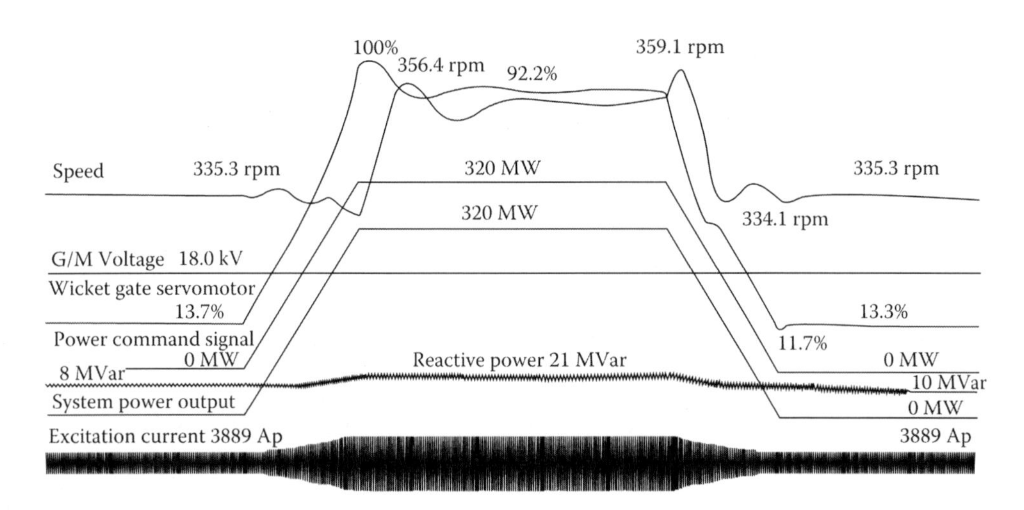

FIGURE 2.40 WRIG (Ohkawachi unit 4) ramp response for generating mode. (From Kawabara, T. et al., *IEEE Trans.*, EC-11(2), 376, 1996.)

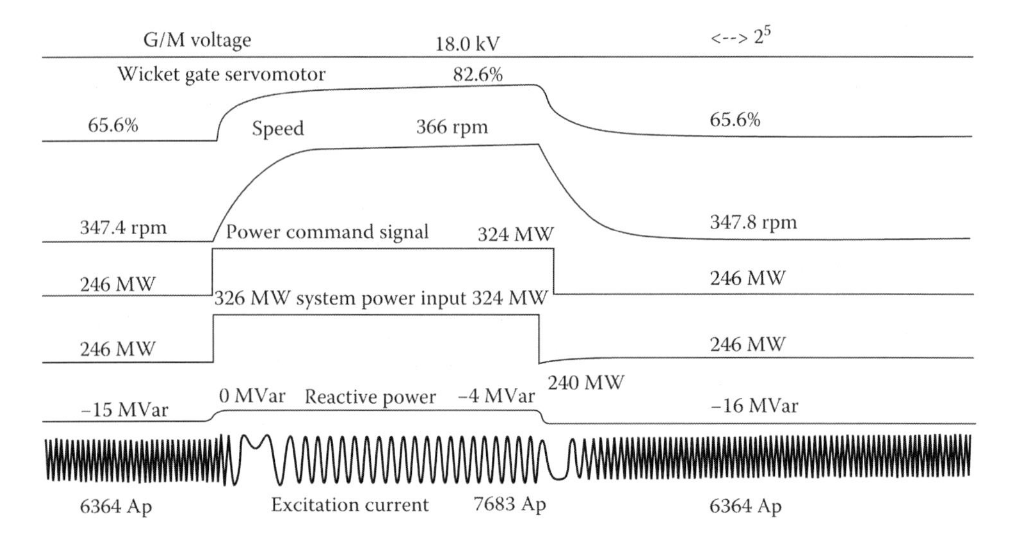

FIGURE 2.41 WRIG (Ohkawachi unit 4) ramp response for pumping (motoring) mode. (From Kawabara, T. et al., *IEEE Trans.*, EC-11(2), 376, 1996.)

2.13 Voltage and Current Low-Frequency Harmonics of WRIG

Current harmonics in a WRIG originate mainly from the following:

- Grid voltage harmonics
- Winding space harmonics
- Switching behavior of the static power converter connected to the rotor

The high-frequency harmonics are left aside here. As the power grid is connected to both the stator (directly) and the rotor (through the AC–AC converter), the voltage harmonics influence the stator and rotor currents. The converter has to handle fundamental power control and also act as a filter for these harmonics.

Alternatively, active filters may be placed between the power grid and the nonlinear loads [23]. In what follows, the low-frequency source current and equations are provided and used as a basis for feed-forward control added to fundamental vector control of a WRIG, to attenuate current harmonics.

The significant voltage harmonics due to the power grid are of the order $6k + 1$ (positive sequence) and $6k - 1$ (negative sequence) in stator coordinates and $\pm 6k$ in stator voltage fundamental vector coordinates:

$$V_s = \hat{V}_{s1}\left(1 + \sum_k \underline{C}_{6k-1} e^{-6jk\omega_1 t} + \sum_k \underline{C}_{6k+1} e^{+6jk\omega_1 t}\right) \tag{2.91}$$

The distorted supply voltage produces current harmonics through the main (sine filter) choke (L_c, R_c) placed in series on the source-side (front end) converter. They are driven by the voltage drop over the main choke:

$$V_{sv} - V_{invv} = R_c I_{s,v} + j\omega_{sv} L_c I_{s,v} \tag{2.92}$$

where V_{sv}, V_{invv} are, respectively, the vectors of harmonics in supply-side converter and source-side output voltages. On the other hand, the nonsinusoidal distribution of WRIG windings (placed in slots) produces rotor flux harmonics of $6k + 1$ orders. In rotor coordinates, they are of the following form:

$$\Psi_r^r = \Psi_{r1}\left(1 + \sum_k \underline{C}_{\Psi 6k-1} e^{-6jk\omega_r t} + \sum_k \underline{C}_{\Psi 6k+1} e^{+6jk\omega_r t}\right) e^{j(\omega_1 - \omega_r)t} \tag{2.93}$$

They produce stator current harmonics of the same order. In stator coordinates,

$$\bar{I}_s^s = I_{s1}\left(1 + \left(\sum_k \underline{C}_{i6k-1} e^{-6jk\omega_r t} + \sum_k \underline{C}_{i6k+1} e^{+6jk\omega_r t}\right) e^{j(\omega_1 - \omega_r)t}\right) \tag{2.94}$$

The amplitude of stator current harmonics thus produced depends heavily on the machine winding type and on power level. They tend to decrease in large power machines but are still nonnegligible.

To compensate for these harmonics of stator current, the machine behavior toward them should be explored. Rewriting the space-phasor equation for the fundamental (Equation 2.15) in stator coordinates ($\omega_b = 0$), for rotor flux $\bar{\Psi}_r$ and stator current $\bar{I}_s$ vectors as variables, we obtain the following:

$$\bar{V}_s - \bar{V}_r \frac{L_m}{L_r} = \left(R_s + \frac{L_m^2}{L_r^2} R_r\right)\bar{I}_s + L_{sc}\frac{d\bar{I}_s}{dt} + \frac{L_m}{L_r}\left(j\omega_r - \frac{R_r}{L_r}\right)\bar{\Psi}_r \tag{2.95}$$

The same equation may be applied for current (voltage) harmonics, provided the influence of frequency on resistances and inductances is considered ($R\nu$, $L_{sc\nu}$). As the rotor flux varies slowly (with the time constant $T_r = L_r/R_r$), the stator voltage harmonics are not followed by $\hat{\Psi}_r$, which then may be eliminated from Equation 2.95:

$$\bar{V}_{s,\nu} - \frac{L_m}{L_\nu}\bar{V}_{r,\nu} \approx (R_\nu + j\nu\omega_1 L_{sc\nu})\bar{I}_{s,\nu} \tag{2.96}$$

On the other hand, harmonics originating from rotor flux are not reflected in the stator voltage, and Equation 2.95 gets the following form:

$$\frac{L_m}{L_r}\left(-j\omega_r + \frac{R_r}{L_s}\right)\overline{\Psi}_{r,\nu} - \frac{L_m}{L_r}\overline{V}_{r,\nu} = (R_{\nu r} + j\nu_r\omega_r L_{scvr})\overline{I}_{s,\nu} \tag{2.97}$$

These rotor-flux-produced stator current harmonics only slightly depend on speed, as speed occurs on both sides of Equation 2.97. However, V_r increases with speed almost linearly in WRIG. Consequently, the compensation voltage, equal and opposite to the left-hand side of Equation 2.97, is proportional to speed. The elimination of stator current harmonics in Equations 2.96 and 2.97 should be done by making the voltage difference on their left side zero by proper compensation. Feed-forward compensation is such a favored method. Advantages of Equations 2.92, 2.96, and 2.97 are taken in such a typical scheme, which are required for each harmonic (Figure 2.42).

The output of the feed-forward controller enters the vector control rotor (machine)-side converter at the rotor voltage references level in synchronous coordinates. It goes without saying that the controller bandwidth has to be increased to cope with these new higher frequency (hundreds of hertz) components. Typical results with 70%–80% compensation of fifth and seventh harmonics are shown in Figure 2.43a

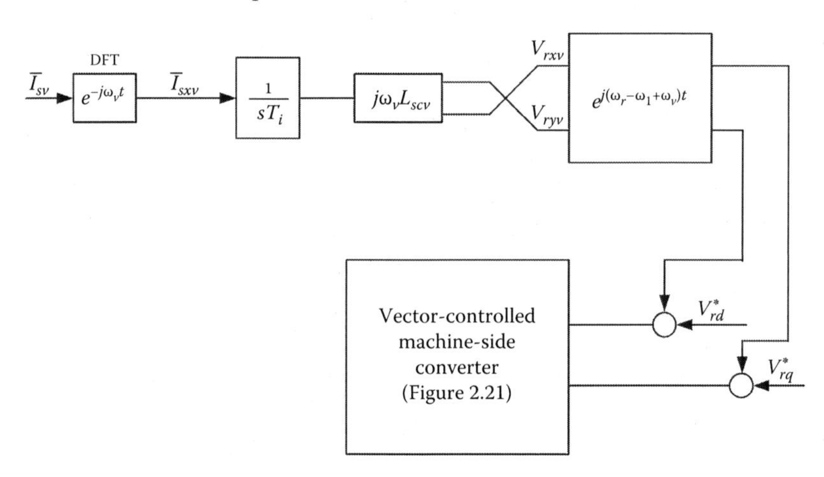

FIGURE 2.42 Feed-forward compensation stator current harmonic ν.

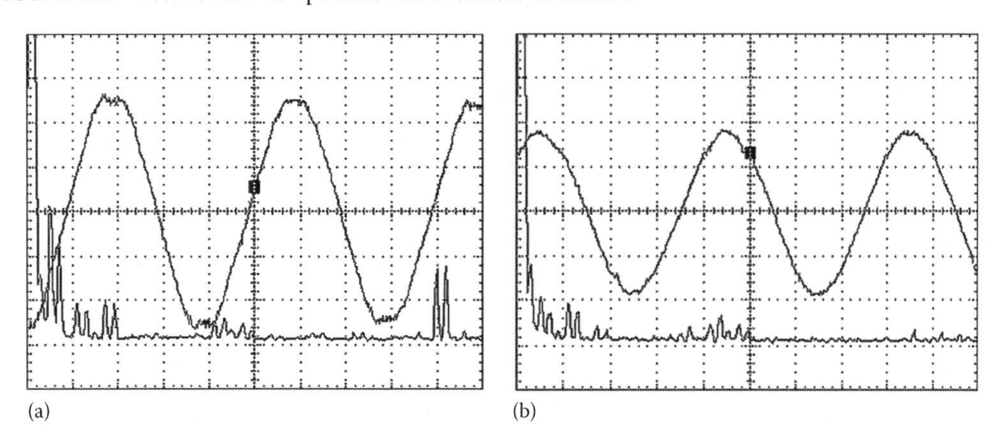

FIGURE 2.43 Stator current harmonic feed-forward compensation: (a) 4 kW WRIG and (b) 2.5 MW WRIG. (From Dittrich, A., Compensation of current harmonics in doubly-fed induction generator system, in: *Record of EPE-2001*, Graz, Austria, 2001, pp. P1–P8.)

for a 4 kW WRIG and in Figure 2.43b for a 2.5 MW WRIG [24]. Though the fifth and seventh current harmonics are larger for the 4 kW machine, their reduction is also worthwhile in the 2.5 MW machine to improve power quality.

Note that converter-produced harmonics are in the hundreds of hertz for thyristor rectifier–current source inverter converter WRIGs and in the kilohertz range for DC voltage link PWM AC–AC converters with IGBTs or IGCTs. They strongly depend on the control and converter switching (PWM) method. And, they are larger and more damaging in DC current link AC–AC converter WRIGs [25]. In DC voltage link PWM AC–AC converters, the main harmonics are around the fixed (if fixed) pulse frequency and its multipliers. Multiple-level voltage PWM AC–AC indirect converters are a good solution for raising the pulse frequency, as seen in the rotor currents.

But essentially, the reduction of converter produced harmonics takes place with the use of the PWM techniques used for the scope. Random switching or randomized switching frequency or pulse positioning are favored methods to reduce the converter-caused line current harmonics. A thorough analysis of various PWM techniques with their effects on stator and rotor current harmonics is given in Reference 26. The common modulation plus triple sine is found to be appropriate. Also, shifting the pulse period of the rotor-side PWM converter by a quarter of a period with respect to the source-side PWM converter produces a reduction of the harmonics spectrum at the double pulse frequency. Passive or active filters may also be used for the scope. More on WRIG transients is available in [27] and [28].

2.14 Ride-Through Control of DFIG under Unbalanced Voltage Sags

As DFIG under symmetric voltage sags has been analyzed previously in this chapter, here special attention is given to ride-through control under unbalanced voltage sags in the power grid.

A pertinent phenomenological description of DFIG behavior to unbalanced voltage sags is given in Reference 29, where also a few fundamental ways to handle it are also illustrated.

Among them is the use of the crowbar for a short interval, around 10 ms or so, to limit the DC link voltage initiation transients and to divert rotor overcurrents due to unbalanced voltage sags [30]. As DFIG is supposed to ride through short transient unbalanced loads, the crowbar short circuit is initiated at a $V_{DC\max}$ and opened at $V_{DC\min}$.

In general, as already pointed out in this chapter, ± vector control of dq rotor currents is used to handle unbalanced voltage sags to prevent large asymmetries in the machine currents and pulsations in the electromagnetic torque and in the DC link voltage.

It was soon realized that meeting all three just mentioned goals is not feasible by intervening only on one of the two parts of the back-to-back converter [31]. The same conclusion was reached either with vector or with direct power control [32].

Skipping the details, comparative results for 50% two-phase voltage sags for a 2 MW DFIG under conventional (symmetric) T1 and both converters ride-through dedicated control, T3, in Figure 2.44a and b [31] in DFIG torque and DC-link voltage are considered conclusive.

Similarly good results have been obtained with dedicated direct power control (DPC) when, again, both RSC and GSC controls are changed, are shown in Figure 2.45 [32].

Though the power pulsations are spectacularly reduced by the improved DPC (Figure 2.45b), the DFIG torque pulsations, though reduced, are still there.

Also, some 150 Hz pulsations occur in the stator currents, just to show that more work is to be done.

Yet another attempt, to deal with these additional problems during unbalanced voltage sags, uses besides positive and negative sequence control, novel current controllers that consist of the standard PI regulator plus a dual-frequency resonant (DFR) compensator tuned at $2\omega_1$ and $6\omega_1$, to regulate the fundamental, fifth and seventh harmonics simultaneously [33].

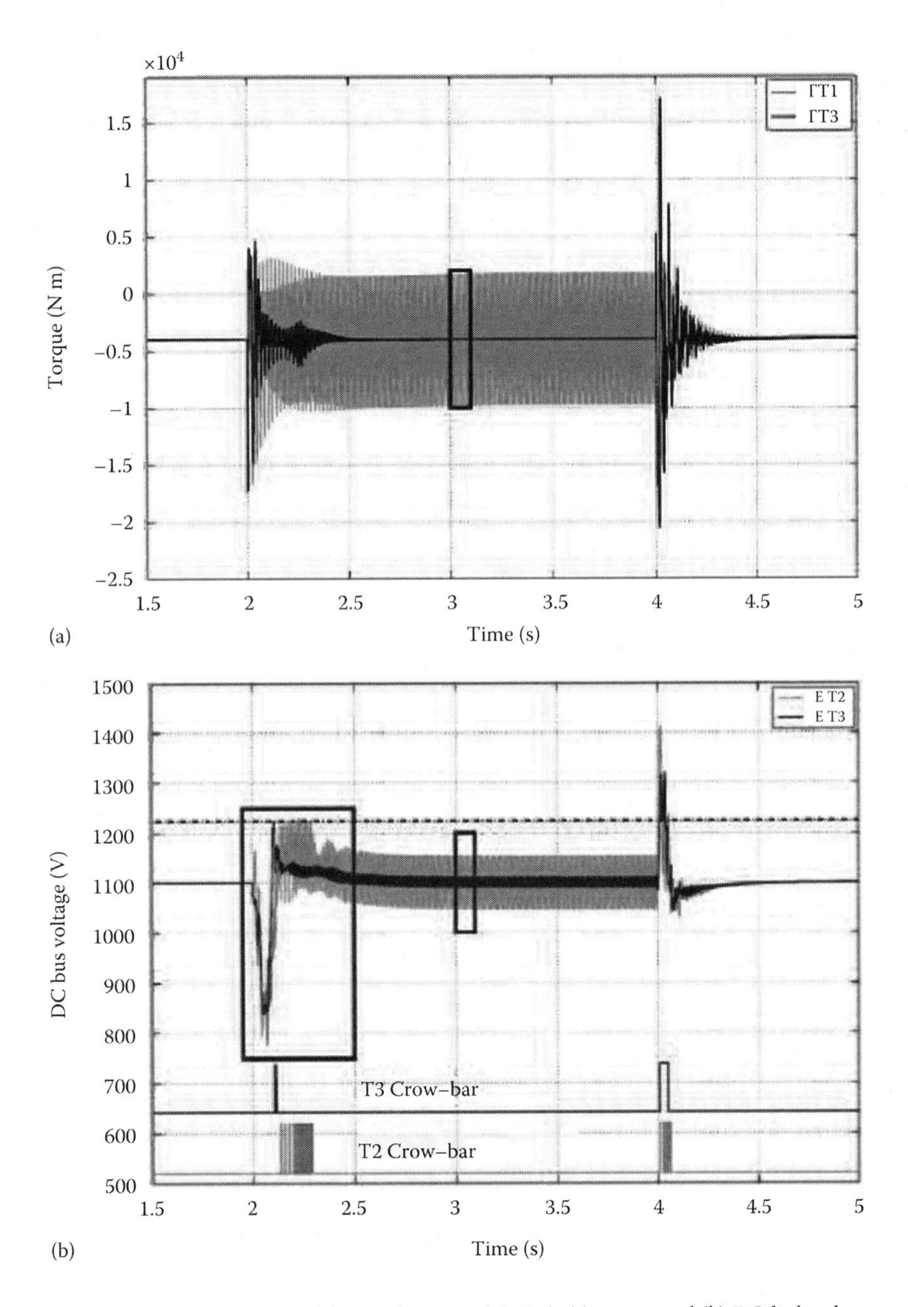

FIGURE 2.44 DFIG through 50% two-phase voltage sag (2 MW): (a) torque and (b) DC-link voltage and dedicated (T3) control. (Adapted after Gomis-Bellmunt, O. et al., *IEEE Trans.*, EC-23(4), 1036, 2008.)

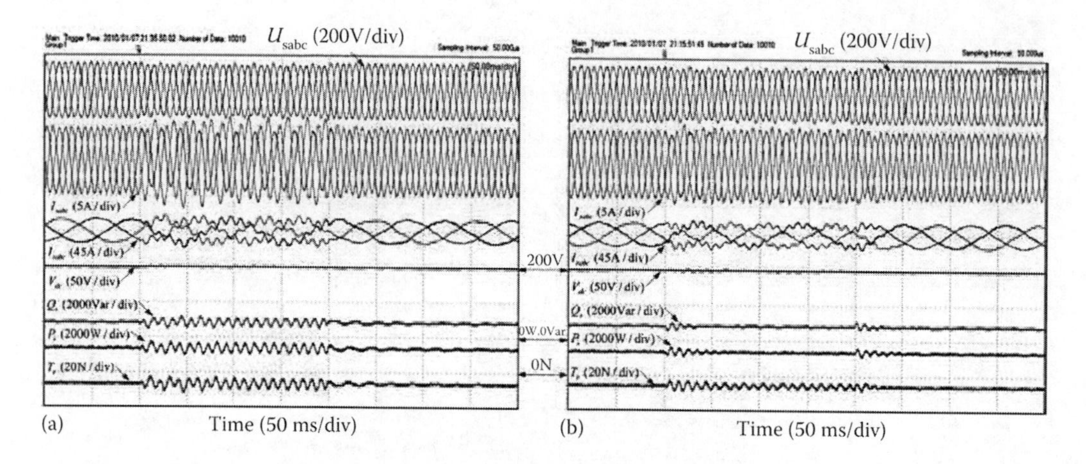

FIGURE 2.45 DPC of DFIG under 200 ms long 80% single phase grid voltage dip: (a) conventional DPC and (b) improved DPC. (Adapted after Nian, H. et al., *IEEE Trans.*, EC-26(3), 976, 2011.)

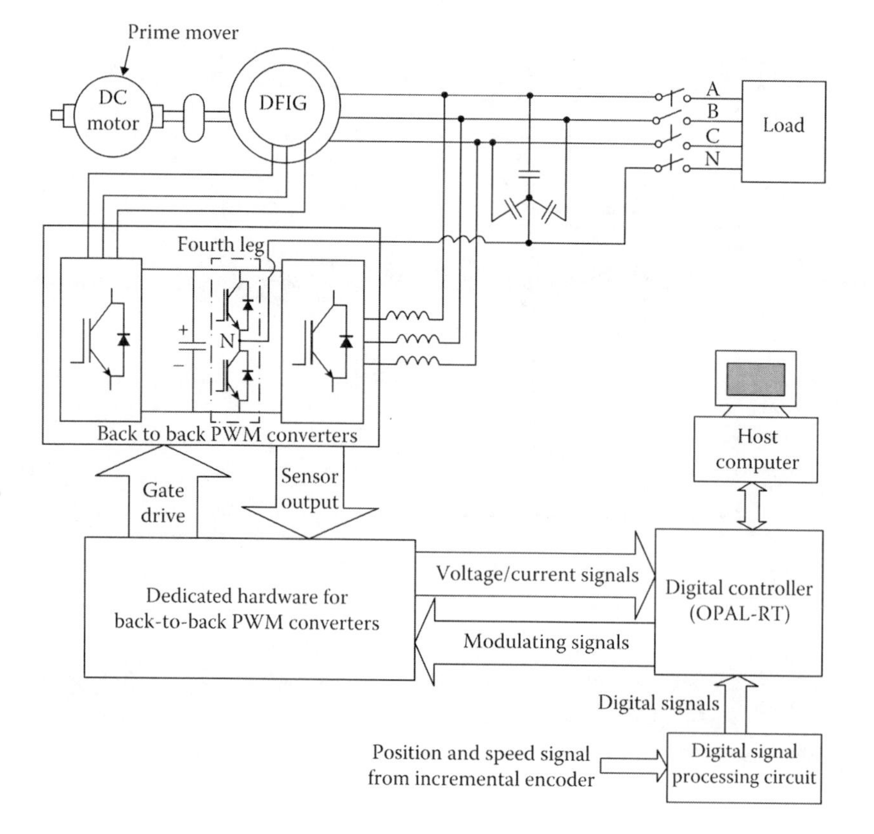

FIGURE 2.46 Stand-alone DFIG system with four-leg stator-side inverter. (Adapted after Pattnaik, M. and Kastha, D., *IEEE Trans.*, IE-60(12), 5506, 2013.)

2.15 Stand-Alone DFIG Control under Unbalanced Nonlinear Loads

Besides the insularization control of DFIG when the power grid connection should be opened, the latter may be used in stand-alone applications, when unbalanced nonlinear loads are typical. Motion sensorless control is preferable for more reliability, but it should be capable to ride through unbalanced and nonlinear loads [34].

Handling single-phase nonlinear (diode rectifier) loads and half-bridge three-phase rectifier loads (with DC zero-sequence current components) requires, apparently, a special scheme that needs an additional lag to the new stator-side connected PWM converter [35]. This leg connects to the null of the stator connected capacitor bank and to the load null point (Figure 2.46).

The case of the one-phase nonlinear load at 2 kW power is illustrated in Figure 2.47, while Figure 2.48 shows the case of the three-phase nonlinear (half-wave rectifier) load at 4.3 kW and supersynchronous speed.

The described control is shown to handle well speed transients also, while sustaining stable (constant) DC and AC output voltage levels [35].

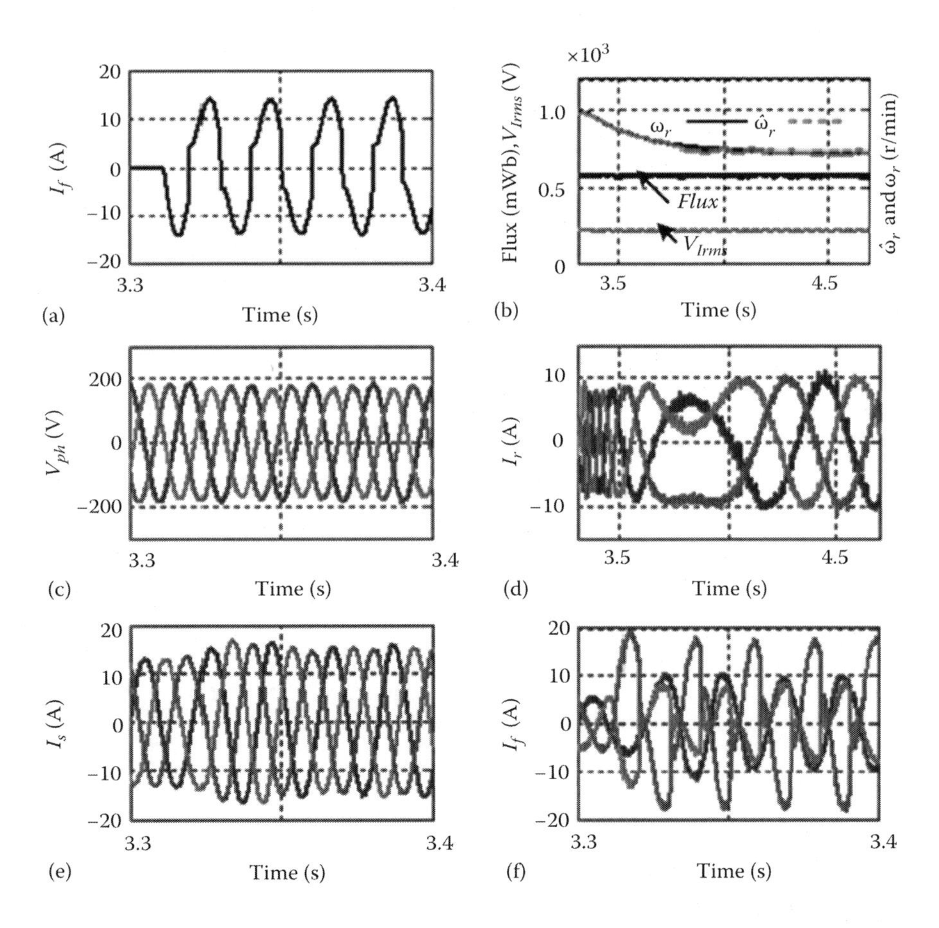

FIGURE 2.47 Experimental transient waveforms of the system with four-leg inverter feeding 2.0 kW one-phase nonlinear load at subsynchronous speed: (a) Load current; (b) rotor speed, stator flux, and rms load line voltage; (c) load phase voltage; (d) rotor current; (e) stator current; and (f) filter current. (Adapted after Pattnaik, M. and Kastha, D., *IEEE Trans.*, IE-60(12), 5506, 2013.)

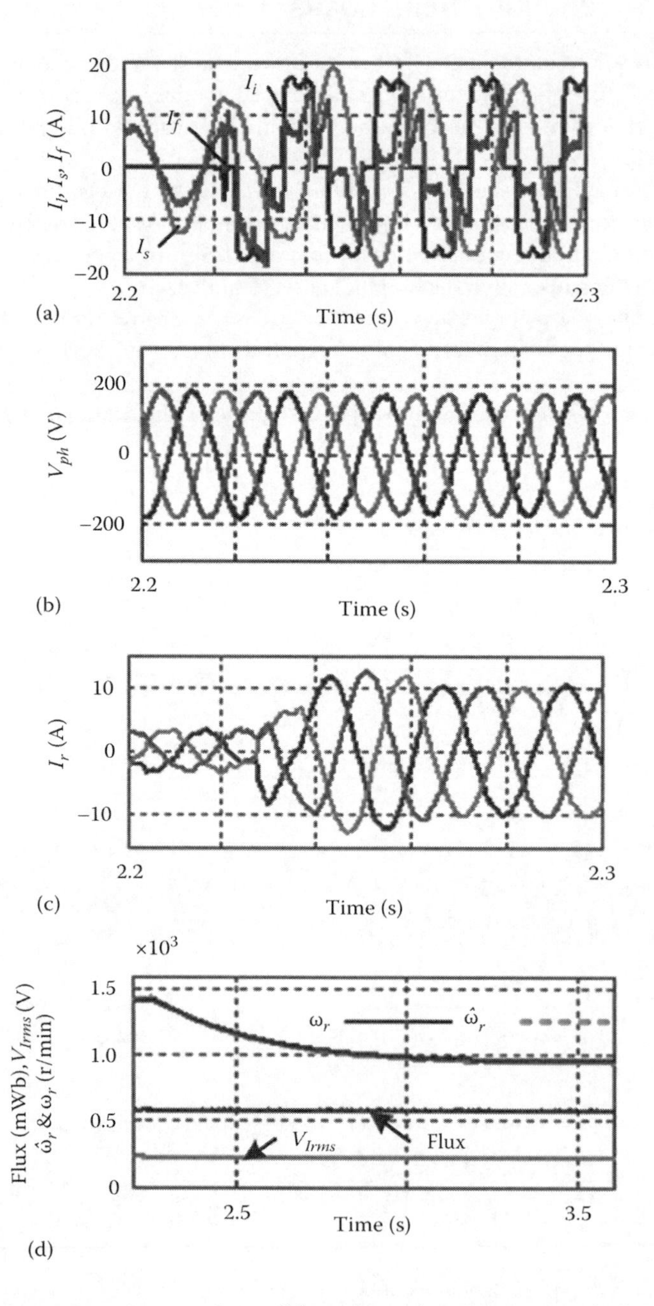

FIGURE 2.48 Experimental transient waveforms of the system with four-leg inverter feeding 4.3 kW three-phase nonlinear load at supersynchronous speed: (a) Inverter, stator, and load currents; (b) load phase voltage; (c) rotor current; and (d) actual and estimated rotor speeds, stator flux, and rms load line voltage. (From Pattnaik, M. and Kastha, D., *IEEE Trans.*, IE-60(12), 5506, 2013.)

2.16 Summary

- WRIGs are also called DFIGs or DOIGs or even doubly fed SGs.
- The uniformly slotted cylindrical laminated stator and rotor cores of WRIG are provided with three-phase AC distributed windings.
- For steady state, the mmfs of the stator and rotor currents both travel at the electrical speed of ω_1, with respect to the stator. However, with electrical rotor speed ω_r, the frequency ω_2 of rotor currents is $\omega_2 = \omega_1 - \omega_r$. Negative ω_2 means negative slip ($S = \omega_2/\omega_1$) and an inverse order of phases on the rotor.

  With a proper power supply connected to the rotor at frequency ω_2 (variable with speed such that $\omega_1 = \omega_r + \omega_2 \approx$ constant), the WRIG may work as a motor and generator subsynchronously ($\omega_r < \omega_1$; $\omega_2 > 0$), supersynchronously ($\omega_r > \omega_1$; $\omega_2 < 0$), and even at synchronism ($\omega_r = \omega_1$, $\omega_2 = 0$). Constant stator voltage and frequency may be secured for variable speed: $\omega_1(1 - |S_{\max}|) < \omega_r < \omega_1(1 + |S_{max}|)$. The larger the speed range, the larger the power rating P_{rN} of the rotor-connected static power converter: $P_{rN} \approx |S_{\max}|P_{sN}$.
- For ±25% slip and speed range, the power rating of the rotor-connected static power converter is around 25% P_{SN}, that is, 25% of stator power. The WRIG may deliver 125% total power at 125% speed, with a stator designed at 100% and 100% speed. The flexibility of a WRIG due to variable speed and the reasonable costs of the converter are the main assets of WRIGs.
- The phase-coordinate model of the WRIG has an eighth order, and some coefficients are rotor-position (time) dependent. It is to be used in special cases only.
- The space-phasor or complex-variable model is particularly suitable for investigating WRIG transients and control. Decomposition along orthogonal axes, spinning at general speed ω_b, leads to the *dq* model. The Park generalized transform relates the phase coordinate to the space-phasor model. The latter has all coefficients independent of rotor position.
- The active power is positive (delivered) in the stator for generating and negative for motoring, for both $S > 0$ and $S < 0$.
- The active power is positive (delivered) in the rotor for motoring (for $S > 0$) and negative (absorbed) for generating. For $S < 0$, the reverse is true: the motor absorbs power through the stator and the rotor, and the generator delivers it through the stator and the rotor.
- With losses neglected, the mechanical power $P_m = P_s + P_r$.
- The space-phasor model of the WRIG, for steady state, in synchronous coordinates, is characterized by DC quantities (voltages, currents, and flux linkages) that make it suitable for control design.
- The space-phasor model is characterized by its equations, space-phasor diagrams, and structural diagrams.
- WRIG transients are to be approached via the space-phasor model. For scalar open-loop control at constant rotor voltage V_r or current I_r, the linearization of the *dq* model leads to a sixth-order equation to determine the complex eigenvalues. The order is reduced to four if the stator resistance R_s is neglected.
- It has, by now, been shown that only rotor current control provides for a stable motor and generator, both undersynchronously ($S > 0$) and supersynchronously ($S < 0$).
- As WRIGs are supplied, in general, in the rotor, by static power converters, and vector or direct power or feedback linearized control is applied, the investigation of transients of the controlled WRIG becomes most relevant.
- Static power converters for the rotor circuit of a WRIG may be classified as follows:
  - DC current link AC–AC converters.
  - DC voltage link AC–AC converters.
  - Direct AC–AC converters (cycloconverters and matrix converters).

  - They all may be built to provide bidirectional power control, both for $S > 0$ and $S < 0$. However, the DC current link AC–AC converter fails to work properly very close to or at synchronism ($\omega_r = \omega_1$). The content in current harmonics depends both on the converter type and on its PWM and control strategies. The DC voltage link AC–AC (back-to-back) converter with IGBTs, GTOs, or IGCTs seems to be the way of the future.
- Vector control of a WRIG refers, separately, to the machine-side converter and to the supply-side converter.
- The vector control of a machine-side converter essentially uses stator active and reactive power closed-loop regulators to set reference rotor d–q current components i_{qr} and i_{dr} in stator flux synchronous coordinates ($\omega_b = \omega_1$). After voltage decoupling, the rotor voltage components V_{dr}^*, V_{dr}^* are Park-transformed into rotor coordinates, to produce the reference rotor voltages V_{ar}^*, V_{br}^*, and V_{cr}^*. A PWM strategy is used to "copy" these patterns.
- Vector control of the source-side converter also works in synchronous coordinates but only if aligned to stator voltage (90° away from the stator flux axis). In essence, the d-axis source-side current i_d is used to control the DC link voltage (active power) through two closed-loop cascaded regulators. The q-axis source-side current i_q is set to provide a certain power factor angle on the source side, through a current regulator. The i_d, i_q components are then Park-transformed in stator coordinates to produce the source-side voltages by the source-side converter.
- Vector control was successfully used for fast active and reactive power control at the power grid and in stand-alone operation.
- Even after a three-phase short circuit, the WRIG recovers swiftly and with small transients, if the reference rotor currents on the machine-side converter are limited by design.
- Besides vector control, for the machine-side converter, direct power control (DPC) was proposed. DPC stems from direct flux and torque control applied to AC drives.
- DPC uses, in principle, hysteresis active and reactive power regulators to directly trigger one (or a sequence of) voltage vector(s) in the machine-side converter. A kind of random PWM is obtained. DPC claims simplicity for fast and robust response in power and implicit motion-sensorless control.
- Vector control requires rotor electrical position θ_{er} and speed ω_r information for implementation. In addition to using electric sensors, to cut costs, motion sensorless control is used.
- The θ_r and ω_r observers are easier to build for a WRIG, as both stator and rotor currents are measurable, though, each in its coordinates. However, by estimating the stator flux, it is easy to use the current model and estimate $\cos\theta_{er}$ and $\sin\theta_{er}$, and then ω_r low-pass filtering is mandatory for both rotor position and rotor speed estimation. Good sensorless operation during synchronization to, and operation at, the power grid was thus demonstrated. Other slip frequency estimators have been proposed recently.
- There are applications where self-starting (and motoring, for pump storage) is required. It is done with the stator short-circuited, while the active power regulator is replaced by a speed regulator, and the reactive power regulator is replaced by constant reference i_{dr}^* current.
- During the synchronization mode, the errors between dq WRIG and power supply voltages are driven to zero by closed-loop regulators that replace the P_r and Q_r regulators. All of these are in the machine-side converter. Setting P_r^*, Q_r^* to zero during synchronization might also work.
- Vector control may be extended to include the negative sequence components and may, thus, enable the handling of asymmetrical power grids, up to zeroing a one-phase current.
- In stand-alone operation, the vector control also works, but the source-side converter has terminal voltage control at a given frequency. The DC link voltage will float with the terminal voltage. The presence of the filter somewhat complicates the control. Smooth passage, from grid to stand-alone operation and back, is feasible.
- Smooth and fast stator active and reactive power control with WRIGs was demonstrated up to 400 MW/unit.

- The static power converter, the distribution of coils in slots in the WRIGs, and the power grid voltage cause (or contain) harmonics.
- The switching harmonics due to PWM in the converter may be attenuated by adequate, quarter of a period delays of pulses on the two sides of the back-to-back converters or through special active filters or by using random PWM.
- The current harmonics due to stator voltage and rotor distributed windings may be compensated one by one by adding their compensating voltages to V_{ds}^{*} and V_{dr}^{*} in the standard machine-side converter vector control scheme. Alternatively, active filters may be used for the scope.
- The power quality of a WRIG can be made really high, and efforts and solutions in that direction are mounting.
- Worldwide research efforts are dedicated to investigating stability in power systems with WRIGs driven by wind or hydraulic turbines [2,36].
- Ride through control for unbalanced voltage sags is being perfected by the day [31–33]; and so is DFIG control for stand-alone applications [31–35].

References

1. I. Boldea and S.A. Nasar, *Induction Machine Handbook*, CRC Press, Boca Raton, FL, 2001, Chapter 13.
2. M.S. Vicatos and J.A. Tegopoulos, Transient state analysis of a doubly fed induction generator under three phase shortcircuit, *IEEE Trans.*, EC-6(1), 1991, 62–68.
3. A. Masmoudi and M.B.A. Kamoun, On the steady state stability comparison between voltage and current control of the doubly fed synchronous machine, In *Proceedings of the IEEE/KTH Power Tech Conference, EMD*, Stockholm, Sweden, 1995, pp. 140–145.
4. M.G. Ioanides, Doubly fed induction machine state variable model and dynamic response, *IEEE Trans.*, EC-6(1), 1991, 55–61.
5. D.P. Gonzaga and Y. Burian Jr., Simulation of a three phase double-fed induction motor (DFIM): A rang of stable operation, In *Proceedings of EPE*, Toronto, Ontario, Canada, 1991.
6. I.F. Soran, The state and transient performance of double-fed asynchronous machine (DFAM), In *Proceedings of EPE*, Firenze, Italy, 1991, pp. 2-395–2-399.
7. I. Cadirci and M. Ermis, Performance evaluation of wind driven DOIG using a hybrid model, *IEEE Trans.*, EC-13(2), 1998, 148–154.
8. E. Akpinar and P. Pillay, Modeling and performance of slip energy recovery induction motor drives, *IEEE Trans.*, EC-5(1), 1990, 203–210.
9. M. Yamamoto and O. Motoyoshi, Active and reactive power control for doubly-fed wound rotor induction generator, *IEEE Trans.*, PE-6(4), 1991, 624–629.
10. P. Bauer, S.W.H. De Hoan, and M.R. Dubois, Wind energy and off shore windparks: State of the art and trends, In *Proceedings of EPE-PEMC*, Dubrovnik and Cavtat, Croatia, 2002, pp. 1–15.
11. R. Rena, J.C. Clare, and G.M. Asher, Doubly-fed induction generator using back to back PWM converters and its application to variable-speed wind-energy generation, *Proc. IEEE*, PA-143(3), 1996, 231–241.
12. I. Serban, F. Blaabjerg, I. Boldea, and Z. Chen, A study of the double-fed wind power generator under power system faults, In *Record of EPE-2003*, Toulouse, France.
13. S. Muller, M. Deicke, and R.W. De Doncker, Adjustable speed generators for wind turbines based on doubly fed machines and 4-quadrant IGBT converters linked to the rotor, In *Record of IEEE-IAS-2000, Annual Meeting*, Roma, Italy, 2000, pp. 2249–2254.
14. I. Boldea and S.A. Nasar, *Electric Drives*, CRC Press, Boca Raton, FL, 1998.
15. R. Datta and V.T. Ranganathan, Direct power control of grid-connected wound rotor induction machine without rotor position sensors, *IEEE Trans.*, PE-16(3), 2001, 390–399.
16. E. Bogalecka and Z. Kvzeminski, Sensorless control of double fed machine for wind power generators, In *Proceedings of EPE-PEMC*, Dubrovnik and Cavtat, Croatia, 2002.

17. J. Bendl, M. Chomat, and L. Schreier, Independent control of positive and negative sequence current components in doubly fed machine, In *Record of ICEM*, Bruges, Belgium, 2002.
18. R. Datta and V.T. Ranganathan, A simple position-sensorless algorithm for rotor-side field-oriented-control of wound-rotor induction machine, *IEEE Trans.*, IE-48(4), 2001, 786–793.
19. L. Morel, H. Goofgroid, H. Mirzgaian, and J.M. Kauffmann, Double-fed induction machine: Converter optimization and field orientated control without position sensor, *Proc. IEEE*, EPA-145(4), 1998, 360–368.
20. U. Radel, D. Navarro, G. Berger, and S. Berg, Sensorless field-oriented control of a slip ring induction generator for a 2.5 MW wind power plant from Mordex Energy Gmbh, In *Record of EPE-2001*, Graz, Austria, 2001, pp. P1–P7.
21. R. Pena, J.C. Clare, and G.M. Asher, A doubly fed induction generator using back to back PWM converters supplying an isolated load from a variable speed turbine, *Proc. IEEE*, EPA-143(5), 1996, 380–387.
22. T. Kawabara, A. Shibuya, and H. Furata, Design and dynamic response characteristics of 400 MW adjustable speed pump storage unit for Ohkawachi Power Station, *IEEE Trans.*, EC-11(2), 1996, 376–384.
23. B. Singh, K. Al-Haddad, and A. Chandra, A review of active filters for power quality improvements, *IEEE Trans.*, IE-46(1), 1999, 960–971.
24. A. Dittrich, Compensation of current harmonics in doubly-fed induction generator system, In *Record of EPE-2001*, Graz, Austria, 2001, pp. P1–P8.
25. M. Ioanides, Separation of transient harmonics produced by double-output induction generators in wind power systems, In *Record of OPTIM-2002*, IEEE-IAS technically sponsored, Poiana Brasov, Romania, 2002.
26. U. Radel and J. Petzoldt, Harmonic analysis of a double-fed induction machine for wind power plants, In *Record of OPTIM 2002*, Poiana Brasov, Romania, 2002.
27. L. Zhang, C. Watthanasarn, and W. Shepherd, Application of a matrix converter for the power control of a variable-speed wind-turbine driving, In *Record of IEEE-IAS 2000 Meeting*, New Orleans, LA, 2000, pp. 906–911.
28. V. Akhmatan, Modelling of variable speed wind turbines with double-fed induction generators in short-term stability investigations, In *Proceedings of Third Workshop on Transmission Networks for Offshore Wind Farms*, Stockholm, Sweden, April 11–12, 2002, pp. 1–23.
29. G. Abad, J. López, M. Rodríguez, L. Marroyo, and G. Iwanski, *Doubly Fed Induction Machine: Modeling and Control for Wind Energy Generation Applications*, IEEE, Wiley, 2011.
30. G. Pannell, D.J. Atkinson, and B. Zahawi, Minimum-threshold crow barfora fault-ride-through grid-code-compliant DFIG wind turbine, *IEEE Trans.*, EC-25(3), 2010, 750–759
31. O. Gomis-Bellmunt, A.O. Junyent-Ferre, A. Sumper, and J. Bergas-Jan, Ride-through control of a doubly fed induction generator under unbalanced voltage sags, *IEEE Trans.*, EC-23(4), 2008, 1036–1045.
32. H. Nian, Y. Song, P. Zhou, and Y. He, Improved direct power controlofa wind turbine driven doubly fed induction generator during transient grid voltage unbalance, *IEEE Trans.*, EC-26(3), 2011, 976–986.
33. H. Xu, J. Hu, and Y. He, Integrated modeling and enhanced control of DFIG under unbalanced and distorted grid voltage conditions, *IEEE Trans.*, EC-27(3), 2012, 725–736.
34. M. Pattnaik and D. Kastha, Adaptive speed observer for a stand-alone DFIG feeding nonlinear and unbalanced loads, *IEEE Trans.*, EC-27(4), 2012, 1018–1026.
35. M. Pattnaik and D. Kastha, Harmonic compensation with zero-sequence load voltage control in a speed-sensorless DFIG-based stand-alone VSCF generating system, *IEEE Trans.*, IE-60(12), 2013, 5506–5514.
36. M.T. Abolhassani, P. Enjeti, and H. Toliyat Integrated doubly fed electric alternator/active filter (IDEA), a viable power quality solution, for wind energy conversion systems, *IEEE Trans.*, EC-23(2), 2008, 642–650.

3

Wound-Rotor Induction Generators: Design and Testing

3.1 Introduction

Wound-rotor induction generators (WRIGs) have been built for power per unit as high as 400 MW/unit in pump storage power plants, to as low as 4.0 MW/unit in wind power plants. Diesel engine or gas-turbine-driven WRIGs for standby or autonomous operation up to 20–40 MW may also be feasible to reduce fuel consumption and pollution for variable loads.

Below 1.5–2 MW/unit, the use of WRIGs is not justifiable in terms of cost versus performance when compared with full-power rating converter synchronous or cage rotor induction generator systems.

The stator-rated voltage increases for power up to 18–20 kV (line voltage, root mean squared [RMS]) at 400 mega volt ampere (MVA). Because of limitations with voltage, for low-cost power converters, the rotor-rated (maximum) voltage occurring at maximum slip currently is about 3.5–4.2 kV (line voltage, RMS) for direct current (DC) voltage link alternating current (AC)–AC pulse-width modulated (PWM) converters with integrated gate-controlled thyristors (IGCTs).

Higher voltage levels are being tested and will be available soon for industrial use, based on multiple-level DC voltage link AC–AC converters made of insulated power cells in series and other high voltage technologies.

Thus far, for the 400 MW WRIGs, the rated rotor current may be in the order of 6500 A; and thus, for $S_{max} = \pm 0.1$, approximately, it would mean 3.6 kV line voltage (RMS) in the rotor. A transformer is necessary to match the 3.6 kV static power converter to the rotor with the 18 kV power source for the stator. The rotor voltage V_r is as follows:

$$V_r = K_{rs} \cdot |S_{max}| \cdot V_s \tag{3.1}$$

For $|S_{max}| = 0.1$, $V_r = 3.6 \cdot 10^3$ V (per phase), $V_s = 18 \cdot 10^{-3}/\sqrt{3}$ V (phase): $K_{rs} = 2/1$. Therefore, the equivalent turn ratio is decisive in the design. In this case, however, a transformer is required to connect the DC voltage link AC–AC converter to the 18 kV local power grid.

For powers in the 1.5–4 MW range, low stator voltages are feasible (690 V line voltage, RMS). The same voltage may be chosen as the maximum rotor voltage V_r, at maximum slip.

For $S_{max} = \pm 0.25$, $V_s = 690/\sqrt{3}$ V, $V_r = V_s$, $K_{rs} = 1/|S_{max}| = 4$. In this case, the rotor currents are significantly lower than the stator currents. A transformer to match the rotor voltage to the stator voltage is not required.

Finally, for WRIGs in the 3–10 MW range to be driven by diesel engines, 3000 (3600) rpm, or gas turbines, stator and rotor voltages in the 3.5–4.2 kV range are feasible. The transformer is again avoided.

Once the stator and rotor rated voltages are calculated and fixed, further design can proceed smoothly. Electromagnetic (emf) and thermomechanical designs are the two types of design. In what follows, we will touch on mainly the emf design.

Even for the emf design, we should distinguish three main operation modes:

1. Generator at a power grid
2. Generator to autonomous load
3. Brushless AC exciter (generator with rotor electric output)

The motor mode is required in applications such as pump storage power plants or even with microhydro or wind turbine prime movers.

The emf design implies a machine model, analytical, numerical, or mixed; one or more objective functions; and an optimization method with a computer program to execute it.

The optimization criteria may include the following:

- Maximum efficiency
- Minimum active material costs
- Net present worth, individual or aggregated

Deterministic, stochastic, and evolutionary optimization methods have been applied to electric machine design [1] (see also Chapter 10). Whatever the optimization method, it is useful to have sound geometrical parameters for the preliminary (or general) design with regard to performance when starting the optimization design process. This is the reason why the general emf design is primarily discussed in what follows. Among the operation modes listed above, the generator at the power grid is the most frequently used, and, thus, it is the one of interest here.

High- and low-voltage stators and rotors are considered to cover the entire power range of WRIGs (from 1 to 400 MW).

The design methodology that follows covers only the essentials. A comprehensive design methodology is beyond the scope of this text.

3.2 Design Specifications: An Example

The rated line voltage of the stator (RMS) (Y) $V_{SN} = 0.38(0.46), 0.69, 4.2(6), \ldots, 18$ kV. The rated stator frequency $f_1 = 50$ (60) Hz. The rated ideal speed $n_{1N} = f_1/p_1 = 3000$ (3600), 1500 (1800). The maximum (ideal) speed $n_{max} = (f_1/p_1)(1 + |S_{max}|)$. The maximum slip $S_{max} = \pm 0.05, \pm 0.1, \pm 0.2, \pm 0.25$. The rated stator power (at unity power factor) $S_{1Ns} = 2$ MW. The rated rotor power (at unity power factor) $S_{Nr} = S_{Ns}|S_{max}| = 0.5$ MW. The rated (maximum) rotor line voltage (Y) is $V_{RN} \leq V_{SN}$. As already discussed, the stator power may reach up to 350 MW (or more) with the rotor delivering maximum power (at maximum speed) of up to 40–50 MW and voltage $V_N^r \leq 4.2(6)$ kV.

Here we will consider the following for our discussion: total 2.5 MW at 690 V, 50 Hz, $V_{RN} = V_{SN} = 690$ V, $S_{max} = \pm 0.25$, and rated ideal speed $n_{1N} = 50/p_1 = 1500$ rpm ($p_1 = 2$).

The emf design factors are as follows:

- Stator (core) design
- Rotor (core) design
- Stator winding design
- Rotor winding design
- Magnetization current computation
- Equivalent circuit parameter computation
- Loss and efficiency computation

3.3 Stator Design

Two main design concepts need to be considered when calculating the stator interior diameter D_{is}: the output coefficient design concept (C_e—Esson coefficient) and the shear rotor stress (f_{xt}) [1]. Here, we will make use of the shear rotor stress concept with $f_{xt} = 1.5$–8 N/cm^2.

The shear rotor stress increases with torque. First, the emf torque has to be estimated, noting that at 2.5 MW, $2p_1 = 4$ poles, the expected rated efficiency $\eta_N > 0.95$.

The total emf power S_{gN} at maximum speed and power is as follows:

$$S_{gN} \approx \frac{(S_{SN} + S_{RN})}{\eta_N} = \frac{(2+0.5)10^6}{0.96} = 2.604 \cdot 10^6 \text{ W} \tag{3.2}$$

The corresponding emf torque T_e is

$$T_e = \frac{S_{gN}}{2\pi(f_1/p_1)(1+|S_{max}|)} = \frac{2.604 \cdot 10^6}{2\pi(50/2)(1+0.25)} = 1.327 \cdot 10^4 \text{ Nm}. \tag{3.3}$$

This torque is at stator rated power and ideal synchronous speed ($S = 0$). The stator interior diameter, D_{is}, based on the shear rotor stress concept, f_{xt}, is

$$D_{is} = \sqrt[3]{\frac{2T_e}{\pi \cdot \lambda \cdot f_{xt}}}; \quad \lambda = \frac{l_i}{D_{is}} \tag{3.4}$$

with l_i equal to the stack length. The stack length ratio $\lambda = 0.2$–1.5, in general. Smaller values correspond to a larger number of poles. With $\lambda = 1.0$ and $f_{xt} = 6$ N/cm² (aiming at high torque density), the stator internal diameter, D_{is}, is as follows:

$$D_{is} = \sqrt[3]{\frac{2 \cdot 1.3270}{\pi \cdot 1 \cdot 6}} = 0.52 \text{ m}. \tag{3.5}$$

At this power level, a unistack stator is used, together with axial air cooling.

The external stator diameter D_{out} based on the maximum air gap flux density per given magnetomotive force (mmf) is approximated in Table 3.1 [1]. Therefore, from Table 3.1, D_{out} is as follows:

$$D_{out} = D_{is} \cdot 1.48 = 0.5200 \cdot 1.48 = 0.796 \text{ m} \tag{3.6}$$

The rated stator current at unity power factor in the stator I_{SN} is

$$I_{SN} = \frac{S_{SN}}{\sqrt{3}V_{SN}} = \frac{2 \cdot 10^6}{\sqrt{3} \cdot 690} = 1675.46 \text{ A} \tag{3.7}$$

The air gap flux density $B_{g1} = 0.75$ T (the fundamental value). On the other hand, the air gap emf E_S per phase is as follows:

$$E_S = \frac{K_E V_{SN}}{\sqrt{3}}; \quad K_E = 0.97 - 0.98 \tag{3.8}$$

$$E_S = \pi\sqrt{2} f_1 W_1 K_{W1a} \cdot \frac{2}{\pi} B_{g1} \cdot \tau \cdot l_i \tag{3.9}$$

TABLE 3.1 Outer to Inner Stator Diameter Ratio

$2p_1$	2	4	6	8	≥10
D_{out}/D_{is}	1.65–1.69	1.46–1.49	1.37–1.40	1.27–1.30	1.24–1.20

where τ is the pole pitch:

$$\tau = \frac{\pi D_{is}}{2p_1} = \frac{\pi \cdot 0.52}{2 \cdot 2} = 0.40 \text{ m} \tag{3.10}$$

K_{W1} is the total winding factor:

$$K_{W1} = \frac{\sin \pi/6}{q_1 \sin \pi/6q_1} \sin\frac{\pi}{2}\frac{y}{\tau}; \quad \frac{2}{3} \le \frac{y}{\tau} \le 1 \tag{3.11}$$

where

q_1 is the number of slot/pole/phase in the stator
y/τ is the stator coil span/pole pitch ratio

As the stator current is rather large, we are inclined to use $a_1 = 2$ current paths in parallel in the stator. With two current paths in parallel, full symmetry of the windings with respect to stator slots may be provided. For the time being, let us adopt $K_{W1} = 0.910$. Consequently, from Equation 3.9, the number of turns per current path W_{1a} is as follows:

$$W_{1a} = \frac{0.97 \cdot 690/\sqrt{3}}{\sqrt{2} \cdot 50 \cdot 0.910 \cdot 2 \cdot 0.75 \cdot 0.52 \cdot 0.4} = 19.32 \text{ turns} \tag{3.12}$$

Adopting $q_1 = 5$ slots/pole/phase, a two-layer winding with two conductors per coil, $n_{c1} = 2$,

$$W_{1a} = \frac{2p_1}{a_1} \cdot q_s \cdot n_{c1} = \frac{2 \cdot 2}{2} \cdot 5 \cdot 2 = 20 \tag{3.13}$$

The final number of turns/coil is $n_{c1} = 2$, $q_1 = 5$, with two symmetrical current paths in parallel. The North and South poles pairs constitute the paths in parallel.

Note that the division of a turn into elementary conductors in parallel with some transposition to reduce skin effects will be discussed later in this chapter.

The stator slot pitch is now computable:

$$\tau_S = \frac{\tau}{3q_1} = \frac{0.40}{3.5} = 0.0266 \text{ m} \tag{3.14}$$

The stator winding factor K_{W1} may now be recalculated if only the y/τ ratio is fixed: $y/\tau = 12/15$, very close to the optimum value, to reduce to almost zero the fifth-order stator mmf space harmonic:

$$K_{W1} = \frac{\sin \pi/6}{5\sin(\pi/6 \cdot 5)} \cdot \sin\frac{\pi}{2} \cdot \frac{4}{5} = 0.9566 \cdot 0.951 = 0.9097 \tag{3.15}$$

This value is very close to the adopted one, and thus, $n_{c1} = 2$ and $q_1 = 5$, $a_1 = 2$ hold. The number of slots N_S is as follows:

$$N_S = 2p_1q_1m = 2 \cdot 2 \cdot 5 \cdot 3 = 60 \text{ slots} \tag{3.16}$$

As expected, the conditions to full symmetry, $N_S/m_1q_1 = 60/(3 \cdot 2) = 10$ = integer, $2p_1/a_1 = 2 \cdot 2/2$ = integer are fulfilled.

The stator conductor cross-section $A_{\cos}$ is as follows:

$$A_{\cos} = \frac{I_{SN}}{a_1 \cdot j_{\cos}} \tag{3.17}$$

With the design current density j_{col} = 6.5 A/mm^2, rather typical for air cooling:

$$A_{\cos} = \frac{1675.46}{2 \cdot 6.5} = 128.88\ \text{mm}^2 \tag{3.18}$$

Open slots in the stator are adopted, but magnetic wedges are used to reduce the air gap flux density pulsations due to slot openings.

In general, the slot width W_S/τ_S = 0.45–0.55. Let us adopt W_S/τ_S = 0.5. The slot width W_S is

$$W_S = \left(\frac{W_S}{\tau_S}\right) \cdot \tau_S = 0.5 \cdot 0.02666 = 0.01333\ \text{m} \tag{3.19}$$

There are two coils (four turns in our case) per slot (Figure 3.1).

As we deal with a low-voltage stator (690 V, line voltage RMS), the total slot filling factor, with rectangular cross-sectional conductors, may be safely considered as $K_{fills} \approx 0.55$.

The useful (above wedge) slot area A_{sn} is

$$A_{sn} = \frac{2n_{c1}A_{\cos}}{K_{fills}} = \frac{2 \cdot 2 \cdot 128.88}{0.55} = 937\ \text{mm}^2 \tag{3.20}$$

The rectangular slot useful height h_{su} is straightforward:

$$h_{su} = \frac{A_{su}}{W_b} = \frac{937}{13.33} = 70.315\ \text{mm} \tag{3.21}$$

The slot aspect ratio h_{su}/W_s = 70.315/13.33 = 5.275 is still acceptable. The wedge height is about $h_{sw} \approx$ 3 mm. Adopting a magnetic wedge leads to the apparent reduction of slot opening from W_s= 13.33 mm to about W_s = 4 mm, as detailed later in this chapter.

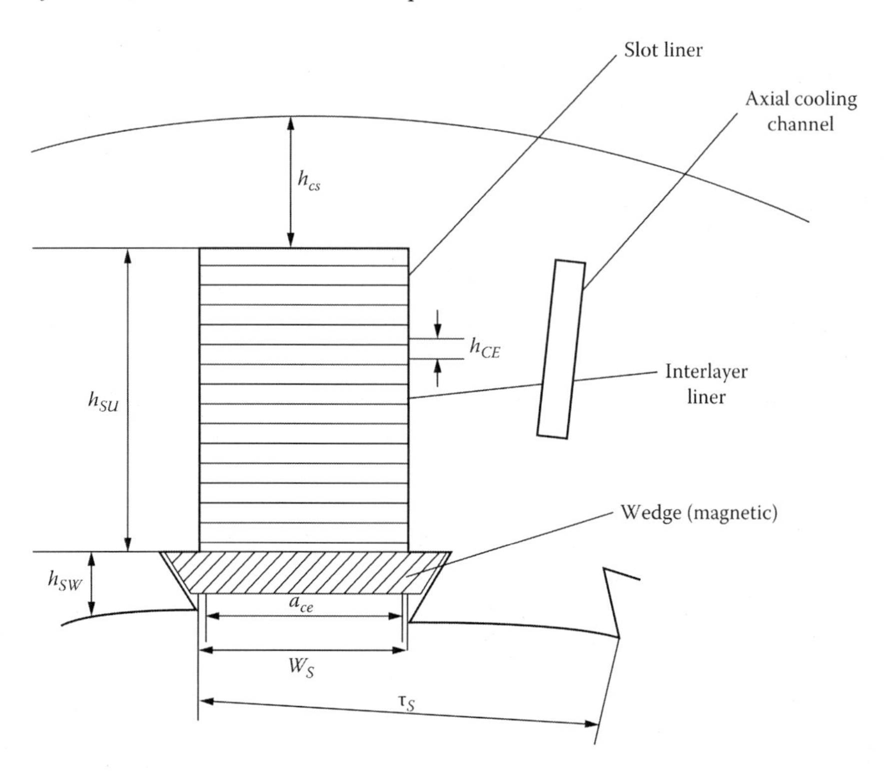

FIGURE 3.1 Stator slotting geometry.

The air gap g is

$$g \approx (0.1+0.012\sqrt[3]{S_{SN}})10^{-3} = (0.1+0.0012\sqrt[3]{2\cdot 10^{6}})10^{-3} = 1.612\text{ mm} \tag{3.22}$$

With a flux density $B_{cs} = 1.55$ T in the stator back iron, the back core stator height h_{cs} (Figure 3.1) may be calculated as follows:

$$h_{cs} = \frac{B_{g1}\tau}{\pi B_{cs}} = \frac{0.75\cdot 0.4}{\pi\cdot 1.5} = 0.0637\text{ m} \tag{3.23}$$

The magnetically required stator outer diameter D_{outm} is

$$\begin{aligned} D_{outm} &= D_{is} + 2(h_{sm} + h_{sw} + h_{cs}) \\ &= 0.52 + 2(0.07315 + 0.003 + 0.0637) = 0.7997 \approx 0.8\text{ m} \end{aligned} \tag{3.24}$$

This is roughly equal to the value calculated from Table 3.1 (Equation 3.6).

The stator core outer diameter may be increased by the double diameter of axial channels for ventilation, which might be added to augment the external cooling by air flowing through the fins of the cast iron frame.

As already inferred, the rectangular axial channels, placed in the upper part of stator teeth (Figure 3.1), can help improve the machine cooling once the ventilator of the shaft is able to flow part of the air through these stator axial channels.

The division of a stator conductor (turn) with a cross-section of 128.8 mm^2 is similar to the case of synchronous generator design in terms of transposition to limit the skin effects. Let us consider four elementary conductors in parallel. Their cross-sectional area $A_{\cos e}$ is as follows:

$$A_{\cos e} = \frac{A_{\cos}}{4} = \frac{128.8}{4} = 32.2\text{ mm}^2 \tag{3.25}$$

Considering only $a_{ce} = 12$ mm, out of the 13.33 mm slot width available for the elementary stator conductor, the height of the elementary conductor h_{ce} is as follows:

$$h_{ce} = \frac{A_{\cos e}}{a_{ce}} = \frac{32.2}{12} \approx 2.68\text{ mm} \tag{3.26}$$

Without transposition and neglecting coil chording the skin effect coefficient, with $m_e = 16$ layers (elementary conductors) in slot, is as follows [1]:

$$K_{Rme} = \varphi(\xi) + \frac{\left(m_e^2 - 1\right)}{3}\psi(\xi) \tag{3.27}$$

$$\begin{aligned} \xi = \beta h_{ce};\quad \beta &= \sqrt{\frac{\omega_1\mu_0\sigma_{co}}{2}\cdot\frac{a_{ce}}{W_s}} \\ &= \sqrt{\frac{2\pi\cdot 50\cdot 4.3\cdot 10^{7}\cdot 1.256\cdot 10^{-6}}{2}\cdot\frac{12}{13.3}} = 0.6964\cdot 10^{2}\text{ m}^{-1} \end{aligned} \tag{3.28}$$

Finally, $\xi = \beta h_{ce} = 0.6964 \cdot 10^{-2} \cdot 2.68 \cdot 10^{-3} = 0.1866$. With,

$$\varphi(\xi) = \xi \frac{(\sinh 2\xi + \sin 2\xi)}{(\cosh 2\xi - \cos 2\xi)} \approx 1.00 \tag{3.29}$$

$$\psi(\xi) = 2\xi \frac{(\sinh \xi - \sin \xi)}{(\cosh \xi + \cos \xi)} \approx 5.55 \cdot 10^{-4} \tag{3.30}$$

From Equation 3.27, K_{Rme} is

$$K_{Rme} = 1.00 + \frac{(16^2 - 1)}{3} \cdot 5.55 \cdot 10^{-4} = 1.04675$$

The existence of four elementary conductors (strands) in parallel leads also to circulating currents. Their effect may be translated into an additional skin effect coefficient K_{rad} [1]:

$$K_{rad} = 4\beta^4 \cdot h_{ce}^4 \left(\frac{l_i}{l_{turn}} \right)^2 n_{cn}^2 \frac{(1 + \cos\gamma)^2}{4} \tag{3.31}$$

where

- l_{turn} is the coil turn length
- n_{cn} is the number of turns per coil
- γ is the phase shift between lower and upper layer currents ($\gamma = 0$ for diametrical coils; it is $(1 - y/\tau)(\pi/2)$ for chording coils)

With $l_i/l_{turn} \approx 0.4$, $n_{cn} = 2$, $h_{ce} = 2.68 \cdot 10^{-3}$ m, and $\beta = 0.6964 \cdot 10^{-2}$ m^{-1} (from Equation 3.28), K_{rad} is as follows:

$$K_{rad} = 4 \cdot (0.6964 \cdot 10^2 \cdot 2.68 \cdot 10^{-3})^4 \cdot 0.4^2 \cdot 2^2 \frac{(1 + \cos 18°)^2}{4} = 0.01847! \tag{3.32}$$

The total skin effect factor K_{Rll} is

$$K_{Rll} = 1 + (K_{rme} - 1)\frac{l_{stack}}{l_{turn}} + K_{rad} = 1 + (1.04675 - 1)0.4 + 0.01847 = 1.03717 \tag{3.33}$$

It seems that at least for this design, no transposition of the four elementary conductors in parallel is required, as the total skin effect winding losses add only 3.717% to the fundamental winding losses.

The stator winding is characterized by the following:

- Four poles
- 60 slots
- $q_1 = 5$ slots/pole/phase
- Coil span/pole pitch $y/\tau = 12/15$
- $a_1 = 2$ symmetrical current paths in parallel
- Four elementary rectangular wires in parallel

With t_1 equal to the largest common divisor of N_s and $p_1 = 2$, there are $N_s/t_1 = 60/2 = 30$ distinct slot emfs. Their star picture is shown in Figure 3.2. They are distributed to phases based on 120° phase shifting after choosing $N_s/2m_1 = 60/(2 \cdot 3) = 10$ arrows for phase a and positive direction (Figure 3.3a and b).

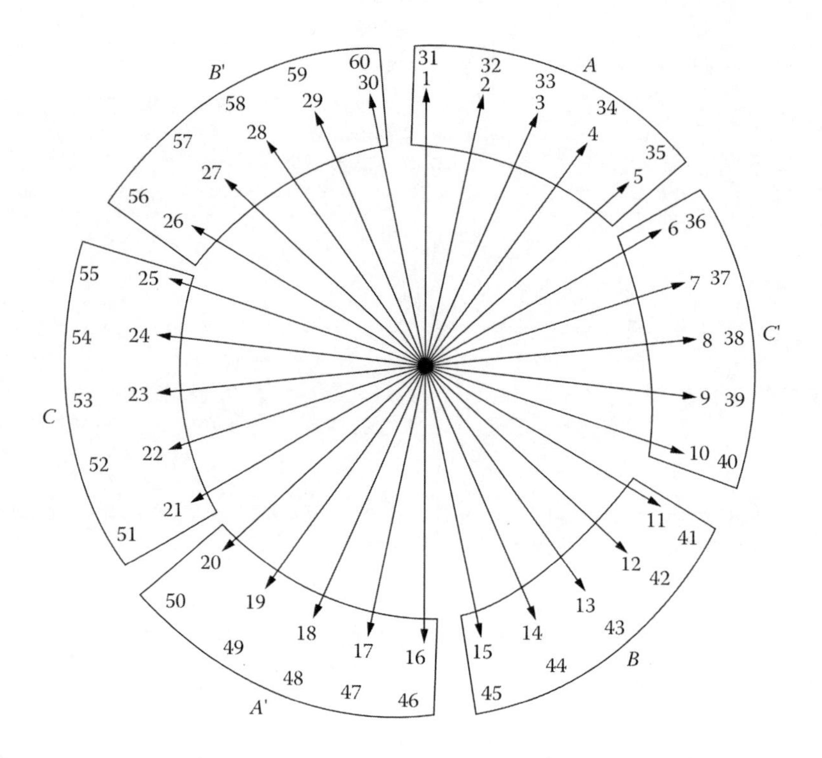

FIGURE 3.2 Slot allocation to phases for $N_S = 60$ slots, $2p_1 = 4$ poles, and $m = 3$ phases.

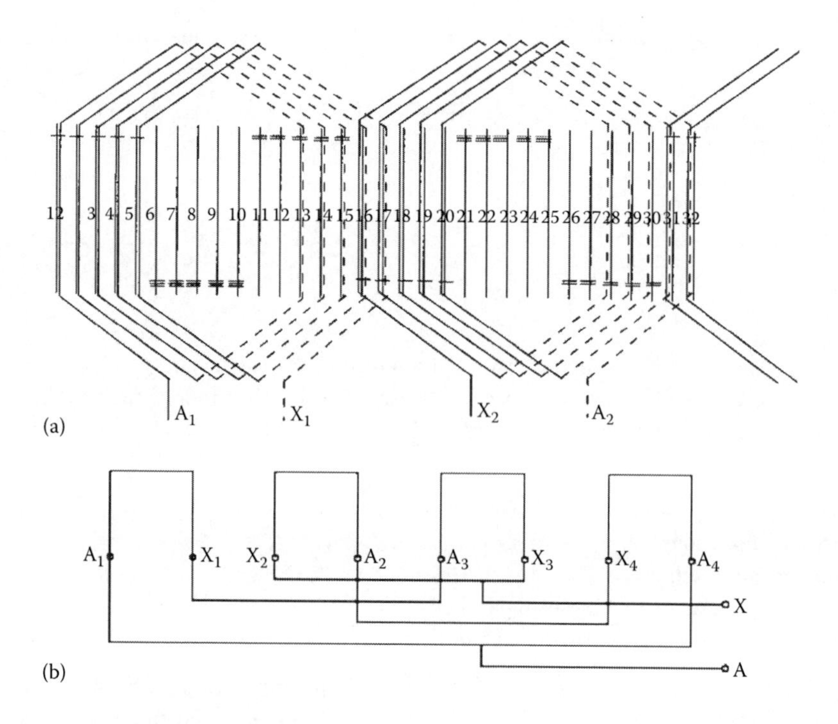

FIGURE 3.3 (a) Half of coils of stator phases A and (b) their connection to form two current paths in parallel.

3.4 Rotor Design

The rotor design is based on the maximum speed (negative slip)/power delivered, P_{RN}, at the corresponding voltage $V_{RN} = V_{SN}$. Besides, the WRIG is designed here for unity power factor in the stator. Therefore, all the reactive power is provided through the rotor. Consequently, the rotor also provides for the magnetization current in the machine.

For $V_{RN} = V_{SN}$ at S_{max}, the turn ratio between rotor and stator K_{rs} is obtained:

$$K_{rs} = \frac{W_2 K_{W2}}{W_1 K_{W1}} = \frac{1}{|S_{max}|} = \frac{1}{0.25} = 4.0 \tag{3.34}$$

Now the stator rated current reduced to the rotor I'_{SN} is as follows:

$$I'_{SN} = \frac{I_{SN}}{K_{rs}} = \frac{1675.46}{4} = 418.865 \text{ A} \tag{3.35}$$

The rated magnetization current depends on the machine power, the number of poles, and so forth. At this point in the design process, we can assign I'_m (the rotor-reduced magnetization current) a per unit (P.U.) value with respect to I_{SN}:

$$\begin{gathered} I'_m = K_m I'_{SN} \\ K_m = 0.1 - 0.30 \end{gathered} \tag{3.36}$$

Let us consider here $K_m = 0.30$. Later on in the design, K_m will be calculated, and then adjustments will be made.

Therefore, the rotor current at maximum slip and rated rotor and stator delivered powers is as follows:

$$I_{RN}^R = \sqrt{I_{SN}'^2 + I_m'^2} = I'_{SN}\sqrt{1 + K_m^2} = 418.865\sqrt{1 + 0.30^2} = 437.30 \text{ A} \tag{3.37}$$

The rotor power factor $\cos \varphi_{2N}$ is

$$\cos\varphi_{2N} = \frac{P_{RN}}{\sqrt{3} V_{RN} I_{RN}^R} = \frac{500{,}000}{\sqrt{3} \cdot 690 \cdot 437.30} = 0.9578 \tag{3.38}$$

It should be noted that the oversizing of the inverter to produce unity power factor (at rated power) in the stator is not very important.

Note that, generally, the reactive power delivered by the stator Q_1 is requested from the rotor circuit as SQ_1.

If massive reactive power delivery from the stator is required, and it is decided to be provided from the rotor-side bidirectional converter, the latter and the rotor windings have to be sized for the scope.

When the source-side power factor in the converter is unity, the whole reactive power delivered by the rotor-side converter is "produced" by the DC link capacitor, which needs to be sized for the scope.

Roughly, with $\cos \varphi_2 = 0.707$, the converter has to be oversized at 150%, while the machine may deliver almost 80% of reactive power through the stator (ideally 100%, but a part is used for machine magnetization).

Adopting $V_{RN} = V_{SN}$ at S_{max} eliminates the need for a transformer between the bidirectional converter and the local power grid at V_{SN}.

Once we have the rotor-to-stator turns ratio and the rated rotor current, the designing of the rotor becomes straightforward.

The equivalent number of rotor turns, W_2K_{W2} (single current path), is as follows:

$$W_2K_{W2} = W_{1a}K_{W1} \cdot K_{rs} = 20 \cdot 0.908 \cdot 4 = 72.64 \text{ turns/phase} \tag{3.39}$$

The rotor number of slots N_R should differ from the stator one, N_S, but they should not be too different from each other.

As the number of slots per pole and phase in the stator $q_1 = 5$, for an integer q_2, we may choose $q_2 = 4$ or 6. We choose $q_2 = 4$. Therefore, the number of rotor slots N_R is as follows:

$$N_R = 2p_1m_1q_2 = 2 \cdot 2 \cdot 3 \cdot 4 = 48 \tag{3.40}$$

With a coil span of $Y_R/\tau = 10/12$, the rotor winding factor (no skewing) becomes

$$K_{W2} = \frac{\sin \pi/6}{q_2 \sin \pi/6q_2} \cdot \sin \frac{\pi}{2} \cdot \frac{y_R}{\tau} = \frac{0.5}{4\sin(\pi/24)} \cdot \sin \frac{\pi}{2} \cdot \frac{10}{12}$$

$$= 0.95766 \cdot 0.9659 = 0.925 \tag{3.41}$$

From Equation 3.39, with Equation 3.41, the number of rotor turns per phase W_2 is

$$W_2 = \frac{W_2K_{W2}}{K_{W2}} = \frac{72.64}{0.925} = 78.53 \approx 80 \tag{3.42}$$

The number of coils per phase is $N_R/m_1 = 48/3 = 16$. With $W_2 = 80$ turns/phase; it follows that each coil will have n_{c2} turns:

$$n_{c2} = \frac{W_2}{(N_R/m_1)} = \frac{80}{16} = 5 \text{ turns/coil} \tag{3.43}$$

Therefore, there are ten turns per slot in two layers and one current path only in the rotor.

Adopting a design current density $j_{con} = 10$ A/mm^2 (special attention to rotor cooling is needed) and, again, a slot filling factor $K_{fill} = 0.55$, the copper conductor cross-section A_{cor} and the slot useful area A_{slotUR} are as follows:

$$A_{cor} = \frac{I_{RN}}{j_{cor}} = \frac{434.3}{10} = 43.4 \text{ mm}^2 \tag{3.44}$$

$$A_{slotUR} = \frac{2n_{c2}A_{cor}}{K_{fill}} = \frac{2 \times 5 \times 43.40}{0.55} = 789.09 \text{ mm}^2 \tag{3.45}$$

The rotor slot pitch τ_R is

$$\tau_R = \frac{\pi(D_{is} - 2g)}{N_R} = \frac{\pi(0.52 - 2 \cdot 0.001612)}{48} = 0.033805 \text{ m} \tag{3.46}$$

Assuming that the rectangular slot occupies 45% of rotor slot pitch, the slot width W_R is

$$W_R = 0.45\tau_R = 0.45 \cdot 0.033805 = 0.0152 \text{ m} \tag{3.47}$$

The useful height h_{RU} (Figure 3.4) is, thus,

$$h_{SU} = \frac{A_{slotUR}}{W_r} = \frac{789.09}{15.20} = 51.913 \text{ mm} \tag{3.48}$$

This is an acceptable (Equation 3.48) value, as $h_{SU}/W_R < 4$.

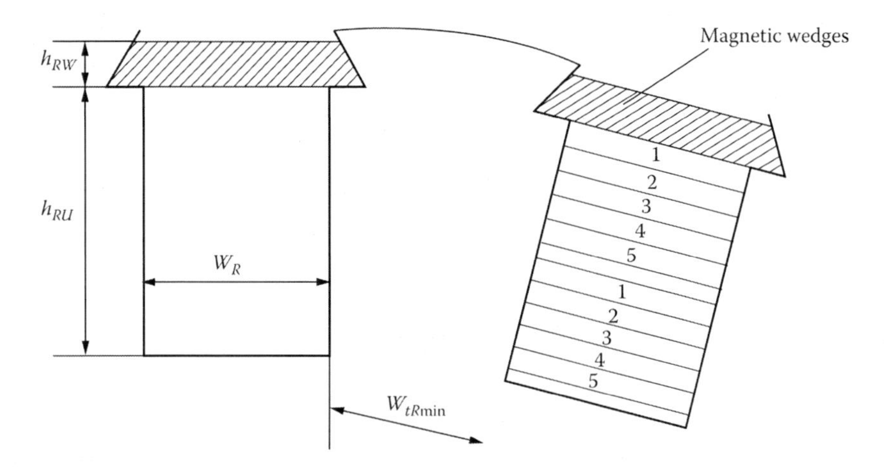

FIGURE 3.4 Rotor slotting.

Open slots have been adopted, but with magnetic wedges ($\mu_r = 4$–5), the actual slot opening is reduced from $W_R = 15.2$ mm to about 3.5 mm, which should be reasonable (in the sense of limiting the Carter coefficient and surface and flux pulsation space harmonics core losses).

We need to verify the maximum flux density, B_{tRmax}, in the rotor teeth at the bottom of the slot:

$$B_{tRmax} = \frac{B_{g1} \cdot \tau_R}{W_{tRmin}} = 0.75 \cdot \frac{33.805}{11.42} = 2.22 \text{ T} \tag{3.49}$$

with

$$\begin{aligned} W_{tRmin} &= \frac{\pi(D_{is} - 2(g + h_{RU} + h_{RW}))}{N_R} - W_R \\ &= \frac{\pi(0.520 - 2(0.001612 + 0.05191) + 0.003)}{48} - 0.0152 \\ &= 11.42 \cdot 10^{-3} \text{ m} \end{aligned} \tag{3.50}$$

Though the maximum rotor tooth flux density is rather large (2.2 T), it should be acceptable, because it influences a short path length.

The rotor back iron radial depth h_{cr} is as follows:

$$h_{cr} = \frac{B_g \cdot \tau}{\pi \cdot B_{cr}} = \frac{0.75 \cdot 0.4}{\pi \cdot 1.6} = 0.0597 \approx 0.06 \text{ m} \tag{3.51}$$

The value of rotor back iron flux density B_{cr} of 1.6 T (larger than in the stator back iron) was adopted, as the length of the back iron flux lines is smaller in the rotor with respect to the stator.

To see how much of it is left for the shaft diameter, let us calculate the magnetic back iron inner diameter D_{ir}:

$$\begin{aligned} D_{ir} &= D_{is} - 2(g + h_{RU} + h_{RW} + h_{cr}) \\ &= 0.52 - 2(0.001612 + 0.05191 + 0.003 + 0.06) = 0.287 \text{ m} \end{aligned} \tag{3.52}$$

This inner rotor core diameter may be reduced by 20 mm (or more) to allow for axial channels—10 mm (or more) in diameter—for axial cooling, and thus, 0.267 m (or slightly less) are left of the shaft. It should be enough for the purpose, as the stack length is $l_i = D_{is} = 0.52$ m.

The rotor winding design is straightforward, with $q_2 = 4$, $2p_1 = 4$ and a single current path. The cross-section of the conductor is 43.20 mm², and thus, even a single rectangular conductor with the width $b_{cr} < W_R \approx 13$ mm and its height $h_{cr} = A_{cor}/b_{cr} = 43.40/13 = 3.338$ mm, will do.

As the maximum frequency in the rotor, $f_{q\max} = f_1^* |S_{\max}| = 50 \cdot 0.25 = 12.5$ Hz, the skin effect will be even smaller than in the stator (which has a similar elementary conductor size); and thus, no transposition seems to be necessary.

The winding end connections have to be tightened properly against centrifugal and electrodynamic forces by adequate nonmagnetic bandages.

The electrical rotor design also contains the slip-rings and brush system design and the shaft and bearings design. These are beyond our scope here. Though debatable, the use of magnetic wedges on the rotor also seems doable, as the maximum peripheral speed is smaller than 50 m/s.

3.5 Magnetization Current

The rated magnetization current was previously assigned a value (30% of I_{SN}). By now, we have the complete geometry of stator and rotor slots cores, and the magnetization mmf may be considered. Let us consider it as produced from the rotor, though it would be the same if computed from the stator.

We start with the given air gap flux density $B_{g1} = 0.75$ T and assume that the magnetic wedge relative permeability $\mu_{RS} = 3$ in the stator and $\mu_{RR} = 5$ in the rotor. Ampere's law along the half of the Γ contour (Figure 3.5), for the main flux, yields the following:

$$F_m = \frac{3W_2K_{W2}I_{R0}\sqrt{2}}{\pi p_1} = (F_{AA'} + F_{AB} + F_{BC} + F_{A'B'} + F_{B'C'}) \tag{3.53}$$

where

$F_{AA'}$ is the air gap mmf
F_{AB} is the stator teeth mmf
F_{BC} is the stator yoke mmf
$F_{A'B'}$ is the rotor tooth mmf
$F_{B'C'}$ is the rotor yoke mmf

The air gap mmf $F_{AA'}$ is as follows:

$$F_{AA'} = gK_C\frac{B_{g1}}{\mu_0}; \quad K_C = K_{C1} \cdot K_{C2} \tag{3.54}$$

K_C is the Carter coefficient:

$$K_{C1,2} = \frac{1}{1 - \gamma_{1,2}g/2\tau_{s,r}}; \quad \gamma_{1,2} = \frac{(2W_{S',R'}/g)^2}{5 + 2W_{S',R'}/g} \tag{3.55}$$

The equivalent slot openings, with magnetic wedges, $W_{S'}$ and $W_{R'}$, are as follows:

$$W_{S'} = W_S/\mu_{RS}$$
$$W_{R'} = W_R/\mu_{RR} \tag{3.56}$$

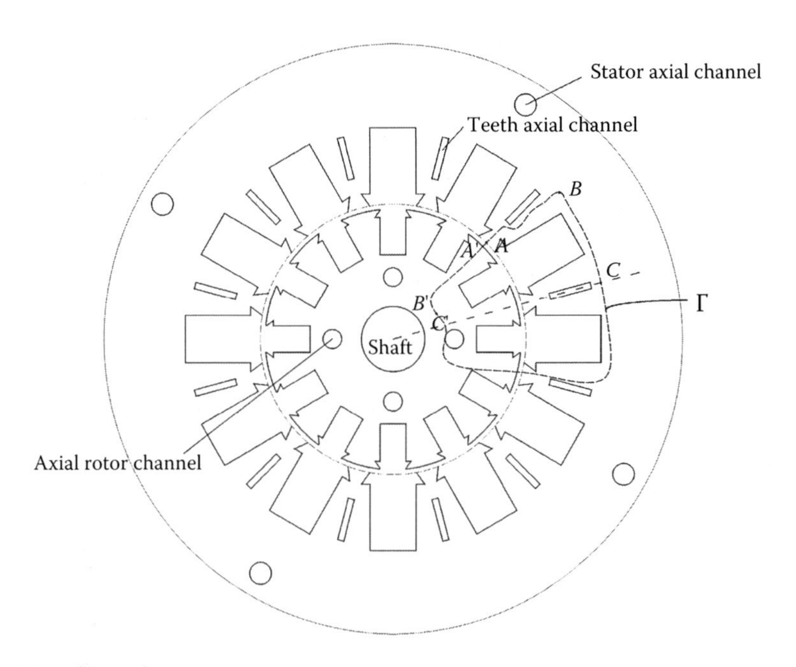

FIGURE 3.5 Main flux path.

where

$\tau_{s,r}$ = 26.6, 33.8 mm
g = 1.612 mm
$W_{S'} = 13.2/3 = 4.066$ mm
$W_{R'} = 15.2/5 = 3.04$ mm (from Equation 3.56 through Equation 3.58)
K_{C1} = 1.0826
K_{C2} = 1.040

Finally,

$$K_C = K_{C1}K_{C2} = 1.0826 \cdot 1.04 \approx 1.126 \tag{3.57}$$

This is a small value that will result, however, in smaller surface and flux pulsation core losses. The small effective slot opening, $W_{S'}$ and $W_{R'}$ will lead to larger slot leakage inductance contributions. This, in turn, will reduce the short circuit currents.

The air gap mmf (from Equation 3.54) is as follows:

$$F_{AA'} = 1.612 \cdot 10^{-3} \cdot 1.126 \frac{0.75}{1.256 \cdot 10^{-6}} = 1083.86 \text{ A turns}$$

The stator and rotor teeth mmf should take into consideration the trapezoidal shape of the teeth and the axial cooling channels in the stator teeth.

We might suppose that, in the stator, due to axial channels, the tooth width is constant and equal to its value at the air gap $W_{ts} = \tau_S - W_S = 26.6 - 13.2 = 13.4$ mm. Therefore, the flux density in the stator tooth B_{ts} is

$$B_{ts} = \frac{B_{g1} \cdot \tau_S}{W_{ts}} = 0.75 \cdot \frac{0.0266}{0.0134} = 1.4888 \text{ T} \tag{3.58}$$

From the magnetization curve of silicon steel (3.5% silicon, 0.5 mm thickness, at 50 Hz; Table 3.2), H_{ts} = 1290 A/m by linear interpolation.

TABLE 3.2 *B–H* Curve for Silicon (3.5%) Steel (0.5 mm Thick) at 50 Hz

B (T)	0.05	0.1	0.15	0.2	0.25	0.3	0.35	0.4	0.45	0.5
H (A/m)	22.8	35	45	49	57	65	70	76	83	90
B (T)	0.55	0.6	0.65	0.7	0.75	0.8	0.85	0.9	0.95	1
H (A/m)	98	106	115	124	135	148	162	177	198	220
B (T)	1.05	1.1	1.15	1.2	1.25	1.3	1.35	1.4	1.45	1.5
H (A/m)	237	273	310	356	417	482	585	760	1,050	1,340
B (T)	1.55	1.6	1.65	1.7	1.75	1.8	1.85	1.9	1.95	2.0
H (A/m)	1760	2460	3460	4800	6160	8270	11,170	15,220	22,000	34,000

Therefore, the stator tooth mmf F_{AB} is

$$F_{AB} = H_{ts} \cdot (h_{su} + h_{sw}) = 1290(0.070357 + 0.003) = 94.62 \text{ A turns} \tag{3.59}$$

The stator back iron flux density was already chosen: $B_{cs} = 1.5$ T.

H_{cs} (from Table 3.2) is $H_{cs} = 1340$ A/m. The average magnetic path length l_{csw} is as follows:

$$(l_{csw})_{BC} \approx \frac{2}{3}\frac{\pi(D_{out} - h_{cs})}{2 \cdot 2p_1} = \frac{2}{3}\frac{\pi(0.780 - 0.0637)}{2 \cdot 2 \cdot 2} = 0.19266 \text{ m} \tag{3.60}$$

Therefore, the stator back iron mmf F_{BC} is

$$F_{BC} \approx H_{cs} \cdot l_{csav} = 1340 \cdot 0.19266 \approx 258.17 \text{ A turns} \tag{3.61}$$

In the rotor teeth, we need to obtain first an average flux density by using the top, middle, and bottom tooth values B_{trt}, B_{tRm}, and B_{tRb}:

$$\begin{aligned} B_{trt} &= B_{g1} \cdot \frac{\tau_R}{W_{tR}} = 0.75 \cdot \frac{0.0338}{(0.0338 - 0.0152)} = 1.3629 \text{ T} \\ B_{tRm} &= B_{g1} \cdot \frac{\tau_R}{W_{tRm}} = 0.75 \cdot \frac{0.0338}{0.01501} = 1.68887 \text{ T} \\ B_{tRb} &= B_{g1} \cdot \frac{\tau_R}{W_{tRmin}} = 0.75 \frac{0.0338}{0.01142} = 2.2 \text{ T} \end{aligned} \tag{3.62}$$

The average value of B_{tR} is as follows:

$$B_{tR} = \frac{B_{tRt} + B_{tRb} + 4B_{tRm}}{6} = \frac{1.3629 + 2.2 + 4 \cdot 1.68887}{6} = 1.7197 \approx 1.72 \text{ T} \tag{3.63}$$

From Table 3.2, $H_{tR} = 5334$ A/m. The rotor tooth mmf $F_{A'B'}$ is, thus,

$$F_{A'B'} = H_{tr} \cdot (h_{RU} + h_{RW}) = 5334(0.0519 + 0.003) = 292.8366 \text{ A turns} \tag{3.64}$$

Finally, for the rotor back iron flux density $B_{CR} = 1.6$ T (already chosen), $H_{cr} = 2460$ A/m, with the average path length l_{CRav} as follows:

$$l_{CRav} = \frac{2}{3}\frac{\pi(D_{shaft} + h_{CR} + 0.01)}{2 \cdot 2p_1} = \frac{2}{3}\pi\frac{(0.267 + 0.01 + 0.06)}{2 \cdot 2 \cdot 2} = 0.088 \text{ m} \tag{3.65}$$

The rotor back iron mmf $F_{B'C'}$,

$$F_{B'C'} = H_{cr} \cdot l_{CRav} = 2460 \cdot 0.088 = 216.92 \text{ A turns} \tag{3.66}$$

The total magnetization mmf per pole F_m is as follows (Equation 3.55):

$$F_m = 1083.86 + 94.62 + 258.17 + 292.8366 + 216.92 = 1946.40 \text{ A turns/pole}$$

It should be noted that the total stator and rotor back iron mmfs are not much different from each other. However, the stator teeth are less saturated than the other iron sections of the core. The uniform saturation of iron is a guarantee that sinusoidal air gap tooth and back iron flux densities distributions are secured. Now, from Equation 3.55, the no-load (magnetization) rotor current I_{R0} might be calculated:

$$I_{R0} = \frac{F_m \cdot \pi \cdot p_1}{3W_2 \cdot K_{W2} \cdot \sqrt{2}} = \frac{1946.4 \cdot \pi \cdot 2}{3 \cdot 80 \cdot 0.925 \cdot 1.41} = 39.05 \text{ A} \tag{3.67}$$

The ratio of I_{R0} to $I_{SN'}$ (Equation 3.35; stator current reduced to rotor), K_m, is, in fact,

$$K_m = \frac{I_{R0}}{I'_{SN}} = \frac{39.05}{418.865} = 0.09322 < 0.3 \tag{3.68}$$

The initial value assigned to K_m was 0.3, so the final value is smaller, leaving room for more saturation in the stator teeth by increasing the size of axial teeth cooling channels (Figure 3.1). Alternatively, the air gap may be increased up to 3 mm if so needed for mechanical reasons.

The total iron saturation factor K_s is

$$K_s = \frac{(F_{AB} + F_{BC} + F_{A'B'} + F_{B'C'})}{F_{AA'}} = \frac{(94.62 + 258.17 + 292.5366 + 216.92)}{1083.66} = 0.79568 \tag{3.69}$$

Therefore, the iron adds to the air gap mmf 79.568% more. This will reduce the magnetization reactance X_m accordingly.

3.6 Reactances and Resistances

The main WRIG parameters are the magnetization reactance X_m, the stator and rotor resistances R_s and R_r, and leakage reactances X_{sl} and X_{rl}, reduced to the stator. The magnetization reactance expression is straightforward [1]:

$$\begin{aligned}
X_m &= \omega_1 \cdot L_m; \\
L_m &= \frac{6\mu_0 (W_{1a} K_{WS})^2 \tau l_i}{\pi^2 p_1 g K_C (1 + K_S)} \\
L_m &= \frac{6 \cdot 1.256 \cdot 10^{-6} (20 \cdot 0.908)^2 \cdot 0.4 \cdot 0.52}{\pi^2 \cdot 2 \cdot 1.612 \cdot 10^{-3} \cdot 1.126 \cdot (1 + 0.79568) \cdot 2} = 8.0428 \cdot 10^{-3} \text{ H} \\
X_m &= 2 \cdot \pi \cdot 50 \cdot 8.0428 \cdot 10^{-3} = 2.5254\ \Omega
\end{aligned} \tag{3.70}$$

The base reactance $X_b = V_{SN}/I_{SN} = 690/(\sqrt{3} \cdot 1675.46) = 0.238\ \Omega$. As expected, $x_m = X_m/X_b = 2.5254/0.238 = 10.610 = I_{SN}/I_{R0}$ from Equation 3.67.

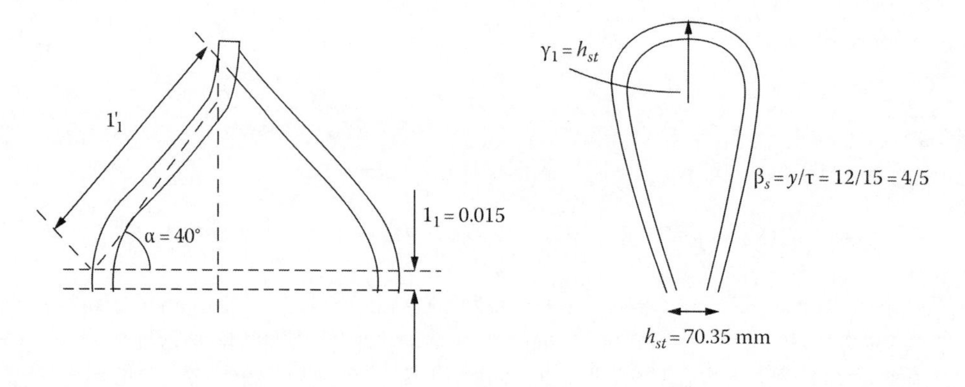

FIGURE 3.6 Stator coil end connection geometry.

The stator and rotor resistances and leakage reactances strongly depend on the end connection geometry (Figure 3.6). The end connection l_{fs} on one machine side, for the stator winding coils, is

$$l_{fs} \approx 2(l_l + l_{l'}) + \pi h_{st} = 2\left(l_l + \frac{\beta_s \tau}{2\cos\alpha}\right) + \pi h_{st}$$

$$= 2\left(0.015 + \frac{0.8 \cdot 0.4}{2\cos 40°}\right) + \pi \cdot 0.07035 = 0.668 \text{ m} \quad (3.71)$$

The stator resistance R_s per phase has the following standard formula (skin effect is negligible as shown earlier):

$$R_S \approx \rho_{co100°} \cdot \frac{W_{1a} 2}{A_{cos}} (l_i + l_{fs}) \cdot \frac{1}{a_1}$$

$$= \frac{1.8 \cdot 10^{-8}}{2}\left(1 + \frac{100 - 20}{272}\right) \cdot \frac{20 \cdot 2(0.52 + 0.668)}{128.88 \cdot 10^{-6}} = 0.429 \cdot 10^{-2}\ \Omega \quad (3.72)$$

The stator leakage reactance X_{sl} is written as follows:

$$X_{sl} = \omega_1 L_{sl}; \quad L_{sl} = \mu_0 (2n_{c1})^2 \cdot l_i (\lambda_S + \lambda_{end} + \lambda_{ds}) \cdot \frac{N_s}{m_1 a_1^2} \quad (3.73)$$

where

$n_{c1} = 2$ turns/coil

$l_i = 0.52$ m (stack length)

$N_s = 60$ slots

$m_1 = 3$ phases

$a_1 = 2$ stator current paths

λ_s, λ_{end}, and λ_{ds} are the slot, end connection, and differential geometrical permeance (nondimensional) coefficients

For the case in point [1],

$$\lambda_s = \frac{h_{su}}{3W_s} + \frac{h_{sw}}{W_s'} = \frac{70.35}{3 \cdot 15.2} + \frac{3}{4.066} = 2.2806 \quad (3.74)$$

$$\lambda_{end} = \frac{0.34 \cdot q_1(l_{fs} - 0.64\beta\tau)}{l_i}$$

$$= \frac{0.34 \cdot 5(0.668 - 0.64 \cdot 0.8 \cdot 0.4)}{0.52} = 1.5143 \quad (3.75)$$

The differential geometrical permeance coefficient λ_{ds} [1] may be calculated as follows:

$$\lambda_{ds} = \frac{0.9 \cdot \tau_s (q_s K_{W1}) \cdot K_{01} \cdot \sigma_{ds}}{K_c g} \quad (3.76)$$

$$K_{01} = 1 - 0.033\left(\frac{W_s^2}{g\tau_s}\right) = 1 - 0.033 \cdot \frac{4.066^2}{1.612 \cdot 26.66} = 0.9873$$

The coefficient σ_{ds} is the ratio between the differential and magnetizing inductance and depends on q_1 and chording ratio β. For $q_1 = 5$ and $\beta = 0.8$ from Figure 3.7 [1], $\sigma_{ds} = 0.0042$.

Finally,

$$\lambda_{ds} = \frac{0.9 \cdot 0.0266(5 \cdot 0.908)^2 \cdot 0.9873 \cdot 0.0042}{1.126 \cdot 1.612 \cdot 10^{-3}} = 1.127$$

The term λ_{ds} generally includes the zigzag leakage flux.

Finally, from Equation 3.73,

$$L_{sl} = 1.256 \cdot 10^{-6}(2 \cdot 2)^2 \cdot 0.52(2.2806 + 1.5143 + 1.127) \cdot \frac{60}{3 \cdot 2^2} = 0.257 \cdot 10^{-3}\ \text{H}$$

As can be seen, $L_{sl}/L_m = 0.257 \cdot 10^{-3}/(8.0428 \cdot 10^{-3}) = 0.03195$.

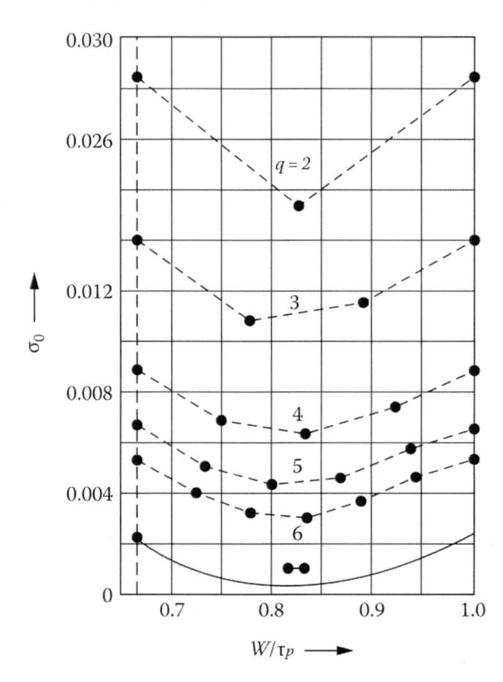

FIGURE 3.7 Differential leakage coefficient σ_d.

$X_{sl} = \omega_1 L_{sl} = 2\pi \cdot 50 \cdot 0.257 \cdot 10^{-3} = 0.0807\ \Omega$. The rotor end connection length l_{fr} may be computed in a similar way as in Equation 3.71:

$$l_{fr} = 2\left(l_l + \frac{\beta_r \tau}{2\cos\alpha}\right) + \pi(h_{RU} + h_{RW})$$

$$= 2\left(0.015 + \frac{10 \cdot 0.4}{12 \cdot 2 \cdot \cos 40^0}\right) + \pi(0.0519 + 0.003) = 0.6375\ \text{m} \tag{3.77}$$

The rotor resistance expression is straightforward:

$$R_R^r = \rho_{co100^\circ} \cdot \frac{W_2 \cdot 2}{A_{cor}}(l_i + l_{fr}) = 1.8 \cdot 10^{-8} \cdot \frac{80 \cdot 2(0.52 + 0.6775)}{43.40 \cdot 10^{-6}} = 7.68 \cdot 10^{-2}\ \Omega \tag{3.78}$$

Equation 3.73 also holds for rotor leakage inductance and reactance:

$$L_{rl}^r = \mu_0 (2n_{c2})^2 \cdot l_i (\lambda_{SR} + \lambda_{endR} + \lambda_{dR}) \cdot \frac{N_R}{m_1} \tag{3.79}$$

$$\lambda_{SR} = \frac{h_{RU}}{3W_R} + \frac{h_{RW}}{W_{R'}} = \frac{51.93}{3 \cdot 15.2} + \frac{3}{3.04} = 2.1256 \tag{3.80}$$

$$\lambda_{endR} = \frac{0.34 q_2 (l_{fr} - 0.64 \beta_R \tau)}{l_i}$$

$$= \frac{0.34 \cdot 4(0.6375 - 0.64 \cdot (5/6) \cdot 0.4)}{0.52}$$

$$= 1.1093 \tag{3.81}$$

$$\lambda_{dR} = \frac{0.9 \tau_R (q_R K_{W2})^2 \cdot K_{o2} \cdot \sigma_{dR}}{K_C g} \tag{3.82}$$

with $\sigma_{dR} = 0.0062$ ($q_2 = 4$, $\beta_R = 5/6$) from Figure 3.7:

$$K_{o2} = 1 - 0.033(W_{R'}^2) g \tau_R = 1 - 0.033\left(\frac{3.04^2}{1.612 \cdot 33.8}\right) = 0.994 \tag{3.83}$$

$$\lambda_{dR} = \frac{0.9 \cdot 33.8(4 \cdot 0.925)^2 \cdot 0.994 \cdot 0.0062}{1.126 \cdot 1.612} = 1.413$$

From Equation 3.79, L_{Rl}^r is

$$L_{Rl}^r = 1.256 \cdot 10^{-6} (2.5)^2 (2.1256 + 1.1093 + 1.413) \cdot \frac{48}{3} = 4.858 \cdot 10^{-3}\ \text{H}$$

$$X_{RL}^r = \omega_1 L_{RL}^r = 2\pi \cdot 50 \cdot 4.858 \cdot 10^{-3} = 1.525\ \Omega$$

Noting that R_R^r, L_{Rl}^r, and X_{Rl}^r are values obtained prior to stator reduction, we may now reduce them to stator values with $K_{RS} = 4$ (turns ratio):

$$R_R = \frac{R_R^r}{K_{RS}^2} = \frac{7.68 \cdot 10^{-2}}{16} = 0.48 \cdot 10^{-2}\ \Omega$$

$$L_{Rl} = \frac{L_{Rl}^r}{K_{RS}^2} = \frac{4.858 \cdot 10^{-3}}{16} = 0.303 \cdot 10^{-3}\ \text{H} \tag{3.84}$$

$$X_{Rl} = \frac{X_{Rl}^r}{K_{RS}^2} = \frac{1.525}{16} = 0.0953\ \Omega$$

Let us now add here the R_S, L_{sl}, X_{sl}, L_m, and X_m to have them all together: $R_s = 0.429 \cdot 10^{-2}\ \Omega$, $L_{sl} = 0.257 \cdot 10^{-3}$ H, $X_{sl} = 0.0807\ \Omega$, $L_m = 8.0428 \cdot 10^{-3}$ H, and $X_m = 2.5254\ \Omega$.

The rotor resistance reduced to the stator is larger than the stator resistance, mainly due to notably larger rated current density (from 6.5 to 10 A/mm^2), despite having shorter end connections.

Because of smaller q_2 than q_1, the differential leakage coefficient is larger in the rotor, which finally leads to a slightly larger rotor leakage reactance than in the stator.

The equivalent circuit may now be used to compute the power flow through the machine for generating or motoring, once the value of slip S and rotor voltage amplitude and phase are set. We leave this to the interested reader. In what follows, the design methodology, however, explores the machine losses to determine the efficiency.

3.7 Electrical Losses and Efficiency

The electrical losses are made of the following:

- Stator winding fundamental losses: p_{cos}
- Rotor winding fundamental losses: p_{cor}
- Stator fundamental core losses: p_{irons}
- Rotor fundamental core losses: p_{ironR}
- Stator surface core losses: p_{iron}^{SS}
- Rotor surface core losses: p_{iron}^{rr}
- Stator flux pulsation losses: p_{iron}^{SP}
- Rotor flux pulsation losses: p_{iron}^{RP}
- Rotor slip-ring and brush losses: p_{srb}

As the skin effect was shown small, it will be considered only by the correction coefficient $K_R = 1.037$, already calculated in Equation 3.33. For the rotor, the skin effect is neglected, as the maximum frequency $S_{max}f_1 = 0.25f_1$. In any case, it is smaller than in the stator, because the rotor slots are not as deep as the stator slots:

$$p_{cos} = 3K_R R_S I_{SN}^2 = 3 \cdot 1.037 \cdot 0.429 \cdot 10^{-2} \cdot 1675^2$$
$$= 37464.52\ \text{W} = 37.47\ \text{kW} \tag{3.85}$$

$$p_{cor} = 3R_R^R I_{RN}^2 = 3 \cdot 7.62 \cdot 10^{-2} \cdot 437^2$$
$$= 43715.47\ \text{W} = 43.715\ \text{kW} \tag{3.86}$$

The slip-ring and brush losses p_{sr} are easy to calculate if the voltage drop along them is given, say $V_{SR} \approx$ 1 V. Consequently,

$$p_{sr} = 3V_{SR} I_R^R = 3 \cdot 1 \cdot 437.3 = 1311.9\ \text{W} = 1.3119\ \text{kW} \tag{3.87}$$

To calculate the stator fundamental core losses, the stator teeth and back iron weights G_{ts} and G_{cs} are needed:

$$G_{ts} = \left\{\frac{\pi}{4}\left[(D_{is}+2(h_{su}+h_{sw}))^2 - D_{is}^2\right] - N_s\cdot(h_{su}+h_{sw})\right\} l_i \gamma_{iron}$$

$$= \left\{\frac{\pi}{4}[(0.52+2(70.315+3)10^{-3})^2] - 60\cdot(70.315+3)10^{-3}\cdot 13.3\cdot 10^{-3}\right\}\cdot 0.52\cdot 7600$$

$$\approx 313.8 \text{ kg} \tag{3.88}$$

$$G_{cs} \approx \pi(D_{out}-h_{cs})\cdot h_{cs}\cdot l_i\cdot \gamma_{iron}$$

$$= \pi(0.8-0.0637)\cdot 0.0637\cdot 0.52\cdot 7600$$

$$= 582 \text{ kg} \tag{3.89}$$

The fundamental core losses in the stator, considering the mechanical machining influence by fudge factors such as $K_t = 1.6$–1.8 and $K_y = 1.3$–1.4, are as follows [1]:

$$P_{irons} \approx p_{10/50}\left(\frac{f_1}{50}\right)^{1.5}\left(K_t B_{ts}^2\cdot G_{ts} + K_y B_{cs}^2 G_{cs}\right) \tag{3.90}$$

With $B_{ts} = 1.488$ T, $B_{cs} = 1.5$ T, $f_1 = 50$ Hz, and $p_{10/50} = 3$ W/kg (losses at 1 T and 50 Hz),

$$P_{irons} = 3\cdot\left(\frac{50}{50}\right)^{1.3}[1.6\cdot(1.488)^2\cdot 313.8 + 1.3\cdot(1.5)^2\cdot 582] = 8442 \text{ W} = 8.442 \text{ kW}$$

The rotor fundamental core losses may be calculated in a similar manner, but with f_1 replaced by Sf_1, and introducing the corresponding weights and flux densities.

As the $S_{max} = 0.25$, even if the lower rotor core weights will be compensated for by the larger flux densities, the rotor fundamental iron losses would be as follows:

$$P_{ironr} < S_{max}^2\cdot P_{irons} = \frac{1}{16}\cdot 8442 = 527.6 \text{ W} = 0.527 \text{ kW} \tag{3.91}$$

The surface and pulsation additional core losses, known as strayload losses, are dependent on the slot opening/air gap ratio in the stator and in the rotor [1]. In our design, magnetic wedges reduce the slot-openings-to-air gap ratio to 4.066/1.612 and 3.04/1.612; thus, the surface additional core losses are reduced. For the same reason, the stator and rotor Carter coefficients K_{c1} and K_{c2} are small ($K_{c1}\cdot K_{c2}$ = 1.126); and therefore, the flux-pulsation additional core losses are also reduced.

Consequently, all additional losses are most probably well within the 0.5% standard value (for detailed calculations, see Reference 1, Chapter 11) of stator rated power:

$$p_{add} = p_{iron}^{SS} + p_{iron}^{SR} + p_{iron}^{SP} + p_{iron}^{RP} = \frac{0.5}{100}\cdot 2{,}000\cdot 10^6 = 10{,}000 \text{ W} = 10 \text{ kW} \tag{3.92}$$

Thus, the total electrical losses Σp_e are

$$\Sigma p_e = p_{cos} + p_{cor} + p_{irons} + p_{ironr} + p_{ad} + p_{sr}$$

$$= 37.212 + 43.715 + 8.442 + 10 + 0.527 + 1.3119 = 101.20 \text{ kW} \tag{3.93}$$

Neglecting the mechanical losses, the "electrical" efficiency η_e is

$$\eta_e = \frac{P_{SN} + P_{RN}}{P_{SN} + P_{RN} + \Sigma p_e} = \frac{(2+0.5)10^6}{(2+0.5)10^6 + 0.101 \cdot 10^6} = 0.9616 \tag{3.94}$$

This is not a very large value, but it was obtained with the machine size reduction in mind.

It is now possible to redo the whole design with smaller f_{xt} (rotor shear stress), and lower current densities to finally increase efficiency for larger size. The length of the stack l_i is also a key parameter to design improvements. Once the above, or similar, design methodology is computerized, then various optimization techniques may be used, based on objective functions of interest, to end up with a satisfactory design (see Reference 1, Chapter 10).

Finite element analysis (FEA) verifications of the local magnetic saturation, core losses, and torque production should be instrumental in validating optimal designs based on even advanced analytical nonlinear models of the machine.

Mechanical and thermal designs are also required, and FEA may play a key role here, but this is beyond the scope of our discussion [2,3]. In addition, uncompensated magnetic radial forces have to be checked, as they tend to be larger in WRIG (due to the absence of rotor cage damping effects) [4].

3.8 Testing of WRIGs

The experimental investigation of WRIGs at the manufacturer's or user's sites is an indispensable tool to validate machine performance.

There are international (and national) standards that deal with the testing of general use induction machines with cage or wound rotors (International Electrotechnical Commission [IEC]-34, National Electrical Manufacturers Association [NEMA] MG1-1994 for large induction machines).

Temperature, losses, efficiency, unbalanced operation, overload capability, dielectric properties of insulation schemes, noise, surge responses, and transients (short circuit) responses are all standardized.

We avoid a description of such tests [5] here, as space would be prohibitive, and the reader could read the standards above (and others) by himself. Therefore, only a short discussion, for guidance, will be presented here.

Testing is performed for performance assessment (losses, efficiency, endurance, noise) or for parameter estimation. The availability of rotor currents for measurements greatly facilitates the testing of WRIGs for performance and for machine parameters. On the other hand, the presence of the bidirectional power flow static converter connected to the rotor circuits, through slip-rings and brushes, poses new problems in terms of current and flux time harmonics and losses.

The IGBT static converters introduce reduced current time harmonics, but measurements are still needed to complement digital simulation results.

In terms of parameters, the rotor and stator are characterized by single circuits with resistances and leakage inductances, besides the magnetization inductance. The latter depends heavily on the level of air gap flux, whereas the leakage inductances decrease slightly with their respective currents.

As for WRIGs, the stator voltage and frequency stay rather constant, and the stator flux $\bar{\Psi}_s$ varies only a little. The air gap flux $\bar{\Psi}_m$

$$\bar{\Psi}_m = \bar{\Psi}_s - L_{sl}\bar{I}_s \tag{3.95}$$

varies a little with load for unity stator power factor:

$$I_s R_s + V_s + j\omega_1 L_{sl} I_s = -j\omega_1 \bar{\Psi}_m \tag{3.96}$$

The magnetic saturation level of the main flux path does not vary much with load. Therefore, L_m does not vary much with load. In addition, unless $I_s/I_{sn} > 2$, the leakage inductances, even with magnetic wedges on slot tops, do not vary notably with respective currents. Therefore, parameters estimation for dealing with the fundamental behavior is greatly simplified in comparison with cage rotor induction motors with rotor skin effect.

On the other hand, the presence of time harmonics, due to the static converter of partial ratings, requires the investigation of these effects by estimating adequate machine parameters of WRIG with respect to them. Online data acquisition of stator and rotor currents and voltages is required for the scope.

The adaptation of tests intended for cage rotor induction machines to WRIGs is rather straightforward; thus, the following may all be performed for WRIGs with even better precision, because the rotor parameters and currents are directly measurable (for details see Reference 1, Chapter 22):

- Loss segregation tests
- Load testing (direct and indirect)
- Machine parameter estimation
- Noise testing methodologies

3.9 Summary

- WRIGs are built for powers in the 1.5–400 MW and more per unit.
- A flexible structure DFIG recent design at 10 MW and 10 rpm, 50 Hz shows promising potential [6].
- With today's static power converters for the rotor, the maximum rotor voltage may go to 4.2–6 kV, but the high-voltage direct current (HVDC) transmission lines techniques may extend it further for very high powers per unit.
- For stator voltages below 6 kV, the rotor voltage at maximum slip will be adapted to be equal to stator voltage; thus, no transformer is required to connect the bidirectional AC–AC converter in the rotor circuit to the local power grid. This way, the rotor current is limited around $|S_{\max}|I_{SN}$; I_{SN} rated stator current.
- If stator unity power factor is desired, the magnetization current is provided in the rotor; the rotor rated current is slightly increased and so is the static converter partial rating. This oversizing is small, however (less than 10%, in general).
- For massive reactive power delivery by the stator, but for unity power factor at the converter supply-side terminals, all the reactive power has to be provided via the capacitor in the DC link, which has to be designed accordingly. As for Q_s reactive power delivered by the stator, only SQ_s has to be produced by the rotor; the oversizing of the rotor connected static power converter still seems to be reasonable.
- The emf design of WRIGs as generators is performed for maximum power at maximum (supersynchronous) speed. The torque is about the same as that for stator rated power at synchronous speed (with DC in the rotor).
- The emf design basically includes stator and rotor core and windings design, magnetization current, circuit parameter losses, and efficiency computation.
- The sizing of WRIGs starts with the computation of stator bore diameter D_{is} based either on Esson's coefficient or on the shear rotor stress concept $f_{xt} \approx 2.5$–8 N/cm^2. The latter path was taken in this chapter. Flux densities in air gap, teeth, and back irons were assigned initial values, together with the current densities for rated currents. (The magnetization current was also assigned an initial P.U. value.)
- Based on these variables, the sizing of the stator and the rotor went smoothly. For the 2.5 MW numerical example, the skin effects in the stator and rotor were proven to be small ($\leq$4%).

- The magnetization current was recalculated and found to be smaller than the initial value, so the design holds; otherwise, the design should have been redone with smaller f_{xt}.
- Rather uniform saturation of various iron parts—teeth and back cores—leads to rather sinusoidal time waveforms of flux density and, thus, to smaller core losses in a heavily saturated design. This was the case for the example in this chapter.
- The computation of resistances and leakage reactances is straightforward. It was shown that the differential leakage flux contribution (due to mmf space harmonics) should not be neglected, as it is important, though $q_1 = 5$, $q_2 = 4$ and proper coil chording is applied both in the stator and in the rotor windings.
- Open slots were adopted for both the stator and the rotor for easing the fabrication and insertion of windings in slots. However, magnetic wedges are mandatory to reduce additional (surface and flux pulsation) core losses and magnetization current.
- Though mechanical design and thermal design are crucial, they fall beyond our scope here, as they are strongly industry-knowledge dependent.
- The testing of WRIGs is standardized mainly for motors. Combining these tests with synchronous generator tests may help generate a set of comprehensive, widely accepted testing technologies for WRIGs.
- Though important, the design of a WRIG as a brushless exciter with rotor output was not pursued in this chapter, mainly due to the limited industrial use of this mode of operation.

References

1. I. Boldea and S.A. Nasar, *Induction Machine Handbook*, CRC Press, Boca Raton, FL, 1998.
2. E. Levi, *Polyphase Motors—A Direct Approach to Their Design*, Wiley Interscience, New York, 1985.
3. K. Vogt, *Electrical Machines: Design of Rotary Machines*, 4th edn., VEB Verlag, Berlin, Germany, 1988 (in German).
4. D.G. Dorrel, Experimental behavior of unbalanced magnetic pull in 3-phase induction motors with eccentric rotors and the relationship with teeth saturation, *IEEE Trans.*, EC-14(3), 1999, 304–309.
5. Institute of Electrical and Electronics Engineers (IEEE), *IEEE Standard Procedure for Polyphase Induction Motors and Generators*, IEEE Standard 112-1996, IEEE Press, New York, 1996.
6. V. Delli Colli, F. Marignetti, and C. Attaianese, Analytical and multiphysics approach to the optimal design of a 10 MW DFIG for direct drive wind turbines, *IEEE Trans.*, IE-59(7), 2012, 2791–2799.

4

Self-Excited Induction Generators

4.1 Introduction

By self-excited induction generators (SEIGs), we mean cage-rotor induction machines (IMs) with shunt (and series) capacitors connected at their terminals for self-excitation.

The shunt capacitors may be constant or varied through power electronics (or step-wise). SEIGs may be built with either a single-phase or three-phase output and may supply alternating current (AC) loads or AC-rectified (direct current [DC]) autonomous loads. We also include here SEIGs connected to the power grid through soft-starters or resistors and having capacitors at their terminals for power factor compensation (or voltage stabilization).

Note that power electronics-controlled cage-rotor induction generators (IGs) for constant voltage and frequency output at variable speed, for autonomous and power grid operation, will be dealt with in Chapter 5.

This chapter will introduce the main schemes for SEIGs and their steady-state and transient performance, with sample results for applications such as wind machines, small hydrogenerators, or generator sets. Both power grid and stand-alone operations and three-phase and single-phase output SEIGs are discussed in this chapter.

4.2 Principle of Cage-Rotor Induction Machine

The cage-rotor IM is the most built and the most used electric machine, mainly as a motor but, recently, as a generator, too.

The cage-rotor induction machine contains cylindrical stator and rotor cores with uniform slots separated by a small air gap (0.3–2 mm, in general).

The stator slots host a three-phase or two-phase AC winding meant to produce a traveling magnetomotive force (mmf). The windings are similar to those described for synchronous generators (SGs) in Chapter 4 of *Synchronous Generators*, or for wound-rotor induction generators (WRIGs) in Chapter 3 of this book. This traveling mmf produces a traveling flux density in the air gap, B_{g10}:

$$B_{g10} = \frac{\mu_0 F_{10}}{g}\cos(\omega_1 t - p_1\theta_r) \tag{4.1}$$

$$F_{10} = \frac{3\sqrt{2}I_{10}W_1K_{W_1}}{\pi p_1} \quad \text{(for three phases)} \tag{4.2}$$

where

- g is the air gap
- θ_r is the rotor position
- p_1 equals the pole pairs

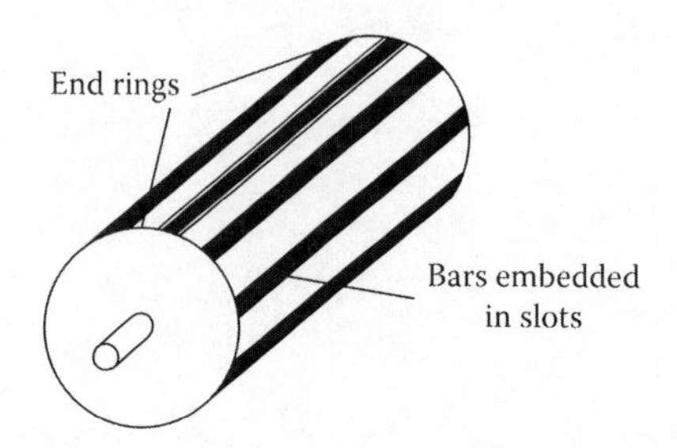

FIGURE 4.1 The cage rotor.

The cage rotor contains aluminum (or copper or brass) bars in slots. They are short-circuited by end-rings with resistances that are, in general, smaller than those of bars (Figure 4.1).

The angular speed of the traveling fields is obtained for the following:

$$\omega_1 t - p_1\theta_r = \text{const.} \tag{4.3}$$

That is, for

$$\frac{d\theta_r}{dt} = \frac{\omega_1}{p_1}; \quad n_1 = \frac{f_1}{p_1} \tag{4.4}$$

The speed n_1 (in revolutions per second, r/s) is the so-called ideal no-load or synchronous speed and is proportional to stator frequency and inversely proportional to the number of pole pairs, p_1.

The traveling field in the air gap induces electromagnetic fields (emfs) in the rotor that rotates at speed n, at frequency f_2:

$$f_2 = \left(\frac{n_1 - n}{n_1}\right) \cdot f_1 = Sf_1$$
$$S = \frac{n_1 - n}{n_1} \tag{4.5}$$

As expected, the emfs induced in the short-circuited rotor bars produce in them AC currents at slip frequency $f_2 = Sf_1$.

Let us now assume that the symmetric rotor cage, which has the property to adapt to almost any number of pole pairs in the stator, may be replaced by an equivalent (fictitious) symmetric three-phase winding (as in WRIGs) that is short-circuited. The traveling air gap field produces symmetric emfs in the fictitious three-phase rotor with frequency that is Sf_1 and with amplitude that is also proportional to slip S:

$$E_2 = SE_1 = S\omega_1 L_m I_m \tag{4.6}$$

where L_m is the magnetization inductance.

E_1 is the stator phase self-induced emf, generally produced by both stator and rotor currents, or by the so-called magnetization current I_m ($\underline{I}_m = \underline{I}_1 + \underline{I}_2$).

The rotor phases may be represented by a leakage inductance L_{2l} and a resistance R_2. Consequently, the rotor current I_2 is as follows:

$$I_2 = \frac{SE_1}{\sqrt{(R_2)^2 + (S\omega_1 L_{2l})^2}} \tag{4.7}$$

The rotor currents interact with the air gap field to produce tangential forces—torque. In Equations 4.6 and 4.7, the rotor winding is reduced to the stator winding based on energy (and loss) equivalence.

Note that the stator phases are also characterized by a resistance R_1 and a leakage inductance L_{1l}; the stator and rotor equations may be written, for steady state, in complex numbers, as for a transformer but with different frequencies in the primary and secondary. Let us consider the generator association of signs for the stator:

$$\begin{aligned} \underline{I}_1(R_1 + j\omega_1 L_{1l}) + \underline{V}_1 &= \underline{E}_1 \\ \underline{I}_r(R_2 + jS\omega_1 L_{2l}) &= S\underline{E}_1 \\ \underline{E}_1 &= -j\omega_1 L_m(\underline{I}_s + \underline{I}_r) \end{aligned} \tag{4.8}$$

Dividing the second expression in Equation 4.8 by S yields the following:

$$\underline{I}_2\left(R_2 + \frac{R_2(1-S)}{S} + j\omega_1 L_{2l}\right) = \underline{E}_1 \tag{4.9}$$

This way, in fact, the frequency of rotor variables becomes ω_1, and it refers to a machine at standstill, but with an additional (fictitious) rotor resistance $R_2(1 - S)/S$. The power dissipated in this resistance equals the mechanical power in the real machine (minus the mechanical losses):

$$T_e \cdot 2\pi n_1(1-S) = 3I_2^2 R_2 \frac{(1-S)}{S} \tag{4.10}$$

Finally,

$$T_e = \frac{3p_1}{\omega_1} I_2^2 \frac{R_2}{S} = 3\frac{p_1}{\omega_1} P_{elm} \tag{4.11}$$

P_{elm} is the so-called electromagnetic power: the total active power that crosses the air gap. Equations 4.8 and 4.9 lead to the standard equivalent circuit of the induction machine (IM) with cage rotor (Figure 4.2).

The core loss resistance R_m is added to account for fundamental core losses located in the stator, as $S \ll 1$, in general. R_m is determined by tests or calculated in the design stage. As can be seen from Equation 4.11, the electromagnetic power P_{elm} is positive (motoring) for $S > 0$ and negative (generating) for $S < 0$. For details on parameter expressions, various losses, parasitic torques, design, and so forth, of cage rotor IMs, see Reference 1.

As can be seen from Figure 4.2, the equivalent (total) reactance of the IM is always inductive, irrespective of slip sign (motor or generator), while the equivalent resistance changes sign for generating.

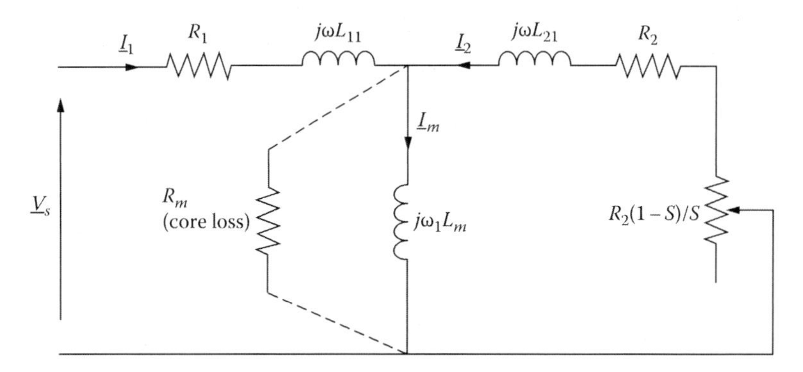

FIGURE 4.2 The cage-rotor induction machine equivalent circuit.

Therefore, the IM takes the reactive power to get magnetized either from the power grid to which it is connected or from a fixed (or controlled) capacitor at terminals. Note that when a full power static converter is placed between the IG and the load (or power grid), the IG is again self-excited by the capacitors in the converter's DC link or from the power grid (if a direct AC–AC converter is used).

As the operation of an IM at the power grid is straightforward ($S < 0$, $\omega_r > \omega_1$), the capacitor-excited IG will be treated here first in some detail.

4.3 Self-Excitation: A Qualitative View

The IG with capacitor excitation is driven by a prime mover with the main power switch open (Figure 4.3a). As the speed increases, due to prime-mover torque, eventually, the no-load terminal voltage increases and settles to a certain value, depending on machine speed, capacitance, and machine parameters.

The equivalent circuit (Figure 4.2) is further simplified by neglecting the stator resistance and leakage inductance and by considering zero slip ($S = 0$: open-rotor circuit) for no-load conditions (Figure 4.3b).

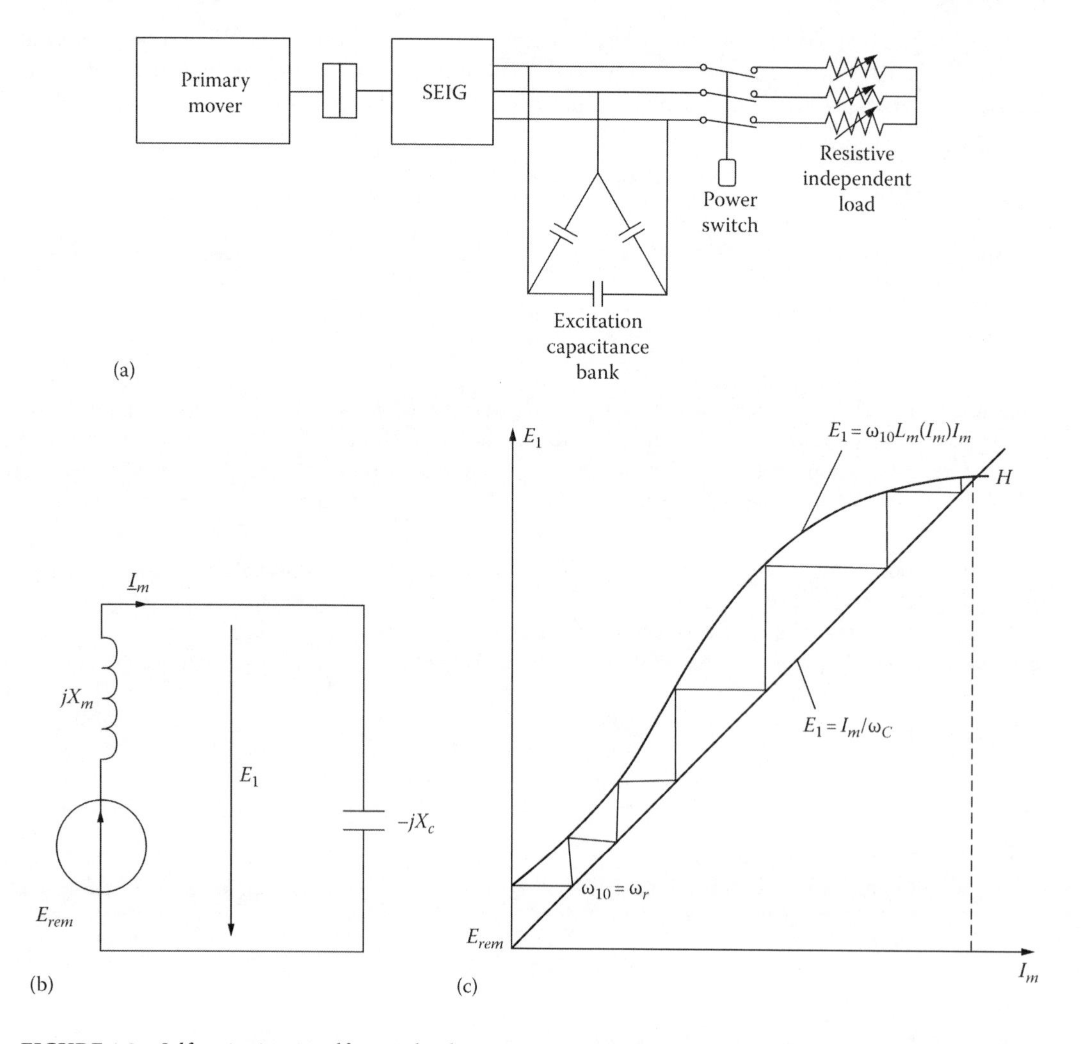

FIGURE 4.3 Self-excitation on self-excited induction generator (SEIG): (a) the general scheme, (b) oversimplified equivalent circuit, and (c) quasi-steady-state self-excitation characteristics.

E_{rem} represents the no-load initial stator voltage (before self-excitation), at frequency $\omega_{10} = \omega_r$, produced by the remnant flux density in the rotor left there from previous operation events.

To initiate the self-excitation process, E_{rem} has to be nonzero.

The magnetization curve of the IG, obtained from typical motor no-load tests, $E_1(I_m)$, has to advance to the nonlinear (saturation) zone in order to firmly intersect the capacitor straight-line voltage characteristic (Figure 4.3c) and, thus, produce the no-load voltage E_1. The process of self-excitation of IG has been known for a long time [2].

The increase in the terminal voltage from V_{rem} to V_{10} unfolds slowly in time (seconds), and Figure 4.3c presents it as a step-wise quasi-steady-state process. It is only a qualitative representation. Once the SEIG is self-excited, the load is connected. If the load is purely resistive, the terminal voltage decreases and so does (slightly) the frequency ω_1 for constant (regulated) prime-mover speed ω_r.

With $\omega_1 < \omega_r$, the SEIG delivers power to the load for negative slip $S < 0$:

$$f_1 = \frac{np_1}{1+|S|}; \quad S < 0 \tag{4.12}$$

The computation of terminal voltage V_1, frequency f_1, stator current I_1 delivered active and reactive powers (efficiency) for given load (speed n), capacitor C, and machine parameters R_1, R_2, L_{1l}, L_{2l}, and $L_m(I_m)$ represents, in fact, the process of obtaining the steady-state performance.

The nonlinear function $L_m(I_m)$—magnetization curve—and the variation of frequency f_1 with load, at constant speed n, make the process mathematically intricate.

4.4 Steady-State Performance of Three-Phase SEIGs

Various analytical (and numerical) methods to calculate the steady-state performance of SEIGs were proposed. They, however, fall into two main categories:

1. Loop impedance models [3]
2. Nodal admittance models [4]

Both models are based on the SEIG equivalent circuit (Figure 4.2) expressed in per unit (P.U.) form for frequency f (P.U.) and speed U (in P.U.) as follows:

$$f = \frac{f_1}{f_{1b}} \qquad U = \frac{np_1}{f_{1b}} \tag{4.13}$$

The base frequency for which all reactances X_{1l}, X_{2l}, and $X_m(I_m)$ are calculated is f_{1b}.

With an R_L, L_L, C_L load, the equivalent circuit in Figure 4.2 with speed and frequency in P.U. terms becomes as shown in Figure 4.4.

The presence of frequency f in the load, the dependence of core loss resistance R_m of frequency f, and the nonlinear dependence on X_m of I_m make the solving of the equivalent circuit difficult. The SEIG plus load show zero total impedance:

$$\begin{aligned} &R_e(\text{IG} + \text{load}) = 0 \\ &X_e(\text{IG} + \text{excitation} - \text{capacitor} + \text{load}) = 0 \end{aligned} \tag{4.14}$$

for self-excitation, under load.

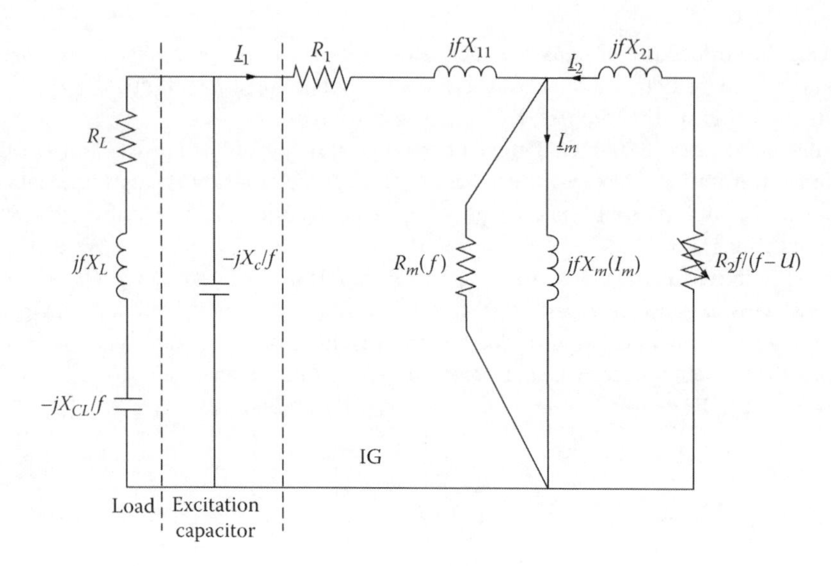

FIGURE 4.4 Self-excited induction generator (SEIG) equivalent circuit in per unit (P.U.) frequency f and speed U.

To solve it simply, the problem is reduced to two unknowns: f (frequency) and X_m for given excitation capacitor, IG(R_1, R_2, X_{1l}, X_{2l}, $X_m[I_m]$), load(R_L, X_L, X_C), and speed U.

Two main impedance approaches to solve Equation 4.14 were developed:

1. High-order polynomial equation (in f) approaches [5]
2. Optimization approaches [6,7]

The high-order polynomial and the optimization method solutions obscure the intuitive understanding of performance sensitivity to various parameters, but they constitute mighty computerized tools.

In References 8 and 9, admittance models that led to a quadratic equation for slip $f - U = S$ (instead of f) for given frequency f were introduced for balanced resistive loads (RLs) without additional simplifying assumptions. A simple iterative method is used to adjust frequency until the desired speed is obtained.

For the sake of simplicity, the admittance model will be used in what follows.

4.4.1 Second-Order Slip Equation Methods

The standard equivalent circuit of Figure 4.4 may be changed by lumping together the IG stator (R_1, jfX_{1l}), the excitation capacitor reactance ($-jX_c/f$), and the load (R_L, jfX_L, $-jX_{CL}/f$) into an equivalent series circuit (R_{1L}, jfX_{1L}) (Figure 4.5). For self-excitation, $X_{1L} \leq 0$:

$$R_{1L} + jf\omega_1 X_{1L} = R_1 + j\omega_1 L_{1l} + \frac{-j(X_c/f)\left(R_L + +jfX_L - j(X_{CL}/f)\right)}{R_L + j\left(fX_L - (X_c/f) - (X_{CL}/f)\right)} \tag{4.15}$$

In general, both R_{1L} and X_{1L} are dependent on frequency f (P.U.), though they get simplified forms if only a resistive or an R_L, X_L (an induction motor) is considered.

For self-excitation, the summation of currents in the node should be zero (with $E_1 \neq 0$):

$$-\underline{I}_1 + \underline{I}_m - \underline{I}_2 = 0 \tag{4.16}$$

or

$$f\underline{E}_1 \cdot \left(\frac{1}{R_{1L} + jfX_{1L}} + \frac{S}{R_2 + jfSX_{2l}} + \frac{1}{jfX_m} \right) = 0 \tag{4.17}$$

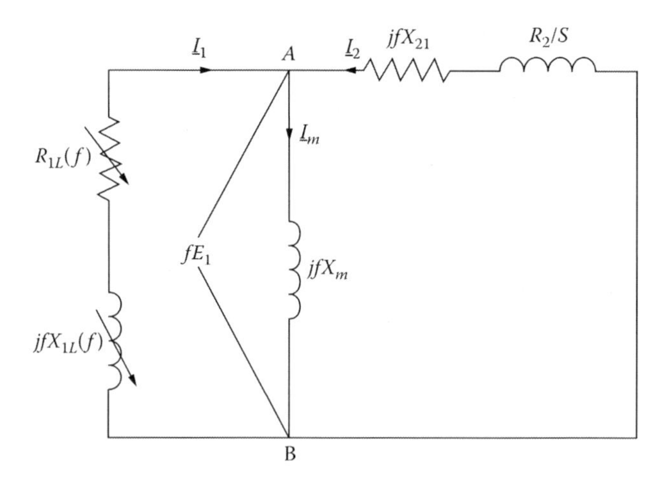

FIGURE 4.5 The nodal equivalent circuit of a self-excited induction generator (SEIG).

The real and imaginary parts in Equation 4.17 have to be zero for self-excitation (it is, in fact, an energy balance condition):

$$\frac{R_{1L}}{R_{1L}^2 + f^2 X_{1L}^2} + \frac{SR_2}{R_2^2 + S^2 f^2 X_{2l}^2} = 0 \tag{4.18}$$

$$\frac{1}{fX_m} - \frac{fX_{1L}}{R_{1L}^2 + f^2 X_{1L}^2} + \frac{SfX_{2l}}{R_2^2 + S^2 f^2 X_{2l}^2} = 0 \tag{4.19}$$

For given frequency f (P.U.), Equation 4.18 remains (for given excitation capacitors, IG parameters, and load), with only one unknown, the slip S:

$$aS^2 + bS + c = 0 \tag{4.20}$$

with

$$\begin{aligned} a &= f^2 X_{2l}^2 R_{1L} \\ b &= R_2\left(R_{1L}^2 + f^2 X_{1L}^2\right) \\ c &= R_{1L} R_2^2 \end{aligned} \tag{4.21}$$

Equations 4.20 and 4.21 have two solutions, but only the smaller one (in amplitude) is really useful. For the larger one, most of the power is consumed into the rotor resistance:

$$S_{1,2} = \frac{-b \pm \sqrt{b^2 - 4ac}}{2a} \tag{4.22}$$

If complex solutions of S are obtained, it means that self-excitation is impossible.

With slip S_1 found from Equation 4.22, for given f, the corresponding P.U. speed $U = f - S$ is determined. With $S = S_1$, from Equation 4.19, the magnetization reactance X_m is calculated as follows:

$$X_m = f^{-1}\left[\frac{fX_{1L}}{R_{1L}^2 + f^2 X_{1L}^2} - \frac{SfX_{2l}}{R_2^2 + S^2 f^2 X_{2l}^2}\right]^{-1} < X_{\max} \tag{4.23}$$

$X_{\max}$ is the maximum (unsaturated) value of the magnetization reactance (at base frequency f_{1b}). With $X_m > X_{\max}$, self-excitation is again impossible.

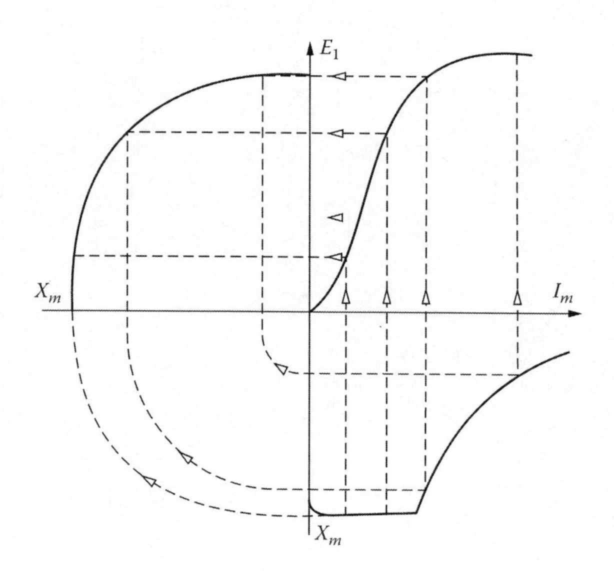

FIGURE 4.6 Magnetization curve at base frequency f_{1b}.

Further on, from no-load motor testing, or from design calculations, the $E_1(I_m)$ or $X_m(I_m) = E_1/I_m$ characteristic will be determined (Figure 4.6).

$E_1(X_m)$ from Figure 4.6 may be curve fitted by mathematical approximations such as the following [10]:

$$E_1 = \omega_{1b} K_1 I_m; \quad I_m < I_0$$
$$E_1 = \omega_{1b}\left[K_1 I_m + \frac{K_2}{d}\tan^{-1}(d(I_m - I_0))\right] \tag{4.24}$$

for $I_m \le I_0$.

The coefficients K_1, K_2, d are calculated to preserve continuity at $I_m = I_0$ in E_1 and in dE_1/dI_1, and they reasonably approximate the entire curve. This particular approximation has a steady decrease in the derivative, and its inverse is readily available:

$$X_m = X_{\max} = K_1\omega_b \quad \text{for } I_m < I_0 \tag{4.25}$$

$$E_1 = X_m\left[I_0 + \frac{1}{d}\tan\left(\frac{E_1\left(1 - X_{\max}/X_m\right)}{dK_2\omega_{1b}}\right)\right]; \quad I_m > I_0 \tag{4.26}$$

Though Equation 4.26 is a transcendent equation, its numerical solution in E_1, for the now calculated X_m (Equation 4.23), is rather straightforward.

Once E_1 is known, the equivalent circuit in Figure 4.5 produces all required variables:

$$\begin{aligned}
&I_2 = \frac{-fE_1}{(R_2/S) + jfX_{2l}}; \quad \underline{I}_m = \frac{E_1}{jX_m}\\
&\underline{I}_1 = \frac{-fE_1}{R_{1L} + jfX_{1L}}; \quad X_{1L} < 0\\
&-\underline{V}_1 = f\underline{E}_1 + (R_1 + jfX_{1l})\underline{I}_1\\
&\underline{I}_C = -V_1 j\frac{1}{fX_c}; \quad X_c = \frac{1}{\omega_1 C}\\
&\underline{I}_L = \underline{I}_1 + \underline{I}_C\\
&\underline{I}_m = \underline{I}_1 + \underline{I}_2
\end{aligned} \tag{4.27}$$

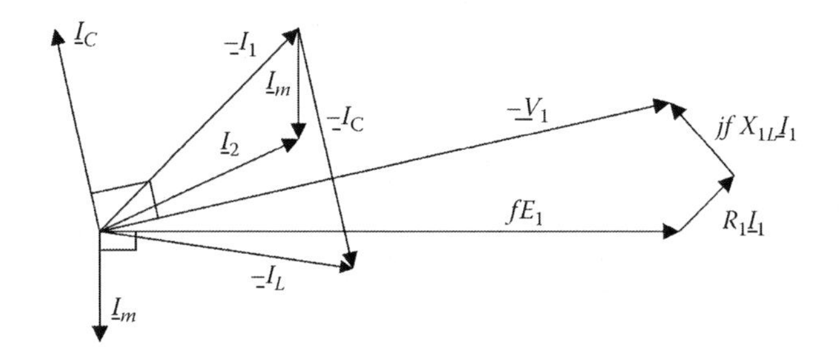

FIGURE 4.7 The phasor diagram.

Let us now draw a general phasor diagram for a typical R_L, L_L load when the load current $\underline{I}_L$ is lagging behind the terminal voltage V_1. Also, note that in Equation 4.27 $\underline{I}_1$ is leading fE_1, because $V_{1L} < 0$ to fulfill the self-excitation conditions.

The phasor diagram starts with fE_1 in the real axis and $(-\underline{I}_1)$ leading it (Figure 4.7). Then, from Equation 4.27 (the third expression), $\underline{V}_1$ is constructed. Also, from Equation 4.27 (the first expression), for $S < 0$, $\underline{I}_2$ is ahead of $\underline{E}_1$. For resistive-inductive load, the capacitor current is in a leading position with respect to terminal voltage $(-\underline{V}_1)$.

The whole computation process described thus far may be computerized; and for given speed U (P.U.), the initial value of f may be taken as $f(1) = U$. After one computation cycle, the slip $S(1)$ is calculated, and the new value $f(2)$ is $f(2) = v + S(1)$. The whole iterative process continues until the frequency error between two successive computation cycles is smaller than a desired value.

It was demonstrated [11] that less than 10 cycles are required, even if the core loss resistance (R_m) would be included. It was also shown that core losses do not modify the machine capability, except for the situation around maximum power delivery.

Once fE_1 is known, power core losses p_{iron} may be calculated as follows:

$$p_{iron} \approx \frac{3(fE_1)^2}{R_m} \tag{4.28}$$

Therefore, the efficiency on SEIG is

$$\eta = \frac{3V_1I_L\cos\varphi_L}{3V_1I_L\cos\varphi_L + 3R_1I_1^2 + 3R_2I_2^2 + p_{iron} + p_{mec} + p_{stray} + p_{cap}} \tag{4.29}$$

where

p_{mec} is the mechanical loss
p_{stray} is the IG stray load loss (Reference 1, Chapter 3)
p_{cap} is the excitation capacitor loss

Typical load voltage V_1 versus load current I_L for unity power factor load ($\varphi_L = 0$), for a 20 kW, 4-pole, 50 Hz, 380 V (Y), 5.4 A star connected machine are shown in Figure 4.8 [11]. The machine parameters are $R_1 = 0.10$ P.U., $X_{1l} = 0.112$ P.U., $R_2 = 0.0736$ P.U., $X_{2l} = 0.1$ P.U., and

$$\begin{aligned} &E_1\,(\text{P.U.}) = 1.345 - 0.203X_m; \quad X_m < 1.728\ \text{P.U.} \\ &E_1\,(\text{P.U.}) = 1.901 - 0.525X_m; \quad 1.728 \le X_m \le 2.259 \\ &E_1\,(\text{P.U.}) = 3.156 - 1.08X_m; \quad 2.259 \le X_m \le 2.446 \\ &E_1\,(\text{P.U.}) = 37.79 - 15.12X_m; \quad 2.446 \le X_m \le 2.48 \\ &0; \quad X_m > 2.48 \\ &R_m = 18.51 + E_1 \times 4.197 \end{aligned} \tag{4.30}$$

The typical collapse of terminal voltage, even at RL ($\varphi_L = 0$), below rated machine current, is evident.

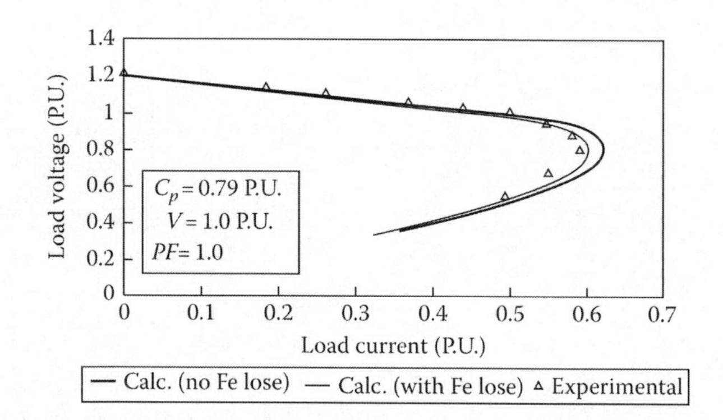

FIGURE 4.8 Voltage vs. load current of self-excited induction generator (SEIG) with shunt self-excitation.

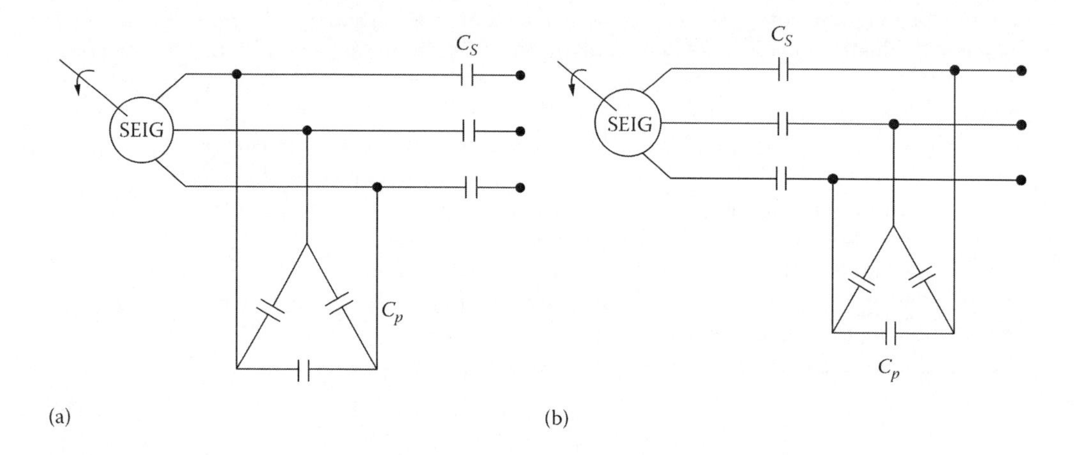

FIGURE 4.9 Series compensation by capacitance: (a) short shunt and (b) long shunt.

4.4.2 SEIGs with Series Capacitance Compensation

In an attempt to increase the load range (in P.U.), series capacitors are added in short-shunt (Figure 4.9a) or long-shunt (Figure 4.9b) connections.

The short shunt was proven superior in extending the stable operation load range for the same capacitance effort. The investigation of both connections can be done by following the iterative method in the previous paragraph by incorporating ($-jX_{cs}/f$) in the load (short shunt) or to the stator leakage reactance fX_{1l} (long shunt). Typical load voltage/current curves with long-shunt and short-shunt compensation, for the same machine, are shown in Figure 4.10 [11], with $K = X_{cs}/X_{cp}$ as the ratio between series and parallel capacitor reactances (Figure 4.10a and b).

Though the voltage collapse was avoided up to rated machine current, the voltage regulation is still noticeable for both connections. Note that the parallel capacitor C_p is larger for the long-shunt connection ($K = C_s/C_p$).

4.5 Performance Sensitivity Analysis

In this analysis, the influence of IG resistances R_1, R_2, leakage reactances X_{1l}, X_{2l}, magnetization reactance X_m, and parallel and series capacitances C_p, C_s on an SEIG's performance is investigated for constant speed (controlled prime mover) and constant head hydroturbine or uncontrolled speed wind turbine.

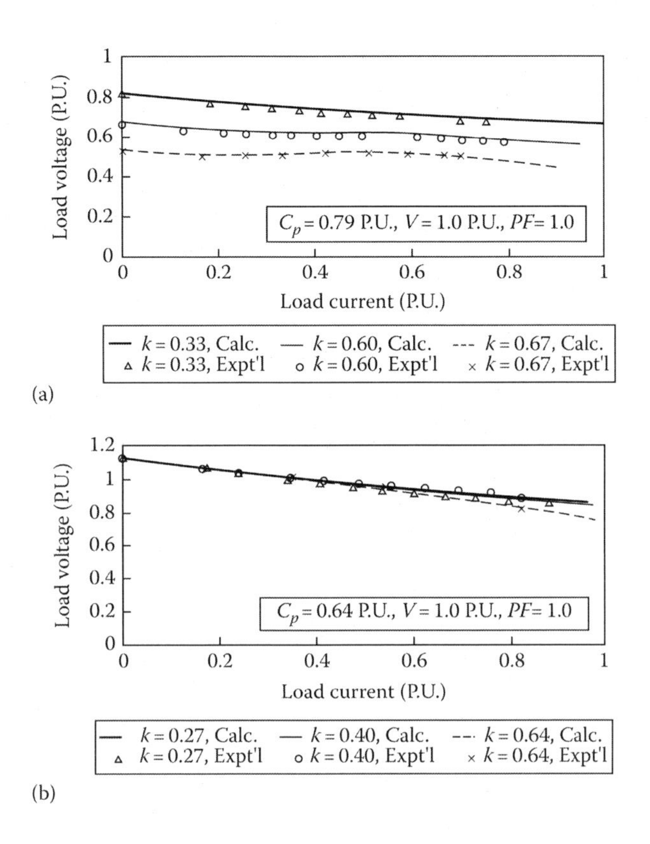

FIGURE 4.10 Load curve of three-phase SEIG with series compensation: (a) long shunt and (b) short shunt.

4.5.1 For Constant Speed

- The no-load voltage increases with the parallel (excitation) capacitance C_p.
- The maximum output power and terminal voltage increase significantly with capacitance C_p.
- For constant load voltage, the required capacitor C_p increases with delivered power.
- When no-load voltage increases, the magnetization reactance X_m decreases, due to advancing magnetic saturation.
- For self-excitation, a necessary but insufficient condition for purely RL (R_L) is (Equation 4.15) $X_{1L} < 0$:

$$fX_{1l} < X_L; \quad X_L = \frac{X_c/f}{1+\left(X_c^2/f^2R_L^2\right)} \tag{4.31}$$

With given capacitor (X_c), Equation 4.31 reduces itself to a minimum load resistance R_L condition:

$$R_L \geq X_C\sqrt{\frac{X_{1l}}{X_c - f^2X_{1l}}} \tag{4.32}$$

The smaller the stator leakage reactance, the better (the smaller the value of R_L). Note again that Equation 4.32 is a necessary but not sufficient condition.

- From Equation 4.22, a real value of S is required for self-excitation: $b^2 > 4ac$. For RL,

$$R_{1L}^2 + f^2 X_{1L}^2 > 2 \cdot R_{1L} \cdot f \cdot X_{2l} \tag{4.33}$$

or simply,

$$fX_{2l} < X_{1L} = -fX_{1l} + \frac{X_c/f}{1+\left(X_c^2/f^2R_L^2\right)} \tag{4.34}$$

Finally,

$$f(X_{1l} + X_{2l}) < \frac{X_c/f}{1+\left(X_c^2/f^2R_L^2\right)} \tag{4.35}$$

and

$$R_L \geq X_c\sqrt{\frac{X_{1l} + X_{2l}}{X_c - (X_{1l} + X_{2l})f^2}} \tag{4.36}$$

- It is very clear that Equation 4.36 is stronger than Equation 4.32 and should be the only one of the two conditions to be considered for practical purposes.
- Smaller short-circuit reactance is better, while at least $X_c > (X_{1l} + X_{2l})f^2$. This corresponds to a large capacitor (perhaps the largest ideal limit).
- The largest slip S (Equation 4.22) is obtained for $b^2 = 4ac$, and for small slips, S_1 is

$$S_{1\max} \approx \frac{-2ac}{2ab} \approx -\frac{c}{b} \tag{4.37}$$

For RL, S_1 becomes

$$S_{1\max} \approx -\frac{R_2}{fX_{2l}} \tag{4.38}$$

Incidentally, this ideal condition (for $S_{1\max}$ the voltage already collapses) corresponds to maximum torque for constant air gap flux (E_1) control in vector-controlled IMs.

4.5.2 For Unregulated Prime Movers

- Along a 40% load resistance variation—around maximum output power—the output power varies only a little.
- Within this rather constant power load range, however, the load voltage drops notably with load.
- Up to maximum power, the frequency and speed decrease with power, while they tend to increase after the maximum power load; a kind of self-stabilization, in this respect, takes place.
- The maximum power depends on $C_p^{2.2}$ and the corresponding load voltage, on $C_p^{0.5}$. This tends to be valid both for three-phase and single-phase SEIGs.
- For no load, the minimum capacitance for self-excitation is inversely proportional to speed squared.
- Under load, the minimum capacitance depends on speed, load impedance, and load power factor.

- When the capacitance is too small to handle the total reactive power (of IG and load) of the SEIG's, the voltage collapses. When the parallel capacitance C_p is too large, the rotor impedance of IG causes de-excitation, and the voltage collapses again. There should be an optimum capacitor between $C_{p\min}$ and $C_{p\max}$ to provide maximum output power for good efficiency.
- The linear (stable) zone of $V_1(I_L)$ curve can be extended notably—that is, larger maximum power with reasonable voltage regulation—by using a series capacitor $C_s(X_{cs})$. A good first guess for C_p corresponds to $X_{cp} \approx 0.7$–0.8 (P.U.) and $X_{cs} \approx 0.4$–$0.6X_{cp}$ (see Reference 12 for more on capacitance selection). Again, there is an optimal series capacitor for a given SEIG to produce minimum average voltage regulation.
- Constant-speed-regulated prime movers lead to notably larger powers delivered by the SEIG for other given data.
- A combination of speed regulation and capacitance variation with load may provide rather constant voltage and frequency for up to rated load [13].
- Capacitance and speed coordinated control, in a simple rugged configuration, to provide voltage and frequency control, with a small droop for the entire power range, is still to be accomplished.

4.6 Pole Changing SEIGs for Variable Speed Operation

There are SEIG applications where the speed varies notably with the load, but a certain frequency and voltage regulation is allowed for. Such a case is wind power. As the wind turbine power varies with cubic speed, a 4/6 pole changing in the stator winding leads to a (4/6) speed variation and, thus, a reduction in power of $(4/6)^3 = 8/27$. This kind of reduction is practically acceptable. Either two windings with different ratings and pole numbers are inserted in the stator slots, or a single winding is connected two ways for the 2 pole pairs: $p_1 < p_1'$ [14,15]. Provided the winding factors K_{W_1} and K'_{W_1} are acceptably high for both pole counts, and the emf space harmonics content is moderate, the dual winding choice seems to be an overall superior solution.

The main condition is to keep the machine saturated for both pole counts. In other words, the air gap flux densities B_{g10}, B'_{g10} on no load for the 2 pole counts, that is, for the two speeds (in the ratio of p_1/p_1'), have to be about the same:

$$\frac{B_{g10}}{B'_{g10}} = \frac{p_1}{p_1'}\frac{V_1}{V_1'}\frac{K'_{W_1}W_1'}{K_{W_1}W_1} \tag{4.39}$$

where

V_1 and V_1' are the phase voltages
W_1 and W_1' are the turns per phase for the p_1 and p_1' pole counts

For usual pole changing windings and various winding connections, we have the following [14]:

$$\begin{aligned} \frac{V_1 W_1'}{V_1' W_1} &= 2; && \text{for } \frac{YY}{Y} \\ &= 2.732; && \text{for } \frac{Y\Delta}{Y} \\ &= \frac{2}{1.73}; && \text{for } \frac{YY}{\Delta} \end{aligned} \tag{4.40}$$

In reality, p_1/p_1' should not be too far away from unity, to avoid needing too large a capacitance, and not too close to unity, to secure a sizable speed regulation range. Therefore, $p_1/p_1' = 1/2$ to $2/3$.

With parallel star/series star combination (Equation 4.40), it is possible to obtain $B_1/B_1' \approx 1$, provided K_{W_1}/K'_{W_1} makes up for the difference between the p_1/p_1' and $V_1W_1'/V_1'W_1$ ratios. Balanced

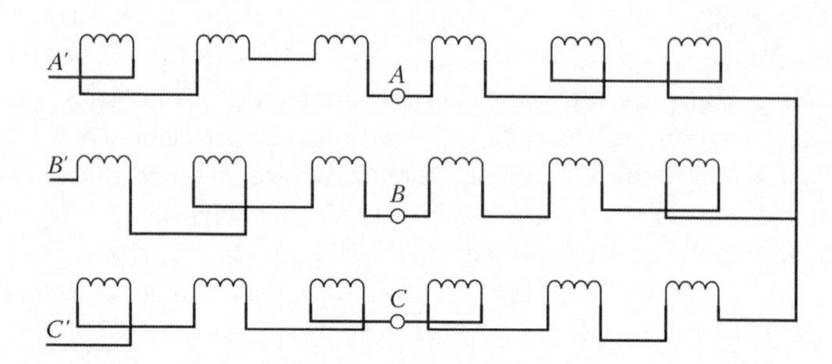

FIGURE 4.11 Stator 4/6 pole combination winding for pole changing self-excited induction generator (SEIG).

three-phase voltages for both pole counts are required. Also, simple pole count switching should be provided to reduce the costs of pole switching.

For a 4/6 pole combination, with 36 slots, the winding in Figure 4.11 [16] provides for $K_{W_1} = 0.831$ and $K'_{W_1} = 0.644$, the coil pitch being six slots. From Equation 4.39, with YY/Y connection,

$$\frac{(B_{g10})_{2p=4}}{(B'_{g10})_{2p=6}} = \frac{4}{6} \cdot \frac{2 \times 0.644}{0.831} \approx 1 \tag{4.41}$$

In all, two standard power switches are required to power either of two winding connections.

The typical torque–speed curves for the 2 pole counts are shown qualitatively in Figure 4.12. The switching from smaller pole count to larger pole count should be made such that the transients before reaching the new steady-state point 0′ are limited.

A larger slip frequency for peak torque seems to be advantageous from this viewpoint, at the price of reduced efficiency. An additional capacitor should also help. It is essential that no-load self-excitation conditions be provided at least for 1 pole count.

Typical voltage versus speed characteristics for the 4/6 pole counts, obtained with a $2p_1 = 4$ pole, 3.7 kW, 50 Hz, 440 V IM under various phase resistance loads are shown in Figure 4.13 [15,16]. A rather wide speed range is visible in Figure 4.13, ideally from 1500 rpm to roughly 800 rpm with a single capacitance. A large voltage variation takes place with speed. However, for the cubic speed/power law of wind turbines, the load resistance $R_L = 55$ Ω/phase for $2p_1 = 4$ poles at $n \approx 1500$ rpm leads to $V_L \approx 420$ V at 3238 W, while for $2p'_1 = 6$, $R_L = 185$ Ω/phase, the SEIG produces 959 W, at the same 420 V and at about 1000 rpm.

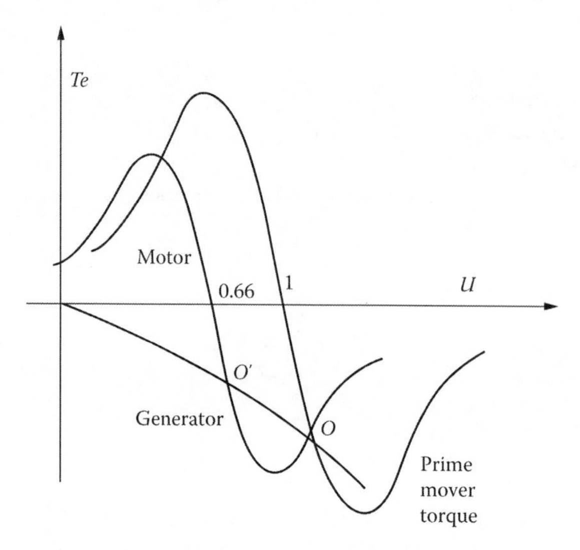

FIGURE 4.12 Torque–speed curves for the 4/6 pole combination.

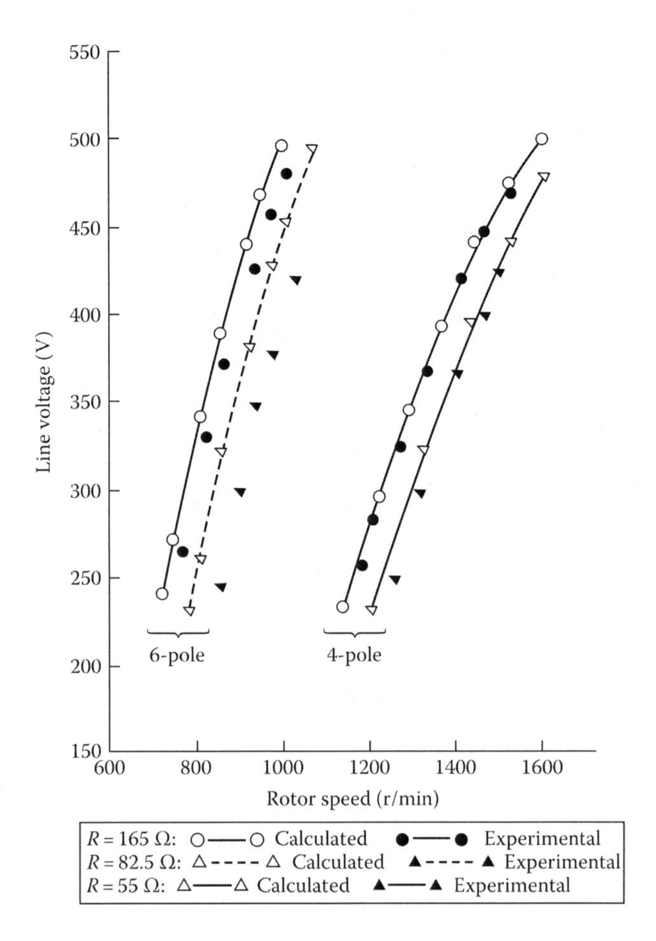

FIGURE 4.13 Line voltage/speed with $C = 100$ μF for a 4/6 pole self-excited induction generator (SEIG).

Space harmonics are present in the machine, with the subharmonics of 2 poles for $2p_1' = 6$ pole connection. In general, pole changing windings with $2p_1' = 3$ K are not balanced for harmonics. Terminal voltage unbalance of less than 2%–3% was observed for $2p_1' = 6$ pole count. Apart from the harmonics problem, the efficiency for the $2p_1' = 6$ pole count is notably lower than that for the $2p_1' = 4$ count.

The reduced cost for variable speed is the main advantage of the pole changing solution, but voltage harmonics, reduced efficiency, and strong transients during pole count switching should also be considered when this solution is applied.

4.7 Unbalanced Operation of Three-Phase SEIGs

Unbalanced phase load impedances or (and) failure of one (or two) capacitor(s) leads to unbalanced operation of SEIGs. Though both Δ and star connections are feasible, only the Δ connection will be treated here in some detail via the method of symmetrical components (Figure 4.14).

The admittance of excitation capacitance (X_c) and load $\left(\underline{Z}_{ab}^L\right)$, Y_{ab} is as follows:

$$\underline{Y}_{ab} = \frac{1}{\underline{Z}_{ab}^L(f)} + j\frac{f}{X_c} \tag{4.42}$$

Y_{bc} and Y_{ca} have similar expressions.

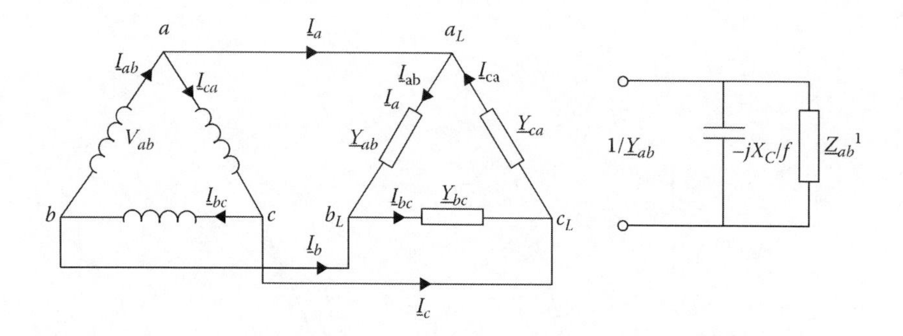

FIGURE 4.14 Δ-Connected self-excited induction generator (SEIG) and load.

To solve for the situation in Figure 4.14, first the assumed unbalanced load impedances are replaced by their symmetrical admittances $\underline{Y}^0, \underline{Y}^+, \underline{Y}^-$:

$$\begin{vmatrix} \underline{Y}^0 \\ \underline{Y}^+ \\ \underline{Y}^- \end{vmatrix} = \frac{1}{3}\begin{vmatrix} 1 & 1 & 1 \\ 1 & a_1 & a_1^2 \\ 1 & a_1^2 & a_1 \end{vmatrix}\begin{vmatrix} \underline{Y}_{ab} \\ \underline{Y}_{bc} \\ \underline{Y}_{ca} \end{vmatrix} \tag{4.43}$$

The total load plus capacitor currents I_{abL}, I_{bcL}, and I_{caL} are also transformed into their symmetric components:

$$\begin{vmatrix} \underline{I}_{abL}^0 \\ \underline{I}_{abL}^+ \\ \underline{I}_{abL}^- \end{vmatrix} = \frac{1}{3}\begin{vmatrix} \underline{Y}^0 & \underline{Y}^- & \underline{Y}^+ \\ \underline{Y}^+ & \underline{Y}^0 & \underline{Y}^- \\ \underline{Y}^- & \underline{Y}^+ & \underline{Y}^0 \end{vmatrix}\begin{vmatrix} \underline{V}_{ab}^0 \\ \underline{V}_{ab}^+ \\ \underline{V}_{ab}^- \end{vmatrix} \tag{4.44}$$

As in the Δ connection, there is no zero-voltage sequence, $\underline{V}_{ab}^0$, and the components I_a^+, I_a^- of line currents are as follows:

$$\begin{aligned} \underline{I}_a^+ &= \underline{I}_{abL}^+ - \underline{I}_{caL}^+ = (1-a)\underline{I}_{abL}^+ = (1-a)\left(\underline{Y}^0 \times \underline{V}_{ab}^+ + \underline{Y}^- \times \underline{V}_{ab}^-\right) \\ \underline{I}_a^- &= \underline{I}_{abL}^- - \underline{I}_{caL}^- = (1-a^2)\underline{I}_{abL}^- = (1-a^2)\left(\underline{Y}^+ \times \underline{V}_{ab}^+ + \underline{Y}^0 \times \underline{V}_{ab}^-\right) \end{aligned} \tag{4.45}$$

On the other hand, the generator current components I_1^+ and I_1^- are as follows:

$$\begin{aligned} \underline{I}_1^+ &= \underline{Y}_G^+ \cdot V_1^+ \\ \underline{I}_1^- &= \underline{Y}_G^- \cdot V_1^- \end{aligned} \tag{4.46}$$

For the load (Equation 4.45), the I_a components I_a^+, I_a^- as seen from the generator, are as follows:

$$\begin{aligned} \underline{I}_a^+ &= (1-a)I_{ab}^+ = -(1-a)V_{ab}^+Y_G^+ \\ \underline{I}_a^- &= (1-a^2)\underline{I}_{ab}^- = -(1-a^2)V_{ab}^-Y_G^- \end{aligned} \tag{4.47}$$

The generator voltages are considered balanced; and by eliminating V_{ab}^+ and V_{ab}^- after equating Equations 4.45 and 4.47, as they refer to the same currents, the self-excitation condition is obtained as follows:

$$\left(\underline{Y}_G^+ + Y_0\right)\left(\underline{Y}_G^- + \underline{Y}_0\right) - \underline{Y}^+\underline{Y}^- = 0 \tag{4.48}$$

Two conditions are contained in Equation 4.48.

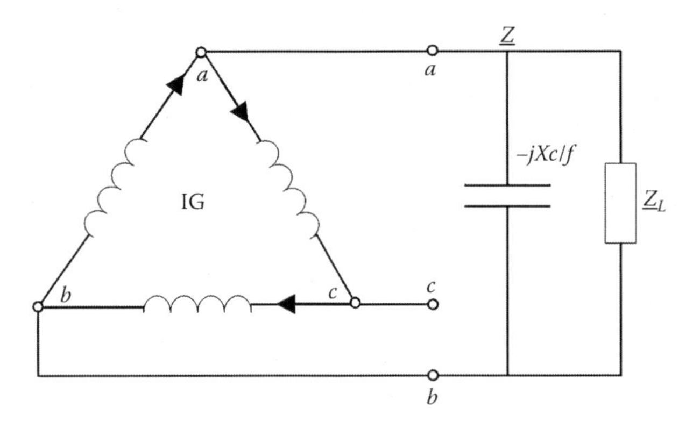

FIGURE 4.15 Δ/Δ-Connection with one load phase open.

The form of Equation 4.48 for balanced operation, that is, for $Y_G^+ = Y_G^- = Y_G, \underline{Y}^+ = Y^- = Y, \underline{Y}_0 = 0$, degenerates into the following:

$$\underline{Y}_G - \underline{Y} = 0 \tag{4.49}$$

It is evident that Equation 4.49 is identical to Equation 4.17, obtained for balanced operation. Let us consider the particular case of one phase open (Figure 4.15).

With a single load,

$$\begin{aligned} \underline{Y}_{ab} &= \underline{Y} = \frac{1}{\underline{Z}} \\ Y_{bc} &= Y_{ca} = 0 \end{aligned} \tag{4.50}$$

Therefore,

$$\underline{Y}^0 = \underline{Y}^+ + \underline{Y}^- = \frac{\underline{Y}}{3} \tag{4.51}$$

The self-excitation condition (Equation 4.48) degenerates into

$$\frac{3}{\underline{Y}} + \frac{1}{\underline{Y}_G^+} + \frac{1}{\underline{Y}_G^-} = 0 \tag{4.52}$$

Equation 4.52 refers to the series connection of positive (+) and negative (−) generator sequence equivalent circuits to $3/\underline{Y} = 3\underline{Z}$ (Figure 4.16).

Other unbalanced situations may be imagined. For example, if phase C opens but the load impedance $\underline{Z}$ remains balanced, the rationale discussed applies again, but with $Y_{ab} = 1.5Y$ and $Y_{bc} = Y_{ca} = 0$ [17].

A short circuit leads to voltage collapse, but at least with induction motor loads, not before notable transients. The computation of performance for unbalanced operation, based on conditions in Equation 4.48, is to be done as for balanced conditions but most probably through optimization methods to solve for frequency f and magnetization reactance X_m simultaneously.

It seems, however, that whenever a zero line current occurs [17], the one-line open magnetization curve is to be applied, at least in the high saturation zone.

Maximum and minimum capacitors for self-excitation for various voltages at given speed and no load may be calculated as for balanced operation. However, efficiency is diminished.

Unbalanced grid operation connections may also lead to unbalanced operation (no capacitor in this case). The absence of capacitors leads to a different approach; but again, the positive and negative components of active and reactive powers are to be calculated.

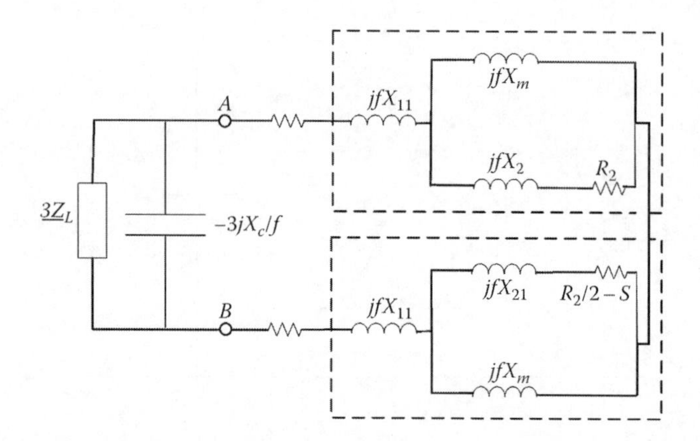

FIGURE 4.16 Δ/Δ-Self-excited induction generator (SEIG) with single capacitor and single load.

4.8 One Phase Open at Power Grid

When the IG is connected to the power grid (Figure 4.17), the latter may sometimes be unbalanced [18].

The Z_G^+, Z_G^- of the IG are easily recognizable in Figure 4.16, while the symmetrical components of the line voltage V_{ab}, V_{bc}, and V_{ca} are as follows:

$$\begin{vmatrix} V^+ \\ V^- \\ V^0 \end{vmatrix} = \frac{1}{3} \begin{vmatrix} 1 & a_1 & a_1^2 \\ 1 & a_1^2 & a_1 \\ 1 & 1 & 1 \end{vmatrix} \begin{vmatrix} V_{ab} \\ V_{bc} \\ V_{ca} \end{vmatrix} \tag{4.53}$$

The equivalent circuit of the IG (positive and negative sequences) connected to the unbalanced power grid (with $V_a^0 = 0$) is shown in Figure 4.18. The input (mechanical) power of IG connected to the power grid P_{input} is as follows:

$$P_{input} = P_{input}^+ + P_{input}^- = -3\left(I_2^+\right)^2 \frac{R_2}{S}(1-S) + 3\left(I_2^-\right)^2 \frac{R_2}{2-S}(1-S) \tag{4.54}$$

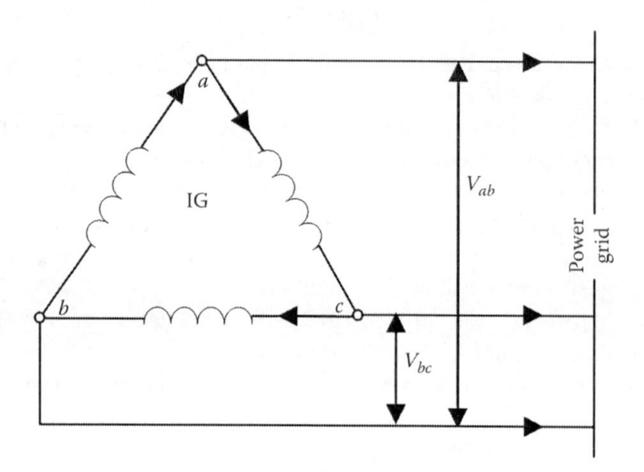

FIGURE 4.17 Induction generator (IG) to power grid.

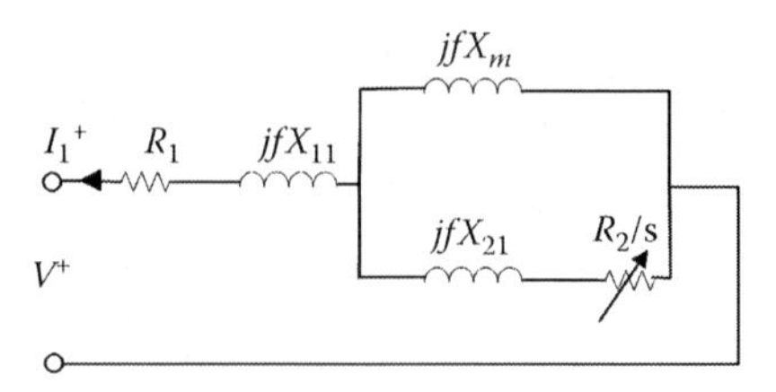

FIGURE 4.18 Sequence equivalent circuit of an induction generator (IG).

The total IG output active and reactive powers are

$$P_{out} = 3\text{Re}\left(\underline{V}^{+}\underline{I_1}^{*+} + \underline{V}^{-}\underline{I_1}^{*-}\right) \tag{4.55}$$

$$Q_{out} = 3\text{Imag}\left(\underline{V}^{+}\underline{I_1}^{*+} + \underline{V}^{-}\underline{I_1}^{*-}\right) \tag{4.56}$$

The efficiency of the IG connected to the grid η is

$$\eta = \frac{P_{out}}{P_{input}} \tag{4.57}$$

While this represents a solution for the general case, a phase open (say *c*) represents a probable practical case:

$$\begin{aligned} \underline{I}_{ab} - I_{ca} &= \underline{I}_a \\ \underline{I}_{bc} - \underline{I}_{ab} &= \underline{I}_b \\ \underline{I}_{ca} - \underline{I}_{bc} &= \underline{I}_c = 0 \end{aligned} \tag{4.58}$$

The symmetrical components of $\underline{I}_{ab}$, $\underline{I}_{ac}$, $\underline{I}_{ca}$ are

$$\begin{aligned} \underline{I}^{+} &= \underline{I}^{-} \\ \underline{I}_{ab} &= 2I^{+} \\ \underline{I}_{bc} &= -\underline{I}^{-} \end{aligned} \tag{4.59}$$

with

$$\underline{V}_{ab} = V^{+} + V^{-} \tag{4.60}$$

and

$$\begin{aligned} \underline{V}^{+} &= \underline{Z}^{+} \times \underline{I_1}^{+} \\ \underline{V}^{-} &= \underline{Z}^{-} \times \underline{I_1}^{-} \end{aligned} \tag{4.61}$$

I^+ is

$$\underline{I}^+ = \frac{\underline{V}_{ab}}{\underline{Z}^+ + \underline{Z}^-} \tag{4.62}$$

Based on the data from Figure 4.18 for Z^+ and Z^-, the input (mechanical) power P_{input} is as follows:

$$\begin{aligned} P_{input} &= -3I_2^2 K_S R_2 \\ K_S &= \frac{2(1-S)^2}{S(2-S)} \\ I_2^2 &= \frac{\underline{V}_{ab}^2}{(R_{total} + K_S R_2)^2 + X_{total}^2} \\ R_{total} &= 2(R_1 + R_2) \\ X_{total} &= 2(X_{1l} + X_{2l}) \end{aligned} \tag{4.63}$$

Now, with V_{ab}, V_{ca}, V_{bc} given, as amplitude and phase, and known machine parameters, ω_1 and given speed ω_r, the slip S may first be calculated as follows:

$$S = \frac{\omega_1 - \omega_r}{\omega_1} \tag{4.64}$$

Then, the values of K_s, I_2, and P_{input} are determined. Then, I_1^+, I_1^- and I_1 and V^+, V^-, P_{out}, and Q_{out} are computed. That is, steady-state performance versus slip may be calculated for various degrees of grid voltage unbalance V^-/V^+, for example. Both the reactive power drawn from the power grid and the IG losses increase with input power.

The increase in grid voltage imbalance leads to increased currents, losses, and power factor reduction for given slip. The current also depends on how the voltage unbalance is produced: by one increased or by one reduced voltage.

A small voltage unbalance has a large impact on the IG currents. For the same total losses, the output power is reduced considerably with respect to balanced operation. For 10% voltage unbalance, it may be allowed to handle only 30%–40% of rated power at an acceptable temperature level.

Derating in direct relationship to voltage unbalance V^-/V^+ ratio is recommended in order to avoid IG overheating.

4.9 Three-Phase SEIG with Single-Phase Output

Single-phase output SEIGs may be approached with a two-phase induction generator as shown in the next paragraph. However, the power density, power pulsations, vibration, and noise are notable. Also, when the power level goes above 2–3 kW, the single-phase IG becomes less attractive and is not available off the shelf.

Using a three-phase IG, with an advanced degree of phase symmetrization, may prove a practical solution for single-phase output, at least above 2–3 kW per unit. Three main connections were proposed:

1. Steinmetz connection (Figure 4.19a) [19]
2. Smith connection (Figure 4.19b) [20]
3. Fukami connection (Figure 4.19c) [21]

All connections are, in principle, capable of providing machine symmetrization at some speed (load) for given frequency and perform with good efficiency and power factor for a notable power range. However, it seems that the Steinmetz connection, augmented with series compensation (C_s), does better with only two capacitors that are wisely used (at high voltage level): about 2.0 P.U. power delivery with limited voltage self-regulation.

Consequently, we will treat here in detail only the Steinmetz connection with series compensation [19].

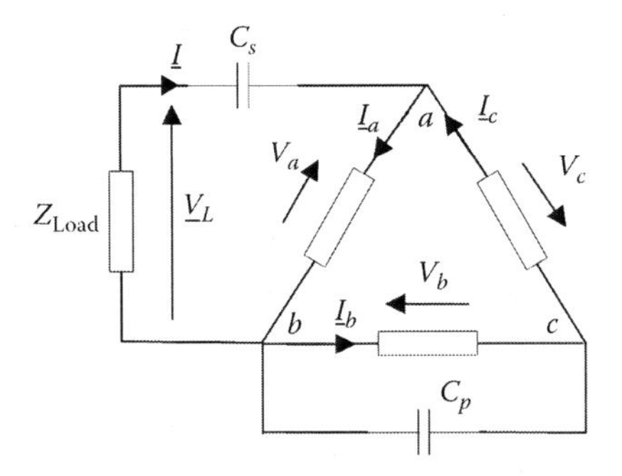

FIGURE 4.19 Three-phase self-excited induction generator (SEIG) with single-phase output: (a) Steinmetz connection, (b) Smith connection, and (c) Fukami connection.

For steady state, again, the method of symmetrical components is applied. The connection in Figure 4.19a suggests the following relationships:

$$\underline{V} = V_a \qquad \underline{V}_a + \underline{V}_b + \underline{V}_c = 0 \tag{4.65}$$

$$\underline{I}_1 = \underline{V}_b \cdot \underline{Y}_{Cp} = \underline{I}_c - \underline{I}_b \tag{4.66}$$

$$\underline{I} = \underline{I}_a - \underline{I}_c \tag{4.67}$$

where Y_{Cp} is the admittance of the parallel capacitance:

$$Y_{Cp} = jX_{Cp}^{-1}f; \quad f \text{ is P.U. frequency} \tag{4.68}$$

The symmetrical components of voltages in Equation 4.53 with Equations 4.65 through 4.67 are as follows:

$$\underline{V}^{+} = \frac{V\sqrt{3}\left(\underline{Y}_G^{-} + (e^{j\pi/6}/\sqrt{3})\underline{Y}_{Cp}\right)}{\underline{Y}_{Cp} + \underline{Y}_G^{+} + \underline{Y}_G^{-}} \qquad \underline{V}^{-} = \frac{V\sqrt{3}\left(\underline{Y}_G^{+} + (e^{-j\pi/6}/\sqrt{3})\underline{Y}_{Cp}\right)}{\underline{Y}_{Cp} + \underline{Y}_G^{+} + \underline{Y}_G^{-}} \tag{4.69}$$

where Y_G^{+} and Y_G^{-} are positive and negative sequence admittances of IG at P.U. frequency f (Figure 4.16):

$$\frac{1}{\underline{Z}_{ab}} = Y_G^{+} \qquad \frac{1}{\underline{Z}_{bc}} = Y_G^{-} \tag{4.70}$$

The equivalent IG impedance at points a, b, $\underline{Z}_{IG}$, is as follows:

$$\underline{Z}_{IG} = \frac{\underline{Z}_G^{+}\underline{Z}_G^{-} + \underline{Z}_G^{+}\underline{Z}_{Cp} + \underline{Z}_G^{-}\underline{Z}_{Cp}}{3\underline{Z}_{Cp} + \underline{Z}_G^{+} + \underline{Z}_G^{-}} \tag{4.71}$$

The sum of impedances $\underline{Z}_{IG}$, $\underline{Z}_{CS}$, and $\underline{Z}_L$ should be zero for self-excitation:

$$\underline{Z}_{IG} + \underline{Z}_L + \underline{Z}_{CS} = 0 \qquad \underline{Z}_{CS} = -j\frac{X_{CS}}{f} \tag{4.72}$$

The load impedance $\underline{Z}_L$ (at P.U. frequency f) is

$$\underline{Z}_L = R_L + j\left(fX_L - \frac{X_{CL}}{f}\right) \tag{4.73}$$

For the impedance model solution, the frequency f and reactance X_m are the unknowns, to be solved for iteratively from Equation 4.72.

The complexity of $\underline{Z}_G^+$ and $\underline{Z}_G^-$ leaves little room for an analytical solution.

It is practical to calculate the amplitude of the complex impedance in Equation 4.72 and to force it to zero, that is, to find a minimum by an optimization method (Hooke–Jeeves, for example [19]):

$$Z(f, X_m) = \sqrt{(R_{IG}(f, X_m) + R_L)^2 + \left(X_{IG}(f, X_m) + fX_L - \frac{X_{CL}}{f} - \frac{X_{cs}}{f}\right)^2} = 0 \tag{4.74}$$

The unsaturated value of X_m, $X_{\max}$ may be taken as an initial value of $X_m(1)$ or as a constraint. Again, the compensation ratio K is defined as $K = X_{cs}/X_{cp}$. As K increases, in general, the voltage reduction with load is decreased. Values of $K \approx 0.3$–0.5 seem to be practical (as for the three-phase balanced SEIG in the previous paragraph).

For a 2.2 kW, 4-pole, 220 V, 50 Hz, 9.4 A IM with $R_1 = 0.0844$ P.U., $X_{1l} = 0.112$ P.U., $R_2 = 0.098$ P.U., $X_{2l} = 0.1$ P.U., $X_m = 2.2$ P.U., and $E_1(X_m)$ as in Equation 4.30, the load voltage versus load current for $PF = 1.0$, $K = 0.34$ speed $U = 1$ is shown in Figure 4.20a and the power in Figure 4.20b [19].

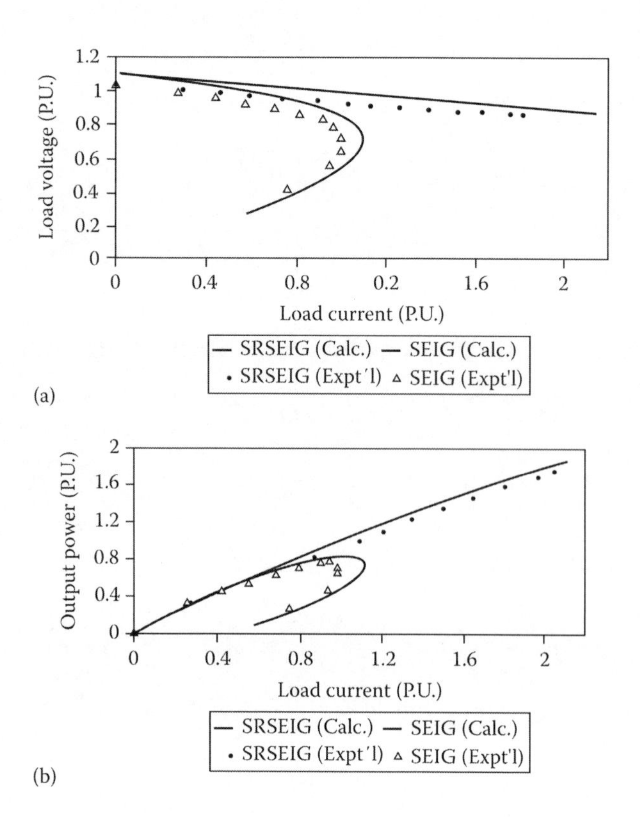

FIGURE 4.20 (a) Voltage and (b) output power vs. load in per unit (P.U.).

The extension of the stability zone due to the series compensation is spectacular. As expected, purely balanced operation is hardly present, but the phase current and voltage unbalances (Figure 4.21a and b) are acceptable. The efficiency is rather large for a wide range of loads (Figure 4.22) [19].

In Figures 4.20 through 4.22, SEIG means without series compensation, and SRSEIG means with series compensation ($K = 0.34$, $C_p = 125\ \mu F$, $C_s = 370\ \mu F$).

As evident from Figures 4.20 through 4.22, the largest phase voltage is below 1.2 P.U., which is an acceptable value.

The phase voltages are almost symmetric at 1.6 P.U. load, while below and above that load, a certain degree of unbalance remains.

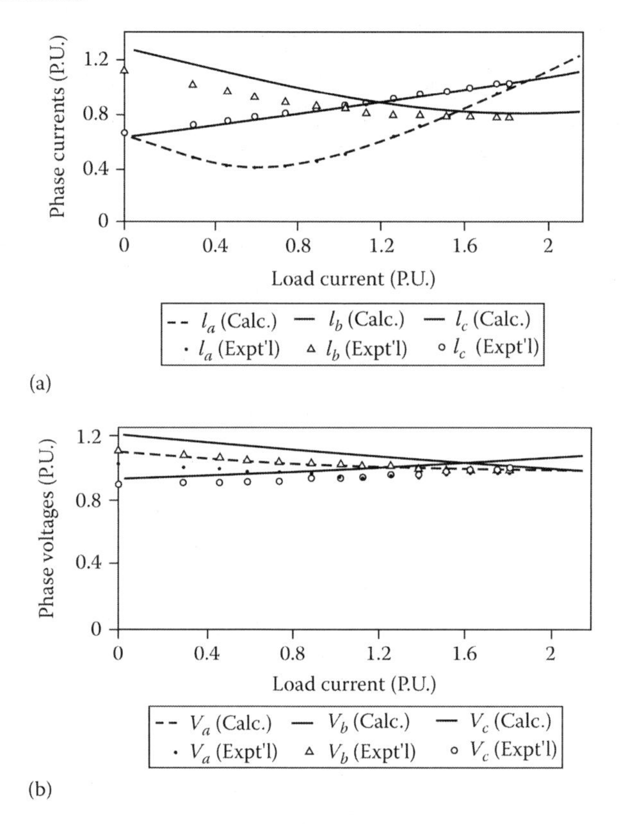

FIGURE 4.21 (a) Phase currents and (b) voltages vs. load in per unit (P.U.) for $PF = 1.0$.

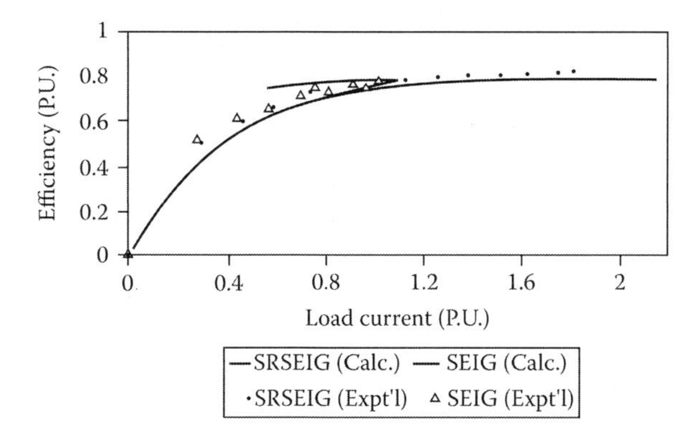

FIGURE 4.22 Efficiency vs. load in per unit (P.U.).

The increase in output power to 2.0 P.U. with the Steinmetz connection and series compensation may make it a practical solution, for powers above 3 kW, where single-phase IMs are not available off the shelf.

4.10 Two-Phase SEIGs with Single-Phase Output

As already mentioned, two-phase IMs may also be used as SEIGs with single-phase output. The self-regulated such configuration contains a self-excitation capacitor C_e placed in parallel with the excitation (additional) stator winding, while the power winding is provided with a single series capacitor C_s or with a long- or short-shunt capacitor C_p (Figure 4.23).

The configuration with only two capacitors C_e and C_s will be considered further here (for the others, see the literature [22–24]), as it seems a good compromise between capacitor costs and performance.

To investigate the steady state, the symmetrical component method is used again. We should note the following:

- The excitation capacitor C_e may be lumped in series with the excitation winding, which is short-circuited ($V_a = 0$).
- The series compensation capacitor C_s may be lumped into the load Z'_L:

$$\underline{Z}'_L(f) = \underline{Z}_L(f) - j\frac{X_{CS}}{f} \tag{4.75}$$

For an R_L, L_L load,

$$\underline{Z}_L(f) = R_L + jfX_L \tag{4.76}$$

As the excitation voltage $V_a = 0$ and $V_m = V_s$, the positive and negative sequence voltages are as follows:

$$\underline{V}_m^+ = \underline{V}_m^- = \frac{\underline{V}_s}{2} = \frac{-\underline{Z}'_L\left(\underline{I}_m^+ + \underline{I}_m^-\right)}{2} \tag{4.77}$$

The equivalent circuit in positive and negative components of the two-phase machine (p. 830 in Reference 1) may be slightly simplified, as the voltage $\underline{V}_m^+$ and $\underline{V}_m^-$ (Equation 4.77) may be written as follows:

$$\begin{aligned} \underline{V}_{AB} &= \underline{V}_m^+ = +\frac{\underline{Z}'_L}{2}\left(\underline{I}_m^+ - \underline{I}_m^-\right) - \underline{Z}'_L\underline{I}_m^+ \\ \underline{V}_{BC} &= \underline{V}_m^- = -\underline{Z}'_L\underline{I}_m^- - \frac{\underline{Z}'_L}{2}\left(\underline{I}_m^+ - \underline{I}_m^-\right) \end{aligned} \tag{4.78}$$

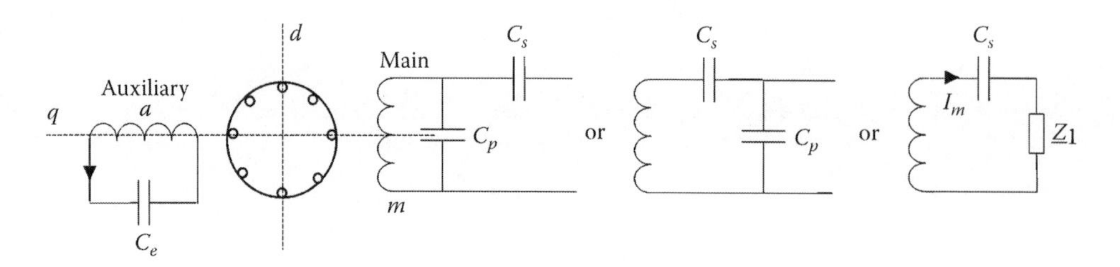

FIGURE 4.23 Two-phase self-excited induction generator (SEIG) configurations with single-phase output.

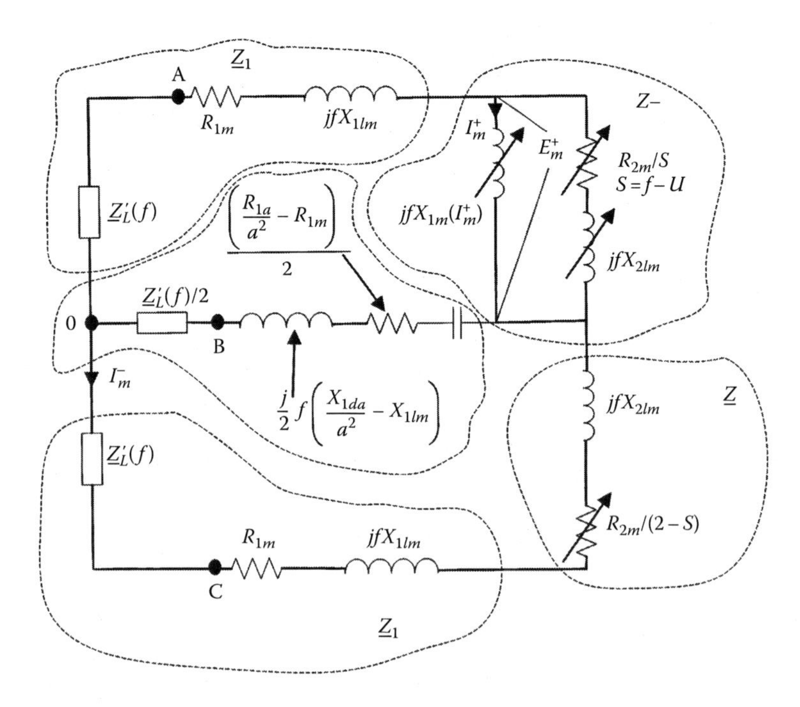

FIGURE 4.24 The equivalent circuit of a two-phase self-excited induction generator (SEIG) with single-phase output.

This way, the complete equivalent circuit of the two-phase machine with unequivalent windings is shown in Figure 4.24.

Note that "a" is the turn ratio between the auxiliary excitation and main power winding. This allows for the design with a different voltage in the excitation winding and also introduces one more variable for symmetrization at a desired load.

The equivalent circuit in Figure 4.24 may be reduced easily to the one in Figure 4.25a and b.

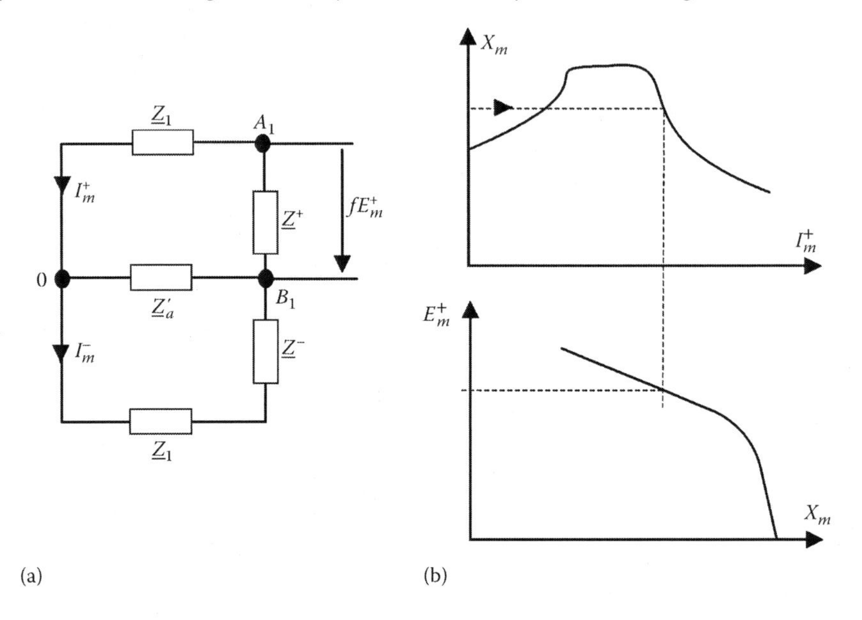

FIGURE 4.25 (a) Simplified equivalent circuit of a two-phase self-excited induction generator (SEIG) and (b) magnetization curves.

The self-excitation condition is evident: the summation of currents in the node 0 has to be zero:

$$\underline{Y}_E = \frac{1}{\underline{Z}_1 + \underline{Z}^+} + \frac{1}{\underline{Z}_1 + \underline{Z}^-} + \frac{1}{\underline{Z}'_a} = 0 \tag{4.79}$$

All parameters in Equation 4.79 depend on frequency f, but only X_m is dependent on I_m^+. Therefore, again, we have two variables. Optimization methods to minimize (make it zero) the amplitude of the admittance in Equation 4.79 would be a fairly straightforward way to solve for both f and X_m with given speed (slip S) and other IG load parameters and capacitors.

The magnetization reactance X_m is a function of the positive sequence magnetization current I_m^+, and the relationship $V_{AB} = E_m^+(I_m^+)$ is to be known from measurements or through computation (in the design stage). Once X_m and f are known, from the magnetization curve $X_m(I_m^+)$, the value of I_m^+ is calculated. Further on, the air gap voltage fE_m^+ is determined:

$$fE_m^+ = V_{A_1B_1} = fX_m I_m^+ \tag{4.80}$$

From now on, the equivalent circuit in Figure 4.25a and b may be used to calculate the following:

$$I_m^+ = fE_m^+ \frac{1}{\underline{Z}_1 + (\underline{Z}'_a \cdot \underline{Z}_1)/(\underline{Z}'_a + \underline{Z}_1)} \tag{4.81}$$

$$\underline{I}_m^- = \underline{I}_m^+ \cdot \frac{\underline{Z}'_a}{\underline{Z}^- + \underline{Z}_1} \tag{4.82}$$

The load current $\underline{I}$ is simply

$$\underline{I} = \underline{I}_m^+ + \underline{I}_m^- \tag{4.83}$$

The excitation current $\underline{I}_a$ is

$$\underline{I}_a = j\left(\underline{I}_m^+ - \underline{I}_m^-\right) \tag{4.84}$$

It is evident that symmetrization is obtained when $\underline{I}_m^- = 0$.

The output electrical power P_{out} is

$$P_{out} = R_L I_m^2 \tag{4.85}$$

The positive sequence rotor current I_{2m}^+ (from Figure 4.24) is as follows:

$$\underline{I}_{2m}^+ = I_m^+ \frac{jfX_{1m}}{(R_{2m}/S) + jf(X_{1m} + X_{2m})} \tag{4.86}$$

Neglecting the mechanical losses, the input (shaft) power P_m is

$$P_m = 2\left[\left(I_{2m}^+\right)^2 \frac{R_{2m}}{S} - \left(I_{2m}^-\right)^2 \frac{R_{2m}}{2-S}\right](1-S) \tag{4.87}$$

The negative sequence power is translated into losses.

To calculate the efficiency, the core losses p_{iron} and the mechanical losses p_{mec} have to be added.

$$\eta = \frac{P_{out}}{P_m + p_{iron} + p_{mec}} \tag{4.88}$$

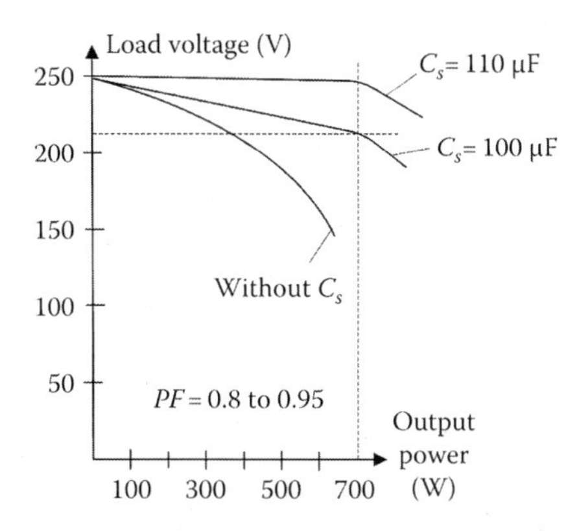

FIGURE 4.26 Load voltage vs. output power.

The generator voltage V is as follows (Equation 4.77):

$$\underline{V}_{Gen} = -Z'_L \underline{I}_m \tag{4.89}$$

The load voltage V_{load} differs from V_{Gen} by the voltage drop along the series capacitor C_s:

$$\underline{V}_{load} = \underline{V}_{Gen} + j\frac{X_c}{f}\underline{I}_m \tag{4.90}$$

This model may be instrumental, after arranging it into a computer program, in calculating the steady-state performance for given speed, capacitors, and load impedance. Changing the load impedance will produce data for steady-state characteristics.

For a 700 W, 50 Hz, $2p_1 = 2$ poles, $V_N = 230$ V machine with $R_{1m} = 3.94\ \Omega$, $R_{sa} = 4.39\ \Omega$, $R_{2m} = 3.36\ \Omega$, $X_{1lm} = X_{2lm} = 5.48\ \Omega$, $X_{1la} = 7.5\ \Omega$, unsaturated $X_{1m} = 70\ \Omega$ and $C_e = 40$ μF, $C_s = 100$ μF, load voltage versus current curves are shown in Figure 4.26 [25].

The beneficial effect of the series compensation by C_s is evident. The symmetrization problem remains, but, beside C_s and C_e, the turns ratio a is a variable to take advantage of in the design optimization process.

For powers below 3 kW, where the two-phase IM is available, its use as a generator still seems to be a practical option in low-cost applications.

4.11 Three-Phase SEIG Transients

SEIGs undergo transients on no load during self-excitation, when electrical load is connected or rejected or varied. When speed is modified or the self-excitation capacitor is varied with load to control the load voltage, transients also occur.

The method of approach to SEIG transients is the dq (space-phasor) model. We will use here the space-phasor model, as it facilitates the accounting of magnetic saturation. The space-phasor model of the IM in stator coordinates (see Chapter 2) is as follows:

$$\begin{aligned} \bar{V}_1 &= -R_1\bar{I}_1 - \frac{d\bar{\Psi}_1}{dt} \\ \bar{\Psi}_1 &= L_{1l}\cdot\bar{I}_1 + \bar{\Psi}_m \end{aligned} \tag{4.91}$$

$$\begin{aligned} 0 &= -R_2\bar{I}_2 - \frac{d\bar{\Psi}_2}{dt} + j\omega_r\bar{\Psi}_2 \\ \bar{\Psi}_2 &= L_{2l}\cdot\bar{I}_2 + \bar{\Psi}_m \end{aligned} \tag{4.92}$$

The association of signs corresponds to generating mode. The magnetizing (air gap) Ψ_m is

$$\begin{aligned} \bar{\Psi}_m &= L_m(I_m)\bar{I}_m \\ \bar{I}_m &= \bar{I}_1 + \bar{I}_2 \end{aligned} \tag{4.93}$$

The $L_m(I_m)$ function is obtained from tests or through computation in the design stage.

The electromagnetic torque T_e is

$$T_e = \frac{3}{2} p_1 \text{Imag}\left(\bar{\Psi}_1 \bar{I}_1^*\right) \tag{4.94}$$

The capacitors connected at the terminals ($C_Y = 3C\Delta$) lead to the voltage equation:

$$\frac{d\bar{V}_1}{dt} = \frac{1}{C_Y} \cdot \bar{I}_C \tag{4.95}$$

The generator motion equation is written as follows:

$$J \frac{d\omega_r}{dt} = T_{prime\ mover} - T_e \tag{4.96}$$

The dynamics of the turbine with its speed control model (if any) that drives the SEIG should be added here.

With the flux linkages as variables and the currents as dummy variables, only the $L_m(I_m)$ function needs to be known to account for magnetic saturation. The decomposition into dq components to deal with "real number variables" is straightforward, as ($\bar{V}_1 = \bar{V}_d + j\bar{V}_q$).

A passive load may be described as follows:

$$\begin{aligned} \bar{V}_1 &= R_L \bar{I}_L + L_L \frac{d\bar{I}_L}{dt} + \bar{V}_{CL} \\ \frac{d\bar{V}_{CL}}{dt} &= \frac{1}{C_{YL}} \bar{I}_L \\ \bar{I}_L &= \bar{I}_1 - \bar{I}_C \end{aligned} \tag{4.97}$$

On the other hand, an induction motor load may be described as follows:

$$\begin{aligned} \bar{V}_1 &= R_{1M} \bar{I}_M + \frac{d\bar{\Psi}_{1M}}{dt} \\ 0 &= R_{2M} \bar{I}_{2M} + \frac{d\bar{\Psi}_{2M}}{dt} - j\omega_{rM} \bar{\Psi}_{2M} \\ \bar{\Psi}_{1M} &= L_{1lM} \bar{I}_{1M} + \bar{\Psi}_{mM} \\ \bar{\Psi}_{2M} &= L_{2lM} \bar{I}_{2M} + \bar{\Psi}_{mM} \\ \bar{\Psi}_{mM} &= L_{mM}\left(\bar{I}_{1M} + \bar{I}_{2M}\right) \\ T_{eM} &= \frac{3}{2} p_1 \text{Imag}\left(\bar{\Psi}_{1M} \bar{I}_{1M}^*\right) \\ J_M \frac{d\omega_{rM}}{dt} &= T_{eM} - T_{Load} \end{aligned} \tag{4.98}$$

The same models could also be used in synchronous coordinates, but then the currents and voltages are transformed back into stator coordinates.

Typical transients for a 1.1 kW, 127/220 V, 8.3/4.8 A line current, 60 Hz, 2-pole machine with $R_1 = 0.078$ P.U., $X_{1l} = X_{2l} = 0.0895$ P.U. are shown in what follows [25].

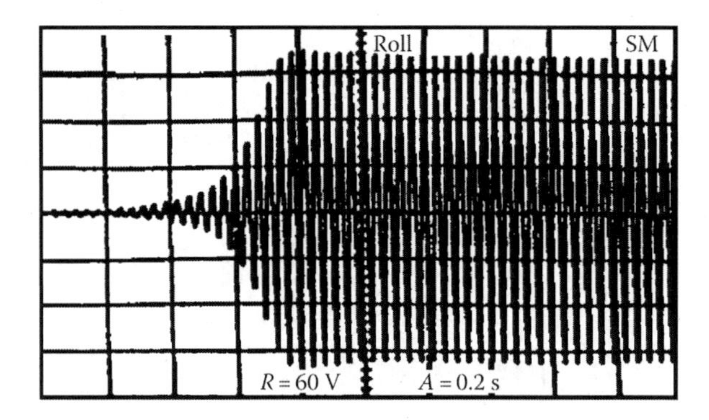

FIGURE 4.27 Voltage buildup at no load.

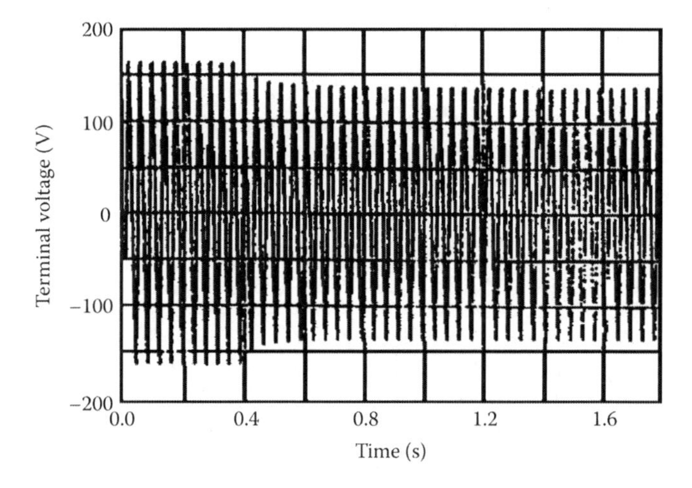

FIGURE 4.28 Sudden connection of resistive load.

The rather slow phase a voltage buildup at no load for $\omega_r = 3600$ rpm and $C_\Delta = 248.7\ \mu F$ is evident in Figure 4.27. Sudden connection of an RL ($R_L = 80\ \Omega$), $t = 0.4$ s shows a mild reduction of phase a voltage with stable response (Figure 4.28), as the load is not too large. A step loading from $R_L = 80$ to $10\ \Omega$ leads to voltage collapse [25]. A notable voltage increase occurs when the RL is rejected (Figure 4.29).

A few remarks are in order:

- Sudden application of load—be it R_L or R_L, L_L—leads to fast voltage reduction at the SEIG terminals, while RC load leads to very small voltage variation.
- The sudden disconnection of the capacitor leads to quick decay of terminal voltage to zero.
- Load rejection leads to stable voltage recovery toward the no-load steady state (for constant speed).
- When the load is greatly increased, the voltage of the machine collapses, as the SEIG de-excites and thus desaturates rapidly.
- A short circuit at SEIG terminals may cause up to 2.0 P.U. voltage surges and 5 P.U. current transients for a short period of around 20 ms before the voltage collapses.
- An SEIG supplying an induction motor load requires large capacitances to cover for the motor reactive power. However, when the motor is disconnected, the large capacitance may produce overvoltages that are too large. It is, thus, practical to divide the capacitances between the SEIG and the induction motor (IM) [26]. For safe starting of an IM, the SEIG-to-IM power ratio should be at least 1/0.6, but, to handle stably up to 100% step mechanical load, the power ratio should be 3(4) to 1.

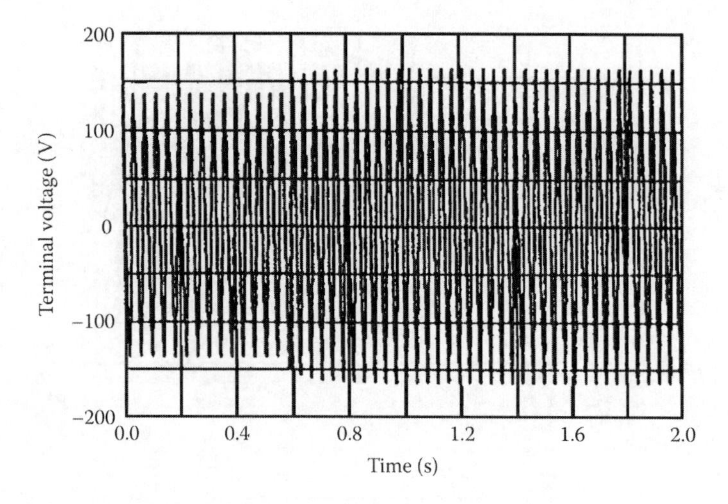

FIGURE 4.29 Resistive load rejection at 0.58 s ($R_L = 80\ \Omega$).

4.12 Parallel Connection of SEIGs

As there are constraints on SEIG ratings per unit for microhydro or wind turbines, several such SEIGs may have to be operated in parallel with a single capacitance bank C (Figure 4.30) for self-excitation.

As one of the main problems with SEIGs is their poor voltage regulation, improvements may be obtained by capacitor variation or (and) speed variation of some SEIGs of the group, as is done for single such generators operating on autonomous load.

For self-excitation, at node 0 (Figure 4.30), the total admittance has to be zero. All generators have the same terminal voltage and frequency f (P.U.).

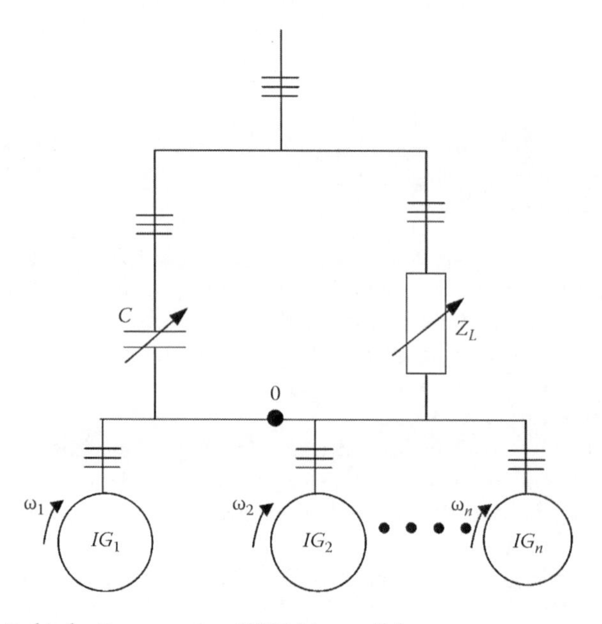

FIGURE 4.30 Self-excited induction generators (SEIGs) in parallel.

On the other hand, the summation of currents in the n generators is as follows:

$$\sum_{i=1}^{n} I_{1i} = V_1\left(\frac{1}{R_L} - \frac{j}{fX_L} + \frac{jf}{X_C}\right) \tag{4.99}$$

In Equation 4.99, an additional variable inductance (X_L) is placed in parallel with the fixed excitation capacitor (X_C) to vary the total equivalent capacitor. Ballast RL is used to control the IG group when the load decreases.

Also, for each generator, from the equivalent circuit,

$$\begin{aligned} \underline{I}_{1i} &= -\frac{\underline{V}_1}{\underline{Z}_{Gi}} \\ \underline{I}_{1i} + \underline{I}_{mi} = \underline{I}_{2i} &= \frac{-f\underline{E}_{1i}}{(R_{2i}/S_i) + jfX_{2i}} \end{aligned} \tag{4.100}$$

The air gap emf $\underline{E}_{1i}$ depends on the magnetization reactance X_{mi}, as already discussed extensively earlier in this chapter.

The self-excitation condition is, thus, from Equation 4.99,

$$\underline{Y}_t = \sum_{i=1}^{n} \frac{1}{\underline{Z}_{Gi}(f)} + \frac{1}{R_L} - \frac{j}{fX_L} + \frac{jf}{X_C} = 0 \tag{4.101}$$

Equations 4.99 through 4.101 provide for $n + 2$ unknowns, X_{mi}, and C and f. Various iterative procedures such as Newton–Raphson's may be used to solve such a problem [27]. The voltage V_1, all parameters, and speeds are given. Initial values of variables are given and then adjusted until good convergence is obtained.

As expected, for given voltage, the required capacitance increases with load. The frequency f drops when the load increases. For a fixed load power, the frequency f increases, as the voltage increases, as for single SEIGs. Also, lagging power factor loads require larger capacitance for given voltage and load power. Increasing the number of identical SEIGs for the same load power tends to require more capacitance at given voltage. The frequency increases in such a case as the load of each SEIG is decreased.

The machine speeds u_1 and u_2 also influence the performance (Figure 4.31) [27]. The capacitance increases with decreasing speed [1–5] and while frequency decreases for given voltage and load power.

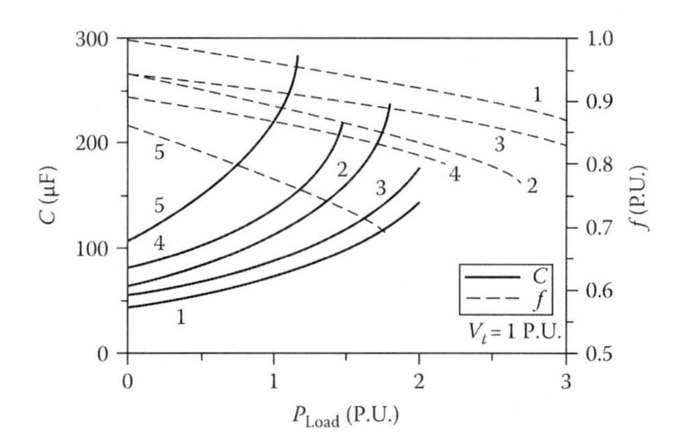

FIGURE 4.31 Influence of self-excited induction generator (SEIG) speeds (in P.U.) on required capacitance C and on frequency f.

The load power factor also influences the required speed for given capacitor and voltage. The speed and frequency increase with load and more so with lagging power factor loads.

Again, as for single SEIGs, voltage control of frequency-insensitive loads may best be accomplished by combining capacitance and speed control. In this case, the frequency variation is also limited.

Note that the parallel operation of an SEIG may be approached through the *dq* model, valid for transients and steady state. The eigenvalue method is then to be applied to predict the system's behavior and to determine the capacitance values [28].

4.13 Direct Connection to Grid Transients in Cage-Rotor Induction Generators

Rigid as they may seem, cage-rotor IGs are still connected, up to some power level per unit, directly to the power grid. Adjustments in its power delivery are made by controlling the turbine speed (torque). The direct connection of a cage-rotor IM to a strong power grid leads, irrespective of machine initial speed, to large current and torque transients. When the IM rating increases, or (and) the local power grid is not so strong, the disturbances produced by such large transients are severe. Moreover, the torque transients are so large that they can, in time, damage the turbine.

If the simplicity and low costs of cage-rotor IGs are to keep this solution in perspective, besides speed (small range) control, the switch-on (off) transients to the power grid have to be drastically reduced. To accomplish such a goal, with limited expense, it seems that either soft-starters or additional resistors should be connected in series with the stator windings for a short period of time (a few seconds). For wind turbines as prime movers, rotor wind and hub wind speeds (in m/s) vary continuously with time (Figure 4.32a) [29]. The wind turbine generator scheme is shown in Figure 4.32b.

In a direct start transient simulation, using the *dq* model, a 4-pole 0.5 MW IG transmission model response (Figure 4.33) shows aerodynamic torque, mechanical torque, turbine rotor speed, and generator speed. Some small oscillations are visible in the turbine rotor speed, but hardly so in the generator rotor speed [29].

For the same IG, accelerated freely by the wind turbine up to 1500 rpm and then directly connected to the grid, the speed, phase current amplitude, and reactive and active power transients for some loading are shown in Figure 4.34 [29]. There is a very short-lived superhigh peak in phase current (up to 10 P.U.) and in reactive and active powers. Then, they all stabilize but still retain some small pulsations due to wind turbine speed pulsations (Figure 4.34).

The power grid was realistically modeled [29]. The connection to the power grid of larger power IGs (say 2 MW) poses problems of voltage and too large current transients. This is why soft-starters were proposed for the scope. They may also automatically disconnect the generator when there is not enough wind power and reconnect again based on the power factor of IG, which varies notably with the slip.

A typical start-up to 1500 rpm and then connection and loading of a 2 MW, 4-pole IG through a soft-starter is illustrated in Figure 4.35 [29]. The current peaks are still above the rated value but are much less than those for direct connection to the grid. The same is true for reactive and active power transients at the price of slower speed stabilization to load (1.8 MW).

A capacitance bank controlled in steps may be used to compensate for the IG reactive power (of about 0.936 MVAR for 1.8 MW) and keep the voltage regulation within limits when load varies in the local grid. Alternatively, an external resistor may be used to reduce the grid connection transients (Figure 4.36) [30]. The resistor connection procedure is shown to perform well with respect to all three connection factors:

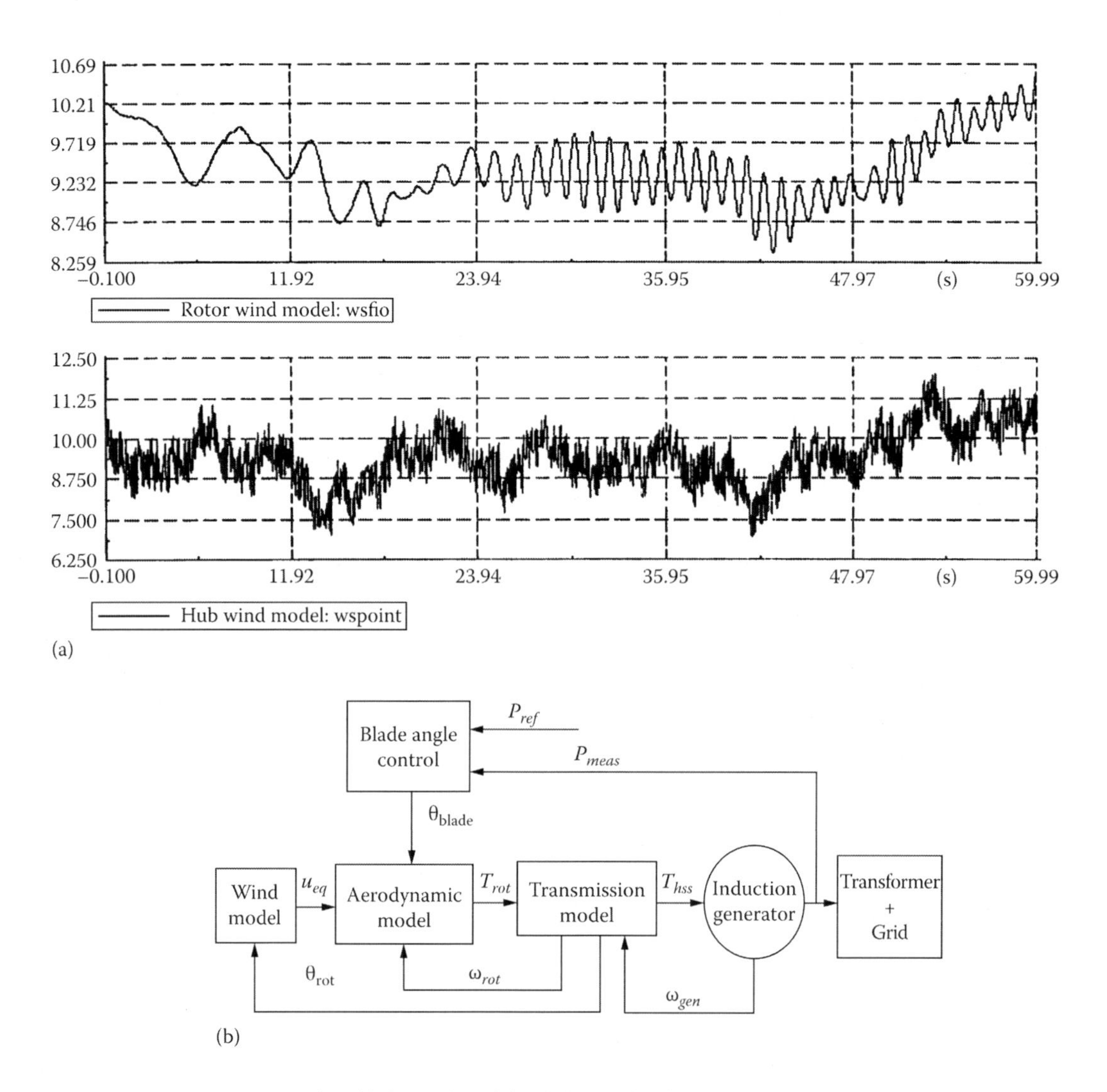

FIGURE 4.32 (a) Rotor wind and hub wind model and (b) wind turbine model.

1. Maximum voltage change factor K_u
2. Maximum current factor K_i
3. Flicker step factor K_f [31]

A voltage change factor of only 4% is now enforced in Europe at the IG connection to the grid.

Active stall wind turbine regulation is standard for the smooth connection of megawatt (MW)-size wind induction generators. A variable slip IG may also be used, when the IG has a wound rotor and a controlled or self-controlled rotor connected additional resistor. Figure 4.37 shows 15 kW, 0.8% slip active-stall regulated IG connection to the grid at no load. The external stator resistance is $R_{ext} = 50R_1 = 1.8\ \Omega$ [30]. A very smooth connection is evident. The costs of the short-lived current external resistance is quite small in comparison with the soft-starter, while fulfilling the smooth connection conditions, as the reactive power requirements are nonzero for a soft-starter.

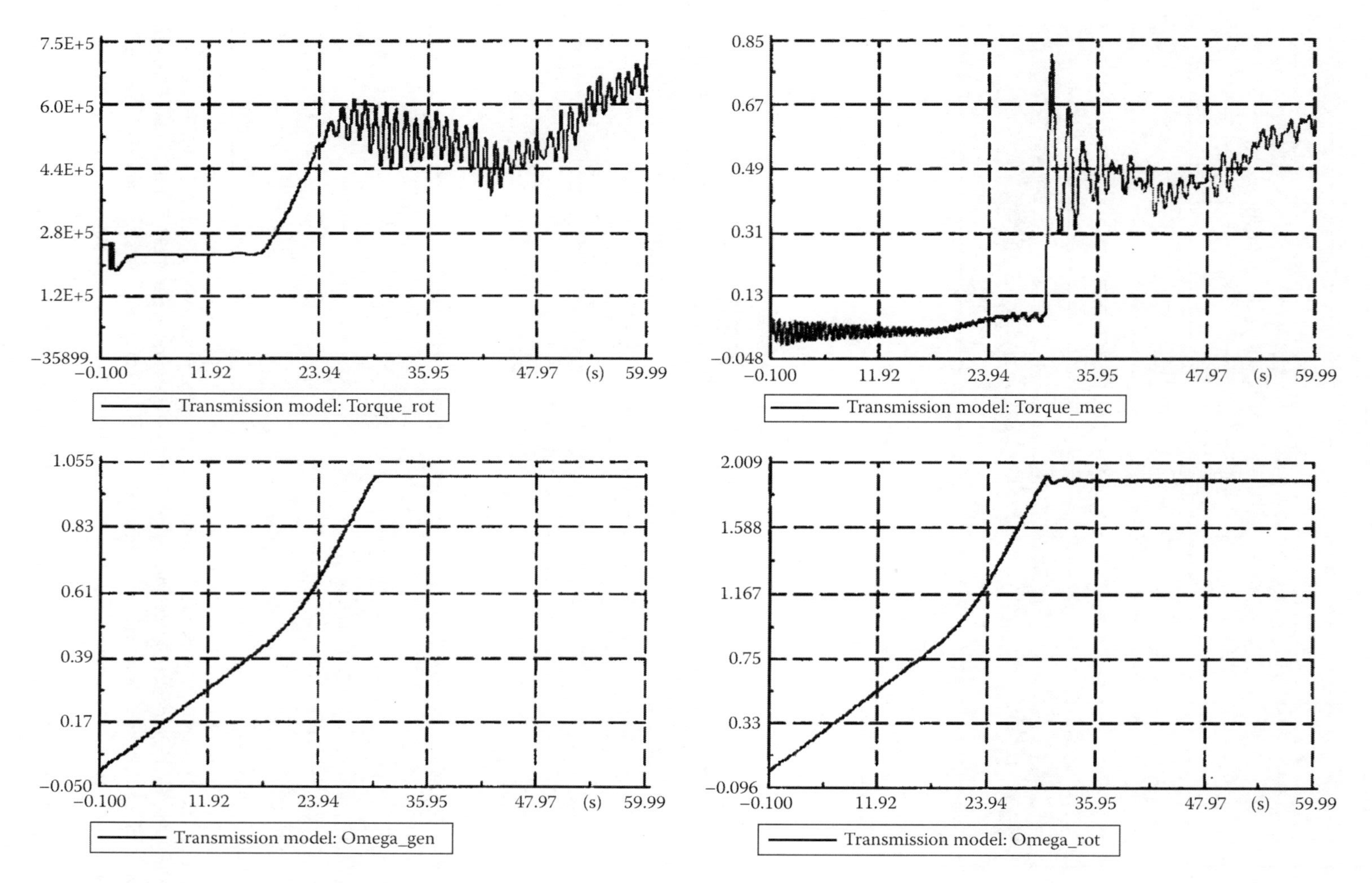

FIGURE 4.33 Transmission model response during start-up.

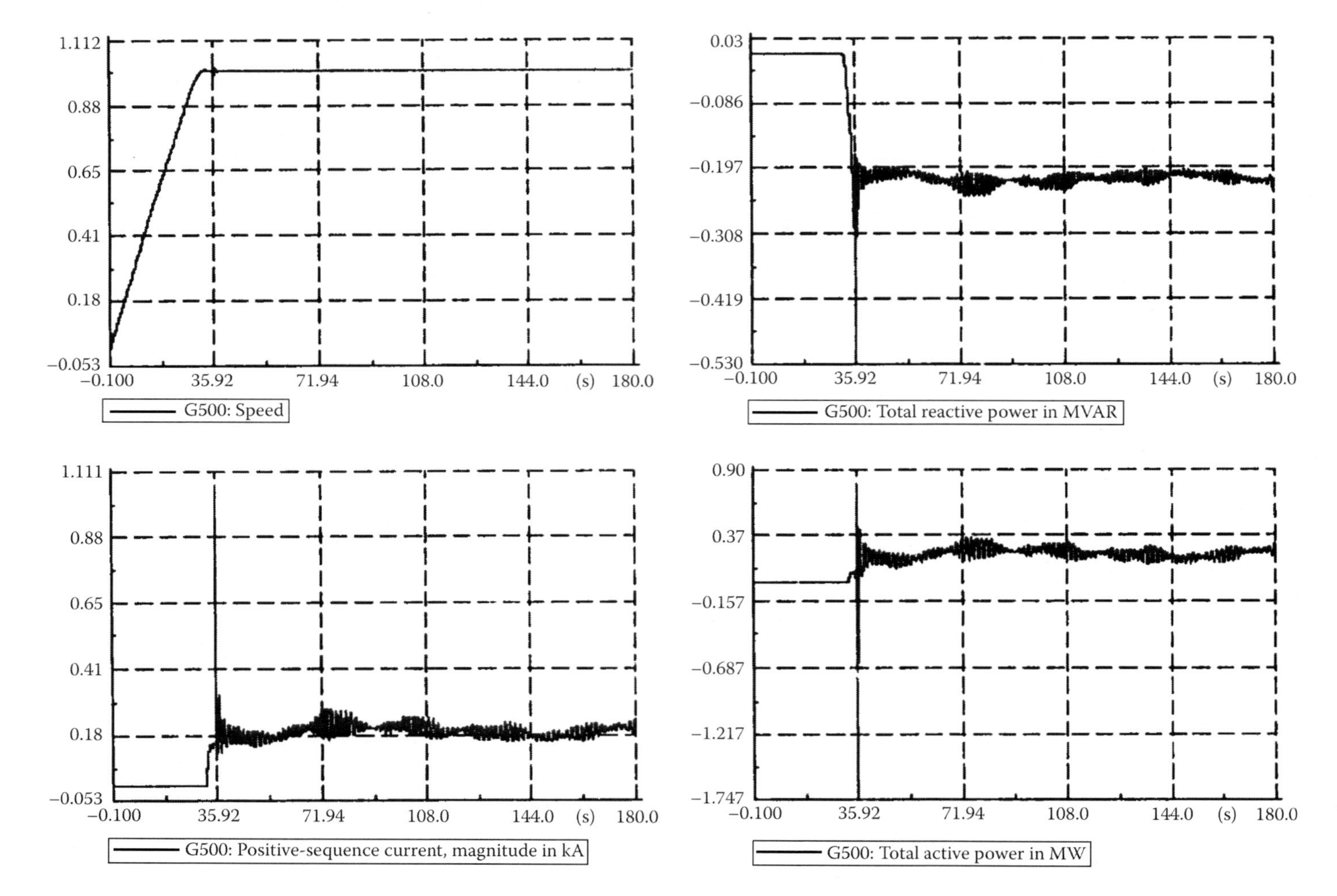

FIGURE 4.34 Direct start-up of a 4-pole, 50 Hz, 0.5 MW induction generator (IG).

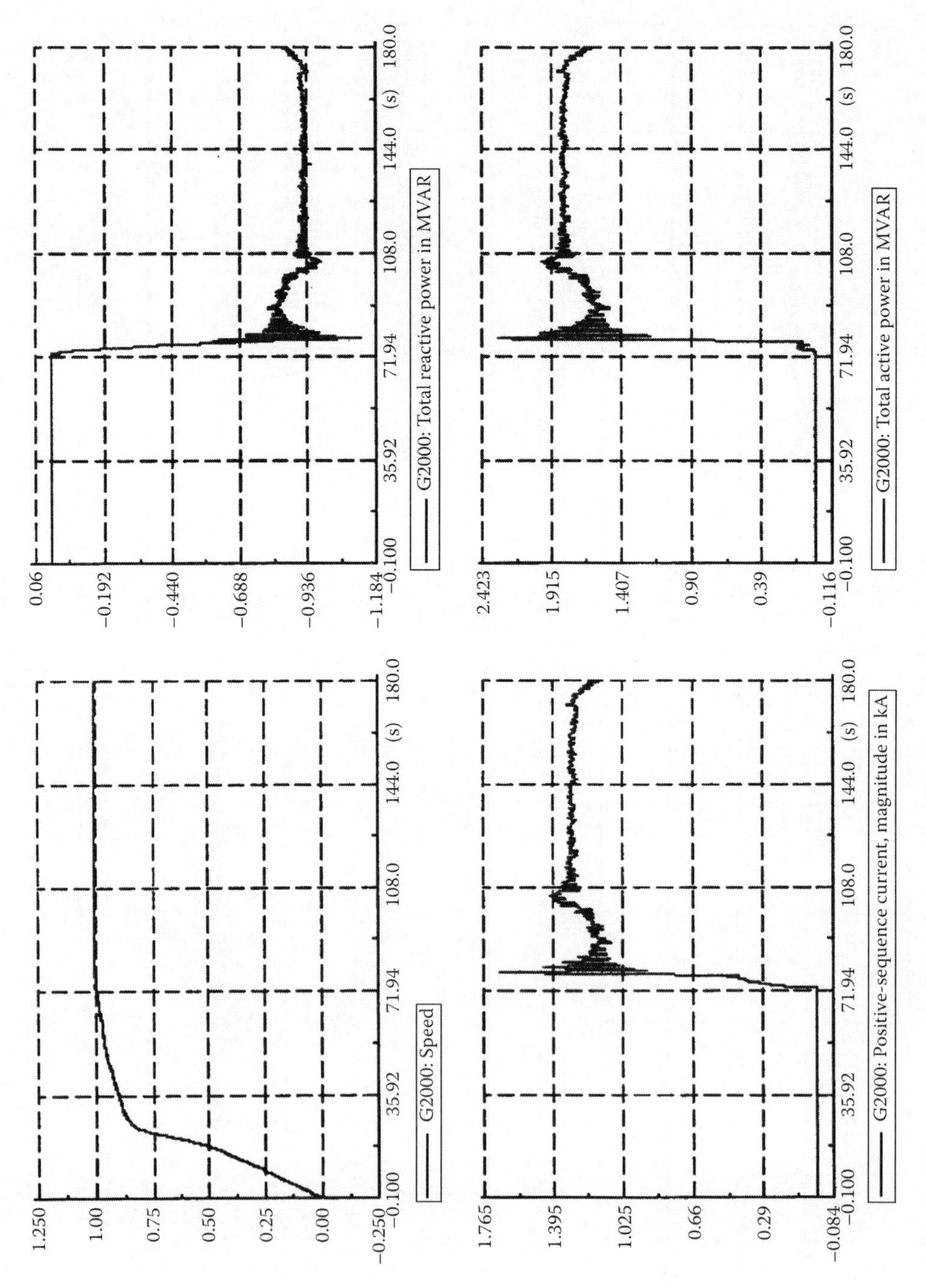

FIGURE 4.35 Start-up and connection via soft-starter to grid at 1500 rpm (2 MV A self-excited induction generator [SEIG]).

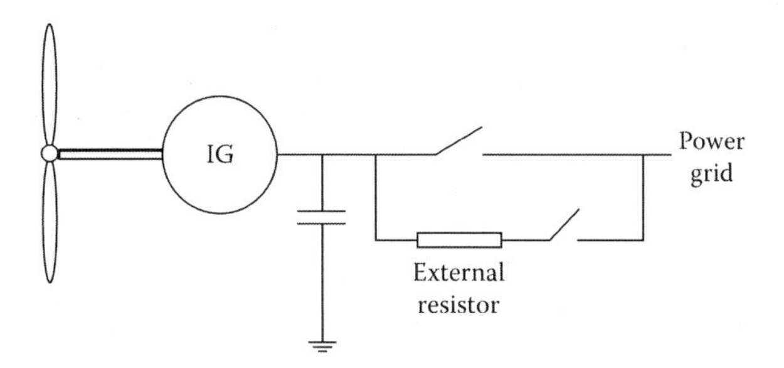

FIGURE 4.36 Connection of induction generators (IGs) with external resistor.

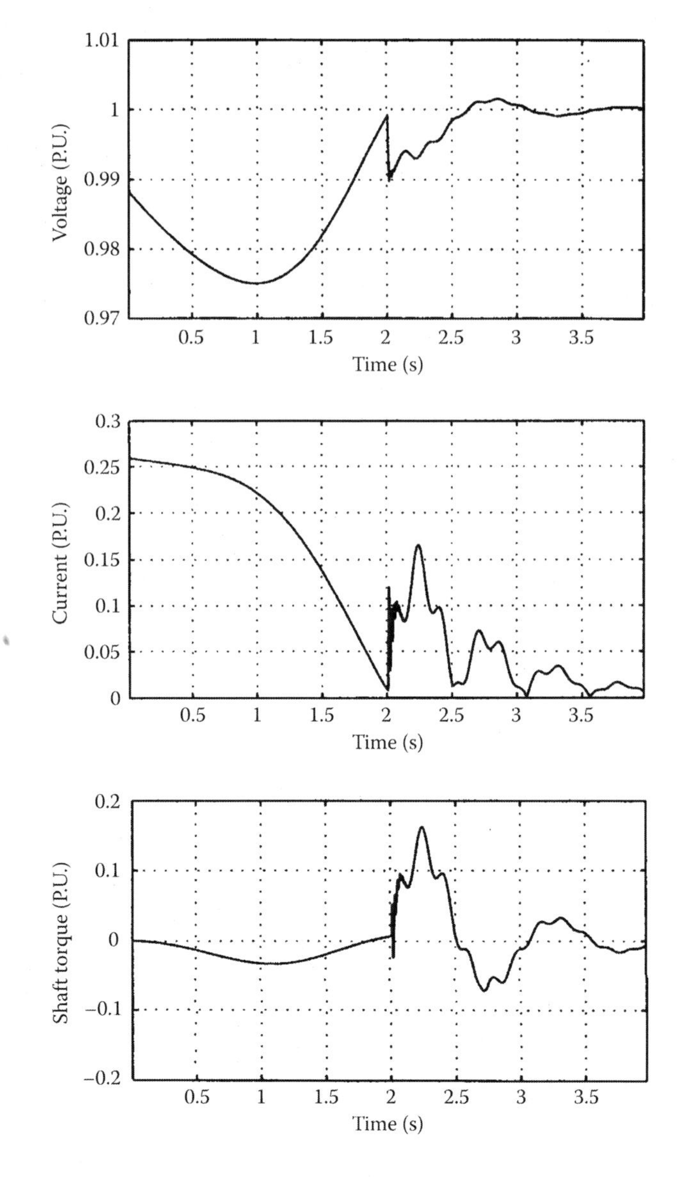

FIGURE 4.37 Voltage, current, and shaft torque vs. time for resistor connection to grid of a 15 kW, 0.8% slip induction generator (IG) with active-stall wind turbine regulation, at no load.

4.14 More on Power Grid Disturbance Transients in Cage-Rotor Induction Generators

IGs connected to the power grid are driven by wind turbines, hydroturbines, diesel engines, and so forth. In most cases, the speed of the IG is larger than the speed of the prime mover; thus, a gearbox transmission is required.

Further on, the prime mover has a number of lumped inertias, elastically coupled to each other. The three blades of a standard wind turbine have their inertias as follows: H_{B1}, H_{B2}, and H_{B3} (Figure 4.38). The hub, the gearbox, and the IG rotor have the inertias H_H, H_{GB}, and H_G. Axes and brakes are integrated with them. Spring stiffness and damping elements are also introduced, while the inputs to the wind turbine model are the aerodynamic torques of the blades T_{B1}, T_{B2}, and T_{B3} and the generator torque T_G (Figure 4.38).

The state-space equations of such a drive train in P.U. are given here for convenience:

$$\frac{d}{dt}\begin{vmatrix}[\theta]\\ [\omega]\end{vmatrix}=\begin{bmatrix}[0] & [I]\\ -[2H]^{-1}[C] & -[2H]^{-1}[D]\end{bmatrix}\begin{bmatrix}[\theta]\\ [\omega]\end{bmatrix}+\begin{bmatrix}[0]\\ [2H]^{-1}\end{bmatrix}[T] \tag{4.102}$$

where

$[\theta]$, $[\omega]$, $[T]$ are 6×1 matrix vectors of positions, angular velocities, and torques
[0] and [1] are 6×6 zero and identity matrices
$[H]$ is the 6×1 matrix of inertias
$[C]$ and $[D]$ are stiffness and damping matrices

$$[C]=\begin{bmatrix}C_{HB} & 0 & 0 & -C_{HB} & 0 & 0\\ 0 & C_{HB} & 0 & -C_{HB} & 0 & 0\\ 0 & 0 & -C_{HB} & -C_{HB} & 0 & 0\\ -C_{HB} & -C_{HB} & -C_{HB} & -C_{HGB}+3HB & -C_{HGB} & 0\\ 0 & 0 & 0 & -C_{HGB} & C_{HGB}+C_{GBG} & -C_{GBG}\\ 0 & 0 & 0 & 0 & -C_{GBG} & C_{GBG}\end{bmatrix} \tag{4.103}$$

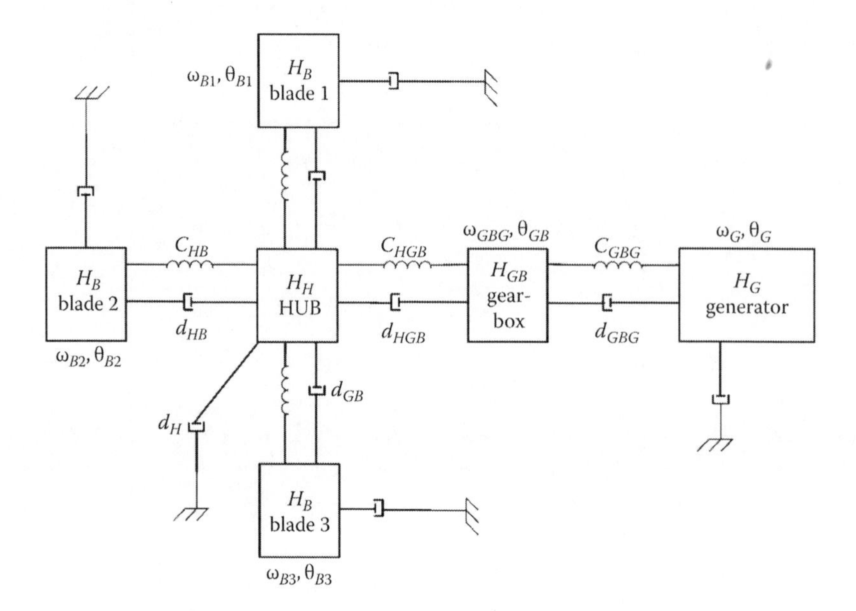

FIGURE 4.38 Wind turbine induction generator (IG) drive train with six inertias.

$$[D]=\begin{bmatrix} D_B+d_{HB} & 0 & 0 & -d_{HB} & & \\ 0 & D_B+d_{HB} & 0 & -d_{HB} & & \\ 0 & 0 & D_B+d_{HB} & -d_{HB} & & \\ -d_{HB} & -d_{HB} & -d_{HB} & \begin{matrix} D_H+d_{HGB} \\ +3d_{HGB} \end{matrix} & -d_{HGB} & 0 \\ 0 & 0 & 0 & -d_{HGB} & (D_{GB}+d_{HGB}+d_{GBG}) & -d_{GBG} \\ 0 & 0 & 0 & 0 & -d_{GBG} & D_G+d_{GBG} \end{bmatrix} \quad (4.104)$$

The IG model for transients is the already described space-phasor model. However, as during some operation modes the IG may end up as self-excited, supplying its own load after disconnection from the power grid, the model should include the magnetic saturation. The model in Equations 4.91 and 4.92 may be decomposed along d and q axes with Ψ_{1d}, Ψ_{1q}, Ψ_{md}, and Ψ_{mq} as variables:

$$\begin{aligned} &s\Psi_{1d} = R_1 I_{1d} + \omega_b \Psi_q + V_d = F_{1d} \\ &s\Psi_{1q} = R_1 I_{1q} - \omega_b \Psi_{1d} + V_q = F_{1q} \\ &A_{dd} s\Psi_{md} + A_{dq} s\Psi_{mq} = -R_2 I_{2d} + (\omega_b - \omega_r)\Psi_{2q} + \frac{X_{2l}}{X_{1l}} F_{1d} \\ &A_{dq} s\Psi_{md} + A_{qq} s\Psi_{mq} = -R_2 I_{2q} - (\omega_b - \omega_r)\Psi_{2d} + \frac{X_{2l}}{X_{1l}} F_{1q} \\ &A_{dd,qq} = X_{2l}\left(\frac{1}{X_{1l}} + \frac{1}{X_{2l}} + \frac{1}{X_m}\right) - X_{2l}\left(\frac{1}{X_m} - \frac{1}{X'_m}\right)\frac{\Psi^2_{md,q}}{\Psi^2_m} \\ &A_{dq,qd} = X_{2l}\left(\frac{1}{X_m} - \frac{1}{X'_m}\right)\frac{\Psi_{md}\Psi_{mq}}{\Psi^2_m} \end{aligned} \quad (4.105)$$

$$\begin{aligned} &\omega_1 L_m = X_m = \omega_1 \frac{\Psi_m(I_m)}{I_m} \\ &\omega_1 L'_m = X'_m = \omega_1 \frac{d\Psi_m}{dI_m}(I_m) \end{aligned} \quad (4.106)$$

$$I_m = \sqrt{(I_{1d}+I_{2d})^2 + (I_{1d}+I_{2d})^2} \quad (4.107)$$

The rotor flux linkages Ψ_{2d} and Ψ_{2q} and the stator and rotor currents are all dummy variables to be eliminated via the flux/current relationships:

$$\begin{aligned} &\Psi_{1d} = -L_1 I_{1d} + L_m I_{2d} \\ &L_1 = L_{1l} + L_m \\ &\Psi_{1q} = -L_1 I_{1q} + L_m I_{2q} \\ &\Psi_{md} = L_m(-I_{1d} + I_{2d}) \\ &\Psi_{mq} = L_m(-I_{1q} + I_{2q}) \\ &\Psi_{2d} = -L_m I_{1d} + L_2 I_{2d} \\ &L_2 = L_{2l} + L_m \\ &\Psi_{2q} = -L_m I_{1q} + L_2 I_{2q} \end{aligned} \quad (4.108)$$

There are four currents, I_{1d}, I_{1q}, I_{2d}, and I_{2q}, and the rotor fluxes Ψ_{2d} and Ψ_{2q} to eliminate from the six expressions of Equation 4.108. Implicitly, the so-called cross-coupling magnetic saturation is accounted for in these equations [32,33]. Note that the stator equations are written for the generator mode association of current signs.

The electromagnetic torque T_e is as follows:

$$T_e = \frac{3}{2} p_1 \left(\Psi_{1d} I_{1d} - \Psi_{1q} I_{1q} \right) \tag{4.109}$$

The power network is, in general, represented by its sequence equivalents.

Among most abnormal operation modes, we consider here the following:

- Three-phase sudden short circuits
- Line-to-line short circuit
- Breaker reclosing before generator defluxing
- One-phase interruption
- Unbalanced system voltages

A three-phase fault at the low-voltage side of the transformer, on a 500 kW IG with a 200 kVA parallel capacitor, connected to a local grid containing an 800 kVA (20/0.69 kV) transformer, a 2 km line, and a circuit breaker at the point of common connection is illustrated in Figure 4.39a through c [34].

About 3 P.U. peak torque at the IG shaft occurs. The propagation of torque through the drive train shows "high fidelity" on the high-speed side, with a more compliant response on the low-speed side (with a 2–3 Hz natural frequency and low damping). The response to a line-to-line fault is similar, but

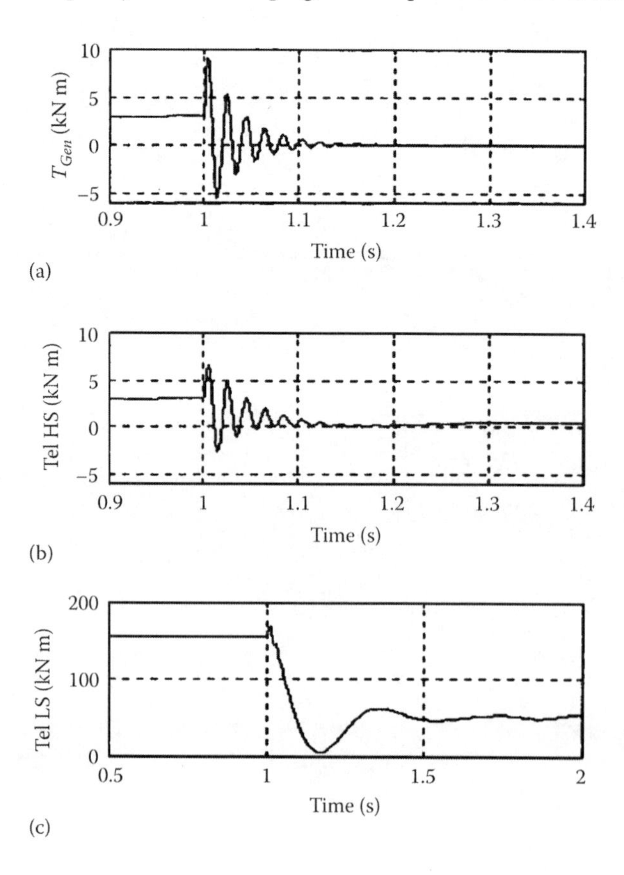

FIGURE 4.39 Three-phase short circuit at the low-voltage busbars: (a) induction generator (IG) torque, (b) high-speed side torque, and (c) low-speed side elasticity torque.

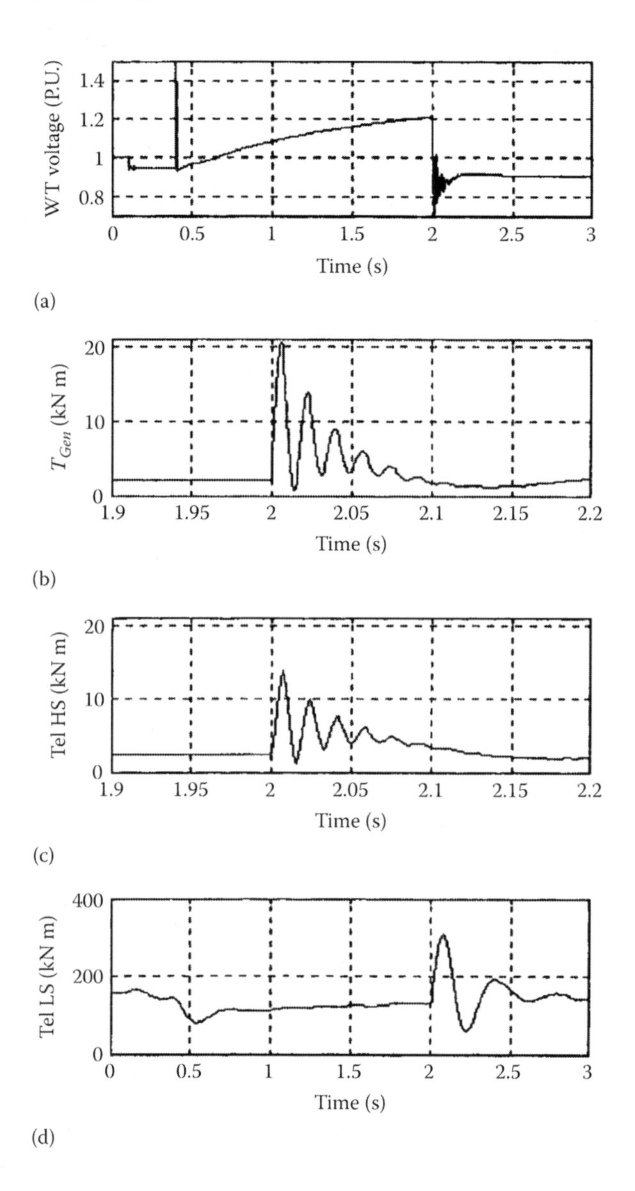

FIGURE 4.40 Breaker reclosing after a remote fault: (a) induction generator (IG) voltage, (b) IG torque, (c) high-speed side torque, and (d) low-speed elasticity torque.

with sustained 100 Hz ($2f_1$) oscillations, as expected. Breaker reclosing, however, produces notably larger torque transients (Figure 4.40a through d) [34]. A remote fault at 0.1 s is cleared at 0.4 s and is followed by a reclosing at 2.0 s.

The voltage builds up to 120% as the IG stator circuits are open and the 200 kVA capacitors remain at IG terminals. The speed (not shown) does the same. The peak IG torque transients reach 7 P.U. values after reclosing. The gearbox "feels" about 4 P.U. torque oscillations. On the low-speed side, the effect is small. The generator and gearbox stresses are considered potentially harmful. Unbalanced voltages in the local power network produce 100 Hz pulsations in torque. These pulsations may have a notable effect on the fatigue life of the gearbox.

We may conclude here that the "constant speed" IG is vulnerable to power grid disturbances due to its "rigid" response in contrast to variable speed IGs. Note that single-phase SEIG induction transients are to be treated again through the *dq* model but in stator coordinates (Chapter 26 in Reference 1).

4.15 Summary

- Induction machines with cage rotor and capacitor excitation are coded as SEIGs.
- SEIGs operate typically alone or in a group, autonomously on their particular loads.
- Cage-rotor IGs may be connected to the power grid also, with capacitor banks connected at terminals to make up for the reactive power required for their magnetization and to contribute to voltage stabilization.
- In SEIGs with capacitor excitation, self-excitation at no load at a frequency f (P.U.) and voltage V_0 (P.U.) is dependent on speed U (P.U.), capacitance C_p at terminals, and on IG parameters. At no load, the slip $S_0 = f_0 - U$ is negative but very small.
- When the SEIG, already self-excited at no load, is loaded, the terminal voltage and frequency vary with load and its power factor. In general, for constant speed and R or R, L load, the load voltage decreases markedly with load. The slip is negative and increases with load.
- For capacitor self-excitation on no load, there should be an initial level of magnetization in the rotor, left from previous operation, and operation at a notable degree of magnetic saturation.
- The computation of frequency and voltage, for given speed, parallel capacitance C_p, load, and IG parameters, may be approached through two main categories of methods: impedance and admittance types.
- Only the admittance methods, for given frequency, magnetization curve, IG and load parameters, and capacitance, may allow for the computation of slip S (and speed $U = f - S$, $S < 0$) from a second-order algebraic equation. Then, iteratively, the frequency f is changed until the final speed reaches the desired value. Only a few iterations are required. The computation of the corresponding magnetization reactance X_m is straightforward. Then, the air gap emf is found from the magnetization curve ($X_m = f(E_1)$). Further on, the terminal voltage, load current, capacitor current, load power, and so forth are calculated without any iteration. For successful self-excitation, the solutions of slip S have to be real numbers, and $X_m < X_{max}$; X_{max} is the unsaturated value of X_m.
- The SEIG exhibits a voltage collapse point on its $V(I)$ curve. To extend the stable (linear) part of the $V(I)$ curve, a series capacitance C_s is added. The short-shunt connection performs better. The optimum ratio $K = X_{cs}/X_{cp} = 0.4$–0.5.
- Performance/parameter sensitivity studies reveal that the smaller the IG leakage reactances X_{1l} and X_{2l} in P.U., the better. Also, prime movers with controlled speed are capable of producing notably more power. As the speed goes up with the electric load power, so does the frequency.
- Combined parallel capacitance C_p and speed control are recommended to keep the voltage regulation within 4%–5% up to full load, with small frequency variations a bonus.
- In applications such as wind machines, where the power decreases with cubic speed, variable speed is desirable when the wind speed decreases notably. Pole-changing SEIGs may handle such situations at low costs. The pole count ratio p_1/p_1' should vary, generally between one half and two thirds. Two separate windings (of different ratings) may be placed in the stator slots and switched on or off at a certain power (speed) level, or a pole-changing winding may be used for the scope.
- Pole-changing windings with 4/6 pole count ratio, for example, need different connections to provide the same voltage, with the same capacitance (at no load) at different speeds. The key issue is to maintain the air gap flux density (magnetic saturation) at about the same level and the winding factors reasonably high. Attention has to be paid to space harmonics, including subharmonics.
- The commutation of pole count has to be made at a speed that is generally below the peak torque situation for the active connection, in order to safeguard stable transients.
- Three-phase SEIGs may perform on unbalanced loads by accident or by necessity. The investigation of such situations for steady state is performed with the symmetrical components method.
- In general, it is found that the efficiency is notably reduced due to negative sequence losses.

- One phase open, with SEIG connected to the power grid, shows a similar reduction in power at higher losses. The derating of an SEIG (IG) is required for long unbalanced operation, in order to meet the rated temperature limit.
- In some situations, single-phase power is required above 2–3 kW, and two-phase IMs are not available off the shelf.
- Three-phase SEIG connections for single-phase output were proposed. The Steinmetz connection (single-capacitance $C+$ in parallel with the second phase, for Δ connection) augmented with series capacitance C_s ($K = X_{cs}/X_{cp} = 0.3–0.6$) was demonstrated to extend the $V(I)$ linear curve up to 2.0 P.U. power at reasonable efficiency. The symmetrization of phases is obtained above rated power at a certain value, but the highest phase voltage does not go above 1.2 P.U.
- For powers below 2–3 kW, two-phase IMs are available off the shelf. They were, thus, proposed for small-power single-phase output. A practical solution contains an excitation capacitor C_e to close the auxiliary winding and a series capacitor C_s in the main (power) winding. Again, the method of symmetrical components is to be used to assess the steady-state performance with magnetic saturation consideration as necessary. Because of the complexity of the two self-excitation equations, an optimization method seems to be of practical to calculate f, C_p, and C_s simultaneously. The Hooke–Jeeves method was proven to be proficient for this endeavor. Again, the series capacitor C_s extends the output power range notably with reasonable voltage self-regulation. Small-power generator sets may take advantage of this inexpensive solution.
- Three-phase SEIG transients occur at self-excitation at no load, load connection and rejection, speed variation, and so forth. The dq model is used to handle the operation modes that are crucial for power quality and protection of the system design. Again, magnetic saturation has to be included in the model, which may be used first in the space-phasor form.
- Among the main results that we mention after investigating SEIG transients is that the sudden short circuit at SEIG terminals leads to 5 P.U. current transients for around 20 ms before the voltage collapses. Also, a safe starting of an IM connected to an SEIG requires only 160% overrating of the SEIG. However, to maintain stable IM operation for 100% step mechanical load, a 300(400)% overrating is needed.
- Load rejection leads to a slow increase in terminal voltage to its steady-state no-load value. This might be harmful if the speed of the prime mover is not regulated, and speed increases notably after load rejection.
- The parallel connection of SEIGs is required, as there are limitations on the prime-mover power (unit) due to local energy resource limitations. In general, a single variable capacitor is used to self-excite such a group. Again, voltage regulation is better if the speed of some IGs of the group is also regulated with load.
- IGs are connected to the power grid experience connecting transients. Recent international standards drastically limit the voltage change factor, the current change factor, and the voltage flick factor for such transients, in order to maintain power quality in the power grid.
- Direct connection of a cage rotor IG to the grid shows very large transients. However, soft-starter connection leads to much lower transients and is recommended for use, specifically for larger power per unit. Alternatively, series resistors may be used for the scope at lower costs but with lower flexibility [35].
- Power grid disturbances (short circuit, breaker reclosing, one phase open, etc.) may produce very high-peak torques in the generator and in the gearbox (if any), 7 P.U. and 3.0 P.U. oscillations, respectively, for breaker reclosing after clearing a distant short circuit. These torque oscillations may reduce the fatigue life of the IG shaft and of the gearbox. The transmission power train is to be modeled as a multiple inertia system with stiffness and damping connection elements.
- Electrical flicker is a measure of the voltage variation that may cause eye disturbance for the consumer. The variation of wind power (speed) in time may induce flicker during continuous

operation. Flicker may also be induced during switchings. For constant speed IGs, this is a particularly sensitive issue [36,37].

- SEIGs in autonomous [38,39] and grid connection [40,41] are in investigation for involved transients and stability as they still are in wide commercial use.
- New fault ride through solutions for SEIG are still proposed to further improve performance [42].

References

1. I. Boldea and S.A. Nasar, *Induction Machine Handbook*, CRC Press, Boca Raton, FL, 2001.
2. E.D. Basset and F.M. Potter, Capacitive excitation of induction generators, *Electrical Eng.*, 54, 1935, 540–545.
3. S.S. Murthy, O.P. Malik, and A.K. Tandon, Analysis of self-excited induction generators, *Proc. IEE*, 129C(7), 1982, 260–265.
4. L. Ouazone and G. McPherson, Analysis of the isolated induction generator, *IEEE Trans.*, IAS-102(8), 1983, 2793–2798.
5. L. Shrider, B. Singh, and C.S. Tha, Towards improvements in the characteristics of selfexcited induction generators, *IEEE Trans.*, EC-8(1), 1993, 40–46.
6. S.P. Singh, B. Singh, and M.P. Jain, A new technique for the analysis of selfexcited induction generator, *EMPS J.*, 23(6), 1995, 647–656.
7. T.F. Chan and L.L. Lai, Steady state analysis and performance of a stand alone three phase induction generator with asymmetrical connected load impedances and excitation capacitances, *IEEE Trans.*, EC-16(4), 2001, 327–333.
8. N. Ammasaigounden, M. Subbiah, and M.R. Krishnamurthy, Wind driven self-excited pole changing induction generator, *Proc. IEE*, 133B(6), 1986, 315–321.
9. K.S. Sandhu and S.K. Jain, Operational aspects of self-excited induction generator using a new model, *EMPS J.*, 27(2), 1999, 169–180.
10. S. Rajakaruna and R. Bonert, A technique for the steady state analysis of a selfexcited induction generator with variable speed, *IEEE Trans.*, EC-8(4), 1993, 757–761.
11. T.F. Chan, Analysis of self-excited induction generators using an iterative method, *IEEE Trans.*, EC-10(3), 1995, 502–507.
12. L. Shridha, B. Singh, C.S. Tha, B.P. Singh, and S.S. Murthy, Selection of capacitors for the self-regulated short shunt self-excited induction generators, *IEEE Trans.*, EC-10(1), 1995, 10–16.
13. E. Suarez and G. Bortolotto, Voltage-frequency control of a self-excited induction generator, *IEEE Trans.*, EC-14(3), 1999, 394–401.
14. P. Chidambaram, K. Achutha, M. Subbiah, and M.R. Krishnamurthy, A new pole changing winding using star/star delta switching, *Proc. IEE*, B-EPA-130(2), 1983, 130–136.
15. R. Parimelalasan, M. Subbiah, and M.R. Krishnamurthy, Design of dual speed single-winding induction motors—A unified approach, Record IEEE-IAS-1976 Annual Meeting, Paper 398, pp. 1071–1079.
16. N. Ammasaigounden, M. Subbiah, and M.R. Krishnamurthy, Wind-driven selfexcited pole-changing induction generators, *Proc. IEE*, 133B(5), 1986, 315–322.
17. A.M. Bahrani, Analysis of selfexcited induction generators under unbalanced conditions, *EMPS J.*, 24(2), 1996, 117–129.
18. A.W. Ghorashi, S.S. Murthy, B.P. Singh, and B. Singh, Analysis of wind-driven grid connected induction generators under unbalanced conditions, *IEEE Trans.*, EC-9(2), 1994, 217–223.
19. T.F. Chan and L.L. Lai, A novel single-phase self-regulated self-excited induction generator using three phase machine, *IEEE Trans.*, EC-16(2), 2001, 204–208.
20. T.F. Chan and L.L. Lai, Single-phase operation of a three-phase induction generator with Smith connection, *IEEE Trans.*, EC-17(1), 2002, 47–54.

21. T. Fukami, Y. Kaburaki, S. Kawahara, and T. Miyamoto, Performance analysis of self-regulated and self-excited single phase induction machine using a three phase machine, *IEEE Trans.*, EC-14(3), 1999, 622–627.
22. Y.H.A. Rahim, I.I. Alolah, and R.I. Al-Mudaiheem, Performance of single phase induction generators, *IEEE Trans.*, EC-8(3), 1993, 389–395.
23. M.H. Salama and P.G. Holmes, Transients and steady-state load performance of stand-alone self-excited induction generator, *Proc. IEE*, EPA-143(1), 1996, 50–58.
24. O. Ojo, Performance of self-excited single-phase induction generators with shunt, short-shunt and long-shunt excitation connections, *IEEE Trans.*, EC-14(1), 1999, 93–100.
25. L. Wang and J.Y. Su, Dynamic performance of an isolated self-excited induction generator under various loading conditions, *IEEE Trans.*, EC-14(1), 1999, 93–100.
26. B. Singh, L. Shridar, and C.B. Iha, Transient analysis of selfexcited generator supplying dynamic load, *EMPS J.*, 27(9), 1999, 941–954.
27. A.H. Al-Bahrani and N.H. Malik, Voltage control of parallel operated self-excited induction generators, *IEEE Trans.*, EC-8(2), 1993, 236–242.
28. C.H. Lee and L. Wang, A novel analysis of parallel operated self-excited induction generators, *IEEE Trans.*, EC-14(2), 1998, 117–123.
29. L. Mihet-Popa, F. Blaabjerg, and I. Boldea, Wind turbine generator modeling and simulation where rotational speed is the controlled variable, *IEEE Trans.*, IA-40(1), 2004, 8–13.
30. T. Thiringer, Grid friendly connected of constant speed wind turbines using external resistors, *IEEE Trans.*, EC-17(4), 2002, 537–542.
31. A. Larsson, Guidelines for grid connection of wind turbines, *Record of 15th International Conference on Electricity Distribution*, Nice, France, June 1–4, 1999.
32. P. Vas, K.E. Hallenius, and J.E. Brown, Cross-saturation in smooth airgap electrical machines, *IEEE Trans.*, EC-1(1), 1986, 103–112.
33. I. Boldea and S.A. Nasar, Unified treatment of core loss and saturation in the DQ models of electrical machines, *Proc. IEE*, 134B(6), 1987, 355–363.
34. S.A. Papathanassiou and M.P. Papadopoulos, Mechanical stresses in fixed speed wind turbines due to network disturbances, *IEEE Trans.*, EC-16(4), 2001, 361–367.
35. A.P. Grilo, A. Sharghi, and W. Freitas, An analytical approach to determine optimal resistance for three-series-resistor method for IG connection, *IEEE Trans.*, EC-23(4), 2008, 1111–1113.
36. A. Larsson, Flicker emission of wind turbines during continuous operation, *IEEE Trans.*, EC-17(1), 2002, 114–118.
37. A. Larsson, Flicker emission of wind turbines caused by switching operations, *IEEE Trans.*, EC-17(1), 2002, 119–123.
38. S.N. Mohato, S.P. Singh, and M.P. Sharma, Capacitors required for maximum power of a selfexcited IG using a three phase machine, *IEEE Trans.*, EC-23(2), 2008, 372–381.
39. L. Wang and P.-Y. Lin, Analysis of a commercial biogas generator system using a gas-engine-induction generator set, *IEEE Trans.*, EC-24(1), 2009, 230–239.
40. Y. Zhang, D. Xie, J. Feng, and R. Wang, Small signal modeling and model analysis of wind turbine based on three-mass shaft model, *EPCS J.*, 42(7), 2014, 693–702.
41. A. Moharana, R.K. Varma, and R. Seethapathy, Modal analysis of IG based wind form connected to series compensated transmission line and line commutated converter HVDC, *EPCS J.*, 42(6), 2014, 612–628.
42. M. Firouzi and G.B. Gharehpetian, Improving FRT capability of fixed speed wind turbine by using bridge-type fault current limiter, *IEEE Trans.*, EC-28(2), 2013, 361–369.

5

Stator-Converter-Controlled Induction Generators

5.1 Introduction

As already discussed at length in Chapter 4, the cage-rotor induction generator (IG) with fixed capacitance self-excitation produces output at slightly variable frequency with load even at regulated (constant) prime-mover speed. Many electrical loads are, however, frequency and voltage sensitive. On the other hand, variable speed generation is needed in many applications to extract more energy from the primary source (wind, water energy, internal combustion engine [ICE]).

Power electronics control in the stator of cage-rotor IG may be used to allow variable speed operation in two main situations:

1. Grid or stand-alone operation with constant alternating current (AC) voltage and frequency output
2. Stand-alone operation with rectifier (direct current [DC]) loads

AC output stator-converter-controlled IGs (SCIGs) require AC–AC (cascaded or direct) full power rating static converters. Nowadays, they are fabricated, for variable speed drives, with ±100% reactive and ±100% active power capabilities by proper design, and they are roughly only 50% more expensive than the stand-alone diode rectifier power voltage source insulated gate bipolar transistor (IGBT) inverters that produce up to 2–3 MW per unit.

The cost of a full power rating converter—with ±100% active and reactive power capabilities—may be justified by the large motoring and generating speed range and reactive power (or voltage) control and flexibility in power grid applications. For reactive loads, the forced-commutated IGBT rectifier and capacitor filter, eventually, with a boost DC–DC converter, is sufficient to control the converter DC voltage output for variable speed.

In applications where the prime-mover power decays sharply with speed (with cubic speed, for example, in wind turbines), it may be sufficient to locate an additional winding in the stator with a larger number of poles $p_1' > p_1$ (say 6/4, 8/6) and use an AC–AC converter for this auxiliary winding, to augment the power of the power winding of the stator that is directly connected to the grid.

The two windings may work simultaneously or successively. The four-quadrant converter operation is very useful to produce additional output above synchronous speed $n_1 = f_1/p_1$. Alternatively, for lower speeds, the auxiliary winding may work alone at variable speed to tap the energy below 25% of the rated power. We will explore in some detail the above configurations.

5.2 Grid-Connected SCIGs: The Control System

SCIGs may be read as SCIGs or cage-rotor IGs. There are two basic schemes:

1. With AC–AC cascaded pulse-width modulator (PWM) converter (Figure 5.1a).
2. With direct AC–AC PWM converter (Figure 5.1b) [1,2].

The configurations with thyristor DC current link AC–AC converter and, respectively, with thyristor cycloconverter seem to be merely of historical interest, as their reactive power drainage and current harmonics content are no longer acceptable in terms of power quality standards.

While the matrix converter is still in advanced laboratory status, the cascaded AC–AC PWM converter is available off the shelf for powers up to 1 MW and more, with up to ±100% reactive power capability. The so-called high-voltage direct current (HVDC) light technology uses, in fact, IGBT multilevel AC–AC cascaded power converters [3], but for higher DC link voltage levels (tens of kilovolts) for DC power transmission.

The vector or direct torque control [4] of cascaded PWM converters was applied for variable speed drives with fast and frequent regenerative braking. In essence, the control is similar to the case of wound-rotor induction generators (WRIGs) with the cascaded converter connected to the rotor (Chapter 2). What differ are the control and the state estimation for the machine-side converter, as IGs here have a cage rotor.

The motor starting is inherent in the control. Speed, torque, or power control of the machine-side converter may be adopted, depending on the prime-mover operation modes.

For a pump-storage small hydropower unit, motor starting and operation at variable speed are required besides generating at variable speed; for a wind turbine unit, no motoring is, in general, required.

The grid-side converter is controlled for constant DC-link voltage and the desired direct reactive power exchange.

5.2.1 Machine-Side PWM Converter Control

To let the control system open for motoring and generating, let us consider that only torque versus speed is performed. In essence, a functional generator produces the desired torque versus speed curve imposed

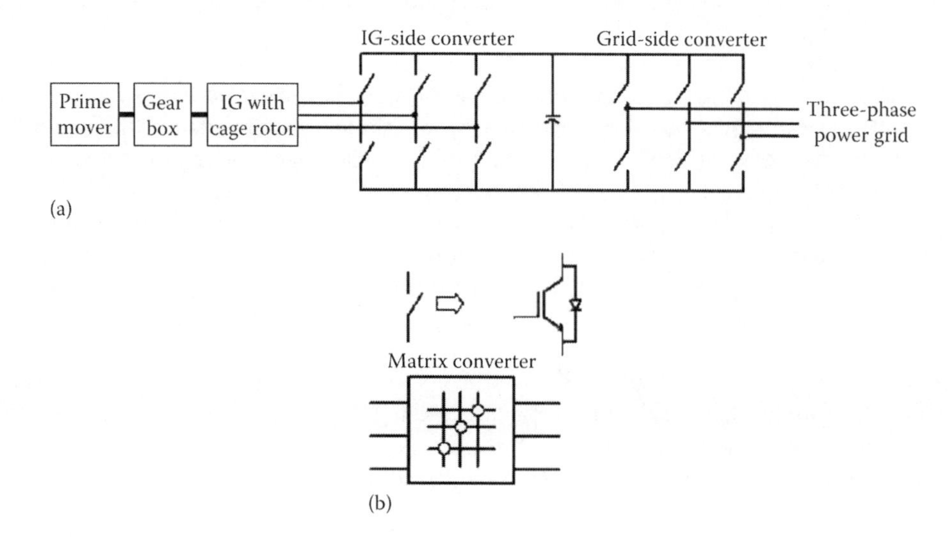

FIGURE 5.1 Grid-connected stator-converter-controlled induction generators (SCIGs): (a) with cascade alternating current (AC)–AC pulse-width modulator (PWM) converter, and (b) with direct (matrix) converter.

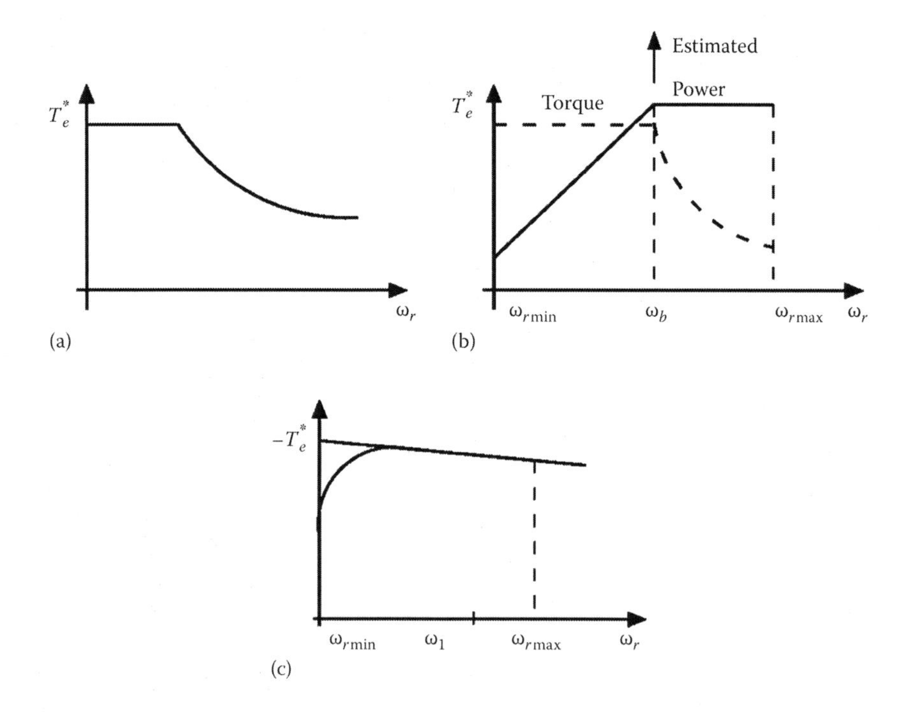

FIGURE 5.2 Typical desired torque/speed curves for (a) motor starting and operation, (b) wind turbines stall regulated, and (c) uncontrolled microhydro turbine.

for the IG (Figure 5.2a through c). For motor starting, the torque versus speed may decrease notably with speed (Figure 5.2a).

In essence, by an a priori applied optimization process, involving the prime-mover characteristics and IG capability, the optimum torque/speed curves are calculated. From now on, positive or negative torque control is performed with the various torque/speed curves stored in tables and called upon according to the operation mode.

For generating, the reference power P^* is set, but then its value is translated into the torque/speed functional reference generator (Figure 5.3).

The direct torque and flux control (DTFC) seems to be inherent to the application once torque control is required. Stator flux (Ψ_s) control is added, and thus, the control system becomes robust and presents fast response. The stator flux functional may also be expressed in terms of flux versus torque to minimize the losses in the IG over the whole-speed and power range. The space-vector modulation (SVM) is added to further reduce the IG current harmonics, converter losses, and noise.

The two main components of DTFC for SCIGs are the state observers and the DTFC–SVM strategy. Vector control strategies perform similarly but, apparently, with slightly larger online computation efforts and higher sensitivity to the machine parameter variation.

5.2.1.1 State Observers for DTFC of SCIGs

DTFC of SCIGs requires stator flux vector instantaneous amplitude and position, torque, and (for motion sensorless control) speed observers.

Essentially, the whole body of knowledge on state observers for sensorless cage-rotor induction motor drives is available for DTFC of SCIGs. The state observers constitute a dynamic subsystem that produces an approximation for the state vector of the actual system. The inputs of the state observer are chosen from the inputs and available outputs of the actual system. Observers for linear systems were first proposed by Luenberger [5]. Then, they were extended for nonlinear systems [6] and by Kalman for stochastic systems [7].

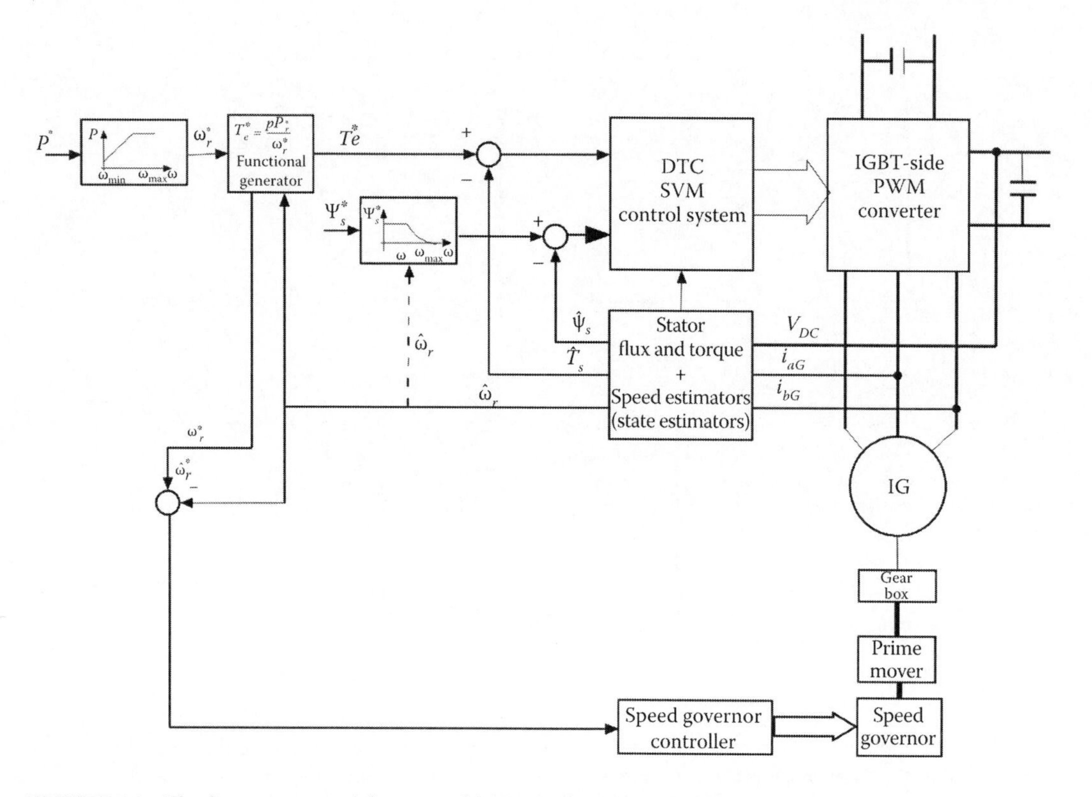

FIGURE 5.3 The direct torque and flux control (DTFC) of machine-side converter.

There are two basic categories of state observers for adjustable speed electric machine control:

1. Full or reduced order linear structure observers for linear systems
2. Nonlinear observers such as the following:
    a. Variable structure
    b. Stochastic or adaptive observers, suitable for uncertain or nonlinear systems
    c. Artificial intelligence observers, such as fuzzy logic, artificial neural networks (ANNs), and genetic algorithms (GAs)

Typical performance indexes for state observers are as follows:

- Accuracy (steady state and transient)
- Robustness
- Convergence quickness
- Behavior at zero, very low, and very high rotor speeds
- Complexity and cost of digital implementation versus performance

For a practical overview on state observers for motion sensorless control, see Reference 8. From the myriad of proposed state observers for motion sensorless control, we treat here only the sliding mode state observers and direct torque and flux controllers, as they demonstrate the following attributes:

- Capable of good accuracy down to 3–5 rpm in speed control and zero speed in torque control mode
- Robust, once the stator resistance is corrected through an estimator
- Capable of working for the entire speed range without making changes in software or hardware

We will treat separately, for convenience, only the flux–torque observer and the speed observer issues. To decouple the flux observer from the speed observer errors, the former is conceived as inherently speed sensorless. The typical form of the sliding-mode flux observer is based on the IM space-phasor model [8]:

$$\frac{d\hat{\bar{\Psi}}_s}{dt} = -R_s\hat{\bar{I}}_s + \hat{\bar{V}}_s + K_1\text{sgn}(\bar{I}_s - \hat{\bar{I}}_s) + K_1'(\bar{I}_s - \hat{\bar{I}}_s) \tag{5.1}$$

$$\frac{d\hat{\bar{\Psi}}_r}{dt} = \frac{L_m}{L_r T_r \sigma}\hat{\bar{\Psi}}_s - \left(\frac{1}{T_r\sigma} + j(\omega_{\psi r} - \omega_r)\right)\hat{\bar{\Psi}}_r + K_2\,\text{sgn}(\bar{I}_s - \hat{\bar{I}}_s) + K_2'(\bar{I}_s - \hat{\bar{I}}_s) \tag{5.2}$$

where

R_s is the stator resistance
L_m is the magnetization inductance
L_r is the total rotor inductance $T_r = L_r/R_r$
R_r is the rotor resistance
ω_r is the rotor speed
$\hat{\bar{V}}_s$ is the stator voltage vector
$\hat{\bar{\Psi}}_s, \hat{\bar{\Psi}}_r$ are the stator and rotor flux vectors

The first equation is written in stator frame ($\omega_b = 0$) and the second in rotor flux frame ($\omega_b = \omega_{\psi_r}$). Consequently, in rotor flux orientation, the term in $j(\omega_{\psi_r} - \omega_r)$ disappears. This is how the state observer becomes inherently speed sensorless. In addition, the stator and the rotor flux observers use combined sliding mode nonlinear and linear feedback terms $S_s = \bar{I}_s - \hat{\bar{I}}_s$: coefficients K_1, K_2, and K_1', K_2'.

To further reduce chattering, the "sgn" functional may be replaced by

$$\text{sat}(x) = \begin{cases} \text{sgn}(x) & \text{if } |x| > h \\ \dfrac{x}{h} & \text{if } |x| < h \end{cases} \tag{5.3}$$

A low-pass filter on the local dynamics of the functional S_s is obtained. To compensate for the inevitable offset in voltage or current measurements, the terms are replaced by

$$K_{1,2}^{\sigma} = \begin{vmatrix} K_1 + (K_{1I}/s) \\ K_2 \end{vmatrix} \tag{5.4}$$

A typical embodiment of such a sliding mode plus linear flux observer is depicted in Figure 5.4 [8].

The reduction in error with offset compensation of a 0.3 V voltage offset is shown in Figure 5.5a and b.

The rotor resistance detuning also produces notable observer errors. A stator and rotor resistance estimation is added:

$$\hat{R}_s = -K_{Rs}\frac{1}{s}(\hat{I}_{s\alpha}(I_{s\beta} - \hat{I}_{s\beta}) + \hat{I}_{s\beta}(I_{s\alpha} - \hat{I}_{s\alpha})) \tag{5.5}$$

$$\hat{R}_r = K_R\hat{R}_s$$

$\hat{I}_s$ is the estimated current vector.

A standard Luenberger current estimator in the stator frame is used for the scope:

$$\frac{d}{dt}\begin{vmatrix} \hat{\bar{\Psi}}_s \\ \hat{\bar{\Psi}}_r \end{vmatrix} = \begin{vmatrix} A_{11} & A_{12} \\ A_{21} & A_{22} \end{vmatrix} \cdot \begin{vmatrix} \hat{\bar{\Psi}}_s \\ \hat{\bar{\Psi}}_r \end{vmatrix} + \begin{vmatrix} 1 \\ 0 \end{vmatrix} \cdot \bar{V} + \begin{vmatrix} K_1 \\ K_2 \end{vmatrix}(\bar{I}_s - \hat{\bar{I}}_s) \tag{5.6}$$

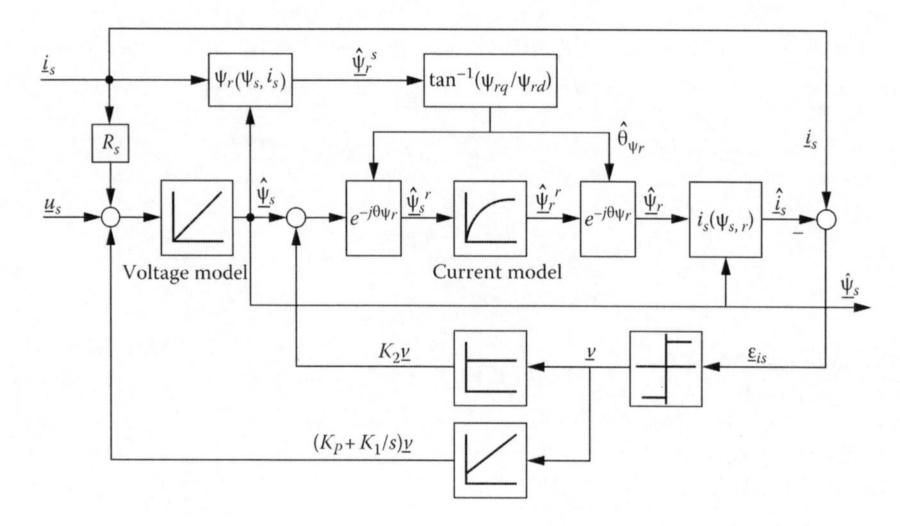

FIGURE 5.4 Inherently speed sensorless combined sliding mode (SM) linear observer with offset compensation.

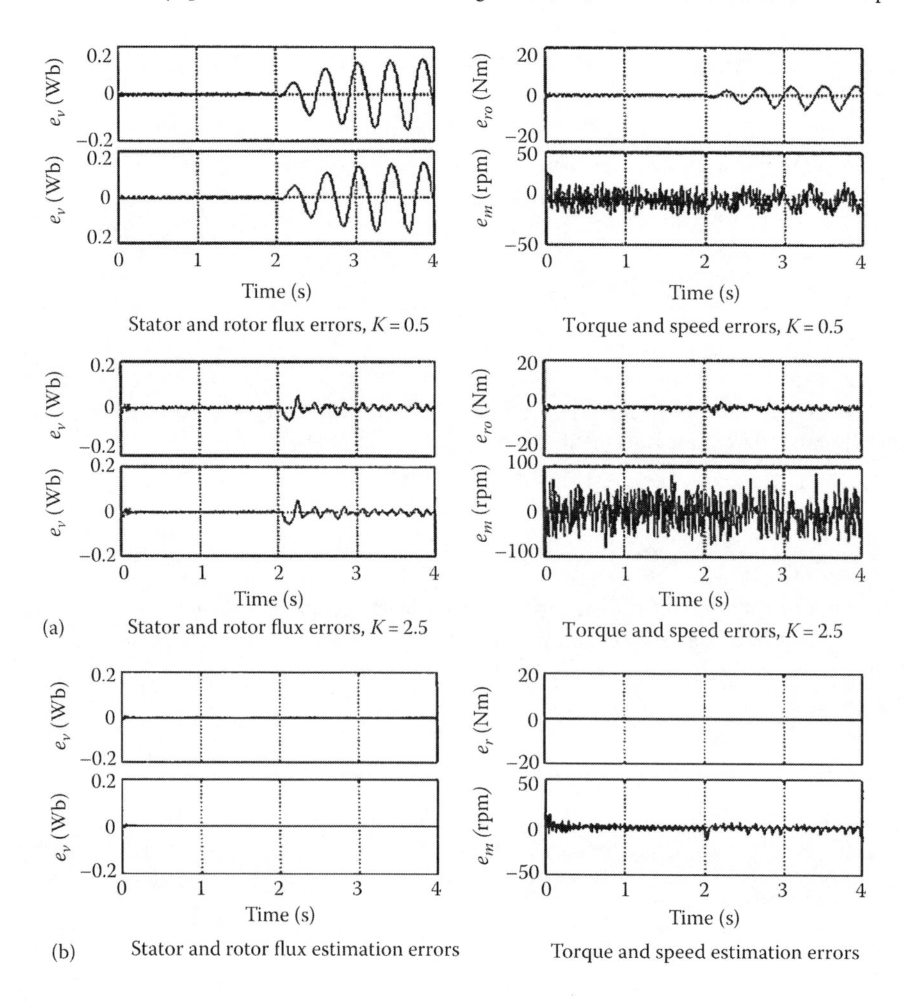

FIGURE 5.5 Voltage offset compensation of 0.3 V in flux, torque, and speed errors: (a) two gains $K = 0.5$, 2.5 and (b) modified SM (5.4) $-K_p = 20$, $K_i = 80$.

$$\hat{I}_s = C_1\hat{\bar{\Psi}}_s + C_2\hat{\bar{\Psi}}_r \tag{5.7}$$

$$A_{11} = -\left(\frac{1}{T_s\sigma} + \frac{(1-\sigma)}{T_r\sigma}\right) A_{12} = \frac{L_m}{L_s L_r \sigma}\left(\frac{1}{T_r} - j\omega_r\right) A_{21} = \frac{L_m}{T_r} A_{22} = -\left(\frac{1}{T_r} - j\omega_r\right) \tag{5.8}$$

$$|C| = \left|\frac{1}{L_s\sigma}, -\frac{L_m}{L_s L_r \sigma}\right| = |C_1, C_2| \tag{5.9}$$

Zero-speed torque-mode operation is shown in Figure 5.6a through f for a 1.1 kW IM with cage rotor [8].

Acceleration to 1500 rpm with the same observer is presented in Figure 5.7a through f. The same observer configuration performs well from 0 to 1500 rpm and more.

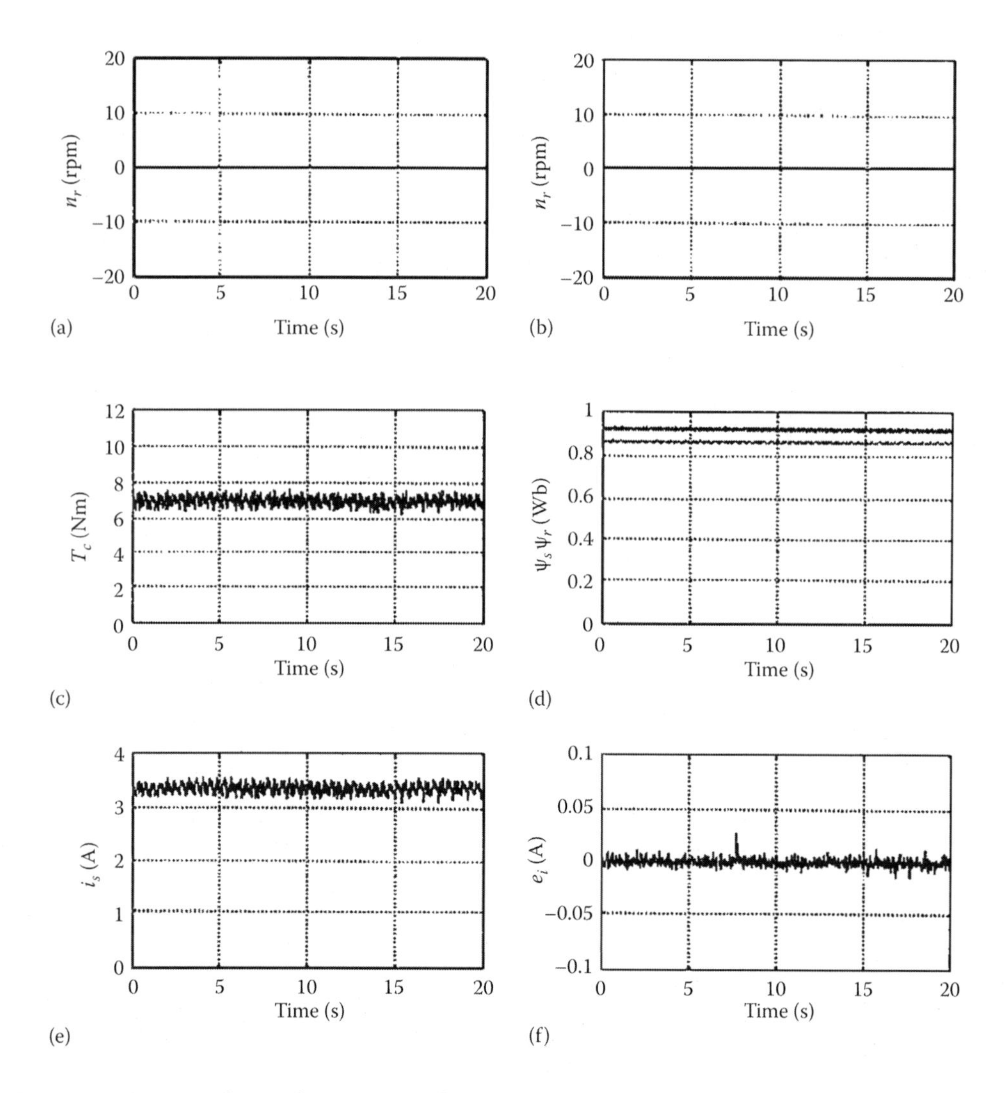

FIGURE 5.6 Zero-speed sensorless torque mode with sliding mode flux observers: (a) estimated rotor speed, (b) measured rotor speed, (c) estimated torque, (d) estimated stator and rotor flux, (e) estimated stator current magnitude, and (f) stator current estimation error. (Adapted from C. Lascu, Direct Torque Control of Sensorless Induction Machine Drives, Ph.D. thesis, University of Politehnica, Timisoara, Romania, 2002.)

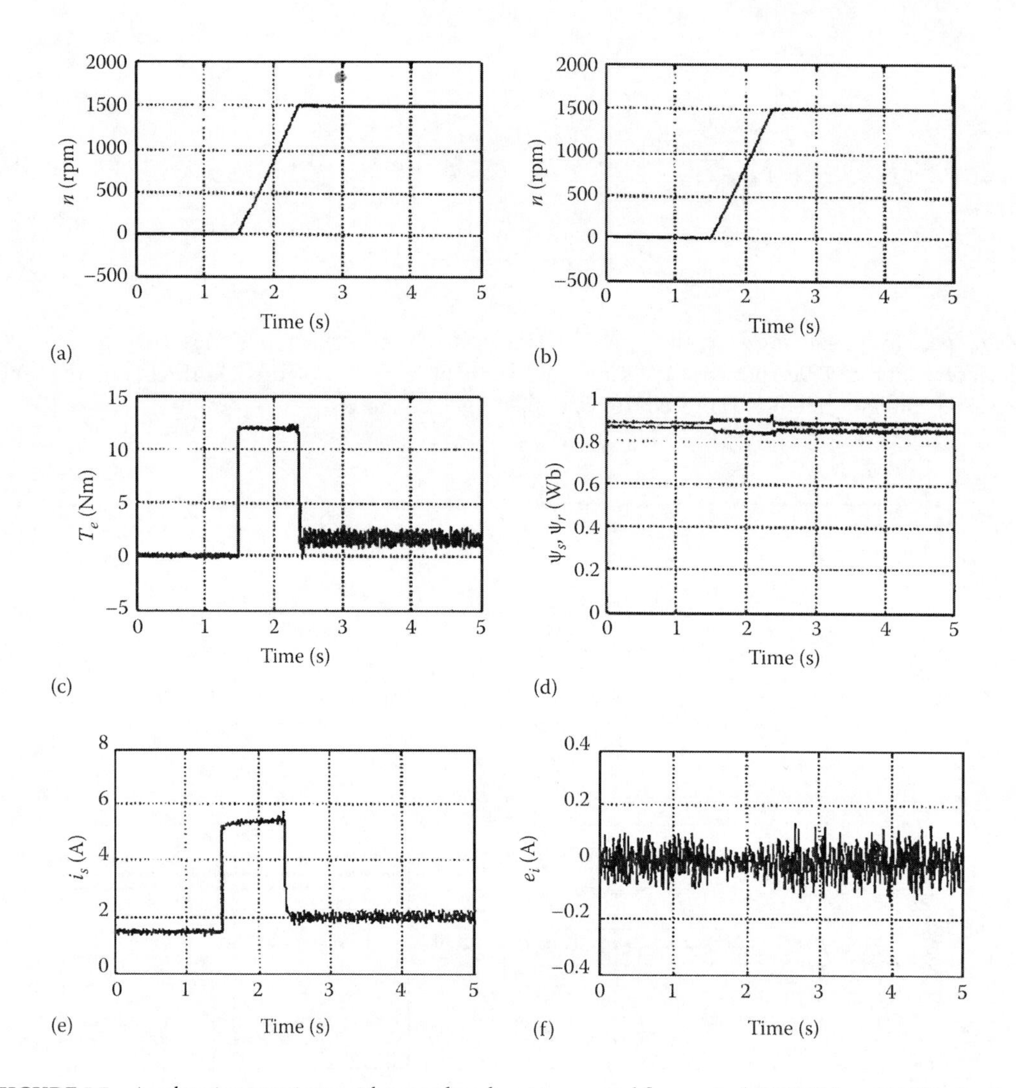

FIGURE 5.7 Acceleration transients with sensorless direct torque and flux control (DTFC) for a 1.1 kW induction machine (IM): (a) estimated rotor speed, (b) measured rotor speed, (c) estimated torque, (d) estimated stator and rotor flux, (e) estimated stator current magnitude, and (f) stator current estimation error.

The speed and torque observers used here are standard, though many others were investigated [8]. As the stator and rotor resistances are corrected online,

$$\hat{\omega}_r = \hat{\omega}_{\Psi r} - \hat{\omega}_{slip} \tag{5.10}$$

$$\hat{\omega}_{\Psi r} = \frac{\hat{\Psi}_{r\beta}(k)\cdot\hat{\Psi}_{r\alpha}(k-1) - \hat{\Psi}_{r\alpha}(k)\cdot\hat{\Psi}_{r\beta}(k-1)}{T_{sa}\left(\hat{\Psi}^2_{r\alpha}(k) + \hat{\Psi}^2_{r\beta}(k)\right)} \tag{5.11}$$

$$\hat{\omega}_{slip} = \hat{T}_e \frac{2\hat{R}_r}{3p_1\hat{\Psi}_r^2} \tag{5.12}$$

with

$$T_e = \frac{3}{2} p_1 \frac{L_m}{L_r} (\hat{\Psi}_{r\alpha}(k)\hat{i}_{s\beta} - \hat{\Psi}_{r\beta}\hat{i}_{s\alpha}(k)) \tag{5.13}$$

The rotor flux vector instantaneous speed $\hat{\omega}_{\Psi_r}$ calculator (Equation 5.11) includes the sampling time T_{sa}.

5.2.1.2 DTFC–SVM Block

As explained in Chapter 2 for the WRIG, DTFC was developed as a fast-response, deadbeat torque and flux control to stand as an alternative implementation to vector control.

The principle of analogy with the DC motor, embedded on vector control, is replaced in DTFC by stator flux direct acceleration or deceleration and increasing or decreasing amplitude. An adequate combination of voltage vectors in the PWM voltage source converter is triggered based on torque error, flux error signs, and stator flux vector position in one of the six, 60° wide sectors of an electrical period (Figure 5.8; [9]).

For torque control alone, speed estimation is not even required, though for speed control it is required. Also, the speed estimator is needed when the prime mover speed control is operated by an SCIG application.

The original DTFC is characterized by the following:

- Fast dynamic response in torque and flux
- Robustness
- No need for current controllers
- Simplicity
- Inherent sensorlessness (no rotor position is required for control)
- High torque and current ripple
- Voltage source inverter (VSI) switching frequency that is variable
- High acoustical noise at low speeds
- Steady-state torque error

To take advantage of DTFC qualities but circumventing its difficulties, the space-vector modulation (SVM) [9] and various implementations of torque and flux regulators were successfully introduced [10]. Variable structure of torque and flux controllers (VSCs) plus SVM solutions are presented here [9–11].

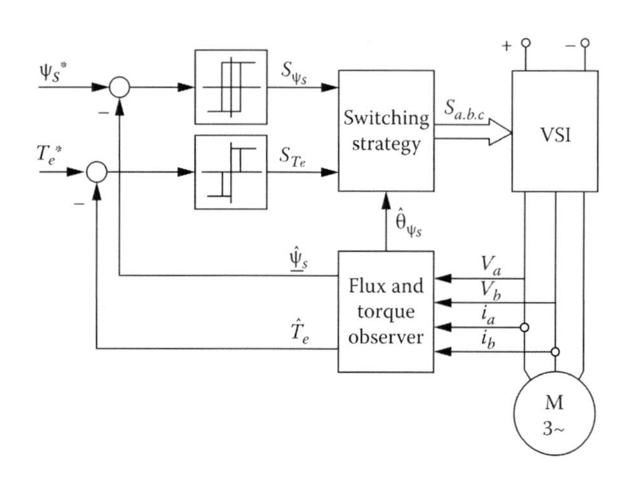

FIGURE 5.8 Original direct torque and flux control (DTFC) of an induction machine (IM).

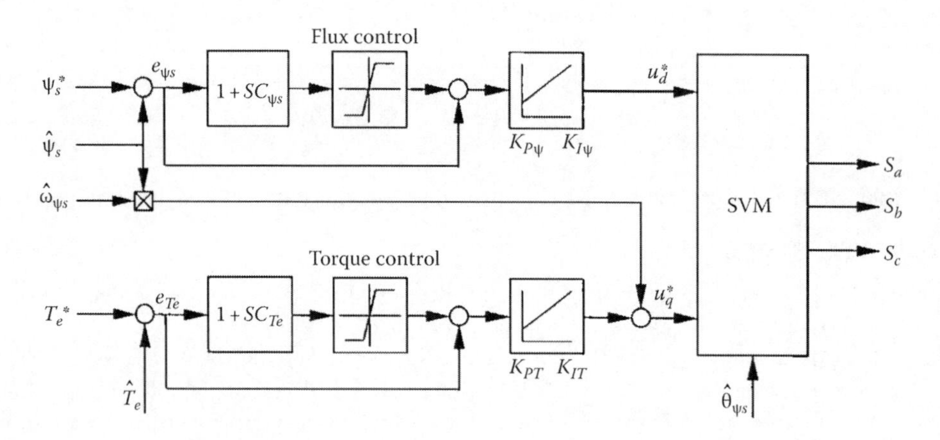

FIGURE 5.9 Linear and SM feedback direct torque and flux control (DTFC).

Basically, the dq voltage components $V_d^* \, V_q^*$ are calculated from flux and torque errors $\varepsilon_{\Psi_s} \varepsilon_{Te}$ by combining proportional integral (PI) with sliding mode control. The sliding mode functional vector is S_s:

$$S_s = S_{\Psi_s} + jS_{Te} = \varepsilon_{\Psi_s} + j\varepsilon_{Te} \tag{5.14}$$

$$V_{sd}^* = \left(K_{P\Psi} + \frac{1}{s} K_{I\Psi} \right) (\varepsilon_{\Psi_s} + K_{Vsc\Psi} \text{sgn}(S_{\Psi_s})) \tag{5.15}$$

$$V_{sd}^* = \left(K_{PTe} + \frac{1}{s} K_{ITe} \right) (\varepsilon_{Te} + K_{VscTe} \text{sgn}(S_{Te})) + \hat{\omega}_{\Psi_s} \hat{\Psi}_s \tag{5.16}$$

Equations 5.15 and 5.16 reveal the combination of PI and nonlinear (sliding mode) regulators and also the motion electromagnetic field (emf) compensation ($\hat{\omega}_{\Psi_s} \hat{\Psi}_s$). Again, the stator flux speed $\hat{\omega}_{\Psi_s}$ (estimated earlier in this paragraph), and not the rotor speed, is required. The block diagram of the stator flux and torque linear and discontinuous (SM) controllers is shown in Figure 5.9.

The reference stator voltages $V_{ds}^* \, V_{qs}^*$ in stator flux coordinates are transformed by the Park transformation:

$$\begin{vmatrix} V_a^* \\ V_b^* \\ V_c^* \end{vmatrix} = \begin{vmatrix} \cos(\theta_{\Psi_s}) & -\sin(\theta_{\Psi_s}) \\ \cos\left(\theta_{\Psi_s} - \frac{2\pi}{3}\right) & -\sin\left(\theta_{\Psi_s} - \frac{2\pi}{3}\right) \\ \cos\left(\theta_{\Psi_s} + \frac{2\pi}{3}\right) & -\sin\left(\theta_{\Psi_s} + \frac{2\pi}{3}\right) \end{vmatrix} \cdot \begin{vmatrix} V_{ds}^* \\ V_{qs}^* \end{vmatrix} \tag{5.17}$$

From now on, an open-loop PWM technique may be used to "construct" the stator voltage waveforms [9,11]. The operation of such an IM sensorless 1.1 kW drive with DTFC–SVM during ±6 rpm speed reversal under full load is indicative of good performance (Figure 5.10a through h).

While DTFC–SVM sensorless control with linear and SM controllers and observers was illustrated for an IM drive, the same may be used directly in the SCIG control system shown in Figure 5.3.

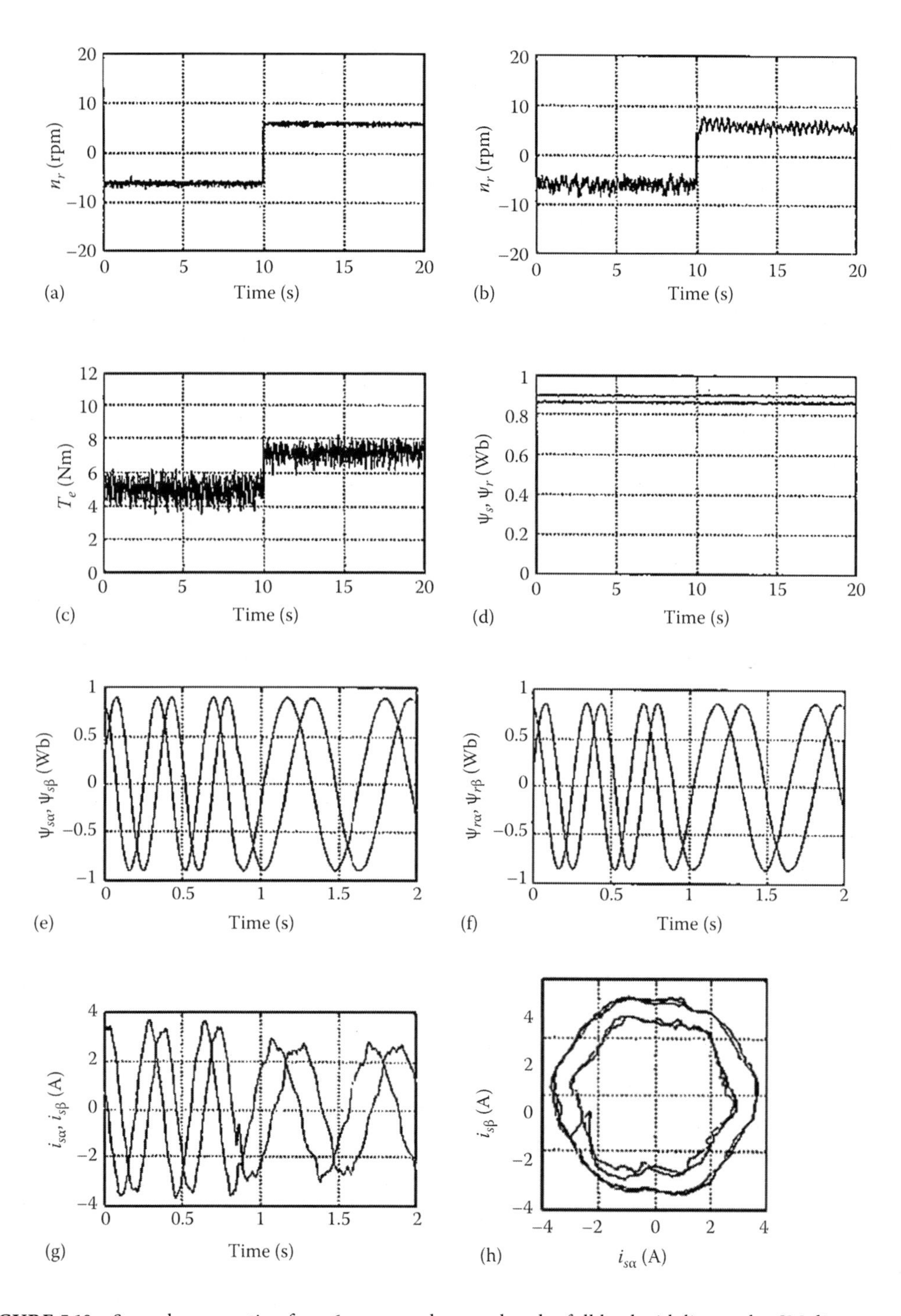

FIGURE 5.10 Sensorless operation for ±6 rpm speed reversal under full load with linear plus SM direct torque and flux control (DTFC)-space vector modulation (SVM) control: (a) estimated rotor speed, (b) measured rotor speed, (c) estimated torque, (d) estimated stator and rotor flux, (e) estimated (α,β) stator flux, (f) estimated (α,β) rotor flux, (g) estimated (α,β) stator current, and (h) estimated stator current trajectory.

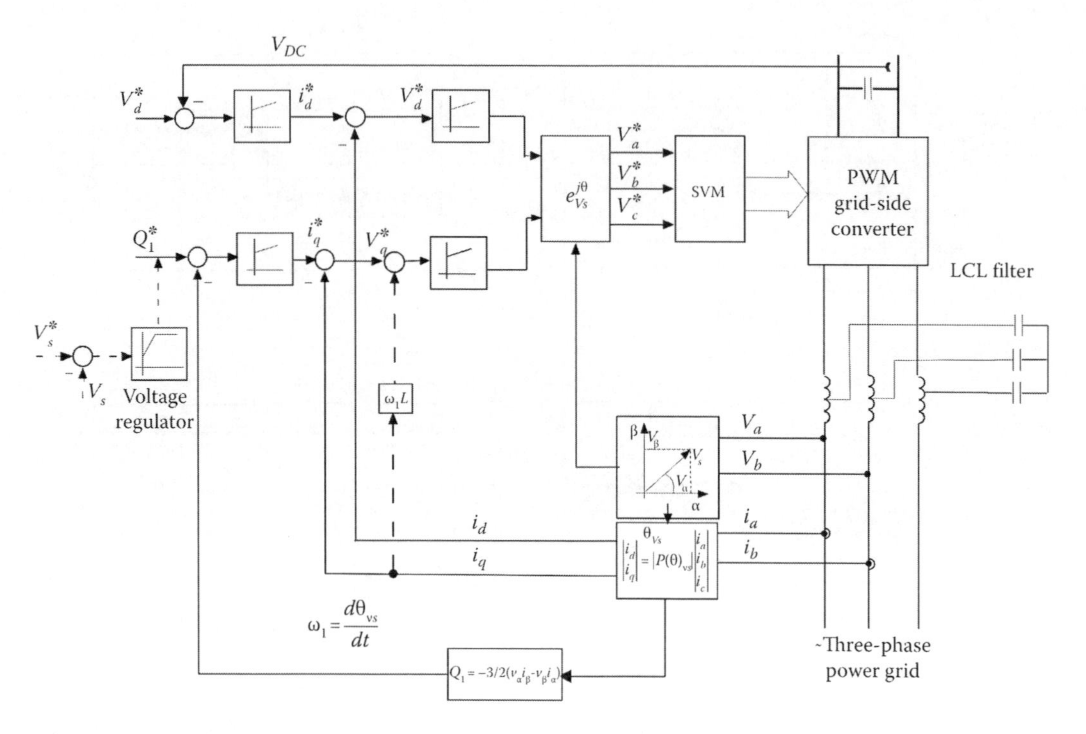

FIGURE 5.11 The grid-side pulse-width modulator (PWM) converter control.

5.2.2 Grid-Side Converter Control

Grid-side converter control is, in general, standard vector control, where DC link voltage control provides for active power from (to) DC link voltage to (from the power grid, while reactive power control provides for reactive power exchange with the power grid) (Figure 5.11). The reactive power exchange with the power grid is, in fact, provided by the oversized DC link capacitor, which also "covers" the IG magnetization.

The active power exchange is controlled through the machine-side converter from (to) the IG. Adequate voltage and capacitance oversizing of the DC link may provide up to ±100% reactive power exchange Q_1^*, which is very useful in the local power grid voltage control and stabilization. Q_1^* may be commanded by the grid voltage error with respect to a desired value. The DC link reference voltage V_{DC}^* is generally kept constant under normal operational circumstances, but it may be reduced in relation to the reactive power requirements.

When an nductance–capacitor–inductance (LCL) filter is introduced between the grid-side converter and the grid, speed decoupling of filter inductance L along the q axis current control is added, to construct V_q^* (Figure 5.11). The measured frequency of power grid voltage ω_1 is required for decoupling, to speed up the response in the presence of the power filter.

5.3 Grid Connection and Four-Quadrant Operation of SCIGs

Consider the case of a microhydro turbine, also capable of pumping operation, that needs variable speed during generating periods to eliminate the speed governor and provide good overall turbine efficiency for variable water heads.

The standard synchronous generator solutions require speed governing in the microhydro turbine for constant speed, to provide constant frequency. Also, the acceleration and the synchronization take time, as they are done by the turbine and are not protected from severe transients.

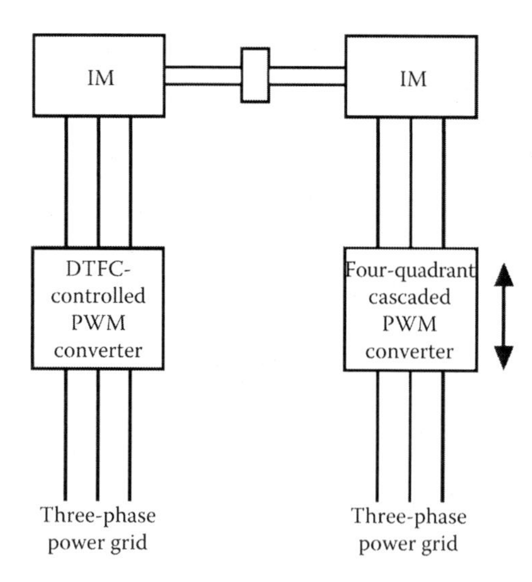

FIGURE 5.12 Testing stator-converter-controlled induction generator (SCIG) to the power grid.

The SCIG, on the other hand, may start with the IG in motoring by fixing a positive torque reference to the machine-side converter to complement the unregulated torque contribution of the turbine, after the water gate is opened. The acceleration is fast, and the "synchronization" sequence is eliminated. All that is needed is to set a negative reference torque (or power P_1^*) to control the system and a positive (or negative) reactive power reference Q_1^* to the grid-side converter. If pumping is required, the positive torque (power) reference is maintained and tailored to speed to best exploit the pump induction motor system up to 20%–50% above base (rated) speed $n_1 = (f_{1b}/p_1)$. For better pumping efficiency, the turbine pump needs more speed than that needed for good turbining.

Experiments were performed on a laboratory system using two 10 kW cage rotor IMs, one playing the role of the turbine and the other the role of the SCIG (Figure 5.12). The 25 kVA four-quadrant cascaded PWM AC–AC converter was an off-the-shelf device intended for variable speed drives with fast regenerative braking of large inertia loads.

The turbine was emulated by a variable speed drive in speed control mode. Starting can be performed either by the "turbine" up to a preset speed or simultaneously by the turbine and the SCIG in the motoring mode.

Steady-state operations at the power grid generating between 0% and 50% reactive power delivery are illustrated in (Figure 5.13a and b).

The power grid current evolution when, for –100% reference torque (generator) at the IG side converter, control input is maintained, and the speed is ramped down by "turbine" control from 1500 to 800 rpm as shown in Figure 5.14.

Rather smooth generating to motoring transients were obtained. Grid current versus voltage waveforms during motoring acceleration (for pumping) at zero reactive power exchange with the power grid are shown in Figure 5.15. Full four-quadrant operation was thus performed.

It goes without saying that "synchronization" has become an irrelevant concept, as it can be done at variable speed. Also, the disconnection from the power grid can be done smoothly via grid-side and machine-side converters. The two converters provide flexibility and opportunities for various actions, should power grid faults occur.

The full rating of a four-quadrant AC–AC cascaded PWM converter turns out to be a performance asset, as it controls the whole power exchanged with the grid: active and reactive. All of this comes at

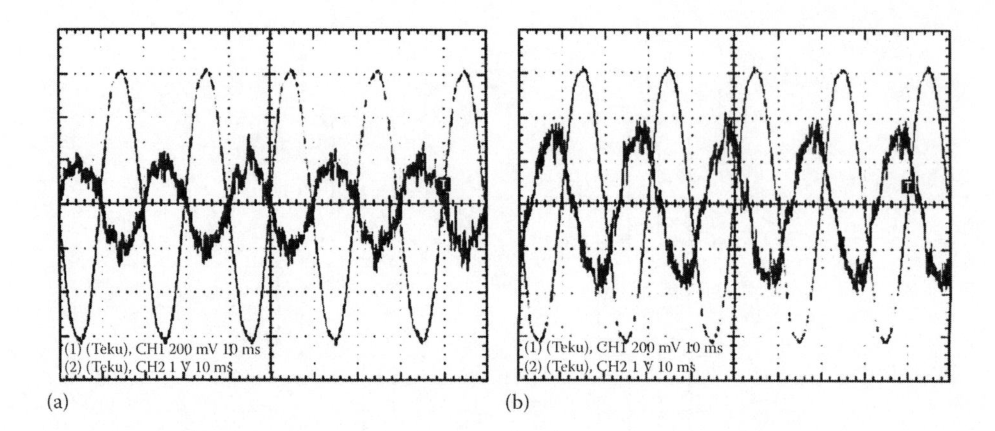

(a) (b)

FIGURE 5.13 Steady-state generating at the power grid (1500 rpm), voltage and current at (a) zero reactive power and at (b) 50% reactive power delivery and 100% torque.

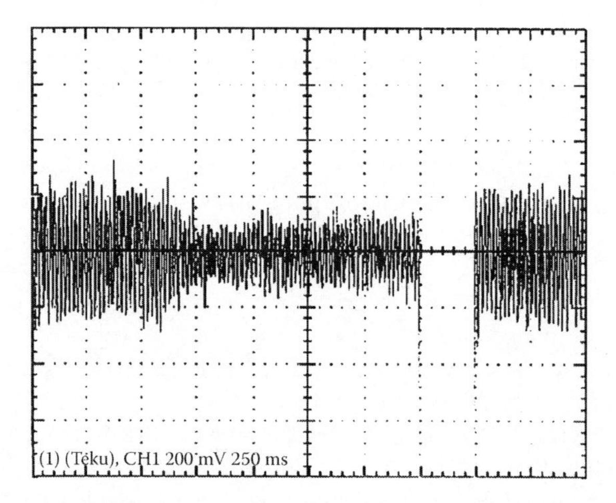

FIGURE 5.14 Grid phase current for generating at 100% torque when speed is ramped down by the turbine control from 1500 to 800 rpm.

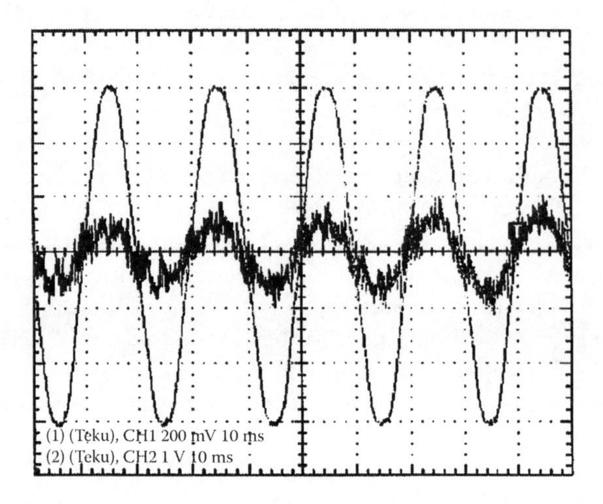

FIGURE 5.15 Grid voltage and current during acceleration for motoring.

higher cost than in WRIGs, where the rating of the four-quadrant cascaded AC–AC PWM converter is 25%–30% of the rated power. The latter, however, has tight control only on ±25% of the power. It should be noted that the commercial four-quadrant PWM IGBT converter used in our experiments and built for drives requires additional LCL filtering between the grid-side converter section and the power grid to improve the current waveforms in order to fully comply with the contemporary strict power quality standards.

Load rejection of SCIG at the power grid with controlled turbine tends to lead to overspeeding, unless a ballast (alternative) load is provided in the DC voltage link. But, this represents a transition from power grid to stand-alone operation, which will be discussed next.

5.4 Stand-Alone Operation of SCIG

In stand-alone operations, the SCIG is considered the only source meant to produce constant voltage and frequency output.

The IG-side converter control may remain the same as that of the grid connection (DTFC, in our earlier presentation). The power (torque) versus speed reference function generator is again based on optimal utilization of the turbine generator and the available primary energy (water or wind energy). The control of the output voltage and frequency and the power balance between the generator and the load powers are performed through the grid-side PWM converter, which needs alteration in comparison with the grid-connected operation mode. Current limitation means, through the control, are required along with voltage feed-forward and filter inductance L decoupling.

A DC voltage limiter V_{dcmax} triggers a DC–DC converter that feeds a resistive load connected to the DC link, to balance the generator to load power. Finally, a battery in the DC link is needed to provide initial DC link voltage to initiate the self-excitation of the IG.

All things considered, a generic control system for stand-alone operation would look like that shown in Figure 5.16. The controlled system in Figure 5.16 is basically an indirect voltage vector with current limiters [12]. Vector current indirect control is also possible. The vector control is performed in load voltage orientation and, at least at the initialization, this voltage is zero, therefore the angle θ_v of the load voltage vector has to first be imposed as follows:

$$\theta_v = \int \omega_1^* dt \tag{5.18}$$

where ω_1^* is the reference load frequency. Moreover, the reference DC voltages V_{df}^* and V_{df}^* have to be carefully correlated, depending on requested active and reactive power loads and on the LCL filter characteristics (to some extent).

As the active power is transmitted through the IG, its control should be placed there, while the reactive power control requirement should govern the V_{qf}^* value. When the load is too large, the V_{DC} controller decreases the reference voltage V_{df}^* by fast action.

Instead of the actual load voltage vector angle in the Park transformation, for open-loop PWM of V_{df}^* and V_{qf}^*, it seems reasonable to use the value from Equation 5.18, as frequency ω_1^* has to be imposed. Full load, applied on 50% load, transient operation of such an 11 kW laboratory system is illustrated in Figure 5.17a and b [12], for constant speed, in terms of DC voltage V_{DC} and V_d (in fact, load voltage amplitude) responses.

The DC voltage recovers quickly at the reference value after a notable drop. And, V_d (load voltage) recovers quickly, but after a peak, possibly due to the LCL filter intervention.

Note that the grid-connected and stand-alone control systems may be integrated and allow for both operation modes and smooth transitions from one to the other [12].

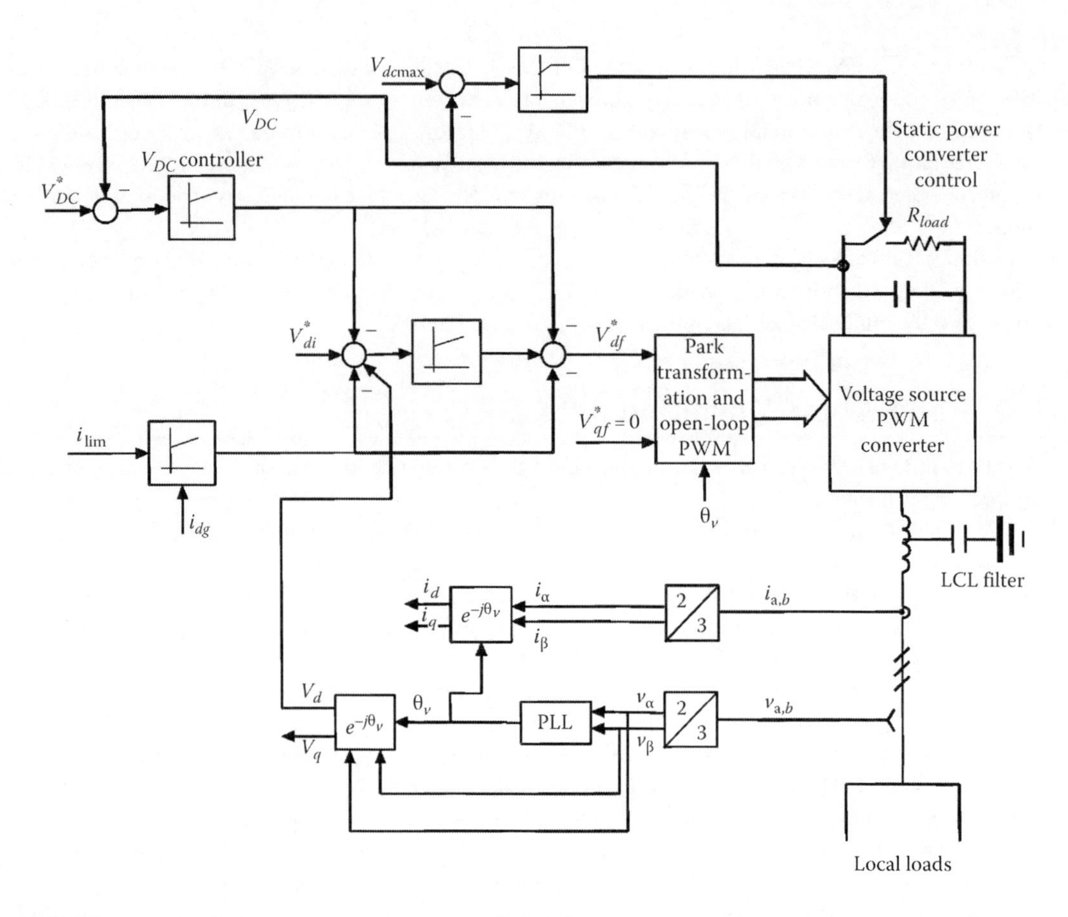

FIGURE 5.16 Stand-alone stator-converter-controlled induction generator (SCIG): load-side PWM converter control system.

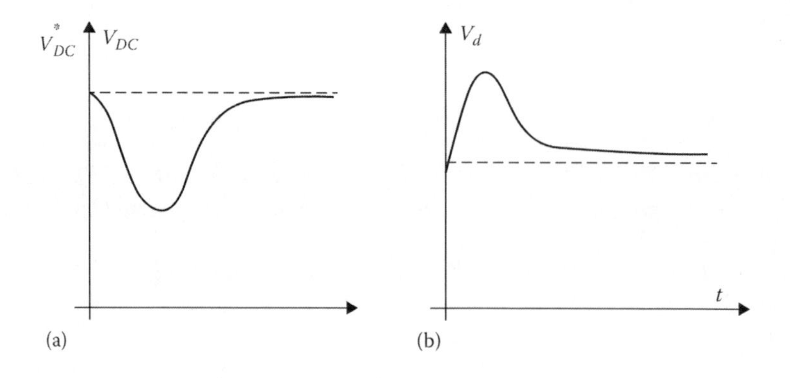

FIGURE 5.17 Full-load application over 50% load: (a) V_{DC} vs. time and (b) V_d vs. time.

5.5 Parallel Operation of SCIGs

Stand-alone SCIGs are to be connected in parallel, and they have to share equitably the load power (active and reactive). Sharing power "equitably" between PWM converters may be accomplished as done for synchronous generators: through voltage and frequency droop curves (Figure 5.18).

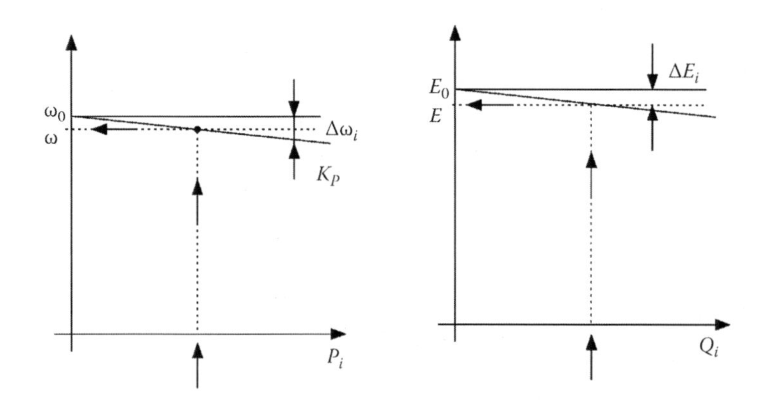

FIGURE 5.18 Frequency and voltage droop curves.

It is well understood that the frequency and voltage in the load are the same for all SCIGs. The active and the reactive power sharing of each SCIG depends on the frequency and on the voltage droops $\Delta\omega_i$, ΔE_i of each component of the group.

If the SCIGs are of different ratings, they have to load in the same proportion to the rated load, in order to ensure maximum output, without overloading too much any of the SCIG components of the group. The load active and reactive power requirements P_iQ_i have to be measured (estimated) from measured voltages V_a, V_b and currents i_ai_b. They then should be divided with corresponding weightings among the SCIGs of the group:

$$P_L = \sum C_{Pi}(P_L)P_{irated} = \sum P_i \tag{5.19}$$

$$Q_L = \sum C_{Qi}(Q_L)Q_{irated} = \sum Q_i \tag{5.20}$$

The weight coefficients C_{Pi} and C_{Qi} depend on P_L and Q_L and on the availability, costs, and risks of various SCIGS, as some may be driven by wind turbines, and some by gas turbines, microhydro turbines, or diesel engines.

The frequency and voltage droops of each SCIG increase with active and reactive power. Therefore, V_{DC}^* and ω_1 should be lowered slightly with power P_iQ_i to allow for load sharing but not compromise the power quality. The transient operation of two single-phase converters in parallel for step power commands is shown in Figure 5.19 [13].

The responses in frequency (Figure 5.19a) reveal clearly the fact that the first converter experiences a reduction in active power (P_1), while the second shows an increase in active power (P_1) (Figure 5.19b). Though the changes in frequencies of the two converters are rather small, they (Figure 5.19a) are decisive for active power sharing. The same rationale applies for changes in voltages.

Going back to the control of each SCIG means that we only need to change the V_d^* and ω_1^* (in θ_v^*), accordingly, to active and reactive power requests from that particular SCIG.

Note that handling unbalanced power grid voltages and unbalanced loads in stand-alone operation may be done through separate control of positive and negative sequences. This way, the voltage dips in the power grid are markedly reduced [14].

5.6 Static Capacitor Exciter Stand-Alone IG for Pumping Systems

Considering the energy storage capacity/water pumping in a reservoir for later use seems to be one of the best ways of using the wind energy, which has a supply that is time dependent (by hour, day, and season). As variable speed is useful, to tap most of the wind energy from cut-in to cut-off wind speeds,

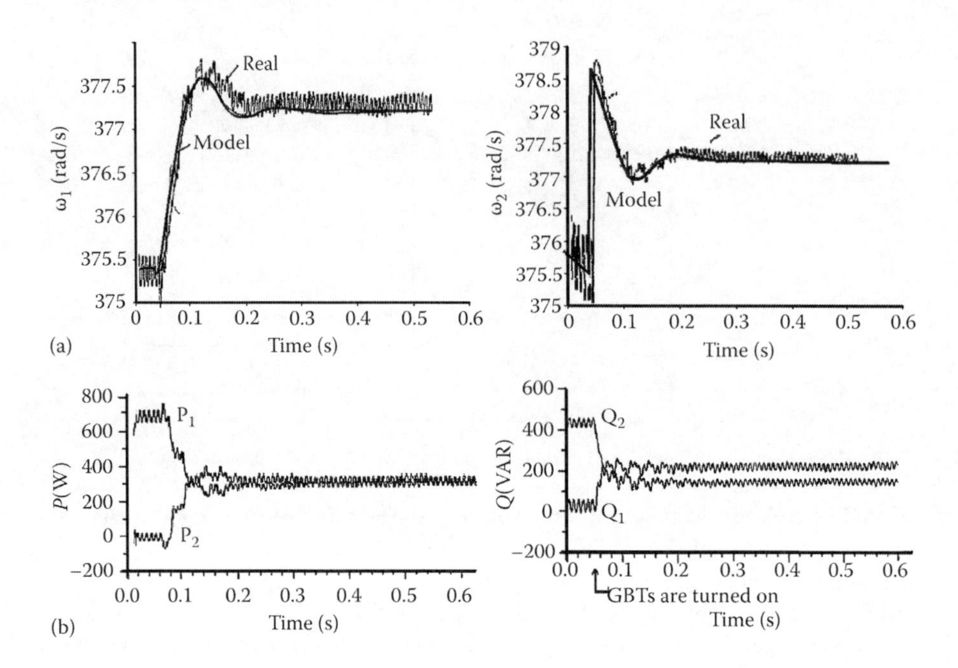

FIGURE 5.19 Two pulse-width modulator (PWM) converters in parallel: (a) frequency and transients and (b) active and reactive power transients (P_1, P_2, Q_1, Q_2).

the frequency of the voltage generated by the IG varies markedly, but the ratio *V/f* does not vary much. For induction-motor-driven pumps, such a situation is adequate. Consequently, the frequency of the IG does not need to be controlled at induction motor terminals. To provide controlled voltage at various speeds, the single value capacitor needed to self-excite the IGs, must be varied. A static capacitor exciter (SCE; Figure 5.20) does just that.

The static exciter handles only the reactive power requirements of both the IG and IM through adequate control [15,16].

For self-excitation, the magnetic flux in the IG has to remain above a certain level, and the output voltage V_G proportional to IG speed. ω_G must be producing a good approximation to that. If speed feedback is not available, a V_G proportional to frequency ω_1 will do the job.

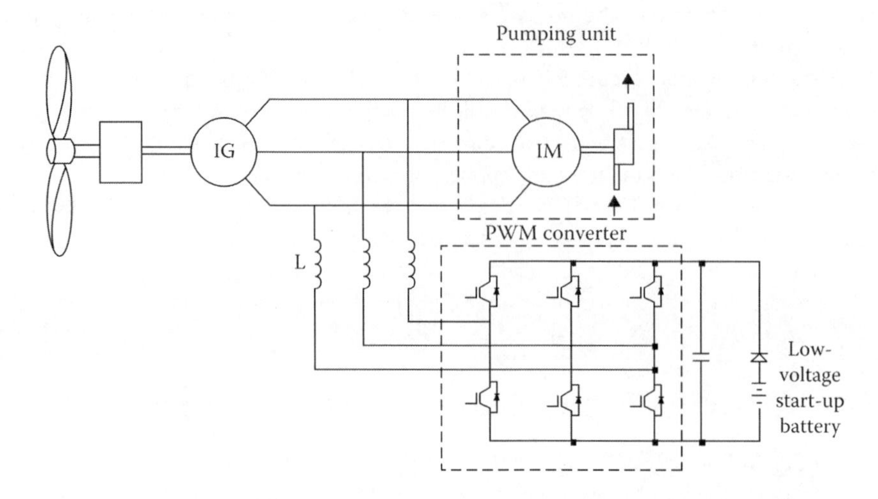

FIGURE 5.20 Static capacitor exciter (SCE) stand-alone induction generator (IG) for water pumping.

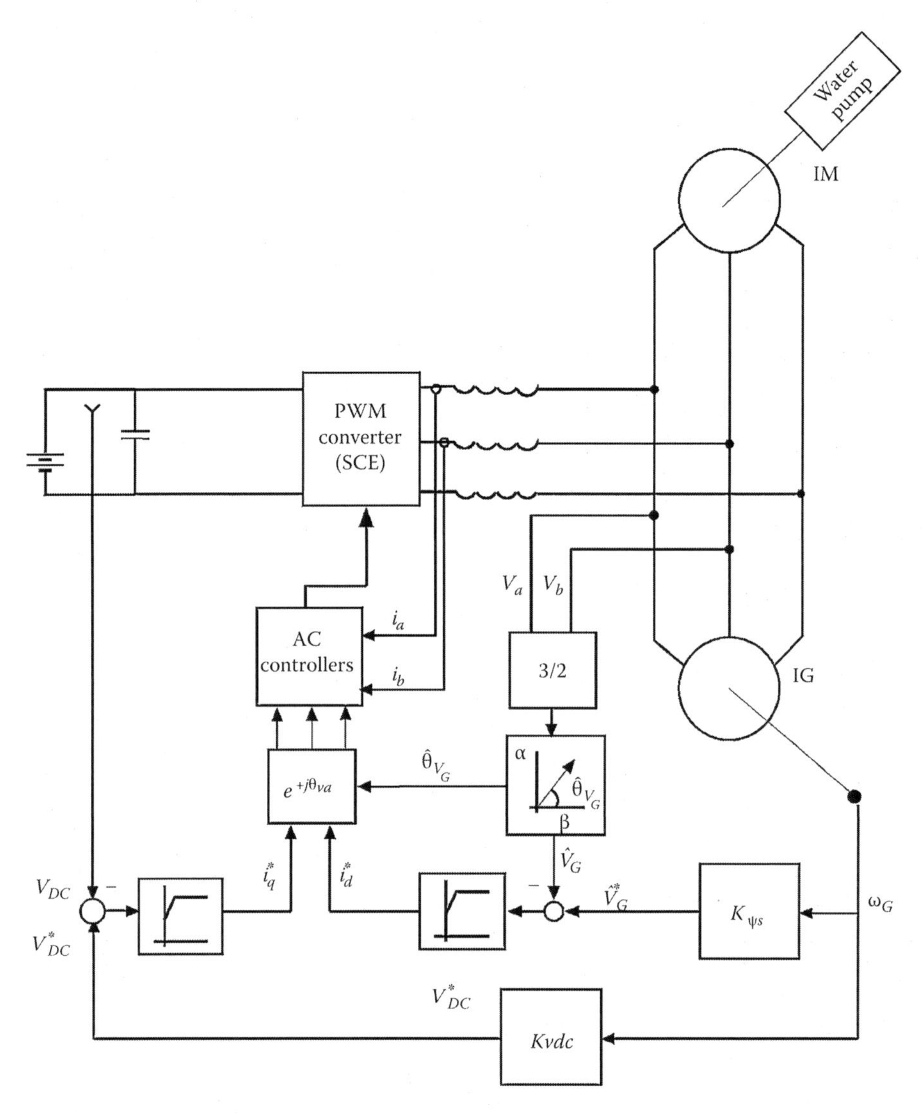

FIGURE 5.21 Vector control of static capacitor exciter (SCE)induction generator (IG) for induction motor (IM)) water pumping system.

A vector control system that keeps both the capacitor voltage and the generator voltage proportional to speed (or frequency) is shown in Figure 5.21.

The stator flux in the IG is close-loop controlled (rather than imposed only) through the generator voltage amplitude regulator along the i_d channel.

The DC capacitor voltage is also PI controlled to correspond to generator voltage proportional to speed.

As the magnetization current in the machine is maintained rather constant through *V*/*f* control, the equivalent value of the capacitor is inversely proportional to speed. The IG stator flux is, thus, constant over a wide range of speeds. The induction motor flux should also be constant with generator speed (*V*/*f* control).

Typical behavior of such a system [16] is depicted in Figure 5.22, where the speed is ramped from 1200 to 1800 rpm and back. The IG and IM flux and capacitor voltage V_{DC} are recorded.

As expected, the response is smooth and stable. The centrifugal characteristics are such that the pumping flow rate Q varies proportionally with the wind speed, until the maximum power ceiling of the stall-regulated wind turbine is reached.

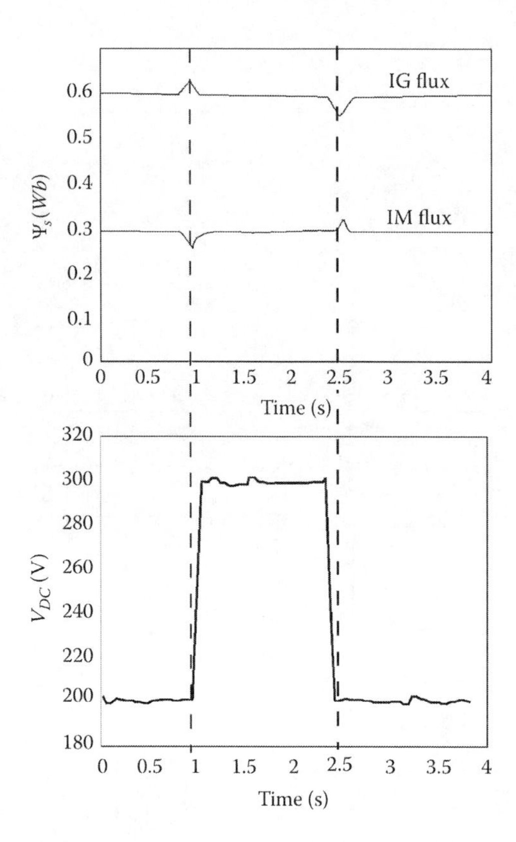

FIGURE 5.22 Stator flux Ψ_s and capacitor voltage V_{DC} transients during speed ramping from 1200 to 1800 rpm and then back to 1200 rpm.

5.7 Operation of SCIGs with DC Voltage-Controlled Output

Variable speed operation, with some stator converter control of cage rotor IGs, may be obtained at reasonable cost if DC voltage controlled output is considered. DC loads with battery backup are a typical application for stand-alone operations.

When the SCIG is part of a group of generators to be paralleled, on a common DC voltage bus, with a single inverter to connect the group to a local power grid or to a cluster of AC loads, the same situation occurs.

Offshore wind farms are a typical example, as is a group of microhydro generators. In principle, the SCIG is self-excited through a static exciter (SCE; Figures 5.20 and 5.21). The IG is designed to produce a voltage that even at maximum speed will need a bit of boosting through the diode rectification and filtration, in order to produce fast-controlled constant DC voltage output (Figure 5.23).

If the IG is built with a rather high-rated power factor (above 0.8), the SCE is designed for partial power (reactive power) ratings (below 0.6 per unit [P.U.]) and, thus, for lower cost.

In essence, as described in a previous paragraph (Figure 5.21), the generator voltage increases proportionally to speed. Consequently, the capacitor rating is minimum. The diode rectifier with capacitance filter provides for the active power input to boost the DC–DC converter with filter. A DC voltage robust regulator regulates the output DC voltage.

The boost converter makes use of an IGBT power switch, for sufficient switching frequency, to reduce the size of the output filter $(f_1)C_{f2}$. The boost converter may also be implemented with three AC power switches (with six thyristors of partial rating) connected to the diode rectifier median point (Figure 5.24) [17].

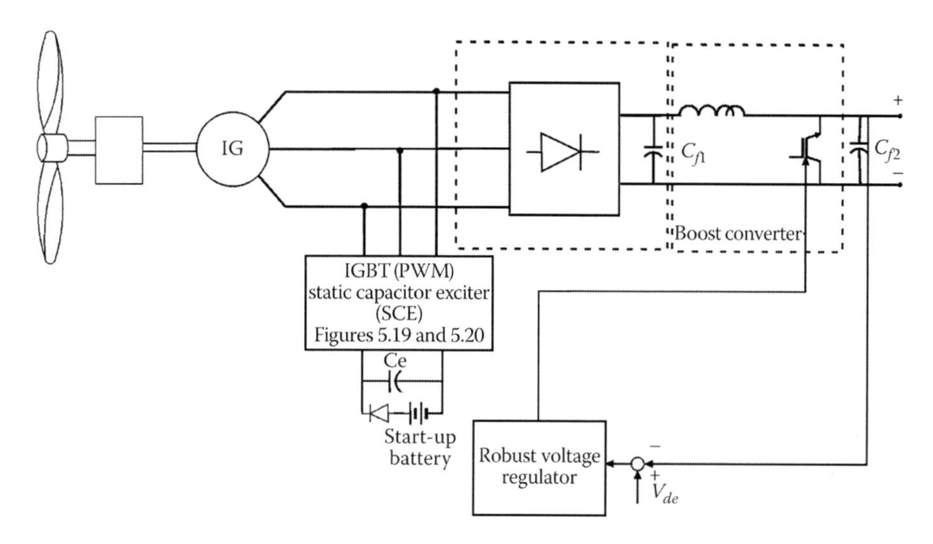

FIGURE 5.23 Static capacitor excited (SCE) induction generator (IG) with controlled direct current (DC) voltage output.

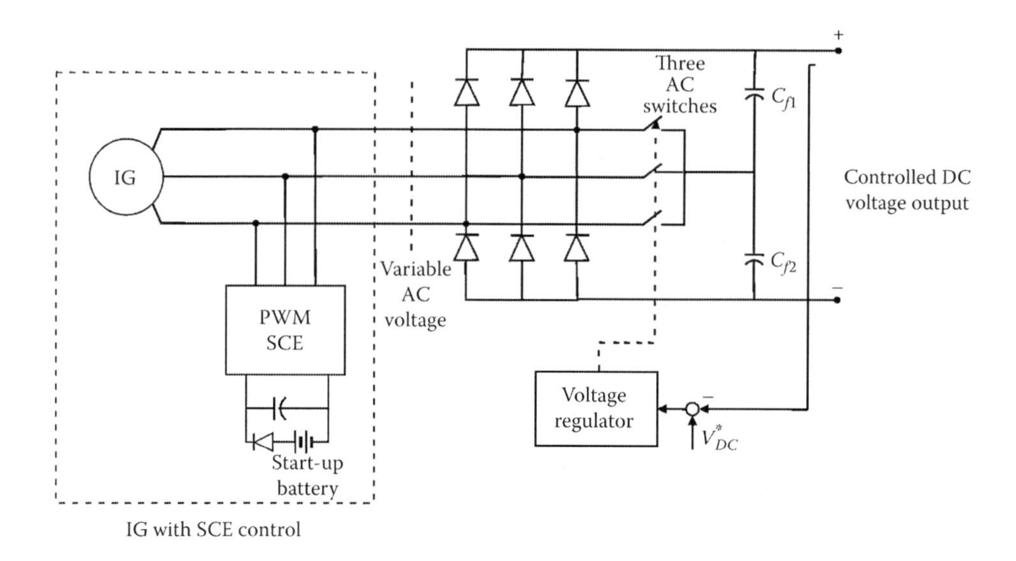

FIGURE 5.24 Static capacitor excited (SCE) induction generator (IG) with median point diode rectifier voltage booster and direct current (DC) controlled output.

The voltage boost is provided by the capacitance C_{f1}, C_{f2} as energy storage elements. The three AC (bidirectional) thyristor switches of partial rating plus the capacitance C_{f1}, C_{f2} are expected to cost less than the single IGBT and the inductor energy storage element.

The diode rectifier generally provides on the input side a unity power factor for the fundamental. Consequently, no reactive power is transferred to the DC side for the scheme in Figure 5.23. The phase delay action of the thyristor switches in the median point voltage booster slightly modifies this situation, but the capacitor energy storage elements C_{f1} and C_{f2} compensate for it. The capacitance required for the same DC voltage output ripple should be larger for the latter case, as the thyristor switching frequency is less.

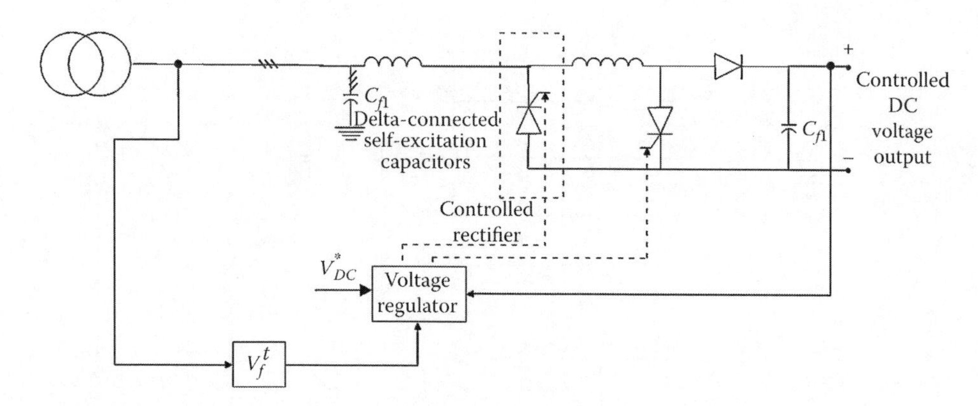

FIGURE 5.25 Self-excited induction generator (SEIG) with controlled rectifier and voltage booster.

It is also possible to use the uncontrolled capacitor self-excited IG, a fully controlled rectifier, to reduce the DC output voltage at higher speed, and an inductor plus an IGBT switch voltage booster with a capacitance filter (Figure 5.25).

The thyristor rectifier lowers the generator voltage, which tends to increase with speed, while the booster increases it to maintain the output voltage constant with speed. The concerted action of rectifier and booster control is to produce fast and stable DC voltage control. At light load, the rectifier absorbs almost only reactive power to produce controlled IG terminal voltage. Operation of the controlled rectifier at large delay angle (almost 90%) is not appropriate for DC load voltage control.

The operation is proper when the boosting stage is at work. Therefore, the minimum reference voltage V_{DC}^* (Figure 5.25) has to be higher than the maximum voltage of the rectifier for 1.0 P.U. IG voltage [18].

As expected, in all of the above schemes, load rejection, for constant speed, does not produce significant overvoltage transients. In the SCE schemes, the failure of the SCE or voltage booster control at maximum speed does not produce severe overvoltage to the diode rectifier, as the IG de-excites quickly in the process. This is not so for the controlled rectifier scheme, where the whole capacitor remains connected at the generator terminals.

Connecting a few such SCE IGs in parallel on the DC voltage bus poses not only the problem of voltage stabilization at a constant value but also the problem of power sharing. Changing slightly the reference voltages V_{DC}^* of various SCE IGs, based on voltage to power droop curves, is appropriate for the scope of our discussion here (Figure 5.26 for two SCE IGs).

The total load power is divided between the SCE IGs in parallel, based on the energy availability and minimum risk or other criteria, accordingly, to voltage versus power droop curves calibrated at commissioning.

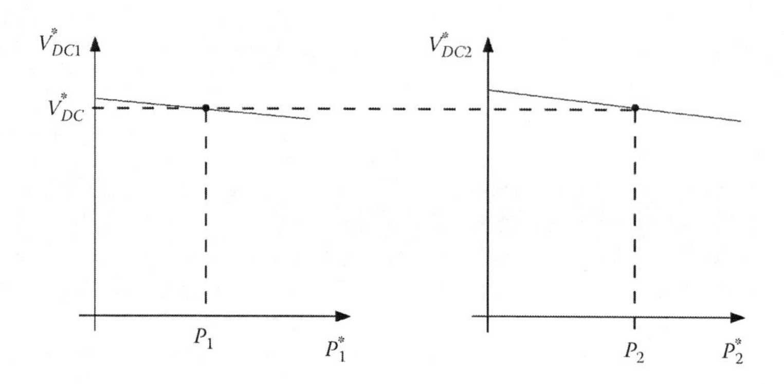

FIGURE 5.26 Direct current (DC) voltage vs. power droop curves.

The power of each SCE IG system, on the DC side, has to be measured, and, eventually, closed-loop voltage droop setting is performed.

Note that a secondary stator winding may be used and supplied through an SCE to improve the characteristics of a self-excited (with fixed capacitors) primary AC output IG [18].

Though some reduction in frequency variation was obtained, the flexibility of the solution is still limited in terms of speed for frequency-sensitive loads. The solution is similar in effect to the SCE IG, but it costs more, as the IG is provided with dual windings.

5.8 Stand-Alone SCIG with AC Output and Low Rating PWM Converter

Thus far, full power rating back-to-back converters have been discussed in relation to SCIG grid operation while for stand-alone operation the lower rating converter with DC capacitor link covered all reactive power of the induction generator for DC output voltage loads. A myriad of topologies with transformers is feasible to handle nonlinear and unbalanced loads [19].

Here, we briefly introduce an even lower rating stator connected PWM converter with DC link capacitor and uncontrolled capacitor bank (c) at IG terminals, while also only AC loads are served, Figure 5.27 [20].

As visible in Figure 5.27, d_q reference currents i_{di}^*, i_{qi}^* are the results of the DC link, V_{DC} voltage controller and IG terminal voltage amplitude V_s controller. An additional machine i_d (magnetization) current regulator is added to avoid excessive magnetic saturation. Finally, i_{ai}, i_{bi}, and i_{ci} (inverter) AC current controllers are used. A low-voltage battery with a diode is used as an auxiliary source to help starting IG after demagnetization. For a speed range of 1.0–1.2 P.U., the inverter rating is only 0.21 P.U. There is no constant frequency assurance for the AC loads, however, in this low-cost scheme.

5.9 Dual Stator Winding for Grid Applications

A dual stator winding cage rotor IG [21,22] may have a main three-phase winding with p_1 pole pairs, designed at 100% rated power, and an auxiliary three-phase winding with p_1' pole pairs ($p_1/p_1' = 2/3, 3/4$) designed at around 25% power rating. Alternatively, such two separate machines may be placed on the same shaft (Figure 5.27a and b).

The 25% rating cascaded (bidirectional power convertor has four-quadrant control capabilities for 25% of rated power of the main winding [generator]) and may be at work when the following conditions are met:

- Load is below 25%, and the main winding (IG) is turned off.
- The main winding (IG) is at work above 100% power, and additional power is available at the prime-mover shaft and should be delivered to the power grid.

The larger number of poles ($p_1 > p'$) makes the auxiliary winding (IG), fed at variable speed, suitable for lower speeds. Also, with $p_1'/p_1 = 2/3, 3/4$ not very different from unity, less than 150% of power grid frequency is needed in the auxiliary winding fed through the AC–AC converter to add power at speeds between synchronous and peak torque speed for main winding.

As motoring and generating are feasible, through the PWM AC–AC converter, the latter may also be used to damp the rotor oscillations due to natural variations (wind gusts, tower shadow blade pulsations, etc.) in wind speed.

As the number of poles of the two windings $2p_1$ and $2p_{1'}$ are different from each other, there is no main flux coupling between them other than through the leakage flux, which may be neglected, to a first approximation.

The two windings are fed with voltages of two different frequencies ω_1 and ω_1', and their rotor currents may have different frequencies: $\omega_2 \neq \omega_2'$. Therefore, the two windings interact with the rotor cage

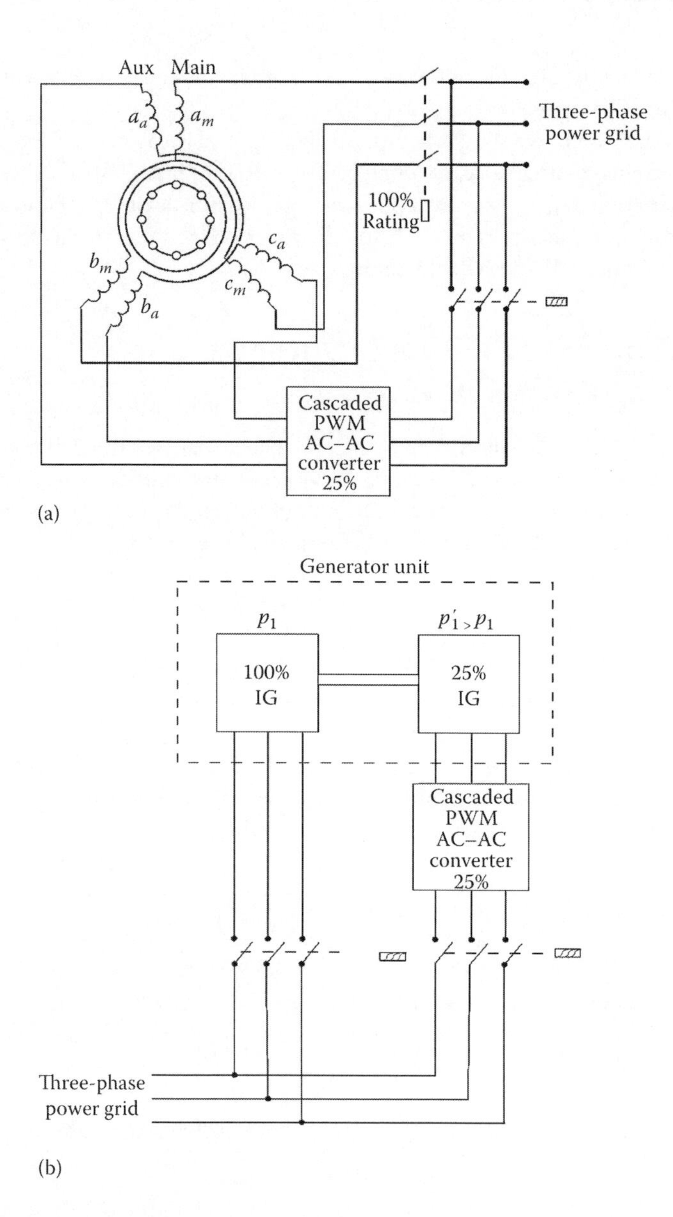

FIGURE 5.27 Dual stator-converter-controlled induction generator (SCIG) for power grid applications: (a) with dual stator winding and (b) with dual induction generator (IG).

independently, and thus, they may be modeled separately, as in the case of two IGs (Figure 5.26) that have the pertinent rotor parameters.

Typical steady-state characteristics, shown in Figure 5.27, may look like those of the pole-changing SEIG in Chapter 4, but the auxiliary winding IG is now capable of four-quadrant smooth operation at a roughly −50% speed range, with notable reactive power exchange capability. The 25% rating was taken as a good compromise between additional costs and performance gains (Figure 5.27). The transients in the power grid may be reduced through the PWM AC–AC converter control.

Note that the dual stator winding concept was proposed for widely different pole counts in variable speed, IM drives with two PWM converter controls in order to provide for more controllable torque at very low speeds [22].

Static converter full power control may also be used for single-phase AC power grids in remote populated areas. As only the load-side PWM converter is changed to a single-phase one, this case will not be treated further here (see Reference 18 for more details).

Cage IG design is similar to motor design for variable speed with special attention paid to losses and overspeeding. For details, see Reference 1, Chapters 14 through 18.

5.10 Twin Stator Winding SCIG with 50% Rating Inverter and Diode Rectifier

In yet another configuration, the two stator windings of an SCIG are designed with same number of poles and eventually different number of turns/phase, but occupying each half of the stator slot area.

One winding is connected to the DC voltage link by an inverter while the other, provided with capacitors at terminals, is connected through a diode rectifier to the common voltage link (Figure 5.28) [23].

Up to a certain speed, only winding 2 is operational in motoring or generating. Above a certain speed, the output voltage of winding 2 is sufficient to open the diode rectifier and to produce additional power in the DC voltage link or AC power in a frequency-insensitive AC load. As the speed increases, so does the frequency in the stator windings and the delivered power increases with frequency and speed (Figure 5.29) [23].

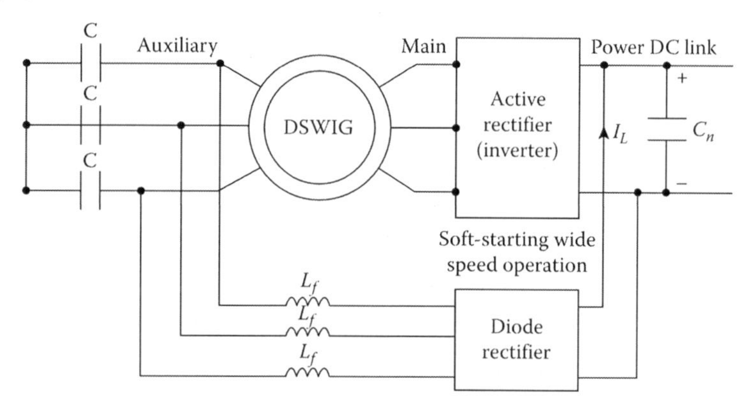

FIGURE 5.28 Torque vs. speed curves of an induction generator (IG) with auxiliary stator winding four-quadrant power converter control.

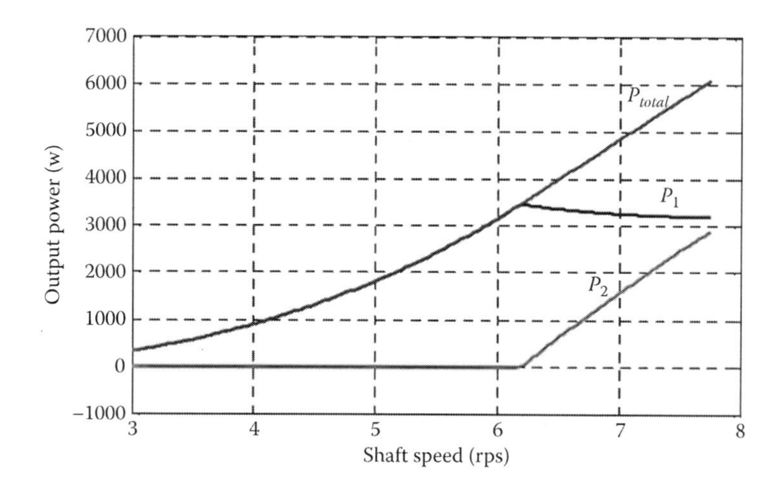

FIGURE 5.29 Output power versus frequency speed at constant DC voltage output

The configuration is not only prone to reduce initial cost but provide for feasible output power control in generating while assisting the starting of the prime mover (ICE, for example) in ship-like generators.

5.11 Dual Stator Winding IG with Nested Cage Rotor

In an effort to build brushless, doubly fed induction generators, decades ago a dual stator winding (with $p_m \neq p_c$ pole pairs) configuration was considered for a rotor with:

- $2p_r = p_1 + p_2$ nested cage poles
- $2p_r = p_1 + p_2$ salient poles

A dual PWM converter is connected to the control winding (p_c pole pairs), while the main winding (p_m pole pairs) is connected directly to the grid (or load in stand-alone mode) (Figure 5.30).

The coupling of the two windings is realized by the nested cage rotor via a harmonic of air gap permeance. As the first air gap permeance harmonic based on the different number of poles in the three windings is at most of half of the constant component, it follows that both windings, besides the useful inductance, will experience a strong leakage inductance due to the large constant component of air gap permeance.

There are two consequences of this phenomenon:

1. Reduction of fundamental winding factor of the two windings to around 50% [24], which means inevitably low torque density
2. A reduction of power factor (and efficiency by consequence) due to the additional air gap-leakage inductance

In contrast to same stator but with high saliency rotor configuration, there are notable rotor nested-cage losses.

Apart from these disadvantages, the machine works similarly to a brushless DFIG with partial rating PWM converter [25].

5.12 Summary

- Cage rotor IGs connected directly to the power grid (Chapter 4) behave rigidly, causing a severe transient both in the power grid and in the IG torque. To soften this behavior and tap most of the prime-mover energy (wind turbines, for example), or reduce fuel consumption (in diesel engines), variable speed generation at constant voltage and frequency power grid is required.

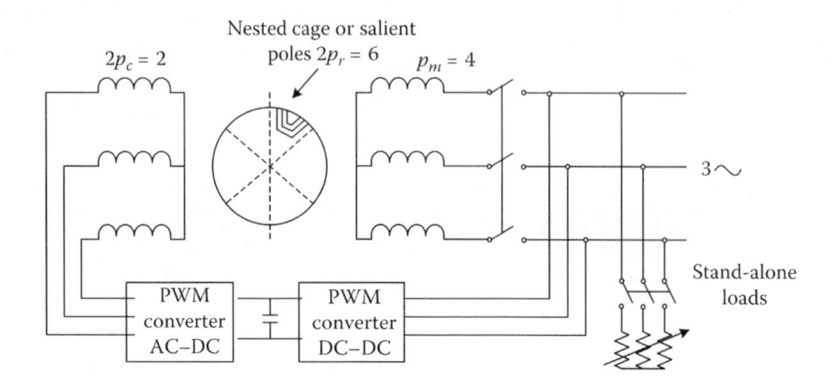

FIGURE 5.30 Dual stator winding IG with nested cage rotor.

- Full power bidirectional (four-quadrant AC–AC) PWM static converters are soft interface between IG and the power grid at variable speed. They are called here SCIGs.
- Four-quadrant PWM static converters may be of cascaded (indirect) type or direct (matrix) type.
- Only the back-to-back voltage source PWM converters are available off the shelf up to 1 megawatt (MW) per unit and more for special orders.
- In the cascaded PWM AC–AC converter, direct torque and flux control strategy may be applied to the machine-side converter, and vector control is adequate for the machine-side converter. The prime-mover optimal use leads, finally, to an almost linear power vs. speed curve, with a limiter. The machine-side converter may be torque-controlled with torque reference calculated from power reference versus speed of wind or hydroturbine or diesel engine.
- DTFC of machine-side converters seems the natural choice. The reference stator flux may be calculated as torque and speed dependent, to reduce IG losses at all speeds.
- An implementation case study, with sliding mode flux observers and torque and flux controllers for sensorless control, is illustrated in this chapter. Space-vector modulation in the machine-side PWM converter is added to reduce current, flux, and torque ripple. The same implementation would work for machine-side PWM converters.
- The grid-side converter is vector controlled with orthogonal axes aligned to grid voltage vector.
- Along axis d (voltage vector position), active power exchange is provided through DC link voltage and d axis current controls that output V_d^*. Along axis q, the grid voltage is regulated, and its regulator output commands the reactive power reference.
- Cascade reactive power and i_q regulators then provide V_q^*.
- After transformation to $V_a^*\ V_b^*\ V_c^*$, the three AC voltages are produced by the inverter through PWM techniques.
- The cascaded AC–AC PWM converter provides for smooth motor starting and then motoring or generating to the power grid. The standard synchronization sequence is fully eliminated. Safe and soft connection and disconnection to the power system are inherently available.
- Up to ±100% reactive power exchange with the power grid is available, which eliminates the external capacitor bank, with its stepwise control so typical and problematic in fixed speed IGs at the power grid.
- Experimental work on a 10 kW unit is solid proof of these claims. The price to pay for this performance is the additional cost of the four-quadrant PWM AC–AC converter.
- Stand-alone operation of SCIGs occurs when during operation at power grid, full load rejection is required or when the system is designed for separate loads.
- In stand-alone operation, the machine-side converter control stays the same as for power grid operation; the load-side converter control changes markedly, from current control type to voltage control type.
- For stand-alone operation, the DC link voltage control imposes on the V_d (load voltage in synchronous coordinates) control, with given frequency. Voltage decoupling of the filter ($\omega L i_q$) might also be added. Good voltage response to sudden load increase was demonstrated.
- Parallel operation of stand-alone SCIGs leads to paralleling of voltage source inverters.
- To provide for desired power sharing between SCIGs of the group ΔV_1 versus P_i, Q_i droops are used as in synchronous generators in power systems [26].
- Static capacitor exciters (SCEs) at IG terminals are a lower cost alternative to control the IG terminal voltage. For variable speed operation, the terminal voltage is kept proportional to speed, as the DC (capacitance) voltage does. For induction motors, this is equivalent to V/f, variable frequency scalar control, ideal for centrifugal water pumps.
- The same SCE solution, but with rectifiers and DC voltage booster, may provide constant DC output voltage, output at variable IG speed. Again, the SCE is vector controlled. IG voltage speed and capacitance voltage are kept proportional to speed (frequency) to maintain large stator flux in the IG, which is needed for self-excitation.

- Load rejection in SCE IG, when all power electronics elements are faulting, does not lead to dangerous over voltages, as the IG de-excites rapidly.
- Dual stator winding IGs or dual IGs, one of 100% power rating and the other of about 25% power rating and slightly larger pole count $2p_1'/2p_1 > 1$, were also investigated.
- The lower power rating winding (machine) is connected to the power grid through a four-quadrant AC–AC PWM converter. The main winding (or IG) is connected directly to the power grid.
- The four-quadrant AC–AC PWM converter may work alone below 25% rated power at lower speed, or it may assist the full power winding (IG) with up to 25% more power, or it may help in damping the prime-mover torque oscillations.
- IG design for variable speed is similar to IM design for variable speed. For more details, see Reference 1, Chapters 14 through 18.

References

1. P.W. Wheeler, J. Rodriguez, J.C. Clare, L. Empringham, and A. Wheinstein, Matrix converters: A technology review, *IEEE Trans.*, IE-49(2), 2002, 276–288.
2. C. Klumpner, P. Nielsen, I. Boldea, and F. Blaaabjerg, A new matrix converter motor (MCM) for industry applications, *IEEE Trans.*, IE-49(2), 2002, 325–335.
3. J. Rodriguez, J.B. Lai, and F.Z. Peng, Multilevel inverters: A survey of topologies, control and applications, *IEEE Trans.*, IE-49(4), 2002, 724–734.
4. I. Boldea and S.A. Nasar, *Electric Drives*, CRC Press, Boca Raton, FL, 1998.
5. D.G. Luenberger, An introduction to observers, *IEEE Trans.*, AC-16(6), 1971, 596–602.
6. E.A. Misawa and J.K. Hedrick, Nonlinear observers—A state of the art survey, *Trans. ASME J. Dyn. Syst., Meas. Control*, 111, September, 1989, 344–352.
7. R.G. Brown and P.Y.C. Huang, *Introduction to Random Signals and Applied Kalman Filtering*, 3rd edn., John Wiley & Sons, New York, 1997.
8. C. Lascu, Direct torque control of sensorless induction machine drives, Ph.D. thesis, University of Politehnica Timisoara, Timisoara, Romania, 2002.
9. C. Lascu, I. Boldea, and F. Blaabjerg, A modified direct torque control for induction motor sensorless drive, *IEEE Trans.*, IA-36(1), 2000, 130–137.
10. I. Boldea, Direct torque and flux control of a.c. drives: A review, Record of EPE–PEMC-2000, Kosice, Slovakia, 2000, pp. 88–97.
11. J. Holtz, Pulse width modulation for electronic power conversion, In *Power Electronics and Variable Speed Drives*, B. Bose, Ed., IEEE Press, Washington, DC, 1997, Chapter 4.
12. R. Teodorescu, F. Blaabjerg, and F. Iov, Control strategy for small standalone wind turbines, Record of PCIM 2003 Power Quality, Nurnberg, Germany, May 20–22, 2003, pp. 201–206.
13. E.A.V. Coelho, P. Cortez, and P.F. Garcia, Small signal stability for parallel connected inverters in standalone a.c. supply systems, *IEEE Trans.*, IA-38(2), 2002, 533–542.
14. G. Saccomando, J. Svensson, and A. Sannino, Improving voltage disturbance rejection for variable speed wind turbines, *IEEE Trans.*, EC-17(3), 2002, 422–428.
15. D.W. Novotny, D.J. Gritter, and G.H. Studmann, Self excitation in inverter driven induction machines, *IEEE Trans.*, PAS-96(4), 1977, 1117–1125.
16. M.S. Miranda, R.O.C. Lyra, and S.R. Silva, An alternative isolated wind electric pumping system using induction machines, *IEEE Trans.*, EC-14(4), 1999, 1611–1616.
17. T.L. Maguire and A.M. Gole, Apparatus for supplying an isolated d.c. load from a variable speed selfexcited induction generator, *IEEE Trans.*, EC-8(3), 1993, 468–475.
18. H. Huang and L. Chang, A new d.c. link boost scheme for IGBT inverters for wind energy extraction, In *Proceedings of the Electrical and Computer Engineering 2000 Canadian Conference*, Vol. 1, Halifax, Nova Scotia, Canada, 2000, pp. 540–544.

19. G.K. Kasal and B. Singh, Voltage and frequency controllers for an asynchronous generator-based isolated wind energy conversion system, *IEEE Trans.*, EC-26(2), 2011, 402–416.
20. G. Garcia and M.I. Valla, Induction generator controller based on instantaneous reactive power theory, *IEEE Trans.*, EC-17(3), 2002, 368–373.
21. O. Ojo and I.E. Davidson, PWM VSI inverter assisted standalone dual stator winding generator, Record of IEEE–IAS 1999, Annual Meeting, 1999.
22. A. Munoz-Garcia and T.A. Lipo, Dual stator winding induction motor drive, Record of IEEE–IAS 1998, Annual Meeting, 1998, Vol. I, pp. 601–608.
23. L.N. Tutelea, I. Boldea, N. Muntean, and S.I. Deaconu, Modeling and performance of novel scheme dual winding cage rotor variable speed induction generator with DC link power delivery, IEEE–ECCE 2014, Pittsburgh, PA.
24. A.M. Knight, R.E. Betz, and D.G. Dorel, Design and analysis of brushless doubly fed reluctance machines, *Trans. IEEE*, IA-49(1), 2013, 50–58.
25. E. Abdi, A. Oraee, S. Abdi, and R.A. Mc Mahon, Design of the brushless DFIG for optimal inverter rating, In *7th IET International Conference on Power Electronics, Machines and Drives (PEMD2014)*, Manchester, England, 2014, pp. 1–6.
26. M.N. Marwali and A. Keyhani, Control of distributed generation systems, parts I + II, *IEEE Trans.*, PE-19(6), 2004, 1541–1561.

6

Automotive Claw-Pole-Rotor Generator Systems

6.1 Introduction

Increasing comfort and safety in cars, trucks, and buses driven by combustion engines requires more electric power to be installed on board [1]. As of now, the claw-pole-rotor generator is the only type of automotive generator used in industry, with total powers per unit up to 5 kW and speeds up to 18,000 rpm.

The solid rotor claw-pole structure with ring-shaped single direct current (DC) excitation coil, though supplied through slip-rings and brushes from the battery on board, has proven simple and reliable, with low cost, low volume, and low excitation power loss.

In general, the claw-pole-rotor generator is a three-phase generator with three or six slots per pole and with 12, 14, 16, or 18 poles, and a diode full power rectifier.

Its main demerit is the rather large losses (low efficiency), around 50% at full power and highest speed. Producing electricity on board with such high losses is no longer acceptable as the electric power requirements per vehicle increase.

Improvements to the claw-pole-rotor (or Lundell) generator design for better efficiency at higher powers per unit are currently under aggressive investigation by both industry and academia, and encouraging results were recently published.

The first commercial mild hybrid electrical vehicle, launched in 2002, uses the Lundell machine as a starter/generator.

This chapter presents both simplified and advanced modeling of the Lundell generator for steady-state and transient performance. Pertinent control and the latest design improvement efforts and results are also included. The chapter ends with the discussion of the starter–generator mode Lundell machine, with its control and design aspects for hybrid electric vehicles.

6.2 Construction and Principle

A cross section of a typical industrial claw-pole-rotor generator is shown in Figure 6.1. It contains the following main parts:

- Uniformly slotted laminated stator iron core
- Three-phase alternating current (AC) winding: typically one layer with $q = 1, 2$ slots per pole per phase, star or delta connection of phases
- Claw-pole rotor made of solid iron parts that surround the ring-shaped DC-fed excitation (single) coil

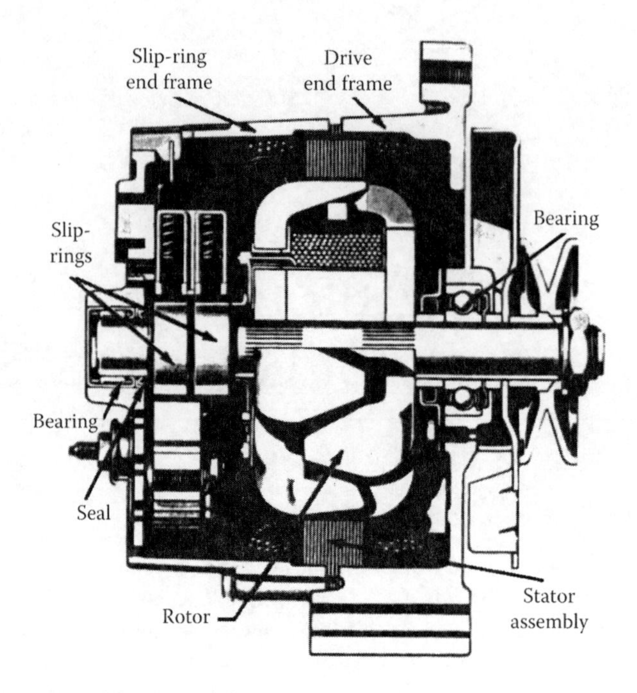

FIGURE 6.1 Claw-pole-rotor (Lundell) generator.

- Copper slip-rings with low voltage drop brushes to transfer power to the DC excitation coil on the rotor
- Bearings and an end frame made of two sides—the slip-ring side and the drive-end side; the generator is driven by the internal combustion engine (ICE) through a belt transmission

The Lundell generator AC output is rectified through a three- or four-legged diode rectifier and connected to the on-board battery (Figure 6.2). Today, 14 V_{DC} batteries are used, but 42 V_{DC} batteries are now adopted as the new standard for automotive application loads [1].

The diodes D_1 through D_6 (D_8) serve the full-power output rectification and are designed for the maximum power of the generator. For large units (for trucks, etc.), three elementary diodes in parallel are mounted on radiator semilegs to comply with the rather high current levels involved (28 V_{DC} batteries are typical for large vehicles).

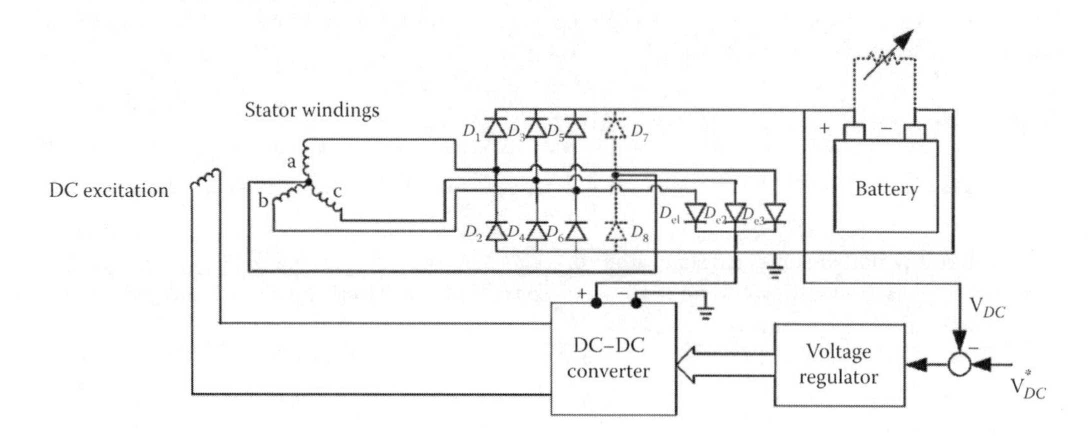

FIGURE 6.2 Typical industrial Lundell generator system.

The excitation coil is supplied from the generator terminals through a half-bridge diode rectifier of low current level (in the 5–20 A interval) and a DC–DC static power converter. A voltage sensor and regulator command the DC–DC converter to keep the voltage of the battery in a certain interval (roughly 12–17 V_{DC}, for 14 V_{DC} batteries) at all times and provide overvoltage and overcurrent protection.

The battery voltage depends on the state of charge (SOC), on its ambient temperature, and on the load level. In designing the generator, the extreme conditions for the battery have to be considered. When the temperature decreases, the battery voltage increases for all states of battery charge. For 100% battery charge, the battery voltage may increase from 13.4 V at 60°C to 16 V at 20°C. For 10% battery charge, the same voltages are 10.75 and 14 V, respectively. Battery age plays an important role in voltage regulation. Comprehensive modeling of the battery is required to exploit it optimally, as the average life of a battery is around 5 years or more for a typical contemporary car.

The generator should provide the same current for the load (or for the battery charge) from the ICE idle speed (700–1000 rpm, in general) onward. The excitation circuit is disconnected when the ICE is shut down, to save in battery life and in fuel consumption.

The stator windings for cars, especially, are the one-layer type, with $q = 1$ slot/pole/phase and with diametrical coils. They are machine-inserted in the slots, and the slot filling factor K_{fill} is modest (around or less than 0.3–0.32).

Only with large power units ($P > 2.5$–3 kW), $q = 2$, when chorded coils are used, to reduce magnetomotive force (mmf) first-space harmonics (fifth and seventh), the distribution and chording factors, for the γth harmonic, $K_{d\gamma}$, $K_{y\gamma}$, are as follows:

$$K_{d\gamma} = \frac{\sin\gamma(\pi/6)}{q\sin\gamma(\pi/6q)}; \quad K_{y\gamma} = \sin\gamma\frac{y}{\tau}\frac{\pi}{2} \tag{6.1}$$

For $q = 1$, chording the coils to $y/\tau = 2/3$, as the only possibility, the mmf fundamental (or power) is as follows:

$$(K_{y1})_{(y/\tau)=(2/3)} = \sin\frac{2}{3}\frac{\pi}{2} = 0.865 \tag{6.2}$$

This is why chording is not generally used for $q = 1$, although the length of coil end turns would be reduced notably, and so would the stator winding losses.

The number of poles is, in general, $2p_1 = 12$, as a compromise between size reduction and increasing iron core loss. Also, $2p_1 = 14, 16, 18$ are used for larger power units (for buses, trucks, etc.).

For completeness, let us include here the typical three-phase, $N_s = 36$ slots, $2p_1 = 12$ poles winding (Figure 6.3a and b). Only one phase is shown in slots. The mmf distributions for sinusoidal currents for $i_{a1} = i_{max} = -2i_{b1} = -2i_{c1}$ and for $i_{a1} = 0$, $i_{b1} = i_{c1} = -i_{max}(\sqrt{3}/2)$ are also shown.

The mmf distribution changes between the extreme shapes as can be seen in Figure 6.3a. With the diode rectifier, the waveform of the stator phase currents changes from quasirectangular (discontinuous) at idle engine speed to quasi-sinusoidal (continuous) waveform at higher speeds (Figure 6.4).

For the eight-diode bridge, the existence of null current leads to a third harmonic in the phase currents.

Therefore, even without considering the magnetic saturation, the mmf space harmonics and phase current time harmonics pose specific problems to the operation of the Lundell machine, which is, otherwise, a salient-pole rotor synchronous machine.

Typical phase voltage and current waveforms at 1500 and 6000 rpm are shown in Figure 6.5a and b, with rectifier and resistive load (no battery).

As the speed increases notably, third harmonics show up in the phase voltage. In addition, the phase shift between the fundamental voltage and current increases from about 1° at 1500 rpm to 9° at 6000 rpm.

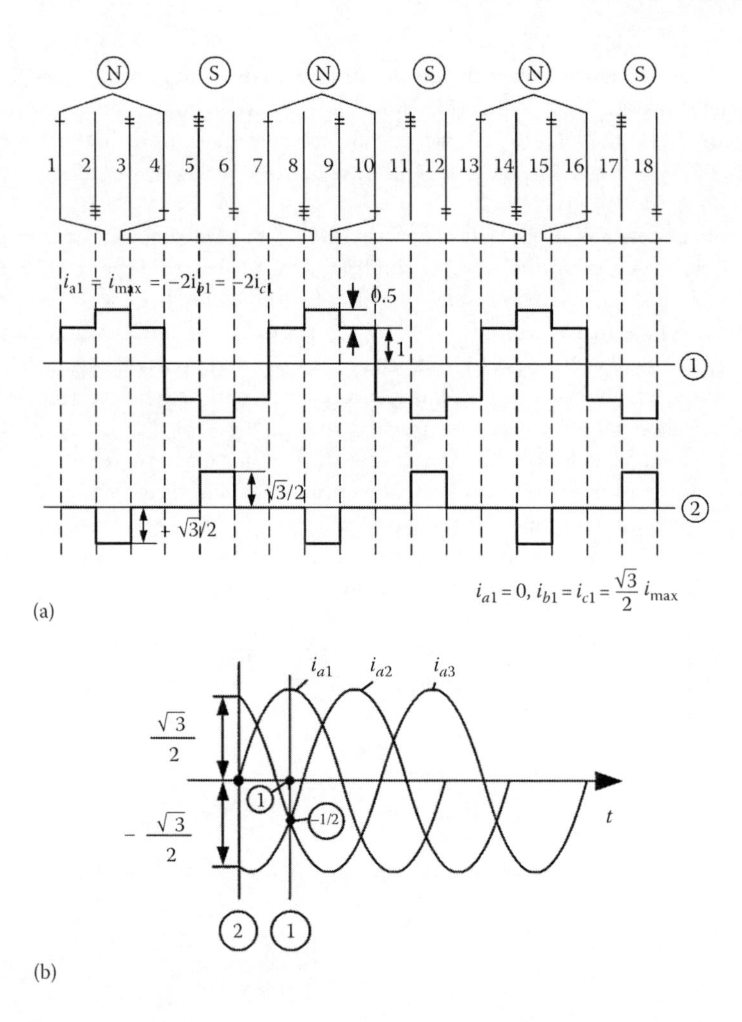

FIGURE 6.3 (a) The 36-slot, 12-pole winding (half of it is shown) and (b) the stator magnetomotive force (mmf) distribution for three and two conducting phases.

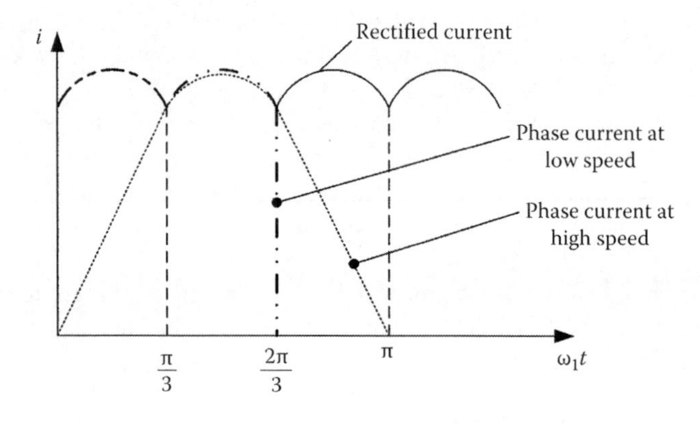

FIGURE 6.4 Ideal shapes of generator currents (star connection).

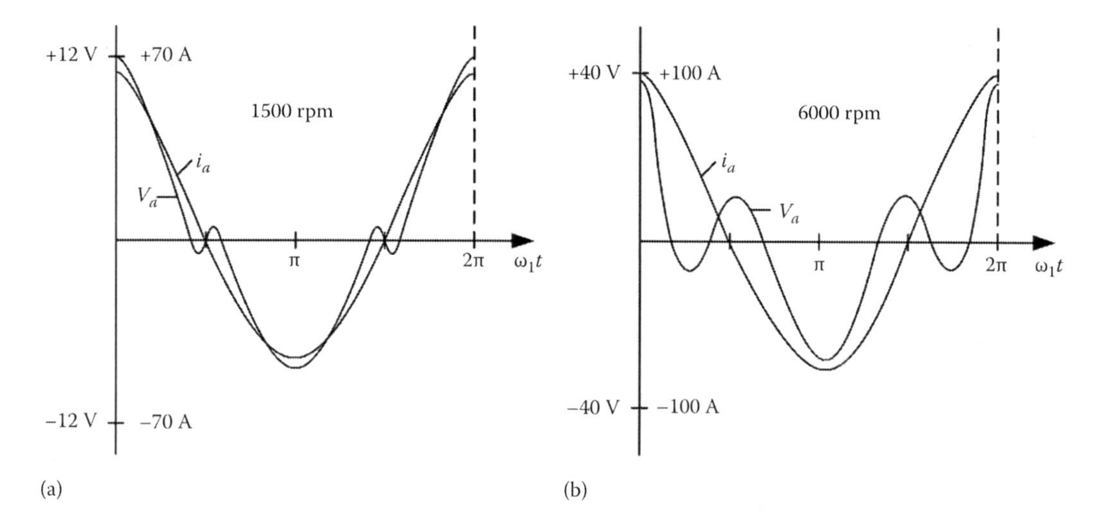

FIGURE 6.5 Phase voltage and current with diode rectifier and resistive load and same field current: (a) 1500 rpm and (b) 6000 rpm.

This latter phase lag is categorically due to the commutation of diodes—three diodes work at a time at high speeds—in corroboration with the machine commutation inductances.

The third harmonic in the phase voltage under load comes from the distribution of the flux density in the air gap. This, in turn, is due to magnetic saturation in corroboration with the $q = 1$ diametrical winding and slot openings.

The trapezoidal shape of the claw poles (Figure 6.6) corresponds to a kind of double skewing, and thus produces some reduction in flux density and radial force harmonics. Noise is reduced this way too, and so is the axial force.

To further reduce the noise, the claw poles have a chamfer (Figure 6.6). The chamfer also reduces the excitation field leakage between neighboring poles.

With τ the pole pitch, the same in the stator and rotor, the claw-pole span in the middle of stator stack, $\alpha_{c\tau}$, is the main design variable for claw poles. The additional geometrical variables are (Figure 6.6) φ_c, W_1, W_2, d_c, and W_c.

In general, the pole angle $\alpha_c = 0.45–0.6$; the claw angle is $\varphi_c = 10°–20°$. The lower values of α_c and φ_c in the above intervals tend to produce higher output with rectifier and battery operation.

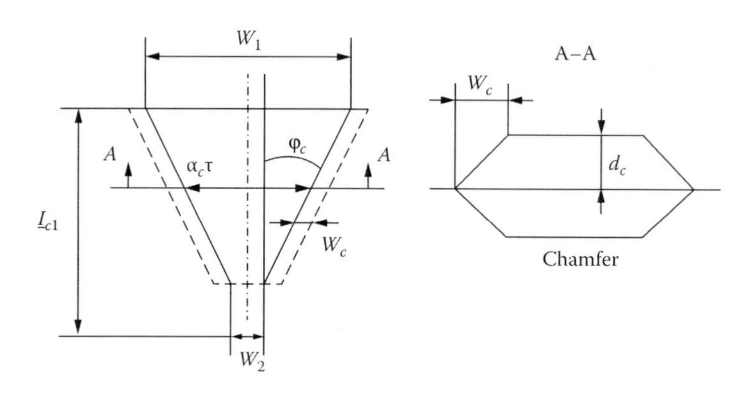

FIGURE 6.6 The rotor claw with chamfer.

Taking note of the above peculiarities of the Lundell generator, a few remarks can be made:

- An analytical constant parameter treatment, with only the fundamental phase voltage and current considered, should be used only with extreme care, as it is prone to large errors due to magnetic saturation at low speeds and due to armature-reaction flux density caused distortion at high speeds.
- Including magnetic saturation only by equivalent saturation factors pertaining to fundamental flux distribution may not produce practical enough third harmonic values at high speeds (in Figure 6.5b: $V_3 = 9.71$ V and $V_1 = 7.21$ V at 6000 rpm).
- For the eight-diode rectifier, the phase voltage third harmonic in the phase current occurs. The null current is three times the third-phase harmonic current.
- In this case, applying the fundamental circuit model brings a bit more realistic results, but still, the third harmonic current has to be calculated separately, and its losses are included when efficiency is determined.

Consequently, a nonlinear, iterative circuit model is required to properly describe Lundell generator performance over a wide range of speeds and loads.

Conversely, a nonlinear field model could be used.

Three-dimensional finite element method (FEM) or magnetic equivalent circuit (MEC) modeling was applied to portray Lundell generator performance—steady state and transients—with remarkable success. The MEC approach, however, requires at least two orders of magnitudes (100 times) less computer time. Both these field methods will be presented in some detail, as applied to the Lundell generator, in what follows.

For voltage control design, however, an equivalent circuit model is required. Even the fundamental electric circuit with magnetic saturation coefficients along the two orthogonal axes should do (after linearization) for such purposes.

Composite FEM circuit models are also a way out of difficulties with Lundell generator steady-state performance computation.

6.3 Magnetic Equivalent Circuit Modeling

The construction of the MEC should start with observation of a typical main magnetic flux path. Such a three-dimensional path is shown in Figure 6.7a and b. It corresponds to no-load conditions (zero stator current). Each flux path section is characterized by a magnetic reluctance:

- Air gap: R_g
- Stator tooth: R_{st}
- Stator yoke: R_{sy}
- Claw, axial: R_{ca}
- Claw, radial: R_{cr}
- Rotor yoke: $2R_{cy}$

Besides, there are at least two essential leakage paths for this main flux line: one tangential and the other axial between claws: R_{ctl} and R_{cal}. In addition, there is the "slot" leakage magnetic reluctance of the ring-shaped DC excitation coil: R_{csl}.

A simplified MEC for no load is shown in Figure 6.7b, where the axial interclaw leakage flux R_{cal} was neglected, for reluctance simplicity.

Analytical expressions for all these magnetic reluctances have to be adopted. While for the stator sections (R_{sy}, R_{st}) and the air gap (R_g) the formulae are rather straightforward (with zero slot openings), for the rotor sections (R_{cr}, R_{ca}, R_{cy}, R_{ctl}), the magnetic reluctance expressions, including magnetic saturation effects, are cumbersome, if good precision is expected, because the cross section of the claw poles varies axially and so does the magnetic permeability in them. For details, see pp. 125–127 in Reference 2.

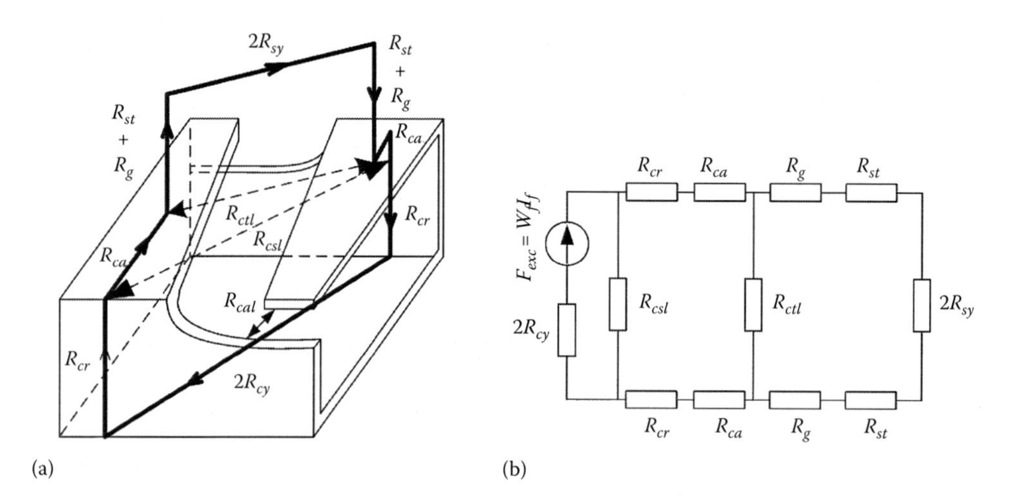

FIGURE 6.7 (a) Typical field magnetomotive force or mmf flux line and (b) simplified magnetic equivalent circuit (MEC).

Acceptable no-load magnetization curves may be obtained this way:

$$E_1(I_f) = \pi\sqrt{2} \cdot n \cdot p_1 \cdot W_1 \cdot \Phi_{p1}(I_f); \quad \Phi_{p1} = B_{g1} \cdot \frac{2}{\pi} \cdot \tau \cdot l_{stack} \tag{6.3}$$

where

Φ_{p1} is the fundamental of stator polar flux ($q = 1$)
E_1 is the root-mean-squared (RMS) value of the electromagnetic force (emf) fundamental
τ is the pole pitch; p_1—pole pairs; w_1—turns/stator phase (one current path)
l_{stack} is the stator stack length
n is the speed in revolution per second (rps)

The relationship between the polar flux Φ_{p1} and the field current has to account for the approximated air gap flux density distribution at no load (Figure 6.8).

The rectangular distribution, calculated with the average rotor claw span $\alpha_p = 0.45$–0.6 (Figure 6.8) leads to an ideal flux density fundamental in the air gap B_{g1} of the following:

$$B_{g1} = \frac{4}{\pi} \cdot B_g \cdot \sin \alpha_p \frac{\pi}{2} = K_{\alpha_p} \cdot B_g \tag{6.4}$$

$$\Phi_{p1} = K_{\alpha_p} \cdot \Phi_p \tag{6.5}$$

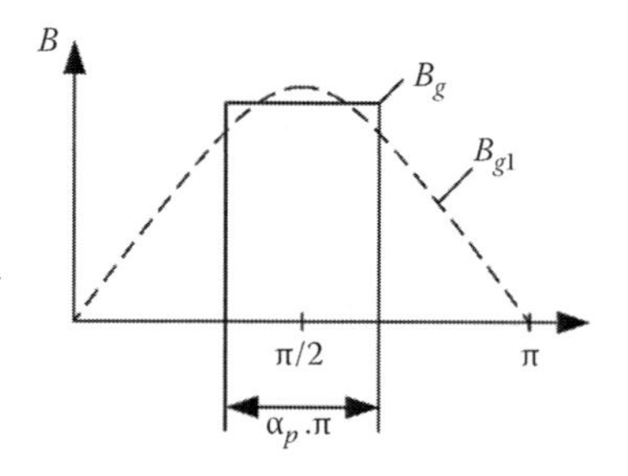

FIGURE 6.8 Ideal no-load air gap flux density.

Reducing the field current I_f to the stator I'_f is also required for the machine fundamental equation:

$$W_f \cdot I_f \cdot K_{\alpha_p} = 2 \cdot F_{10}; \quad F_{10} = \frac{3 \cdot W_1 \cdot I'_f \cdot \sqrt{2}}{\pi \cdot p_1} \tag{6.6}$$

F_{10} is the mmf fundamental per stator pole of a three-phase current I'_f that produces the same air gap field as the DC field current I_f:

$$I'_f = I_f \cdot K_{if}; \quad K_{if} = \frac{W_f \cdot K_{\alpha p}}{(6 \cdot W_1 \cdot \sqrt{2})/(\pi \cdot p_1)} \tag{6.7}$$

In general, to limit the DC excitation losses, the air gap is kept small (g = 0.35–0.5 mm); but to reduce the machine reactances and volume, the magnetic circuit is saturated for rated field current. Also, the solid-iron claw poles serve as a damper winding during transients and during commutation in the rectifier.

However, air gap field harmonics produce additional eddy currents in the claw poles even during steady state. These additional losses increase with speed up to a point and pose a severe limitation on machine output.

In the above approximations, the stator slot openings were neglected. To a first approximation, they may be considered by increasing the air gap magnetic reluctance R_g per pole by the known Carter coefficient—$K_c > 1$ [3]:

$$R_g \approx \frac{1}{\mu_0} \cdot \frac{g \cdot K_c}{\alpha_p \cdot \tau \cdot l_{stack}} \tag{6.8}$$

The Carter coefficient produces a rough approximation that is also global in the sense that the tangential actual distribution, strongly disturbed by the stator slot openings, is not visible.

For no load, however, the emf measured harmonics are smaller than 10%, even for rated field current, when heavy magnetic saturation occurs, which tends to "induce" a third harmonic in the emf. This saturation-produced third harmonic may be interpreted as if the inverse air gap function contains a second harmonic, besides the constant value.

While the above MEC is suitable for no load, a more complicated one, to take care of local magnetic saturation, an inverse air gap function variation with rotor position and along axial direction (due to the tapering shape of claws) is needed. Such a comprehensive model is presented in Reference 4, where the on-load operation is directly considered.

It is a three-dimensional model in the sense that the permeance of air gap varies along axial and tangential directions, while the stator and rotor permeances vary along radial and tangential directions. Axially and in the radial plane, the areas are divided into a few sections of almost uniform (but different) permeability (in iron). The more the elements are considered, the better the precision, but the computation time increases notably. The magnetic permeance approximations are, in general, analytical functions of rotor position (of time).

Each slot and tooth is modeled by one (or two) element(s). Axially, the machine is divided into a few sections. The connection of phases is arbitrary.

With a few hundred elements, a comprehensive MEC can be built [5], with the voltage–current relationship added for completeness.

Remarkably good agreement between calculated and measured phase and neutral (for eight-diode rectifier) current is obtained with a few minutes of time on a laptop computer [4].

In the same time, the distribution of air gap flux density along circumferential directions is properly, though approximately, simulated.

6.4 Three-Dimensional Finite Element Method Modeling

Though requiring much more computer time—intensive, three-dimensional—the FEM can capture the shear complexity of the flux lines in a Lundell generator at load, including magnetic saturation, claw-pole tapping, claw chamfering, and slot openings [6].

The periodicity conditions allow us to simulate only one pole (Figure 6.9). This way, the computation time for the 3D FEM becomes more reasonable, though still high (hours for complete steady-state characteristics at a given speed).

Typical radial air gap and flux density distributions obtained through 3D FEM are shown in Figure 6.10a and b [6].

The influence of slot openings is evident. The strong departure of the air gap flux density from a sinusoid is due to the low number of slots (three) per pole and due to magnetic saturation (over the whole pole at no load, and over half of it on load).

On load, the stator mmf at 6000 rpm is only slightly dephased with respect to the rotor longitudinal (pole) axis, as indicated by the severe demagnetization in axis d (at 180° in Figure 6.10b). Still, there is a rather high peak of air gap flux density at about 140° electrical, as expected.

The distortion in the total (resultant) emf is due to this mix of causes: three slots per pole and local magnetic saturation, dependent on load current and speed. The influence of speed comes into play through the stator mmf wave angle with the rotor poles, which tends to decrease to low values at high speed when the diode rectifier battery imposes active-power-only transfer from generator at all speeds. The small power factor angle increasing with speed, from 0° to 9° to 10°, is due to the reactive power "consumption" of machine inductances during diode commutation.

FEM application needs the values of the phase currents for a given rotor position. However, to calculate these values, a circuit model of the machine, including the diode rectifier and the battery, is necessary. An iterative circuit model adopted through results from FEM is the obvious choice for the scope.

Calculating the operation point—for given speed, DC load (resistance), and field current—may be done by first approximating the relationship between the fundamental phase current and voltage V_1, I_1 and their DC battery correspondents V_{DC}, I_{DC}.

With zero losses in the diode rectifier,

$$
\begin{aligned}
I_{DC} &= K_i \cdot I_1; \quad K_i = \frac{3\sqrt{3}}{\pi} \approx 1.35 \\
V_{DC} &= \frac{3V_1 \cdot I_1}{I_{DC}} = K_V \cdot V_1; \quad K_V \approx \frac{\pi}{\sqrt{2}} = 2.22
\end{aligned}
\tag{6.9}
$$

These approximations hold for the six-diode rectifier and star phase connection.

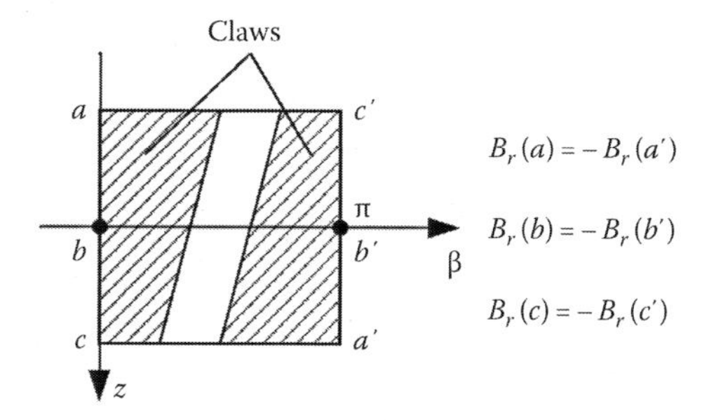

FIGURE 6.9 Symmetry conditions.

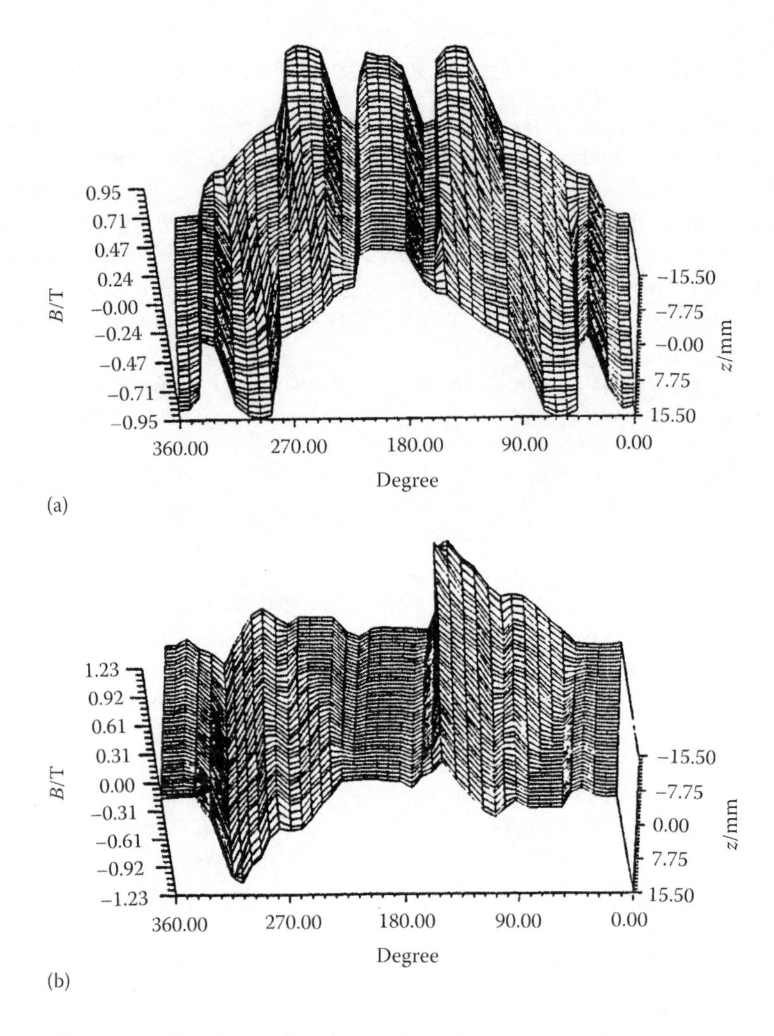

FIGURE 6.10 Radial air gap flux density distribution (Z is the axial variable): (a) on no-load and (b) at 6000 on-load.

For AC loads, the situation is simpler, as the load resistance and reactance are given, and we need to calculate the output voltage and current fundamentals.

The phase current is considered essentially sinusoidal (star connection) for simplicity, though, at low speeds, it tends to be rectangular—discontinuous.

The phasor diagram may be used in the iterative procedure. The dq model equations—valid for fundamental quantities only—of the Lundell (synchronous) machine are as follows:

$$\underline{I}_1 R_1 + \underline{V}_1 = \underline{E}_1 - jX_d \underline{I}_{d1} - jX_q \underline{I}_{q1}; \quad \underline{I}_1 = \underline{I}_d + \underline{I}_q \tag{6.10}$$

$$\underline{E}_1 = -jX_{dm} \underline{I}_f'; \quad X_d = X_{dm} + X_{1l}; \quad X_q = X_{qm} + X_{1l} \tag{6.11}$$

X_{dm}, X_{qm}, and X_{1l} are, respectively, magnetization synchronous and leakage stator reactances in the dq model. (See Chapter 4 in *Synchronous Generators* on synchronous generators, steady state.)

For a small power factor angle, the phasor diagram corresponding to Equations 6.10 and 6.11 is illustrated in Figure 6.11. The small ($\varphi_1 < 0$) power factor angle takes into account the effect of diode commutation on reactive power absorption from the generator.

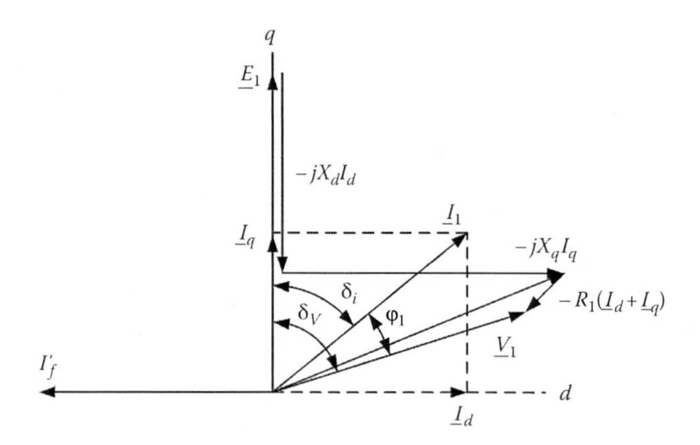

FIGURE 6.11 The phasor diagram.

The iterative computation process starts with given phase current $\underline{I}_1$ and current power angle δ_i. (δ_i increases with speed, in general, for constant DC output voltage.)

Using FEM, the air gap flux distribution with its fundamental and space harmonics is determined. Then, the fundamental flux per pole Φ_{p1} in the air gap is obtained. Also, its position angle δ_V with respect to q rotor axis is found. This is, in fact, the standard power (voltage) angle. The air gap fluxes in axes d and q are thus,

$$\Psi_{dm} \approx \Phi_{p1} \cdot W_1 \cdot \cos\delta_V \tag{6.12}$$

$$\Psi_{qm} \approx \Phi_{p1} \cdot W_1 \cdot \sin\delta_V \tag{6.13}$$

But, the dq components of current I_1 are known, as I_1, δ_i are known:

$$I_{d1} = I_1 \cdot \sin\delta_i \tag{6.14}$$

$$I_{q1} = I_1 \cdot \cos\delta_i \tag{6.15}$$

Consequently, the magnetization reactances as affected by magnetic saturation are calculated:

$$X_{dm} = \frac{\Psi_{dm}}{I_d + I'_f} \tag{6.16}$$

$$X_{qm} = \frac{\Psi_{qm}}{I_q} \tag{6.17}$$

The field current, reduced to the stator, I'_f, was defined earlier in this chapter as a function of actual field current, with its value also given. The stator leakage reactance X_{sl} is considered to be known.

The given phase current I_1 is based on the existence of a DC current I_{DC} (Equation 6.8): $I_{DC} = K_i \cdot I_1 \approx 1.35\ I_1$. For given power and neglected diode rectifier losses, given load means given voltage (or load resistance) and current: V_{DC} and I_{DC}. This way, from Equation 6.8, the phase voltage fundamental (RMS) value V_{10} is, in fact, computed. The speed n is known, and thus, $\omega_1 = 2\pi p_1 n$ is given.

With the values of Ψ_{dm} and Ψ_{qm} from FEM, the phasor diagram "produces" the new phase voltage V'_1:

$$V'_1 = \sqrt{(\Psi_{dm}\omega_1 + X_{1l}I_d + R_1 I_q)^2 + (\Psi_{qm}\omega_1 + X_{1l}I_q - R_1 I_d)^2} \tag{6.18}$$

After the first computation cycle $V_1' \neq V_{10}$; the computation cycle should restart, with a new value of the current power angle δ_i, but with the same I_1:

$$\delta_i(K+1) = \delta_i(K) \cdot \left(1 - C_c \cdot \frac{V_1(K) - V_1(K-1)}{V_{10}}\right) \tag{6.19}$$

The coefficient C_c is chosen by trial and error to provide fast convergence, while $V_1(K)$, $V_1(K-1)$ are the calculated phase voltage fundamentals in the K and $K-1$ computation cycles, respectively. If, after a certain number of iterations, convergence is not met, it means that the value of I_1 has to be reduced.

This way, families of saturation curves $\Psi_{dm}(i_{d1}+i_f', i_{q1})$, $\Psi_{dm}(i_{d1}+i_f', i_{q1})$ may be obtained, to be used later for curve fitting by analytical approximations, for a complete modeling of the machine (as in Chapter 4 in *Synchronous Generators*).

Typical I_{DC} versus speed curves for given battery voltage V_{DC}, obtained through procedures as above, are shown in Figure 6.12 for three different 14 V_{DC} Lundell automotive generators of increasing power.

At the idle engine speed, the current that may be injected in the load, with battery backup, is limited and may constitute a design value in per unit (P.U.). Above 3000 rpm, almost full current is produced. The close to unity power factor in the generator with diode rectifier limits the maximum power (current) that the machine is capable of, for given DC voltage and speed, but the diode rectifier is safe and inexpensive.

To capture the third harmonic in the phase voltages V_{3o}, air gap flux density distribution for final current angle δ_i (and I_1) is decomposed in harmonics, and the third harmonic is detected as amplitude and phase B_{g3}, ε_3, with respect to the fundamental. Therefore, the third harmonic total emf V_{3o} (RMS value) is as follows:

$$V_{3o} = 3 \cdot \Phi_{p3} \cdot \omega_1 \cdot \frac{1}{\sqrt{2}} \cdot W_1; \quad \Phi_{p3} = \frac{2}{\pi} \cdot \frac{\tau}{3} \cdot l_{stack} \cdot B_{g3} \tag{6.20}$$

The phase lag between V_1 and V_{3o} is again ε_3, approximately.

This way, the total phase waveform is obtained as follows:

$$V_a(t) = V_1\sqrt{2}\cos\omega_1 t + V_{3o}\sqrt{2}\cos(3\omega_1 t - \varepsilon_3) \tag{6.21}$$

A comparison of $V_a(t)$ calculated and from tests showed an acceptable agreement Figure 6.13 [9].

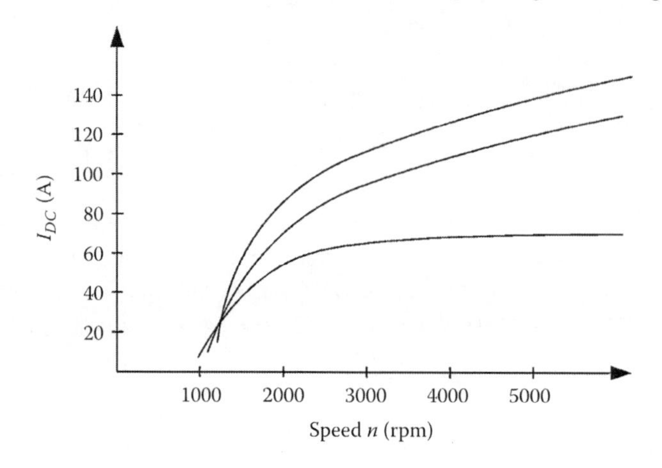

FIGURE 6.12 Typical DC output vs. speed for constant battery voltage.

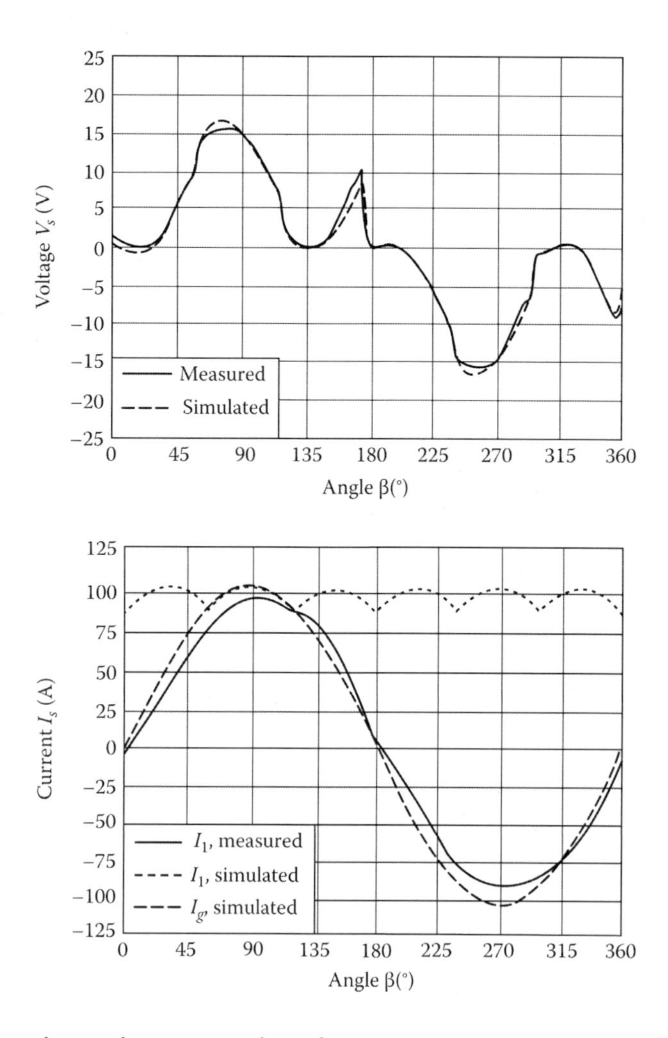

FIGURE 6.13 Phase voltage and current typical waveforms.

An alternative way to evidentiate the voltage time harmonics consists of indirectly using the FEM to calculate the machine characteristics via an inverse P.U. air gap function $g/g(\beta)$ (obtained from FEM) to define the phase air gap self-inductance and mutual inductance. Unfortunately, this function should account for saturation, which depends on speed, current I_1, its power angle δ_i, and I_f [6]:

$$\left(\frac{g_0}{g}\right) = [K_1 + K_2(\cos\beta - 2\omega_1 t + \alpha_2) + K_4\cos(4\beta - 4\omega_1 t + \alpha_4)](1 + \cos 6\beta)$$

$$L_g = \frac{4\mu_0 \cdot W_1^2 \cdot l_{stack} \cdot \tau}{\pi^2 g_0}; \quad L_{abg} = -\frac{L_{aag}}{3}$$

$$L_2 = L_g \cdot \frac{\sqrt{3}}{\pi} \cdot K_2; \quad L_4 = L_g \cdot \frac{\sqrt{3}}{\pi} \cdot K_4 \tag{6.22}$$

$$L_{ag}(t) = L_g \cdot K_1 + L_2\cos(2\omega_1 t + \alpha_2) + L_4\cos(4\omega_1 t + \alpha_4)$$

$$L_{abg}(t) = -\frac{L_g}{3}K_1 + L_2\cos\left(2\omega_1 t + \frac{\pi}{3} + \alpha_2\right) + L_4\cos\left(4\omega_1 t - \frac{\pi}{3} + \alpha_4\right)$$

The coefficients K_1 ($K_1 < 1$), K_2, K_4 account for magnetic saturation and slot opening effects. And, they depend heavily on speed and load conditions.

Defining this way phase a self-inductance $L_{aa}(t) = L_{1l} + L_{ag}(t)$; a, b phase mutual inductance $L_{ab}(t) = L_{abg}(t)$; and, in a similar way, the other inductances for sinusoidal current, the steady-state (constant speed and load resistance) performance may be calculated without the d–q model. That is, in phase coordinates,

$$\left|i_{a,b,c,f}\right| \cdot \left|R_1\right| + \left|V_{a,b,c}\right| = -\frac{d\left|\Psi_{a,b,c}\right|}{dt}; \quad \left|\Psi_{a,b,c}\right| = \left|L_{a,b,c,f}(t)\right| \cdot \begin{vmatrix} i_a \\ i_b \\ i_c \\ i'_f \end{vmatrix} \tag{6.23}$$

Typical results obtained with this method are shown in Figure 6.13 [7].

The highly distorted phase voltage and sinusoidal current waveforms are self-evident.

Note that the presence of booster diodes (the fourth leg of diodes connected to the machine null point) was not yet considered for star connection. Therefore, for most speeds, the phase current is quasi-sinusoidal. For a comprehensive study of a stator excited Lundell generator by FEM, see Reference 8.

6.5 Losses, Efficiency, and Power Factor

Losses occur in the Lundell machine both in the stator and in the rotor:

- Stator iron losses: p_{is}
- Stator copper losses: p_{Cos}
- Rotor copper losses: p_{Cor}
- Rotor claw harmonics losses: p_{claw}
- Mechanical losses: p_{mec}
- Diode rectifier losses: p_{diode}

A complete study of these loss components is presented in Reference 9 with ample experimental results on industrial Lundell generators.

Noting that, for the star connection of phases, the phase current is quasi-sinusoidal, except for speeds close to ICE idling speed (700–1100 rpm), the winding loss calculation is straightforward:

$$p_{Cos} = 3R_1 \cdot I_1^2 \tag{6.24}$$

As the field current has small ripple at six times the stator frequency (speed), we may calculate the DC excitation (rotor copper) losses easily:

$$p_{Cor} = R_f \cdot (I_f)^2_{average} \tag{6.25}$$

Mechanical losses may be approximated by analytical expressions, but measurements are mandatory for a good approximation of them.

The diode rectifier losses, neglected thus far, may not be left out in efficiency calculations, as for 14 V_{DC} battery systems they are about as large as the stator copper losses:

$$p_{diode} = 3(\Delta V_{diode} + R_{diode} I_1) I_1 \tag{6.26}$$

where

ΔV_{diode} is the constant voltage drop along a diode

R_{diode} is the equivalent diode resistance

If the diode rectifier loss and the actual power factor cos φ_1 (close to unity but not equal to) in the machine are considered, the power balance between the machine and the diode rectifier is altered to the following:

$$V_1 = \frac{V_{DC}I_{DC} + p_{diode}}{3I_1\cos\varphi_1}; \quad K_V = \frac{V_{DC}}{V_1}$$
$$K_i = \frac{I_{DC}}{I_1} = \frac{3\sqrt{3}}{\pi}\cdot\cos\varphi_1 = 1.35\cdot\cos\varphi_1 \tag{6.27}$$

For better precision, Equation 6.27 should replace Equation 6.9. In this latter case, K_i varies a little and so does K_V, with speed (cos φ_1) and current.

The power factor may be attempted based on the idea that it is related to the reactive power Q_{con1} "absorbed" by the commutation inductances of the machine. For sinusoidal current, Q_{con1} is

$$Q_{con1} = 3\cdot\omega_1\cdot L_c\cdot I_1^2 \tag{6.28}$$

L_c is, in fact, an average of subtransient inductances of the generator at the d and q axes:

$$L_c \approx \frac{L_d'' + L_q''}{2} < L_d, L_q \tag{6.29}$$

The solid-iron claw poles represent a not-so-strong but still important damper cage on the rotor. The computation of L_d'', L_q'' is complicated, and measurements are preferable if comprehensive 3D analytical or numerical (FEM) eddy current models are not used. The power factor is, approximately,

$$\cos\varphi_1 \approx \sqrt{1-\left(\frac{Q_{con1}}{3\cdot V_1\cdot I_1}\right)^2} = \sqrt{1-\left(\frac{\omega\cdot L_c\cdot I_1}{V_1}\right)^2} \tag{6.30}$$

The smaller L_c is, the better.

The power factor decreases slightly with speed (ω_1) for the same V_1 and I_1, as proven by tests (see the literature [6,9]).

The claw-pole eddy current losses p_{claw} are produced by the air gap field space harmonics, for sinusoidal stator currents. Claw eddy current losses occur during both no-load and on-load. With no-load, the stator slot openings modulate the excitation-only produced air gap field, and its pulsations are "seen" by the claws as flux density variations. The frequency of these eddy currents corresponds basically to the first slot harmonic and, for $q = 1$, it leads to $6\omega_1$ frequency.

For load conditions, the stator mmf produces space harmonic fields in the air gap, and the fifth and seventh orders are the most important (for $q = 1$). The slot openings appear to augment these harmonics and produce claw-pole eddy currents, as for the no-load situation. As resultant field on load is burdened with local magnetic saturation, the situation becomes further complicated.

A practical analytical model for claw-pole eddy current losses p_{claw} is still not available. Alternatively, when neglecting the eddy currents reaction field, the 3D FEM or MEC models could produce the flux density in all elements, investigated in successive time decrements. Then both the stator iron loss p_{is} and p_{claw} may be calculated through advanced analytical formulas [10].

In these formulas, the loss coefficient in p_{claw} should correspond to solid iron, while for the stator core, it should refer to laminations.

The rotor claw-pole losses tend to increase with speed up to a point and then level out, while the stator iron losses slowly decrease with speed after a peak. The stator current has an increasingly demagnetization effect as speed increases because of the need for unity power factor demanded by the diode rectifier.

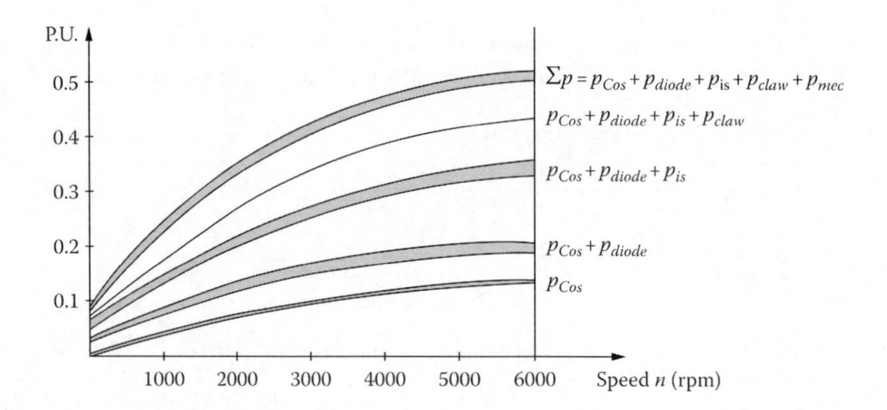

FIGURE 6.14 Loss components vs. speed for a Lundell generator.

A qualitative view of machine losses variation with speed for a given DC voltage battery is shown in Figure 6.14. For a detailed study on losses, see Reference 9.

Mechanical, diode, winding, and claw-pole losses are of the same order magnitudes, above 2000 rpm.

Reducing the claw-pole eddy currents by, eventually, using soft magnetic powder material for the claws would notably improve the high-speed performance and power capability of the Lundell generator, provided the mechanical ruggedness is somehow preserved.

Increasing the DC bus (battery) voltage from 14 to 42 V is another step forward, in the sense that the diode rectifier loss in the P.U. is decreased notably, because the voltage drop on the diode will count less.

Using thinner laminations would notably decrease the stator iron losses, and smoothing the rotor and stator surfaces would reduce the mechanical losses.

Currently, Lundell generators are designed for low volume and costs at the penalty of down to 50% efficiency at high speed and full power. Bold steps in design are required to keep the existing power/volume while increasing the efficiency at moderate cost increases. Such recent attempts are dealt with in what follows.

6.6 Design Improvement Steps

The modeling tools of the previous paragraph are the foundation for design optimization. A few design improvement steps are discussed:

- Claw-pole geometry optimization
- Booster diodes
- Additional permanent magnets (PMs)
- Increase of the number of poles
- Winding tapping
- Claw-pole damper
- Controlled rectifier
- Winding-type effect on noise level

6.6.1 Claw-Pole Geometry

As shown in Figure 6.6, the claw-pole geometry may be defined by a few parameters, such as claw-pole span α_c (nondimensional), claw-pole angle φ_c, pole pitch τ, claw-pole depth d_c, and the chamfer width W_c.

W_1/τ, W_2/τ—the maximum and minimum claw-pole widths ratios—may be calculated, with α_c, τ, and claw-pole angle φ_c given, as dummy variables.

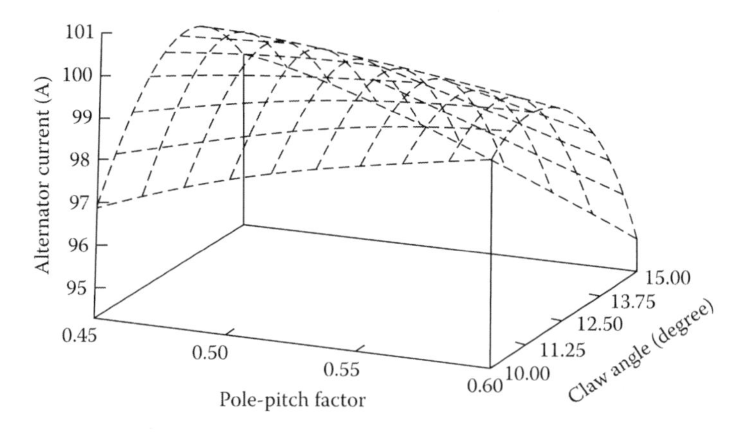

FIGURE 6.15 Generator DC vs. pole pitch α_c coefficient and φ_c.

This small number of independent variables, with all other machine parameters kept constant, makes a direct FEM local optimization feasible through a few tens of runs in a practical time interval. Typical results are shown in Figure 6.15 [7].

The penalty function is the DC current, and, for $\alpha_c \approx 0.45$ and $\varphi_c = 12.5°$, it is maximum.

6.6.2 Booster Diode Effects

A four-legged diode rectifier (Figure 6.16) is sometimes used to tap the third voltage harmonic power. To deliver power through the third harmonic, the latter has to exist both in phase voltage and in current. The connection of the machine null point to diode booster point *o* leads to nonzero third harmonic phase currents. However, the null current is discontinuous when the third harmonic voltage does not surpass the DC battery voltage, that is, at low speeds.

The presence of the third harmonics power above a certain speed boosts the DC current delivered by the generator for constant battery voltage (Figure 6.17). The effect is notable only at high speeds, where, in general, it is not so much needed. The additional third harmonic current produces stator copper and stator core and claw-pole losses.

At least for constant speed synchronous generators with discontinuous current, at 50 (60) Hz, it was proven that the extra power with booster diodes offsets the increase in losses, and an increase in efficiency of about 1% is obtained [11], together with 10%–15% more power.

For automotive applications, the advantages of booster diodes are considered less practical, as they do not come at low speeds when more power is badly needed.

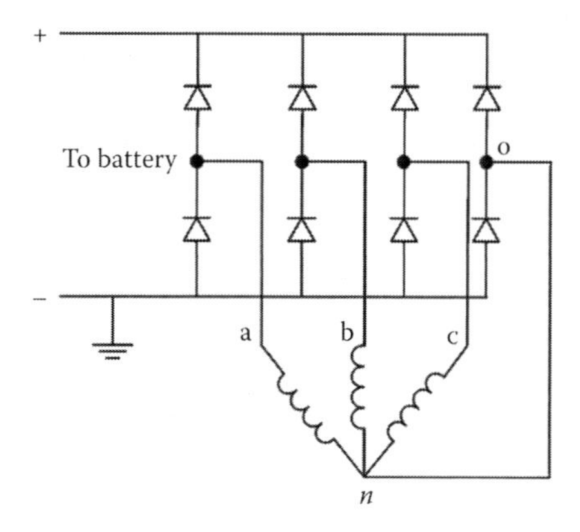

FIGURE 6.16 Diode rectifier with power booster diodes.

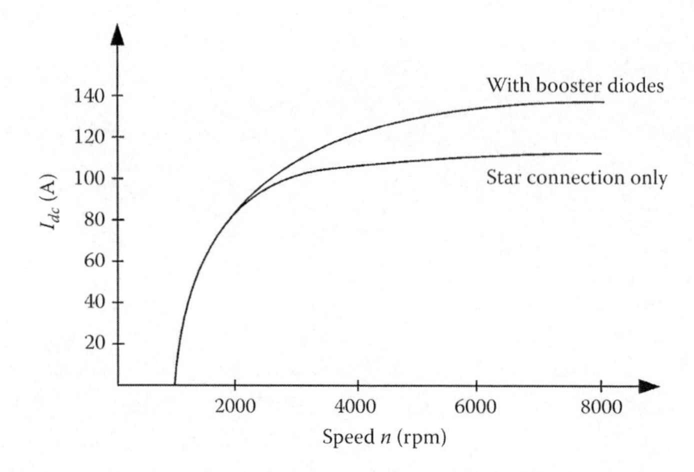

FIGURE 6.17 Booster diode effect on DC vs. speed.

6.6.3 Assisting Permanent Magnets

Axially magnetized PMs may be added beside the coil (Figure 6.18a), on the surface of the claws (Figure 6.18b), and between the claws (Figure 6.18c). In the third case, the PMs act to destroy the interclaw leakage flux of excitation coil. The first two solutions are meant to produce more excitation flux for lower excitation losses. Unfortunately, at low speeds, the influence on the DC output current is small due to magnetic saturation (Figure 6.19).

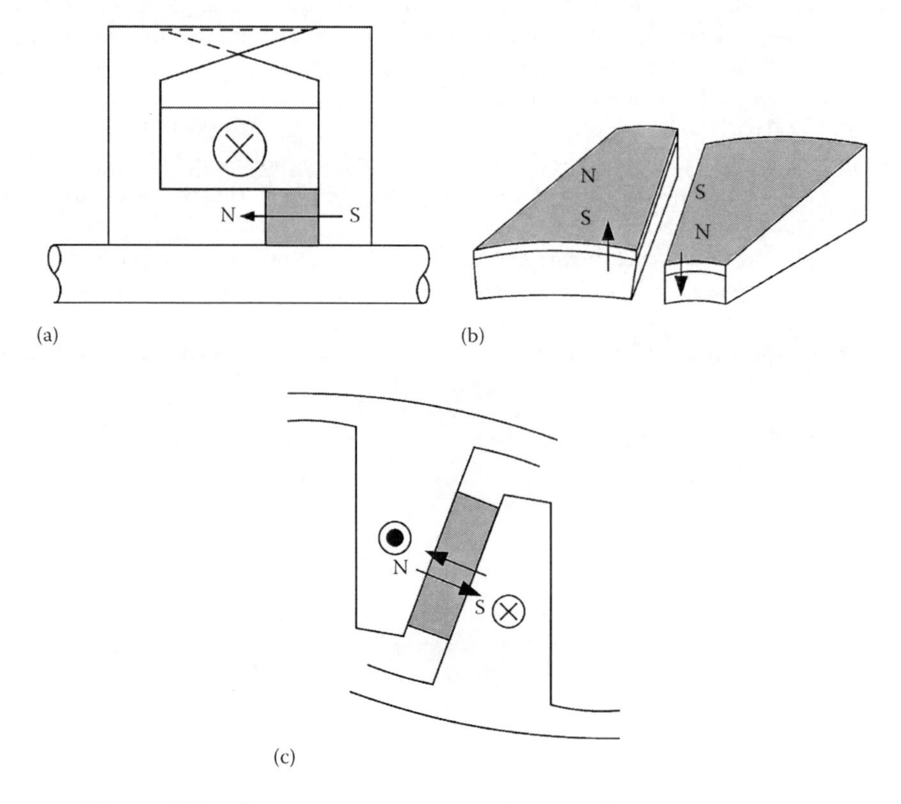

FIGURE 6.18 Placing additional permanent magnets (PMs) for given stator stack: (a) PMs on the shaft, (b) PMs on the claw surface, and (c) PMs between claws.

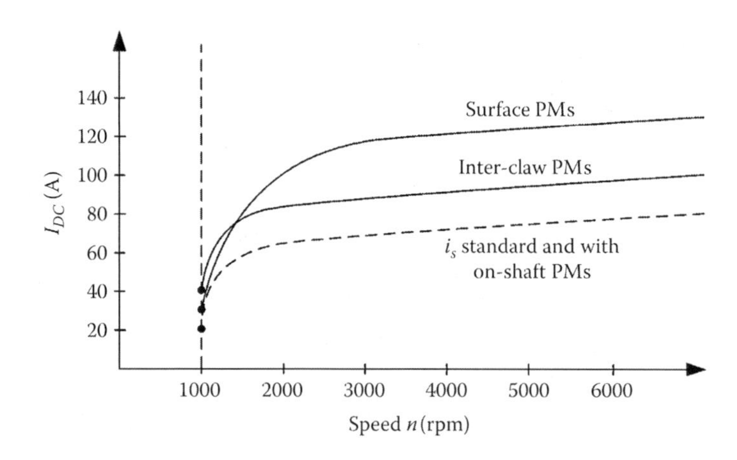

FIGURE 6.19 Influence of additional PMs on DC output for the same stator.

Due to high leakage flux, the PMs on the shaft produce little effect at all speeds. The surface PMs produce notably larger DC output current but, unfortunately, only at higher speeds.

The PMs placed between the claws destroy the excitation leakage flux in the rotor and, thus, enhance the main flux for all speeds. The improvement occurs at all speeds (Figure 6.19) [7].

Though the interclaw PMs, eventually made of ferrites, seem to be an attractive solution, their mechanical placement and integrity are not easy to secure.

6.6.4 Increasing the Number of Poles

Increasing the pole number, for given stator and rotor diameters and lengths, and at the same total stator slot area, seems to be another way of notably increasing the output. The reason is the torque multiplication effect of increasing the "speed" of flux variation.

When the number of poles goes up from $2p = 12$ (the standard) to 14 and 16, the DC output current steadily increases, especially above 2000 rpm [7].

Unfortunately, the frequency $f_1 = p_1 \cdot n$ also increases with pole number. Consequently, the iron losses tend to increase, especially in the stator teeth. Additionally, the claw eddy current losses tend to increase. Skin effect losses in the stator conductors increase also.

For larger power Lundell generators—used for trucks or buses—such a solution may prove to be better. Still, the effect at low speed is not so important.

6.6.5 Winding Tapping (Reconfiguration)

It is a known fact that at low speeds the Lundell generator is not capable of producing enough DC due to insufficient emf, as the number of turns is kept low to secure a good cross section of copper in the coils and so as to handle large currents at high speeds. Increasing the number of turns per phase (or per current path) at low speeds while reducing it at high speeds is the way out of this difficulty.

Switching the machine connection from star to delta or tapping the stator winding are two alternatives to produce increased output. Halving (tapping) the winding reduces twice the stator resistance (Figure 6.20).

To produce large output at low (idle engine) speed, the number of turns per phase has to be increased: $W_1' > W_1$. If the total area of slots remains constant, the $W_1 \cdot I_1$ should be the same for a given current

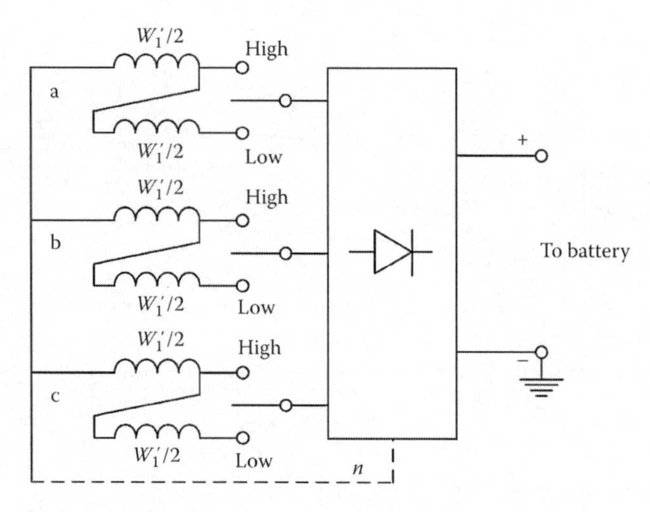

FIGURE 6.20 Winding tapping (halving).

density. However, an increase in W_1 (to W_1') should be accompanied by a reduction of copper wire cross section; thus, higher current density at higher (or same) current is required:

$$W_1 \cdot A_{Co} \approx W_1' \cdot A_{Co}' \tag{6.31}$$

with W_1, W_1', A_{Co}, and A_{Co}' equal to the initial and modified total number of turns and copper wire cross section, for the tapped winding design. Increasing W_1 and W_1' when the winding tapping is applied leads to higher maximum emf at idle engine speed. This is essential progress.

Neglecting the machine saliency ($X_d = X_q = X_s$), the phasor diagram in Figure 6.11, at unity power factor, takes shape in Figure 6.21.

Only the fundamental components of generator phase voltage and current are considered for simplicity.

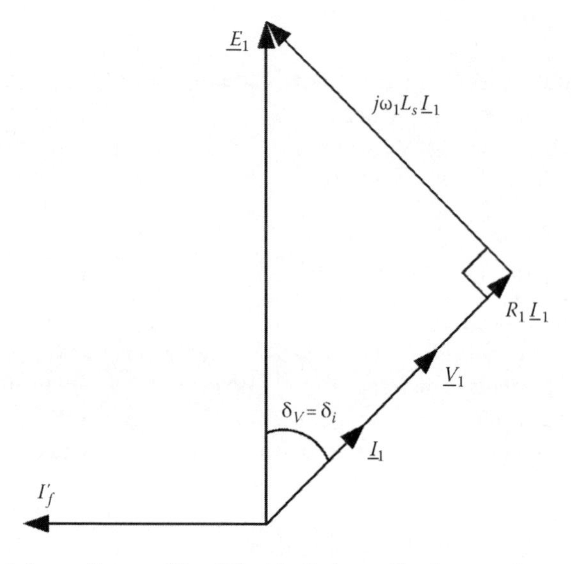

FIGURE 6.21 Simplified phasor diagram ($X_d = X_q$) with diode rectifier ($\cos \varphi_1 = 1$).

The stator resistance R_1 and R_1' (initial and new) are as follows:

$$R_1 = \rho_{Co} \cdot l_c \cdot \frac{W_1}{A_{Co}}; \quad R_1' = \rho_{Co} \cdot l_c \cdot \frac{W_1'}{A_{Co}'} \tag{6.32}$$

with l_c as coil length.

The reactance X_s is

$$\begin{aligned} X_s &\approx \omega_1 \cdot L_s = \omega_1 \cdot K_L \cdot W_1^2 \\ X_s' &= \omega_1 \cdot K_L \cdot W_1'^2 \end{aligned} \tag{6.33}$$

The emfs are

$$\begin{aligned} E_1 &= K_e \cdot \omega_1 \cdot W_1 \\ E_1' &= K_e \cdot \omega_1 \cdot W_1' \end{aligned} \tag{6.34}$$

From the phasor diagram, the following voltage relationship is obtained:

$$E_1 = \sqrt{(\omega_1 \cdot L_s \cdot I_1)^2 + (V_1 + R_1 \cdot I_1)^2} \tag{6.35}$$

$$E_1' = \sqrt{(\omega_1 \cdot L_s' \cdot I_1')^2 + (V_1 + R_1' \cdot I_1')^2} \tag{6.36}$$

The problem is to yield a higher current I_1 at idle engine speed ($\omega_{1\min}$) for the same battery voltage with a larger number of turns, that is, larger L_s and R_s: $L_s' > L_s$ and $R_s' > R_s$.

For simplicity, let us neglect the $R_1 I_1$ terms to obtain the following:

$$I_1^* \approx \frac{\sqrt{E_1^{*2} - V_1^{*2}}}{\omega_1 \cdot L_s} = \frac{\sqrt{K_e^2 \cdot \omega_1^2 \cdot W_1'^2 - V_1^2}}{\omega_1 \cdot K_e \cdot W_1'^2} \tag{6.37}$$

It is evident from Equation 6.35 that there is an optimum number of turns that produces the maximum current:

$$W_{1opt} = \frac{V_1 \cdot \sqrt{2}}{K_e \cdot \omega_1} \tag{6.38}$$

Equation 6.38 spells out the fact that for maximum current output at given speed, the number of turns should vary inversely proportional to speed so that the emf is $\sqrt{2}$ times the phase voltage ($E_{1opt} = K_e \cdot \omega_1 \cdot W_{1opt} = V_1 \cdot \sqrt{2}$). If the initial design is fixed at W_{1opt}, it should remain thus, and nothing is gained at low speeds. At high speeds, the halving of the winding will produce more DC output current for slightly more losses. However, if for the existing design $W_1 < W_{1opt}$, then it should be brought to W1opt while reducing, accordingly, the cross section of copper wire.

The DC output current at idle engine speed will increase to its maximum value:

$$I_{\max} \approx \frac{V_1}{\omega_1 \cdot L_s} \tag{6.39}$$

At high speeds, with winding halving, more current will be obtained but, again, with slightly more copper losses.

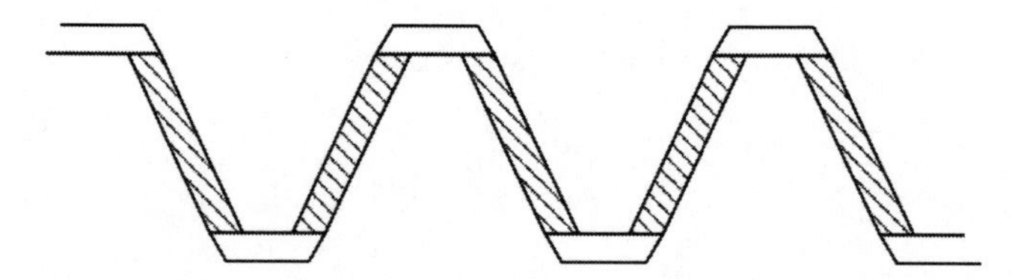

FIGURE 6.22 Aluminum damper pieces.

6.6.6 Claw-Pole Damper

Yet another way of increasing the DC output current at all speeds consists of decreasing the commutation inductance L_c of the machine and, thus, reducing the diode commutation process reactive power.

But, the commutation inductance $L_c = (L''_d + L''_q)/2$ relies on the damper effect of the rotor solid claw poles. An additional damper placed between the claws in the form of solid aluminum plates should do (Figure 6.22). If the electrical contact between the aluminum dampers and the solid iron of claws is good, their combination will act more like a one-piece conductor for the space harmonics field of the stator. A stronger damper winding is obtained.

A practical technology for tightly sticking the aluminum dampers between the claw poles is yet to be developed, but decreasing the commutation (subtransient) inductance of the machine seems to be a practical way to increase output.

Increasing the air gap would also help in terms of obtaining more output from the same volume but for higher losses in the excitation winding. Also, if the air gap is increased too much, the interclaw excitation leakage field tends to increase. Preliminary test results presented in Reference 12 seem to substantiate the aluminum damper's advantages.

6.6.7 Controlled Rectifier

Yet another way of increasing the output at low speeds is by using a controlled thyristor rectifier instead of the diode rectifier. This way, the unity power factor condition is eliminated, and reactive power exchange capability with the battery is established. The delivered power is approximately

$$P = \frac{3 \cdot E_1 \cdot V_1 \cdot \sin \delta_V}{X_s} \tag{6.40}$$

The voltage power angle δ_V of the machine could be raised close to 90° to produce more output [13] (Figure 6.23a and b).

For the same emf E_1 in Figure 6.23a and b, more current is feasible at given voltage V_1 and speed (X_s), as the $X_s \cdot I_1$ vector changes the location in the rectangular triangle.

An increase in DC output current of almost 50% is reported in Reference 13.

Further, the introduction of the controlled rectifier may be accompanied by winding reconfiguration (Figure 6.24), when the controlled rectifier works on the full winding at low speed, while a (additional) diode rectifier works at high speed, on half the winding, to combine the advantages of both methods [13].

The switches A and B are closed only for low speed. The speed for opening switch A is a matter of compromise and depends on the winding tap point, which, in general, may be placed at 1/2 or 1/3 P.U. distance from the null point. The tap point should be easily accessible. Reference 13 reports not only improved DC output current at low and especially at high speeds but also an increase in efficiency.

The added complexity and costs of equipments are the prices to pay for a 5%–10% increase in generator–rectifier system efficiency (from 47% to 57%).

A switched diode rectifier may also be used to boost the generator voltage at low speeds and buck it at high speeds, in order to increase the output of an existing claw-pole-rotor alternator.

Note that the design problems of the Lundell alternator do not end here. Designers should continue with noise and vibration and thermal and mechanical redesigns [14].

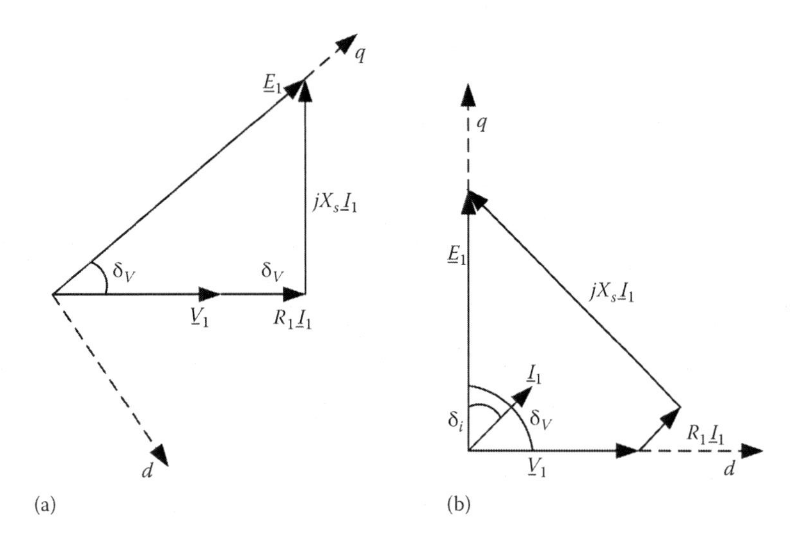

FIGURE 6.23 Simplified phasor diagrams ($X_d = X_q = X_s$): (a) with diode rectifier $\delta_V < 90°$ and (b) with controlled rectifier ($\delta_V < 90°$).

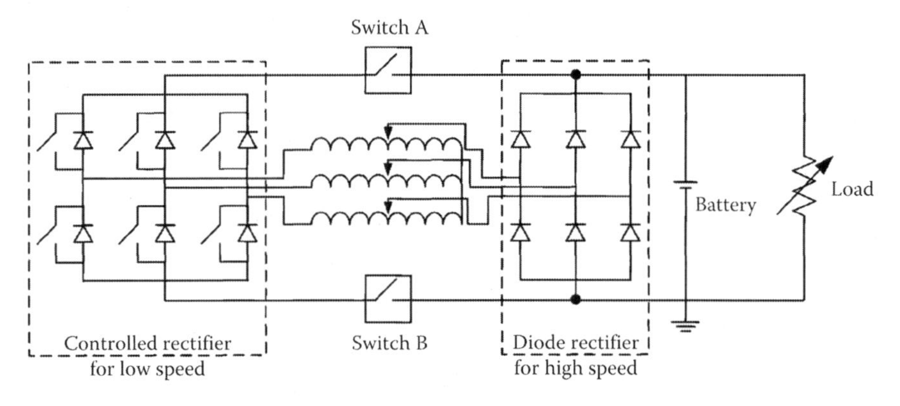

FIGURE 6.24 Winding reconfiguration with controlled rectifier.

6.7 Lundell Starter/Generator for Hybrid Vehicles

Recently, the Lundell machine was used on the first commercial mild hybrid electrical vehicle. Its use as a starter presupposes inverter (frequency) control and, thus, implicitly controls rectification capability during generating mode. The design accent will be placed on motoring mode with verifications for the generating mode.

The efficiency problem becomes very important, besides size, both during starting assistance of the ICE, and during driving the air-conditioning system when the ICE is shut down in traffic jams. The energy extracted from the battery is decreased, and thus, increases in gas mileage and battery life are obtained (Figure 6.25) [15].

The power requirements of air conditioners increase to a peak 2.5 kW (for a luxury car) at 15,000 rpm, and the peak torque is about constant, up to maximum speed. Through an electric clutch, placed on the ICE shaft, the single, ribbed, belt transmission allows for the starter/generator to engage the ICE when the latter is on and, respectively, the air conditioner, power steering, and water pumps (auxiliaries), otherwise. The conventional starter is still there for first (morning) start.

Mild hybrid vehicles rely on the contribution of the starter/generator supplied from the 42 V_{DC} battery. Three 14 V_{DC} batteries in series, with a typical 100 A maximum current absorption (for a number

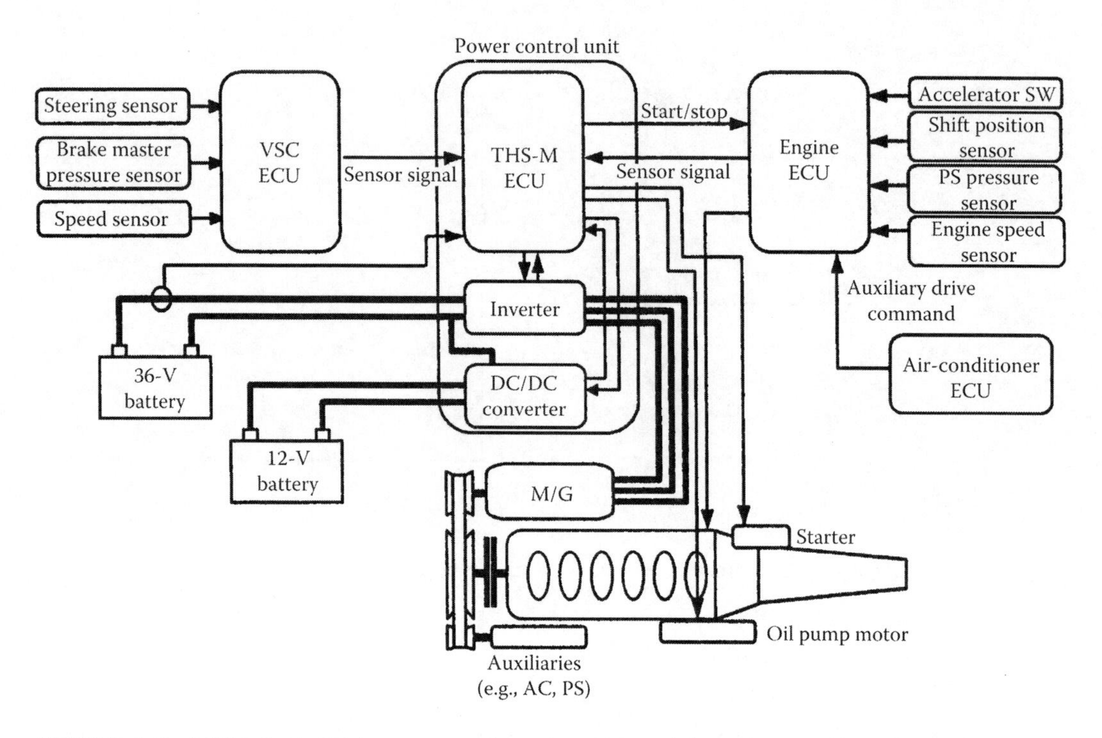

FIGURE 6.25 Mild hybrid vehicle system configuration. (Adapted from Teratani, T. et al., Development of Toyota mild hybrid systems (THS-M) with 42 V power net, Records of IEEE–IEMDC–2003, Madison, WI, 2003.)

of cycles corresponding to 5 years of regular driving), provide for a moderate cost battery. The battery also serves all the auxiliary equipment on board a medium-to-large car. A 14 V_{DC} bus is built through a special DC–DC converter to serve the low-voltage loads.

The starter–alternator design should be worked on with the following objectives in mind:

- Maximum motoring torque for starting and assisting the ICE at low speed, and for driving the ICE air conditioner compressor load when the ICE is off
- Maximum generating power versus speed
- Observation of the battery state and reduction of the losses in the system to obtain a reasonable battery life
- Minimum possible machine volume and pulse-width modulator (PWM) converter costs

As a result, the main specifications for the Lundell starter–generator in Reference 15 are as follows:

- Rated voltage: 36 V (42 V_{DC} bus)
- Rated output: motoring, 3 kW; generating, 3.5 kW
- Maximum torque: 56 Nm at 300 rpm
- Permissible maximum speed: 15,000 rpm
- Air cooling

Typical battery specifications are as follows:

- Capacity: 20 Ah valve-regulated lead acid battery
- Weight: 27 kg
- Volume: 9.21 L
- Starting performance: 6.1 kW for 1 s
- Auxiliary drive: 2.1 kW
- Regenerative performance: 3.5 kW for 5 s

There are three ways of saving fuel with the starter–generator system:

1. Starting assistance of ICE
2. Energy saving during "idle stop," when the ICE is shut down at traffic lights or in traffic jams
3. Regenerative electric braking of the vehicle as much as the battery recharging permits, before bringing passenger's discomfort.

A sophisticated control system is needed to perform these actions, and a 40% fuel savings in town driving and 15% for overall driving was reported with such a system that now costs probably about $1500 and fills within the volumes of an existing car.

Here, we will pay attention to Lundell starter–generator control through the PWM inverter, as coordinated excitation and armature control are required for optimum performance in the four main operation modes:

1. Vehicle stopping (idle stop): the Lundell machine drives the air conditioner, power steering, and other auxiliaries
2. Starting: after initial starting by the starter, the Lundell machine restarts the ICE
3. Normal driving: generating as needed by the battery monitored state
4. Deceleration: regenerative braking to recharge the battery

An intelligent power metal-oxide semiconductor field-effect transistor (MOSFET) module PWM inverter is used.

The Lundell machine may be either vector- or direct torque and flux-controlled (DTFC) [16,17]. In what follows, we will dwell on DTFC, as it leads to an inherently more robust control. The rotor position is required for speed calculation and in the flux observers but not for coordinate transformation in the control. Also, DTFC may be adapted easily for the generating operation mode.

Essentially, for motoring, the Lundell machine should be controlled at unity power factor to minimize the machine losses and inverter kilovoltamperes. The generator excitation-only control may be used with the MOSFETs inhibited in the converter, where only the diodes are working.

The torque remains the main controlled variable during motoring—for starting the ICE operation mode. For auxiliary motoring (idle stop operation mode), speed control may be used by adding an external speed control loop.

As at low speed, calculations of the power factor are not very robust, a feed-forward field current calculator is added. The stator flux reference level is adapted to speed for generating and motoring (during idle stop).

The *dq* model of the Lundell machine is imperative in conceiving the DTFC system for the former. The *dq* model of the SM is as follows [16]:

$$\begin{aligned}
&\bar{\imath}_s R_s - \bar{V}_s = -\frac{\partial \bar{\Psi}_s}{\partial t} + j\omega_r \bar{\Psi}_s \\
&\bar{\imath}_s = i_d + j i_q; \quad \bar{V}_s = V_d + jV_q; \quad \bar{\Psi}_s = \Psi_d + j\Psi_q \\
&\Psi_d = L_{sl} i_d + \Psi_{dm}; \quad \Psi_{dm} = L_{dm} i_{dm}; \quad i_{dm} = i_d + i_{dr} + i'_f \\
&\Psi_q = L_{sl} i_q + \Psi_{qm}; \quad \Psi_{qm} = L_{qm} i_{qm}; \quad i_{qm} = i_q + i_{qr} \\
&i'_f R'_f - V_f = -\frac{\partial \Psi'_f}{\partial t}; \quad \Psi'_f = L'_{fl} i'_f + \Psi_{dm}; \quad R_s = R_1\text{—phase resistance} \\
&i_{dr} R_{dr} = -\frac{\partial \Psi_{dr}}{\partial t}; \quad \Psi_{dr} = L_{sl} i_{dr} + \Psi_{dm} \\
&i_{qr} R_{qr} = -\frac{\partial \Psi_{qr}}{\partial t}; \quad \Psi_{qr} = L_{sl} i_{qr} + \Psi_{qm} \\
&T_e = \frac{3}{2} p_1 \, \mathrm{Re}al\left(j\psi_s i_s^* \right) = \frac{3}{2} p_1 (\Psi_d i_q - \Psi_q i_d)
\end{aligned} \tag{6.41}$$

i_{dr} and i_{qr} are the rotor damper (claw) currents reduced to the stator.

For steady state, in rotor coordinates, $(\partial/\partial t) = 0$ and, for unity power factor, the space-lag angle between the stator voltage and current vectors (V_s, I_s) is zero, and thus, the stator flux vector is $\overline{\Psi}_s$ and is located at 90° (electrical) with respect to $\overline{V}_s$:

$$\overline{i}_{so}R_s - \overline{V}_{so} = -j\omega_r\overline{\Psi}_{so} \tag{6.42}$$

For $\cos\varphi_1 = 1$,

$$\begin{aligned} &i_{so}R_s - V_{so} = -\omega_r\Psi_{so}; \quad \Psi_{so} = L_{sl}\cdot i_{do} + L_{dm}(i_{do} + i'_{fo}) + j(L_{sl} + L_{qm})\cdot i_{qo} \\ &T_e = \frac{3}{2}p_1\cdot\Psi_{so}\cdot i_{so} \\ &i'_{fo} = \frac{V'_f}{R'_f}, \quad i_{dro} = i_{qro} = 0 \end{aligned} \tag{6.43}$$

The vector diagram of the Lundell machine for $\cos\varphi_1 = 1$ (and steady state) is shown in Figure 6.26. It resembles the phasor diagrams dealt with earlier in this chapter; but in the time delay angle, there are now space angles of the same value, and one vector represents the whole machine and not only a phase.

The vector diagram allows for computation of field current i^*_{fo} for given stator flux Ψ^*_{so} and torque T^*_{eo}. From the torque expression (Equation 6.43), the stator current i^*_{so} is as follows:

$$i^*_{so} = \frac{2}{3}\cdot\frac{T^*_{eo}}{p_1\cdot\Psi^*_{so}} \tag{6.44}$$

From Figure 6.26,

$$\begin{aligned} \cos\delta^*_\psi &= \frac{i^*_{qo}}{i_{so}} \\ \sin\delta^*_\psi &= \frac{L_q i^*_{qo}}{\Psi^*_{so}} \end{aligned} \tag{6.45}$$

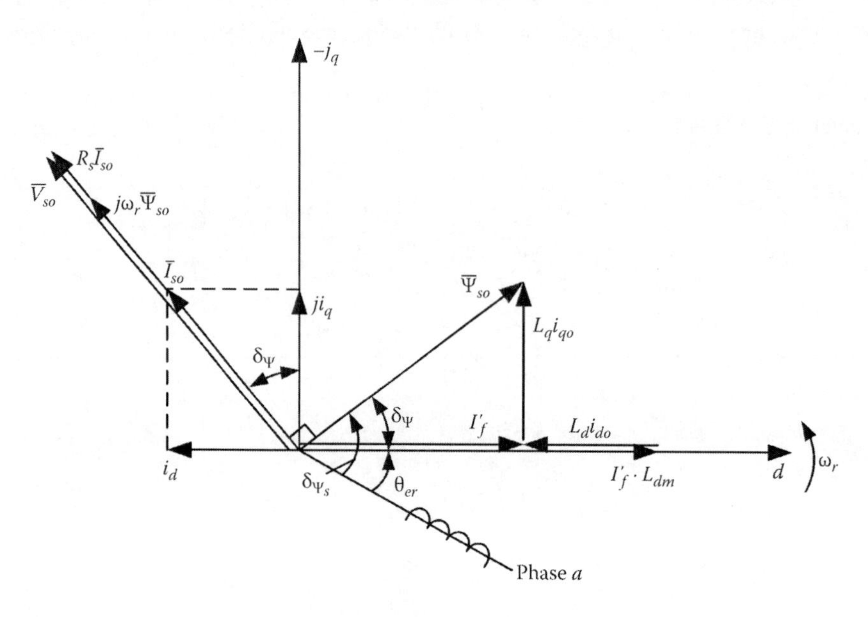

FIGURE 6.26 Vector diagram for $\cos\varphi_1 = 1$ and motoring.

Consequently,

$$\tan\delta_{\Psi}^{*} = \frac{L_q \cdot i_{so}^{*}}{\Psi_{so}^{*}} \tag{6.46}$$

with δ_{Ψ}^{*} known, the required field current $i_{fo}'^{*}$ is obtained:

$$i_{fo}'^{*} = \frac{\Psi_{so}^{*}\cos\delta_{\Psi}^{*} + L_d i_{so}^{*}\sin\delta_{\Psi}^{*}}{L_{dm}} \tag{6.47}$$

Also, with δ_{Ψ}^{*} known, the stator flux desired position with respect to phase a in the stator $\theta_{\Psi_s}^{*}$ is as follows:

$$\theta_{\Psi_s}^{*} = \delta_{\Psi}^{*} + \theta_{er} \tag{6.48}$$

The rotor position is measured through a rugged (magnetic) encoder or it is estimated.

For firm driving around zero speed, an encoder is recommended, though robust flux estimators at zero speed for torque control are now available.

The PWM inverter (Figure 6.27a) produces six nonzero and two zero-voltage vectors: V_1, …, V_6, V_0, V_7 (Figure 6.27b).

The voltage vector $\overline{V}_1$ "acts" along phase a because phase a is connected in series with phases b and c in parallel.

The vector definition, in stator coordinates, is

$$V_s = \frac{2}{3}\left(v_a + v_b e^{j(2\pi/3)} + v_c e^{-j(2\pi/3)}\right) \tag{6.49}$$

For $\overline{V}_1 : v_a - v_b = V_{DC}$, $v_b = v_c$, and $v_a + v_b + v_c = 0$, and thus, $v_a = (2/3)\ V_{DC}$, $v_b = v_c = -(1/3)\ V_{DC}$. For the other voltage vectors, similar expressions are obtained.

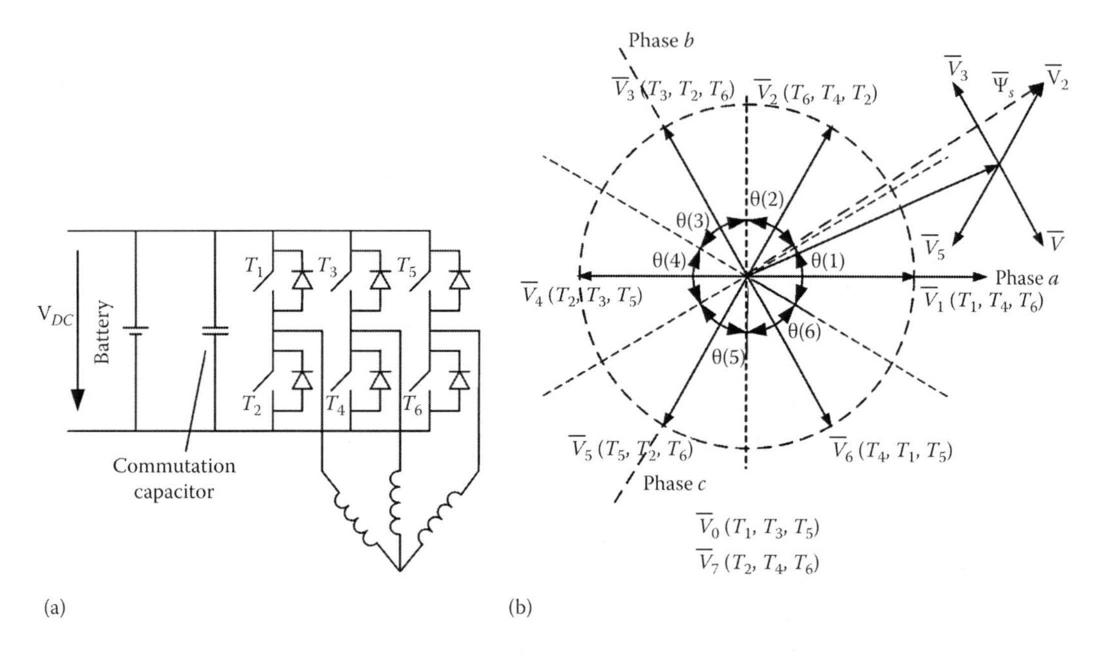

FIGURE 6.27 (a) Typical voltage-source pulse-width modulator (PWM) inverter and (b) its voltage vectors.

In DTFC, the stator flux $\overline{\Psi}_s(\Psi_s, \delta_{\Psi_s})$ and the torque T_e are estimated and compared to their desired values Ψ_s^* and T_e^* to find the errors ε_{Ψ_s} and ε_{T_e}:

$$\begin{aligned}\varepsilon_{\Psi_s} &= \Psi_s^* - \Psi_s \\ \varepsilon_{T_e} &= T_e^* - T_e\end{aligned} \tag{6.50}$$

Together with the stator flux, δ_{Ψ_s}, ε_{Ψ_s}, and ε_{T_e} (values and sign) form the variables by which a single-voltage vector (or a given sequence of them) is triggered in the converter. No coordinate transformation is needed for control.

Although the stator flux vector position angle δ_{Ψ_s} may be precisely found, only its location in one of the six (60° wide) sectors in Figure 6.27b is needed.

The method for choosing the adequate voltage vector for given stator flux vector $\theta(i)$ and torque and flux errors comes from the stator equation in stator coordinates:

$$\bar{i}_{ss}R_s - \overline{V}_{ss} = -\frac{d\overline{\Psi}_{ss}}{dt} \tag{6.51}$$

if $R_s \approx 0$,

$$\overline{\Psi}_{ss} - \overline{\Psi}_{ss}^0 \approx \int_0^T \overline{V}_{ss} dt = \overline{V}(i) \cdot T \tag{6.52}$$

The flux vector variation falls along the applied voltage vector direction.

Also, for increasing motoring torque, the flux vector has to be accelerated in the direction of motion. It should be decelerated for motoring torque reduction.

For the flux vector in Section 1 ($\theta(1)$), to increase the flux amplitude and increase the torque, $\overline{V}_2$ should be turned on; to increase the torque, but decrease the flux, $\overline{V}_3$ is chosen; to decrease the torque and increase flux, $\overline{V}_6$ is needed; while to decrease torque and decrease flux, $\overline{V}_5$ is applied. A unique table of commutations is thus feasible (Table 6.1) (p. 232 in Reference 16).

The flux and torque error value of 1 means positive value, while –1 means negative value. To improve the stator current waveform and reduce torque pulsations, instead of each voltage vector in Table 6.1, a combination of the two neighboring vectors and a zero vector, or a more sophisticated regular combination, may be implemented. This is space-vector modulation (SVM) in a kind of random "expression."

Although the 1, –1, 0 outputs of flux and torque errors suggest hysteresis torque and flux regulators, more advanced (proportion integral [PI], sliding mode) regulators were recently used with remarkable success [18].

The general control scheme for motoring is as in Figure 6.28. In general, full flux is needed for full torque at low speeds. The eventual errors in the machine parameters, as influences on i_f^*, are corrected by the power factor regulator.

A four-quadrant DC–DC converter seems best to control the field current in order to obtain very fast field current response. This way, overvoltages are avoided, and the torque response is faster. The stator

TABLE 6.1 Direct Torque and Flux Control (DTFC) Switching Table

		$\theta(i)$					
ε_{Ψ_s}	ε_{T_e}	$\theta(1)$	$\theta(2)$	$\theta(3)$	$\theta(4)$	$\theta(5)$	$\theta(6)$
1	1	V_2	V_3	V_4	V_5	V_6	V_1
1	–1	V_6	V_1	V_2	V_3	V_4	V_5
0	1	V_0	V_7	V_0	V_7	V_0	V_7
0	–1	V_0	V_7	V_0	V_7	V_0	V_7
–1	1	V_3	V_4	V_5	V_6	V_1	V_2
–1	–1	V_5	V_6	V_1	V_2	V_3	V_4

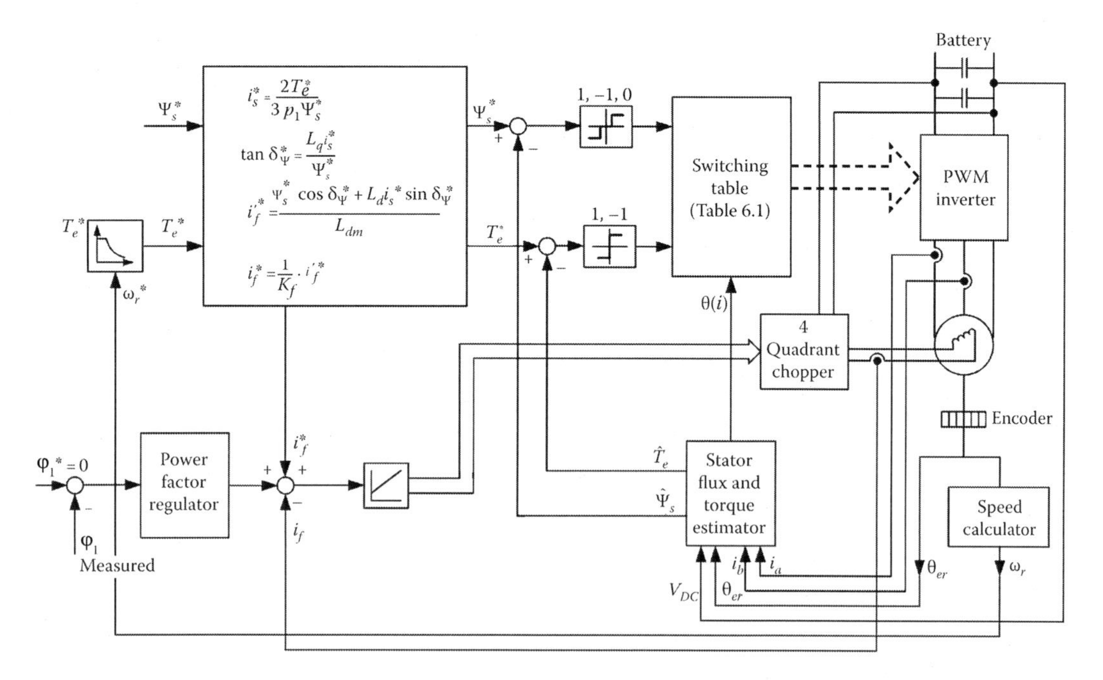

FIGURE 6.28 Motoring direct torque and flux control (DTFC) of Lundell machine.

flux and torque estimator configurations are a matter of choice from many possibilities [16]. Here, a combined voltage–current model is used, as position feedback $\overline{\theta}_{er}$ is available (Figure 6.29a and b).

The PI compensator drives to zero the error between the voltage and current models, providing for current model domination at low frequencies (speeds). In general, only the DC voltage is measured, but for the 42 V_{DC} system, care must be exercised at very low speeds to correct for the stator resistance variation and for MOSFET voltage drops in the inverter. The AC voltages are constructed from the knowledge of the active voltage in the inverter and the DC voltage, as discussed earlier in this paragraph.

The errors in parameters for current model have an effect only at low speed, where the current model is dominant. The trustworthy encoder feedback should secure fast and reliable torque response of the system from zero speed. A robust position and speed observer from zero speed is another possibility. The reference torque T_e^* should be determined in a torque sharing program based on speed, total drivetrain torque requirements, battery stage, comfort, and so forth. A speed close loop may be added to provide reference torque of the air conditioner during idle stop driving.

For the generating mode, it is possible to use full control of the converter as for motoring but with the reference flux tailored with speed:

$$\Psi_s^* \approx \frac{V_{1\max}}{\omega_r}(1.03-1.05) < \psi_{s\max}^* \tag{6.53}$$

The torque reference T_e^* should now be negative and may be calculated based on braking paddle position during braking, with a limitation based on the battery energy acceptability.

During normal driving, based on battery voltage and its SOC and load requirements, a certain power P_e^* is required. A power close loop may be added to generate the negative torque reference after division by speed.

These two options are depicted in Figure 6.30.

It is also possible to totally inhibit the MOSFET in the inverter during generating modes and let the diodes do the rectification, with field current control only. In this latter case, the torque reference (in Figure 6.30) should act directly (after scaling) as excitation current reference i_f^* (Figure 6.28).

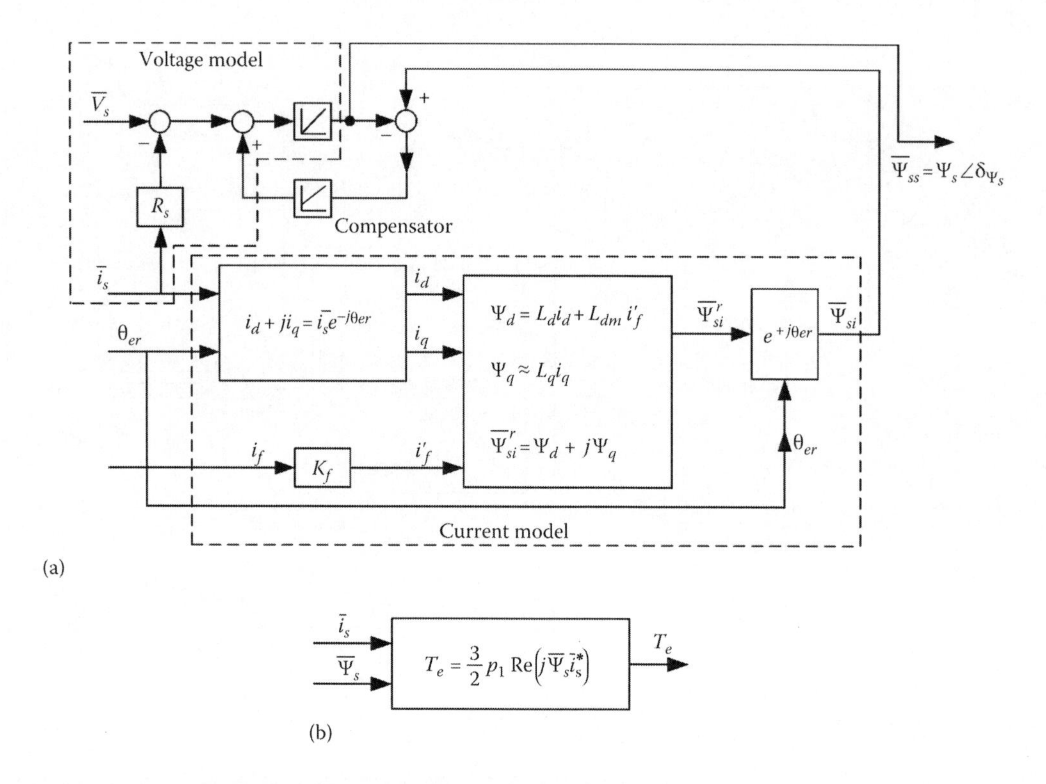

FIGURE 6.29 Typical (a) flux observer and (b) torque calculator.

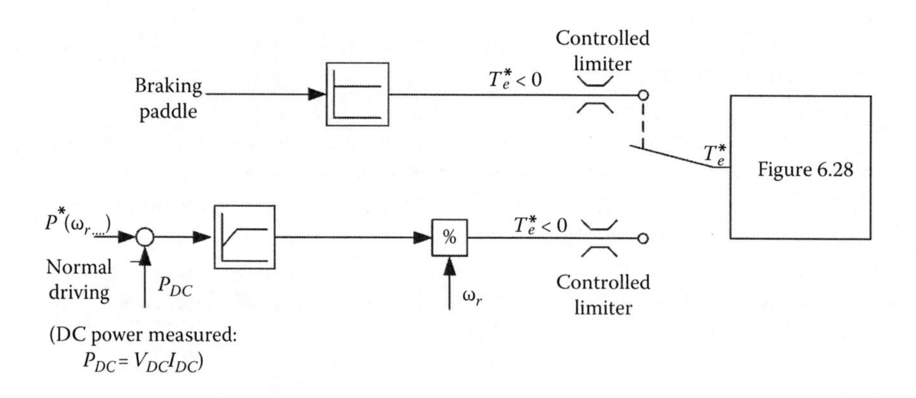

FIGURE 6.30 Generator control settings.

As out-of-town driving may be predominant, generating is required a lot, and standard generating control (by field current) might be the way to obtain greater reliability of the control system.

Note that the rather low efficiency in existing Lundell standard generator designs has to be improved notably for the starter–generator in mild hybrid cars. Allowing for more volume and using more copper in the windings, which then have to be manufactured and inserted in a special manner, and higher DC—voltage seem to be the evident ways to better performance.

The use of the Lundell machine in mild hybrid vehicles is motivated by the fact that it is a proven technology, and it has very good generating characteristics over a wide speed range, which are necessary characteristics in the application. The exclusive field current control during generating is another essential advantage of the claw-pole-rotor starter–alternator.

6.8 IPM Claw-Pole Alternator System for More Vehicle Braking Energy Recuperation: A Case Study

Abstract

This paragraph aims at demonstrating that by using PMs between the rotor poles, larger air gap and raised voltage, the hereby called interpolar magnets (IPMs) claw-pole alternator (CPA) (Figure 6.31) is suitable up to 100 kW peak power for more vehicle braking energy recuperation or propulsion support in HEVs. The main contributions of this paragraph may be divided into a comprehensive 3D magnetic circuit for IPM-CPA with magnetic saturation and skin effect considered and its validation through 3D-FEM and some experimental results; optimal design method and code for IPM-CPA; system simulation model and code for a proposed Li-ion battery backed 42 V_{DC} "hidden" power bus for 8 kW peak power of vehicle braking energy recuperation; decoupled control of 42 V_{DC} Li-ion battery and DC–DC converter control of 14 V_{DC} loads is demonstrated for up to 5 kW load levels. The encouraging results obtained thus far are hoped to be a sounder basis toward more braking energy recuperation on automobiles for further gas mileage improvements with moderate initial costs.

6.8.1 3D Nonlinear Magnetic Circuit Model

The optimal design with experimental validation of the IPM-CPA is based on the MEC and Hooke–Jeeves modified algorithm. In the MEC method, each region of the machine is represented by a permeance whose value depends on local geometrical quantities and on the local geometry and/or flux through the region.

The mathematical model of a claw-pole machine is complicated because the paths of the magnetic flux are 3D. Each flux path section is characterized by a magnetic reluctance.

For all these magnetic reluctances, analytical expressions have been adopted. The MEC of the IPM-CPA (Figure 6.32) is a three-dimensional model in the sense that the permeances of air gap, of stator and of rotor vary along radial, axial, and tangential directions. The presence of IPMs on rotor is also considered in the MEC.

The MEC model is used for the optimization of the machine using a nonlinear iterative method based on the linearization of the B–H curve.

The dynamic characteristics of the machine can be studied by rotating the rotor mesh with respect to the stator. When the rotor is rotated the radial and axial air gap reluctances, together with the axial leakage paths between the claw-poles and the plates, are modified accordingly.

The details of MEC composition are skipped here to save space.

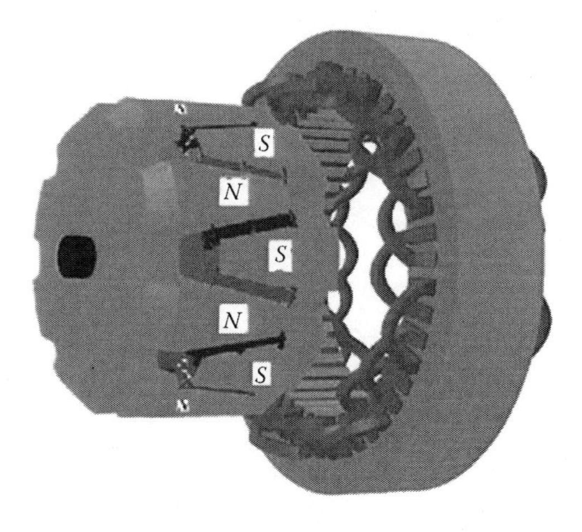

FIGURE 6.31 The geometry of the IPM-CPA (from 3D FEM analysis).

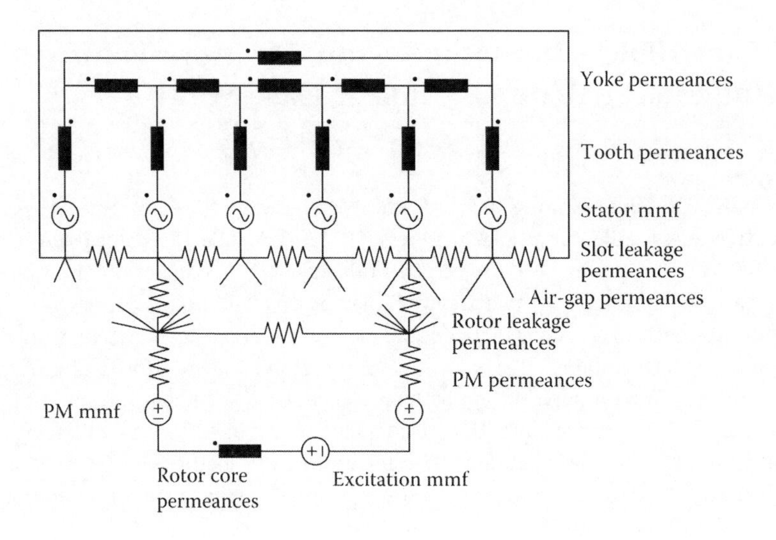

FIGURE 6.32 Multiple MEC of the IPM-CPA (one pole pair).

6.8.1.1 Evaluation Design Calibration

The described MEC model had to be checked before proceeding to optimization design, finite element analysis and control.

For this purpose, a 14 V_{DC}, 2.5 kW Bosch CPA without IPMs has been used. A good agreement between simulation and experiment can be seen both in Figure 6.33 (where the efficiency has been plotted) and in Figure 6.34 (where the maximum DC current has been plotted). Based on these results, we concluded that the MEC model is precise enough to continue the analysis of CPA through it.

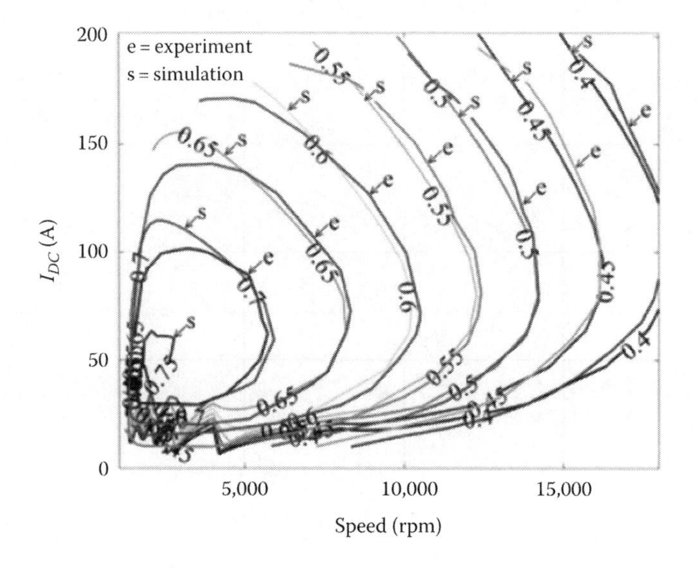

FIGURE 6.33 Efficiency of a 2.5 kW, 14 V_{DC} CPA without IPMs.

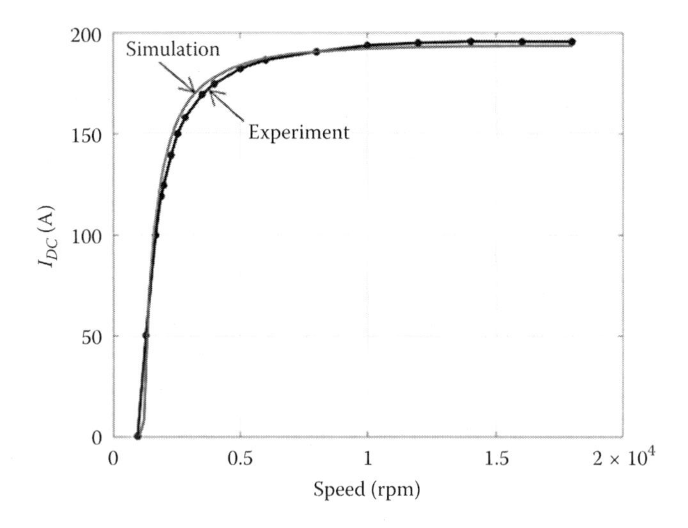

FIGURE 6.34 Maximum DC current of a 2.5 kW, 14 V_{DC} CPA without IPMs.

6.8.2 Optimal Design: Method, Code, and Sample Results with Prototype Test Results

The modified Hooke–Jeeves optimization algorithm is used in order to allow constrained system optimization using external penalty functions.

The complete analytical model has been written in MATLAB®, being structured in m-files.

The objective function represents the total (initial plus losses) cost and has the most important parameters as variables. This function will be minimized using the Hooke–Jeeves algorithm.

The optimized variables are grouped in the vector

$$\overline{X} = [sDo, sDi, rDci, lc, shy, stw, sMs, hag, poles, sb, ap, rpl, wrd, lrc, rdcl, rph, Iextmax, kipm] \quad (6.54)$$

The 18 optimization variables are explained in Table 6.2.

TABLE 6.2 Optimization Variables of the IPM-CPA

sDo	Stator outer diameter
sDi	Stator inner diameter
rDci	Rotor coil inner diameter
l_c	Stator core length
shy	Stator core yoke width
stw	Stator tooth width
sMs	Mouth of stator slot
hag	Air gap length
poles	Number of poles
sb	Turns per coil
ap	Polar cover
rpl	Rotor pole length
wrd	Width of rotor end disk
lrc	Rotor core length
rdcl	Rotor core length variation due disk
rph	Rotor pole height
Iexmax	Maximum (A turns) do field coil
kippm	Inter pole PM: 0—no IPM

For the 3.2 kW rated power and 8 kW peak power at 42 V_{DC}, the values (in mm) of their initial dimensions are:

$$\overline{X} = [151,106,54,37,7,3,2.8,0.8,16,8,0.82,37,14.4,54.8,5.4,10,2800,0.5] \tag{6.55}$$

This optimization vector varies in a range, between $\overline{X}_{\min}$ and $\overline{X}_{\max}$ ($\overline{X}_{\min} < \overline{X} < \overline{X}_{\max}$).

The chosen lower and upper limits are as follows:

$$\overline{X}_{\min} = [100,75,35,25,3,3,2.5,0.3,4,2,0.6,15,7,25,14,500,0] \tag{6.56}$$

$$\overline{X}_{\max} = [330,230,130,80,20,20,8,2,32,20,0.9,80,40,150,18,20,5500,0.9] \tag{6.57}$$

The initial step vector (initial variation of the optimization variable) $d\overline{X}$ and the minimum step vector (final variation of the optimization variable) $d\overline{X}_{\min}$ are as follows:

$$d\overline{X} = [16,16,10,5,2,1,0.5,0.4,4,2,0.1,5,2,10,2,2,500,0.1] \tag{6.58}$$

$$d\overline{X}_{\min} = [0.5,0.5,0.5,0.5,0.1,0.1,0.1,0.05,2,1,0.01,0.1,0.1,0.1,0.1,0.1,20,0.01] \tag{6.59}$$

The program achieves a minimum of the objective function after around 70 steps. The resulting optimized vector (from which the 3D-FEM analysis will start) is as follows:

$$\overline{X}_0 = [164,107,53.5,34.5,12.8,3,3.1,0.4,16,8,0.81,34.3,13.4,70.5,4.2,10,2780,0.49] \tag{6.60}$$

The optimization reveals that a maximum braking output power of 8 kW is achievable into the frame of a 3.2 kW, 14 V_{DC} CPA, after it is redesigned for 42 V_{DC} (Figure 6.35). The efficiency of the IPM-CPA has increased by 5% by optimization (a very practical initial design variable vector was chosen), from 80% to 85% (Figure 6.35). This is 10% higher than the efficiency achieved in a 2.5 kW, 14 V_{DC} alternator (without IPMs).

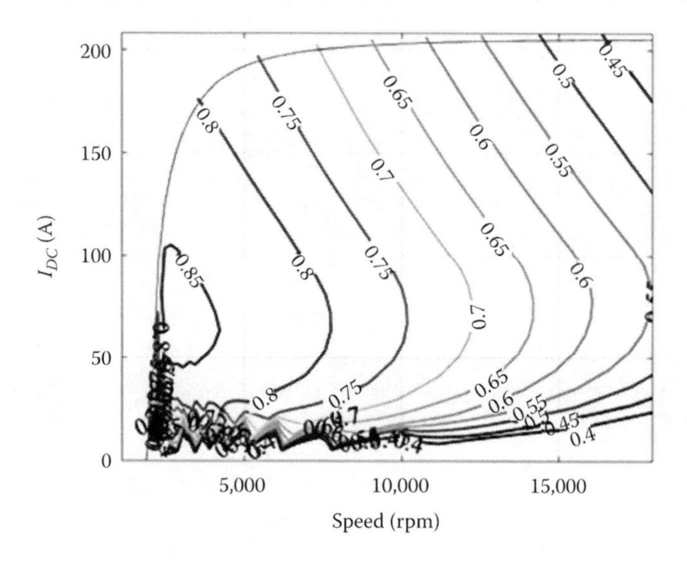

FIGURE 6.35 Efficiency of the optimized IPM-CPA at 42 V_{DC}.

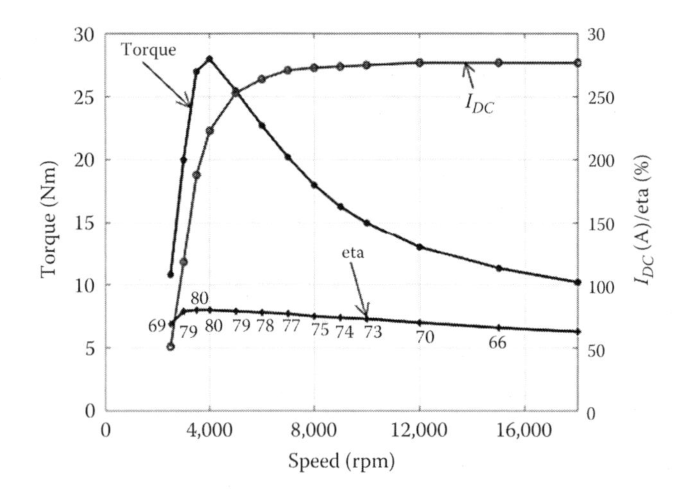

FIGURE 6.36 Experimental results of a retrofitted Bosch 3.2 kW, 42 V_{DC} IPM-CPA.

An experimental 42 V_{DC} IPM-CPA has been built by Robert Bosch GmbH within the FP7 EE-VERT 2009–2011 European program in which the authors were involved. The experimental results (Figure 6.36) show a maximum efficiency of 80% and a maximum braking output power of 11 kW.

6.8.3 3D-FEM Analysis

FEM is a valuable tool in the final stage design verifications, where it is used to check the saturation level and air gap magnetic flux density waveform at steady state. Due to the high number of elements compared to MEC, FEM offers higher accuracy. The magnetic flux flows through the claw-poles and rotor yoke in z-direction, therefore 3D FEM analysis is mandatory.

The variable that is solved in an electromagnetic FE problem is the magnetic potential at the nodes of the elements. All important quantities can be obtained by applying differential operators to magnetic vector potential.

The FE model of the 16 poles IPM-CPA (Figure 6.31) was implemented in a commercial software package by means of the built-in script programming. In the FEA, the demagnetizing curve of the NdFeB IPM (VAC 677 type) was taken into account. They have a remnant flux density Br (at 20°C) equal to 1.13 (T) and a coercive field strength (at 20°C) of 860 kA/m.

All the dimensions were imported from the MEC optimal design of the IPM-CPA ($\bar{X}_0$).

Studying Figure 6.37 we can gain better understanding of the magnetic flux density in the air gap of an IPM-CPA. Only at axial distance $z = 0$ (in the middle) the magnetic flux density has a symmetrical variation when we move from one pole to another.

The influence of slot opening in the magnetic flux density at no-load operation is evident. The strong departure of the air gap flux density distribution from a sinusoidal waveform is due to the low number of slots per pole (3), and to magnetic saturation.

The three main harmonics of the magnetic flux density in the air gap of the IPM-CPA are presented in Figure 6.38. In addition, the harmonic spectrum of the CPA (which was obtained based on the air gap magnetic flux density variation) is shown in Figure 6.39.

The air gap magnetic flux density harmonics (Figures 6.38 and 6.39) cause emf's harmonics "pollution." In addition, the air gap field harmonics produce additional eddy currents in the claw poles. These losses increase with speed.

To find the flux through a stator coil, a surface corresponding to the coil geometry was defined. The script rotates the mobile part with a step of 1.5 mechanical degrees over one pole, analyses the model and

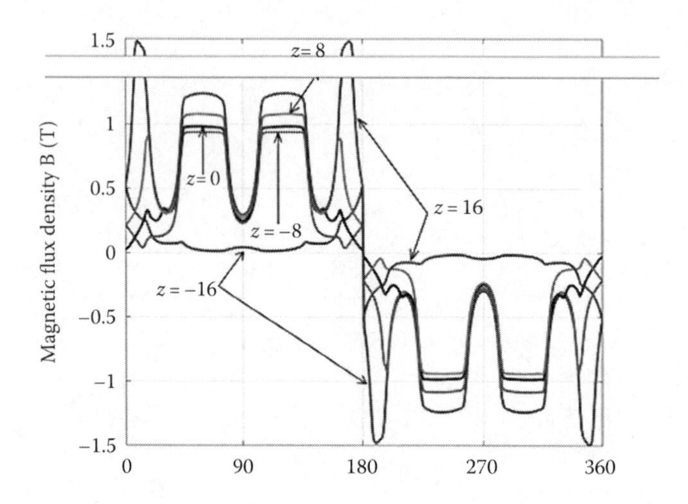

FIGURE 6.37 Distribution of the magnetic flux density at no-load-IPM-CPA (DC field mmf = 1500 At) at different axial lengths: $z = -16; -8; 0; 8; 16$ mm.

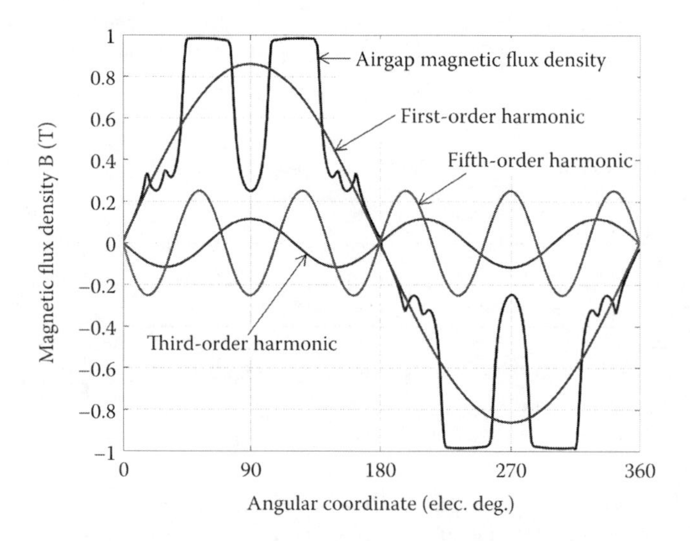

FIGURE 6.38 The three main harmonics of the air gap magnetic flux density (at axial distance $z = 0$); IPM-CPA.

computes the integral of magnetic flux density over the defined surface. At the same mmf, a 40% growth in air gap flux appears when IPMs are used (Figure 6.40), due to the fact that the flux between rotor poles is "destroyed"—redirected, and thus the useful flux is enhanced.

Given the Faraday law: derivating the flux with respect to the mechanical angle and multiplying it by speed (and all the other transformation constants), we obtain the emf induced at no-load in a phase (Figure 6.41). The emf induced only by the IPM at 3000 rpm is about 3 V (peak value).

This value assures us that even at maximum alternator speed (18,000 rpm) the no-load voltage (18 V peak, per phase), at zero field current presents no threat to power electronics equipment.

The torque was computed using Maxwell stress tensor. At constant stator currents: $I_a = I_{max}$, $I_b = -I_{max}/2$, $I_c = -I_{max}/2$, the rotor is rotated along a pole with a step of one mechanical (eight electrical) degree. The maximum stator current is the one from optimization design, $I_{max} = 223$ A at DC field mmf of 2800 A-turns. This gives us the total torque in Figure 6.42.

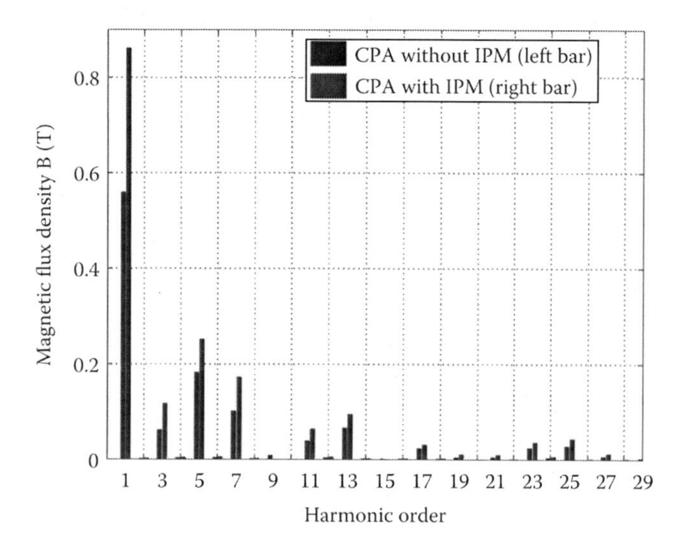

FIGURE 6.39 Harmonic content of the CPAs air gap magnetic flux density with and without IPM (at no load).

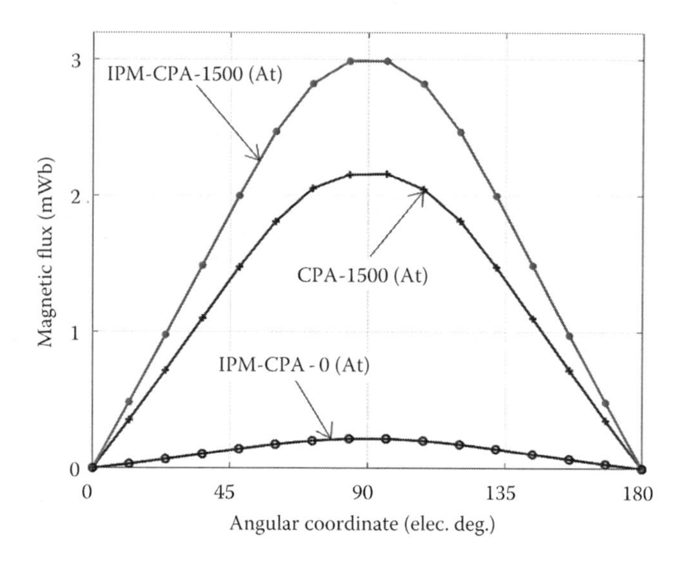

FIGURE 6.40 Flux in a coil at zero load.

The camel-back shape of the total torque is due the "cogging" torque (Figure 6.42), which appears because of the interaction between the stator teeth and magnetized rotor, when the field current and IPMs are present.

The torque pulsations in total torque (Figure 6.43) were obtained by modifying the three sinusoidal stator currents: I_a, I_b, and I_c with respect to the angular shifting of the rotor for a given *dq* power angle and initial rotor angle of 104° (13° × 8°) (Figure 6.42)—which should correspond to unity power factor operation due to diode rectifier constraint. The average value of the torque of the CPA with IPM is 7.8 (Nm), while the average torque value for the CPA without IPM is about 5.1 (Nm). More than 40% torque increase is obtained by PMs. (The Bosch modified IPM-CPA in Figure 6.36 is not exactly our optimized design, which explains its higher power range, but confirms the rising torque trend.)

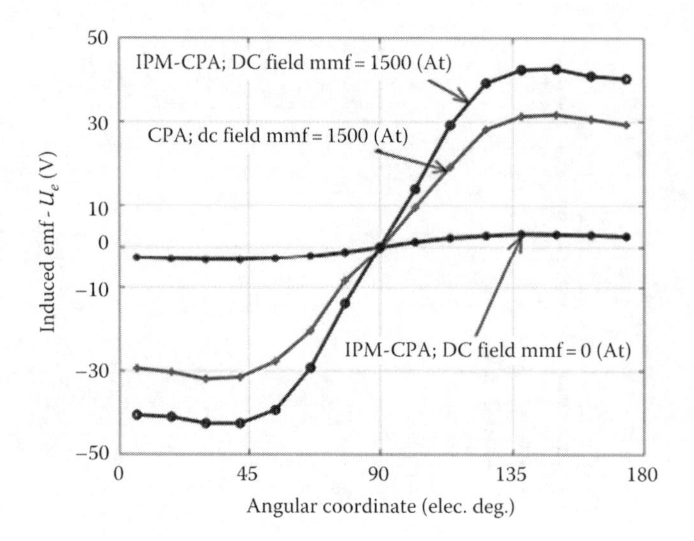

FIGURE 6.41 Computed emf in a stator phase at 3000 rpm.

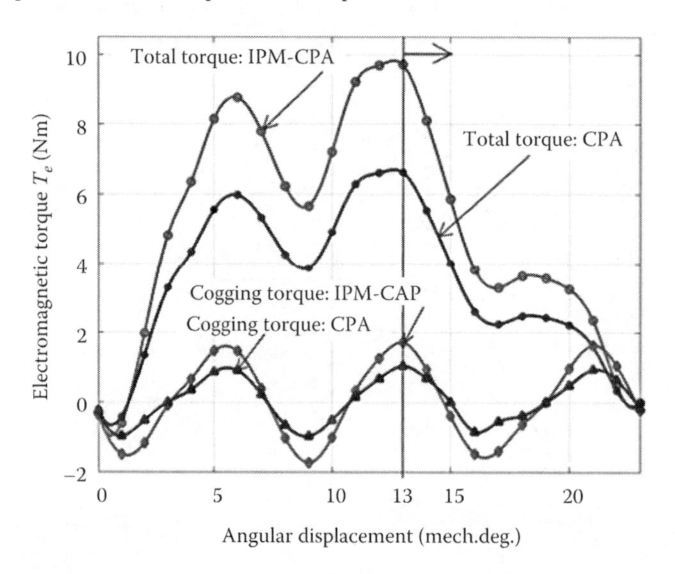

FIGURE 6.42 Total and cogging torque at $I_{max} = I_a = -2\ I_b = -2\ I_c = 200$ A.

6.8.4 Vehicle Braking Energy Recuperation Scheme and Its Control

The proposed vehicle power net (Figure 6.44) for vehicle braking energy recuperation contains the electrically excited CPA, with its embedded diode rectifier, the storage Li-ion battery with high voltage loads, the lead acid battery with classical low voltage loads and two DC–DC quadrant converters, one to interconnect the 42 V_{DC} bus with the 14 V_{DC} bus and the second to supply the excitation circuit of the alternator. This architecture has the advantage of preserving the actual power bus structure for existing loads and at the same time, to offer a larger voltage for new high power consumers as air-conditioners and other new electrical actuators.

The power converter supplies only the 14 V_{DC} loads so its cost and power level and losses are reduced. The 14 V_{DC} DC bus experiences less influence of the engine speed variation and, consequently, its level is more stable. The energy management block is controlling the alternator excitation current in order

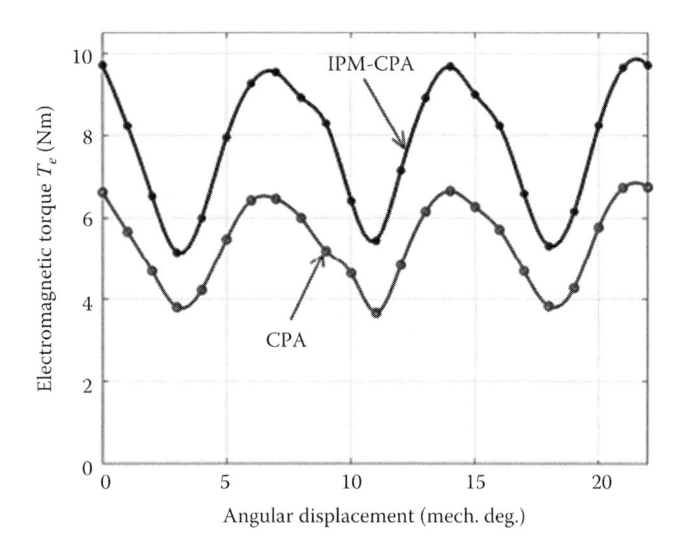

FIGURE 6.43 Torque pulsation of the CPA.

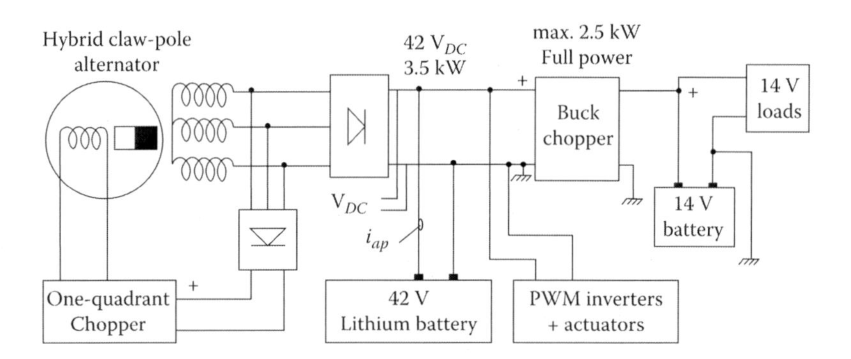

FIGURE 6.44 Vehicle power net.

to extract the maximum calculated power of 8–11 kW, sufficient in vehicle urban driving to reduce fuel consumption by 8%–10%, if the battery SOC is smaller than unity; respectively, to produce power between zero and maximum load required (according to battery SOC) when the vehicle is not braking (Figure 6.45).

6.8.4.1 Dynamic Model of the Proposed System

The MATLAB-Simulink® model of power net is reflecting its structure in Figure 6.45, containing the alternator with embedded rectifier model, storage battery model, load model, excitation circuit model, and the energy management block. The alternator speed, that is in direct relation with the vehicle engine speed, is the input variable of the simulations.

The main element of the power net is the new alternator capable to produce maximum power (during the vehicle braking) which should be installed in less than 0.1 s because the duration of one braking period is only a few seconds. The model of alternator is based on the *dq* equations that handle DC variables in steady state. Usually, the *dq* variables (currents and voltage) are linked to the phase components through the Park transformation and then, through the rectifier, to the DC bus. The alternator current is rather sinusoidal at working frequency (400–800 Hz for 3000–6000 rpm) despite the diode rectifier; and consequently, the dominant power transfer is through the fundamental harmonic.

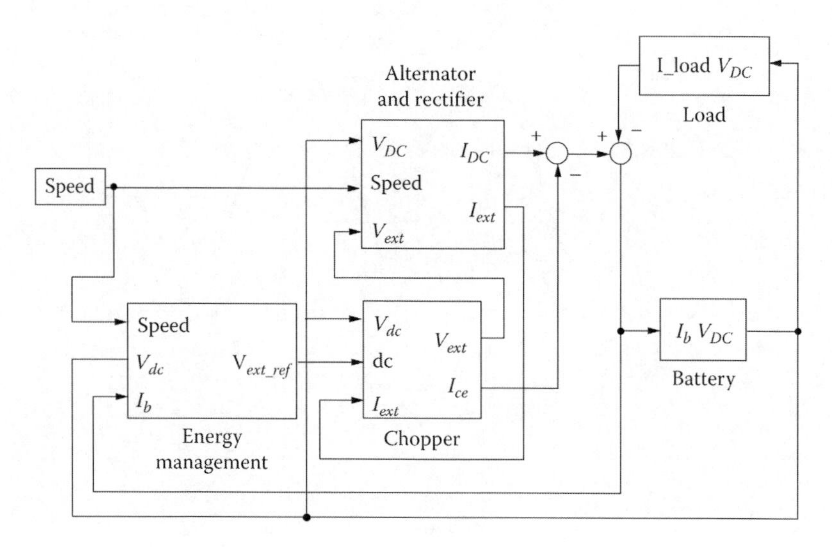

FIGURE 6.45 Power net Simulink model.

A simple algebraic equation could be used between phase voltage and DC bus voltage, respectively, phase's current and DC bus current. The alternator fundamental voltage is considered in phase with its current and, using the relation between phase voltage/currents and DC bus voltage/currents, it is possible to use directly the reduced value of the DC voltage (Equation 6.61) in the *dq* model, and also to compute directly the DC current from *dq* currents (Equation 6.63).

$$V_m = k_v(V_{dc} - 2V_d) \tag{6.61}$$

where

V_m is the *d*, *q* model voltage vector magnitude
V_{DC} is the DC voltage
V_d is the direct voltage drop on the diode
k_v is the DC voltage reduction factor (2)

$$k_v = \begin{cases} \frac{2}{\pi} & \text{for star winding connection} \\ \frac{2}{\pi}\sqrt{3} & \text{for delta winding connection} \end{cases} \tag{6.62}$$

$$I_{DC} = k_i\sqrt{I_d^2 + I_q^2} \tag{6.63}$$

where

I_{DC} is the DC alternator current
I_d is the *d* axis current component
I_q is the *q* axis current component
k_i is the stator current reduction factor (4)

$$k_i = \begin{cases} \frac{3}{\pi} & \text{for star winding connection} \\ \frac{3}{\pi}\sqrt{3} & \text{for delta winding connection} \end{cases} \tag{6.64}$$

Permanent magnets are placed between claw poles in order to reduce the excitation field fringing. A small PM flux appears in d axis in the same direction with the excitation flux, but the main benefit of PMs is the improvement of excitation characteristics. Finally, the alternator dq voltage equations are presented in Equations 6.65 and 6.66 while the excitation equation, reduced to the stator winding, appears in Equation 6.67. The transformer coupling between stator d axis and excitation flux appears also between I_d, I_{el} time derivatives (Equation 6.68).

$$\frac{d\Psi_d}{dt} = -V_m \sin(\theta) - k_s R_s I_d + \omega_e L_q I_q \tag{6.65}$$

$$\frac{d\Psi_q}{dt} = -V_m \cos(\theta) - k_s R_s I_q + \omega_e(\Psi_{PM} + L_{md} I_e) - \omega_e L_q I_q \tag{6.66}$$

$$\frac{d\Psi_{el}}{dt} = V_{el} - R_{el} I_{el} \tag{6.67}$$

$$\frac{d}{dt}\begin{pmatrix} I_d \\ I_{el} \end{pmatrix} = \begin{pmatrix} L_d & -L_{md} \\ -L_{md} & L_{el} \end{pmatrix}^{-1} \cdot \begin{pmatrix} \frac{d}{dt}\Psi_d \\ \frac{d}{dt}\Psi_{el} \end{pmatrix} \tag{6.68}$$

where

Ψ_d is the d axes flux
Ψ_q is the q axes flux
Ψ_{el} is the excitation stator reduced flux
R_s is the DC stator winding resistance
k_s is the skin effect factor
L_d is the d axes inductance
L_q is the q axes inductance
L_{md} is the d axes main inductance
V_{el} is the excitation reduced voltage
I_{el} is the excitation stator reduced current
R_{el} is the excitation stator reduced resistance
L_{el} is the excitation stator reduced inductance
ω_e is the electric speed
θ is the internal angle (between voltage vector and q axis)

The currents are computed by integrating their derivatives from Equations 6.68 and 6.66, but the integration results are limited to positive values to consider the embedded rectifier in the stator and the one quadrant chopper in the excitation. When either the excitation current or d axis current is zero (smaller than a positive given error in the simulation), decoupled circuits are considered (Equation 6.69):

$$\frac{d}{dt}\begin{pmatrix} I_d \\ I_{el} \end{pmatrix} = \begin{pmatrix} \frac{1}{L_d} & 0 \\ 0 & \frac{1}{L_{el}} \end{pmatrix} \begin{pmatrix} \frac{d}{dt}\Psi_d \\ \frac{d}{dt}\Psi_{el} \end{pmatrix} \tag{6.69}$$

The internal angle θ is the current vector angle with axis q (the voltages are in phase with currents) if either I_d or I_q is different from zero; else, θ is the angle of total rotor produced (transformer and rotating) induced voltage (Equation 6.70) with axis q. This "trick" is required to start the diode rectifier action when I_d and $I_q = 0$.

$$\theta = \begin{cases} \arg\tan(I_d, I_q) & \text{if } I_d \neq 0 \vee I_q \neq 0 \\ \arg\tan(V_{e1} - R_{e1}I_{e1},\ \omega_e(L_{md}I_{e1} + \Psi_{PM})) & \text{if } I_d = 0 \wedge I_q = 0 \end{cases} \quad (6.70)$$

A simplified empirical skin effect Equation 6.71 is proposed considering that the electric speed, ω_{sk}; leads to almost doubling the AC resistance compared with DC resistance at high speeds. The equation has the merit to consider the skin effect dependence with the frequency at 1.5 exponent at very large frequency, while a single parameter, ω_{sk}, should be provided. It shows a good agreement with the standard hyperbolic functions that are considering the slot geometry and conductor diameter, parameters that are not generally available for commercial alternators.

$$k_s = 1 + k_{sk}\frac{\omega_e^{2.5}}{\omega_e + \omega_{sk}}; \quad k_{sk} = 2\omega_{sk}^{-1.5} \quad (6.71)$$

The model is computing also the shaft torque that could be used as input for other blocks in complex vehicle dynamic simulation or to compute the alternator efficiency. The shaft torque, T_S, Equation 6.72 has three components: the electromagnetic torque associated with the power transmitted to stator winding, the electromagnetic torque associated with the iron losses, and mechanical losses torque.

$$T_s = \frac{3}{2}\left[p(\Psi_d I_q - \Psi_q I_d) + p\omega_e \frac{\Psi_d^2 + \Psi_q^2}{R_{Fe}}\right] + T_{f0} + k_f n^2 \quad (6.72)$$

where

- p is the number of pole pairs
- R_{Fe} is the equivalent iron losses resistance
- T_{f0} is the friction torque
- k_f is the coefficient of ventilation torque
- n is the mechanical speed in rpm

The discontinuous alternator current mode is not considered in this model because it appears only at low speeds and has low influence on the energy balance.

6.8.4.2 42 V_{DC} Storage Battery Model

The Li-ion 42 V_{DC} storage battery is an important power net component, and a simple battery model, considering the internal constant resistance and a nonlinear battery emf voltage dependence with battery SOC, is proposed here:

$$V_b = R_b I_b + V_{be} \quad (6.73)$$

$$V_{be} = \begin{cases} V_0 \dfrac{Q}{Q_0} & \text{for } 0 < Q \leq Q_0 \\ V_0 + (V_{bn0} - V_0)\dfrac{Q - Q_0}{Q_n - Q_0} & \text{for } Q_0 < Q \leq Q_n \end{cases} \quad (6.74)$$

where

- R_b is the battery resistance
- I_b is the battery current
- V_{be} is the battery emf voltage
- V_{bn} is the battery rated emf voltage
- V_0 is the emf voltage for discharged battery
- Q is the actual battery electric charge
- Q_n is the battery rated charge
- Q_0 is the battery residual charge

The battery charge is computed by integrating the current and is limited between zero and rated charge as shown in Equation 6.75 where $Q(t_k)$, $Q(t_{k-1})$ is charge at time t_k, respectively, at t_{k-1} and dt is an infinitesimal time difference between t_k and t_{k-1}.

$$Q(t_k)=\begin{cases}0 & \text{if } Q(t_{k-1})+I_b dt \le 0\\ Q(t_{k-1})+I_b dt & \text{if } 0<Q(t_{k-1})+I_b dt<Q_n\\ Q_n & \text{if } Q_n \le Q(t_{k-1})+I_b dt\end{cases} \tag{6.75}$$

The real battery behavior is more complex, with temperature dependences, aging, and history of charge and discharge cycles but, for our scope here, a simple model is enough accurate.

6.8.4.3 Control Strategy

The load power is assumed to be constant and the load current is computed considering the DC variable voltage. A low pass filter on the load current simulates the time constant of the load power converter. The energy management block is organized on two hierarchical levels: the battery voltage control and battery current control.

The battery actual emf voltage is computed considering the DC voltage, battery current, and battery resistance. The control is based on a monotonous relation between battery emf voltage and battery SOC. The battery voltage control block (Figure 6.46) is the key of the energy recovering and storage energy management and it consists in a variable structure control. The speed derivative is computed; and if it is smaller than a certain negative value, then the emf battery voltage reference is set at maximum value, $V_{b\max}$, otherwise it is set to an optimal value, V_{bopt}, small enough to secure energy recovering from

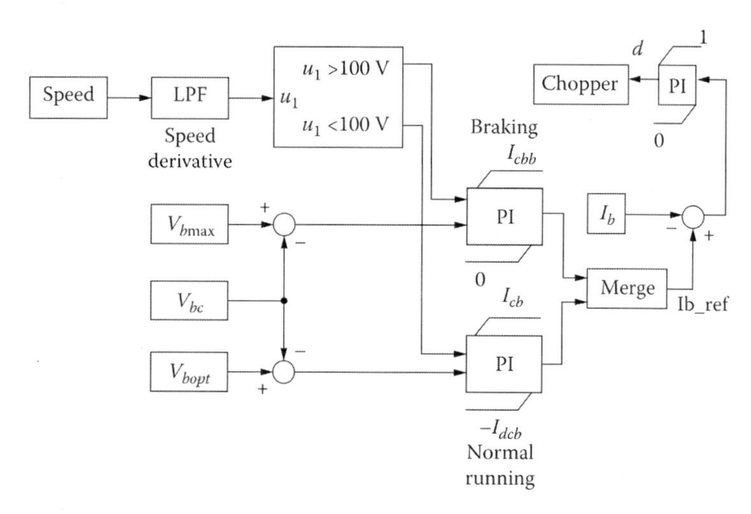

FIGURE 6.46 Battery voltage and current control.

several repetitive braking events, but high enough to supply the vehicle loads a certain time interval when the engine is stopped. In a real vehicle a signal from the braking pedal or from a sudden release of acceleration pedal could be used to change the battery reference voltage in the voltage controller. Both controllers are PI controllers with limitations but with different parameters. The "Braking" PI controller (Figure 6.46) has a larger gain and its upper current limit is set equal to maximum battery charge current and its minimum is zero, while the "Normal running" PI controller (Figure 6.46) has a smaller gain, smaller charger current, and the lower limit is negative and it means maximum battery discharging current. The internal states of controllers are frozen when inactive and when the control structure is changed, the active controller starts from the last internal state.

The battery current controller (not shown in a graph) is also a PI controller with limitation and a large frequency band (Figure 6.45). Its output between zero and unity (in P.U.) represents the duty cycle for the excitation chopper. The chopper model is computing the excitation voltage, V_{ex}, and the required current from excitation, I_{ch}, considering the DC voltage, voltage drop on the chopper device and excitation current from the alternator model:

$$V_{ex} = d \cdot (V_{DC} - \Delta V_{ch}) \tag{6.76}$$

$$I_{ch} = d \cdot I_{ex} \tag{6.77}$$

6.8.4.4 Simulation Results

The scope of simulations is to show the controlled alternator capabilities to change quickly its states to follow the required dynamics for braking energy recovery. To this scope, a repetitive alternator speed input signal was considered with an acceleration time from 3000 to 6000 rpm, running time at 6000 rpm, deceleration time at 3000 rpm and running time at 3000 rpm (Figure 6.47a).

The alternator parameters are $p = 8$ pole pairs, $R_s = 0.0416\ \Omega$, $R_{ex} = 7.07\ \Omega$, $k_{ex} = 28.2$, $L_{md} = 0.26$ mH, $L_{mq} = 0.21$ mH, $L_{se} = 0.8$ H, $\Psi_{PM} = 0.4$ mWb, $f_{skr} = 1003$ Hz, $V_d = 0.65$ V, $R_{fe} = 5\ \Omega$, winding connection—delta, $T_{f0} = 0.17$ Nm, $k_f = 10^{-9}$ Nm/rpm^2.

The battery parameters are $C_n = 64$ Ah, $V_n = 42$ V, $V_0 = 36$ V (10% charged), $R_b = 0.05\ \Omega$.

The control parameters are $V_{b\max} = 42$ V, $V_{bopt} = 39$ V, $I_{cb} = 6.5$ A (slow charging battery current), $I_{cbb} =$ 120 A(maximum battery charge current), $I_{dcb} = 20$ A (maximum battery discharge current).

In the first simulation the load is set at 2.5 kW and the battery SOC is 0.8. The power net evolution is presented for 120 s through several key variables as alternator shaft speed Figure 6.47a, excitation Figure 6.47b, alternator I_a and battery current I_b, Figure 6.47c, battery terminal voltage and battery emf, Figure 6.47d and e.

Three acceleration–deceleration pulses occur during the investigated period. The initial states of controllers are set to zero. They reach the quasi steady state after the first acceleration deceleration cycle, so the behavior of the system is different during the first cycle and then it will be repeated identical for thousands of cycle until the battery emf reaches its reference (39 V, assumed to be optimal).

The limitation of the discharge current made the battery emf regulation process slow but it distributed uniformly the recovery energy usage between deceleration periods and limits the battery aging. A larger value of discharge current lets the system to consume the recovery energy immediately, accelerates the battery discharging process but it will increase the battery aging.

Next simulations (Figure 6.48), which consider a battery initial SOC equal to 0.4, show that the battery charging process is also very slow, so there is no reason to accelerate the discharge process. This behavior is a consequence of larger battery capacity that was chosen considering the maximum permissive recharge current. This is unlike the criterion for super-capacitors sizing, should they be used instead of Li-ion battery, which should be based on storage capacity.

The previous simulations show that the alternator has reserve power and it could produce quickly the required power. At the process scale there is no observable difference between reference battery current from voltage controller and real battery current (Figures 6.47 and 6.48).

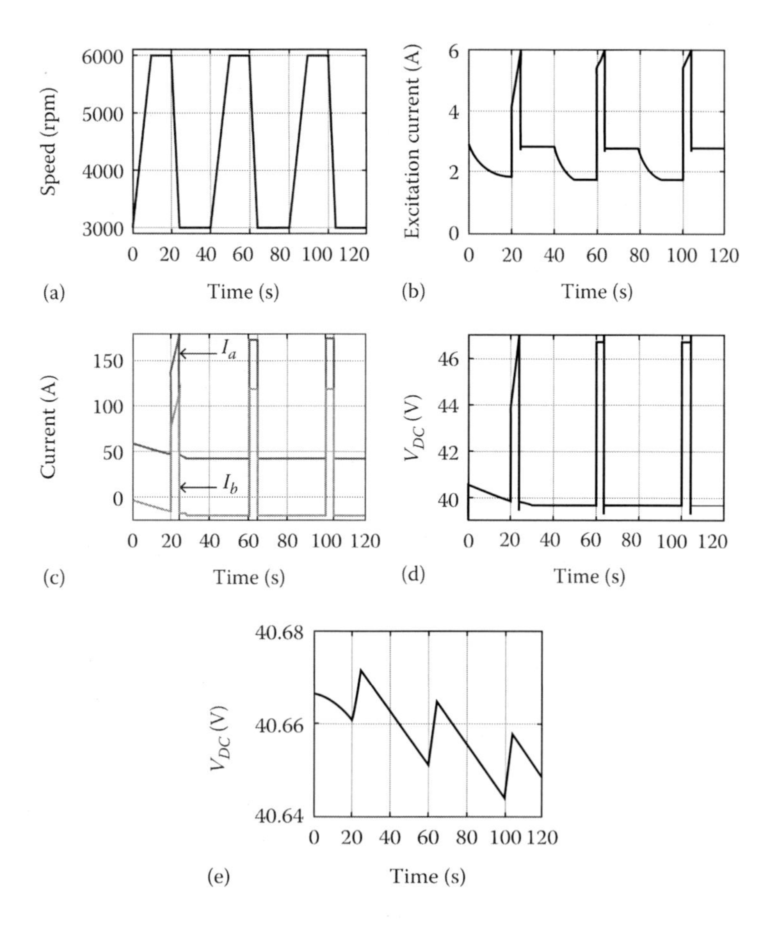

FIGURE 6.47 Parameters variation at SOC = 0.8, P_{load} = 2.5 kW.

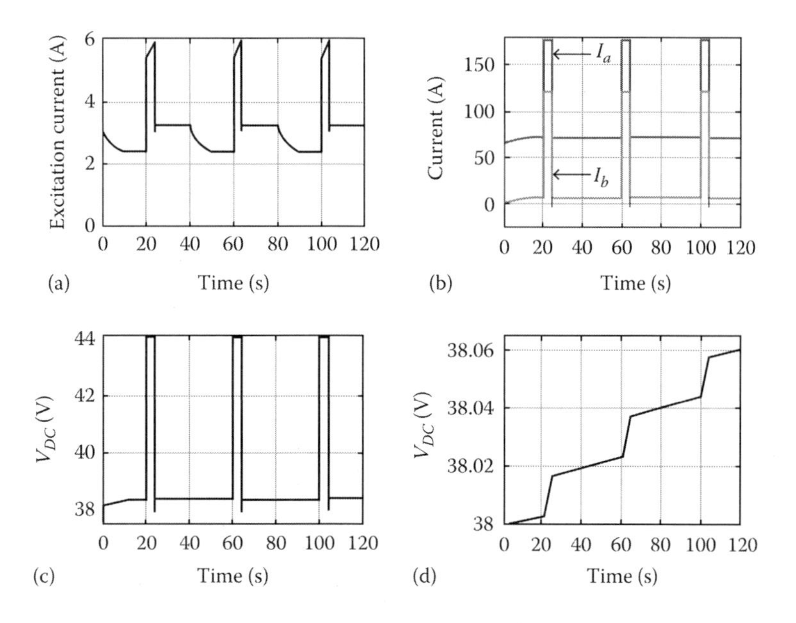

FIGURE 6.48 Parameters variation at SOC = 0.4, P_{load} = 2.5 kW.

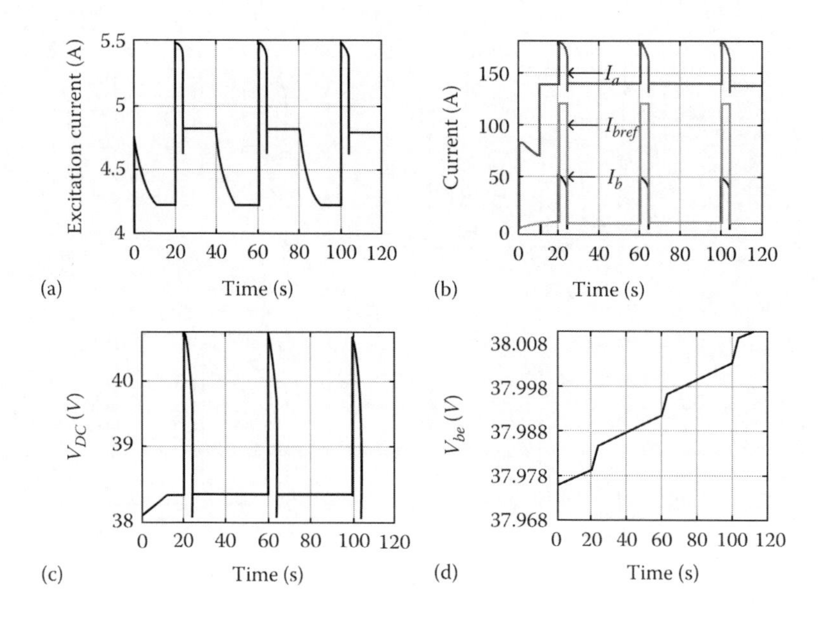

FIGURE 6.49 Parameter variation at SOC = 0.4, P_{load} = 5 kW.

The following simulations (Figure 6.49) show the system behavior when the load is increased to 5 kW and the battery initial SOC is 0.4. The battery terminal low voltage limits the maximum value of the excitation current, and the actual recharging current during the braking is not able to follow the reference current, Figure 6.49b, and the recover energy is smaller. This is proved by lower torque despite of larger power requirement.

Better results are obtained if the battery initial voltage and SOC is larger (SOC = 0.8), as they are shown in Figure 6.50. A little larger terminal voltage makes the difference. This simulation results prove that a battery storage energy system is a better solution than a super-capacitor whose terminal voltage

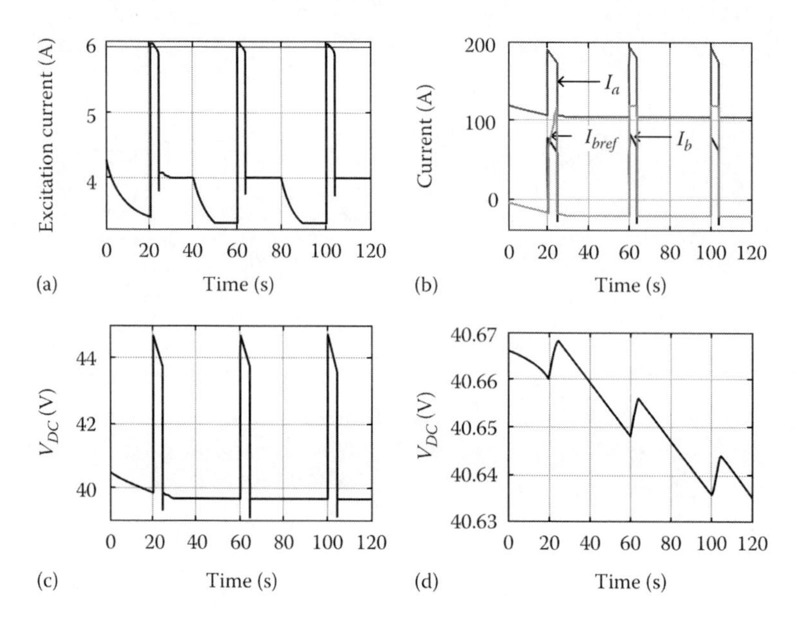

FIGURE 6.50 Parameter variation, SOC = 0.8, P_{load} = 5 kW.

TABLE 6.3 Dimensions Evolution of the IPM-CPA

	Power (kW)				
	5	10	25	50	100
U (V)	40	42	250	500	500
η (%) at $6 \cdot 10^3$ (rpm)	85	89	93	95	95
Total mass (kg)	7.5	13	43	59	62
hag (mm)	0.4	0.85	0.85	0.9	1.2
sDo (mm)	163	195	300	370	375
sDi (mm)	107	145	230	300	308
l_c (mm)	35	47	52	52	49

variation has to be much larger (50% to tap 75% of its stored energy). This would trigger the necessity of higher initial voltage to produce enough large electric power during vehicle braking.

6.8.5 Extension of IPM Alternator Utilization up to 100 kW Systems

Based on the same optimization program used in the analysis of the 8 kW braking power IPM-CPA, the possibility of obtaining CPAs up to 100 kW will also be investigated here. The study is made on the IPM-CPA with the following powers: 5, 10, 25, 50, and 100 kW. The idea is that at this range of powers, the claw-pole machine could be used also as a starter/alternator with full inverter control. The evolution of the main dimensions can be seen in Table 6.3 [20].

It seems that the IPM-CPA is fully competitive with alternative electric propulsion systems up to 100 kW on automobiles.

6.9 Summary

- Electric power needs on-board ICE vehicles (cars, trucks, buses) is on the rise.
- The Lundell machine with single ring-shaped coil multipolar rotor-controlled excitation and diode rectifier output is the only generator technology on board road vehicles today. It is basically a synchronous generator.
- The power and volume of the Lundell machine are becoming a problem with power per unit on the rise. The low efficiency of the system (about 50%) at highest speed is already a big issue, as it hampers the gas mileage.
- The improvement of Lundell machine performance (efficiency, power, and volume) for reasonable extra costs per kilowatt of installed power is the key design issue for securing some future for this generator in automotive applications.
- The power per volume specifications led to a high number of poles (12 in general).
- To reduce DC excitation losses for a 12-pole excitation, the magnetically homopolar rotor configuration leads to a very good solution.
- The same large number of poles and winding manufacturing cost limitations prompted the use of a single-layer three-phase winding on the stator with three slots per pole ($q = 1$). Only in large power units $q = 2$ (six slots per pole).
- The three slots per pole choice leads to important low-order space harmonics in the air gap permeance with a dominant emf third harmonic that increases with speed but only up to 10%.
- During on-load operation, the three slots per pole, slot openings, and claw-pole geometry concur to produce a much stronger third harmonic in the air gap (resultant) emf, especially at high speeds for star connection of phases.

- For star connection of phases, which is the dominant solution, the diode rectifier leads to discontinuous (trapezoidal shape) phase currents at low speed and to sinusoidal phase currents above 1500 rpm, in general.
- Only when a fourth leg (connected to the null point) in the diode rectifier is added does a notable third harmonic reoccur in the phase currents. The third harmonic in the phase voltage is reduced, however.
- In preliminary modeling, for transients, only the fundamental components of voltage and current may be considered. Even in this case, the equivalent magnetic saturation influence on L_d and L_q inductances should be considered.
- Comprehensive (3D) MECs or FEMs are required for a pertinent description of steady-state performance of the Lundell machine with diode rectifier and battery backup DC loads. The 3D trajectory of the main flux paths reclaims such a treatment.
- The power factor angle (for the fundamental components) φ_1 increases from 1° at low speeds to up to 10° at high speeds due to diode commutation-caused reactive power "consumption" in the Lundell machine.
- The maximum DC output current increases steadily with speed for a given battery voltage.
- The four-legged diode rectifier leads to notable increases in the DC output current only at high speeds when it is not badly needed.
- The loss components in the Lundell machine are stator copper loss, stator iron loss, diode rectifier loss, rotor claw-pole eddy current loss, DC excitation loss, and mechanical loss. Only stator core loss and excitation loss tend to decrease above a certain speed. All the other components increase notably with speed.
- At 14 V_{DC}, the diode rectifier losses are about as large as the stator copper losses. Increasing the DC bus voltage to 42 V_{DC} is a way to notably decrease the relative importance of the rectifier loss, besides decreasing the wire size in the stator coils and the power equipment cable cross section (and costs).
- Claw-pole geometry can be optimized for maximum DC output current, and a claw-pole span coefficient of around 0.45 in the axial middle of the stack was found to be best.
- Adding PMs to boost performance for all speeds (DC output current) was proven to be effective only if the former are placed between the rotor claw to reduce the excitation flux leakage.
- Increasing the number of poles from 12 to 18, for the same stator and rotor overall size and stator slot total area, leads to a steady increase in maximum DC output current for given voltage, especially at high speeds. Losses and cooling aspects should be closely monitored.
- Winding reconfiguration (tapping or star to delta), with a larger number of turns in series at low speeds and a smaller number of turns at high speeds, causes an increase in output for all speeds. Again, losses and cooling aspects should be considered carefully.
- Enforcing the damper capability of the solid-iron claw poles on the rotor by placing aluminum plates between the claws also produces some DC output current increases for all speeds, because the commutation (subtransient) inductance L_c of the machine is reduced.
- Combining winding tapping with controlled rectifier (for low speeds) and the diode rectifier for high speeds produces remarkably higher output at all speeds. At low speeds, the controlled rectifier allows the machine to be partly magnetized from the battery, and thus, voltage power angle δ_V may increase up to almost 90° to extract the maximum available power from the Lundell machine. Adding a single IGBT DC–DC converter to the diode rectifier may produce output increases almost similar to those obtained with a controlled rectifier [19].
- The Lundell machine is also suitable as a starter–generator in mild hybrid vehicles with a 42 V_{DC} power bus, even further, to 500 V_{DC}, 100 kW, with inverter control [20].
- Motoring is required for starting the vehicle and at idle stop (ICE shutdown) to drive the air conditioner's compressor and other auxiliaries through a single ribbed-belt transmission.
- Generating is required during normal car driving and during vehicle braking.

- An 8%–10% fuel consumption reduction for in town driving has been obtained with increased (to 8 kW) alternator power and efficiency (80%) with increased air gap, interclaw PMs and a 42 V_{DC} "hidden" DC bus [21].
- A DTFC scheme is introduced for motoring, and—after adequate new settings—for generating. Also, the DTFC scheme may be adapted for field-current-only control during generating, to increase reliability.
- Further efforts to increase the Lundell machine performance in terms of power, volume, higher speeds, efficiency, noise, vibration, and cooling are underway both for standard and mild hybrid vehicles [20,21].

References

1. I.G. Kassakian, H.-C. Wolf, J.M. Miller, and C.J. Hurton, Automotive electrical systems circa 2005, *IEEE Spectrum*, 33(8), August 1996.
2. G. Hennenberger and J.A. Viorel, *Variable Reluctance Electrical Machine*, Shaker Verlag, Aachen, Germany, 2001.
3. I. Boldea and S.A. Nasar, *Induction Machine Handbook*, CRC Press, Boca Raton, FL, 2001.
4. V. Ostovic, J.M. Miller, V. Garg, R.D. Schultz, and S. Swales, A magnetic equivalent circuit based performance computation of a Lundell alternator, In *Record of IEEE-IAS 1998, Annual Meeting*, 1998, vol. 1, pp. 335–340.
5. V. Ostovic, *Dynamics of a Saturated Electric Machine*, Springer-Verlag, New York, 1989.
6. R. Block and G. Hennenberger, Numerical calculation and simulation of a claw-pole alternator, Record of ICEM-1992, 1992, vol. 1, pp. 127–131.
7. G. Hennenberger, S. Küppers, and I. Ramesohl, Sensorless permanent-magnet brushless D.C. and synchronous motor drives, *Electromotion*, 3(4), 1996, 165–177.
8. R. Wang and N.A. Demerdash, Computation of load performance and other parameters of extra-high-speed modified Lundell alternator from 3D-FE magnetic field solutions, *IEEE Trans.*, EC-7(2), 1992, 342–361.
9. S. Küppers, Numerical method for the study and design of A.C. Claw-pole generator, PhD thesis, RWTH, Aachen, Germany, 1996 (in German).
10. K. Xamazaki, Torque and efficiency calculation of an interior PM motor considering harmonic iron losses of both stator and rotor, *IEEE Trans.*, MAG 39(3), 2003, 1460–1463.
11. A. Munoz-Garcia and D.W. Novotny, Utilization of third harmonic-induced-voltages in PM generators, In *Record of IEEE-IAS 1996, Annual Meeting*, 1996, vol. 1, pp. 525–532.
12. A. Mailat, D. Stoia, and D. Ilea, Vehicle alternator with improved performance, Record of Electromotion-2001, Bologna, Italy, 2001, vol. 2, pp. 511–514.
13. F. Liang, J. Miller, and X. Xu, A vehicle electric power generation system with improved power and efficiency, *IEEE Trans.*, IA-35(6), 1991, 1341–1346.
14. C. Kahler and G. Henneberger, Calculation of the mechanical and acoustic behavior of a claw pole alternator in a double and single star connection, In *Records of Electromotion-2001 Symposium*, Bologna, Italy, 2001, vol. 2, pp. 553–558.
15. T. Teratani, K. Kurarnoki, H. Nakao, T. Tachibana, K. Yagi, and S. Abae, Development of Toyota mild hybrid systems (THS-M) with 42 V power net, Records of IEEE-IEMDC-2003, Madison, WI, 2003, pp. 3–10.
16. I. Boldea and S.N. Nasar, *Electric Drives*, CRC Press, Boca Raton, FL, 1998.
17. O. Pyrhönen, Analysis and control of excitation, fields weakening and stability in direct torque controlled electrically excited synchronous motor drives, PhD dissertation, Lapeenranta University of Technology, Lappeenranta, Finland, 1998.

18. C. Lascu, I. Boldea, and F. Blaabjerg, Very low speed sensorless control of induction motor drives without signal injection, In *Record of IEEE–IEMDC–2003 Conference*, Madison, WI, 2003, pp. 1395–1401.
19. D.J. Perreault and V. Caliskan, Automotive power generation and control, *IEEE Trans.*, PE-19(3), 2004, 618–630.
20. B. Simo, D. Ursu, L. Tutelea, and I. Boldea, Automotive generators/motors with claw pole PM—less, NeFeB or ferrite IPM rotors for 10 kW and 100 kW at 6 krpm: Optimal design performance and vector control dynamics, Record of IEEE-IEVC 2014, Firenze, Italy, 2014.
21. L. Tutelea, D. Ursu, and I. Boldea, IPM claw-pole alternator system for more vehicle braking energy recuperation, www.jee.ro, 12, 2012.

7

Induction Starter/Alternators for Electric Hybrid Vehicles

7.1 Electric Hybrid Vehicle Configuration

In this book, EHV stands for electric hybrid vehicle. EHV constitutes an aggressive novel technology aimed at improving comfort, gas mileage, and environmental performance of road vehicles [1,2].

The degree of "electrification" in a vehicle may be defined by the electric fraction $\%E$ [3]:

$$\%E = \frac{\text{Peak electric power}}{\text{Peak electric power} + \text{peak ICE power}} = \frac{P_{(el)}}{P_{(el)} + P_{(ICE)}} \tag{7.1}$$

For a mild hybrid car with battery soft-replenishing, $\%E$ is lower than 40% in town driving. It may reach up to 70% when the battery is replenished from the power grid daily (plug-in). $\%E$ becomes 100% for fully electric vehicles, with fuel cells or batteries or inertial batteries (flywheels) as the energy storage system.

The larger the electric fraction $\%E$, the lower the internal combustion engine (ICE) rating (it is zero for a fully electric vehicle).

The electrification of vehicles is approached through a plethora of system configurations that may be broken down as illustrated in Figure 7.1a through d.

In series hybrids (Figure 7.1a), the full-size ICE drives an electric generator on the vehicle that then produces electric energy for all tasks, from the electric drives to auxiliaries and battery recharging.

In parallel hybrids (Figure 7.1b and c), the downsized ICE is started by the starter/alternator that then assists in propulsion at low-to-medium speeds and, respectively, works as a generator to feed the electrical loads and recharge the battery.

In fully electric vehicles (Figure 7.1d), a large high-voltage battery, recharged from the power grid once every day, supplies all electric drives used for vehicle propulsion. It also contains a 42 V_{DC} battery that supplies the auxiliaries. The latter battery is recharged from the main battery through a dedicated direct current (DC)–DC converter.

In mixed hybrids (Figure 7.1d), two or more motor/generators are used, for example, to electrically drive the front and the rear wheels (Figure 7.2) [3]. Toyota Prius EHV includes a planetary gear and two electric drives to keep the ICE close to the sweet point (maximum efficiency) for most vehicle speeds.

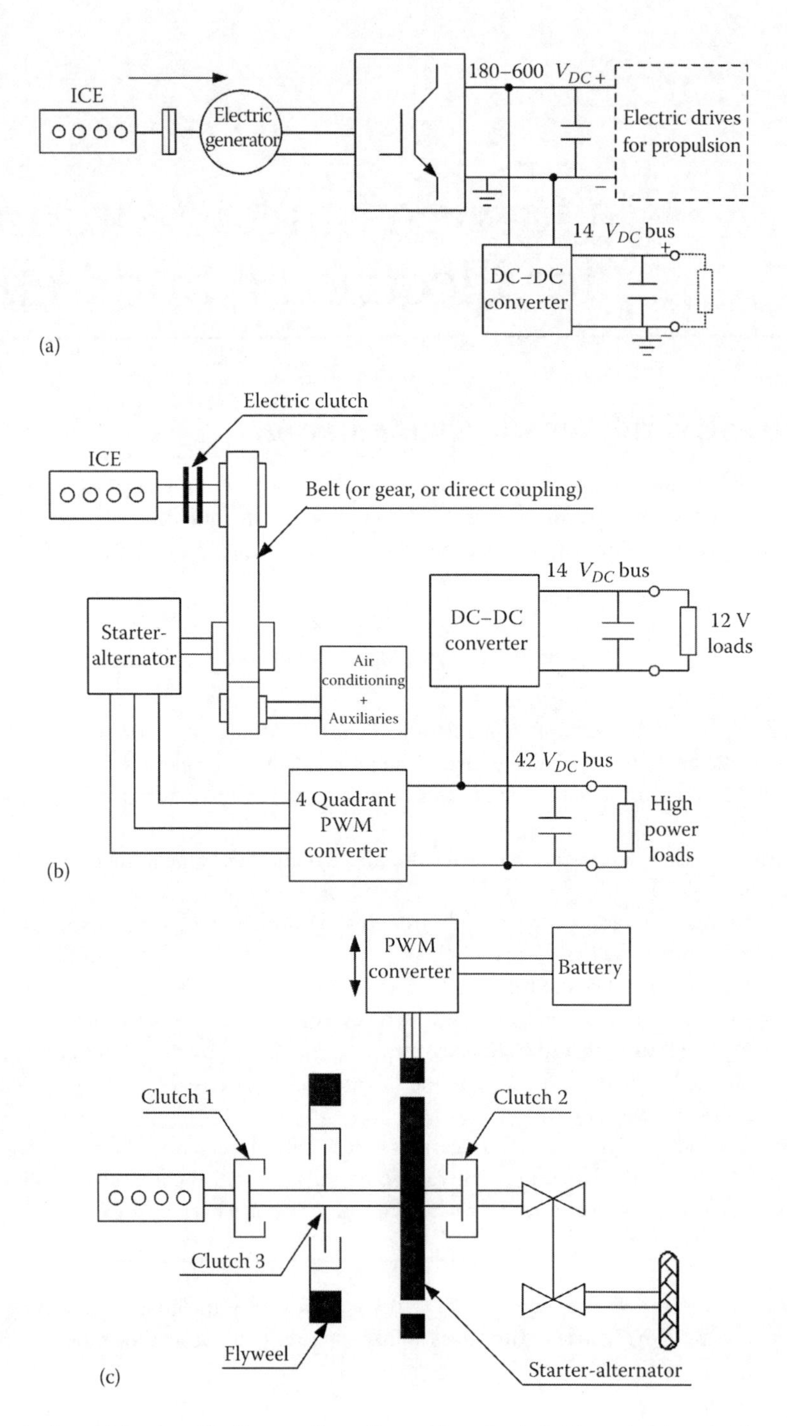

FIGURE 7.1 Basic vehicle electrification configurations: (a) series hybrids and (b) and (c) parallel hybrids. (*Continued*)

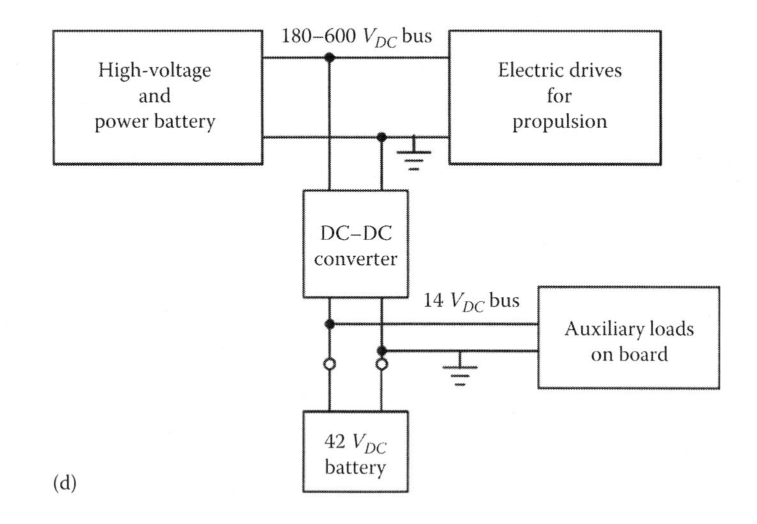

FIGURE 7.1 (*Continued*) Basic vehicle electrification configurations: (d) electric hybrids.

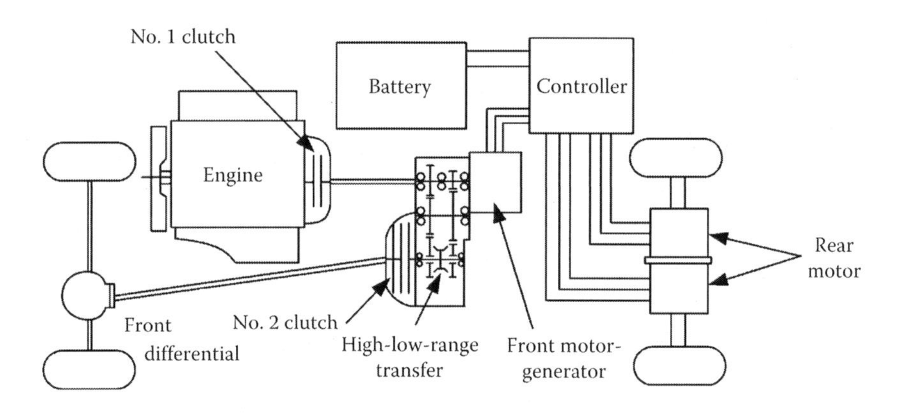

FIGURE 7.2 Mixed hybrid with front and rear motor/generator.

In all hybrid (and electric) vehicles, the air-conditioning and some auxiliaries should remain on duty during idle stop. Idle stop means stopping the engine or electric driving during halts in traffic jams or at traffic stoplights (Figure 7.1b).

Also, depending on the electric fraction %E and vehicle size (car, bus, and truck), the specifications vary within a large range.

The placing of the electrical machine on the ICE shaft or its coupling with an additional transmission (belt or gear) is essential in the design of the starter/alternator, as the peak torque will depend on this transmission ratio $k_{El} \geq 1$.

The pulse-width modulator (PWM) converter peak kilovolt ampere (kVA) rating depends heavily on the starter/alternator design, as the peak current required for peak torque "defines" the converter costs for given battery voltage.

Defining pertinent specifications and design optimization multiobjectives for the starter/alternator (motor/generator) system is of utmost importance. Such an endeavor, as for Toyota Prius EHV, with dual electric drives, is beyond our scope here.

7.2 Essential Specifications

Essential specifications for starter/alternators (motor/generator), typical on mild EHVs, are considered here to be the following:

- Starter/alternator functions
- ICE to starter/generator transmission ratio k_{El}
- Peak torque versus speed for motoring
- Peak generator power (torque) versus speed
- "Battery" voltage
- Battery self- or independent-replenishing method

7.2.1 Peak Torque (Motoring) and Power (Generating)

The peak torque for motoring is defined as the engine starting torque at 20°C and varies between 120 and 300 Nm for cars, but it may reach 1200 Nm for buses. This peak torque level has to be sustained up to n_b = 250–400 rpm for mild hybrids and up to n_b = 1000–1200 rpm for full hybrids. Above base speed n_b, a constant peak power, up to maximum speed n_{max}, has to be provided:

$$P_{ekm} = T_{ekm} \cdot 2\pi \cdot n_b \tag{7.2}$$

The larger the n_{max}/n_b, the larger the contribution of propulsion and its impact on fuel consumption reduction in city driving. This ratio n_{max}/n_b in motoring may range from 3:1 to 6:1. The larger the better for HEV performance, but this comes at the price of stator/alternator and (or) PWM converter oversizing.

A large constant peak power range imposes a few design solutions for which the whole system—starter/alternator, battery and power converter—costs, size, and losses all have to be simultaneously considered.

A multiratio mechanical transmission reduces the constant power speed range of the starter/alternator and allows for a smaller size (volume) electrical machine at the price of the additional costs for a more complex transmission.

Typical motoring torque/speed and generating peak power/speed requirements for a mild hybrid (42 V_{DC}, I_{peak} < 170 A) small car are shown in Figure 7.3 [4–9].

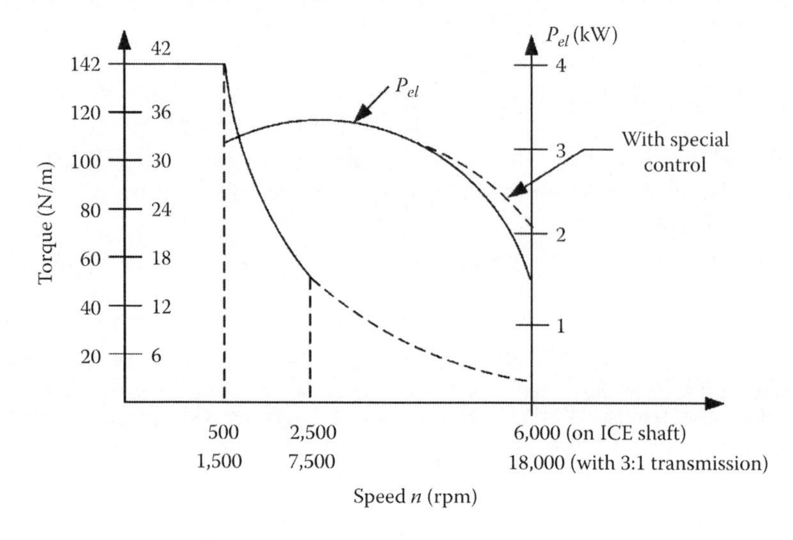

FIGURE 7.3 Potential mild hybrid car peak torque and speed motoring and power and speed generating envelopes.

The limit of 170 A on the battery is based on the acceptable voltage drop (losses) in the 42 V_{DC} battery pack made of three standard lead acid car batteries in series.

The generating power limit is based on the battery receptivity and PWM-converter-controlled starter/generator limits, to safely deliver power at high speed with limited battery overvoltage.

Without winding changeover (reconfiguration), a 3/1 constant power speed range is possible without machine or converter strong oversizing. A 4/1 constant power speed range (n_{max}/n_b = 4/1) is visible in Figure 7.4, and thus oversizing and winding changeover are required. Above 2500 rpm, in Figure 7.4, the driving power decreases due to "lack" of voltage in the battery to sustain it.

The specific propulsion requirements in a fully electric car with standard multispeed transmission are shown in Figure 7.4.

For an electric city bus, 75 kW of electric propulsion is considered in Figure 7.5. A single-stage 6.22/1 gear reduction ratio is mentioned in the literature [10].

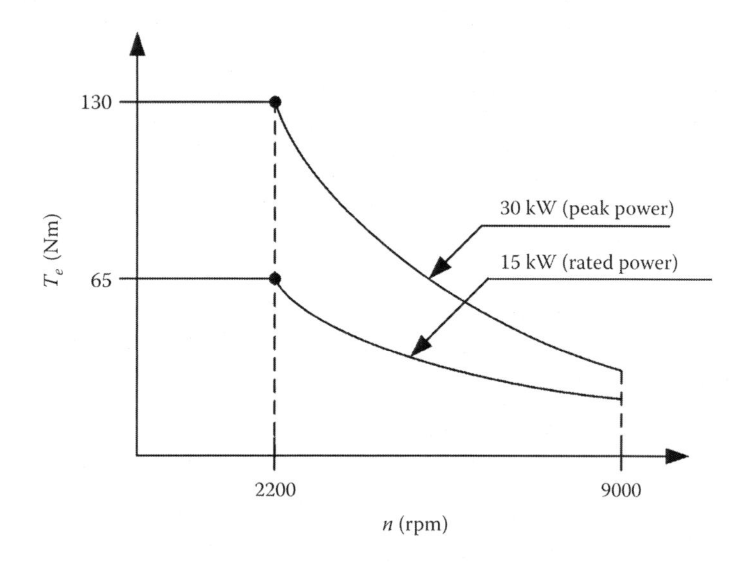

FIGURE 7.4 Typical motoring torque and speed envelopes for a fully electric car.

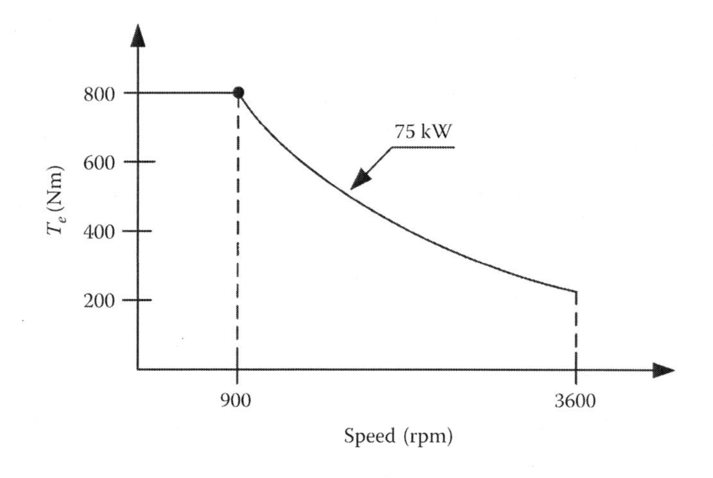

FIGURE 7.5 Typical rated torque and speed envelopes for a fully electric city bus.

7.2.2 Battery Parameters and Characteristics

The main battery parameters are as follows:

- Battery capacity: Q
- Discharge rate: Q/h
- State of charge: SOC
- State of discharge: SOD
- Depth of discharge: DOD

The amount of free electrical charge generated to the active battery material at the negative electrode, ready to be consumed by the positive electrode, is called the battery capacity Q.

Q is measured in ampere hours (Ah); 1 Ah = 3600 C; 1 C is the charge transferred by 1 A in 1 s.

The theoretical capacity Q_T is as follows:

$$Q_T = x \cdot n \cdot F; \quad F = L \cdot e_0 \tag{7.3}$$

where

x is the number of moles of reactant for complete discharge
n is the number of electrons released by the negative electrode during discharge
$L = 6.022 \cdot 10^{23}$ is the number of atoms per mole (Avogadro's constant)
$e_0 = 1.601 \cdot 10^{-19}$ C is the electron charge (F is the Faraday constant)

$$Q_T = 27.8 \cdot x \cdot n \,(\text{Ah}) \tag{7.4}$$

With a number of cells in series, the capacity of a cell equals the capacity of the battery.

The discharge current is called the discharge rate Q (Ah)/h, where h is the discharge rate in hours. If a 200 Ah battery discharges in half an hour, the discharge rate is 400 A.

The SOC represents the battery capacity at the present time:

$$\text{SOC}(t) = Q_T - \int_0^t i(\tau)d\tau \tag{7.5}$$

The SOD(t) represents the charge already drawn from a fully charged battery:

$$\text{SOD}(t) = \int_0^t i(\tau)d\tau = Q_T - \text{SOC}(t) \tag{7.6}$$

The DOD is the per unit (P.U.) battery discharge:

$$\text{DOD}(t) = \frac{\text{SOD}(t)}{Q_T} = \frac{\int_0^t i(\tau)d\tau}{Q_T} \tag{7.7}$$

A deep discharge takes place when DOD >0.8 (80%).

Adequate modeling of the battery is essential in starter/alternator design, as the battery voltage varies with temperature, SOC (SOD); and its recharging process sets limits on the electric energy recovered in the generating mode.

The simplest battery model contains an electromagnetic force (emf), E_v, and an internal resistance R_i that are both dependent on battery SOC and temperature θ (Figure 7.6) [11].

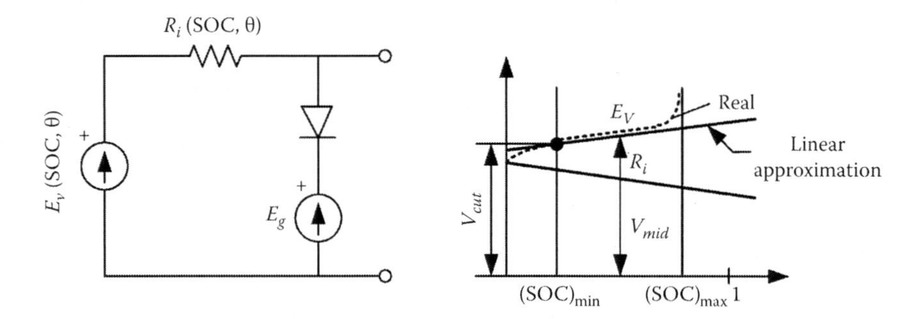

FIGURE 7.6 Simplified battery model.

The battery emf E_v increases with the SOC (and decreases with the SOD), while the internal resistance does the opposite.

When the battery is deeply discharged (SOC_{min}), the voltage tends to drop steeply. This is the cut voltage V_{cut} beyond which the battery should not, in general, be used. The practical capacity is thus,

$$Q_P = \int_0^{t_{cut}} i(t)dt < Q_T \tag{7.8}$$

When the battery is started, the constant discharge current should also be specified.

The emf E_v decreases when the temperature increases. There is also a step increase in E_v when the SOC is high (Figure 7.6). This partially explains, for example, the denomination of 14 (42) V_{DC} batteries when, on load, they work as 12 (36) V_{DC}.

The electrical energy extracted from the battery W_b is as follows:

$$W_b = \int_0^{t_{cut}} V \cdot I\, dt \tag{7.9}$$

With constant discharging current, the total battery voltage versus time looks as shown in Figure 7.7. The discharge time to V_{cut} is larger if the discharge current is smaller.

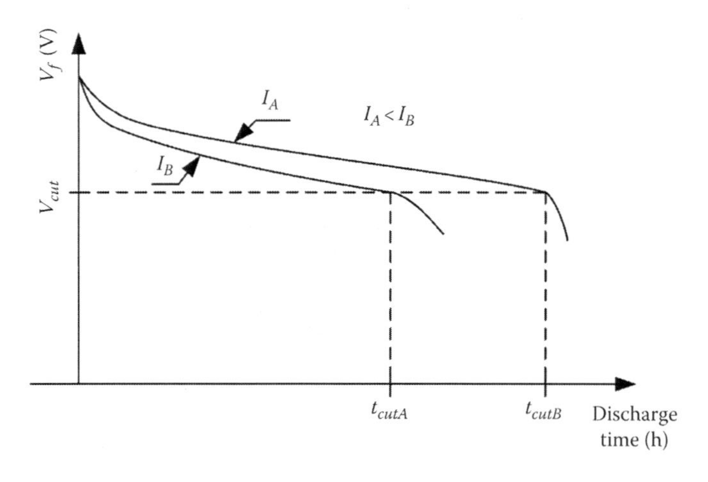

FIGURE 7.7 Voltage per time for constant current discharge.

For constant discharge current (I) the cut-time t_{cut} is offered by Peukert's equation [2]:

$$t_{cut} = \frac{C_c}{I^{n_c}} \tag{7.10}$$

with C_c and n_c as constants.

The specific energy (Wh/kg) is the discharge energy W_b per battery weight. For lead-acid batteries, the specific energy is around 50 Wh/kg at $Q/3$ rate.

Other batteries have higher energy densities, but their costs per watt-hour tend to be higher, in general.

The battery power $P(t)$ is as follows:

$$P(t) = V_t \cdot i = (E_v - R_i i) \cdot i \tag{7.11}$$

For constant E_v, the maximum power $P_{t\max}$ occurs, as known, for $R_{load} = R_i$:

$$P_{t\max} = \frac{E_v^2}{4 \cdot R_i} \tag{7.12}$$

The rated instantaneous power P_{ti} is the maximum power deliverable for a short discharge time without damage, while instantaneous power P_{tc} corresponds to large discharge intervals and no damage to the battery.

The specific power is P_t/M_B (W/kg). The lead acid battery may deliver a maximum of 280–400 W/kg at DOD = 80%. Other batteries may produce less.

Table 7.1 presents typical characteristics of a few batteries for EHVs.

Note that fuel cells, inertial (flywheels) batteries, and supercapacitors may act as alternative energy storage systems on vehicles. They are characterized by smaller energy density but higher power density (2 kW/kg) [12]. The fuel cells tend to have smaller efficiency (60%–70%), while inertial and supercapacitor batteries have higher efficiency.

Super-high-speed inertial (flywheel) batteries, in vacuum, with (eventually) magnetic bearing, should surpass the batteries in all ways, including the possession of a long life and a 2–3 min recharge time. Inertial batteries contain a super-high-speed generator/motor for recharging, controlled through a bidirectional PWM static power converter. This subject will be treated in the chapter on super-high-speed generators (Chapter 10).

In view of the many EHV schemes and ratings, in the following paragraphs, we will concentrate on various aspects of induction machine modeling, and design for variable speed and control for starter/generator applications. In other words, providing for a comprehensive tool for EHV designers is the goal here.

A numerical example emphasizes the essentials and gives a feeling of the magnitude of this topic.

TABLE 7.1 Typical Characteristics of a Few Batteries for Electric Hybrid Vehicles (EHVs)

Battery	Wh/kg	W/kg	Efficiency %	Cycle Life	Cost $/kWh
Lead acid	35–50	150–400	80	500–1000	100–150
Nickel–cadmium	30–50	100–150	75	1000–2000	250–350
Nickel–metal-hydride	60–80	200–300	70	1000–2000	200–350
Aluminum–air	200–300	100	50	—	—
Zinc–air	100–220	30–80	60	500	90–120
Sodium–sulfur	150–240	240	85	1000	200–350
Sodium–nickel–chloride	90–120	140–160	80	1000	250–350
Lithium–polymer	150–200	350	—	1000	150
Lithium–ion	80–130	200–300	95	1000	200

Source: Adapted from Husain, I., *Electric and Hybrid Vehicles*, CRC Press, Boca Raton, FL, 2003.

7.3 Topology Aspects of Induction Starter/Alternator

The operations of induction starter/alternator (ISA) are characterized by the following:

- A thermally harsh environment
- Limited volume (weight)
- Requirements for high efficiency and power factor
- Maximum speed above 6000 rpm
- Bidirectional converter supply to and from a DC bus voltage source (battery); battery voltages vary by about (or more than) 30%, depending on SOC, load, and ambient temperature; the DC voltage goes up with power rating from 42 V_{DC} to 600 V_{DC}

These operating conditions lead to some specifications in ISA design and control.

First, squirrel-cage rotors are to be used. In addition, the high speed corroborated with low volume leads to a large number of poles. As the maximum speed goes up, the number of poles should go down to limit the maximum fundamental frequency f_1 to 500–600 Hz. The frequency limitation is prompted by reasonable skin effect winding, iron losses and by moderate PWM converter switching losses and costs.

The number of poles is larger when the rotor diameter (and peak torque) is larger, as what counts is the ratio of the pole pitch τ to air gap g, to provide a reasonable magnetization current (power factor).

In general, $2p_1 = 8$–12 for $n_{max} < 6{,}000$ rpm and $2p_1 = 4$–6 for $n_{max} > 12{,}000$ rpm.

Higher than 6000 rpm speeds are typical when the belt or gear transmission with $k_e = 2/1$ to 3/1 is used to couple the ISA to the ICE.

Maximum fundamental frequencies of $f_1 = 500$–600 Hz lead to rotor frequencies of $f_2 = 5$–6 Hz, even for a slip $S = 0.01$ ($f_2 = S \cdot f_1$).

Skin effect has to be limited in both the stator and the rotor because of the large fundamental maximum frequencies and due to higher current time harmonics incumbent to PWM converter supplies. In addition, to keep the losses down, while very large torque densities are required, the rotor resistance has to be reduced by design.

The stator slots should be semiclosed and rectangular or trapezoidal. Rotor slot shapes with low skin effect should be chosen (Figure 7.8a through d).

The U-bridge closed rotor slot (Figure 7.8d) is supposed to reduce the surface and flux pulsation losses in the outer part of the rotor cage. Unfortunately, this merit is counteracted by a larger slot leakage inductance. Finally, breakdown torque is reduced this way.

Using copper instead of aluminum may help in reducing the cage losses, despite the fact that the skin effect tends to increase due to higher conductivity of copper in comparison with aluminum.

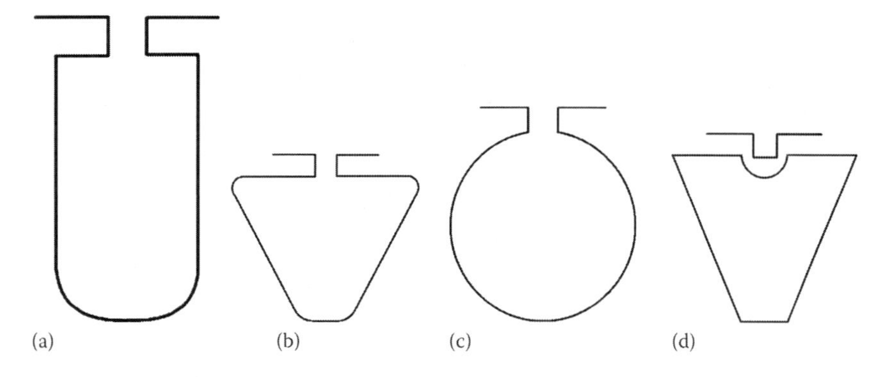

FIGURE 7.8 Suggested induction starter/alternator (ISA) rotor slots: (a) rectangular, (b) trapezoidal, (c) circular, and (d) U-bridge closed slot.

Additionally, smaller cross-sectional rotor bars would allow more room for rotor teeth, leading to reduced rotor core saturation. Insulating the copper bars from slot walls may also prove useful in reducing the inter-bar rotor current losses.

With $2p_1$ = 8–12 poles, in most cases, the number of slots per pole and phase q_1 = 2, 3. Only for $2p_1$ = 4–6 poles, does q_1 = 3–5.

Chorded coils in the stator are required to reduce the fifth and seventh magnetomotive force (mmf) space harmonics with their rotor core surfaces and cage losses.

Skewing is also an option in reducing the first slot harmonics $\upsilon = 6q \pm 1$ with their rotor core surface and pulsation additional losses. When the number of rotor slots N_r is chosen to be smaller than the number of stator slots N_s ($N_r < N_s$):

$$0.8 < \frac{N_r}{N_s} < 1 \tag{7.13}$$

and N_s, N_r combination restrictions are observed for reducing the parasitic torque harmonics and noise (Reference 13, Chapter 11); no skewing is required, even if the rotor cage is not isolated from the rotor core.

To reduce the end-ring leakage permeance, the end-rings should be placed at a distance with respect to the core.

When the ISA is placed on the engine shaft, the rotor diameter D_{in} should be higher than a given value, as the respective inner space is required for cooling.

The environment in such a direct coupling has an ambient temperature $T_{amb} \approx$ 110°C–130°C. Therefore, even with Class $F(H)$ insulation material, the winding over temperature $\Delta\theta_{amb} \le$ 60°C–70°C. The rotor temperature, for prolonged full generating, may reach up to 240°C. Forced cooling may be needed to fulfill such conditions, especially with the ISA designed for low volume at the peak torque.

The peak current density in the stator conductors may reach values around or even larger than 30 A/mm^2.

Further on, the degree of magnetic saturation in ISA, for peak torque, should be high for the same reason. A peak air gap flux density fundamental of 1.0–1.15 T is common for ISA.

Allowing uniform though advanced magnetic saturation in the stator and rotor teeth and yokes seems to be the key to small air gap space harmonics in the air gap flux density (Reference 13, p. 125). Small space harmonics in the air gap flux density lead to small such harmonics in the teeth and yokes. Therefore, in fact, the iron core harmonics losses due to space harmonics are reduced this way, and so are the torque pulsations, vibration, and noise.

In terms of analysis, the gain is exceptional. The phasor and space-phasor (dq) model could be used by simply using the magnetization curve of the machine, calculated (and measured) on no-load to adopt the machine's main field inductance.

For the peak current, to produce limited slot-leakage tangential tooth-top flux density B_{tt} (Figure 7.9a through c), the slot opening W_{os} is kept at 5–6g (g is the air gap).

Also, for open and semiopen slots, the slot opening W_{os} should be "reduced" to W'_{os}, making use of a magnetic wedge with a relative permeability $\mu_{rw} = 2$–4, $W'_{os} = W_{ls}/\mu_{rw}$.

Allowing for $W'_{os}/g > 6$ would imply a notably larger magnetization current and larger teeth flux pulsation that induces large rotor surface and harmonics cage losses.

Open slots allow for the machine placing of coils in slots and allow for cuts to be made in manufacturing costs and time. The optimum air gap is a trade-off between magnetization current and rotor surface harmonics iron losses. In general, for the peak torque in the range from 40 to 1200 Nm, the air gap may vary between 0.5 and 1.0 mm.

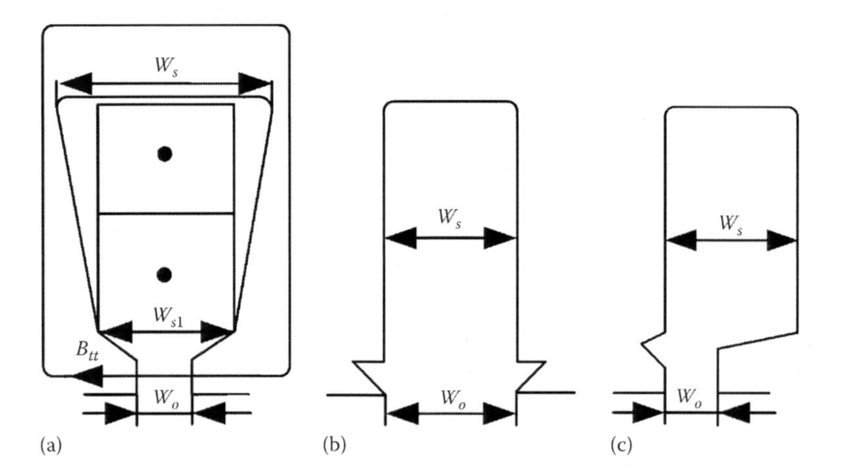

FIGURE 7.9 Typical stator slots for induction starter/alternator (ISA): (a) semiclosed trapezoidal, (b) open rectangular, and (c) semiopen rectangular.

7.4 ISA Space-Phasor Model and Characteristics

As already pointed out in the previous paragraph, uniform and heavy magnetic saturation (by design) of the stator and rotor teeth and yokes leads to a close to sinusoidal air gap flux distribution.

Consequently, the space-phasor model, typically used in modeling induction motor drives [14], may be applied even for calculating torque, provided the magnetization curve,

$$\overline{\Psi}_m(I_m) = L_m(I_m) \cdot \overline{I}_m \tag{7.14}$$

is known either from calculations or from tests. $\overline{\Psi}_m$ is the air gap flux linkage space phasor, and $\overline{I}_m$ is the air gap flux magnetization space-phasor current:

$$\overline{I}_m = \overline{I}_s + \overline{I}_r \tag{7.15}$$

$\overline{I}_s$ is the stator space-phasor current (stator current vector)
$\overline{I}_r$ is the rotor current vector

$$\overline{I}_s^b = \frac{2}{3} \cdot \left(i_a + i_b \cdot e^{j\cdot(2\pi/3)} + i_c \cdot e^{-j\cdot(2\pi/3)} \right) \cdot e^{-j\theta_s^b} \tag{7.16}$$

Equation 7.16 represents the so-called Park complex transformation, with θ_s^b equal to the position angle of the orthogonal axis of coordinate system, with respect to phase a axis in the stator. The same transformation is valid for the stator voltage and flux vectors.

The stator and rotor flux $\overline{\Psi}_s, \overline{\Psi}_r$ are, simply,

$$\begin{aligned} \overline{\Psi}_s &= L_{sl} \cdot \overline{I}_s + \overline{\Psi}_m \\ \overline{\Psi}_r &= L_{rl} \cdot \overline{I}_r + \overline{\Psi}_m \end{aligned} \tag{7.17}$$

L_{sl} and L_{rl} are the leakage inductances, both reduced to the stator, and so are $\overline{I}_r$ and $\overline{\Psi}_r$.

The stator voltage vector equation in stator coordinates is as follows:

$$\bar{I}_s^s \cdot R_s - \bar{V}_s^s = -\frac{d\bar{\Psi}_s^s}{dt} \tag{7.18}$$

And for the rotor, in rotor coordinates,

$$\bar{I}_r^r \cdot R_r = -\frac{d\bar{\Psi}_r^r}{dt} \tag{7.19}$$

For general coordinates,

$$\begin{aligned} &\bar{I}_r^b = \bar{I}_r^r \cdot e^{-j(\theta_s^b - \theta_{er})} \bar{\Psi}_r^b = \bar{\Psi}_r^r \cdot e^{-j(\theta_s^b - \theta_{er})} \\ &\bar{I}_s^b = \bar{I}_s^s \cdot e^{-j\theta_s^b} \bar{\Psi}_s^b = \bar{\Psi}_s^s \cdot e^{-j\theta_s^b} \\ &\bar{V}_s^b = \bar{V}_s^s \cdot e^{-j\theta_s^b} \end{aligned} \tag{7.20}$$

θ_{er} is the rotor position with respect to stator phase a in electrical degrees ($\theta_{er} = p_1 \cdot \theta_r$).

Equations 7.18 and 7.19 thus become

$$\begin{aligned} &\bar{I}_s \cdot R_s - \bar{V}_s = -\frac{\partial \bar{\Psi}_s}{\partial t} - j \cdot \omega_b \cdot \bar{\Psi}_s \omega_b = \frac{d\theta_s^b}{dt} \\ &\bar{I}_r \cdot R_r = -\frac{\partial \bar{\Psi}_r}{\partial t} - j \cdot (\omega_b - \omega_r) \cdot \bar{\Psi}_r \omega_r = \frac{d\theta_{er}}{dt} = p_1 \cdot \Omega_r \end{aligned} \tag{7.21}$$

The superscript b has been dropped in Equation 7.21, which is now both written in general coordinates that rotate at speed ω_b.

For steady-state stator voltages $V_{a,b,c}$,

$$V_{a,b,c} = V_1 \cdot \sqrt{2} \cdot \cos\left(\omega_1 t - (i-1) \cdot \frac{2\pi}{3}\right) \tag{7.22}$$

After applying the Park complex transformation of Equation 7.16, we obtain

$$\bar{V}_s = V_1 \cdot \sqrt{2} \cdot \cos(\omega_1 - \omega_b) \cdot t - j \cdot V_1 \cdot \sqrt{2} \cdot \sin(\omega_1 - \omega_b) \cdot t \tag{7.23}$$

Therefore, the frequency of the voltage vector applied to the induction machine at steady state ($\omega_1 - \omega_b$) depends on the speed of the orthogonal reference, ω_b.

Three reference system speeds are most used:

1. Stator coordinates: $\omega_b = 0$
2. Rotor coordinates: $\omega_b = \omega_r$
3. Synchronous coordinates: $\omega_b = \omega_1$

Synchronous coordinates are used for machine control simulation for vector (flux-oriented) control (FOC) implementation, while stator coordinates are used for direct torque and flux control (DTFC). All three values of ω_b are used for building state observers for FOC or DTFC.

For synchronous coordinates, steady state means zero frequency in Equation 7.23, as $\omega_b = \omega_1$.

Constant rotor flux in Equation 7.21 means $\partial \bar{\Psi}_r / \partial t = 0$ and constitutes the basis for vector control. Let us consider synchronous coordinates: $\omega_b = \omega_1$. This means that only the amplitude of the rotor flux vector has to be maintained constant, to eliminate rotor electrical transients. It was proven that constant rotor flux control leads to the fastest torque transients in the machine fed through a controlled current source. DTFC does almost the same but in stator coordinates.

Eliminating the stator flux and rotor current in Equation 7.21, using Equations 7.15 and 7.17, and replacing $\partial/\partial t$ with s (Laplace operator), yield:

$$\begin{aligned}
&\overline{I}_s \cdot (R_s + (j \cdot \omega_1 + s) \cdot L_{sc}) - \overline{V}_s = -(s + j \cdot \omega_1) \cdot \frac{L_m}{L_r} \cdot \overline{\Psi}_r; \\
&L_{sc} = L_{sl} + L_m \\
&\overline{I}_s \cdot R_r = (R_r + (s + j \cdot S \cdot \omega_1) \cdot L_r) \cdot \frac{\Psi_r}{L_m}; \\
&L_r = L_{rl} + L_m \\
&L_{sc} = L_s - \frac{L_m^2}{L_r} \\
&S = \frac{\omega_1 - \omega_r}{\omega_1} \text{—the slip}
\end{aligned} \tag{7.24}$$

The torque is

$$T_e = \frac{3}{2} \cdot p_1 \cdot \text{real}\left(j \cdot \overline{\Psi}_s \cdot \overline{I}_s^*\right) = \frac{3}{2} \cdot p_1 \cdot L_m \cdot \text{Real}\left(j \cdot \overline{I}_s \cdot \overline{I}_r^*\right)$$

For constant rotor flux, $s = 0$ in the second expression of Equation 7.24. Also, for steady state $s = 0$ in front of $\overline{\Psi}_{ro}$ in the first equation, synchronous coordinates

$$\overline{I}_{so} \cdot (R_s + j \cdot \omega_1 \cdot L_{sc}) - \overline{V}_{so} = -j \cdot \omega_1 \cdot \frac{L_m}{L_r} \cdot \overline{\Psi}_{ro} \tag{7.25}$$

$$\overline{I}_{so} = \frac{\Psi_r}{L_m} + j \cdot S \cdot \omega_1 \cdot T_r \cdot \frac{\Psi_r}{L_m} = I_M + j \cdot I_T; \tag{7.26}$$

$$T_r = \frac{L_r}{R_r} \tag{7.27}$$

For constant rotor flux, the stator flux vector $\overline{\Psi}_s$ is as follows:

$$\overline{\Psi}_s = L_s \cdot I_M + j \cdot L_{sc} \cdot I_T \Psi_r = L_m \cdot I_M \tag{7.28}$$

and the torque T_e is

$$T_e = \frac{3}{2} \cdot p_1 \cdot (L_s - L_{sc}) \cdot I_M \cdot I_T; \quad I_r = -j\frac{S \cdot \omega_1 \cdot \Psi_r}{R_r} \tag{7.29}$$

A few remarks are in order after this theoretical "marathon":

- The torque expression of the induction machine for constant rotor flux (amplitude only in synchronous coordinates) resembles that of a synchronous reluctance machine, where the d-axis inductance is the no-load inductance L_s, and the q-axis inductance is the short-circuit inductance L_{sc}. The higher the difference $L_s - L_{sc}$, the larger the torque for given current.

- For constant rotor flux, the stator current vector has two components: a flux one I_M aligned with the rotor flux vector and the torque one I_T, 90° ahead in the direction of motion for motoring ($S > 0$) and behind ($S < 0$) for generating.
- For constant rotor flux, the stator flux vector has two components (Equation 7.28) produced by the I_M and I_T stator current components.

Equations 7.25 through 7.29 may be represented in a vector diagram as in Figure 7.10a and b:

- For motoring (and braking)—$S > 0$—the stator current vector $\overline{I}_{so}$ is ahead of stator flux vector $\overline{\Psi}_{so}$ in the direction of motion. The opposite is true for generating.
- Voltage, current, and flux vectors power angles $\delta_V, \delta_I, \delta_{\Psi_{s,r}}$ may be defined. For motoring, the stator flux vector $\overline{\Psi}_{so}$ is ahead of the rotor flux vector $\overline{\Psi}_{ro}$ along the direction of motion, and it is lagging for generating.
- The core losses are not yet considered, but for a first approximation, they depend only on ω_1, $S\omega_1$ and Ψ_s or Ψ_m. At low speed, the core losses are small, as ω_1 is small, though full flux is injected in the machine. At high speed, the frequency ω_1 is high, but the flux Ψ_s is small due to voltage limitation. When the winding or pole number changeover occurs at the changeover speed, the core losses are suddenly increased, as the flux Ψ_s is brought to the highest level, again limited only by magnetic saturation. The stator flux Ψ_s surpasses the air gap flux Ψ_m only by a few percent, in general (due to stator leakage flux $L_{sl}\overline{I}_s$).
- The air gap flux Ψ_m magnetization current $\overline{I}_m$ is as follows:

$$\overline{I}_m = \overline{I}_s + \overline{I}_r = I_M + j \cdot I_T \cdot \left(1 - \frac{L_m}{L_r}\right) = \frac{\Psi_r}{L_m} \cdot \left(1 + j \cdot \frac{S \cdot \omega_1 \cdot L_{rl}}{R_r}\right) \tag{7.30}$$

$$\overline{\Psi}_m = \Psi_r - L_{rl} \cdot \overline{I}_r = \Psi_r \cdot \left(1 + j \cdot \frac{S \cdot \omega_1 \cdot L_{rl}}{R_r}\right) \tag{7.31}$$

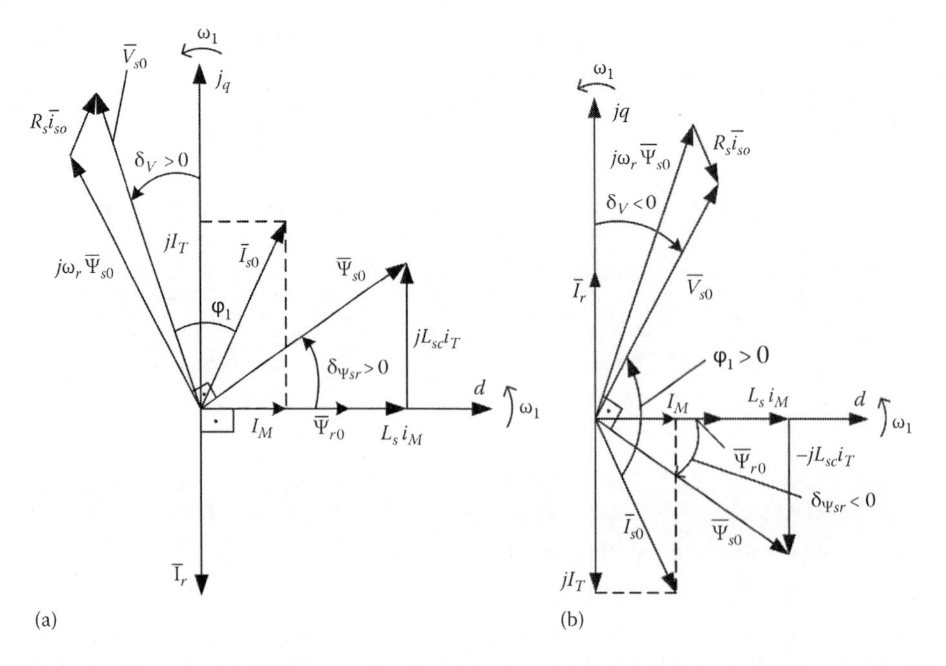

FIGURE 7.10 Induction starter/alternator (ISA) vector diagram (steady state in synchronous coordinates): (a) motoring and (b) generating.

- It should be noted that once the rotor flux Ψ_r and the torque values are set, the amplitude of the stator current I_s, the slip frequency value $S\omega_1$, the air-gap flux Ψ_m, and the stator flux Ψ_s amplitudes are all assigned certain values, depending also on the machine parameters $R_r, L_m(\Psi_m)$ and L_{rl}, L_{sl}.
- The magnetization inductance L_m depends on the air-gap flux only—$L_m(\Psi_m)$ or $L_m(I_m)$—while rotor temperature changes the rotor resistance R_r. The stator and rotor leakage inductances are considered constant, in general.
- At low speed, for peak torque, as the core losses are small, the maximum torque per current principle should be applied:

$$I_s = \sqrt{I_M^2 + I_T^2}$$

$$T_e \approx \frac{3}{2} \cdot p_1 \cdot (L_m - L_{rl}) \cdot I_M \cdot I_T \tag{7.32}$$

$$\frac{\partial T_e}{\partial I_M} = 0$$

Only if L_m = constant, albeit heavily saturated,

$$I_M = I_T = \frac{I_{ski}}{\sqrt{2}} \tag{7.33}$$

The peak torque for given current is, thus,

$$T_{eki} = \frac{3}{2} \cdot p_1 \cdot (L_m - L_{rl}) \cdot \frac{I_{ski}^2}{2}$$

$$(S_{w1})_{ki} \cdot \frac{L_r}{R_r} = 1 \tag{7.34}$$

$$\overline{I}_{mki} = \frac{I_{ski}}{\sqrt{2}} \cdot \left(1 + j \cdot \frac{L_{rl}}{L_r}\right)$$

The copper losses for the peak torque T_{eki}, p_{Coki}, are as follows:

$$p_{Coki} = \frac{3}{2} \cdot I_{ski}^2 \cdot \left(R_s + R_r \cdot \left(\frac{L_m}{L_r}\right)^2 \cdot \frac{1}{2}\right) \tag{7.35}$$

It is now evident that L_m (and L_r) depend predominantly on the magnetization current of the main flux, I_{mki}.

As the short-time-lived peak torque should be large, so too will be the peak current. Consequently, $I_{mki} = (I_{ski}/\sqrt{2})$ (1.02–1.1) should be very large.

Current densities up to 30–40 A/mm^2 are admitted for peak torque short durations and adequate cooling.

Heavy oversaturation of the machine is thus mandatory, for very large peak torque, though $L_m(I_{mki})$ decreases due to heavy saturation.

An optimum saturation level for given geometry and torque (for given volume) is to be obtained.

The ideal current power angle δI = 45°; but due to variable saturation, it departs from 45° to notably larger values, as L_m decreases with saturation to obtain maximum torque for given stator current (Figure 7.11).

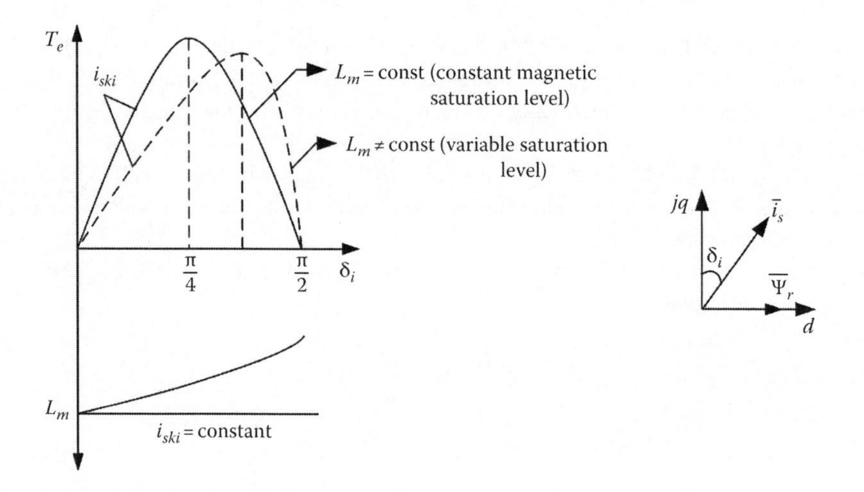

FIGURE 7.11 Torque vs. current power angle δ_i with constant and variable saturation.

The magnetic saturation curve may be calculated or measured and then curve-fitted. A standard approximation is as follows [15]:

$$\Psi_m = \frac{L_o \cdot I_m}{1 + I_m / I_{mo}} + L_{oo} \cdot I_m = L_m \cdot I_m \tag{7.36}$$

Therefore,

$$L_m(I_m) = \frac{L_o}{1 + I_m / I_{mo}} + L_{oo} \tag{7.37}$$

The constants L_o, L_{oo}, I_{mo} are to be determined to best fit the analytical or finite element field model of the saturated machine over the entire magnetization current range.

- As speed goes up, the machine cannot keep the peak torque past a certain speed, called the base speed ω_{rb}, for which full output voltage of the PWM converter is reached. In general, full flux is considered for base speed, full voltage, and peak torque. If the flux is reduced, the base speed may be increased at the cost of larger current. For $I_{M_{ki}} = I_{s_{ki}} / \sqrt{2} = I_{T_{ki}}$ the stator flux $\Psi_{s_{ki}}$ is as follows:

$$\Psi_{s_{ki}} = L_s \cdot \frac{I_{s_{ki}}}{\sqrt{2}} \pm j \cdot L_{sc} \cdot \frac{I_{s_{ki}}}{\sqrt{2}}; \quad \bar{I}_{s_{ki}} = \frac{I_{s_{ki}}}{\sqrt{2}} \cdot (1 \pm j) \tag{7.38}$$

with a positive sign (+) for motoring and a negative sign (−) for generating.

The stator voltage is simply

$$\begin{aligned} \bar{V}_{s_{ki}} &= R_s \cdot \bar{I}_{s_{ki}} + j \cdot \omega_1 \cdot \bar{\Psi}_{s_{ki}} \\ &= R_s \cdot \frac{I_{s_{ki}}}{\sqrt{2}} \mp \omega_1 \cdot L_{sc} \cdot \frac{I_{s_{ki}}}{\sqrt{2}} + j \cdot \left(\pm R_s \cdot \frac{I_{s_{ki}}}{\sqrt{2}} + \omega_1 \cdot L_s \cdot \frac{I_{s_{ki}}}{\sqrt{2}} \right) \end{aligned} \tag{7.39}$$

$$\begin{aligned} V_{s_{ki}} &= \frac{I_{s_{ki}}}{\sqrt{2}} \sqrt{(R_s \mp \omega_{1b} \cdot L_{sc})^2 + (R_s \mp \omega_{1b} \cdot L_s)^2} \\ \omega_{rb} &= \mp \frac{Rr}{Lr} + \omega_{1b} \end{aligned} \tag{7.40}$$

The maximum stator voltage vector fundamental $V_{s_{ki}}$ is dependent mainly on the battery voltage V_{DC} and, to a smaller degree, on the type of PWM strategy that is applied. In general,

$$V_{s_{ki}} = \frac{4}{\pi} \cdot \frac{V_{DC}}{\sqrt{3}} \cdot k_{PWM} \quad \text{(peak value per phase)} \tag{7.41}$$

The coefficient k_{PWM} depends on the PWM strategy, but in general is in the interval of 0.9–0.96.

It is important to mention here again that the battery voltage varies with the SOD (SOC), ambient temperature, and load.

There might be a 30%–50% difference between maximum and minimum value of battery voltage, and to secure the peak torque for the worst case, it might prove to be practical to use a voltage boost bidirectional DC–DC converter in front of a PWM converter, instead of oversizing the PWM converter to comply with peak torque current demand at the lowest battery voltage.

Above base speed, the current angle should be increased, sacrificing more current for flux reduction.

The best exploitation of voltage corresponds to maximum torque per flux:

$$\begin{aligned}
&\Psi_s = \sqrt{(L_s \cdot I_M)^2 + (L_{sc} \cdot I_T)^2} \\
&T_e = \frac{3}{2} \cdot p_1 \cdot (L_s - L_{sc}) \cdot I_M \cdot I_T; \quad I_T = I_M \cdot S \cdot \omega_1 \cdot \frac{L_r}{R_r} \\
&\frac{\partial T_e}{\partial I_M} = 0
\end{aligned} \tag{7.42}$$

Finally, for peak torque at given flux:

$$L_s \cdot I_{Mk\Psi} = L_{sc} \cdot I_{Tk\Psi} = \frac{\Psi_s}{\sqrt{2}} \tag{7.43}$$

and

$$T_e = \frac{3}{2} \cdot p_1 \cdot (L_s - L_{sc}) \cdot \frac{\Psi_s^2}{2 \cdot L_s \cdot L_{sc}}; \quad (\omega_1 S)_{k\Psi} = \frac{R_r}{L_r} \cdot \frac{L_s}{L_{sc}} \tag{7.44}$$

Approximately, for given voltage,

$$\Psi_s \approx \frac{V_{s_{ki}} k_V}{\omega_1}; \quad k_V = 0.95 - 0.97 \tag{7.45}$$

Therefore, the flux level decreases inversely proportional to frequency. The slip frequency $(\omega_1 S)_{k\Psi}$ for maximum torque per flux, though constant, is rather large.

The current angle $(\delta_i)_{k\Psi}$ is, from Equation 7.44,

$$(\delta_i)_{k\Psi} = \tan^{-1}\left(\frac{I_M}{I_T}\right) = \tan^{-1}\left(\frac{L_{sc}}{L_s}\right) \ll \frac{\pi}{4} \tag{7.46}$$

The current power angle (with axis q) for maximum torque per flux is much smaller than 45° in this case.

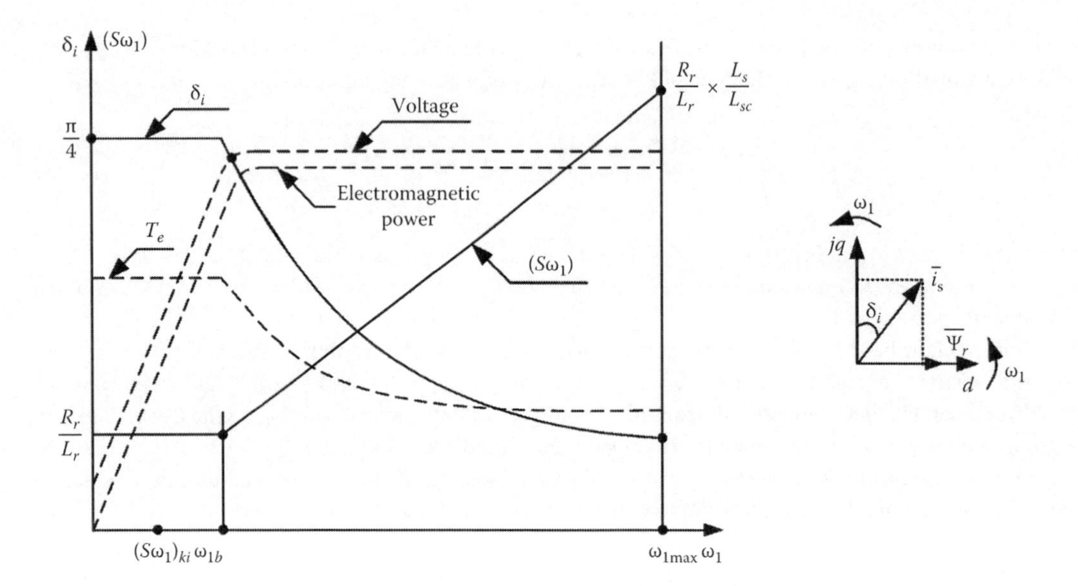

FIGURE 7.12 Current power angle δ_i and slip frequency $S\omega_1$ vs. stator frequency ω_1 for constant torque and then constant power and voltage.

The maximum torque per flux condition (and $\delta_{ik}\Psi$) should be used in the design to define the *maximum speed* for constant power, either motoring or generating.

Beyond this frequency (speed)—$\omega_{1\max} = \omega_{r\max} \pm (S\omega_1)$—producing constant power is not feasible, but some power is available, though at higher losses per Newton meter of torque.

It is, thus, clear that the current power angle δ_i, if controlled, should be $\delta_i = (\delta_i)_{ki} \approx (\pi/4)$ up to base speed and continuously decreasing above this speed down to $\delta_i = (\delta_i)_{ki} = \tan^{-1}(L_{sc}/L_r)$ for maximum constant power speed. In addition, the slip frequency increases from $(S\omega_1)_{ki} = (R_r/L_r)$ to $(S\omega_1)_{k\Psi} = (R_r/L_r)\cdot(L_s/L_{sc})$ (Figure 7.12).

As a conclusion of the above rationale, we mention that once the peak torque T_{ek}, base frequency ω_{1b}, and maximum frequency $\omega_{1\max}$ (for constant power) are fixed, the main machine design options are made.

For constant electromagnetic power P_{elm}:

$$P_{elm} = \frac{T_e\omega_1}{p_1} \tag{7.47}$$

and constant voltage, from ω_{1b} to $\omega_{1\max}$, the reference flux and torque may be calculated easily from Equations 7.44 and 7.47. Then, from Equations 7.42 through 7.43, the required values of stator current orthogonal components I_M and I_T (and δ_i) and $S\omega_1$ may be calculated.

Such typical calculations for 30 kW constant power are shown in Figure 7.13 [9].

It should be noted that, though the constant torque is provided up to 2000 rpm, the voltage V_{sb} is lower than V_{ski} (the maximum value). This means oversizing the converter (in current), but it provided the requested 4:1 constant power speed range without winding reconfiguration.

Once the above theoretical tool is in place, the machine magnetization curve and parameters are known, other steady-state characteristics, such as machine losses and current power angle versus speed, may be determined.

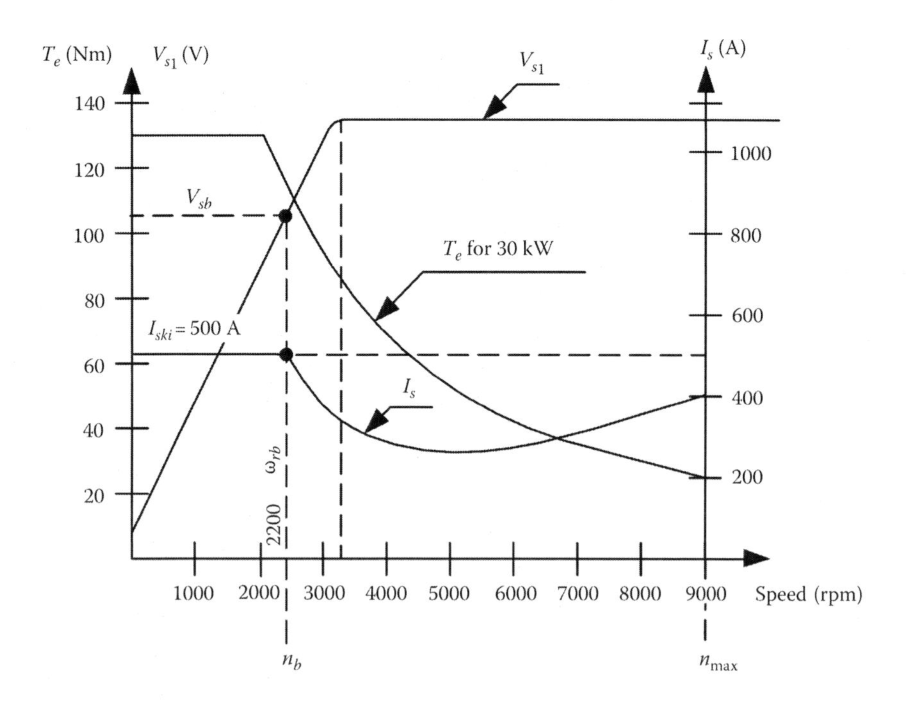

FIGURE 7.13 Voltage and current vs. speed envelopes for constant torque (up to 2000 rpm) and constant power (30 kW) from 2200 to 9000 rpm.

7.5 Vector Control of ISA

A generic vector control system for ISA is shown in Figure 7.14. It has the following main components:

- The reference rotor flux Ψ_r^* calculator (for handling both motoring and generating)
- The reference torque T_e^* calculator (for motoring $[T_e^* > 0]$ and generating $[T_e^* < 0]$)
- The stator space-vector current component calculator, based on Equation 7.42, is adapted to constant torque and constant power conditions as required for motoring and generating below and above base speed, for battery recharging, and for regenerative breaking
- The vector rotator that transforms the currents vector from rotor flux to stator coordinates
- The closed-loop PWM system based on alternating current (AC) regulators

It should be noted that the state of the acceleration and brake pedals and of the battery and the speed have to be considered in the reference flux and torque calculators, in order to harmonize the driver's motion expectations with energy conversion optimal flow on-board.

In a DC vector system, the rotor flux position θ_{Ψ_r}, rotor flux, and speed may all be estimated [14], and thus a motion-sensorless system may be built. The vehicle has to start firmly, even from a stop on a slope; thus firm and fast torque responses are required from zero speed. Only for cruise control is an external speed regulator added.

Therefore, all sensorless systems have to provide safe estimation of θ_{Ψ_r} and Ψ_r at zero speed. This is how signal injection solutions became so important for sensorless ISA control [16].

The signal injection observer of θ_{Ψ_r}, Ψ_r, and ω_r has to be dropped above a certain speed due to large losses in the machine, inverter voltage usage reduction, and hardware and software time and costs limitations. The transition between the two observers has to be smooth [16].

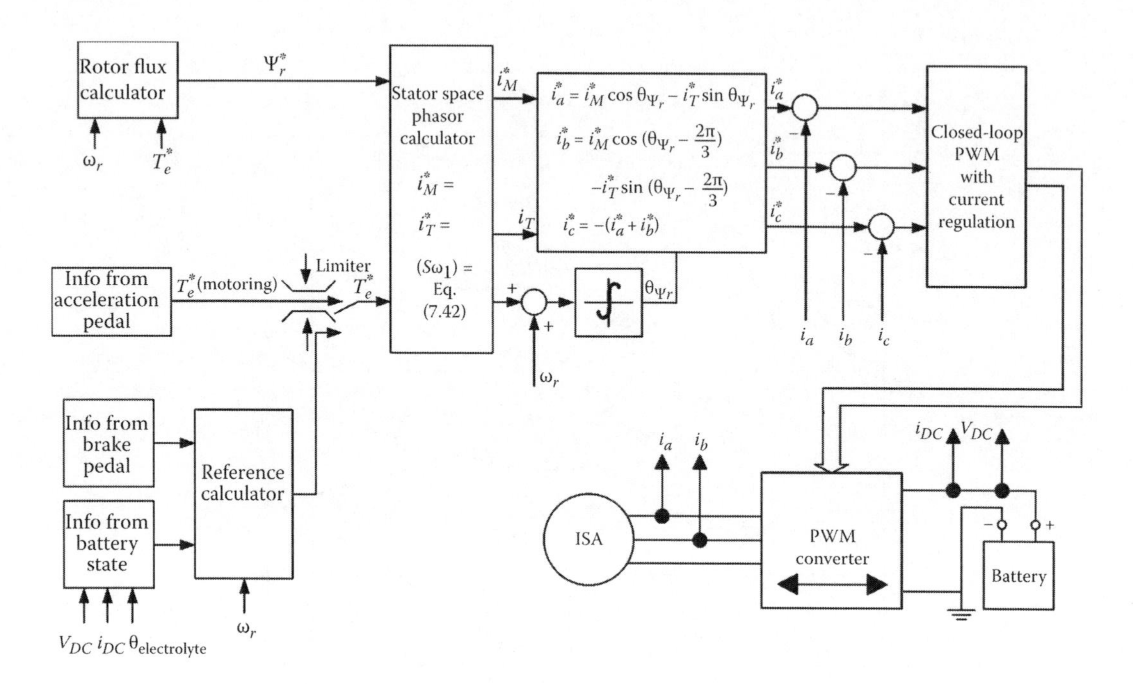

FIGURE 7.14 Generic (indirect AC) vector control system for induction starter/alternators (ISAs).

The above indirect current vector control scheme, Figure 7.14, has the merit that it works from zero speed in the torque mode, but adaptation for rotor resistance R_r and for magnetic saturation have to be added.

If a speed sensor is available on an EHV, the indirect current vector control with rotor resistance adaptation and saturation consideration constitutes a practical solution for the application.

7.6 DTFC of ISA

DTFC provides closed-loop control of flux and torque that directly triggers the adequate voltage vector in the converter. To reduce torque and current ripple, a regular sequence of neighboring voltage vectors with a certain timing is needed. See Chapter 6 for the basic DTFC. A general scheme for DTFC for ISA is shown in Figure 7.15.

For DTFC, the AC (or DC) current regulators are replaced with DC stator flux and torque regulators. In addition, a state observer that calculates stator flux amplitude Ψ_s, and position angle θ_{Ψ_s} (in stator coordinates), and then calculates the torque, is required. In motion sensorless configurations, a speed observer is also needed.

As speed control was not imperative at very low speed, sensorless DTFC without signal injection was proven to produce fast and safe torque response at zero speed [17] (Figure 7.16a and b). Variable structure (sliding mode) control was used both in state observer and in the torque and speed regulators.

The sliding mode flux observer [17], shown in Figure 7.17, combines the voltage model in stator coordinates and the current model in rotor flux coordinates to eliminate the speed estimation interference (errors) in the observers. The two models are connected through a sliding mode current error corrector (the current is also estimated, not only measured). This way, the current model prevails at low speeds and the voltage model prevails at high speeds.

A stator resistance corrector is needed for precision control at very low speeds. The rotor resistance is considered proportional to that of the stator:

$$\hat{R}_r = R_{so} - k_{RS} \cdot (\hat{\Psi}_{rd} \cdot (i_{sq} - \hat{i}_{sq}) - \hat{\Psi}_{rq} \cdot (i_{sd} - \hat{i}_{sd})) \tag{7.48}$$

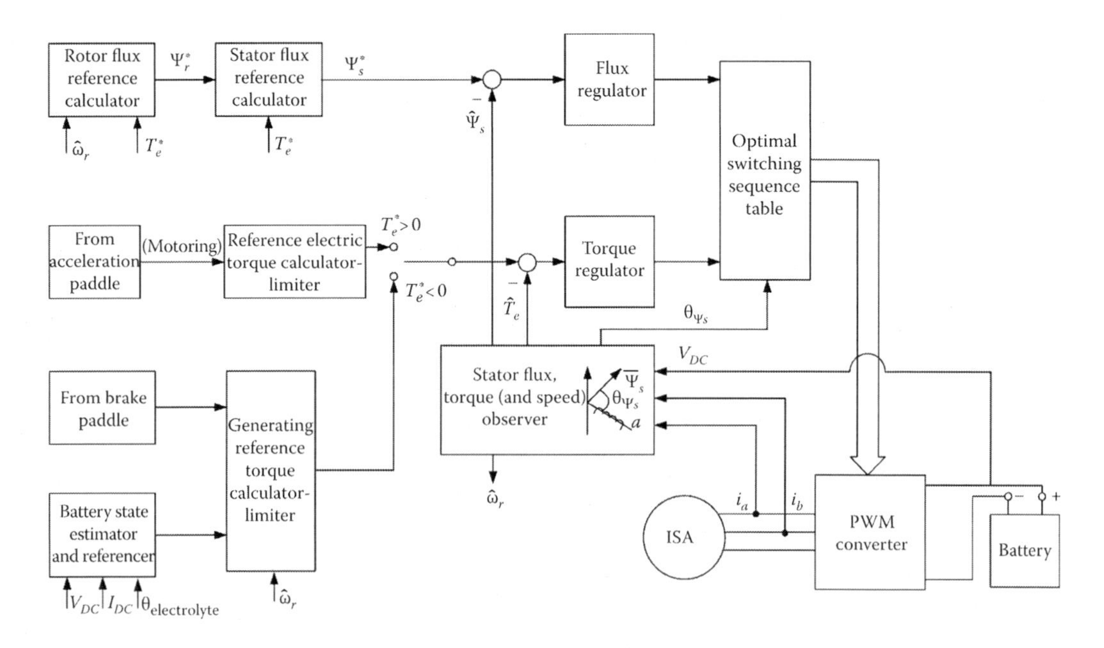

FIGURE 7.15 Direct torque and flux control (DTFC) of induction starter/alternator (ISA).

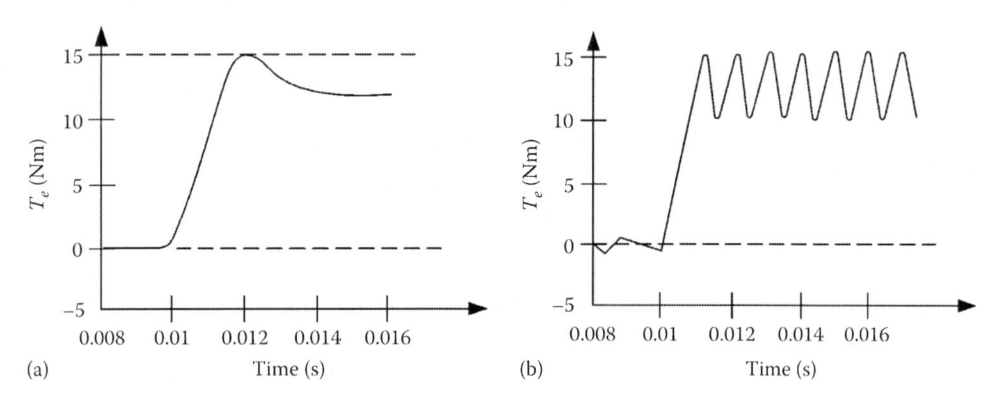

FIGURE 7.16 Step torque response of motion-sensorless direct torque and flux control (DTFC) at zero speed without signal injection with (a) sliding mode control and space-vector modulation and (b) classical DTFC. (From Lascu, C. et al., Very low speed sensorless variable structure control of induction machine, in *Record of IEEE-IEMDC-2003 Conference*, Madison, WI, 2003.)

$$\hat{R}_r = k_{sr} \cdot \hat{R}_s \cdot \frac{R_{ro}}{R_{so}} \tag{7.49}$$

i_s is the measured

$\hat{i}_s$ the estimated current vector (Figure 7.17)

The stator resistance adaptation greatly reduces the flux estimation errors [17].

The torque calculator is straightforward:

$$\tilde{T}_e = \frac{3}{2} \cdot p_1 \cdot \text{Real}(j\hat{\overline{\Psi}}_s \cdot \bar{i}_s) \tag{7.50}$$

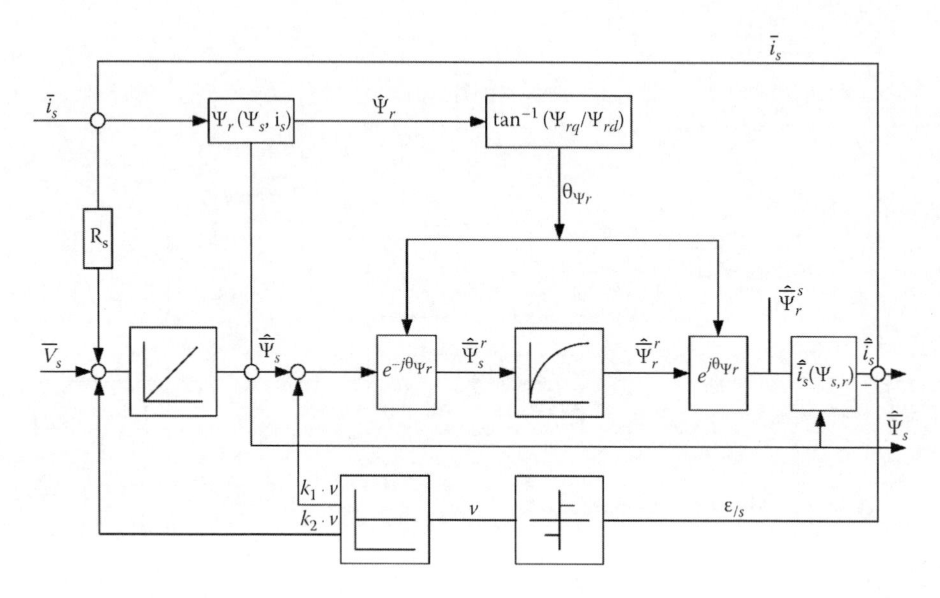

FIGURE 7.17 Sliding mode flux observer.

A few solutions for the speed observer may be applied for the scope [14]; but as speed control of ISA is not required at low speeds, a standard solution may be appropriate for the case:

$$\hat{\omega}_r = \hat{\omega}_{\Psi_r} - (S\hat{\omega}_1) \tag{7.51}$$

$$\hat{\omega}_{\Psi_r} = \frac{\text{Real}(\hat{\bar{\Psi}}_r \cdot j(d\hat{\bar{\Psi}}_r/dt))}{|\hat{\Psi}_r|^2} \tag{7.52}$$

$$(S\hat{\omega}_1) = \frac{2}{3}\frac{\hat{R}_r}{p_1}\frac{\hat{T}_e}{|\hat{\Psi}_r|^2} \tag{7.53}$$

The rotor speed comes as the difference between the rotor flux vector speed and the slip speed. An exact value of rotor resistance $\hat{R}_r$ is required for good precision. This justifies the rotor resistance adaptation, as in the flux estimator. The rotor resistance interferes only in the current model (i.e., at low speeds).

The control hardware and software do not change with speed.

Note that while both vector control and DTFC are capable of similar dynamics and steady-state performance, DTFC seems slightly superior when direct torque control is needed and motion-sensorless control is preferred.

7.7 ISA Design Issues for Variable Speed

There are a few design peculiarities to ISA design for variable speed. We have already mentioned a few in the previous paragraphs. Here we present them in a more systematic manner.

7.7.1 Power and Voltage Derating

There is a rich body of knowledge on induction machine design (mostly for the motor operation mode) for constant voltage and frequency [13].

Epson's constant C_0 (W/m^3), as defined by past experience, is an available starting point in most standard designs. To apply it to ISA, we first have to reduce C_0 to account for additional core and winding time harmonics losses. Then we have to increase it for peak torque requirements in constraint volume.

With today's insulated gate bipolar transistor (IGBT) PWM voltage source converters for DC battery voltage above 200 V, and power MOSFETs for less than 200 V_{DC} batteries, the converter derating of ISA is $P_{derat} = 0.08–0.12$.

The Epson's constant is of little value for ISA designed for speeds above 6000 rpm, belt-driven, as little experience was gained on the subject.

Voltage derating is due to PWM converter voltage drops. It amounts to 0.04–0.06 P.U. for above 200 V_{DC} batteries voltage, but it may go well above these values for 42 V_{DC} batteries.

In designs that are tightly volume constrained, such as ISA, it may be more appropriate to use as a design starting point the specific tangential peak rotor force density (shear stress) f_t (N/cm^2).

Peak values from 4 to almost 12 N/cm^2 may be achieved with current densities ranging from 10 to 40 A/mm^2. Naturally, forced cooling is generally necessary for ISAs.

As the battery voltage varies from V_{dcmin} to V_{dcmax} by 30% or more, the design may be appropriate for average-rated V_{DC}, with verifications on performance for minimum battery voltage.

7.7.2 Increasing Efficiency

Increasing efficiency is important to ISA to save energy on board vehicles. Volume constraint is contradictory to high efficiency, and trade-offs are required.

While volume constraints lead inevitably to increased fundamental winding and core losses, there are ways to reduce the additional (stray load) losses due to space and time harmonics.

Space harmonics are mainly due to stator and rotor slotting and magnetic saturation, but time harmonics are mainly due to PWM converter supply (Reference 13, Chapter 11 on losses).

A few suggestions are presented here:

- Adopt a large number q_1 of slots/pole/phase, if possible, in order to increase the order of the first space slot harmonic ($6q_1 \pm 1$).
- Compare thoroughly, designs with different pole counts for given specifications.
- In long stack designs, use insulated or at least noninsulated rotor cage bars with high bar-contact resistance in skewed rotors, to reduce interbar current losses.
- Use $0.8 < N_r/N_s < 1$ (N_s, N_r stator and rotor slot count) in order to reduce the differential leakage inductance of the first slot harmonics pair $6q_1 \pm 1$ and, thus, reduce interbar rotor current. Skewing may not be needed in this case, provided the parasitic synchronous torques are within limits.
- Skewing is necessary for $q_1 = 1, 2$; even a fractionary winding, with all coils in series, and free of subharmonics ($q = 1(1/2)$), may be tried in order to reduce stray load losses.
- Thin (less than 0.5 mm thick) laminations are to be used in ISA design where the fundamental frequency is above 300 Hz, to reduce all core losses.
- Use chorded coils to reduce end turns (and losses) and reduce the first phase belt harmonics (5,7) parasitic asynchronous torque.
- Carefully increase the air gap to reduce additional surface losses, but check on power factor reduction (peak current increases).
- Re-turn rotor surface to prevent lamination short circuits that may produce notably higher rotor surface additional core losses.
- Use sharp stamping tools and special thermal lamination treatment to reduce fundamental frequency core losses [18] above 300 Hz.
- Use copper rotor bars whenever possible to reduce the rotor cage losses and rotor slot size.

7.7.3 Increasing the Breakdown Torque

Large breakdown torque by design is required when a more than 2:1 constant power speed range is desired. This is the case of ISA, where ratios ω_{max}/ω_{rb} above 4:1 are typical and up to 10(12):1 would be desirable. Breakdown to rated torque T_{ek}/T_{es} ratios in induction machines is in the 2.5–3.5 range. The natural constant power ideal speed range coincides with the T_{ek}/T_{eb} ratio. When

$$\frac{\omega_{r\max}}{\omega_{rb}} > \frac{T_{ek}}{T_{eb}} \tag{7.54}$$

machine oversizing and other means are required.

Still, a high T_{ek}/T_{eb} ratio is desirable without compromising too much efficiency and power factor. As the breakdown torque T_{ek} may be approximated to ($R_s = 0$),

$$T_{ek} \approx \frac{3}{2}\cdot p_1 \cdot \left(\frac{V_s}{\omega_1}\right)^2 \cdot \frac{1}{2L_{sc}} \tag{7.55}$$

reducing the short circuit inductance L_{sc} of ISA is the key to higher breakdown torque.

The two components—stator and rotor—of-circuit (leakage) inductance are as follows (Reference 13, Chapter 6):

$$L_{sl} = 2\mu_o l_{stack}\cdot \frac{w_1^2}{p_1 \cdot q_1}\cdot(\lambda_{ss}+\lambda_{zs}+\lambda_{ds}+\lambda_{end}) \tag{7.56}$$

$$L_{rl} = 4ml_{stack}\cdot\left(\frac{w_1\cdot k_{w1}}{N_r}\right)^2\cdot 2\mu_o(\lambda_b+\lambda_{er}+\lambda_{zr}+\lambda_{dr}+\lambda_{skew}) \tag{7.57}$$

with

λ_{ss}, λ_b the stator (rotor) slot permeance coefficients
λ_{zs}, λ_{zr} the stator (rotor) zigzag permeance coefficients
λ_{ds}, λ_{dr} the stator (rotor) differential leakage coefficients
λ_{end}, λ_{er} the stator (rotor) end connection (ring) leakage coefficients
λ_{skew} the rotor skewing leakage coefficient
w_1 the turns per phase
p_1 the pole pairs

With so many terms in Equations 7.56 and 7.57, an easy way to reduce L_{sc} is not self-evident.

However, the main parameters that influence L_{sc} are as follows:

- Pole count: $2p_1$
- Stack length/pole pitch ratio: l_{stack}/τ
- Slot/tooth width ratio: $b_{ss,sr}/b_{t_s,t_r}$
- Stator slots/pole/phase: q_1
- Rotor slots/pole pair: N_r/p_1
- Slot opening per air gap: $w_{os,or}/g$
- Stator and rotor slot aspect ratio: $h_{ss,sr}/b_{ss,sr}$
- Air flux density: B_{g1}
- Stator (rotor) peak torque current density: j_{sk}, j_{rk}

Parameter sensitivity analyses showed that $2p_1 = 4, 6$ are best suited for speeds above 6000 rpm and $2p_1 = 8, 10, 12$ for speeds below 6000 rpm when wide constant power speed range conjugated with high peak torque at standstill is needed in volume constraint designs such as ISAs. The differential leakage tends to decrease with increased q_1 slots/pole/phase and decreased slot aspect ratios $h_{ss,sr}/b_{ss,sr} < 3.5–4$. However, this tends to increase the machine volume or oversaturates the teeth.

Longer core stacks l_{stack} may allow for a smaller stator bore diameter and, thus, shorter end connection in the stator windings. Slightly less end-connection leakage inductance is obtained this way. Only in high-speed ISAs with four poles does such a strategy seem practical [18].

Skewing, if avoided, as explained in the previous paragraph, may contribute to L_{rl} (and, thus, L_{sc}) reduction. Special stator windings with four layers were proven to reduce, in a few cases, by 30%, the stator leakage inductance [19]. However, this occurred at the price of added winding manufacturing costs.

7.7.4 Additional Measures for Wide Constant Power Range

Beyond the natural constant power speed range (T_{ek}/T_{eb} ratio: 2.5–3.5) illustrated in Figure 7.18a, machine oversizing is required, as the base torque T_{eb} has to go below rated (continuous) torque T_{er} (Figure 7.18b).

When the constant power speed range is larger than 4:1, machine oversizing will not do it alone, and winding reconfiguration or pole count changing is required.

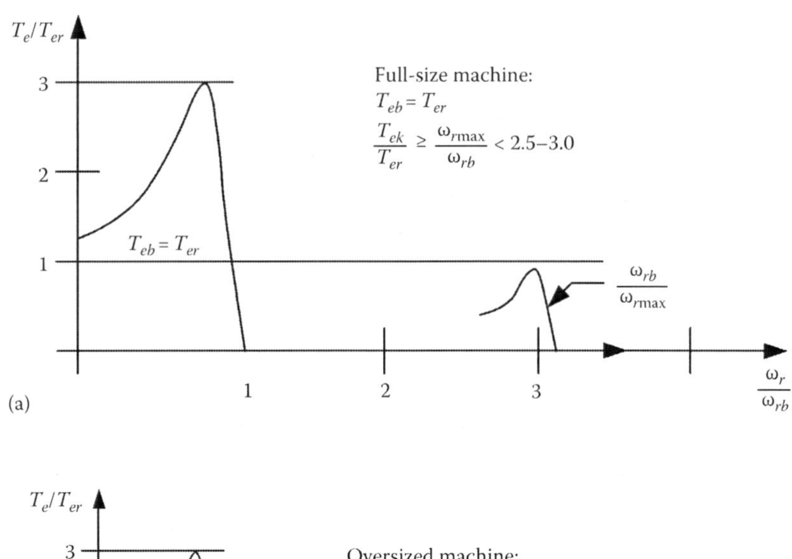

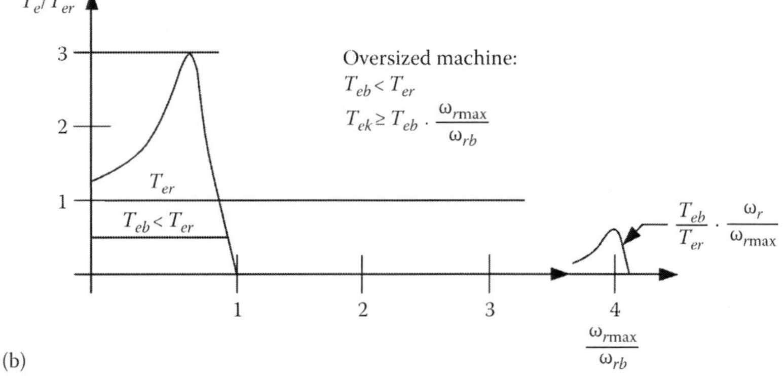

FIGURE 7.18 Natural constant power speed range: (a) full-size machine and (b) oversized machine.

7.7.4.1 Winding Reconfiguration

Consider a wide constant power speed range $c\omega = \omega_{r\max}/\omega_{rb} > 4$ and allow for a torque reserve at all speeds above rated (base) torque requirements (Figure 7.19):

$$\frac{T_{ek}}{T_{eb}} = c_{bT} > 1$$

$$\frac{T_{eMK}}{T_{eM}} = c_{MT} > 1 \tag{7.58}$$

$$\frac{T_{eM}}{T_{eb}} \approx \frac{\omega_{rb}}{\omega_{r\max}} = \frac{1}{c_\omega} \ll 1$$

As in this case $T_{ek}/T_{eb} < \omega_{r\max}/\omega_{rb}$, the phase voltage required to provide the wide constant power range should not reach its maximum value at base speed but somewhere below it.

We may consider T_{ek} and T_{eMK} as breakdown torques; thus,

$$T_{ek} \approx \frac{3}{2}\cdot p_1 \cdot \left(\frac{V_b}{\omega_b}\right)^2 \cdot \frac{1}{2L_{sc}} T_{eMK} = \frac{3}{2}\cdot p_1 \cdot \left(\frac{V_{\max}}{\omega_{\max}}\right)^2 \cdot \frac{1}{2L_{sc}} \tag{7.59}$$

Consequently,

$$\left(\frac{V_{\max}}{V_b}\right)^2 = \left(\frac{\omega_{\max}}{\omega_b}\right)^2 \cdot \frac{T_{eMK}}{T_{ek}} = \frac{c_\omega \cdot c_{MT}}{c_{bT}} \tag{7.60}$$

For

$$\frac{\omega_{\max}}{\omega_b} = 4, \quad c_{bT} = 3, \quad c_{MT} = 1.5, \quad \frac{V_{\max}}{V_b} = 1.41.$$

Consequently, a 41% reduction in the voltage has to be allowed at base speed. Accordingly, the same increase in the phase current is to be allowed for. Consequently, the PWM converter has to be oversized by 1.41 times

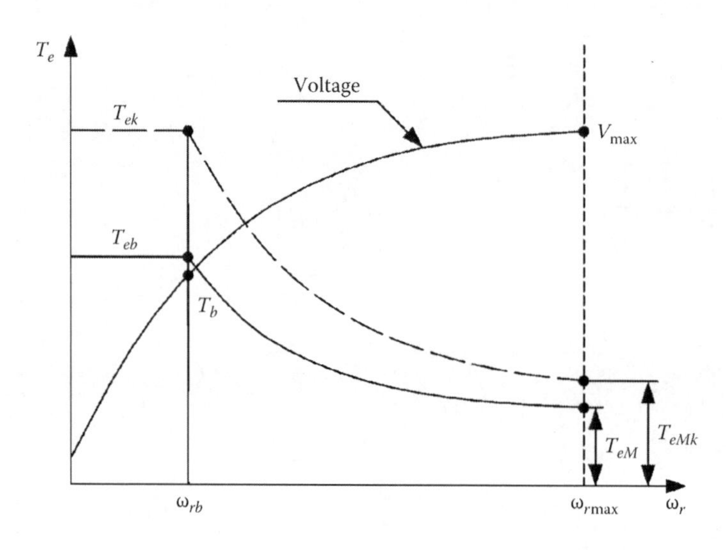

FIGURE 7.19 Phase voltage increasing for extended constant power speed range.

in current. The machine is to be oversized in this situation only if it is not capable of a 4:1 peak torque/rated torque ratio, which is the case in most situations. Realistic values of the above torques may be calculated by making use of induction machine standard equivalent circuits. A further increase in the constant power speed range requires winding changeover. The obvious choice is from star to delta connection (Figure 7.20).

Only a dual action power electromagnetic switch is required to perform the star to delta switch. During this changeover, the PWM converter control should be inhibited. If the time required for the changeover is less than 0.1 s, it will hardly be felt by the EHV power system.

When the winding connection switches from star to delta, it is as if the phase voltage was increased $\sqrt{3}/1$ times.

Consequently, the constant power speed range, proportional to voltage squared, because the torque is proportional to voltage squared, is increased three times:

$$\frac{\omega_{\max_{delta}}}{\omega_{\max_{star}}} = \left[\frac{(V_{phase})_{delta}}{(V_{phase})_{star}}\right]^2 = \frac{3}{1} \quad (7.61)$$

With this 3:1 ratio, the method is efficient for the scope and is in industrial operation in some spindle drives. It should work as well for EHV.

Winding tapping would bring a similar increase in voltage, but only part of the machine winding would remain at work at high speeds. Therefore, the overheating of the machine due to increased copper losses has to be attended for properly.

Therefore, it seems that the star to delta switching changeover is the best option.

Note that inverter pole changing in the ratio of 2:1 with a Dahlander winding may produce an almost 2:1 constant power speed range by using two twin PWM converters for the two half-windings. For one pole count, the currents in the two converters are in phase, and for the other they are opposite. The transition may be smooth [20], but still at the price of two converters, even designed at half the rating of a single one, because full battery voltage is available for both of them.

Note that for the smaller pole count, the winding factor goes down to 0.6–0.7 from 0.86 to 0.933, and its end-connection leakage inductance and resistance are larger than they should be in a dedicated winding. This fact leads to a notable reduction of the actual constant power speed range increase ratio: from 2 to probably less than 1.5.

At a higher ratio (3:1 for example), a pole count changing winding with high performance is not apparently available in an easy-to-manufacture-and-connect winding, therefore this latter approach seems to be less practical for industrial purposes [21].

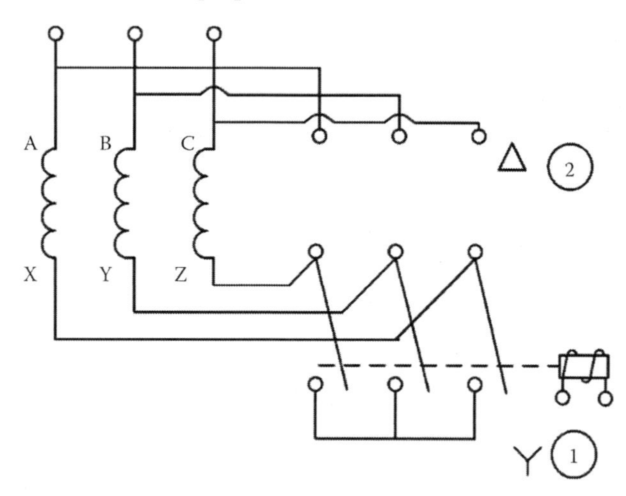

FIGURE 7.20 Star (Y) to delta (Δ) winding changeover.

Example 7.1

Consider a belt-driven ISA. The belt transmission ratio is 3:1, and the ISA has to deliver the peak torque of 140 Nm up to 500 rpm and the constant peak power (7.326 kW) up to 6000 rpm (engine speed). Only 7.326/3 = 2.432 kW is required from ISA for continuous generating from 500 to 6000 rpm: the breakdown to rated torque ratio of the machine $c_{bT} = T_{bk}/T_{eb} = 3.0$.

It is requested that we calculate the PWM converter current oversizing and machine peak torque oversizing to perform the $c_{bT} = \omega_{max}/\omega_{eb} = 6000/500 = 12/1$ constant power speed range.

Solution

The star to delta switching can produce a theoretical 3/1 speed range increase. Therefore, it will be placed at $c_{bT}/3 = 4.0$ base speed. That is, the star to delta switching will take place at 4.0 times the base speed (Figure 7.21).

Consider a zero reserve of torque at maximum engine speed $c_{MT} = 1$. Also, at base speed, $c_{bt} = 3.0$ and $\omega_{\max_{star}}/\omega_b = 4/1$.

Then, from Equation 7.60, applied at $4\omega_b$ for delta connection, and at ω_b for star connection, the respective voltage ratios are calculated:

$$\left(\frac{V_{M_{star/delta}}}{V_{rated}}\right)_B = \sqrt{\frac{c_{bT}}{c_{\omega_{star/delta}} \cdot c_{MT}}} = \sqrt{\frac{3}{3 \cdot 1}} = 1$$

$$\left(\frac{V_{rated}}{V_b}\right)_A = \left(\frac{V_{rated}}{V_{M_{star/delta}}}\right) \cdot \sqrt{\frac{\omega_b}{\omega_{\max_{star}}} \cdot \frac{c_{bT}}{c_{MT}}} = 1 \cdot \sqrt{\frac{3}{4 \cdot 1}} = 0.866$$

For the same power factor, the design current I_b' at base speed (point A in Figure 7.21) is increased with respect to that of base (max) voltage, I_b, by

$$\frac{I_b'}{I_b} = \frac{V_b}{V_{rated}} = \frac{1}{0.866} = 1.1547$$

The same ratio is valid for the peak torque current ratios in the two cases.

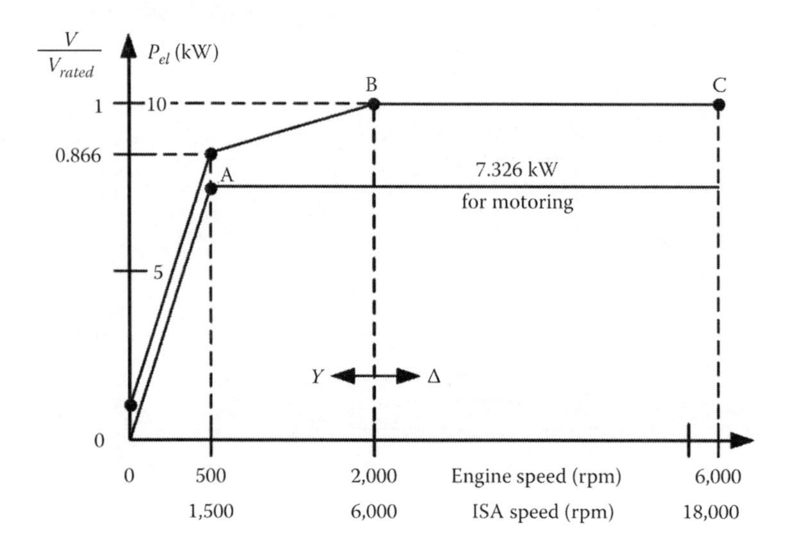

FIGURE 7.21 Voltage for wide speed constant peak power motoring of an induction starter/alternator (ISA) with Y and Δ winding switching.

The converter oversizing is equal to current overloading, which is only 15.47% in our case. However, the machine oversizing in torque is equal to the constant power speed ratio for star connection (4/1) to peak/rated torque ratio (3/1); that is, 4/3 = 1.33, or 33%.

As the machine costs are notably lower than the PWM converter costs for given kilovolt ampere (especially for battery voltages $V_{DC} < 200$ V), the above division of oversizing seems to be adequate.

As in our case, for generator mode, three times less peak power is required, there should be no problem in obtaining it with proper control. Even star connection would do for generating in our case, but performing star to delta and delta to star switching at a given speed may prove more practical to implement in the control, in order to limit the wearing-off of the electromagnetic power switches.

With the exception of generating limitations at high speed, attributed here in part to control insufficiencies, very good performance in torque/speed envelopes was proven in References 6 and 15 for full-size ISAs for mild and, respectively, full hybrid electric cars.

Still, the rather high rotor winding losses at low speed for peak torque constitute a severe obstacle to ISA industrialization.

In Reference 22, an urban electric vehicle with ISA is presented with test results for 14 kW constant power at basically a 3:1 speed range.

A design comparison between two electric vehicles with ISA, one for in-wheel placement (400 rpm, rated speed) and one with a 10/1 gear ratio, at 5 kW, was performed considering the losses in the ISA and in the converter for a 3:1 constant power speed range, for urban- and country-drive standard European cycles.

With stator flux adapted to torque, a notable energy savings for in-town drive is obtained. The geared ISA is shown to peak slightly superior in terms of losses (or range) at 16.3 kg weight (ISA plus gearbox) versus 40 kg for the in-wheel configuration [23].

Yet another such comparison design of direct-driven and transmission-driven ISA for a mild hybrid vehicle (at 42 V_{DC}) for 6.5 kW generating from 1000 to 6500 rpm engine speed, and 350 Nm crankshaft starting torque is given in Reference 24. Theory and some test results are offered. Winding changeover is used, as 6.5/1 constant power speed range is required.

The epicyclic gear is operational only in the motor mode. This way, the ISA peak torque is reduced. A 12-pole ISA is adopted to reduce the slip frequency and losses. The power factor increases, and thus the peak current is reduced. The battery requirements and PWM converter ratings are also reduced.

For the gearless solution, the star to delta switch of the winding is used to reach the large constant power speed range in an 18-pole ISA configuration.

About 6.5 N/cm^2 shear rotor stress for the peak torque is adopted in designs.

Both solutions have their definite merits.

Note that a comprehensive design methodology of an ISA, complete for specifications, should also consider the PWM converter and battery losses and costs in order to provide an industrial design. For simplified PWM loss modeling, with and without DC voltage booster, see Reference 9.

Blending the standard induction machine design body of analytical knowledge with finite element method (FEM) verifications (see Reference 13 for details) for the ISA special operational conditions should lead to an industrial multiobjective optimization design of the ISA system. Such a grueling task is beyond our space and scope here.

For comparisons of ISA and PMSA continuous and short duty (...) power/speed capabilities, for full voltage at base speed designs, without winding reconfiguration, see Reference 25.

7.8 Summary

- "Electrification" of road vehicles is under aggressive development today with some dynamic industrial markets in place.
- The degree of electrification is defined by the electric fraction %E, the ratio between peak electric propulsion power and peak total power of the vehicle.

- With $\%E < (70–80)\%$, the vehicles are called hybrids; with $\%E = 100\%$, they are called electric vehicles.
- High peak starting torques and wide constraint power speed ranges (up to 12:1) are required in EHVs under constraint volume and costs, but with low total system losses, to cut fuel consumption and increase battery life.
- Given its ruggedness and low cost, the cage-rotor induction machine seems a natural choice for starter/alternators on EHVs. They are called here induction starter/alternators (ISAs).
- In mild hybrids, the starter/alternator is to be designed for low battery voltage: the 42 V_{DC} bus, in general. It uses a PWM power converter interface to deliver controlled torque for the entire speed range. It is also limited in power to (in general) a maximum of 6–10 kW.
- Full hybrids have ISAs that deliver above 10 kW of power for both motoring and generating and require high-voltage battery packs (180 V_{DC} and more). ISAs assist the driving substantially, at low speeds (up to 2000 rpm engine speed) and still notably above this speed. Powers of 30–50 kW are typical for full hybrid cars and above 75 kW for hybrid city buses. Electric vehicles with batteries, fuel cells or inertial batteries (flywheels) require even more power, as electric propulsion acts alone.
- The battery voltage is a very important design factor for ISA, and its modeling is crucial to determine the battery receptivity during its normal recharging (at high speed) or during regenerative braking by ISA.
- Battery capacity Q, discharge rate Q/h, SOC, SOD, and DOD, are among the main parameters in battery modeling. Linear models, with variable emf (e_V) and internal resistance R_i are typical.
- The number of charge–discharge cycles, the specific energy (Wh/kg) and specific peak power (W/kg), and efficiency characterize the battery performance for EHVs.
- An actual lead-acid battery is characterized by about 50 Wh/kg, 400 W/kg, 80% efficiency, 500–1000 cycles at $150/kWh. Newer batteries such as lithium polymer batteries are credited in tests with 150–200 Wh/kg, 350 W/kg, 1000 cycles at $150/kWh. Further developments are needed to dethrone the lead-acid batteries, with valve-regulated electrolyte, from road vehicles, because they are an established technology.
- Super-high-speed flywheels (in vacuum) with magnetic suspension and special rotor composite materials have been tested up to 1 km/s peripheral speed, when the 50 Wh/kg of lead-acid battery is surpassed at 2000 W/kg and 2–3 min "recharging." The recharging of the flywheels is performed through the generator/motor on its shaft and its oversized PWM power interface converter.
- Given the challenging torque, efficiency, power factor, speed range, under constraint volume, and costs of the ISA system, quite a few design particularities characterize them.
- As in any variable speed system, the performance and costs prevail. This way, the fundamental frequency in the ISA is limited, with silicon lamination core, from 500 to 600 Hz, to also reduce the PWM converter switching losses.
- Depending on the usage (or not) of the transmission between the ISA and the ICE engine (in hybrid vehicles) and between ISA and wheels in electric vehicles, the maximum speed of ISA varies from 2,000 rpm in city buses to 18,000 rpm in the mild hybrids with belted ISAs.
- Consequently, ISAs placed on the ICE shaft are designed with 8, 10, or 12 poles, while the belted ones have four or six poles.
- Skin effect is not desirable, and this restricts the shape of the rotor slot.
- The high peak temperature in the rotor (240°C or more) tends to suggest using copper bars in the rotor cage.
- Advanced magnetic saturation is needed to reach high peak torque density at low speeds and for start. Uniform oversaturation of stator and rotor teeth and yokes reduces the air gap flux density space harmonics. Therefore, provided the magnetization curve of ISA is known from analytical or FEM models, the equivalent circuit and the space-phasor model may be used in the ISA design, performance, computation, and control.

- Constant rotor flux (rotor flux orientation)—FOC—allows for fastest transient torque of ISA in rotor flux coordinates.
- For constant rotor flux amplitude in synchronous coordinates ($\omega_b = \omega_1$), the ISA acts as a high saliency reluctance synchronous machine; the virtual saliency is "produced" by the rotor current vector, which is at 90° with respect to the rotor flux vector. The d-axis inductance is the no-load inductance L_s, and the q inductance is the short-circuit inductance L_{sc}. The d-axis current is the rotor flux current i_M, and the torque current, along axis q, is i_T.
- For peak torque production, the maximum torque/current criterion is used. Ideally, $i_{MK} = i_{TK} = i_{SK}/\sqrt{2}$, but due to variable (heavy) magnetic saturation, $i_{MK} > i_{TK}$. This should be embedded in the FOC below base speed.
- For maximum speed constant power production, the maximum torque per available stator flux criterion is used when $i_M = i_T \cdot L_{sc}/L_s$.
- The dq current angle $\delta_i = \tan^{-1}(i_M/i_T)$ varies from large values at low speed to low values at high speeds.
- Indirect vector control works naturally for zero speed, with magnetic saturation and rotor resistance adaptation. In the presence of a speed sensor, it seems to be a practical control strategy for ISA.
- The rotor flux and torque reference calculators mitigate the driver motion expectations and energy management on board (with battery state as crucial).
- Still, the online computation effort is high.
- DTFC—used commercially by some manufacturers of electric drives—is also feasible for ISA. DTFC uses the rotor flux and torque reference calculators, but it calculates the required stator flux reference and then close-loop regulates the stator flux and torque to trigger a preset sequence in the PWM converter.
- Instead of vector rotation and sophisticated PWM modulation with current regulators (eventually with emf compensation), typical to FOC, DTFC needs a flux observer and a torque calculator based on measured voltage and current.
- As the speed estimation may be obtained out of the flux observer, a motion-sensorless control system is inherent to DTFC for ISA, where torque control is predominantly used. Only during cruise control is slow speed regulation performed above a certain speed.
- Fast and safe torque response at zero speed, in motion-sensorless DTFC, was reported without signal injection. Operation at zero speed, sensorless, with signal injection is implicit, but injection has to be abandoned above a small speed, for a different strategy, to cover the rest of the large speed range.
- A few design aspects of ISA require special attention to produce competitive results:
  - Power and voltage derating due to PWM converter
  - Design battery DC voltage
  - Ways to decrease stray load losses
  - Increasing the breakdown torque (by decreasing L_{sc})
  - Winding reconfiguration for constant power speed range extension: a 3:1 extension is obtained by star to delta switching; up to 12:1 total constant power speed range could be obtained by some motor and converter oversizing
- In principle, ISA is capable of complying with the challenging EHV requirements, except for the increase in current and losses at high speeds. In terms of PWM converter peak kilovolt ampere (rating), the ISA performs moderately, though overall better than the surface-permanent magnet rotor brushless machine.
- A system optimization design, rather than the ISA design optimization, is the way to a competitive solution with ISA.
- This chapter represents a mere introduction to such a daunting task.

References

1. S.A. Nasar and L.E. Unnewher, *Electrical Vehicle Technology*, John Wiley & Sons, New York, 1981.
2. I. Husain, *Electric and Hybrid Vehicles*, CRC Press, Boca Raton, FL, 2003.
3. T. Yanase, N. Takeda, Y. Suzuki, and S. Imai, Evolution of energy efficiency in series-parallel hybrid vehicle, Record of World EHV-18, Berlin, Germany, 2001.
4. J.M. Miller, V.R. Stefanovic, and E. Levi, Progresses for starter-alternator systems in automotive applications, Record of EPE–PEMC-2002, Dubrovnik and Cavtat, Croatia, 2002.
5. E. Spijker, A. Saibertz, R. Busch, D. Kak, and D. Pelergus, An ISG with dual voltage power net stretching the technology boundaries for higher fuel economy, *Record of EVS-18 World Congress 2001*, Berlin, Germany, 2001.
6. S. Chen, B. Lequesne, R.R. Henry, Y. Xuo, and J. Ronning, Design and testing of belt-driven induction starter-generator, *IEEE Trans.*, IA-38(6), 2002, 1525–1532.
7. T. Teratani, K. Kuramochi, H. Nakao, T. Tachibana, K. Yagi, and S. Abou, Development of Toyota mild hybrid system (THS-M) with 42V PowerNet, Record of IEEE–IEMDC-2003, Madison, WI, 2003, pp. 3–10.
8. I. Boldea, L. Tutelea, and C.I. Pitic, PM-assisted reluctance synchronous motor-generator (PM-RSM) for mild hybrid vehicles, Record of OPTIM-2002, Poiana Brasov, Romania, 2002, vol. 2, pp. 383–388.
9. A. Vagati, A. Fratta, P. Cugliehmi, G. Franchi, and F. Villata, Comparison of AC motor based drives for electric vehicle, In *Record of PCIM-1999, Intelligent Motion Conference*, Nurnberg, Germany, 1999, pp. 173–181.
10. A. Lange, W.R. Canders, F. Laube, and H. Mosebach, Comparison of different drive systems for a 75 kW electrical vehicle drive, Record of ICEM-2000, Espoo, Finland, 2000, vol. 2, pp. 1308–1312.
11. S. Barsali, M. Ceraolo, and A. Possenti, Techniques to control the electricity generation in a series hybrid electrical vehicle, *IEEE Trans.*, EC-17(2), 2002, 260–266.
12. D. Rand, R. Wood, and R.M. Dell, *Batteries for Electric Vehicle*, John Wiley & Sons, New York, 1998.
13. I. Boldea and S.A. Nasar, *Induction Machine Handbook*, CRC Press, Boca Raton, FL, 2001.
14. I. Boldea and S.A. Nasar, *Electric Drives*, CRC Press, Boca Raton, FL, 1998.
15. P.J. McCleer, J.M. Miller, A.R. Gale, M.W. Degner, and F. Leonardi, Nonlinear model and momentary performance capability of a cage rotor induction machine used as an automotive combined starter-alternator, *IEEE Trans.*, IA-37(3), 2001, 840–846.
16. F. Briz, M.W. Degner, A. Diez, and R.D. Lorenz, Static and dynamic behavior of saturation-induced saliences and their effect on carrier signal-based sensorless AC drives, *IEEE Trans.*, IA-38(3), 2002, 670–678.
17. C. Lascu, I. Boldea, and F. Blaabjerg, Very low speed sensorless variable structure control of induction machine, In *Record of IEEE–IEMDC-2003 Conference*, Madison, WI, 2003.
18. W.L. Soong, G.B. Kliman, N.R. Johnson, R. White, and J. Miller, Novel high speed induction motor for a commercial centrifugal compressor, In *Record of IEEE–IAS-1999 Annual Meeting*, 1999, vol. 1, pp. 494–501.
19. L. Küng, K. Reichert, and A. Vezzini, Winding arrangement for multipolar squirrel cage induction machine with low leakage inductance, Record of ICEM-1998, Istanbul, Turkey, 1998, vol. 1, pp. 353–358.
20. M. Osana and T.A. Lipo, A new inverter control scheme for induction motor drives requiring wide speed range, In *Record of IEEE–IAS-1995 Annual Meeting*, 1995, vol. 1, pp. 350–355.
21. C. Zang, I. Yang, X. Chen, and Y. Gue, A new design principle for pole changing winding—The three equation principle, *EMPS J.*, 22(2), 1994, 1859–1865.
22. D. Casadei, C. Rossi, G. Serra, and A. Tani, Performance analysis of an electric vehicle driven by SFVC induction motor drive, Record of Electromotion-2001, Bologna, Italy, 2001, vol. 1, pp. 141–146.

23. L. Tutelea, E. Ritchie, and I. Boldea, Induction machine design with and without mechanical transmission for electrical vehicle drives, Record of Electromotion-2001, Bologna, Italy, 2001, vol. 1, pp. 275–280.
24. R. Nutchler, Two concepts of starter-generator-machine for 8 to 12 cylinder combustion engine, In *Record of IEEE–IEMDC-2003 Conference*, Madison, WI, 2003, vol. 1, pp. 194–199.
25. G. Pellegrino, A. Vagati, B. Boazzo, and P. Guglielmi, Comparison of induction and PM synchronous motor drives for EV applications including design examples, *IEEE Trans.*, IA-48(6), 2012, 2322–2332.

8

Permanent-Magnet-Assisted Reluctance Synchronous Starter/Alternators for Electric Hybrid Vehicles

8.1 Introduction

The permanent-magnet-assisted reluctance synchronous machine (PM-RSM), also called the interior PM (IPM) synchronous machine, with high magnetic saliency was proven to be competitive, price-wise (Figure 8.1) [1], and superior in terms of total losses (Figure 8.2) [2] for starter/alternator automobile applications where a large constant power speed range $\omega_{max}/\omega_b > 3$–4 is required.

The cost comparisons show the PM-RSM starter alternator system, including power electronics control, to be notably less expensive than the surface PMSM or switched reluctance machine system at the same output. It is comparable with the cost of the induction machine system. In terms of total machine plus power electronics losses, the PM-RSM is slightly superior even to the surface PMSM (Figure 8.2), and notably superior to the induction machine system, all designed for the same machine volume, at 30 kW [2].

It was also demonstrated that PM-RSM [2] is capable of larger constant peak power speed range than the surface PM synchronous, or the induction, or the switched reluctance machine of the same volume.

In essence, both the lower cost and the wide constant power speed range are explained by the combined action of PMs and the high magnetic saliency torque to reduce the peak current for peak torque at low speed and reduce flux/torque at high speeds. The use of ferrites in PM-RSM has been proven recently rather competitive if the PM demagnetization is avoided by careful design, up to most extreme operation mode.

Starter/alternators for automobile applications are forced to operate at a constant power speed range $\omega_{max}/\omega_b > 3$–4 and up to 12–1. The higher the interval (without notably oversizing the machine or the converter) constant power speed range, the better.

This is how the PM-RSM becomes a tough competitor for electric hybrid vehicles (EHVs). The larger the ω_{max}/ω_b, the smaller the PM contribution to torque.

In what follows, we will treat the main topological aspects, field distribution, and parameters by finite element method (FEM), lumped parameter modeling of saturated PM-RSM, core loss models, design issues for wide ω_{max}/ω_b ratios, and system models for dynamics and vector flux-oriented control (FOC) and direct torque and flux control (DTFC) with and without position control feedback.

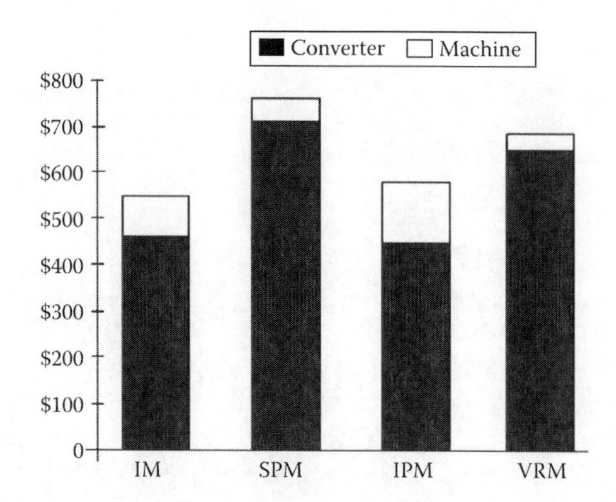

FIGURE 8.1 Cost comparisons for four alternators at 6 kW and 42 V_{DC}.

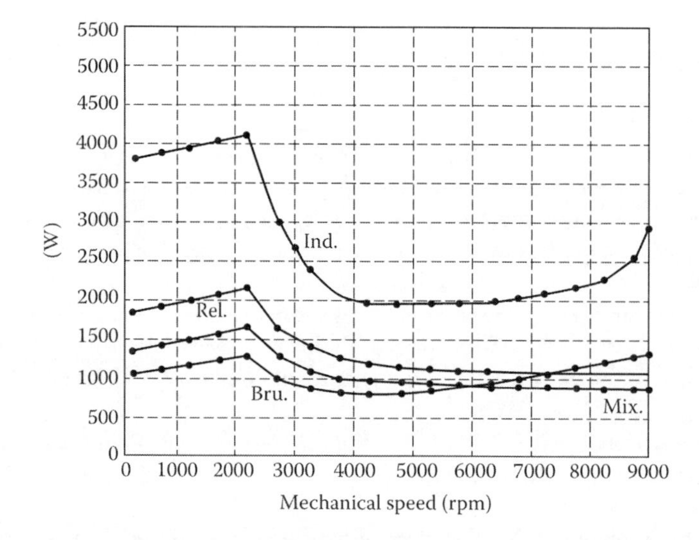

FIGURE 8.2 Loss comparisons between four starter motor-generator systems at 30 kW. (From Vagati, A. et al., Comparisons of AC motor based drives for electric vehicle application, Record of PCIM-1999, Intelligent Motion, Europe, 1999, pp. 173–181.)

8.2 Topologies of PM-RSM

The PM-RSM is, in fact, a multilayer flux barrier and PM rotor SM—an interior multilayer PM rotor machine, in other words. The standard IPM has only one PM rotor layer per pole, and the PM contribution to torque is predominant (Figure 8.3a). In PM-RSMs, the high magnetic saliency created by the multiple flux barriers in the rotor makes reluctance torque predominant at low speeds when the highest torque is required.

The stator core of the PM-RSM is provided with uniform slots that host a distributed ($q > 2$) three-phase winding with chorded coils. The rotor core may be built of conventional (transverse) laminations

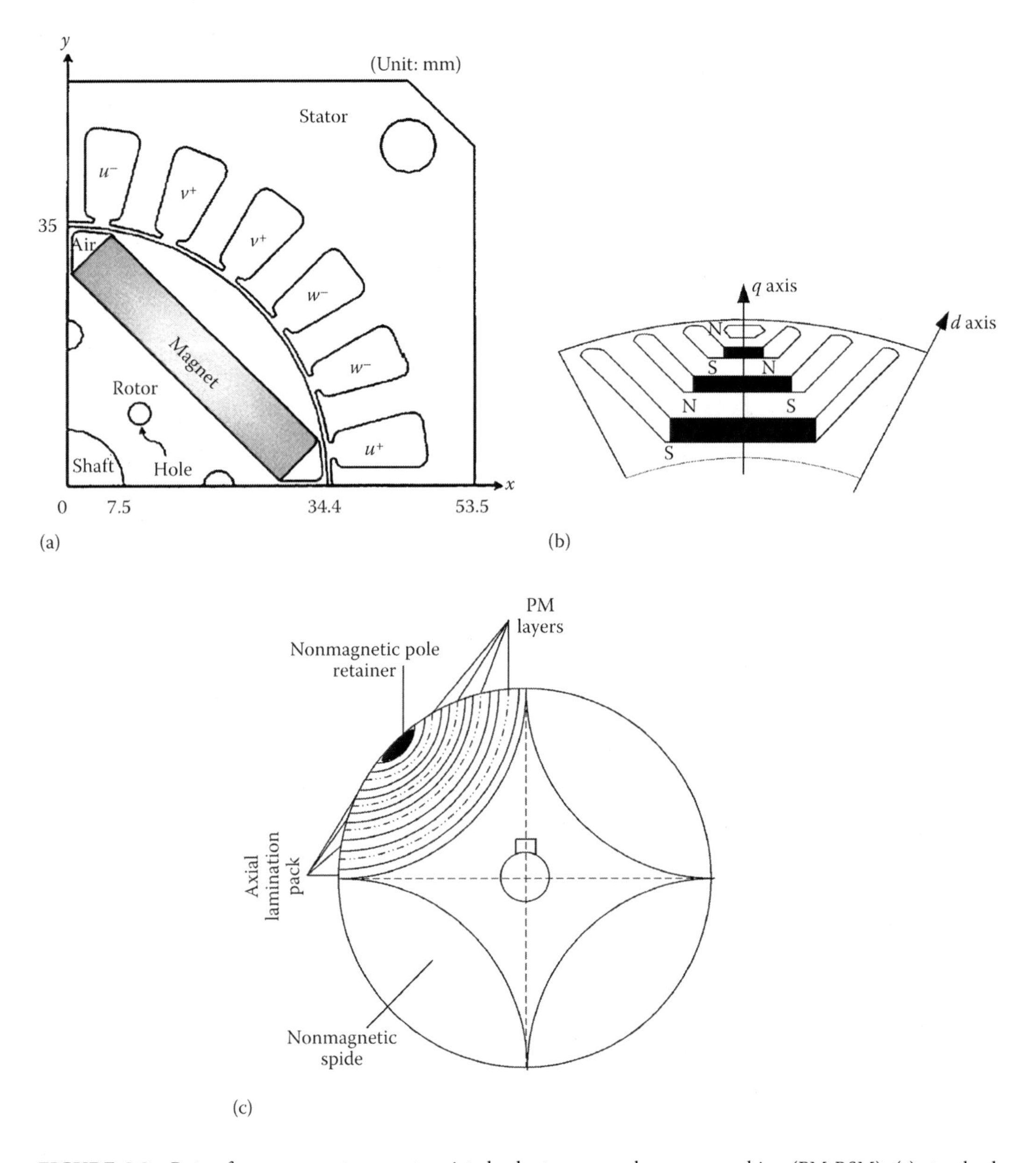

FIGURE 8.3 Rotor for permanent-magnet-assisted reluctance synchronous machine (PM-RSM): (a) standard interior permanent magnet (IPM), (b) with transverse laminations, and (c) with axial laminations.

with stamped multiple flux barriers per pole, filled with PM layers (Figure 8.3b), or from axial laminations with multiple PM layers per pole (Figure 8.3c).

The transverse lamination rotor with multiple flux barriers requires magnetic bridges to leave the lamination in one piece and provide enough mechanical resilience up to maximum design speed.

It turns out that mechanical speed limitations, due to centrifugal forces, basically, lead to magnetic bridges that are shorter (tangentially), while their thickness varies radially from 0.6 to 1 (1.2) mm minimum (Figure 8.4) [3]. The shorter flux bridges tend to result in larger magnetic permeance

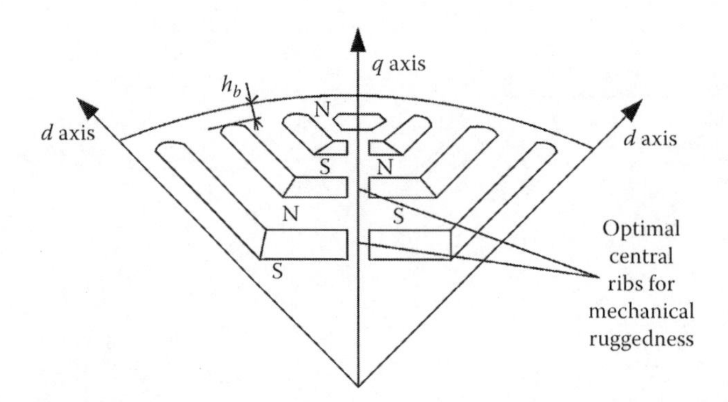

FIGURE 8.4 Mechanically acceptable flux barrier geometry.

along the q axis (which is not desirable) but to larger permeance along the d axis also (which is desirable), and the L_{dm} to L_{qm} difference may be maintained acceptably large. L_{dm}/L_{qm} ratios up to 10/1 may be obtained this way (L_{dm} and L_{qm} are the magnetization inductances). Placing the PMs on the bottom of the flux barriers (Figure 8.3b) has the advantage that the former are more immune to demagnetization at peak current (or torque). However, it has the disadvantage that the PM flux leakage is larger, therefore more PM weight is required. Moreover, when the PMs are placed on the lateral part of the flux barriers, the reverse is true. Depending on the overload specifications and constant power speed range, the PMs may be placed either on the bottom or on the lateral parts of the flux barriers.

The axially laminated rotor (Figure 8.3c) [4,5] is made of overlaid layers of axial laminations and PM flexible ribbons for each pole. A nonmagnetic rotor spider sustains the poles, and a nonmagnetic pole retains them to the spider with nonmagnetic bolts.

There are no flux bridges between the axial lamination layers; thus, L_{qm} is notably smaller. In addition, the PM flux loss (leakage) tangentially in the rotor is smaller. There are, however, three problems associated with the axially laminated anisotropic (ALA) rotor:

- The manufacturing is not standard.
- The mechanical rigidity is not high, therefore the maximum peripheral speed is limited to, perhaps, 50 m/s.
- The flux harmonics due to the open flux barriers produce flux pulsations in the rotor, and the q-axis armature reaction space harmonics create magnetic flux perpendicular to axial lamination plane. Both of these phenomena produce additional losses in the rotor, as the air gap is necessarily low to ensure high magnetic saliency: L_{dm}/L_{qm}.

Two or three radial slits in the rotor core (laminations) and the making of the PM of three (four) pieces axially will reduce those losses to reasonable values [4]. Saliencies $L_{dm}/L_{qm} = 25$ under saturated conditions were obtained with a two-pole 1.5 kW machine [4].

The magnetic saliency ratio L_{dm}/L_{qm} refers here only to the main flux path. The leakage inductance L_{sl}, when included, makes the ratio L_d/L_q smaller:

$$\frac{L_d}{L_q} = \frac{L_{dm} + L_{sl}}{L_{qm} + L_{sl}} < \frac{L_{dm}}{L_{qm}} \tag{8.1}$$

Increasing the number of PM layers/pole leads to increasing the L_{dm}/L_{qm} and $L_{dm} - L_{qm}$ but, above four PM layers (flux barriers) per half-pole, the improvements are minimal. Two to four flux barriers, depending on the rotor diameter and on the number of poles, seem to suffice [6].

Modification of the geometry of flux barriers, mainly their relative thickness, the shape of the flux bridges, together with the number of layers (two, three, or four), and their positioning, is the tool to improve the L_{dm}/L_{qm} ratio and the $1-(L_{qm}/L_{dm})$ saliency index, and thus improve performance [7]. The higher the L_{dm}/L_{qm}, the larger the power factor and the constant power speed range, as the torque is proportional to the $1-(L_{qm}/L_{dm})$ factor.

On the other hand, the lower the L_{qm}, the higher the short-circuit current. Thus high magnetic saliency (low L_{qm}/L_{dm}) means large short-circuit current levels.

8.3 Finite Element Analysis

The complex topology of the rotor, combined with the small air gap, stator slot openings, and the distributed stator windings, makes a finite element analysis of flux distribution in the machine a practical way of grasping some essential knowledge about the PM-RSM capability in terms of torque for given geometry and stator magnetomotive force (mmf).

8.3.1 Flux Distribution

Consider the geometry given in Table 8.1 [8]. The stator–rotor structure is shown in Figure 8.5a, where the flux lines for zero stator currents are shown. As expected, the flux lines "spring" perpendicularly from the PMs and then spread between the flux barriers radially in the rotor [8] and then through the stator teeth.

The PM air gap flux density distribution in Figure 8.5b shows the presence of both multiple PM layers and the stator slot openings. The average value is about 0.3 T, while the distribution is similar to that of no-load air gap flux density in induction machines.

Then, current of given amplitude $I\sqrt{2}$ is injected in phase A with $(-I/\sqrt{2})$ in phases B and C. This way, the stator mmf distribution axis falls along the phase A axis. The rotor axis d (of highest permeance) is then moved from the mmf axis by increasing mechanical angle.

Admitting a certain current density and slot filling factor, the slot mmf may be found without knowing the number of turns per coil in the two-layer, eight-poles, three-phase winding.

The calculation of torque, through FEM, is done for various maximum slot mmfs and angles, until the maximum torque reaches the target of 140 Nm as a certain power angle $\delta_{id} = \tan^{-1}(I_q/I_d)$ (Figure 8.6a and b).

The flux lines for the peak torque conditions, with $\delta_{id} = 46°$ (electrical degrees), are shown in Figure 8.7. The air gap flux density, for the same situation, shown in Figure 8.8, exhibits the strong influence of the stator mmf contribution (see Figure 8.4) to air gap flux density. The question arises as to if, in such conditions, the PMs are not demagnetized. The flux density "getting out" from the lowest barrier PM (Figure 8.3b) indicates clearly that the minimum flux density is 0.42 T and thus, the PMs are safe (Figure 8.9).

TABLE 8.1 Geometry of a 140 nm Peak Torque Permanent-Magnet-Assisted Reluctance Synchronous Machine (PM-RSM)

Parameter	Value
Stator outer diameter	245.2 mm
Rotor outer diameter	174.2 mm
Rotor inner diameter	113 mm
Air gap	0.4 mm
Stack length	68 mm
PM parameter at 20°C	$B_T = 0.87$ T
	$H = 0.66$ kA/m

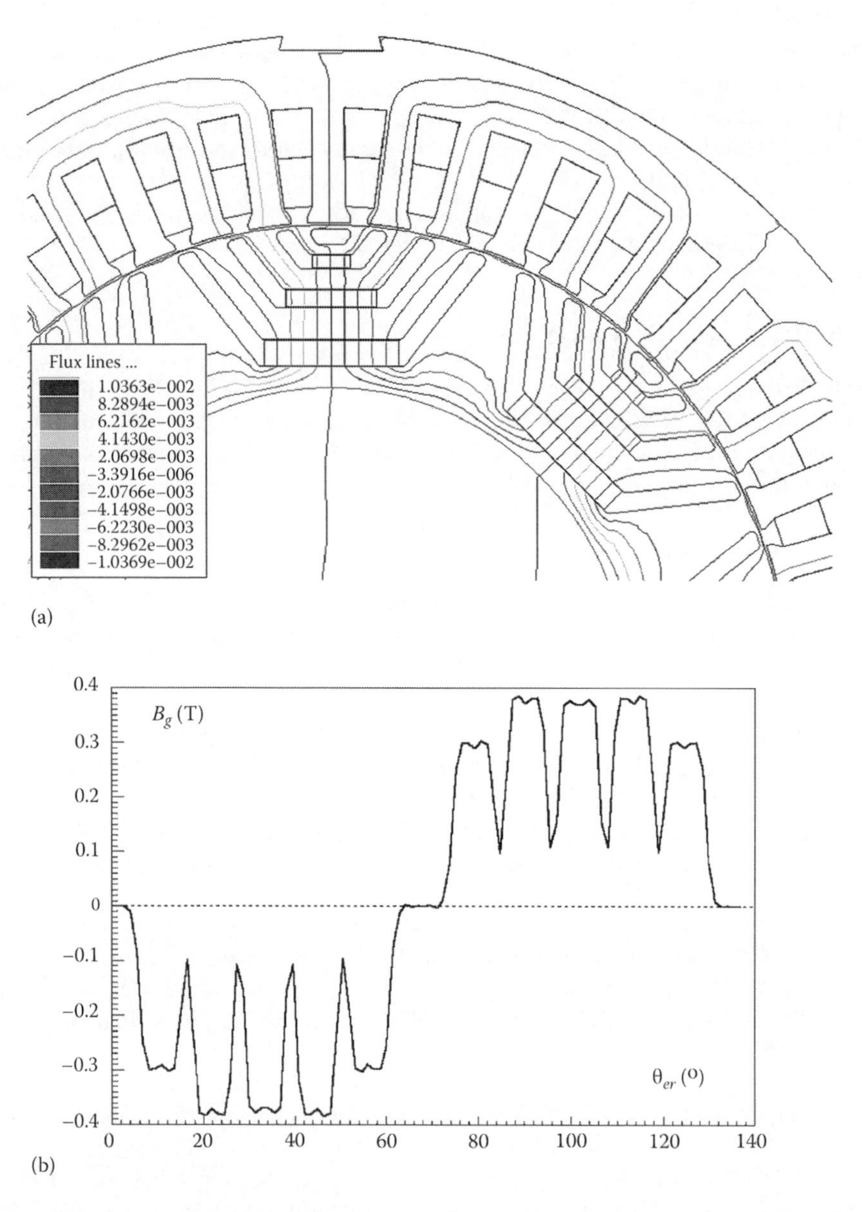

FIGURE 8.5 (a) Permanent magnet (PM) flux lines (zero stator current) and (b) air gap flux density distribution.

The computation of flux density distribution in the stator teeth, in the middle or at their bottom [8], shows that the magnetic saturation may reach 1.85 T in the teeth and 1.65 T in the back iron, in this particular design. The resultant air gap flux for the maximum torque depicted in Figure 8.10 is not far from a sinusoid waveform.

8.3.2 *dq* Inductances

The above analysis indicates that even for rather high electric and magnetic loadings, the air gap flux distribution still keeps close to a sinusoid. Consequently, the orthogonal axes (*dq*) circuit model still leads, at least for preliminary design or control purposes, to practical results. It goes without saying that the varying magnetic saturation has to be considered in the *dq* model. In addition,

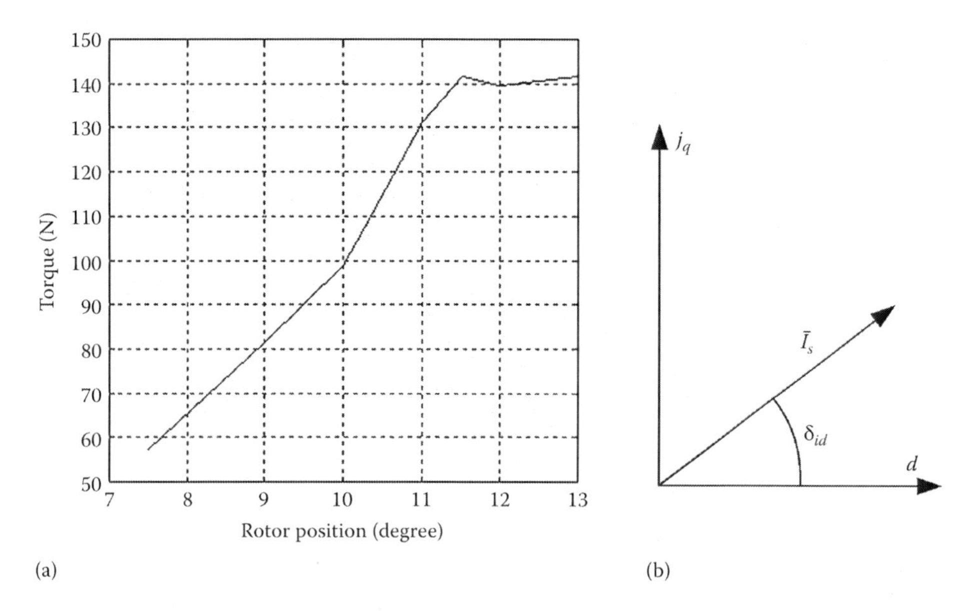

FIGURE 8.6 (a) The finite element method (FEM) calculated torque vs. rotor position for $n_c I\sqrt{2} = 500$ (Aturns/coil) in phase A, and $-n_c I/\sqrt{2}$ in phases B and C ($j_{peak} = 17.66$ A/mm²) and (b) power angle δ_{id}.

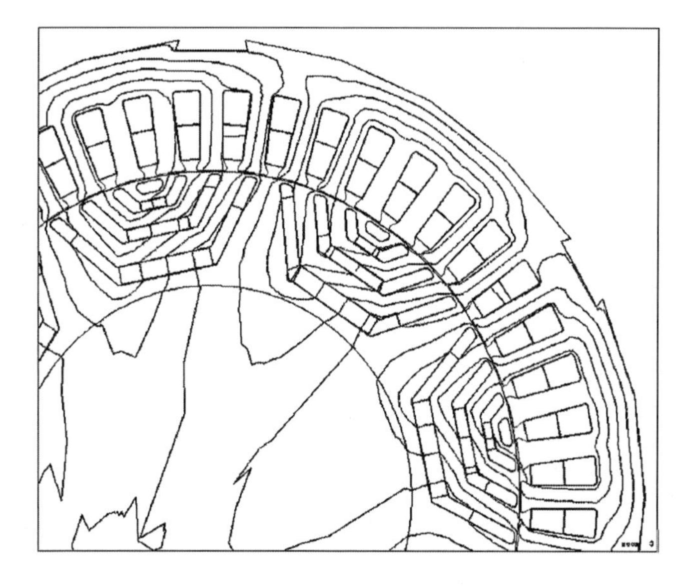

FIGURE 8.7 The flux lines for the maximum torque (140 Nm).

through FEM, we may proceed and calculate, for given stator current, *dq* current load angle, δ_{id}: the flux linkage in phases A, B, C: Ψ_A, Ψ_B, Ψ_C.

Based on these values, the *dq* model flux linkages may be calculated:

$$\Psi_d + j\Psi_q = \frac{2}{3}\left(\Psi_A e^{-j\delta_{id}} + \Psi_B e^{-j(\delta_{id}-(2\pi/3))} + \Psi_C e^{-j(\delta_{id}+(2\pi/3))}\right)$$
$$i_d = I\sqrt{2}\cos\delta_{id}, \quad i_q = -I\sqrt{2}\sin\delta_{id} \tag{8.2}$$

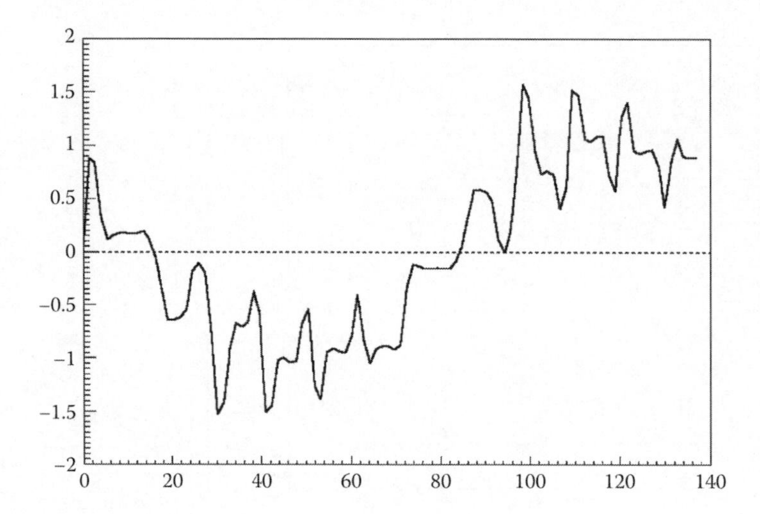

FIGURE 8.8 Permanent magnet (PM) air gap flux density at peak torque.

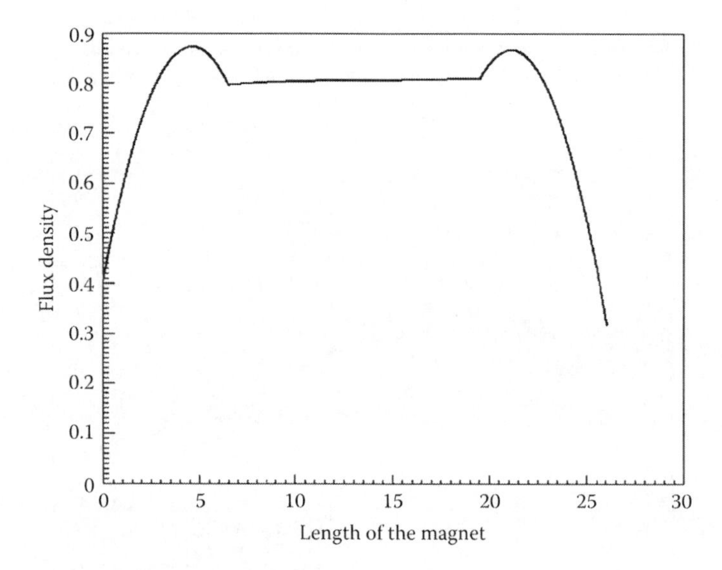

FIGURE 8.9 Permanent magnet (PM) surface flux density at peak torque.

Now, the flux linkages Ψ_{dm}, Ψ_{qm} are

$$\begin{aligned}
\Psi_{dm} &= L_{dm} i_d, \quad L_{dm}(i_d, i_q) = \frac{\Psi_{dm}}{i_d} \\
\Psi_{qm} &= L_{qm} i_q - \Psi_{PMq}, \quad L_{qm}(i_d, i_q) = \frac{\Psi_{qm} + \Psi_{PMq}}{i_q}
\end{aligned} \tag{8.3}$$

Ψ_{PM_q} is the flux produced by the PMs in phase A when aligned to rotor axis q, for zero stator currents. This way it is possible to draw, via FEM, a family of curves $\Psi_{dm}(i_{dm}, i_{qm}), \Psi_{qm}(i_{qm}, i_{dm})$ (Figure 8.11a and b) [9].

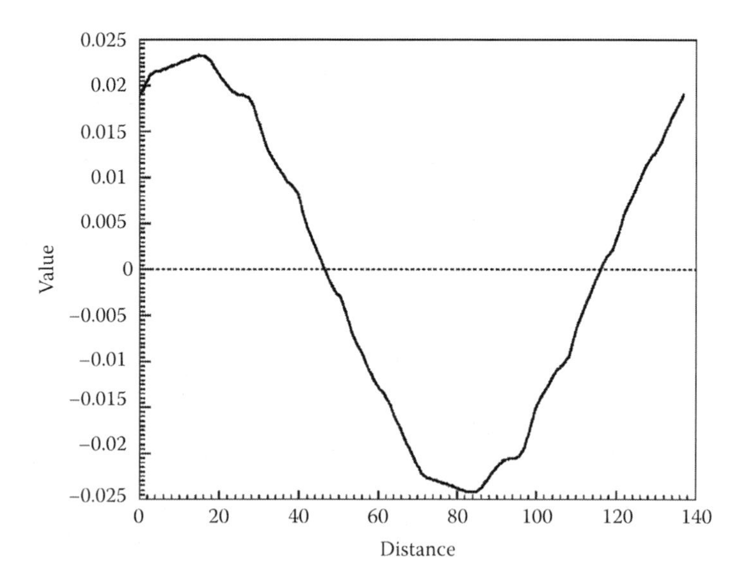

FIGURE 8.10 The air gap flux for the maximum torque.

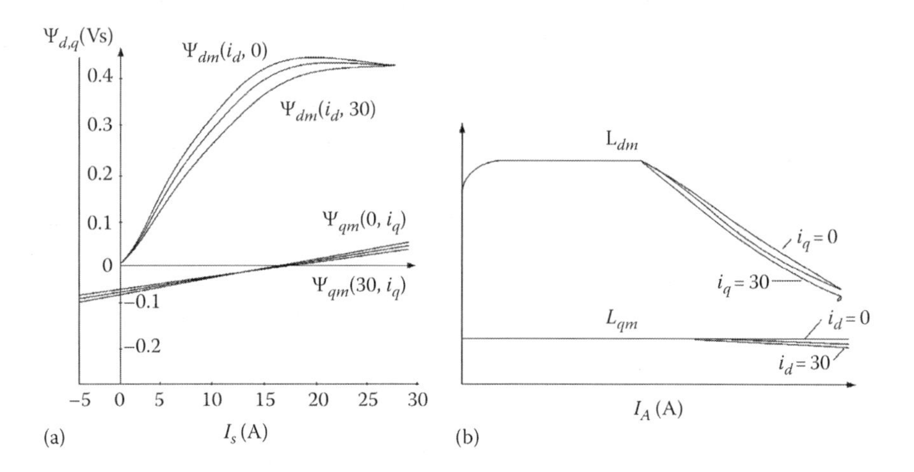

FIGURE 8.11 (a) d–q axes fluxes vs. current families and (b) corresponding L_{dm} and L_{qm}.

Figure 8.11a and b shows that the magnetic saturation is notable, especially in axis d, and though it exists, the cross-coupling magnetic saturation is not generally important [10,11] for PM-RSM. For extremely high electric and magnetic loadings with peak current densities of 30 A/mm^2, the situation might change; thus, direct operation with the family of dq flux/current curves may prove necessary, via curve fitting.

However, for most cases, $\Psi_d(i_d)$ curve suffices as L_{qm} = const:

$$\begin{aligned} \Psi_{dm} &= L_{dm}(i_d)\cdot i_d, \quad \Psi'_d = \Psi_{dm} + L_{sl}i_d \\ \Psi_{qm} &= L_{qm}(i_q)\cdot i_q - \Psi_{PM_q}, \quad \Psi_q = \Psi_{qm} + L_{sl}i_q \end{aligned} \tag{8.4}$$

with, for example,

$$L_{dm} = L_{dm0} - \beta(I_d - I_{ds}) - \gamma(I_d - I_{ds})^2 \quad (8.5)$$

Note that the q-axis magnetization inductance L_{qmo} may depend on i_q, especially at small values of currents, as the iron bridges above the flux barriers may not be fully saturated at zero stator current. In addition, the two-dimensional (2D) FEM analysis might not be satisfactory in predicting L_{qmo} or Ψ_{MPq}, especially in short core stacks when notable axial flux leakage may occur. Three-dimensional (3D) FEM is recommended to avoid the underestimation of L_{qmo}. Also noted on 2D FEM analyzer is the variation (pulsation) of L_{qmo} with rotor position at slot frequency. This phenomenon is confirmed by tests.

8.3.3 Cogging Torque

The torque versus rotor position developed at zero stator current is called the cogging torque. The presence of stator N_S slot openings and the N_r saturated flux bridges in the rotor with PMs placed in the flux barriers is likely to create the variation of stored magnetic energy in the air gap with rotor position. If we consider N_s and N_r as the number of stator and rotor "poles," then the cogging torque fundamental number of periods per mechanical revolution $N_{cogging}$ is

$$N_{cogging} = \text{lowest common multiplier (LCM) of } N_S \text{ and } N_R$$

The higher the $N_{cogging}$, the smaller the cogging torque.

For $q_1 = 2$ and 4 flux barriers per pole in the rotor and $2p_1 = 8$ poles, $N_S = 2p_1qm_1 = 2 \cdot 4 \cdot 2 \cdot 3 = 48$ and $N_r = 2 \cdot p_1 \cdot 4 = 32, N_{cogging} = 96$, which is a pretty large number. The situation is not practical with the same machine and six flux barriers per pole, however, when $N_r = N_S = 48 = N_{cogging}$. With seven flux bridges per pole, however, $N_r = 7 \cdot 8 = 56$, and $N_{cogging} = 8 \cdot 6 \cdot 7 = 336$, a more favorable design is obtained (Figure 8.12). For the example in this paragraph, the cogging torque obtained through 2D FEM is shown in Figure 8.13.

The very large number of periods in the cogging torque and its rather low value (percent of peak torque of 140 Nm) are evident. Numerous methods of reducing cogging torque in IPM machines—with one PM piece and flux barrier per pole—were proposed. Using FEM to verify them is usual [1], but analytical approaches were also introduced [12] with good results. In essence, the single flux barrier radial side angle and other geometrical parameters of the latter are modified in an orderly manner (direct geometry optimization) to minimize cogging torque while limiting the electromagnetic force (emf) and main torque reductions. Generalizing such methods for multiple flux barrier rotors is more complicated and still to come.

Usual methods of reducing cogging torque, such as stator fractionary windings (e.g., $q = 3/2$) or stator slots skewing by up to one slot pitch, are also at hand. But, avoiding the undesirable stator slots N_S rotor flux bridges (ribs), N_r, combinations with small common multiplier are the modality to avoid torque pulsations (five or seven flux bridges per pole are seen here as practical solutions).

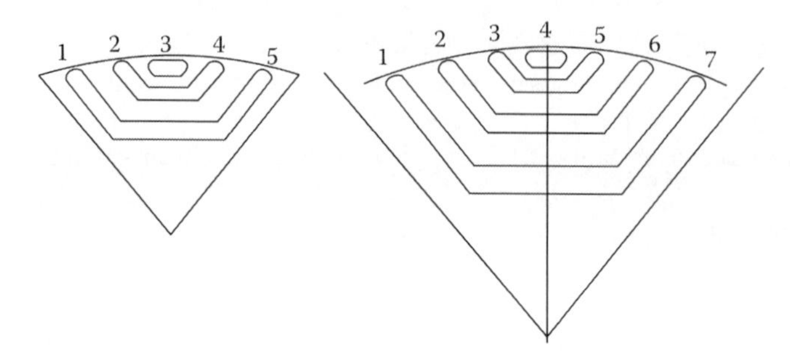

FIGURE 8.12 Rotor poles with 5, 7 flux bridges.

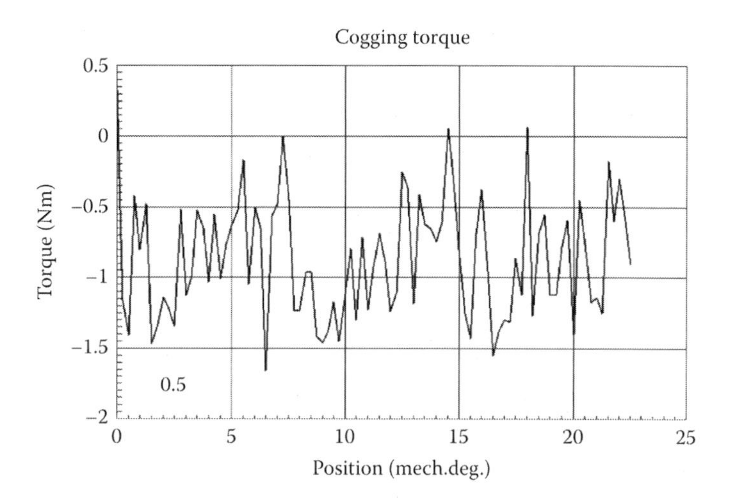

FIGURE 8.13 Cogging torque vs. rotor position.

8.3.4 Core Losses Computation by FEM

The advanced magnetic saturation at peak torque, and small speeds, and the lower but harmonic-rich flux density in various stator and rotor core zones at high speeds make the computation of core losses a difficult task. Fundamental and space and time harmonics losses are to be considered. Rotor core losses are not to be neglected. On top of that, PM eddy current losses at higher speeds (frequency) are also worthy of consideration. It seems that FEM is indispensable for such a task. In Reference 13, an FEM model for core losses is proposed for the single flux barrier (PM) per pole rotor. Still, the PM eddy current losses are not considered. The procedure could be generalized for the PM-RSM without major obstacles.

In essence, the flux density in the various finite elements is calculated N times for a period, and its radial and tangential components B_r, B_θ are determined. Then, the flux density variations are divided by the time step Δt and used to determine the eddy current and hysteresis losses in the stator and in the rotor [13]:

$$P_{iron} = K_{eddy} \cdot \frac{\gamma_{iron}}{2\pi^2} iron \frac{1}{N} \sum_{k=1}^{N} \left[\left(\frac{B_r^{k+1} - B_r^k}{\Delta t} \right)^2 + \left(\frac{B_\theta^{k+1} - B_\theta^k}{\Delta t} \right)^2 \right] dV$$

$$+ K_{hys} \cdot \frac{\gamma_{iron}}{T} \sum_{i=1}^{NE} \frac{\Delta V_i}{2} \left(\sum_{j=1}^{N_{pr}^i} \left(B_{mr}^{ij} \right)^2 + \sum_{j=1}^{N_{p\theta}^i} \left(B_{m\theta}^{ij} \right)^2 \right) \quad (8.6)$$

where

γ_{iron} is the specific iron weight
K_{eddy}, K_{hys} are the specific eddy current and hysteresis coefficients, to be determined on the Epstein frame in variable frequency tests, after dividing losses with frequency (Figure 8.14)
N is the number of ΔT time steps for period T
NE is the number of finite elements
ΔV_i is the volume of finite element
$B_{mr}^{ij}, B_{m\theta}^{ij}$ are the amplitudes of each hysteresis loop

The first term in Equation 8.6 represents the eddy current losses, while the second term is related to hysteresis losses. The eddy current terms may be extended to the PM body zones with minor adaptions. However, using three to five PM pieces in the axial direction leads to a drastic reduction of PM eddy current losses.

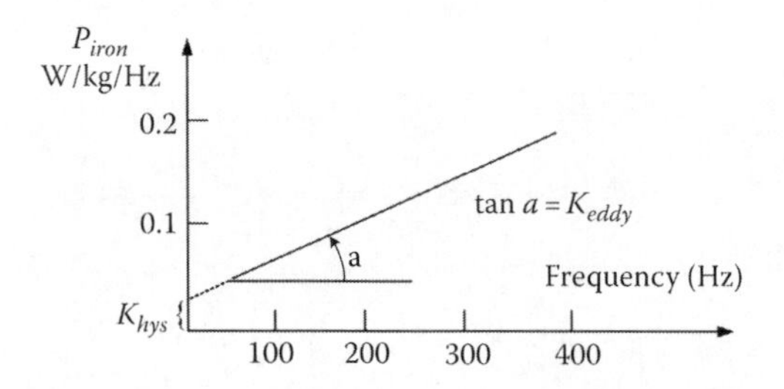

FIGURE 8.14 Iron losses/weight/frequency vs. frequency by the Epstein frame.

Note that this way, for given stator current amplitude, current power angle δ_{id}, and speed, the period of the current is explored for the fundamental. However, the current time harmonics also have to be considered.

The amount of computation time to explore core losses for the entire spectrum of torque and speed, for motoring and generating, becomes large. An equivalent core resistance R_C, even if slightly variable with frequency, may be defined:

$$\frac{1}{R_C} = \frac{1}{R_{eddy}} + \frac{1}{\omega R_{hys}} \tag{8.7}$$

R_{eddy} corresponds to eddy current and is constant. A frequency-dependent term (different in the stator than in the rotor) has to be added, together with a constant R_{hys} term, to represent the hysteresis losses in the dq model. The term R_C acts in parallel to the total stator flux (emf) to represent the core losses:

$$R_C = \frac{3}{2}\frac{(\omega_1 \Psi_s)^2}{P_{iron}} \tag{8.8}$$

in the dq model.

A typical breakdown of iron losses at two different speeds for a 2.5 Nm peak torque, four-poles IPM machine (with one flux barrier per pole) is shown in Figure 8.15 [13]. It should be noted that the stator current produces notable rotor iron losses due to stator-mmf and slot-opening space harmonics. For higher speeds, as expected in starter/alternators, rotor core losses become important. Good efficiency correlation up to 6000 rpm (200 Hz) is claimed with this method in Reference 13.

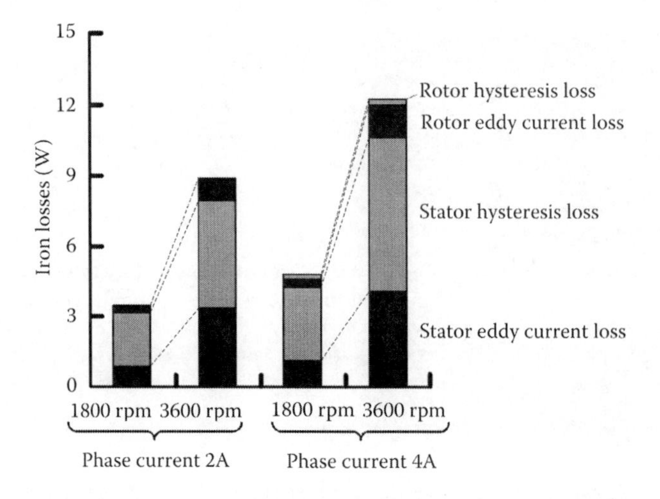

FIGURE 8.15 Calculated iron losses, via finite element method (FEM). (From Yamazaki, K., *IEEE Trans.*, MAG-39(3), 1460, 2003.)

8.4 *dq* Model of PM-RSM

As inferred from FEA, in a professionally designed PM-RSM:

- Magnetic saturation manifests itself along the *d* and *q* axes separately, and there is more along the *d* axis.
- Despite magnetic saturation, the PM and total flux in stator phases with distributed windings varies sinusoidally in time, for sinusoidal currents and constant speed.
- Stator and rotor core losses may be represented through a core resistance R_C that is slightly dependent on frequency and "acts" in parallel to the total (stator) flux produced emf: $\omega_1\Psi_S$.
- The damping effect of rotor core losses is neglected. In such conditions, the *dq* model can be applied directly to the PM-RSM. First, the core losses are neglected, for simplicity, and then they are added to the model for completeness. As PM-RSM is, in fact, a salient-pole synchronous machine, the *dq* model has to be attached to the rotor reference system [4]:

$$\bar{V}_S = R_S\bar{I}_S + j\omega_r\bar{\Psi}_S + \frac{d\bar{\Psi}_S}{dt} \tag{8.9}$$

$$\begin{aligned} \bar{\Psi}_S &= \Psi_d + j\Psi_q \\ \Psi_d &= L_d(i_d)\cdot i_d \\ \Psi_q &= L_{q0}i_q - \Psi_{PMq} \end{aligned} \tag{8.10}$$

$$T_e = \frac{3}{2}p(\Psi_d i_q - \Psi_q i_d) = \frac{3}{2}p(\Psi_{PM_q} + (L_d - L_q)i_q)i_d \tag{8.11}$$

$$\frac{J}{p}\frac{d\omega_r}{dt} = T_e - T_{load} \tag{8.12}$$

Equations 8.9 through 8.12 are written for the motoring mode with positive currents and powers. For the generator mode, the torque has to be negative, and thus, i_d has to change sign. In addition, the T_{load} becomes negative, as it turns into the negative driving torque of the prime mover. The active power is negative for generating in Equations 8.9 through 8.12.

The relationships between stator phase quantities and the *dq* model variables are, as known [4],

$$\bar{V}_S = \frac{2}{3}\left(V_a + V_b e^{j(2\pi/3)} + V_c e^{-j(2\pi/3)}\right)e^{-j\theta_{er}}; \quad \frac{d\theta_{er}}{dt} = \omega_r \tag{8.13}$$

The same transformation is valid for currents and flux linkages. For steady state, with sinusoidal phase voltages of frequency $\omega_1 = \omega_{r0}$,

$$V_{a,b,c} = V\sqrt{2}\cos\left((\omega_{r0}t + \gamma_o) - (i-1)\frac{2\pi}{3}\right) \tag{8.14}$$

$$\bar{V}_{s0} = V\sqrt{2}(\cos\gamma_0 - j\sin\gamma_0) \tag{8.15}$$

Consequently, steady state, in rotor coordinates, corresponds to direct current (DC) voltages, currents, and flux linkages. Therefore, $d/dt = 0$ for steady state in Equation 8.9, and $d\omega_r/dt = 0$ in Equation 8.12.

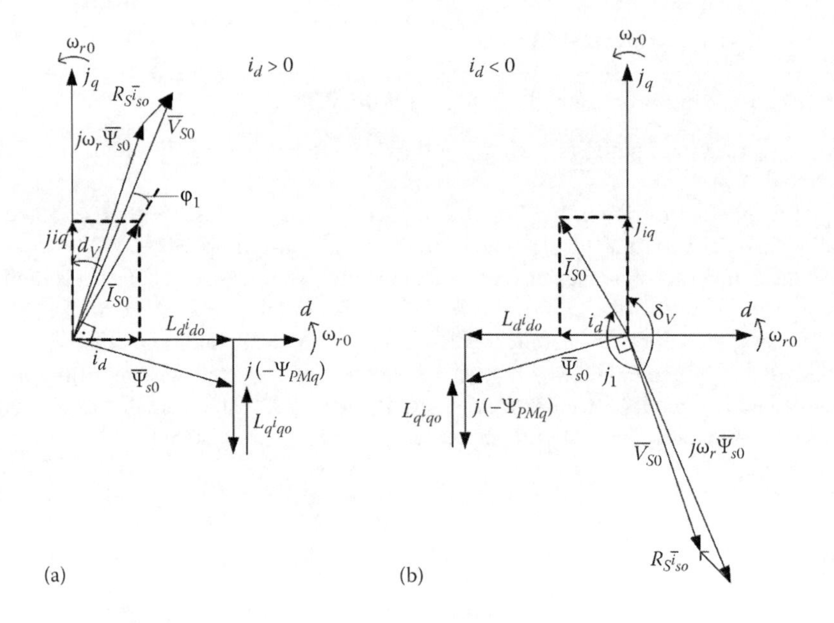

FIGURE 8.16 Vector diagrams of permanent-magnet-assisted reluctance synchronous machine (PM-RSM): (a) motoring and (b) generating.

The vector diagrams for motoring ($i_d > 0$) and generating ($i_d < 0$) are shown in Figure 8.16a and b, respectively, for the trigonometrical positive direction of motion. The flux vector is behind stator current for motoring and ahead for generating. From the vector diagrams,

$$i_d = I_S \sin(\delta_V + \varphi_1) \tag{8.16}$$

$$i_q = I_S \cos(\delta_V + \varphi_1) \tag{8.17}$$

It should be noted that the q-axis current is demagnetizing (defluxing) the machine, while i_d is the main torque current now. From the vector diagram,

$$\begin{aligned} V_d + jV_q &= \omega_r(L_{q0}i_{q0} - \Psi_{PMq}) - R_s i_{d0} + j(R_s i_{q0} + \omega_r L_d(i_d)i_d) \\ V_s &= \sqrt{V_d^2 + V_q^2} \end{aligned} \tag{8.18}$$

$$V_S^2 = (\omega_r(L_{q0}i_{q0} - \Psi_{PMq}) - R_s i_{d0})^2 + (R_s i_{q0} + \omega_r L_d(i_d)i_d)^2 \tag{8.19}$$

Again, $i_{d0} > 0$ for motoring and $i_{d0} < 0$ for generating while $i_{q0} > 0$ for both operation modes, in general.

For given torque T_{e0}, stator vector voltage V_{s0}, and speed, Equations 8.19 and 8.11 provide the values of i_d and i_q. An iterative procedure is required because of the nonlinearity of the equations and due to magnetic saturation ($L_d(i_{d0})$).

The efficiency and power factor may now be calculated for given torque, speed, and stator voltage:

$$\eta_e = \frac{T_e\omega_r/p}{T_e\omega_r/p + (3/2)R_s\left(i_{d0}^2 + i_{q0}^2\right)}, \text{ for motoring} \tag{8.20}$$

$$\cos\varphi_1 = \frac{|T_e\omega_r/p|}{\eta_e\cdot(3/2)V_{s0}\sqrt{i_{d0}^2+i_{q0}^2}} \tag{8.21}$$

Note that only the stator winding losses have been considered in Equations 8.20 and 8.21. Finally, from Equation 8.16, the voltage power angle may be obtained. Alternatively, the steady-state performance calculation may start with given V_{s0}, ω_r, δ_v:

$$\begin{aligned} V_{d0} &= V_{s0}\sin\delta_V \quad 90° > \delta_V > 0, \quad \text{for motoring} \\ V_{q0} &= V_{s0}\cos\delta_V \quad 90° < \delta_V, \quad \text{for generating} \end{aligned} \tag{8.22}$$

Then,

$$\begin{aligned} V_{d0} &= \omega_r(L_{q0}i_{q0} - \Psi_{PMq}) - R_s i_{d0} \\ V_{q0} &= \omega_r L_d(i_{d0})i_{d0} + R_s i_{q0} \end{aligned} \tag{8.23}$$

With the magnetic saturation level defined (L_d = constant), Equations 8.22 and 8.23 yield the currents i_{d0} and i_{q0} analytically. Then the torque, efficiency, and power factor may all be computed from Equations 8.11, 8.20, and 8.21. If the stator resistance R_s is neglected,

$$i_{d0} = \frac{V_{s0}\cos\delta_V}{\omega_r L_d}; \quad \delta_V > 90° \quad \text{for generating} \tag{8.24}$$

$$i_{q0} = \frac{V_{s0}\sin\delta_V + \omega_r\Psi_{PMq}}{\omega_r L_q} \tag{8.25}$$

$$(\cos\varphi_1)_{R_S=0} = \frac{|(3/2)(\Psi_{PMq} + (L_d - L_q)i_{q0})i_{d0}|\omega_r}{(3/2)V_{s0}\sqrt{i_{d0}^2+i_{q0}^2}} \tag{8.26}$$

Example 8.1

Consider a PM-RSM with the constant parameters L_d = 200 mH, L_q = 60 mH, Ψ_{PMq} = 0.215 Wb $2p_1$ = 4 poles, rated voltage 220 V per phase. Neglecting the stator resistance, calculate the steady-state torque, current, and power factor versus voltage power angle δ_V for motoring at 1500 rpm and a stator voltage of 220 V (RMS). Increase PM flux twice, and check the performance at δ_V = 60°.

Solution

The voltage vector amplitude V_{s0} is (Equation 8.15):

$$V_{s0} = V\sqrt{2} = 220\sqrt{2} = 310.2\ \text{V}$$

Now, we just follow Equations 8.24 through 8.26 for a few values of δ_V, and we obtain the results in Table 8.2.

Note that the emf (PM-induced voltage) E_{s0} = 67.51 V, while V_{s0} = 310.2 V. This means that the PM contribution is, in such conditions, rather small. As the saliency ratio L_d/L_q = 200/60 = 3.33 only, it is no wonder that the power factor does not rise above 0.686. The maximum torque is moved from the 45° that characterizes the pure RSM to somewhere between 45° and 60°.

TABLE 8.2 PM–RSM Characteristics

Voltage Power Angle V (°)	I_{do} (A)	I_{qo} (A)	I_{so} (A)	T_{eo} (nm)	$\cos_1$
0	0	20.048	20.048	0	0
15	1.278	19.48	19.5	11.29	0.1942
30	2.469	17.858	18.028	20.28	0.37744
45	3.4922	15.22	15.62	24.58	0.5276
60	4.2825	11.816	12.498	24.01	0.678
75	4.77	7.844	9.18	18.79	0.686
90	4.939	3.583	6.10	10.619	0.5837

With a stronger (doubled) PM contribution, the torque is increased at δ_V = 60° from 24.01 to 33.22 Nm, while the power factor is only slightly improved as, still, E_{s0}/V_{s0} = 135 V/310.2 V.

When the speed increases further, the power factor is likely to be notably increased for constant (given) voltage, as E_{s0} approaches or even surpasses V_{s0} by, in general, at most 50% at highest admissible speed, to avoid voltage oversizing of pulse-width modulated (PWM) converter power switches.

The torque versus voltage power angle is shown in Figure 8.17, for zero losses, constant voltage, and speed.

The expression of torque versus voltage power angle δ_V, for zero stator resistance, is derived simply from Equations 8.24 through 8.26:

$$T_e = \frac{3}{4} p \left(\frac{V_{s0}}{\omega_{r0}}\right)^2 \left(\frac{1}{L_q} - \frac{1}{L_d}\right) \sin 2\delta_V + \frac{3}{2} p \frac{V_{s0}}{\omega_{r0}} \frac{\Psi_{PMq}}{L_q} \sin \delta_V \tag{8.1.1}$$

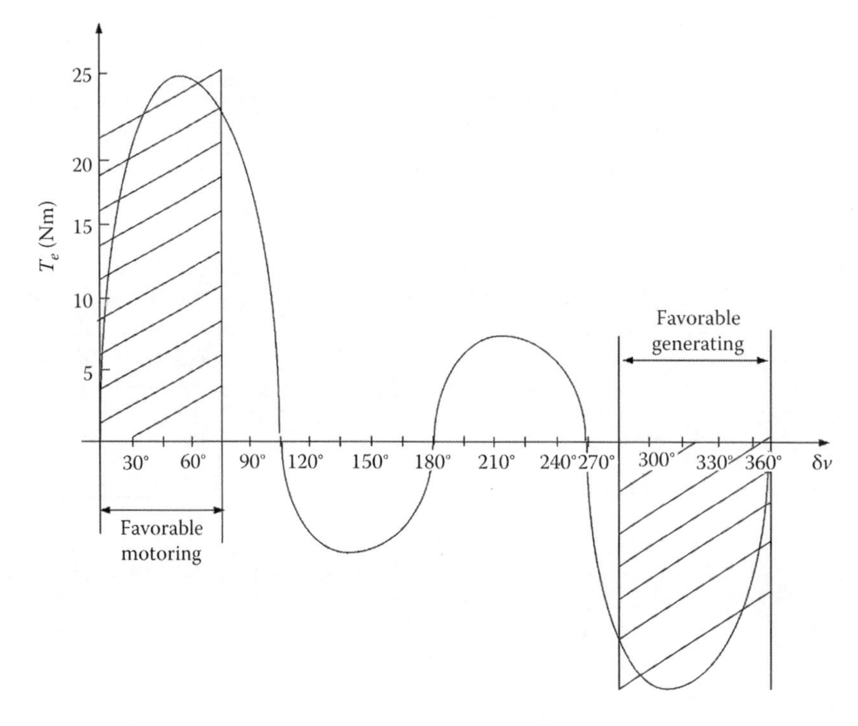

FIGURE 8.17 Typical permanent-magnet-assisted reluctance synchronous machine (PM-RSM) torque vs. voltage power angle curve for constant voltage and speed.

The constant voltage performance is important during motoring and generating inside the constant power speed range, where the PWM converter has to operate at full voltage. However, operation at constant voltage for a wide range of speeds encounters widely variable magnetic saturation and current conditions; thus, it has to be carefully watched through adequate control. In addition, the current should be limited to the peak design value. At low speeds, the stator resistance may not be neglected. The voltage power angle should vary with speed to produce the required torque for minimum current up to base speed ω_b and for minimum flux Ψ_s at high speeds. The dq model equations may be written for transients as follows:

$$V_d = R_s i_d + \frac{d}{dt}(L_d(i_d)i_d) - \omega_r(L_q i_q - \Psi_{PMq}) \tag{8.1.2}$$

$$V_q = R_s i_q + \frac{d}{dt}(L_q i_q) + \omega_r L_d(i_d)i_d \tag{8.1.3}$$

Additionally,

$$\frac{d}{dt}(L_d(i_d)i_d) = \left(L_d(i_d) + i\frac{dL_d}{di_d} \right)\frac{di_d}{dt} = L_d^t(i_d)\frac{di_d}{dt} \tag{8.1.4}$$

$L_d^t(i_d)$ is the transient inductance along axis d, while $L_d(i_d)$ is the normal (rotational) inductance. In general, $L_d^t < L_d$ for saturated core conditions, and $L_d^t = L_d$ under unsaturated conditions. At very low values of current, again, $L_d^t < L_d$ due to the early inflexion in the core magnetization curve.

In case the magnetic saturation along axis q is also considered,

$$\frac{d}{dt}(L_q i_q) = \left(L_q + i_q\frac{dL_q}{di_q} \right)\frac{di_q}{dt} = L_q^t(i_q)\frac{di_q}{dt} \tag{8.1.5}$$

Equations 8.1.2 through 8.1.14 and 8.27 lead to the general equivalent circuits for transients shown in Figure 8.18a for axis d and Figure 8.18b for axis q.

The core losses resistance R_C is "planted across the motion induced emf" and strictly considers the core loss in the stator during steady state. To account for rotor core and PM eddy current losses during transients or due to harmonics, a special additional circuit may be added in parallel to L_{ds}^t and L_{qs}^t terms, as a kind of weak damper case in the rotor (Figure 8.18a and b).

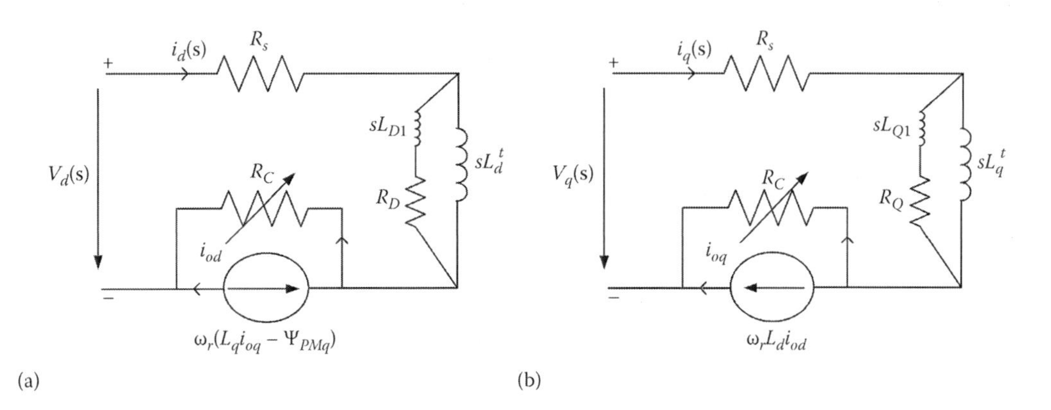

FIGURE 8.18 Permanent-magnet-assisted reluctance synchronous machine (PM-RSM) equivalent circuits for transients (a) for axis d and (b) for axis q.

The equivalent damper winding parameters may be determined from standstill DC current decay or frequency tests as done for synchronous machines. For transients, R_C may be neglected ($R_C = \infty$). The addition of the core loss resistance R_C, in the absence of the damper cages ($R_D = R_Q = 0$, $L_{Dl} = L_{Ql} = 0$), changes, to some extent, the equations of the *dq* model to the following:

$$V_d = R_s i_d + L_d^t \frac{di_d}{dt} - \omega_r (L_q i_q - \Psi_{PMq}) \quad (8.1.6)$$

$$V_q = R_s i_q + L_q^t \frac{di_q}{dt} + \omega_r L_d i_d \quad (8.1.7)$$

$$i_{0d} = i_d + \frac{\omega_r (L_q i_{0q} - \Psi_{PMq})}{R_C} \quad (8.1.8)$$

$$i_{0q} = i_q - \frac{\omega_r L_d i_{0d}}{R_C} \quad (8.1.9)$$

The torque is as follows:

$$T_e = \frac{3}{2} p (\Psi_{PMq} + (L_d - L_q) i_{0q}) i_{0d} \quad (8.1.10)$$

Again, for steady state $(d/dt) = 0$. The core and windings losses are as follows:

$$P_{iron} = \frac{3}{2} \frac{\omega_r \left[(L_q i_{0q} - \Psi_{PMq})^2 + L_d^2 i_{0d}^2 \right]}{R_C} \quad (8.1.11)$$

$$P_{copper} = \frac{3}{2} R_S \left(i_d^2 + i_q^2 \right) \quad (8.1.12)$$

The efficiency η and power factor $\cos \varphi_1$ become thus:

$$\eta = \frac{(3/2)(V_d i_d + V_q i_q) - P_{iron} - P_{copper} - P_{mec}}{(3/2)(V_d i_d + V_q i_q)} \quad (8.1.13)$$

$$\cos \varphi_1 = \frac{(V_d i_d + V_q i_q)}{\sqrt{V_d^2 + V_q^2} \cdot \sqrt{\left(i_d^2 + i_q^2 \right)}} \quad (8.1.14)$$

P_{mec} are the mechanical losses. Additional losses are lumped either into the copper or into core losses. The complete model is to be used for realistic performance computation of PM-RSM.

Once we accept this, we run into the difficulty of accounting for magnetic saturation functions $L_d(i_d)$ and $L_q(i_q)$, based on which $L_d^t(i_d)$, $L_q^t(i_q)$ may be calculated.

For steady state, only the $L_d(i_d)$ and $L_q(i_q)$ functions are needed, and a simple iterative computation procedure may be developed to extract the performance for given current i_{s0}, ω_r for various values of the current power angle $\delta_{i_q} = \tan^{-1} I_d / I_q$; or for given voltage V_{s0} and ω_r for various values of the voltage power angle $\delta_{V_q} = \tan^{-1} V_q / V_d$ ($\delta_V = \delta_{V_q}$ as defined previously in this paragraph; Figure 8.19).

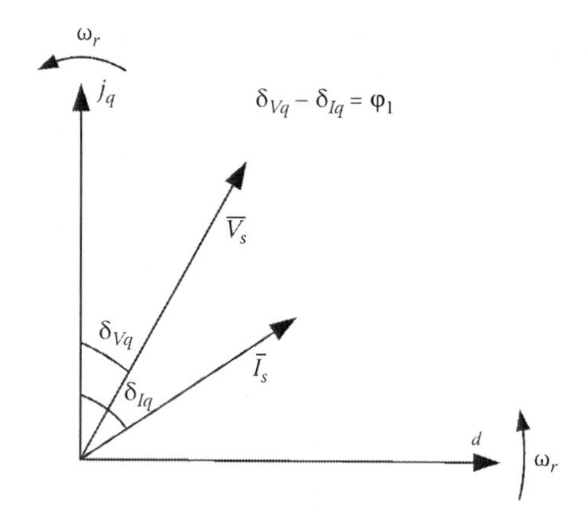

FIGURE 8.19 Control power angles to assess constant voltage (or constant current) and speed performance.

8.5 Steady-State Operation at No Load and Symmetric Short Circuit

8.5.1 Generator No-Load

Generator no-load operation ($i_d = i_{q0} = 0$) at steady state, Equations 8.1.6 through 8.1.9, becomes as follows:

$$i_{0d} = \frac{\omega_r(L_q i_{0q} - \Psi_{PMq})}{R_C} \tag{8.27}$$

$$i_{0q} = -\frac{\omega_r L_d i_{0d}}{R_C} \tag{8.28}$$

Therefore,

$$i_{0d} = -\frac{\Psi_{PMq}\omega_r \cdot R_C}{R_C^2 + \omega_r^2 L_d I_q} < 0; \quad i_{0q} > 0 \tag{8.29}$$

$$V_{d00} = -\omega_r(L_q i_{q0} - \Psi_{PMq}) \tag{8.30}$$

$$V_{q00} = \omega_r L_d i_{0d} \tag{8.31}$$

when $R_C = \infty$ (zero iron losses), $V_{d00} = +\omega_r \Psi_{PMq}, V_{q00} = 0$, and $i_{0d} = i_{0q} = 0$, as expected.

8.5.2 Symmetrical Short Circuit

For steady-state symmetrical short circuit $V_d = V_q = 0$, and thus,

$$0 = R_s\left(i'_{0d} - \frac{\omega_r\left(L_q i'_{0q} - \Psi_{PMq}\right)}{R_C}\right) - \omega_r\left(L_q i'_{0q} - \Psi_{PMq}\right) \tag{8.32}$$

$$0 = R_s\left(i'_{0q} + \frac{\omega_r L_d i'_{0d}}{R_C}\right) + \omega_r L_d i'_{0d} \tag{8.33}$$

$$i_{dsc} = i'_{0d} - \frac{\omega_r L_{0q} i_{oq} - \Psi_{PMq}}{R_C} \quad (8.34)$$

$$i_{qsc} = i'_{0q} + \frac{\omega_r L_d i_{0d}}{R_C} \quad (8.35)$$

Especially at high speeds (frequencies), the core losses are notable and may not be neglected. The torque at symmetrical steady-state short circuit may still be calculated using Equation 8.1.10. With i'_{0d} and i'_{0q} easy to extract from Equations 8.32 and 8.33, i_{dsc} and i_{qsc} are directly obtained from Equations 8.34 and 8.35. Equations 8.32 and 8.33 may be written as follows:

$$R_s i'_{0d} - \omega_r L_q \left(1 + \frac{R_s}{R_C}\right) i'_{0q} = -\omega_r \Psi_{PMq} \left(1 + \frac{R_s}{R_C}\right) \quad (8.36)$$

$$\omega_r L_d \left(1 + \frac{R_s}{R_C}\right) i'_{0d} + R_s i_{0q} = 0 \quad (8.37)$$

Solving for i'_{0q} and i'_{0d} in Equation 8.36 is again straightforward. The electromagnetic power ($T_{esc}\omega_r/p_1$) is transformed into losses of copper and iron. When these losses are neglected ($R_s = 0$, $R_C = \infty$),

$$i'_{od} = i_{dsc} = 0, \quad i'_{oq} = \frac{\Psi_{PMq}}{L_q} = i_{qsc} = i_{sc3i}$$

Therefore, the ideal short-circuit current is proportional to the PM flux linkage Ψ_{PMq} and inversely proportional to L_q ($L_q < L_d$). The higher the Ψ_{PMq} and the lower L_q (higher saliency L_d/L_q) are, the higher will be the short circuit current. In most wide constant power speed range design,

$$\Psi_{PMq} \approx L_q i_{srated} \quad (8.38)$$

For ideal no-load motoring—zero electromagnetic torque—and zero mechanical losses, from Equation 8.1.10, $i_{od} = 0$, and from Equations 8.1.6 and 8.1.7 at steady state ($d/dt = 0$),

$$i_d = -\frac{\omega_r (L_q i_{0q} - \Psi_{PMq})}{R_C} \quad (8.39)$$

$$V_d = -\frac{R_s}{R_C} \omega_r (L_q i_{0q} - \Psi_{PMq}) - \omega_r (L_q i_{0q} - \Psi_{PMq})$$

$$V_q = R_s i_{0q}; \quad i_q = i_{0q} \quad (8.40)$$

$$V_s^2 = V_d^2 + V_q^2$$

The three expressions in Equation 8.40 serve to calculate first i_{0q} and then V_d and V_q, and finally, with Equation 8.39, i_d and i_{q0}.

Example 8.2

For the PM-RSM in Example 8.1, calculate the no-load generator voltage V_{so} at 3000 rpm and the short-circuit symmetric current at the same speed for zero losses in the machine.

Solution

For the generator under no load, the terminal voltage V_{d00} (Equation 8.30) is as follows:

$$V_{d00} = E_{s0} = \omega_r \Psi_{PMq} = 2\pi \frac{3000 \times 2}{60} \times 0.215 = 135 \text{ V(peak voltage per phase)}$$

For the same speed, the ideal symmetrical short circuit $i_{qsc} = i_{sc3}$ (Equation 8.37) is as follows:

$$i_{sc3} = \frac{\Psi_{PMq}}{L_q} = \frac{0.215}{0.06} = 3.5833 \text{ A}$$

With zero losses, the short-circuit torque is zero.

8.6 Design Aspects for Wide Speed Range Constant Power Operation

In automotive applications, a wide speed range constant power operation is required. Speed range (ω_{max}/ω_b) up to 10 (12) would be desirable mostly for generating, but also for motoring, in the same cases. The typical torque/speed and voltage/speed envelopes are given in Figure 8.20.

The base speed ω_b is defined as the maximum speed for which the peak (starting) torque is produced at maximum converter voltage V_{smax} under the condition of maximum torque/current criterion. With wide speed range for constant power ($F > 2$–2.5), oversizing of either the PM-RSM or of the PWM converter or of both is required.

In general, machine oversizing (overrating) is measured in overtorque, while converter oversizing is expressed in current oversizing at the maximum ideal converter line natural peak voltage fundamental, V_{smax}, given for all designs:

$$V_b = V_{smax} = \frac{2}{\pi} V_{DC} \tag{8.41}$$

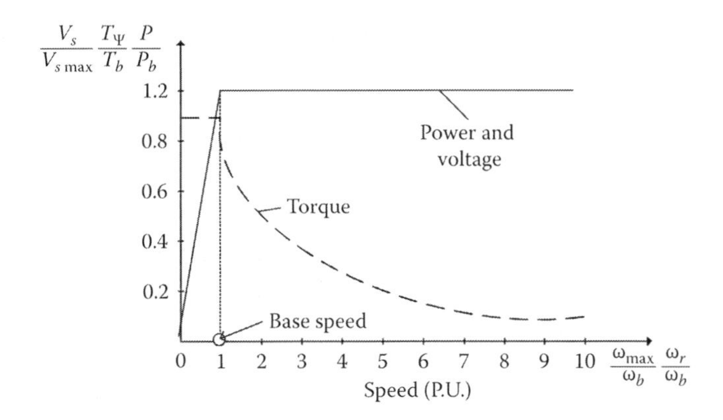

FIGURE 8.20 Torque, power, and, voltage vs. speed.

To investigate the oversizing aspects, a normalization system provides the required degree of generality. The base electrical quantities are as follows:

- V_b is the maximum machine voltage
- I_b is the maximum machine phase current I_b
- ω_b is the base speed that corresponds to the speed at which the machine first reaches V_b at I_b for maximum torque/current criterion
- T_b is the base torque that is produced at I_b for maximum torque/current criterion
- Ψ_b is the base flux: $\Psi_b = (V_b/\omega_b)$

To simplify the analysis, let us neglect losses in the machine at steady state. From Equations 8.9 through 8.12,

$$\begin{aligned} \overline{V}_{s0} &= j\omega_b \overline{\Psi}_{s0} \\ \overline{\Psi}_{s0} &= L_d i_d + j(L_{q0} i_q - \Psi_{PMq}) \\ I_b &= \sqrt{i_d^2 + i_q^2} \\ T_e &= \frac{3}{2} p_1 (\Psi_{PMq} + (L_d - L_q) i_q) i_d \end{aligned} \quad (8.42)$$

Let us neglect magnetic saturation variation and consider L_d= constant. The maximum torque for given current I_b is explored in what follows.

The condition $(dT_e/di_q) = 0$ for given stator current I_b, yields the following:

$$2(L_d - L_q) i_q^2 + i_q \Psi_{PMq} - (L_d - L_q) I_b^2 = 0 \quad (8.43)$$

$$I_{qb} = \frac{-\Psi_{PMq} + \sqrt{\Psi_{PMq}^2 + 8(L_d - L_q)^2 I_b^2}}{4(L_d - L_q)} \quad (8.44)$$

$$I_{db} = \sqrt{I_b^2 - I_{qb}^2} \quad (8.45)$$

Finally,

$$T_b = \frac{3}{2} p_1 (\Psi_{PMq} + (L_d - L_q) i_{qb}) i_{db} \quad (8.46)$$

In addition, the (T_e/I_s) function may be maximized when the maximum Newton meter per ampere of current is obtained.

Note that when magnetic saturation is considered, the above solution does not hold true. I_{qb} tends to be higher than in Equation 8.44. It should be noted that the rated torque T_n would correspond to V_b and I_b and the unity power factor:

$$T_n = \frac{3}{2} \frac{p}{\omega_b} V_b I_b \neq T_b \quad (8.47)$$

The base and rated torque are different in this normalization, as the power factor of the machine is not unity in general. Therefore, $T_b/T_n < 1$. The base speed ω_b, for given V_b and T_b (i_{db} and i_{qb}) has to be calculated from Equation 8.42:

$$\omega_b = \frac{V_b}{\sqrt{(L_d I_{db})^2 + (L_q I_{qb} - \Psi_{PMq})^2}} \quad (8.48)$$

For the maximum speed at constant power ω_{max} and V_b, the machine is to reach the condition of maximum torque per given flux Ψ_{smin}:

$$\Psi_{smin} \approx \frac{V_b}{\omega_{max}} = \frac{V_b}{\omega_b \cdot F} = \frac{\Psi_b}{F} \tag{8.49}$$

Therefore, from Equation 8.42, with $(dT_e/di_q) = 0$:

$$\Psi_{smin} = \sqrt{(L_d i_d)^2 + (L_q i_q - \Psi_{PMq})^2} \tag{8.50}$$

In general, magnetic saturation is negligible at maximum speed, thus finally,

$$2(L_d - L_q)L_q^2 i_q^2 - (3L_d - 4L_q)L_q \Psi_{PMq} i_q - (L_d - L_q)\Psi_s^2 - L_d \Psi_{PMq}^2 = 0 \tag{8.51}$$

$$(i_q)_{\left(\begin{smallmatrix}\omega_{max} \\ \Psi_{smin}\end{smallmatrix}\right)} = \frac{(3L_d - 4L_q)L_q \Psi_{PMq} + \sqrt{\begin{matrix}(3L_d - 4L_q)^2 L_q^2 \Psi_{PMq}^2 \\ +8(L_d - L_q)L_q^2 \cdot \left[(L_d - L_q)\Psi_{smin}^2 + L_d \Psi_{PMq}^2\right]\end{matrix}}}{4(L_d - L_q)L_q^2} \tag{8.52}$$

Knowing ω_{max}, V_b, that is, Ψ_{smin}, the $(i_q)\omega_{max}$, $(i_d)\omega_{max}$ pair is obtained. Then, the corresponding torque is calculated. Finally, the stator current maximum speed $\omega_{max}(I_s)$ is obtained. If $(T_e)^* \ \omega_{max}/p \leq P_b$ and (I_s) $\omega_{max} \leq I_b$, then the machine is capable of handling the respective constant speed power range from ω_b to ω_{max}; if not, oversizing is required. A complete discussion on machine and inverter oversizing for wide F range is given in Reference 14. Thus far, we did not mention any design constraint for wide constant power speed range. Reference 15 contains such a thorough analysis. The main conclusion is that the best designs, if not otherwise constrained, for wide constant power speed range, are obtained if the ideal no-load motoring (zero torque) speed ω_0 is infinite:

$$(\omega_o)_{i_d=0} \cong \frac{V_b}{L_q i_b - \Psi_{PMq}} \cong \infty \tag{8.53}$$

Therefore,

$$\Psi_{PMq} = L_q I_b \tag{8.54}$$

In per unit (P.U.), it means that $\Psi_{PMq} = L_q$. In other words, the flux in the machine at zero torque should be zero. As part of L_q is the leakage inductance, no apparent danger of PM demagnetization occurs from this mathematical cancellation.

It was proven [14] that Ψ_{PMq} (P.U.) and the L_d/L_q ratio remain the independent parameters in the above-presented normalization system. Typical maximum ideal power (zero losses) vs. speed for $L_d/L_q = \zeta$ and three different Ψ_{PMq} (P.U.) and Ψ_{PMq}/L_q ratios are shown in Figure 8.21 [14].

The maximum power is obtained at maximum (I_b) current up to base speed ω_b and at I_b current values (or smaller) above ω_b for base voltage V_b. Figure 8.21 contains fairly general results that warrant the following remarks:

- For $\Psi_{PMq}/L_q = 1$ (P.U.), the P.U. power increases steadily toward unity at infinite speed.
- For $\Psi_{PMq}/L_q = 1$, at base speed, the power in P.U. is less than the base power.
- For Ψ_{PMq} large and $\Psi_{PMq}/L_q > 1$, the power peaks quickly and drops rapidly with speed but is large at base speed.

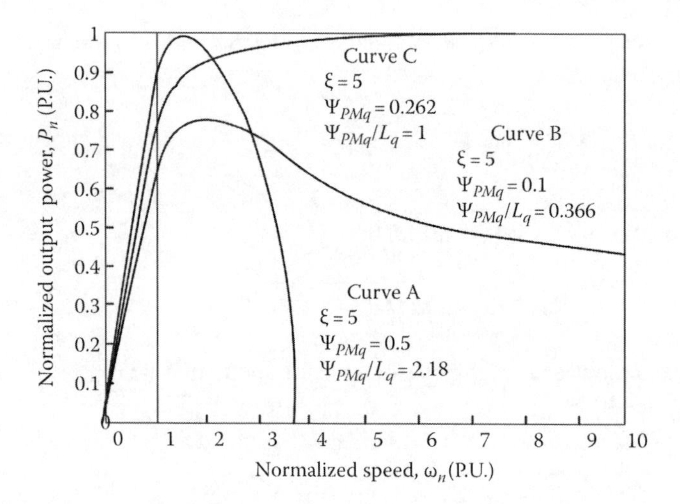

FIGURE 8.21 Maximum power vs. speed for three permanent-magnet-assisted reluctance synchronous machines (PM-RSMs).

- Even designs with small Ψ_{PMq} and thus, $\Psi_{PMq}/L_q < 1$ are poor for wide speed ranges, as the level of constant power that can be maintained is smaller.
- A near optimal design for $F = \omega_{max}/\omega = 10/1$ should depart somewhat from the $\Psi_{PMq}/L_q = 1$ condition to yield more torque at base speed (Figure 8.22) for a higher constant power level [14].
- For $F = 10$, ω_{min} and ω_{max} are left to float; and thus, $\omega_{min} > \omega_b$ allows for more power at minimum speed. At $\omega_{min} > \omega_b$ the inverter already works at maximum voltage. This leads to machine oversizing. The machine oversizing is defined in torque terms (Figure 8.22):

$$t_{over} = \left(\frac{T_b}{T_1} - 1\right) \times 100\% \tag{8.55}$$

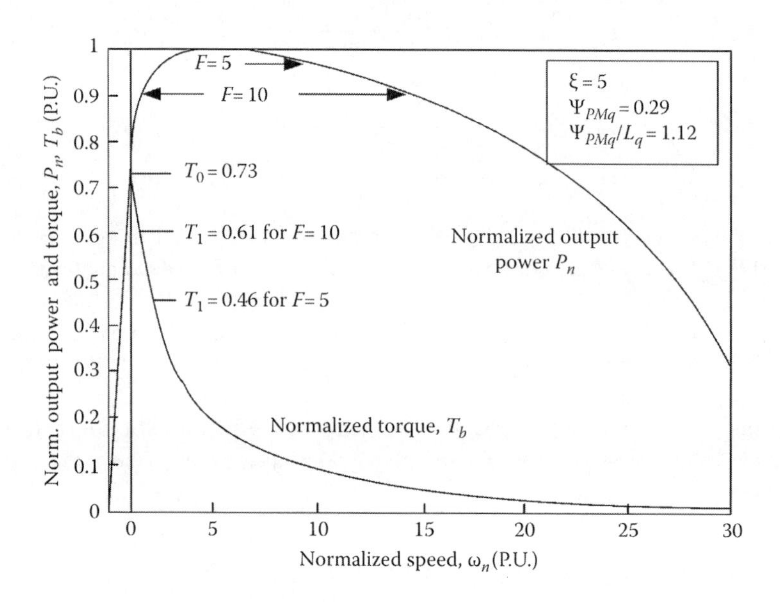

FIGURE 8.22 Near-optimal design for constant power range.

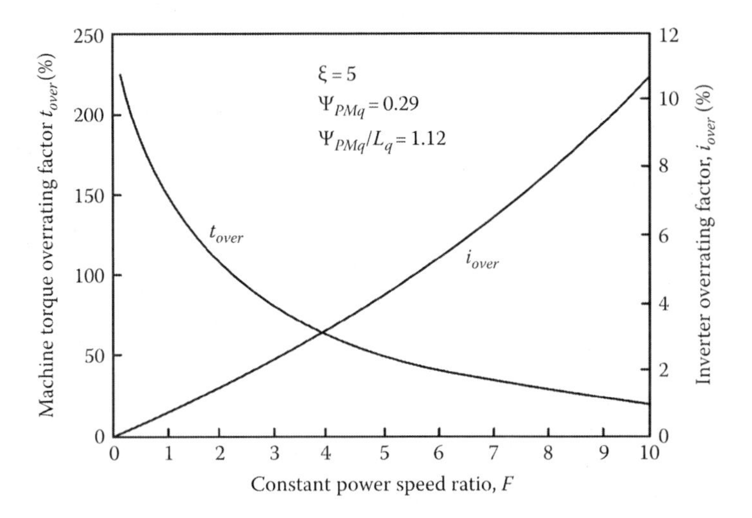

FIGURE 8.23 Oversizing of machine (t_{over}) and of inverter (i_{over}).

The inverter current oversizing i_{over} is as follows:

$$i_{over} = \left(\frac{1}{T_1} - 1\right) \times 100\% \tag{8.56}$$

For the design in Figure 8.22, for $F = 10$, $t_{over} = 20\%$, $i_{over} = 11\%$. A complete illustration of t_{over} and I_{over} for $L_d/L_q = 5$, $\Psi_{PMq}/L_q = 1.12$ is shown in Figure 8.23 [14].

Smaller inverter oversizing and a larger machine oversizing are preferred when system cost is paramount. Machine volume constraints under the vehiclehood might reduce the advantage of such a choice. A higher speed machine with a transmission might do it in such a case. Though instrumental in securing wide constant power speed range, choosing $\omega_{min} > \omega_b$ leads to a design that reaches full voltage in the midst of its constant torque range. It implies saturation of the current regulators and thus, impedes on torque control. In contrast, choosing $\omega_{min}/\omega_b < 1$, it means that $P\omega_{max} > (P_1)\omega_{min}$. This time, the inverter oversizing is $i_{over} = ((1/(T_1)_{\omega_{min}}) - 1) \times 100\%$. The machine is now fully utilized ($t_{over} = 0$), and the current regulators will not saturate. The price to pay for full machine utilization is a higher inverter overrating ($i_{over} = ((1/T_1) - 1) \times 100\% = 40\%$ in Figure 8.24). For $\omega_{min} < \omega_b$, $i_{over} > 40\%$ for $F = 10$, in all designs. For lower values of F, i_{over} may be lower than 40%.

There are few essential constraints that limit the rather large constant power speed range of PM-RSM. The most important are as follows:

- emf at maximum speed
- Uncontrolled generation elimination
- Limiting the short circuit current (torque)

The emf at maximum speed E_{smax} is as follows:

$$e_{smax} = \frac{E_{smax}}{V_b} = \frac{\omega_{max}\Psi_{PMq}}{V_b} \tag{8.57}$$

In general, $e_{smax} \leq 150\%$, to avoid voltage overrating of the converter. This constraint may become severe for $F = \omega_{max}/\omega_b = 10$, as $\Psi_{PMq} = 0.15$ (P.U.), which is a small value. High saliency ratio $L_d/L_q > 5$ would save such a design, eventually.

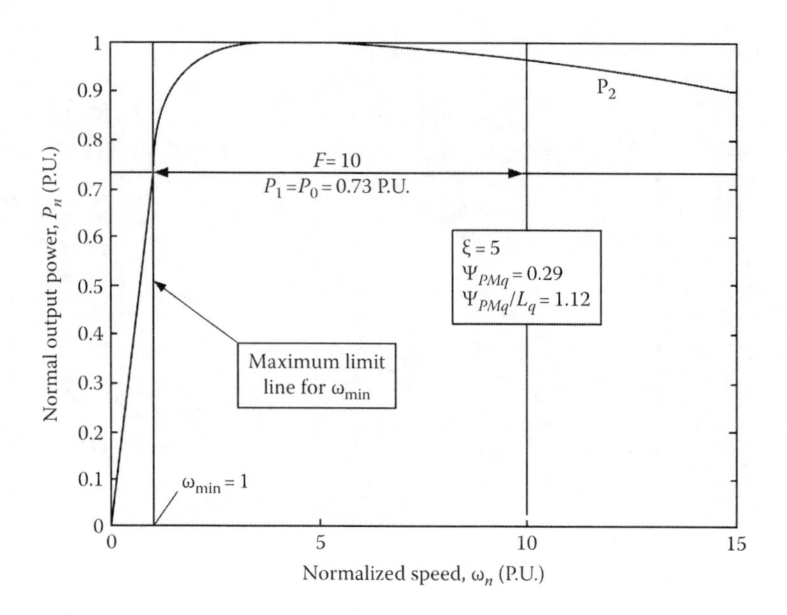

FIGURE 8.24 Design for $\omega_{min}/\omega_b \leq 1$; $t_{over} = 0$, $i_{over} = 40\%$.

The symmetrical ideal short-circuit current i_{sc3} (Equation 8.37) is as follows:

$$i_{sc3} = \frac{\Psi_{PMq}}{L_q} \tag{8.58}$$

In close to optimal designs, however, $\Psi_{PMq}/l_q = 1$, and thus, the symmetrical short-circuit current becomes equal to the base (peak torque) current I_b. Unfortunately, in asymmetrical short circuits, the current is notably larger than $i_{sc3} \cong I_b$. The three-phase fault may occur unintentionally by gating all inverter switches or through DC bus short circuit, when all six freewheeling diodes are conducting. Asymmetrical short circuit through the failure of a single inverter switch in a closed position behaves as single-phase short circuit at the machine terminals. Much higher currents than for symmetrical short circuit occur within 6 ms in two of the three phases that also exhibit DC components. High pulsating torque values are expected, as in any synchronous machine. As a protective means from these high transients, the healthy phases may be opened or the machine may be driven into symmetrical short circuit, when the overcurrents are acceptable ($i_{sc3} \cong i_b$, for $\Psi_{PMq}/L_q \cong 1$). A pertinent analysis of this problem is presented in Reference 16. Finally, uncontrolled generator (UCG) operation may occur at high speeds when gating is removed from all the power switches. The machine acts as a generator with uncontrolled rectifier output. The problem is that due to $L_d/L_q \gg 1$, the machine continues to generate power even when the speed decreases in UCG below the speed that corresponds to $\omega_r\Psi_{PMq} = V_b$. It was shown [17] that, in order to avoid UCG, the following condition is to be met: $\omega_{max} < \omega_{UCG}$.

With

$$F = \frac{\omega_{UCG}}{\omega_b} = 2\frac{\sqrt{(L_d/L_q)-1}}{\Psi_{PMq}(L_d/L_q)} \tag{8.59}$$

For our previous close to optimal design $\Psi_{PMq} = 0.29$, $L_d/L_q = 5$, $F = \omega_{UCG}/\omega_b = 2.75$.

Only by adopting much smaller values of Ψ_{PMq} and a higher saliency ratio, a larger constant power speed range may be produced, at the cost of high inverter oversizing. The maximum emf constraint, for the case in point, fulfilling $\Psi_{PMq} = 0.15$ (P.U.) for $F = 10$, fails to produce an L_d/L_q ratio

TABLE 8.3 Typical Specifications of a Starter/Alternator (ISA) for Cars

Parameter	Value
Starting torque at crankshaft (with or without a gear to ISA)	140 nm
Electric generating power: at crankshaft at 0.6 krpm	2–4 kW
Electric generating power: at crankshaft at 6 krpm	3–6 kW
Maximum starting current density	20–50 A/mm^2
Maximum generating current density	10–20 A/mm^2
Nominal bus voltage	42 V_{DC}
System efficiency at full load and 1500 rpm	75%
Minimum rotor burst speed	10–18 krpm
Maximum electromagnetic field (emf) at 6 krpm	24 V_{ph} (rms)

(even a very large one) in Equation 8.59. Only for $\Psi_{PMq} = 0.1$ for $F = 10$ does $L_d/L_q = 2$ produce a solution to Equation 8.59. Unfortunately, this is so small that the machine torque density is unacceptable. It seems that elimination of UCG in large constant power speed range design ($F > 2.75$), with PM-RSM, is difficult to achieve unless the PM flux is very small and saliency is very high. But then, all the merits of PM-RSM are diminished drastically as the machine turns into an RSM.

UCG operation has to be handled by some other protective means, as the constraint $\omega_{max} < \omega_{UCG}$ for its elimination does not look practical for industrial designs. As discussed, though the PM-RSM seems capable of producing 10:1 constant power speed range with small machine oversizing and 40% converter oversizing, it fails to comply with 150% maximum speed emf conditions and especially fails to avoid UCG at high speeds. Designing PM-RSM for a 4:1 constant power speed range and using a star to delta winding switching, as done for the induction machine seems to be the way to go. Machine oversizing and converter oversizing and the total system losses are about 30%–40% less than for an induction starter alternator designed for the same values. Magnetic saturation reduces the saliency ratio L_d/L_q with all the consequences that follow [18]. Typical specifications for integrated starter alternators (ISA) for cars are shown in Table 8.3.

Offset coupling of ISA to the crankshaft may be preferred. With a 3:1 gear (belt) ratio, the PM-RSM, with an overall slightly expensive design, may be obtained with unlimited emf, as the peak current is lower in the 2:1 ratio. Consequently, the converter voltage oversizing is compensated for the current undersizing [19]. Tripling the speed for geared PM-RSM, however, may impose more severe mechanical constraints on the design.

8.7 Power Electronics for PM-RSM for Automotive Applications

As the PM-RSM in automotive applications undergoes cyclic energy exchanges with a battery, a PWM, voltage source inverter capable of bidirectional power flow is required (Figure 8.25). The capacitor filter helps reduce the current transients in the battery during PWM converter switching sequences. The power switches have to be metal-oxide semiconductor field-effect transistors (MOSFETs) in 42 V_{DC} battery systems with voltage boost via an optional DC–DC converter (Figure 8.25). Introducing the DC–DC converter for voltage boost allows for design of the PM-RSM and the PWM converter for higher voltage, low current ratings. In addition, the DC voltage V_{DC} is made variable and the PWM converter may work extensively in the six-pulse mode when the commutation losses are reduced and the maximum modulation index is increased by 5% in comparison with standard PWM techniques (without overmodulation).

There are additional losses in the DC–DC converter. However, the possibility of using insulated gate bipolar transistor (IGBTs) in the higher-voltage PWM inverter leads to cost reductions in comparison with MOSFETs, but there are the additional costs of the DC–DC converter that have to be added.

The large battery voltage variations of ±20% or more may be absorbed by the DC–DC converter, allowing for designing the PWM inverter at least at maximum battery voltage, V_{bmax}.

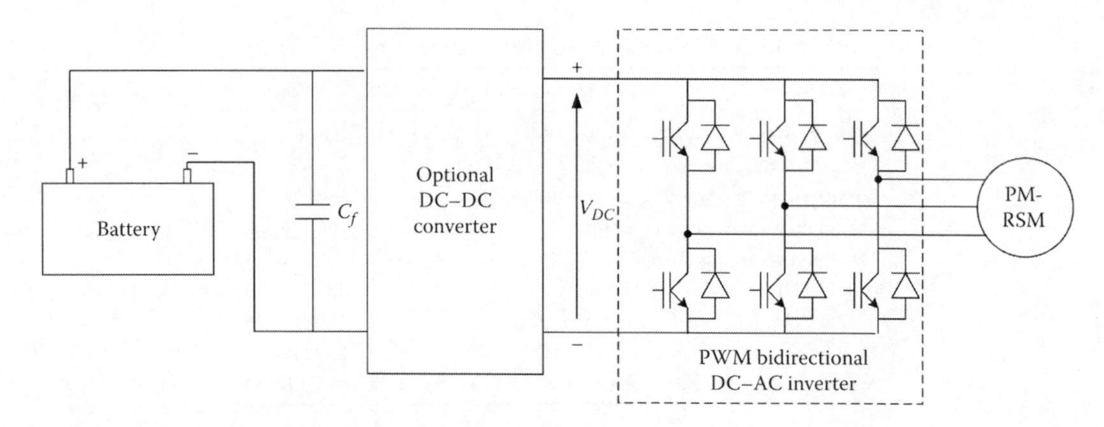

FIGURE 8.25 Typical pulse-width modulator (PWM) converter for permanent-magnet-assisted reluctance synchronous machine (PM-RSM).

The typical kVA_{ref} of the PWM three-phase inverter to deliver given active power is as follows:

$$kVA_{ref} = \frac{\sqrt{3}}{2} V_{\max} I_{\max} \tag{8.60}$$

where

$I_{\max}$ is the peak phase current value
$V_{\max}$ is the peak line voltage value

The inverter component size A_{inv} is defined through the $I_{dim} = I_{\max}$ and V_{dim}:

$$A_{inv} = 6 V_{dim} I_{dim} \tag{8.61}$$

In general, $V_{\max} < V_{dim}$. When a DC–DC converter is present, a safe value of $V_{\max}$ for 600 V IGBTs would be $V'_{\max} = V'_{dc} = 400\ V_{DC}$, with $V'_{\max} = V_{b\min} = 125\ V_{DC}$ when the DC converter is missing.

For a 42 V_{DC} system, MOSFETs are to be used, and again, if a DC–DC converter is present, $V''_{\max} \cong 3\ V_{b\min}$. ($V_{b\min}$ is the minimum battery voltage.) The ratio between the inverter size and delivered mechanical power in motoring is as follows [2]:

$$\frac{A_{inv}}{P_n} = \frac{6 V_{dim} I_{dim}}{P_n} = 4\sqrt{3} \frac{V_{dim}}{V_{max}} \cdot \frac{kVA_{ref}}{P_n} \tag{8.62}$$

The DC–DC converter size is defined as follows:

$$A_{DC-DC} = 2 V_{dim} I_{b\max} \tag{8.63}$$

where $I_{b\max}$ is the maximum battery current.

Adapting a total (system) efficiency η_{tot} yields the following:

$$\frac{A_{DC-DC}}{P_n} = \frac{2}{\eta_{tot}} \frac{V_{d\min}}{V_{b\min}} \tag{8.64}$$

The ratio between the inverter plus DC–DC converter to "inverter-only" size is

$$\frac{A_{DC-DC} + A_{inv}}{A_{inv}} = \frac{V_{b\min}}{V'_{\max}} + \frac{1}{2\sqrt{3}\eta_{tot}} \cdot \frac{P_n}{kVA_{ref}} \tag{8.65}$$

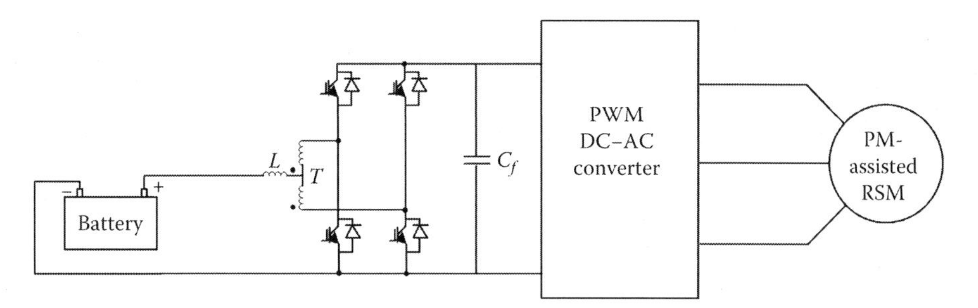

FIGURE 8.26 H-bridge direct current (DC)–DC boost bidirectional converter with transformer and inductance ($T + L$).

The ratio P_n/kVA_{ref} varies between 0.25 and 0.6, while $(1/2\sqrt{3}\eta_{tot})$ is roughly 0.25 ÷ 0.35. For $1/2\sqrt{3}\eta_{tot} = 0.34$ and $P_n/kVA_{ref} = 0.25$, the second term in Equation 8.64 is 0.2; thus, even at $V_{b\min}/V'_{\max} < 0.8$, the total size of the DC–DC PWM converter will be lower than that of the PWM inverter at lower voltage, acting alone, and so will be the costs if the additional power switch control devices costs are small.

Even if the $V'_{dc} = V_{b\max}$ (maximum battery voltage), it seems worth adding the DC–DC converter in terms of initial silicon costs. The DC–DC boost converter includes inductances for energy storage and voltage boost, but the size of the capacitive filter in front of the PWM inverter is reduced [2] and so is the current ripple in the battery. A typical H-bridge DC–DC boost converter with inductance and transformer is shown in Figure 8.26 [2].

Another key issue is the total loss in power electronics. Converter losses are made of conduction and switching losses. For an IGBT inverter leg (two switches),

$$\begin{aligned}(P_{cond})_{leg} &= \left(V_I(I)_{avg} + \frac{\Delta V_R}{I_{rated}}(I)^2_{RMS}\right)\\(P_{sw})_{leg} &= \beta_{sw}V^I_{DC}(I)_{avg}\end{aligned} \tag{8.66}$$

The switching losses per lag in the PWM inverter for six-pulse mode (hexagonal, voltage 5% overmodulation) are as follows [2]:

$$\begin{aligned}P_{sw6pulse} &= \frac{\sin|\varphi|}{2}P_{swPWM}; \quad |\varphi| > \frac{\pi}{6}\\P_{swHpulse} &= \left(\frac{2-\sqrt{3}\cos|\varphi|}{2}\right)P_{swPWM}; \quad |\varphi| < \frac{\pi}{6}\end{aligned} \tag{8.67}$$

The power factor angle φ influences the commutation losses, as known. Typical values of the above coefficients for IGBTs are $V_I = 0.9$ V (IGBT voltage drop), $\Delta V_R = 1.1$ V (diode voltage drop), and $\beta_{sw} = 0.042$.

It was demonstrated that total losses in the power electronics with boost DC–DC converters, for 30 kW, $V_{b\min} = 125\ V_{DC}$ ($V_b = 180\ V_{DC}$), $V_{\max} = 400$ V [2], are smaller than those of the PWM inverter connected directly to the battery terminals. And again, the PM-RSM is better in terms of total losses—machine plus power electronics—than all the other solutions for uphill or flat-terrain driving. In addition, the kVA_{ref} of a PM-assisted RSM converter is smaller, and thus, the cost of power electronics is smaller.

8.8 Control of PM-RSM for EHV

Essentially, as for induction starter/alternators (Chapter 7), the control of PM-RSM in EHVs does torque control from zero speed and has to operate with a large constant power speed range. It has to accommodate the driver's motion expectations while observing the best possible energy transfer to and from the battery. And all this in starter (motoring) generating—for regenerative breaking and for battery recharge—and again in motoring the air conditioner or other auxiliaries during frequent engine idle stops.

As for the induction machine, FOC or direct torque and flux control (DTFC) are both feasible strategies. In addition, the control system may use motion feedback (a position sensor) or may be motion sensorless (without motion feedback).

The control system has to operate safely and nonhesitantly in torque control—for motoring—from zero speed. Therefore, in contrast to the induction machine, the initial rotor position has to be estimated first, in motion sensorless control. Typical control schemes for FOC and DTFC with motion (position) feedback are shown in Figures 8.27 and 8.28 and respectively, in Figures 8.29 and 8.30.

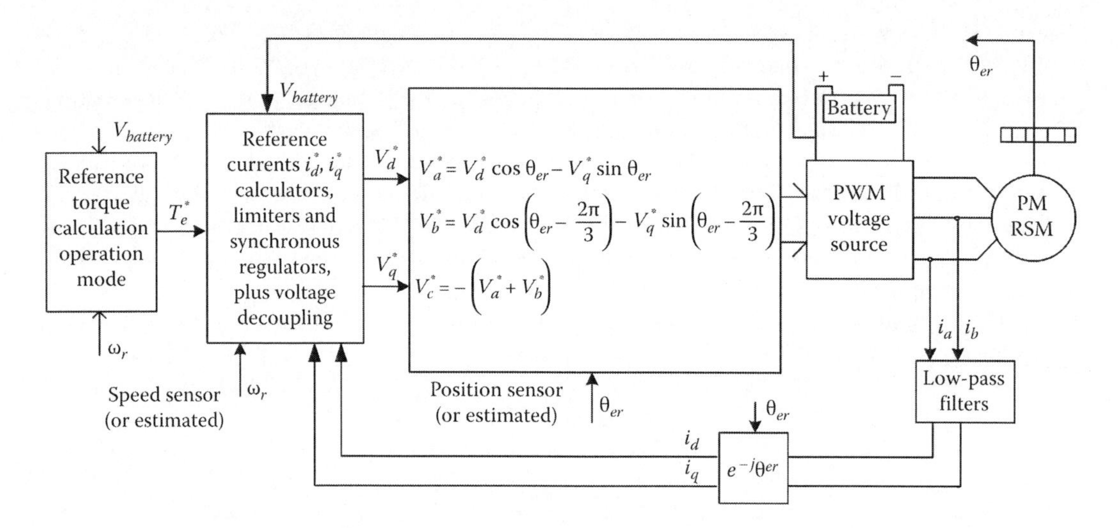

FIGURE 8.27 Indirect current voltage vector control with synchronous current regulators.

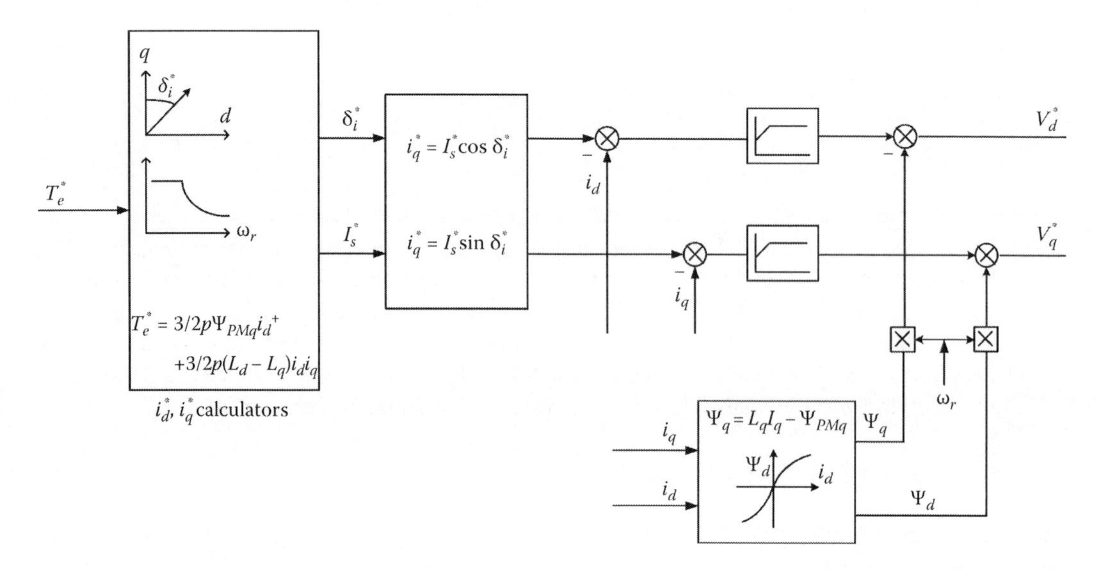

FIGURE 8.28 Reference currents i_d^*, i_q^* calculator and synchronous regulators with voltage decoupling.

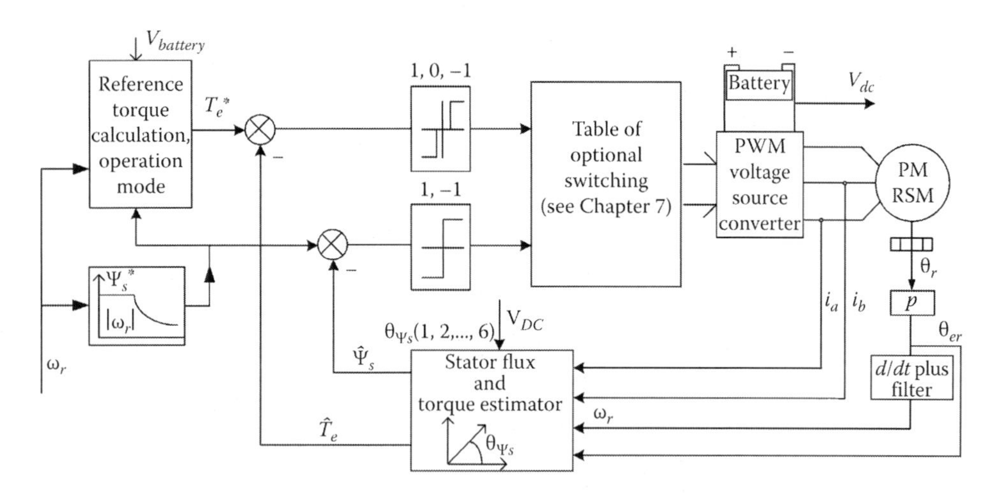

FIGURE 8.29 Direct torque and flux control (DTFC) of permanent-magnet-assisted reluctance synchronous machine (PM-RSM) with position feedback.

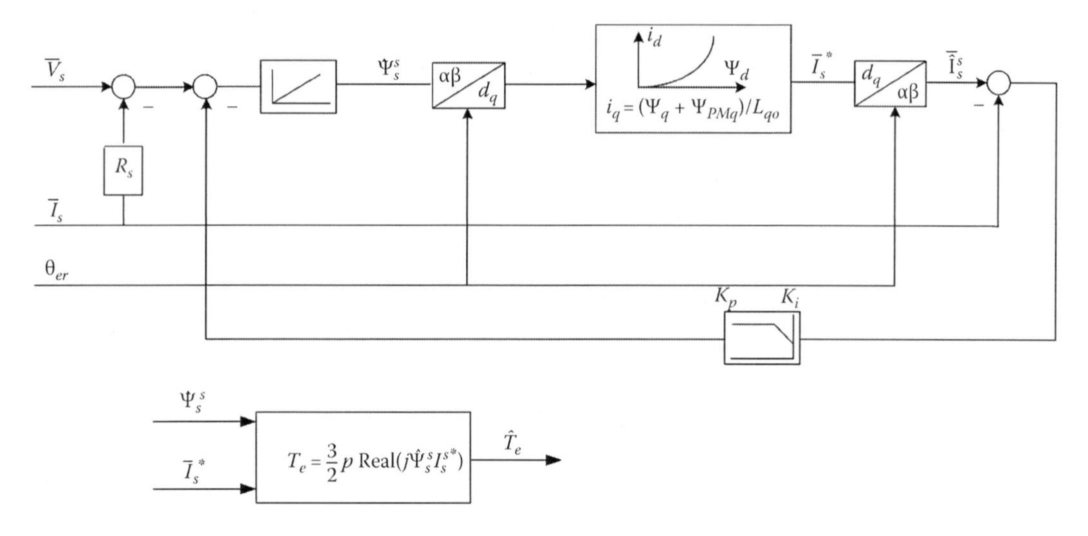

FIGURE 8.30 Combined voltage-current model observers for stator flux and torque.

As explained earlier, in EHV applications, torque control prevails, and thus, even for generating, the reference torque (negative) is to be used. The operation modes are as follows:

- Starting (motoring)
- Regenerative braking (generating)
- Battery slow recharging (generation)
- Motoring during engine idle stop

The battery state characterized here by the battery voltage only (but more complex in general) has to be considered in limiting the reference torque. In addition, the speed will set limits on the available torque, as the inverter output voltage is limited. The torque limitation with speed may be done many ways, but from maximum torque per current (up to base speed) to maximum torque/stator flux at maximum speed, the current d–q angle may be varied as planned offline (Figure 8.28). Only the motoring is treated in Figure 8.28 for i_d^*, i_q^* calculators.

Expressions similar to those in Figure 8.28 may be derived for generating, but for (eventually) more flux in the machine (for given torque). The voltage decoupling is required to secure better control at high speeds, where the synchronous current regulators have less room for adequate control due to the influence of motion-induced voltages.

The presence of an absolute position sensor encoder provides for safe start from any initial position and allows for speed calculation down to less than 1 rpm through adequate digital filtering. The precalculation of optimum current angle δ_i^* as a function of speed (eventually even of reference torque) by gradually advancing it with speed, based on maximum torque/flux condition, simplifies the control notably, as the current amplitude i_s^* and dq angle δ_i^* are found unequivocally once the reference torque and the actual speed are known.

A DTFC system for PM-RSM with position feedback is shown in Figure 8.29. Provided the stator flux and torque estimates are fast, robust, and precise and work from zero to maximum speed, DTFC provides for more robust control with about the same computation effort. The absence of reference online calculators of i_d^*, i_q^* and of the two vector rotators in DTFC is somewhat compensated for by the occurrence of the stator flux and torque estimator.

Stator flux estimation, with position feedback, is straightforward, though many schemes may, in principle, be applied, from model-reference adaptive-system (MRAS) to Kalman filters and so forth. A combined voltage-current model state observer is shown in Figure 8.30. In essence, the voltage model is corrected through the current estimator's error:

$$\bar{\Psi}_s^s = \int (\bar{V}_s - R_s\bar{I}_s + V_{comp})dt$$
$$V_{comp} = \left(K_p + \frac{K_i}{s}\right)\left(I_s^s - \bar{I}_s^s\right) \quad (8.68)$$

The magnetic saturation influence in axis d is considered. The design of the PI controller is done such that at low speeds, the current model acts practically alone. Typical values of K_p and K_I are $K_p = 22$–33 rad/s, $K_i = 40$–90 (rad/s)^2.

Inverter nonlinearities occur at very low speed; but as the current model is dominant, they do not have a very important influence.

8.9 State Observers without Signal Injection for Motion Sensorless Control

In the absence of position feedback, in indirect FOC schemes, only the rotor position θ_{er} and speed ω_r have to be estimated [4]. They have to be used, however, two times for vector rotation and once in the voltage decoupler (ω_r). The problem is that large errors, due to slow observer response or low immunity to machine parameter detuning, may notably influence the field orientation, and the fast, precise, torque control is lost.

In contrast, for DTFC schemes, rotor position is not required in the control, but speed ω_r, the stator flux Ψ_s, and torque T_e have to be estimated. The stator flux estimator, in general, requires the estimated rotor position. However, its influence is notable only at low speeds, when the current model is predominant (Figure 8.30). A straightforward way of calculating the rotor position for the flux observer purposes would be as follows (Figure 8.31a and b):

$$\hat{\theta}_{er} = \hat{\theta}_{\Psi_s} + \delta_{\Psi_s}, \quad \text{for motoring}$$
$$\hat{\theta}_{er} = \theta_{\Psi_s} - \delta_{\Psi_s} - \pi, \quad \text{for generating} \quad (8.69)$$

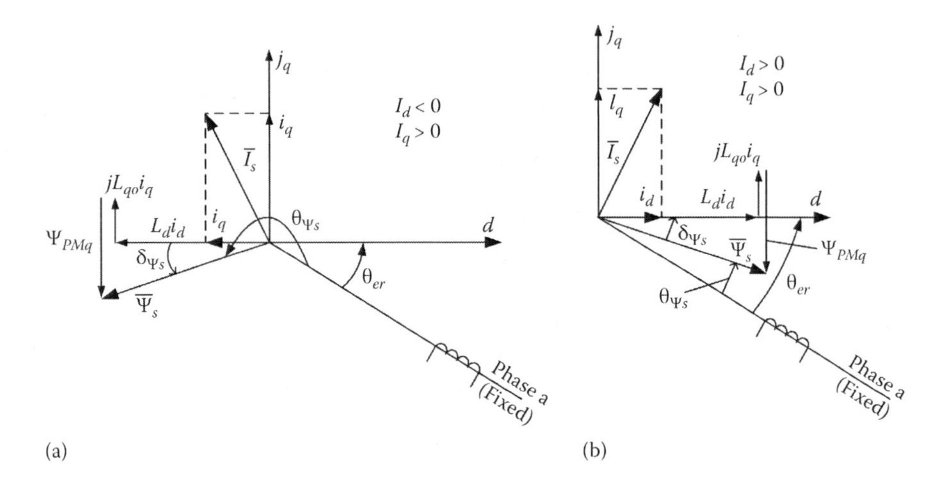

FIGURE 8.31 Rotor position θ_{er} in relation to $\delta\psi_s$ and $\theta\psi_s$: (a) generating and (b) motoring.

$$\bar{\theta}_{\Psi_S} = \sin^{-1}\left(\frac{\Psi^s_{s\alpha}}{\left|\bar{\Psi}_s\right|}\right); \quad \Psi_d = \Psi_s \cos\delta_{\Psi_s}\Psi_q = \Psi_s \sin\delta_{\Psi_s} \tag{8.70}$$

$$T_e = \frac{3}{2}p\left(\Psi_{PMq} + (L_d(\Psi_d - L_{q0})\frac{\Psi_s \sin\delta_{\Psi_S}}{L_{q0}}\right)\frac{\Psi_s \cos\delta_{\Psi_s}}{L_d(\Psi_d)} \tag{8.71}$$

The stator flux and torque observer in Figure 8.30 may be augmented with Equations 8.69 through 8.71 to include the rotor position θ_{er} calculator. Equation 8.71 may be simplified, as δ_{Ψ_s} is, in general, a small angle (below 30° in most cases), and $\cos\delta_{\Psi_s} \approx 1$:

$$|\hat{\delta}_{\Psi_s}| \approx \left(\frac{-2\left|\hat{T}_e\right|}{3p}\frac{L_d(\hat{\Psi}_s)}{\hat{\Psi}_s^2} + \frac{\Psi_{PMq}}{\Psi_s}\frac{L_d(\Psi_s)}{L_q}\right)\frac{1}{((L_d(\Psi_s))/L_{q0})-1} \tag{8.72}$$

This way, the online computation effort is reduced drastically. Such a position estimator θ_{er} (calculator) is acceptable for DTFC schemes, as θ_{er} appears only in the flux observer. This is not so for FOC schemes, where θ_{er} is crucial for field orientation.

Still, the speed has to be estimated. Again, as precise speed control is not needed, the estimated speed is used for adjusting the reference flux Ψ_s^* and reference torque T_e^*. At low speeds, this information is not important, as Ψ_s^* generally remains constant. To a first approximation, then, the steady-state speed estimation may be adopted. That is, the stator flux vector speed is used instead of $\hat{\omega}_r$:

$$\hat{\omega}_{\Psi_S} \approx \hat{\omega}_r = \frac{\Psi_{s\beta}(k)\Psi_{s\alpha}(k-1) - \Psi_{s\alpha}(k)\Psi_{s\beta}(k-1)}{T_{sample}\left(\Psi_{s\alpha}^2(k) + \Psi_{s\beta}^2(k)\right)} \tag{8.73}$$

After low-pass filtration, such a signal should be enough for torque and flux references. Therefore, for sensorless DTFC of PM-RSM, with torque (not speed) control at zero speed, the flux and torque estimators need rotor position and speed calculators that are fast but not so accurate.

The question remains if such a fundamental excitation (no signal injection) state observer can provide nonhesitant start and fast torque control at zero speed. At least for the induction machines, it did.

Many other state observers for FOC or DTFC of IPM synchronous machines without signal injection were proposed. Luenberger, MRAS, and Kalman filter observers all use the stator voltage vector equation (voltage model) and the current model, and the flux current relationship in *dq* rotor coordinates. Sliding mode such observers were proven to yield good stator flux rotor position and speed estimation down to very low speeds (a few rpm) [20]. However, robust fast torque sensorless control of PM-RSM at zero speed, without signal injection, is still due to be proven by tests. And this is so due to the need to estimate first at zero speed, the initial rotor position. This is how the signal injection state observers for PM-RSM come into play.

8.10 Signal Injection Rotor Position Observers

Signal injection state observers are based on machine magnetic saliency tracking. Back-emf methods do not use signal injection but cannot work at very low speeds, or at zero speed. These methods are used to detect rotor position, and then from the latter, the rotor speed is estimated through adequate filtering. Saliency tracking by signal injection works well from zero speed, but it limits available DC link voltage at high speeds. A combination of the saliency tracking—to 10% maximum speed—and back-emf tracking above 10% of maximum speed was recently proven experimentally as being very good from 0 to 5000 rpm in 70 kW IPM synchronous machine control [21] with ±100% torque and 150 m/s ramp control at 100 and at 5000 rpm.

Saliency tracking methods are used to estimate absolute rotor position. A voltage (current) signal at a carrier frequency ω_c above 500 Hz is injected in rotor d–q coordinates in the FOC. The response in current is used to track the spatial saliency and thus the rotor position. The injected signals (in voltage) are of the following form:

$$\bar{V}_{dq}^{c} = V_c \cos \omega_c t - jV_c \sin \omega_c t \tag{8.74}$$

The machine response in current, separated through adequate filtering, is as follows:

$$\bar{I}_{dq}^{c} = I_p \sin \omega_i t - I_n \cos(h\theta_{er} - \omega_i t) - j(I_p \cos \omega_i t - I_n \sin(h\theta_{er} - \omega_i t)) \tag{8.75}$$

I_p and I_n are positive- and negative-sequence carrier currents, while h is the order of the saliency spatial harmonic

θ_{er} is the rotor position

Only the negative sequence current contains the rotor position information. In addition, in general, only one spatial harmonic (h) is considered. The negative-sequence current is processed through heterodyning to produce a signal that is proportional to the rotor position error. This $\Delta\theta_{er}$ (error) signal is then used as input to a Luenberger observer that produces parameter-insensitive zero lag position and speed estimates [22–24].

For the back-emf methods, used for high speeds, the method in Section 8.9 is typical. The basic structure of such a hybrid sensorless FOC, with saliency plus back-emf position and speed tracking, is shown in Figure 8.32 [21]. The switching from low-speed position estimation to high-speed position estimation is done when the two have the same output, around 10%–15% of maximum speed, to leave room for full voltage utilization by the voltage source inverter. Position estimation results at 100 and at 5000 rpm, for a 70 kW IPM-SM, obtained with the above combined control method, is shown in Figures 8.33 and 8.34 [21].

Torque ranging of ±100% is experienced with apparently flawless position estimation. Still, for safe torque control at zero speed, the initial rotor position is required. This is not directly possible with the above method.

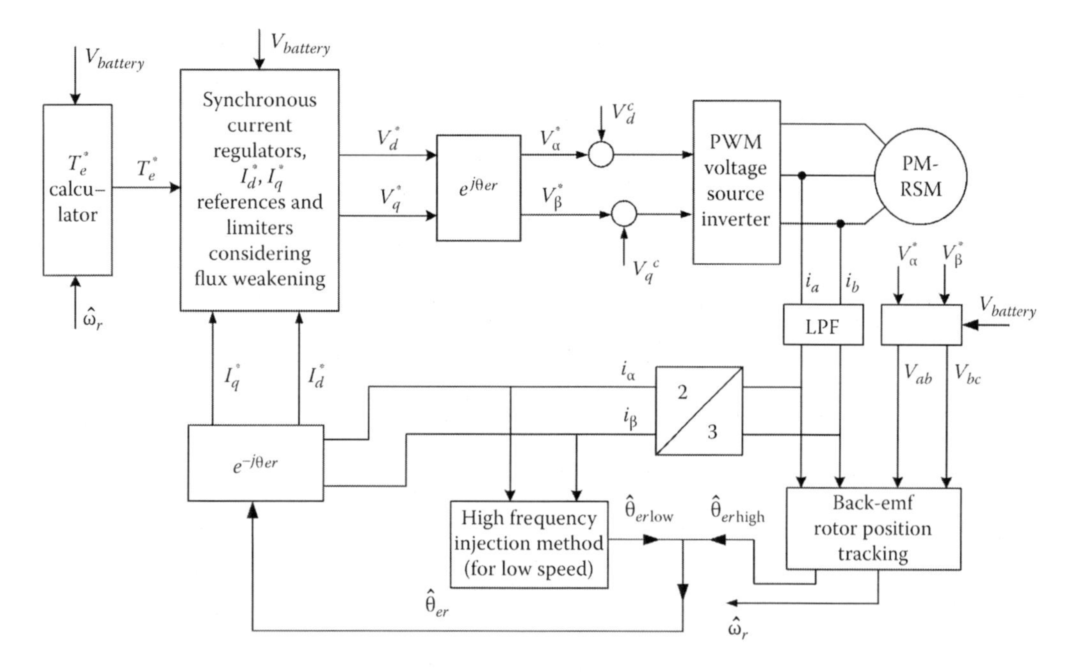

FIGURE 8.32 Sensorless flux-oriented control (FOC) of permanent-magnet-assisted reluctance synchronous machine (PM-RSM) with signal injection plus back electromagnetic force (emf) rotor position estimation.

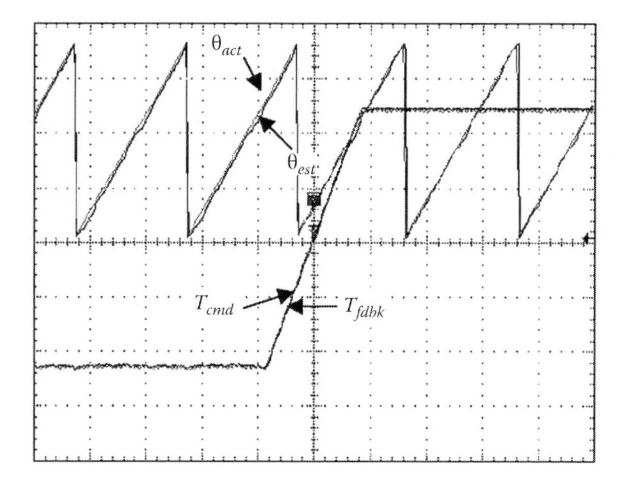

FIGURE 8.33 Dynamic performance of injection at 100 rpm. (Adapted from Patel, M. et al., Encoderless IPM drive system for EH-HEV propulsion applications, Record of EPE-2001, Graz, Austria, 2001.)

8.11 Initial and Low-Speed Rotor Position Tracking

A simple way to detect initial rotor position is to use three proximity Hall sensors placed in the axes of the three phases, to detect the PM flux position. The resolution of such a coarse position tracker is only 180/3 = 60° (electrical). For DTFC, such an error may be acceptable, as the method still indicates correctly the first voltage vector in the table of optimal switchings to be triggered in the inverter. For FOC, better precision is required to secure safe smooth torque for starting. The additional wiring for the Hall sensors may lead to failure. Pulse signal [25] or sinusoidal carrier signal [23–29] methods were proposed in the past decade for initial position tracking in IPM-SMs.

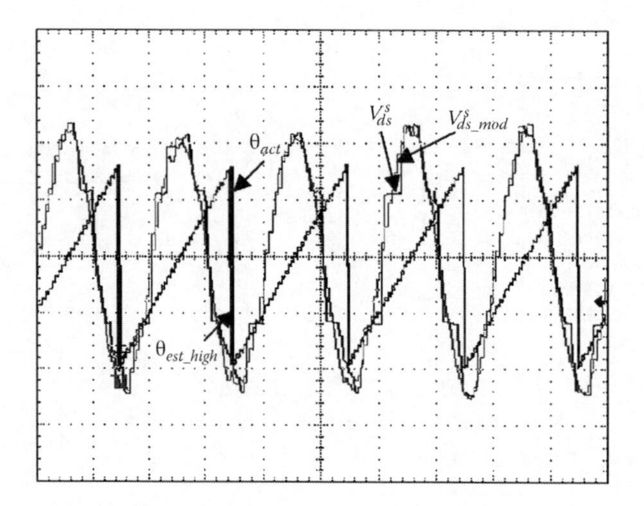

FIGURE 8.34 Comparison of actual and estimated rotor position at 5000 rpm in six-step mode under full-load condition using BEMF method. (Adapted from Patel, M. et al., Encoderless IPM drive system for EH-HEV propulsion applications, Record of EPE-2001, Graz, Austria, 2001.)

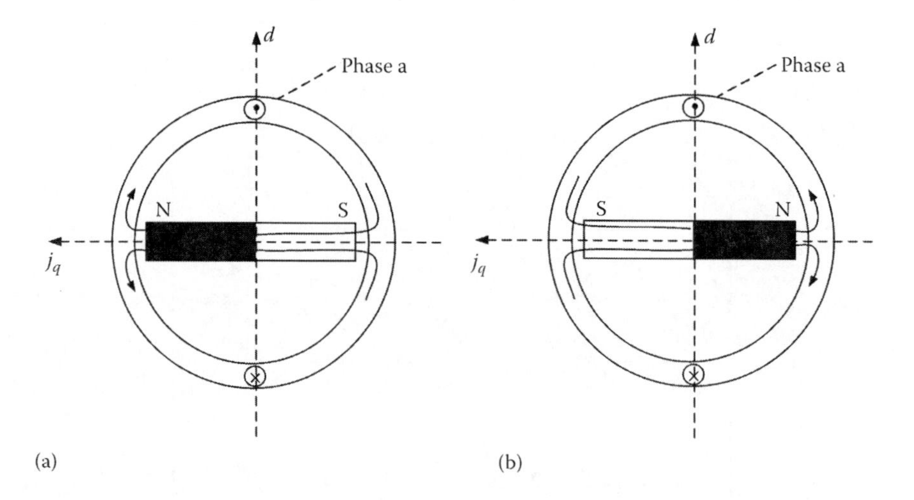

FIGURE 8.35 Magnetic saturation in the q axis of a permanent-magnet-assisted reluctance synchronous machine (PM-RSM). (a) Demagnetization and (b) magnetization.

The slight magnetic saturation due to the PM "North Pole" presence is used to find the PM polarity and thus, settle for the actual rotor initial position (Figure 8.35). Consequently, magnetic saturation has to be considered in the PM-RSM model. The voltage model vector equation of PM-RSM is as follows:

$$\bar{V}_S = R_S\bar{i}_S + \frac{d\bar{\Psi}_r}{dt} + j\omega_r\bar{\Psi}_S \tag{8.76}$$

$$\bar{\Psi}_S = \Psi_d + j\Psi_q \tag{8.77}$$

Magnetic saturation is considered separately for the two axes:

$$i_d + ji_q = R_{d0}\lambda_d + R_{d1}\lambda_d^2 + R_{d2}\lambda_d^3 + j[R_{q0}(\lambda_q - \lambda_{PMq}) + R_{q1}(\lambda_q - \lambda_{PMq})^2] \tag{8.78}$$

Only two terms were retained in axis q, as the magnetic saturation here is less important, as demonstrated earlier in this chapter. Along axis d, magnetic saturation is heavy for the large starting torque. The magnetic saturation functions are mere qualitative, and are used to discriminate the machine current response waveform when a sinusoidal voltage signal injection is applied. A sinusoidal voltage injection method is illustrated in what follows [30].

The injected signal $V^c_{\alpha\beta}$ in stator coordinates is as follows:

$$V^c_{\alpha\beta} = V_c e^{j\omega_i t} \tag{8.79}$$

In rotor coordinates,

$$V^c_{dq} = V^c_{\alpha\beta} e^{-j\theta_{er}} \tag{8.80}$$

Neglecting the resistive voltage drop,

$$\bar{\Psi}^c_{dq} = \int \bar{V}^c_{dq}\, dt \tag{8.81}$$

Then, using Equation 8.78, the injected currents in stator coordinates are as follows:

$$\bar{I}^c_{\alpha\beta} = \bar{I}_{cp1} e^{j(\omega_c t-(\pi/2))} + \bar{I}_{cn1} e^{j(-\omega_c t+2\theta_{er}+\phi_{1n})} + \bar{I}_{cp2} e^{j(2\omega_c t-\theta_{er}+\phi_{2p})} + \bar{I}_{cn2} e^{j(-2\omega_c t+3\theta_{er}+\phi_{2n})} \tag{8.82}$$

The first- and second-time harmonic components are illustrated in Equation 8.82. Only the last three terms contain rotor position information at the second, first, and respectively, third spatial harmonic. The measured carrier current spectrum at standstill in stator coordinates (Equation 8.82) is shown in Figure 8.36 [30]. The negative-sequence first-time harmonic (second-space harmonic) current—the second term in Equation 8.82—contains the largest signal and seems to be most appropriate for initial position detection.

In Reference 30, the I_{cn1} and I_{cn2} terms in Equation 8.82 are used to estimate the position error $\theta_{er} - \hat{\theta}_{er}$, after frame transformations from stationary to ω_c and $2\omega_c$ reference frame speeds, and after taking their dot product:

$$\bar{I}^{-\theta_c}_{dq2\theta_{er}} = I_{cn1} e^{j(2\theta_{er}+\phi_{1n})}; \quad \bar{I}^{-2\theta_c}_{dq3\theta_{er}} = I_{cn2} e^{j(3\theta_{er}+\phi_{2n})} \tag{8.83}$$

$$\bar{I}^s_{dq\theta_{er}} = I^{*-\theta_c}_{dq2\theta_{er}} \cdot I^{-2\theta_c}_{dq3\theta_{er}} = I_{cn1} I_{cn2} e^{j(\theta_{er}+\phi_{1n}-\phi_{2n})} \tag{8.84}$$

A first method to extract the rotor position error from Equation 8.84 is by the vector cross-product with its unit vector:

$$\varepsilon = \hat{U} I^{*s}_{dq\theta_{er}} \times \bar{I}^s_{dq\theta_{er}} = \sin(\theta_{er} - \hat{\theta}_{er}) \tag{8.85}$$

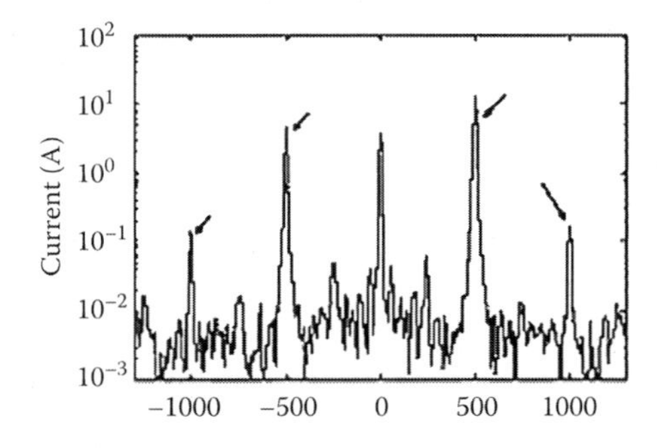

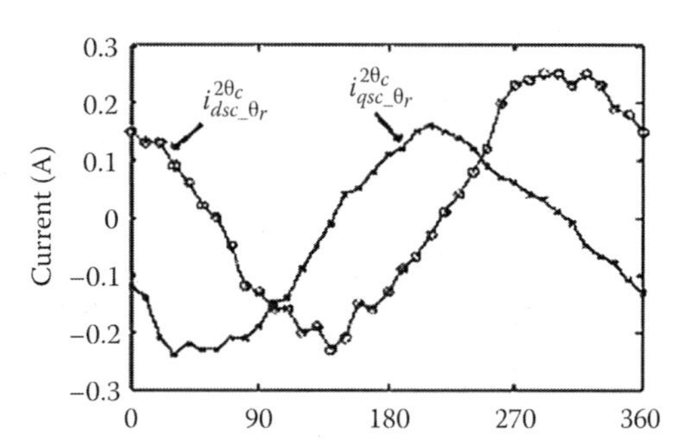

FIGURE 8.36 Measured carrier current spectrum.

No magnet polarity detection is required [30]. The block diagram of the initial position tracking observer, based on this principle, is shown in Figure 8.37. Experimental results in Reference 30 show remarkable initial position tracking precision at 60°, 120°, 240°, and 300° (electrical), with a convergence time of 25 m/s, for f_c = 500 Hz (Figure 8.38). Even faster convergence (10 m/s) is reported for a derivative method with additional magnet polarity compensation (Figure 8.38) [30].

As the above method can be applied as it is at small speeds, also it may be used successfully in combination with the back-emf method for FOC (Figure 8.32 to cover the entire speed range from standstill).

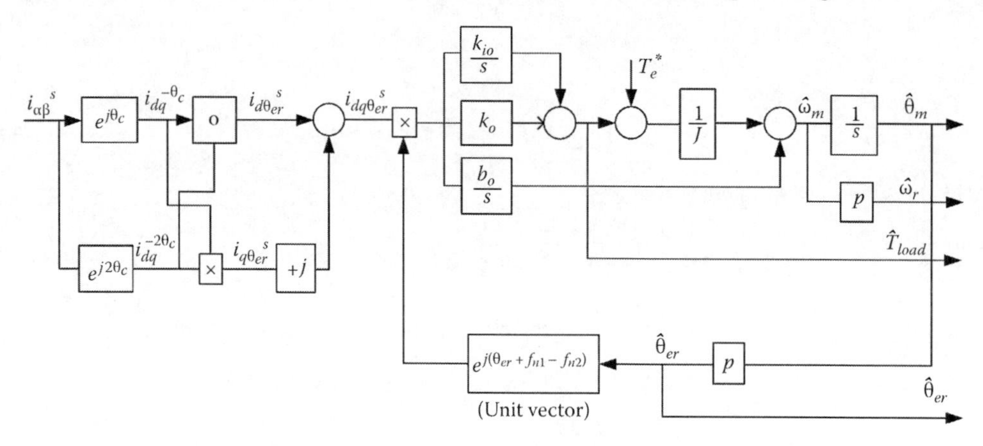

FIGURE 8.37 Fundamental rotor position and speed tracking observer.

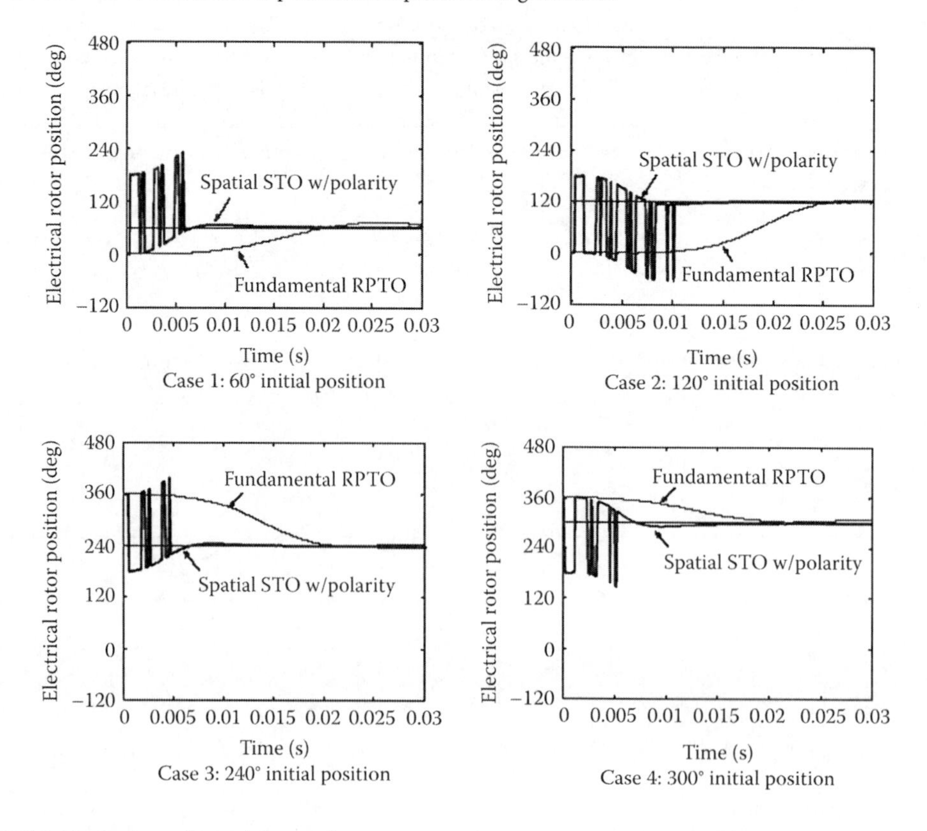

FIGURE 8.38 Estimated initial electrical rotor position. (From Kim, H. et al., Initial rotor position estimation for an integrated starter alternator IPM synchronous machine, Record of EPE-2003, Toulouse, France, 2003.)

The same procedure may be used in DTFC to detect θ_{er} for the use in the flux observer only (Figure 8.30). Recently the active flux concept ($\bar{\Psi}_d^a = \bar{\Psi}_S - L_d\bar{I}_S$), always along rotor axis in SMs, has been proposed to identify directly the rotor position, including signal injection for initial rotor position [34].

8.12 50/100 kW, 1350–7000 rpm (600 Nm Peak Torque, 40 kg) PM-Assisted Reluctance Synchronous Motor/Generator for HEV: A Case Study [31]

Abstract

The aim of this paragraph is to introduce—by general analytical nonlinear and then optimal design with vector control—a hybrid electric and electric vehicles (HEVs/EVs) electric propulsion system for 50/100 kW, 1350–7000 rpm (600 Nm peak torque/40 kg) at above 91% overall efficiency 300 V_{DC} battery for a peak phase current of 520 A. After searching quite a few alternative electric machines the low weight (40 kg) NdFeB (1.1 T) PM–Reluctance synchronous machine (SM) with vector control was developed in detail. Finite element validation of flux density, torque, inductances and nonreconfigurable vector control for the entire peak torque (power) very challenging envelope substantiate a moderate-cost high performance HEV (EV) drive.

8.12.1 Introduction

A 50/100 kW (peak) power 1350–7000 rpm PM-RSM drive system is designed in this paragraph, to produce a peak of 600 Nm for less than 40 kg of active materials (copper, laminations and PMs) for an efficiency overall above 91%–92%, for a limited no load voltage at max. speed, which eliminates an over voltage protection system. A general, nonlinear magnetic circuit-based model is developed and used for a general design to meet the critical torque/speed peak requirements for a 300 V_{DC} battery voltage. Then, an optimal methodology based on Hooke–Jeeves modified algorithm and on the overall lifetime-estimated system cost function is developed and applied; a few percent improvement in efficiency is obtained this way over the initial design for a penalty function with limited 40 kg of active materials weight.

Note: While looking for an electric machine drive system to fulfill the above power/speed peak envelope, for 300 V_{DC} (Figure 8.39) and less than 550 A battery current, the IM, SRM, the 12/14 SPMSM solutions have been tried, but all failed in our particular attempts.

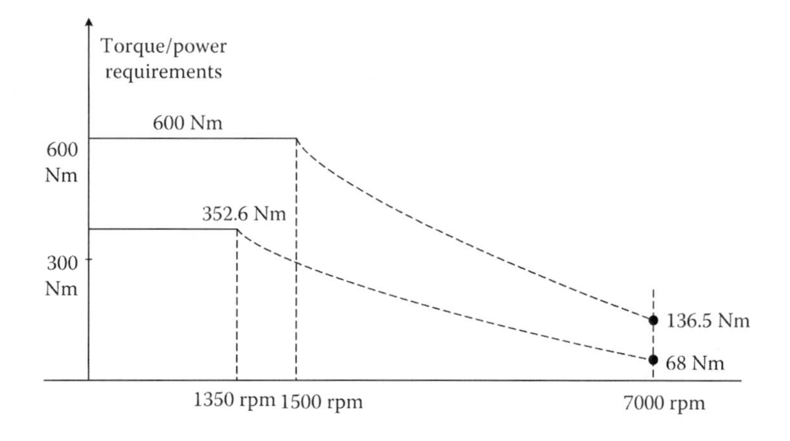

FIGURE 8.39 HEV (EV) torque (power)/speed requirements.

8.12.2 General Design Summary and Results

To start the general design, a few experience-based decisions are made. They are related to max outer stator diameter D_{os} = 285 m, total stator length from 130 mm, $2p_1$ = 8 poles to secure f_{1max} = 466.44 Hz, allowing us to use 600 V IGBTs in a 300 V_{DC} (average) battery system.

To cut the inverter kVA rating and increase torque density, Hiperco 50 (0.25 mm thick lamination) is used both for stator and rotor cores. A 1.1 T peak air gap flux density is allowed for, to secure the 600 Nm peak torque with the 285 mm outer stator and 100 mm stator stack length Hiperco laminations. They are used on aircraft and are believed more suitable mechanically also, up to bursting speed (10,000 rpm: 100 m/s).

8.12.2.1 Stator Core Geometry

With eight poles q = 2 (3) stator slots/pole/phase, we adopt q = 2, y/τ = 5/6 (chorded coil two-layer winding) and a slot pitch τ_s = 12 mm. Consequently, the pole pitch τ is $\tau = 3q\tau_s$ = 72 mm and the interior stator diameter D_{is} is 183.4 mm. With B_{g1max} = 1.1 T peak air gap flux density, the stator back core radial height h_{cs} is 12 mm.

The stator lamination geometry is illustrated in Figure 8.40.

For the B_{g1max} = 1.1 T the required mmf expression is as follows:

$$B_{g1max} = \frac{3\mu_0 \cdot W_1 I_{dak} K_{w1}}{\pi p g K_c (1 + K_{sk})} = 1.1\,\mathrm{T} \tag{8.86}$$

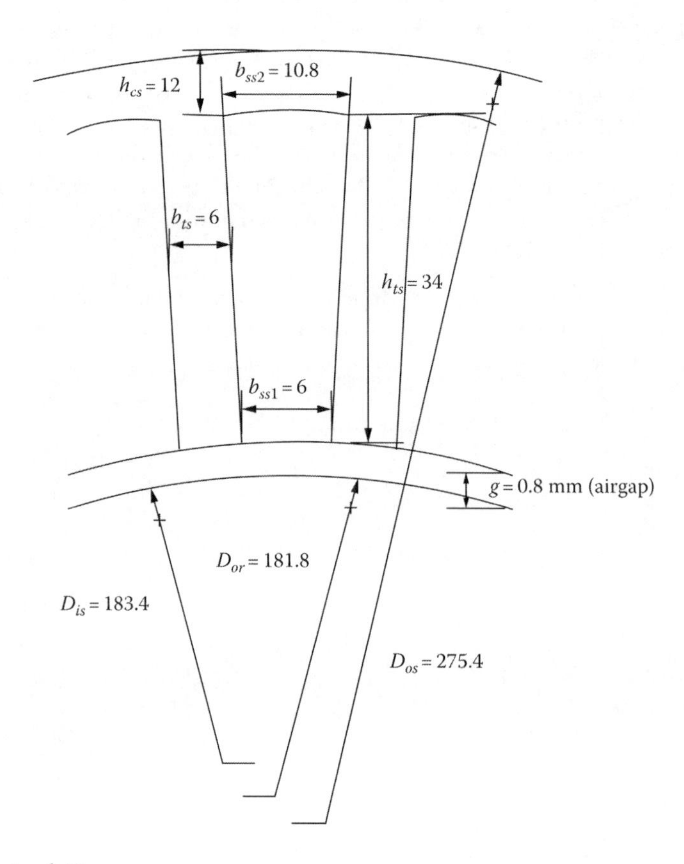

FIGURE 8.40 Stator slotting.

We consider $a = 4$ (not 8) current paths–each pole pair represents a current path-to reduce the circulating currents between current paths. Therefore,

$$W_1 I_{dak} = \frac{W_1}{a} I_{dk}; \quad I_{dk} = a I_{dak} \tag{8.87}$$

where W_1 is total turns per phase: $W_1 = 2pqn_c$ and n_c-turns per coil, with air gap $g = 0.8 \cdot 10^{-3}$ m, $K_c = 1.756$, $K_{sk} = 1.600$, $K_{w1} = 0.933$, from Equation 8.86: $W_1 \cdot I_{dak} = 8.842 \cdot 10^3$ A/turns.

8.12.2.2 Number of Turns per Coil n_c

The number of turns per coil is adopted to limit the no load voltage at 7000 rpm (E_{max}) to 450 V_{DC}.

Consequently, the PM flux linkage per phase (peak value) Ψ_{PM1q} is:

$$\omega_{1\max} \cdot \Psi_{PM1q} = \left(V_{DC} \cdot \frac{2\sqrt{3}}{\pi} \cdot \frac{1}{\sqrt{3}} \right) \cdot 0.95 = 272.25 \text{ V} \tag{8.88}$$

But the Ψ_{PM1q} expression is as follows:

$$\Psi_{PM1q} = B_{gPMav} \cdot l_{stack} \cdot K_{\omega 1} \cdot \frac{W_1}{a}; \quad \omega_{1\max} = 2\pi \cdot 466.66 \text{ rad/s} \tag{8.89}$$

$$\frac{W_1}{a} = 2qn_c \approx 44 \text{ (turns per current path)}; \quad B_{gPMav} = 0.3152 \text{ T}$$

Therefore, $n_c = 11$ turns per coil.

The peak d-axis current per phase is $I_{dk} = a\, I_{dak} = 201$ A.

The d-axis air gap flux linkage is: $\Psi_{dmk} = \frac{2}{\pi} B_{g1\max} \tau l_{stack} K_{w1} \frac{W_1}{a} = 0.207$ Wb, thus the magnetizing inductance L_{dmk} is $L_{dmk} = \frac{\Psi_{dmk}}{I_{dk}} = 1.03 \cdot 10^{-3}$H.

8.12.2.3 The Stator Leakage Inductance L_{s1} and L_{dm}/L_{qm} Requirements

$$L_{s1} = 2\mu_0 \frac{W_1^2 l_{stack}}{pqa^2} \left(\lambda_{slot} + \lambda_t + \lambda_{end} + \lambda_{diff} \right) \tag{8.90}$$

Finally, $L_{s1} = 0.129 \cdot 10^{-3}$ H.

The stator Carter coefficient is as follows:

$$K_{cs} = \frac{1}{1 - \gamma (g/\tau_s)} = 1.756; \quad \gamma = \frac{(b_{ss1}/g)^2}{5 + (b_{ss1}/g)} = 6.46 \tag{8.91}$$

The rotor produced Carter coefficient (due to saturated flux barrier bridges) is almost unity.

By trial and error, a peak torque saliency $L_d/L_q = 4.5$ is required:

$$\frac{L_d}{L_q} = \frac{L_{dmk} + L_{s1}}{L_{qm} + L_{s1}} = \frac{1 + (L_{s1}/L_{dmk})}{(L_{qm}/L_{dmk}) + (L_{s1}/L_{dmk})} = \frac{1 + (0.129 \cdot 10^{-3}/1.03 \cdot 10^{-3})}{(L_{qm}/L_{dmk}) + (0.129 \cdot 10^{-3}/1.03 \cdot 10^{-3})} \tag{8.92}$$

Therefore $L_{qm}/L_{dmk} = 0.125$ which results in $L_{dmk}/L_{qm} = 8$, for magnetizing inductances L_{dmk} and L_{qmk}.

8.12.2.4 Rotor Lamination Design

The rather large L_{dmk}/L_{qm} ratio implies three to four flux barriers/half pole and a total "additional" q-axis air gap:

$$\frac{g_{qa}}{g} = K_c(1+K_{sk})\frac{L_{dmk}}{L_{qm}} - 1 = 21.47 \tag{8.93}$$

Now, with three flux barriers each 6.2 mm thick, we obtain $g_{qaeff}/g = 23.5$ (seven total flux barriers/pole, Figure 8.41).

Careful FEM verification for peak torque will tell the complete truth about this rotor design (Figure 8.41).

$$\alpha_3 < \alpha_2; \quad b_{b3} \geq b_{b2} \geq b_{b1}; \quad \alpha_1 = \alpha_2 = \alpha_0 = \frac{\alpha_3 + \alpha_3}{2}; \quad b_{bav} \approx 0.8 \text{ mm}$$

$$l_{b1} < l_{b2} < l_{b3}; \quad l_{b1} + l_{b2} + l_{b3} = 18.6 \text{ mm}; \quad h_{PM1} = h_{PM2} = h_{PM3} \approx 6.62 \text{ mm}$$

Medium-grade NeFeB or $SmCo_5$ with $B_r = 1.1$ T should be suitable to produce the average PM air gap flux density of $B_{gPMav} = 0.3125$ T.

8.12.2.5 Peak Torque Production

For peak torque production, the flux in axis q is more than canceled:

$$T_{ek} = \frac{3}{2}p_1[L_{dk}I_{dk}I_{qk} + (\Psi_{PMq} - L_qI_{qk})I_{dk}] \tag{8.94}$$

$$L_{dk} = L_{dmk} + L_{s1} = 1.159\cdot 10^{-3} \text{ H}; \quad L_q = \frac{L_{dk}}{4.5} = 0.2575\cdot 10^{-3} \text{ H}$$

By trial and error, we find that for $I_{qk} = 470$ A:

$$T_{ek} = 623 \text{ Nm}; \quad \Psi_{PM1q} = \frac{\omega_{1\max}\Psi_{PM1q}}{\omega_{1\max}} = 0.0929 \text{ Wb}; \quad \Psi_{PM1q} = \frac{\omega_{1\max}\Psi_{PM1q}}{\omega_{1\max}} = 0.0929 \text{ Wb}$$

The slight over cancellation of PM flux does not mean PM demagnetization because $L_{s1} \approx L_{qm}$ (50% of the stator flux in axis q is stator leakage flux).

The placement of PMs only on flux barrier bottoms avoids their easy demagnetization as shown later by FEM.

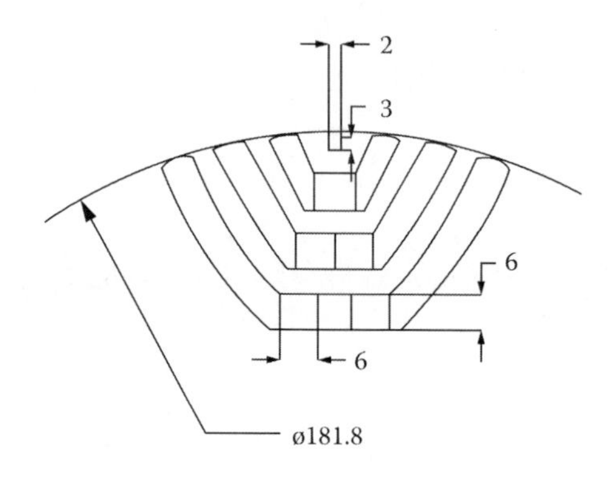

FIGURE 8.41 Rotor pole geometry.

Now the peak stator current I_{sk} is as follows:

$$I_{sk} = \sqrt{I_{dk}^2 + I_{qk}^2} = 511.17 \text{ A} \tag{8.95}$$

Adding 5% more for mechanical and iron losses leads to about 520 A peak phase current for 600 Nm torque.

8.12.2.6 Slot Area/Peak Current Density/Stator Resistance R_s

The current density at 600 Nm is $j_{Co\max} = \dfrac{(2n_c I_{sk}/a\sqrt{2})}{A_{slotu} k_{fill}} = 14.916 \text{ A/mm}^2$, thereby the stator phase resistance R_s is

$$R_s = \rho_{Co} \frac{2n_c q}{a} \frac{l_{coil}}{(I_{sk}/a j_{Co\max}\sqrt{2})} = 2.002 \cdot 10^{-2}\ \Omega \tag{8.96}$$

The copper losses at peak torque (600 Nm) are

$$p_{Costart} = \frac{3}{2} R_s I_{sk}^2 = \frac{3}{2} \cdot 2.002 \cdot 10^{-2} \cdot 511.17^2 = 7841 \text{ W}$$

For performance at 600 Nm and 1350 rpm (90 Hz) still, I_{dk} = 201 A, I_{qk} = 470 A and the *dq* voltage components are

$$\begin{aligned} V_d &= R_s I_{dk} - (L_q I_{qk} - \Psi_{PMq})\omega_r = -11.8 \text{ V} \\ V_q &= R_s I_q + \omega_1 L_{dk} I_{dk} = 141.08 \text{ V} \\ V_s &= \sqrt{V_d^2 + V_q^2} = 141.57 \text{ V} < 180 \text{ V} \end{aligned} \tag{8.97}$$

The 180 V is the maximum peak phase voltage available with 300 V_{DC}. The power factor angle ϕ_1 (neglecting the core loss) is:

$$\phi_1 = -\tan^{-1}\left(\frac{V_d}{V_q}\right) + \tan^{-1}\left(\frac{I_d}{I_q}\right) = 4.78 + 23.15 = 28^0; \quad \cos\phi = 0.883 \tag{8.98}$$

8.12.2.7 Weights of Active Materials

Stator core weight is 15.6 kg

The copper weight is 11.05 kg

The rotor total gross weight is 10.6 kg

Total active material weight:

$$G_a = G_{cores} + G_{copper} + G_{rotor} = 37.25 \text{ kg} \tag{8.99}$$

The 40 kg rounding off may be necessary after FEM verifications for peak torque (600 Nm).

This should be considered a reasonable weight as, for peak torque, it produces 600 Nm/40 kg = 15 Nm/kg.

With 8 W/kg for Hiperco 50 (0.25 mm thick lamination) at 2 T and 50 Hz, the total stator core losses are roughly:

$$p_{cores} = 2.2 p_{Hiperco\,2T\,50Hz} \cdot \left(\frac{B}{2.0}\right)^2 \left(\frac{f}{50}\right)^2 G_{cores} = 1000 \text{ W} \tag{8.100}$$

The coefficient 2.2 includes machining additional losses and harmonics losses.

Allocating 1000 W for mechanical losses, the machine efficiency at 600 Nm and 1350 rpm (90 Hz) is: $\eta_{600\text{ Nm; 1350 rpm; 90 Hz}} = 90\%$.

It should be noted that at 1350 rpm and 600 Nm the stator voltage $V_s = 141.6$ V, less than the maximum available of 180 V at 300 V_{DC} (battery voltage).

8.12.2.8 Performance at 100 kW and 7000 rpm

For 100 kW, approximately the torque at 7000 rpm is

$$(T_{ek})_{7000\text{ rpm}} = \frac{P_{max}}{2\pi n_{max}} = 136.5\text{ Nm} \tag{8.101}$$

We most probably need the maximum available voltage:

$$V_{s\max} = V_{DC} \cdot \frac{2}{\pi} \cdot 0.95 = 181.5 \approx 180\text{ V} \tag{8.102}$$

The level of stator flux admitted in the machine:

$$\Psi_{s7000\text{ rpm}} = \frac{V_{s\max} \cdot 0.95}{2\pi \cdot f_{max}} = 0.05835\text{ Wb} \tag{8.103}$$

We have to get an I_d and I_q pair capable of delivering 136.5 Nm for 0.05835 Wb of stator flux.

$$\Psi_s^2 = (L_d I_d)^2 + (L_q I_q - \Psi_{PMq})^2 \tag{8.104}$$

First, we have to consider that the machine desaturates heavily, so $L_{dk} = 1.159 \cdot 10^3$ becomes now $1.7 \cdot 10^{-3}$ (saturation factor decreases from 1.6 to 1.0908, which is quite reasonable). Adopting $I_q = 400$ A:

$T_e = 136.5 = (3/2) \cdot 4 \cdot (L_d I_q + \Psi_{PMq} - L_q I_q) \cdot I_d$ with $I_q = 400$ A, $I_d = 33$ A, $L_d = 1.7 \cdot 10^{-3}$ H, $L_q = 0.2575 \cdot 10^{-3}$ H Therefore,

$$(I_s)_{100\text{ kW;7000 rpm}} = \sqrt{I_d^2 + I_q^2} = 401\text{ A} \tag{8.105}$$

The dq voltage components are as follows:

$$\begin{aligned} V_d &= R_s I_d - \omega_{1\max} \Psi_q = -23\text{ V} \\ \Psi_q &= L_q I_q - \Psi_{PMq} = 0.101\text{ Wb} \\ V_q &= R_s I_q + \omega_{1\max} L_d I_d = 172.48\text{ V} \\ V_s &= \sqrt{V_d^2 + V_q^2} = 174\text{ V} < 180\text{ V} \end{aligned} \tag{8.106}$$

Note: There is still a voltage reserve, but it is small.

The power factor angle:

$$\phi_1 = -\tan^{-1}\left(\frac{V_d}{V_q}\right) + \tan^{-1}\left(\frac{I_d}{I_q}\right) = 10.88°; \quad \cos\phi = 0.982\text{ (lagging)} \tag{8.107}$$

The copper losses:

$$p_{copper} = \frac{3}{2} \cdot R_s I_s^2 = 4872.2\text{ W} \tag{8.108}$$

The core losses:

$$p_{core} = p_{\substack{core\,600\,\mathrm{Nm}\\1350\,\mathrm{rpm}}} \cdot \left(\frac{V_{s7000\,\mathrm{rpm}}}{V_{s1350\,\mathrm{rpm}}}\right)^2 = 1510\ \mathrm{W} \tag{8.109}$$

Therefore, the efficiency at 7000 rpm and 100 kW (with $p_{mec} = 1.5$ kW) is as follows:

$$\eta_{\substack{100\,\mathrm{kW}\\7000\,\mathrm{rpm}}} = \frac{100 \cdot 10^3}{100 \cdot 10^3 + 4872.2 + 1510 + 1500} \approx 0.926 \tag{8.110}$$

8.12.2.9 Performance at 50 kW, 7000 rpm, and 1350 rpm

The available stator flux is still 0.05834 Wb.

As for peak power (100 kW) for 1350 and 7000 rpm the voltage limit was not surpassed, for 50 kW and same speed limits, for sure, the machine can perform as well. Detailed results reveal approximate 91% efficiency at 1350 and at 7000 rpm.

8.12.2.10 Equivalent Circuit

The equivalent circuit in axes d and q, allowing for core loss and magnetic saturation, are presented in Figure 8.42.

$$R_{core} \approx \frac{3}{2}\left(\frac{V_s^2}{p_{core}}\right)_{1350\,\mathrm{rpm}(90\,\mathrm{Hz})} = \frac{3}{2} \cdot \frac{141.6^2}{1000} = 30.7525\ \Omega \tag{8.111}$$

With: $L_{dk} = 1.159 \cdot 10^{-3}$ H, $L_{dmk} = 1.03 \cdot 10^{-3}$ H at $I_{dk} = 201$ A; $nL_{dmu} = 1.7 \cdot 10^{-3}$ H = ct. for $I_d = 28.8$ A; $L_{qm} = 0.1275 \cdot 10^{-3}$ H = ct., $L_{sl} = 0.129 \cdot 10^{-3}$ H; $L_{dm} = L_{dmu}/(1 + bI_d)$; $L_d = L_{dm} + L_{sl}$; $b = 3.234 \cdot 10^{-3}$

$$L_{dm}^t = L_{dm} + \frac{\partial L_{dm}}{(1 + bI_d)^2}; \quad T_e = \frac{3}{2} \cdot p_1 \cdot [L_d I_q + (\Psi_{PMq} - L_q I_q)] I_d \tag{8.112}$$

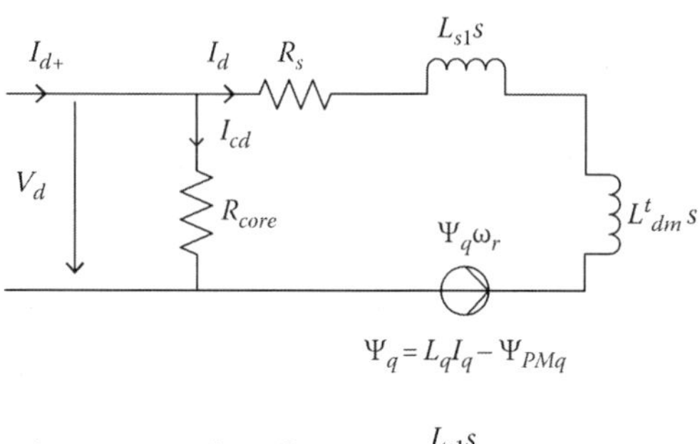

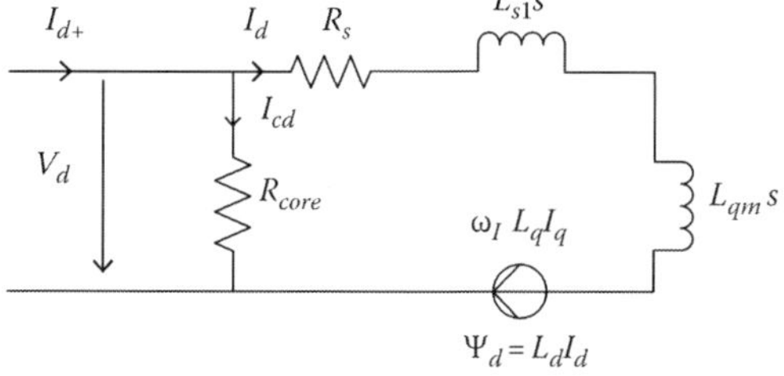

FIGURE 8.42 Equivalent circuits of PM-RSM.

8.12.3 Optimal Design Methodology and Results

8.12.3.1 IPMSM—Analytical Model

A nonlinear analytical magnetic circuit model of IPMSM was developed and implemented in MATLAB® code in order to compute the motor features.

The same nonlinear analytical model is used in optimal design (Figures 8.43 and 8.44).

The input data are introduced using an input file that contains the main design requests and also the several computed geometric dimensions.

8.12.3.2 Optimal Design of IPMSM

The optimal design is based on the analytical model and on the modified Hooke–Jeeves optimization algorithm. The optimized algorithm is divided in a few steps and we mention here some of them:

a. Choose the constant dimensions and the optimized variables, which are grouped in the $\overline{X}$ vector. In our case $\overline{X} = (\text{sDi}\ \ \text{sW3}\ \ sMs\ \ \text{shOA}\ \ \text{klpm}\ \ \text{kmbs}\ \ \text{Br2Bg})^T$

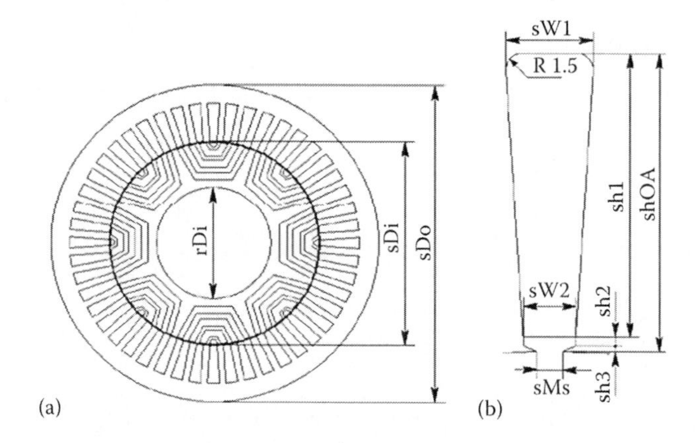

FIGURE 8.43 (a) Cross section of IPMSM and (b) stator slot.

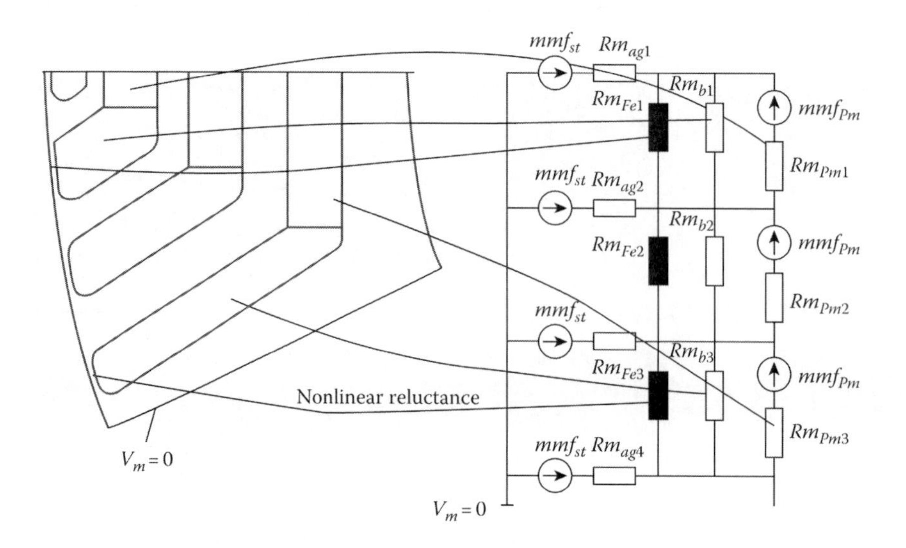

FIGURE 8.44 q-axis equivalent magnetic circuit.

b. Find the technological limitations, geometrical constraints and choose the optimized variable range, $X_{min} \leq X \leq X_{max}$, where $\overline{g}_1(\overline{X}) = \overline{0}$; $\underline{g}_2(\overline{X}) \leq \overline{0}$ are vector functions defined on variable space.
c. Choose the objective scalar function, $f_1(X), f_1 : \Re^n \to \Re$. The constraints could be included in the objective function considering the penalty function $f_p(\overline{X})$ and $f_p : \Re^{p+q} \to \Re_+$.

$$f_p = \sum_{i=1}^{p+q} f_{pi}(g_i(\overline{X})) \quad \text{with } f_{pi} = \begin{cases} 0 \quad \text{if } 1 \leq i \leq p \quad \text{and} \quad g_i(\overline{X}) = 0 \\ 0 \quad \text{if } p+1 \leq i \quad \text{and} \quad g_i(\overline{X}) \leq 0 \\ \text{monotonic positive } \forall \text{ other wise} \end{cases} \tag{8.113}$$

Finally the objective function is: $f(\overline{X}) = f_1(\overline{X}) + f_p(\overline{X})$.

d. Skipping other details, the initial costs are the active material costs:

$$C_i = c_{Cu}m_{Cu} + c_{lam}m_{sFe} + c_{Fe}m_{rFe} + c_{PM}m_{PM} + c_a m_t \tag{8.114}$$

where
C_{Cu} is the copper price assumed 10 USD/kg
C_{lam} is the lamination price assumed 5 USD/kg
C_{Fe} is the rotor iron price assumed 5 USD/kg
C_{PM} is the permanent magnet price assumed 50 USD/kg
C_a is the extra price per total mass m_t
m_{Cu} is the windings mass
m_{sFe} is the stator lamination mass
m_{rFe} is the rotor iron mass
m_{PM} is the permanent magnet mass

The energy losses cost expression is as follows:

$$c_e = P_n\left(\frac{1}{\eta(\overline{X})} - 1\right) \cdot t_1 \cdot p_e \tag{8.115}$$

where
P_n is the motor rated power
η_n is the rated efficiency
t_1 is the operation time, energy price is assumed to be 0.1 USD/kWh

The cost of PM demagnetization is as follows:

$$c_{pdm} = \max\left(0, \quad \frac{\theta_{pk} - \theta_{ad}}{\theta_{ad}} - 1\right) \cdot k_{dm} \cdot c_i \tag{8.116}$$

where
θ_{pk} is the peak magnetomotive force produces by stator current
θ_{ad} is the admissible magnetomotive force of permanent magnet
k_{dm} is the a proportional constant

There are two different optimization objective functions: cycle life cost optimization and active material cost optimization. Further on, the cycle life cost optimization is described and results are shown (Figures 8.45 through 8.48).

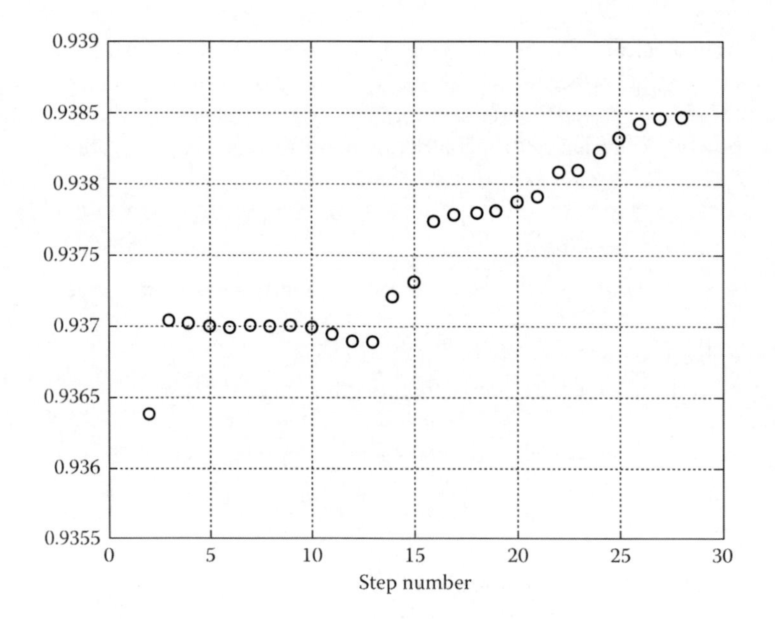

FIGURE 8.45 Electric efficiency evolution.

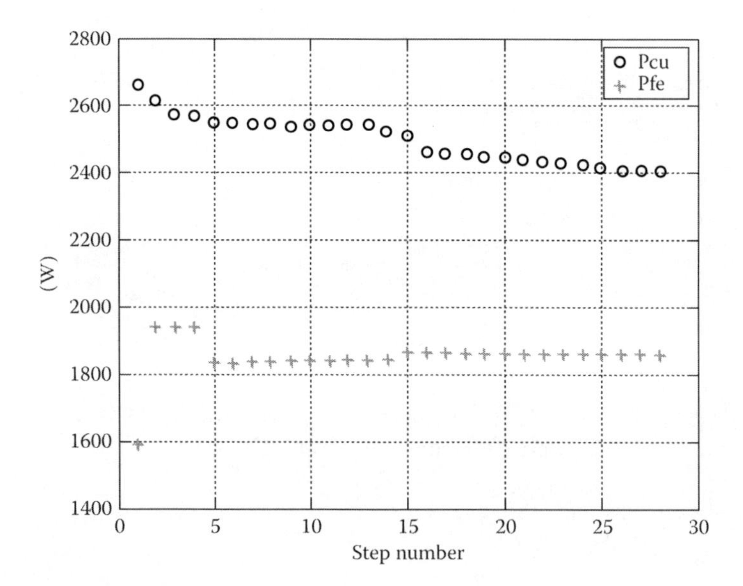

FIGURE 8.46 Losses evolution.

The case study refers to a super high torque density machine that explains the 1.2 T airgap flux density at peak torque (600 Nm). The cycle cost optimization (which includes operation at 50 kW for 15,000 h) objective function has led to initial material cost of €406 for 39.48 kg and an electrical efficiency of 0.93847. The initial material cost optimization has led to €382 initial cost, 38.435 kg, and 0.93347 electrical efficiency. Unfortunately for this latter case, the peak current needed to produce 600 Nm was 600 A (peak value/phase) instead of 520 A as for the case of cycle life cost optimization. Since the cost of the converter is notably higher than that of the motor, it means that the results from cycle cost optimization should be adopted as the global optimum.

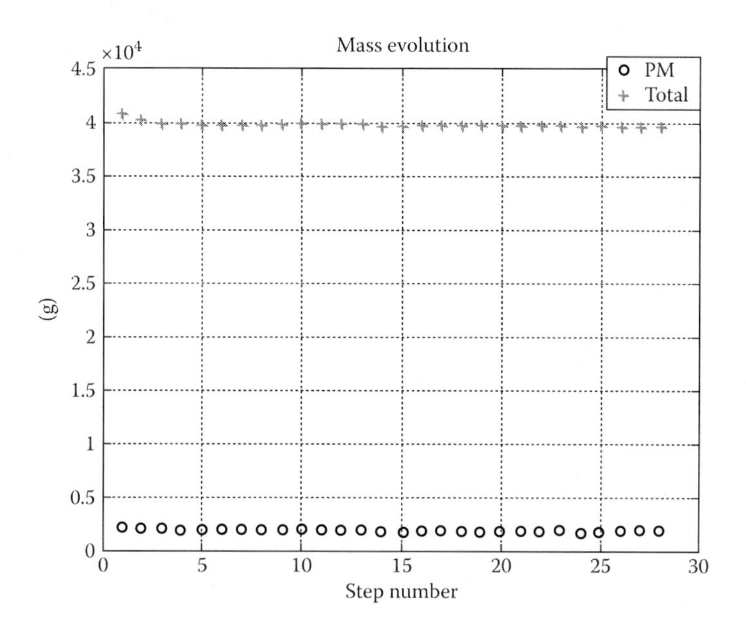

FIGURE 8.47 Active material weight evolution.

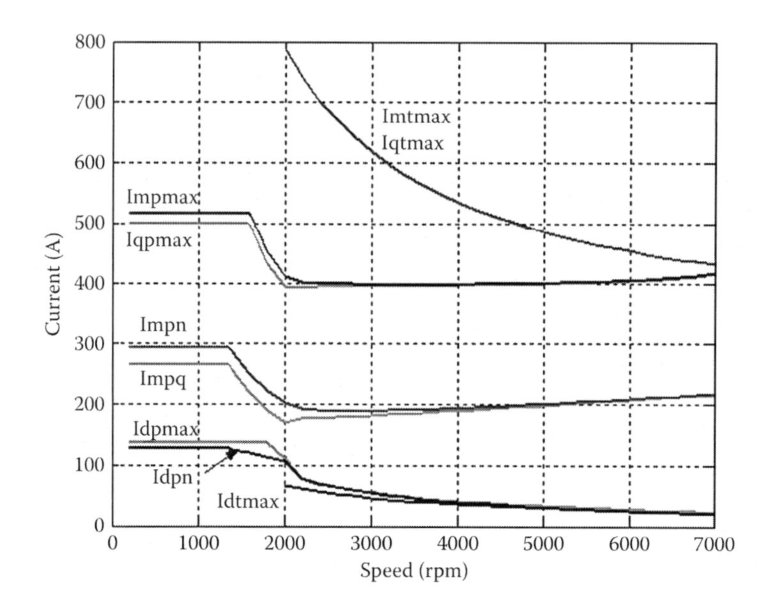

FIGURE 8.48 Current/speed curves family.

The results in the optimization case study have started with a good initial, general, design data; this explains why the optimization study did not change dramatically the machine geometry or performance.

8.12.4 FEM Validation without and with Rotor Segmentation

For cycle life optimization design case study shown in Section IV, FEM verifications are presented in what follows (Figures 8.49 through 8.54).

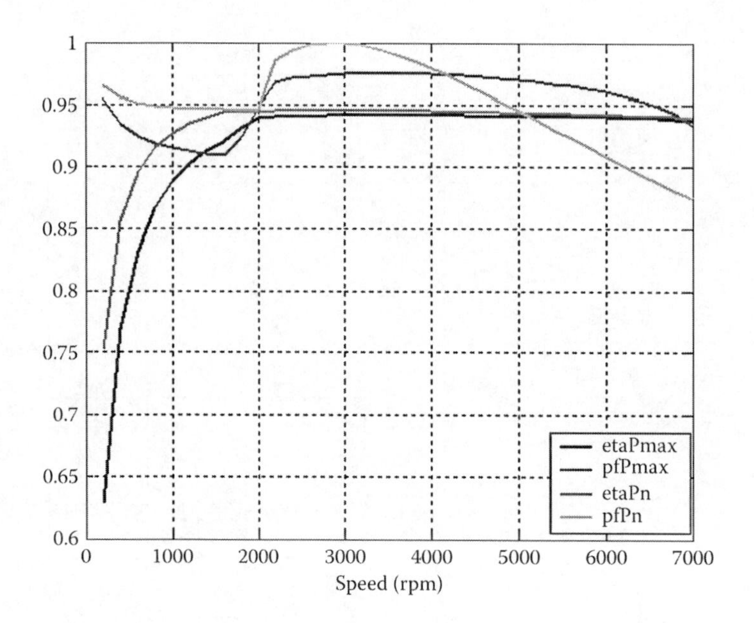

FIGURE 8.49 Efficiency and power factor for optimal cycle cost.

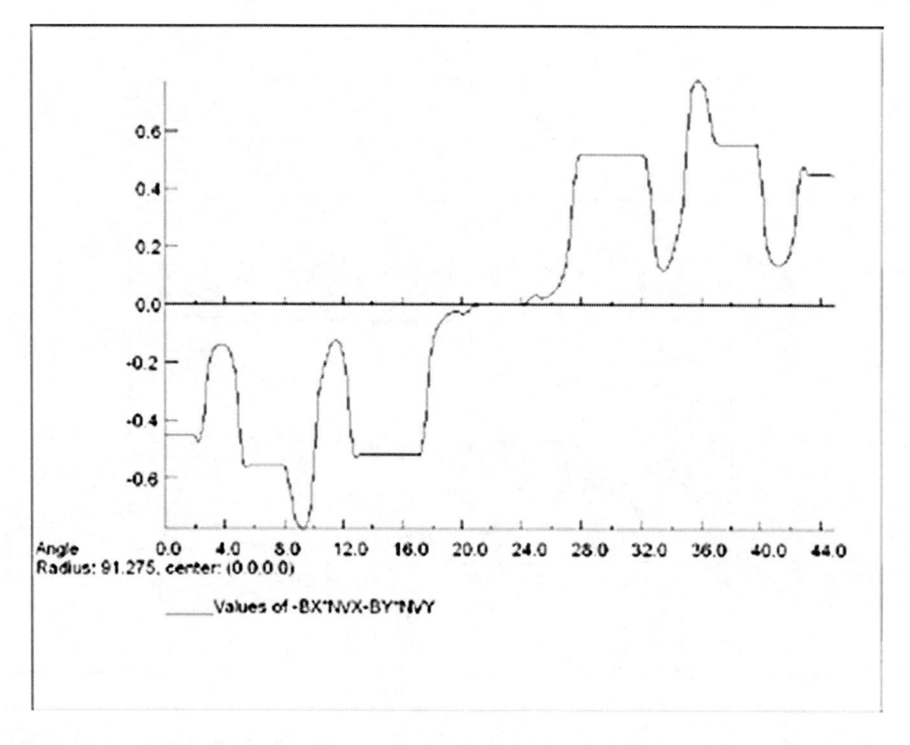

FIGURE 8.50 PM-airgap flux density (normal component) at $I_d = 0$ A, $I_q = 0$ A.

FEM validates within 5% the emf at maximum speed (450 V_{DC}) at 7000 rpm. The flux density in the air gap versus I_d current, d-axis magnetization inductances L_{dm}, flux densities critical values are validated by FEM within 10% despite of heavy magnetic saturation at peak torque.

The q-axis magnetization inductance $L_{qm}(L_{qm} < L_{dm})$, in presence of multiple flux barriers and magnetic saturation of axis d was higher by up to 25% in FEM. Because L_{dm} was also higher in FEM, the peak average

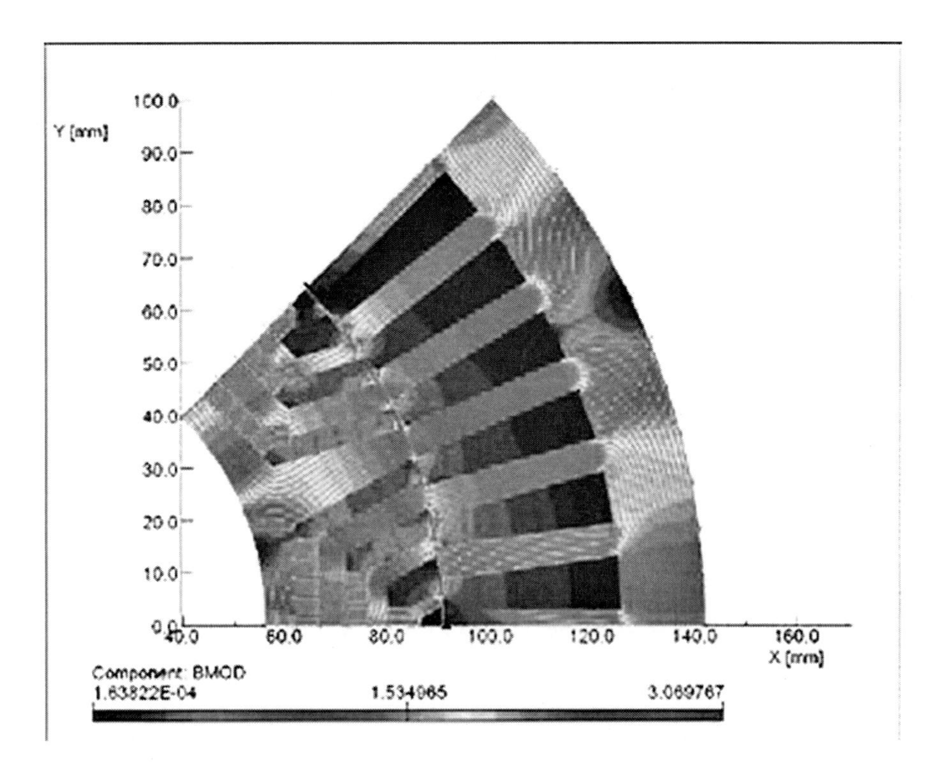

FIGURE 8.51 Flux density distribution and field line at maximum torque, initial position.

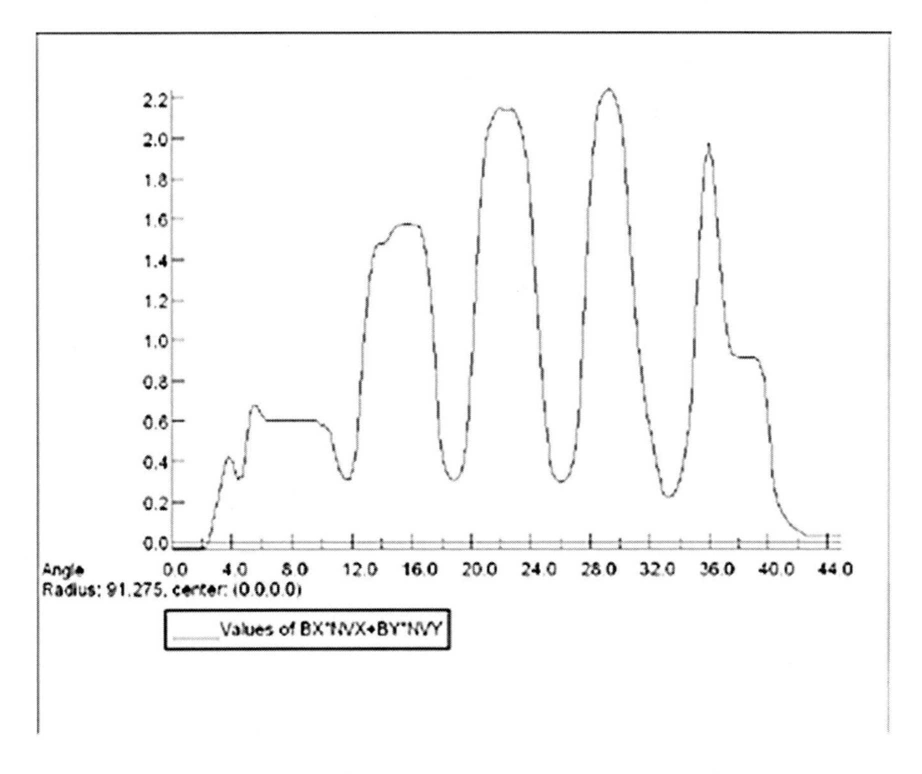

FIGURE 8.52 Airgap flux density (normal component) at maximum torque, initial position.

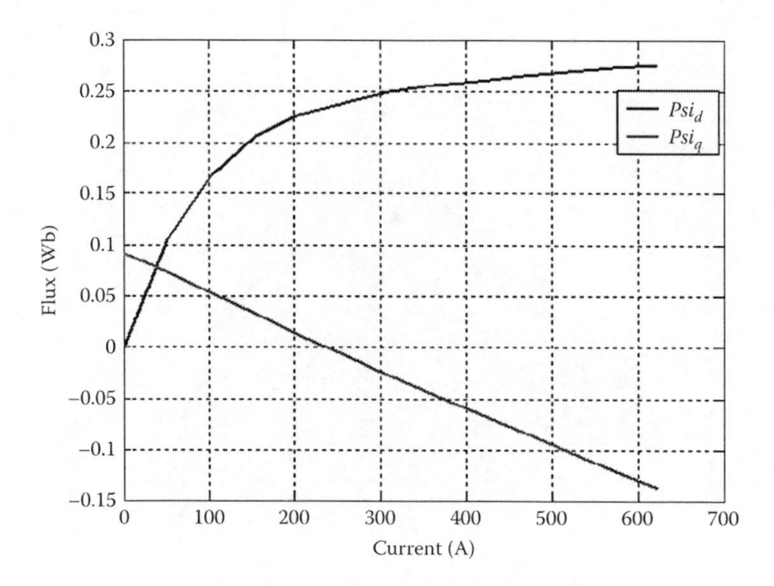

FIGURE 8.53 Linkage flux vs. current.

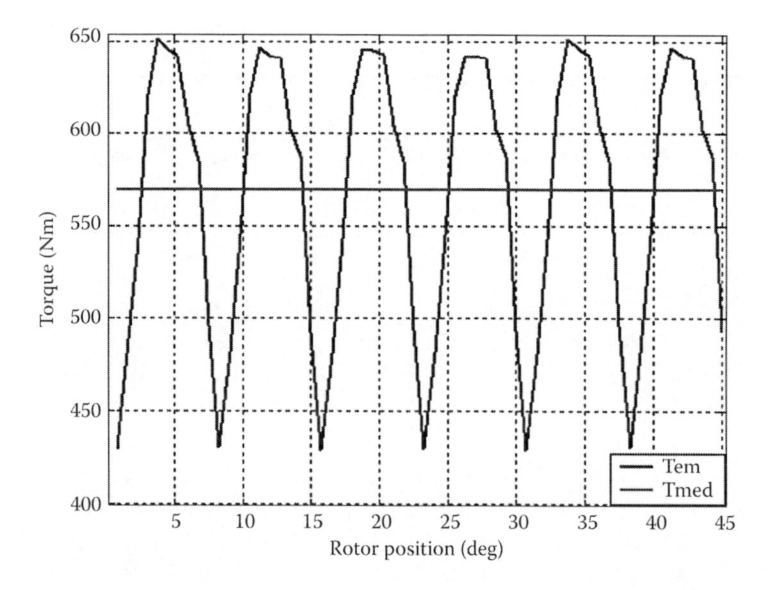

FIGURE 8.54 Torque vs. position.

torque was still obtained with an error of 5%. By making the rotor of two 2.75° axially shifted parts the peak average torque is brought to 590 Nm (almost the target 600 Nm) with torque pulsation reduced to less than 3% (Figure 8.54). The large torque pulsations as can be seen in the figure are due to heavy magnetic saturation, rather large stator slot opening (above 3 mm for 0.6 mm airgap); this was necessary to reach the 600 Nm peak torque with a restricted geometry, given through specifications; and due to long iron bridges in the rotor.

8.12.5 Dynamic Model and Vector Control Performance Validation

The dynamic *dq* model proposed in the dedicated MATLAB computer code (Figures 8.55 and 8.56) accounts for magnetic saturation and uses the parameter expressions from the analytical model. Also, the maximum torque per current curve and the axis magnetization curve are used in the *vector control*

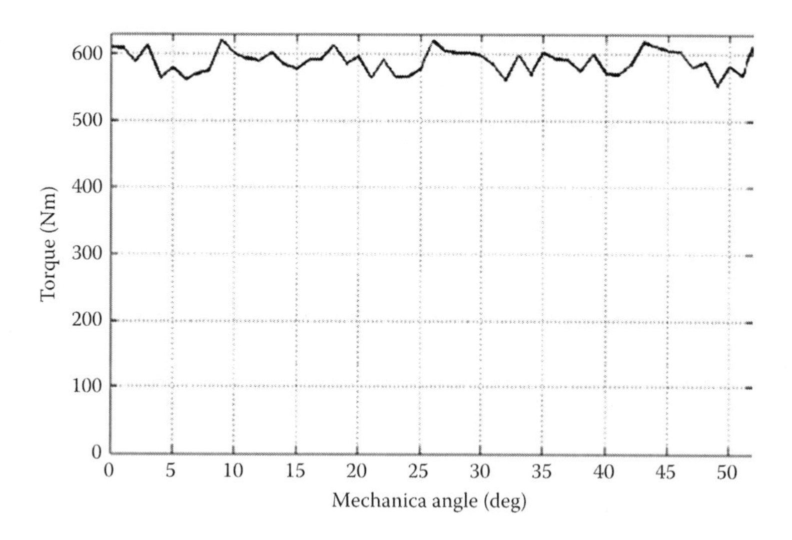

FIGURE 8.55 Torque pulsations for 520 A peak current and $\delta_q = 30°$ with Hiperco50 cores and shifted-segment rotor.

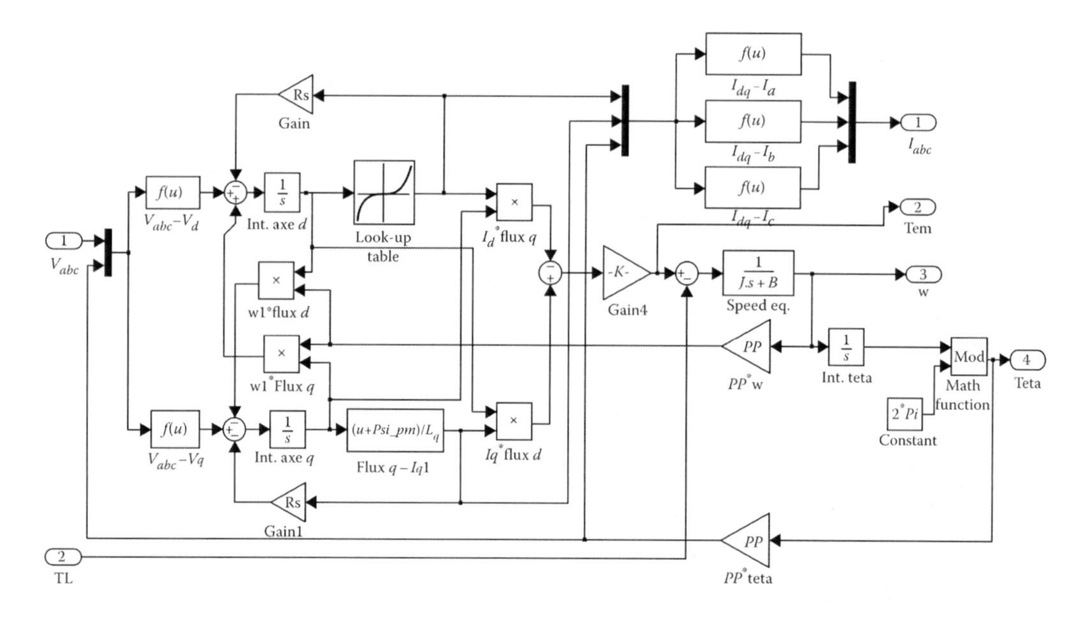

FIGURE 8.56 IPMSM dynamic model.

which, in fact, starts with maximum torque/current control whenever enough voltage is available and drifts automatically towards maximum torque/flux when full voltage use is necessary (*above base speed*). Therefore, flux weakening is implicit in the proposed vector control, accounting for magnetic saturation wide variation along axis *d*.

To produce 600 Nm peak torque at 520 A peak value/phase for the less than 40 kg active material machine, it was necessary to overcompensate a little the PM flux (without demagnetizing the PMs, because the leakage flux of stator in *q* axis is notable). In general this is considered an extreme measure and is not common for less demanding torque/weight specifications.

The transient and steady-state behavior of the proposed vector control system, as illustrated in Figures 8.57 and 8.58, above and below base speed, for maximum (100 kW) and continuous (50 kW) power up to

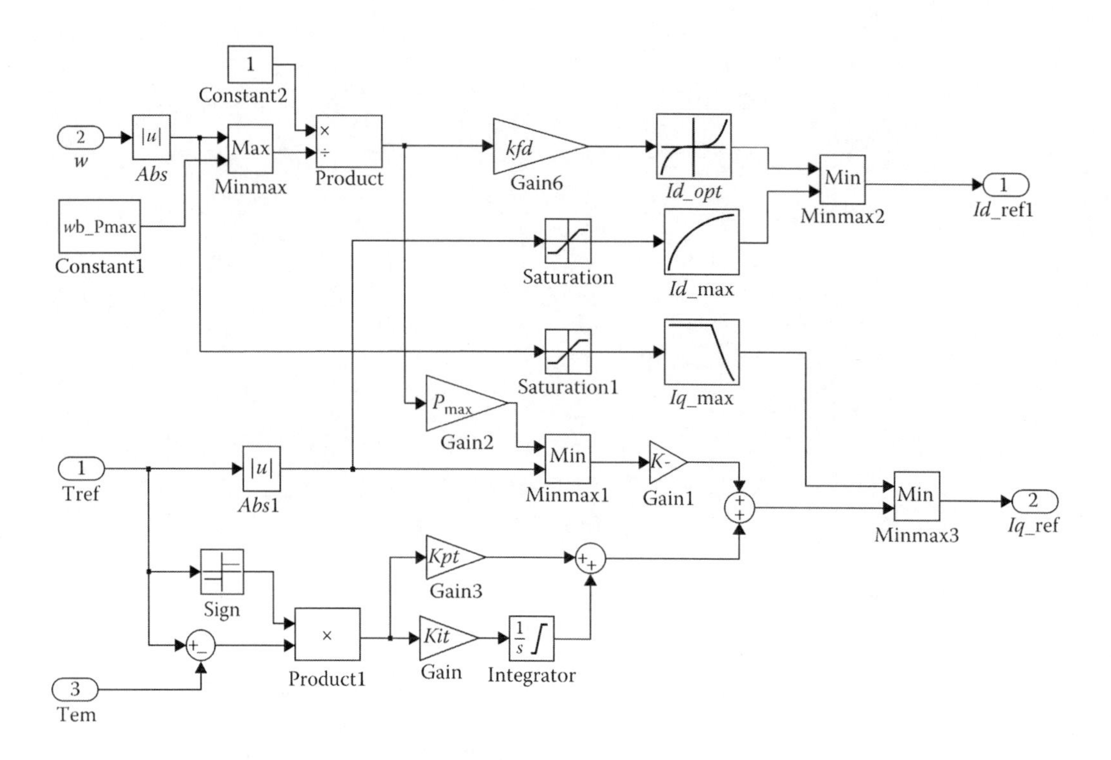

FIGURE 8.57 The references (flux and torque current) block.

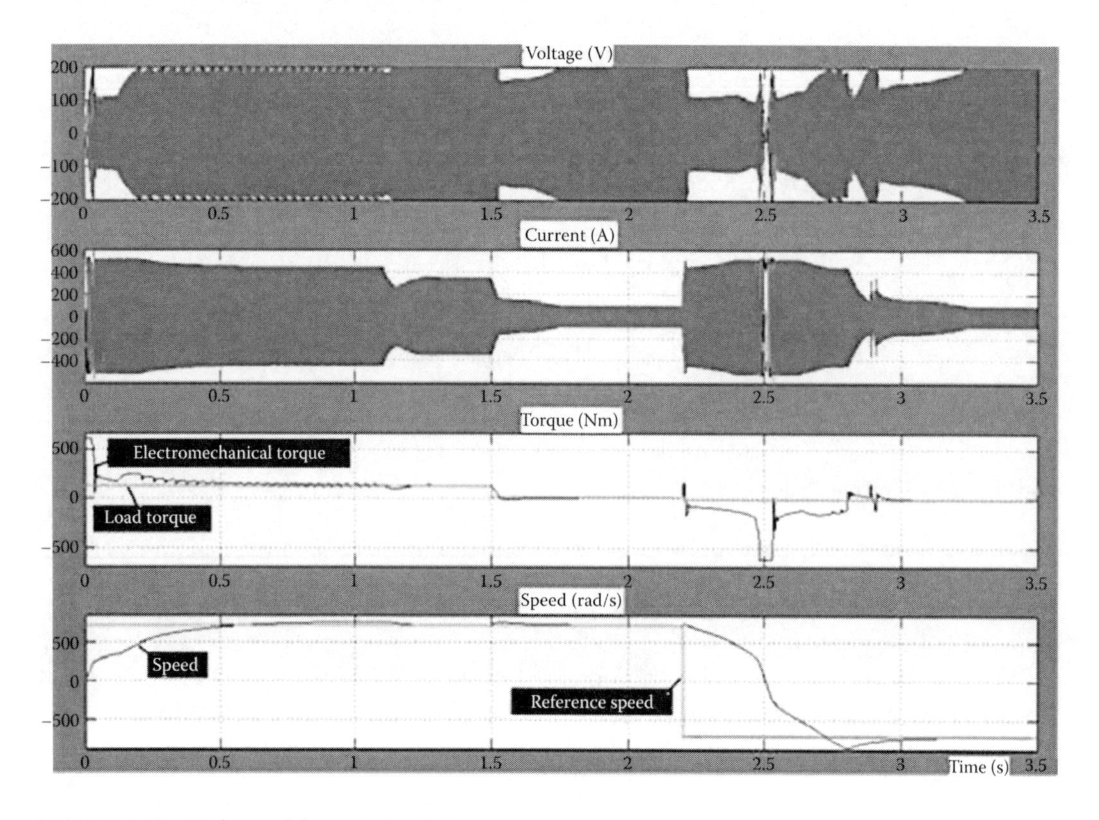

FIGURE 8.58 High-speed dynamic simulations.

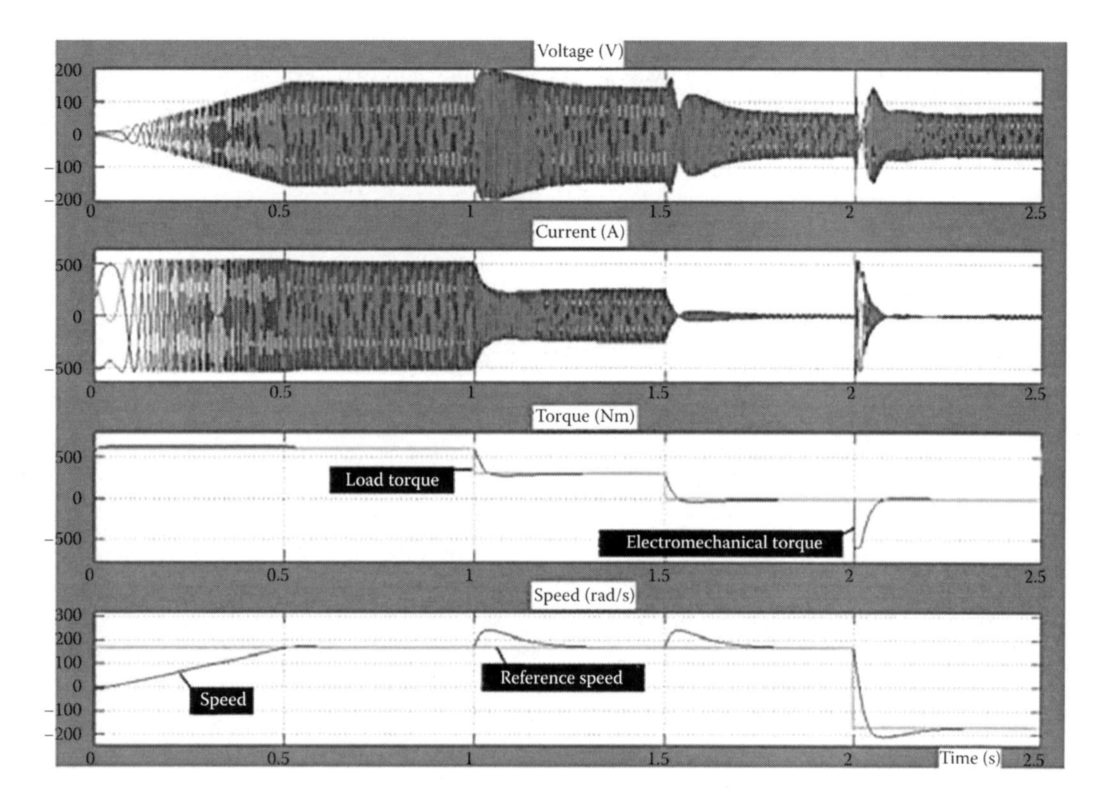

FIGURE 8.59 Low-speed dynamic simulations.

maximum speed (7000 rpm), shows that the drive is fully capable to meet the very challenging torque/speed envelopes. Stable and fast torque and speed response have been obtained (Figure 8.59).

8.13 Summary

- A high magnetic saliency ($L_d/L_q > 3$) RSM with multilayer interior PMs per rotor poles in axis q is called PM-assisted RSM or PM-RSM. It may also be called a special interior PM synchronous machine.
- The reluctance torque is predominant in comparison with the PM interaction torque, in peak torque production.
- Even low remnant flux density PMs ($B_r < 0.8$ T) and thus, lower-cost PMs may be used.
- In automotive applications, typical specifications ask for high start torque up to base speed ω_b, and then a wide constant power speed range $F = \omega_{max}/\omega_b > 4/1$ is required.
- The unique combination of reluctance and PM interaction torque makes the PM-RSM particularly suitable for a wide constant power speed range both in terms of system costs and losses.
- Typical rotor topologies contain two to four (five) flux barriers per half pole, which contains parallelepiped-shaped moderate (low)-cost PMs. The rotor core is made of conventional laminations. The saliency ratio $L_{dm}/L_{qm} \approx 3$–10, but it greatly varies with magnetic saturation level.
- Each flux barrier ends toward air gap with a flux bridge. The bridge thickness varies between 0.5 and 1.5 mm. The smaller the bridge, the better, in terms of rotor saliency L_{dm}/L_{qm}; but alternatively, the flux bridge thickness and length are limited by the mechanical stress allowed before rotor deformation at maximum speed and torque occurred.
- The total saliency is $L_d/L_q = 3$–7 for PM-RSM below 100 kW (peak torque below 400 Nm) when two to four flux barriers per half-pole are used. The stator windings leakage inductance reduces L_d/L_q to values smaller than L_{dm}/L_{qm}.

- FEM direct geometrical optimization of the rotor pole structure may lead to a uniform distribution of PM flux density in the air gap, that is, to a rather sinusoidal emf for a distributed stator winding with $q = 2,3$ slots per pole and phase.
- In addition, a prime number of flux bridges per pole of five or seven, leads to low cogging torque and pulsating torque with sinusoidal currents, even without stator (rotor) skewing.
- Even for heavy saturation levels, the air gap resultant flux linkage per phase is sinusoidal if magnetic saturation is uniform in the stator and rotor teeth and yokes.
- FEM analysis brings out the fact that magnetic saturation moves the maximum torque for given (high) current toward smaller i_d and larger i_q.
- The sinusoidal air gap flux per phase under load allows for full usage of *dq* models of PM-RSM for parameter, steady-state, transient, and control performance estimation.
- FEM analysis shows that magnetic saturation in axis q (where the PMs are located) is almost negligible, while it is important in axis d, especially for high torque operation; cross-coupling magnetic saturation may, thus, be neglected in many practical cases. Saturation in axis d is accounted for by a nonlinear $i_d(\Psi_d)$ function.
- Core losses in the PM-RSM occur both in the stator and the rotor. Only space harmonics in the air gap total flux distribution produce rotor core losses under steady state. Stator current time harmonics produce core losses both in the stator and the rotor. Both space and time harmonics of the air gap field produce eddy currents in the PMs on the rotor, especially at high speeds.
- From FEM static field analysis, the core losses may be calculated. An equivalent, slightly frequency-dependent, core resistance R_C is defined for use in the *dq* model.
- The *dq* model flux linkages $\Psi_d = L_d(i_d)i_d$, $\Psi_q = L_q i_q - \Psi_{PMq}$ lead to a resultant torque that is proportional to i_d. When torque changes sign (from motoring to generating), it is i_d rather than i_q that changes sign.
- The PMs in axis q produce a flux that opposes $L_q i_q$; that is, a demagnetizing armature reaction occurs. The current i_q should not change polarity.
- The PMs in axis q contribute to additional torque for less Ψ_q flux and thus, to high power factor in the machine.
- Generator no-load with nonzero core losses makes the *dq* output current $i_d = i_q = 0$ but the interior currents i_{0d}, $i_{0q} \neq 0$.
- With zero current losses ($R_C = \infty$), all *dq* currents in the *dq* model at no load are zero, and $V_{d00} = \omega_{r\Psi PMq}$ and $V_{q00} = 0$.
- The symmetrical steady-state short-circuit current, when $R_S \neq 0$, $R_C \neq \infty$, has two components i_{dsc3} and i_{qsc3}, and a corresponding torque. With zero losses, the ideal symmetrical short-circuit current $i_{sc3} = i_{qs0} = \Psi_{PMq/}L_q$.
- In most wide constant power speed range designs, $\Psi_{PMq} = L_q I_{srated}$ and thus, symmetrical short circuit is acceptable.
- Unsymmetrical short circuit leads to larger than i_{sc} currents with DC components that have to be treated, eventually, through quick, intentional, triggering of all power switches in the inverter to degenerate in symmetrical short circuit. This way, the inverter is protected.
- The design of PM-RSM for a wide speed range has to compromise machine (torque) and inverter (current) oversizing t_{over}, i_{over}. After smart normalization, only Ψ_{PMq} and L_d/L_q in P.U. remain as independent variables. In essence, design close to $\Psi_{PMq}/(L_q I_{rated}) \approx 1$ in (P.U.) leads to wide constant power speed range.
- With respect to base speed ω_b (full torque at full voltage for maximum torque per current), the minimum speed $\omega_{\min}$ may be chosen either above or under it. For given constant power speed range F, choosing a $\omega_{\min}/\omega_b > 1$ smaller machine and some inverter oversizing are required, but the emf at maximum speed limitation at 150% rules out such a situation with designs for $\Psi_{PMq}/(L_q{}^*I_{rated}) \approx 1$ P.U. In addition, the full voltage at $\omega_b < \omega_{\min}$ may lead to current regulation saturation. For $\omega_{\min}/\omega_b < 1$, full machine utilization is reached, but inverter current oversizing is larger.

- Uncontrolled generation (UCG) occurs even when the speed decreases to ω_{UCG}, below that for which $\omega_r \Psi_{PMq} < V_b$ and the PWM inverter power switches are turned off and diodes act alone. To avoid UCG, the maximum speed ω_{max} should be lower than ω_{UCG}:

$$\omega_{max} < \omega_{UCG} = \frac{2\sqrt{((L_d/L_q)-1)}}{\Psi_{PMq}(L_d/L_q)} * \omega_b; \quad \Psi_{PMq} \text{ in P.U.}$$

- It turns out that, for $F > 4$, only for very small Ψ_{PMq} values and low L_d/L_q ratios, UCG may be eliminated. Essentially, to avoid an unpractical design, UCG has to be eliminated through control (protection) means.
- For EHVs, the PM-RSM starter/alternator is connected to the battery through a PWM inverter. As the battery voltage varies with ±25%, it turns out that a voltage booster, through a DC–DC converter, may be beneficial. Even if the voltage booster acts from V_{bmin} to V_{bmax} of battery, the total ratings and silicon costs of a DC–DC converter plus PWM inverter are smaller than those for the inverter alone at V_{bmin}. Moreover, the total system losses in the machine plus converters are smaller with a DC voltage booster.
- In EHVs, torque, rather than speed, control is required both for motoring and generating.
- FOC and DTFC strategies are feasible.
- The control may be performed with or without a position encoder.
- The control has to produce first reference *dq* currents i_d^*, i_q^* versus torque and speed functions for motoring and generating in FOC, or reference flux vs. speed and torque for DTFC.
- Varying the *dq* current angle with speed from maximum torque per current criterion, below base speed, to maximum torque per flux criterion, at maximum speed, seems a good way for FOC.
- For DTFC, a flux and torque observer is required.
- The position feedback is used two times for vector rotation in FOC, while only in the flux observer for DTFC. For motion-sensorless control, position observers, capable of working from zero speed, are required.
- Initial position estimation is required with FOC, for safe starting.
- Signal injection rotor position observers were proven capable of safe zero and low-speed good performance. In combination with back-emf rotor motion observers, the whole speed range is covered for FOC [32–34].
- For DTFC, as only torque control at zero speed is required, even a rotor position and flux observer without signal injection may be adequate for the whole speed range, with initial position estimation [34].
- New features such as wide constant power speed range—at unity power factor is obtained with biaxial excitation in the rotor (PMs and DC) [35].

References

1. E.C. Lovelace, T.M. Jahns, J.L. Kirtley Jr., and J.H. Lang, An interior PM starter/alternator for automotive applications, Record of ICEM-1998, Istanbul, Turkey, 1998, vol. 3, pp. 1802–1808.
2. A. Vagati, A. Fratta, P. Gugliehmi, G. Franchi, and F. Villata, Comparisons of AC motor based drives for electric vehicle application, Record of PCIM-1999, Intelligent Motion, Europe, 1999, pp. 173–181.
3. E.C. Lovelace, T.M. Jahns, T.A. Keim, and J.H. Lang, Mechanical design considerations for conventionally-laminated high-speed interior PM synchronous machine rotors, Record of IEEE-IEMDC-2001, 2001, pp. 163–169.
4. I. Boldea, *Reluctance Synchronous Machines and Drives*, Oxford University Press, Oxford, U.K., 1996.

5. W.L. Soong, D.A. Staton, and T.J.E. Miller, Design of new axially-laminated interior permanent magnet motor, Record of IEE–IAS, *Annual Meeting*, 1993, pp. 27–36.
6. Y. Honda, Rotor design optimization of a multilayer interior PM synchronous motor, *Proc. IEEE*, EPA-145(2), 1998, 119–124.
7. A. Vagati, A. Canova, M. Chiampi, M. Pastorelli, and M. Repeto, Design refinement of synchronous reluctance motors through finite element analysis, *IEEE Trans.*, IA-36(4), 2000, 1094–1102.
8. I. Boldea, L. Tutelea, and C.I. Pitic, PM assisted reluctance synchronous motor/generator (PM-RSM) for mild hybrid vehicle, Record of OPTIM-2002, Poiana Brasov, Romania, 2002, pp. 383–388.
9. P. Guglielmi, G.M. Pellegrino, G. Griffero, S. Mieddu, G. Girando, and A. Vagati, Conversion concept for advanced autonomy and reliability scooter, Record of IEEE–IECON, 2003, pp. 2883–2888.
10. E.C. Lovelace, T. Jahns, T. Keim, J. Lang, D. Wentzloff, F. Leonardi, J. Miller, and P. McCleer, Design and experimental verification of a direct-drive interior PM synchronous machine using a saturable parameter model, Record of IEEE–IAS-2002, *Annual Meeting*, 2002.
11. D.H. Kim, I.H. Park, J.H. Lee, and C.E. Kim, Optimal shape design of iron core to reduce cogging torque of IPM motor, *IEEE Trans.*, MAG-39(3), 2003, 1456–1459.
12. G.H. Kang, J.P. Hong, and G.T. Kim, Analysis of cogging torque in interior permanent magnet motor by analytical method, *J. KIEE*, 11-B(1), 2001, 1–8.
13. K. Yamazaki, Torque and efficiency calculation of an interior permanent magnet motor considering harmonic iron losses of both stator and rotor, *IEEE Trans.*, MAG-39(3), 2003, 1460–1463.
14. T.A. Jahns, Component rating requirements for vehicle constant power operation of IPM synchronous machine drives, Record of IEEE–IAS-2000, *Annual Meeting*, Rome, Italy, 2000, pp. 1697–1704.
15. R.F. Schiferl and T.A. Lipo, Power capabilities of salient pole permanent magnet synchronous motor variable speed applications, *IEEE Trans.*, IA-26(1), 1990, 115–123.
16. B.A. Welchko, T.A. Jahns, W.C. Long, and J.M. Nagashima, IPM synchronous machine drive response to symmetrical and asymmetrical shortcircuit faults, *IEEE Trans.*, EC-28(2), 2003, 291–298.
17. T.M. Jahns, Uncontrolled generator operation of interior PM synchronous machine following high-speed inverter shutdown, Record of IEEE–IAS, 1998.
18. W.L. Soong, N. Ertugrul, E.C. Lovelace, and T.M. Jahns, Investigation of interior permanent magnet off-set coupled automotive integrated starter/alternator, Record of IEEE–IAS-2001, *Annual Meeting*, 2001.
19. I. Boldea, L. Ianosi, and F. Blaabjerg, A modified direct torque control (DTC) of reluctance synchronous motor sensorless drive, *EMPS J.*, 28(1), 2000, 115–128.
20. G.D. Andreescu, Robust sliding mode based observer for sensorless control of PMSM drives, Record of EPE–PEMC-1998, Prague, Czech Republic, 1998, vol. 6, pp. 172–177.
21. M. Patel, T. O'Meara, J. Nagashinia, and R.D. Lorenz, Encoderless IPM drive system for EH-HEV propulsion applications, Record of EPE-2001, Graz, Austria, 2001.
22. M.M. Degner and R.D. Lorenz, Wide band width flux position and velocity estimation in AC machines at any speed (including zero) using high multiple saliences, Record of EPE-1997, Trondheim, Norway, 1997, pp. 1530–1535.
23. M.M. Corely and R.D. Lorenz, Rotor position and velocity estimation for a permanent magnet synchronous machine at standstill and high speed, *IEEE Trans.*, IA-34(4), 1998, 784–789.
24. T. Aihara, A. Tobu, T. Yanase, A. Mashimoto, and K. Endo, Sensorless torque control of salient pole synchronous motor at zero speed operation, *IEEE Trans.*, PE-14(1), 1999, 202–208.
25. M. Schroedl, Sensorless control of AC machines at low speed and standstill based on the INFORM method, Record of IEEE–IAS-1996, *Annual Meeting*, 1996, vol. 1, pp. 270–277.
26. A. Consoli, G. Scarcella, and A. Testa, Sensorless control of PM synchronous motors at zero speed, Record of IEEE–IAS-1999, *Annual Meeting*, 1999, vol. 2, pp. 1033–1040.

27. T. Noguchi, K. Yamada, S. Kondo, and I. Takahashi, Initial rotor position estimation of sensorless PM synchronous motor with no sensitivity to armature resistance, *IEEE Trans.*, IE-45(1), 1998, 118–125.
28. D.W. Chung, J.K. Kang, and S.K. Sul, Initial rotor position detection of PMSM at standstill without rotational transducer, Record of IEEE–IEMDC-1999, 1999, pp. 785–787.
29. J.I. Ha, K. Ide, T. Sawa, and S.K. Sul, Sensorless position control and initial position estimation of an interior permanent magnet motor, Record of IEEE–IAS-2001, 2001, pp. 2607–2613.
30. H. Kim, K.K. Huh, H. Harke, J. Wai, R.D. Lorenz, and T. Jahns, Initial rotor position estimation for an integrated starter alternator IPM synchronous machine, Record of EPE-2003, Toulouse, France, 2003.
31. L. Tutelea, A. Moldovan (Popa), and I. Boldea, 50/100 kW, 1350–7000 rpm (600 Nm, 40 kg) PM-assisted reluctance synchronous machine: Optimal design with FEM validation and vector control, Record of IEEE xplore OPTIM, Bran, Romania, 2014.
32. Y. Yeong, R.D. Lorenz, T.M. Jahns, and S.K. Sul, Initial rotor position estimation of interior PMSM using carrier-frequency injection methods, Record of IEEE–IEMDC-2003, 2001, vol. 2, pp. 1218–1223.
33. M. Linke, R. Kennel, and J. Holtz, Sensorless speed and position control of synchronous machines using alternating carrier injection, Record of IEEE–IEMDC-2003, 2003, vol. 2, pp. 1211–1217.
34. S.C. Agarlita, I. Boldea, and F. Blaabjerg, High frequency-injection-assisted active-flux based sensorless vector control of RSMs with experiments from zero speed, *IEEE Trans.*, IA-48(6), 2012, 1931–1939.
35. V. Coroban-Schramel, I. Boldea, G.D. Andreescu, and F. Blaabjerg, Active-flux based motion sensorless vector control of biaxial-excitation generator/motor for automobiles (BEGA), *IEEE Trans.*, IA-47(2), 2011, 812–819.

9

Switched Reluctance Generators and Their Control

9.1 Introduction

Switched reluctance generators (SRGs) are double-saliency electric machines with nonoverlapping stator multiphase windings and passive rotors. They may also be assimilated with stepper motors with position-controlled pulsed currents. Multiphase configurations are required for smooth power delivery and eventual self-starting and motoring, if the application requires it.

SRGs were investigated mainly for variable speed operation as starter/generators on hybrid electric vehicles (HEVs) and as power generators on aircraft and for wind energy conversion. They may also be considered for super-high-speed gas turbine generators from kilowatt to megawatt (MW) power per unit.

As SRGs lack permanent magnets (PMs) or rotor windings, they are of low cost, easy to manufacture, and can operate at high speeds and in high-temperature environments.

In vehicular applications, an SRG is required to perform over a wide speed range to comply with the internal combustion engine (ICE) that drives it. For wind energy conversion, limited speed range is needed to extract additional wind energy at lower mechanical stress in the system.

Aware of the very rich literature on SRMs [1,2], we will treat in this chapter the following aspects deemed representative:

- Practical topologies and principles of operation
- Characteristics for performance evaluation
- Design for wide constant power range
- Converters for SRG
- Control of SRG as starter/generator with and without motion sensors

The existence of a handful of companies that fabricate and dispatch SRMs [3] and vigorous recent proposals of SRGs as starters/alternators for automobiles and aircraft (up to 250 kW per unit) seem sufficient reason to pursue the SRG study within a separate chapter such as this one.

9.2 Practical Topologies and Principles of Operation

A primitive single-phase SRG(M) configuration with two stator and two rotor poles is shown in Figure 9.1a and b. It illustrates the principle of reluctance machine, where torque is produced through magnetic anisotropy. The stored magnetic energy (W_e) or coenergy (W_c) varies with rotor position to produce torque:

$$T_e = \left(\frac{\partial W_c(i,\theta_r)}{\partial \theta_r}\right)_{i=\text{cons}} = -\left(\frac{\partial W_c(\Psi,\theta_r)}{\partial \theta_r}\right)_{\Psi=\text{cons}} \tag{9.1}$$

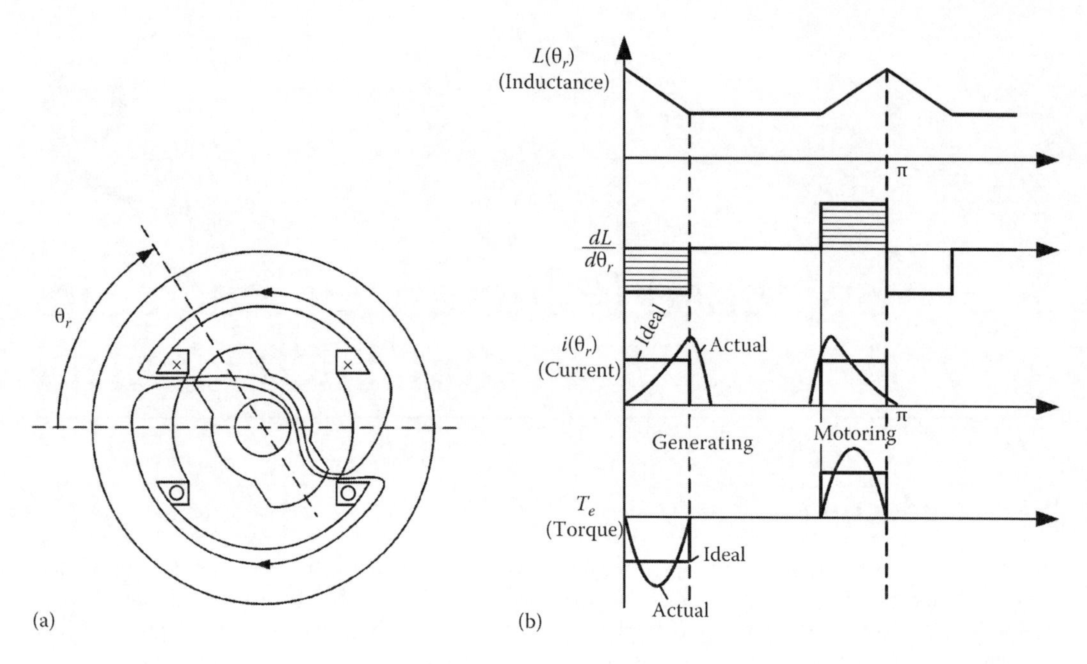

FIGURE 9.1 Primitive switched reluctance generator (SRG)(M) with (a) two stator and rotor poles and (b) ideal waveforms.

$$W_c = \int_0^i \Psi\, di; \quad W_e = \int_0^\Psi i\, d\Psi \tag{9.2}$$

In the absence of magnetic saturation,

$$\Psi = L(\theta_r) \cdot i \tag{9.3}$$

and thus,

$$W_e = W_c = \frac{1}{2} L(\theta_r) \cdot i^2 \tag{9.4}$$

If, further on, we suppose that the phase inductance varies linearly with rotor position, from its maximum to its minimum value, for constant current pulse, the torque is constant over the active rotor position range:

$$T_e = \frac{1}{2} i^2 \frac{dL(\theta_r)}{d\theta_r} = \frac{c(\theta_r)}{2} i^2 = \text{const} \tag{9.5}$$

As $c > 0$ for motoring and $c < 0$ for generating, it is evident that the polarity of the current is not relevant for torque production. It is also clear that because there are N_r poles per rotor, there will be N_r energy cycles for motoring or generating per mechanical revolution per phase.

For single-phase configurations, large pulsations in torque are inevitable, and, as a motor, self-starting from any rotor position is impossible without additional topology changes (a stator parking PM or rotor pole asymmetric air gap).

Actual current pulses (Figure 9.1b) may be made to rectangular shape at low speeds through adequate direct current (DC) output voltage chopping. At high speeds, single pulse operation is inevitable because the electromagnetic force (emf) surpasses the input voltage (Figure 9.1b).

While the instantaneous torque may be calculated from Equation 9.1, the average torque per phase may be determined from the total energy per cycle W_{mec}, multiplied by the number of cycles for revolution, $m \cdot N_r$, and divided by 2π radians:

$$(T_{ave})_{single\ phase} = \frac{m \cdot N_r \cdot W_{mec}}{2\pi}; \quad m\text{-phases} \tag{9.6}$$

The energy per cycle emerges from the family of flux/current/position $\Psi(i, \theta_r)$ curves (Figure 9.2).

As shown in Figure 9.2, magnetic saturation plays an important role in average torque production.

The energy conversion ratio (ECR) is as follows:

$$ECR = \frac{W_{mec}}{W_{mec} + W_{mag}} \geq 0.5 \tag{9.7}$$

For an SRG with the same maximum flux and peak current but larger air gap, when unsaturated, the ECR is around 0.5. For smaller air gap and saturated SRGs, it is larger (Figure 9.2).

SRGs require notably smaller air gaps than PM machines to reach magnetic saturation at small currents. The same small air gap, however, leads to torque pulsations, notable vibration, and noise problems, due to large local radial forces.

Three- and four-phase configurations have become commercial for SRM drives due to their self-starting capability from any rotor position and torque (power) pulsation reduction opportunities through adequate current/position profiling. The basic 6/4, 8/6 three-phase and, respectively, four-phase topologies are shown in Figure 9.3a and b.

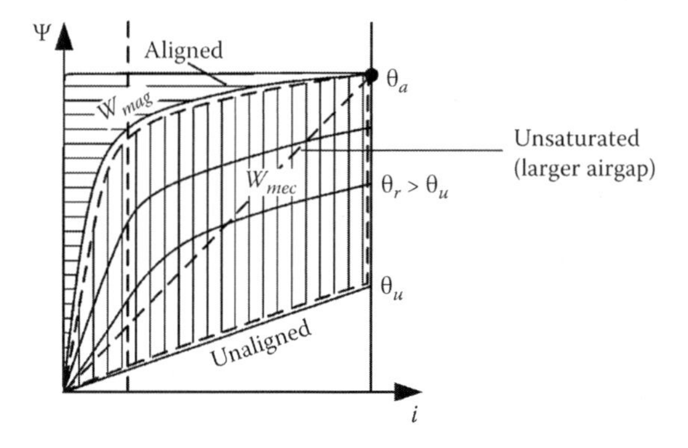

FIGURE 9.2 Magnetization curves.

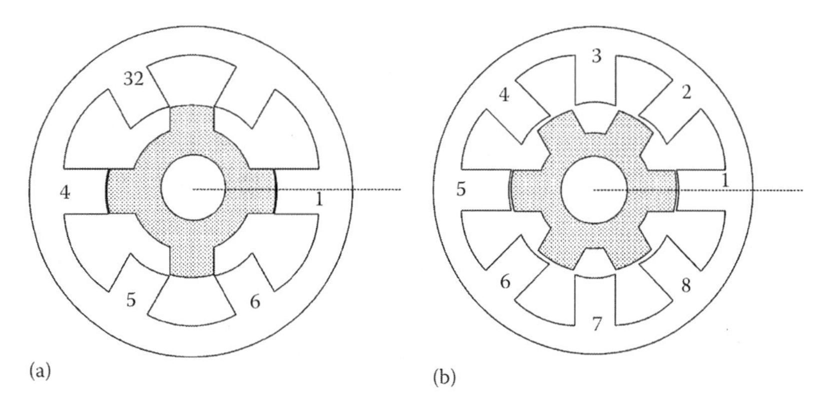

FIGURE 9.3 (a) Three- and (b) four-phase switched reluctance generators (SRGs) motors (Ms) with 6/4 and 8/6 stator/rotor pole combinations.

Ideal phase inductances versus rotor position for the three- and four-phase machines are shown, respectively, in Figure 9.4a and b. It should be noted that with the three-phase machines, only one phase produces positive (or negative) torque at a time ($dL/d\theta_r \neq 0$), while two phases are active at all times in the four-phase machine. Low torque pulsations through adequate current waveform control with phase torque sharing are, thus, more feasible with four phases. To increase the frequency of the pulsations, the number of rotor and stator poles should be increased.

However, the energy conversion tends to deteriorate above a certain number of poles, for given rotor (stator) outer diameter, due to flux fringing and increased rotor core losses. In general, three or four phases are used, but the number of stator and rotor poles may be increased so as to have more such units per stator periphery:

$$N_s/N_r = 6/4, 12/8, 18/12, 24/16, \ldots$$
$$N_s/N_r = 8/6, 16/12, 24/18, \ldots, 32/24 \quad (9.8)$$

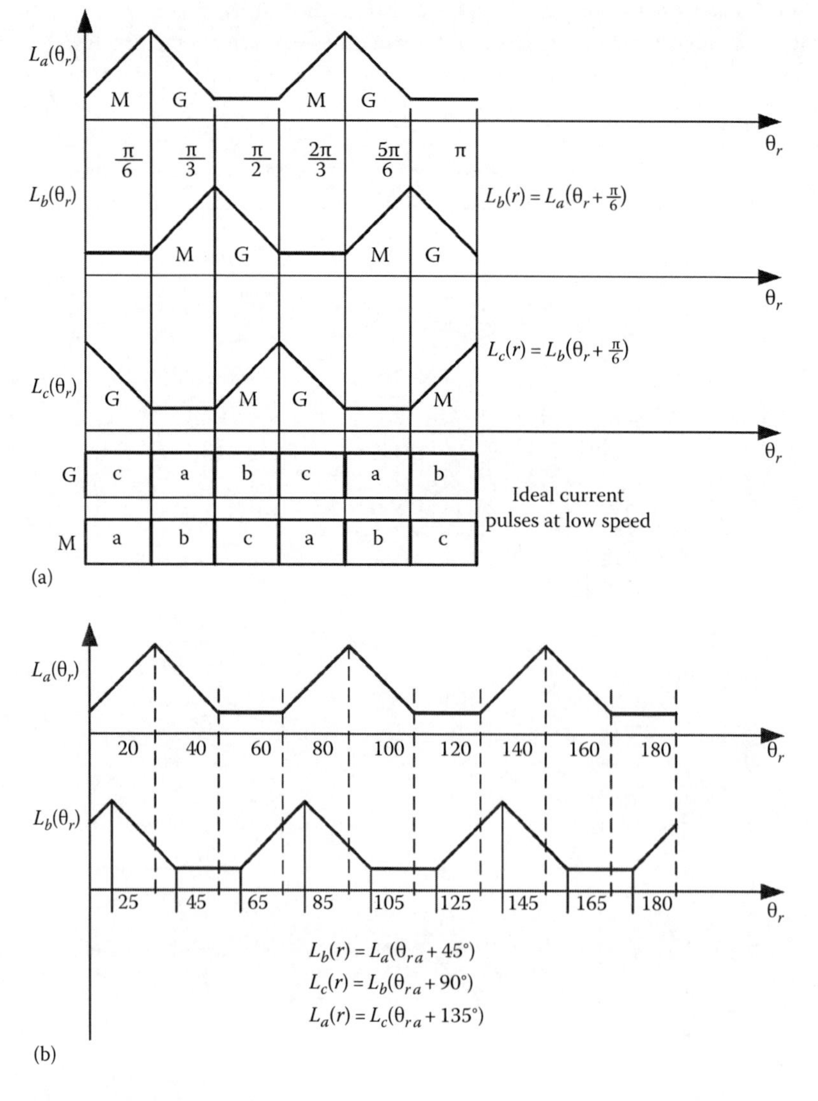

FIGURE 9.4 Ideal phase inductance/position dependence for the (a) three-phase and (b) four-phase switched reluctance generators (SRGs) motors (Ms).

An even number of stator sections (pole pairs) is appropriate when dual-output (two-channel) SRG operation is required.

To reduce the interaction flux between the two sections, the sequence of phase pole polarities along the rotor periphery (for a three-phase 12/8 pole combination) should be N N S S and *not* N S N S standard sequence. However, in this case, the flux in the rotor and stator yokes of the two channels adds up, and thus thicker yokes are required. The standard sequence per phase N S N S is less expensive, though more interference between channels is expected.

Returning to the principle of operation, we should note that for motoring, each phase should be connected when the phase inductance is minimal and constant, because, in this case, the emf is zero at any speed, and thus the phase voltage equation reduces to the following:

$$V_{DC} = R \cdot i + L_u \frac{di}{dt} \tag{9.9}$$

Neglecting the phase resistance voltage drop, the flux accumulated in the phase, Ψ, at constant speed n (rps) is

$$\Psi = \int L_u \frac{di}{dt} dt = \int V_{DC}\, dt \approx V_{DC} \frac{\theta_W}{2\pi n} \tag{9.10}$$

Here, θ_W is the mechanical dwell (conduction) angle of the phase when one voltage pulse is applied. The maximum (ideal) value of θ_W is π/N_r.

In most designs, the value of the maximum flux Ψ_{max} is used to calculate the base speed of the SRG:

$$2\pi n_b = \frac{V_{DC}\theta_W}{\Psi_{max}} \tag{9.11}$$

θ_W is in mechanical radians. This is, in fact, equivalent to the ideal standard condition that the emf equals phase voltage for constant current and zero phase resistance.

The single voltage pulse operation, characteristic of high speeds, is illustrated in Figure 9.5a through d (upper part). The low-speed operation appears in the lower part of Figure 9.5a (current chopping). While current chopping is typical for motoring below base speed, generating is performed, in general, above base speed in the single voltage pulse mode. In special cases, to reduce regenerative braking torque, a pulse-width modulator (PWM) may also be used for the generator mode.

To vary the generator output, only the angles θ_{on} and θ_c may be varied, in general. The turn-on angle θ_{on} may be advanced at high speeds, for both motoring and generating, to produce more torque (or power). The negative voltage pulses refer to the so-called hard switching, where both controlled power switches T_1 and T_2 (Figure 9.5b) are turned off at the same time, and the free-wheeling diodes become active.

The energy cycle is traveled from A to B for motoring and from B to A for generating (Figure 9.5c). The PWMs of voltage effects are shown in Figure 9.5d.

Increasing the energy cycle area W_{mec} means increasing the torque. This is possible by adding a diametrical DC-fed coil to move the energy cycle to the right in Figure 9.5d. In essence, the machine is no longer totally defluxed after each energy cycle. Alternatively, the continuous current control in a phase would lead to similar results, though at the price of additional losses, in both cases.

Besides torque density and losses, which refer essentially to machine size and goodness, the kilowatt to peak kilovolt ampere (kW/peak kVA) ratio defines the ratings of the static converter needed to control the SRG(M).

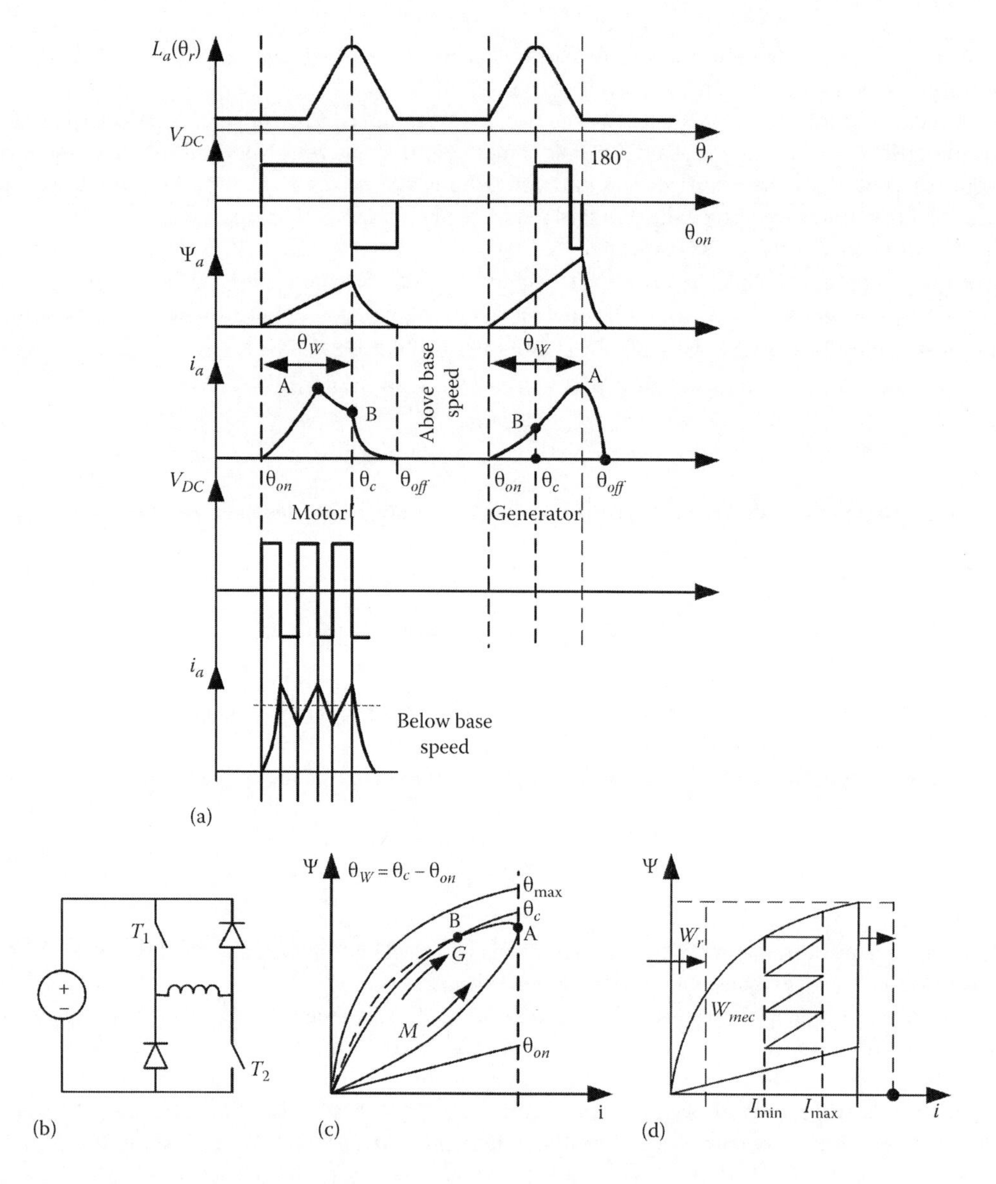

FIGURE 9.5 High-speed single voltage pulse and low-speed pulse-width modulator (PWM) voltage pulse operation: (a) waveforms, (b) phase converter, (c) single pulse energy cycle, and (d) energy cycle with PWM.

9.2.1 kW/Peak kVA Ratio

It was shown [1] that the kW/peak kVA ratio for SRG(M) is as follows:

$$\text{kW/peak kVA} \approx \frac{\alpha_s \cdot N_r \cdot Q}{8\pi} \tag{9.12}$$

where $\alpha_s = 0.4–0.5$ is the stator pole ratio and Q is as follows:

$$Q \approx C_{m,g}\left(2 - \frac{C_{m,g}}{C_s}\right) \tag{9.13}$$

C_m is the ratio between the active dwell angle θ_{wu} and the stator pole span angle $\beta_s \cdot C_m = 1$ only at zero speed, and then it decreases with speed and reaches values of 0.6–0.7 at base speed.

For the generator mode,

$$C_g = \left(1 - \frac{\theta_{wu}}{\beta_s}\right) \tag{9.14}$$

where θ_{wu} is the angle from phase turnoff (after magnetization), when active power delivery starts. Again, $C_g = 0.6–0.7$ should be considered acceptable.

The coefficient C_s [1] is as follows:

$$C_s = \frac{\lambda_u - 1}{\lambda_u \cdot \sigma - 1}; \quad \lambda_u = \frac{L_a^u}{L_u} \approx 4\text{–}10; \quad \sigma = \frac{L_a^s}{L_a^u} = 0.25\text{–}0.4 \tag{9.15}$$

where

- L_a^u is the aligned unsaturated inductance per phase
- L_u is the unaligned inductance
- L_a^s is the aligned saturated inductance

The peak power S of the switches in the SRG(M) converter is as follows:

$$S = 2 \cdot m_1 \cdot V_{DC} \cdot I_{peak} \tag{9.16}$$

For an inverter-fed IM (or alternating current [AC] machine),

$$\text{kW/peak kVA} \approx \frac{3 \cdot V_{DC} \cdot I_{peak} \times PF}{\pi \cdot K \cdot 6 \cdot V_{DC} \cdot I_{peak}} = \frac{3 \times PF}{\pi \cdot 6 \cdot K} \tag{9.17}$$

where K is the ratio between peak current waveform value and its fundamental peak value. For the six-pulse mode of the PWM converter $K = 1.1–1.15$. PF is the power factor for the fundamental.

Example 9.1

Consider a 6/4 three-phase SRG(M) with $\sigma = 0.3$, $\lambda_u = L_a^u / L_a = 8$, $C_g = 0.8$, $\alpha_s = 0.45$, and calculate the kW/peak kVA ratio. Compare it with an induction machine (IM) drive with $PF = 0.81$ and $K = 1.12$.

From Equation 9.15,

$$C_s = \frac{\lambda_u - 1}{\lambda_u \sigma - 1} = \frac{8 - 1}{8 \cdot 0.3 - 1} = 0.5$$

The value of Q comes from Equation 9.13:

$$Q \approx C_g \cdot \left(2 - \frac{C_g}{C_s}\right) = 0.8 \cdot \left(2 - \frac{0.8}{5}\right) = 1.472$$

Finally, from Equation 9.12,

$$(\text{kW/peak kVA})_{RSG} = \frac{0.45 \cdot 4 \cdot 1.472}{8 \cdot \pi} \cong 0.1055$$

For the IM drive (Equation 9.17),

$$(\text{kW/peak kVA})_{IM} = \frac{3 \cdot 0.85}{6 \cdot \pi \cdot 1.12} = 0.1208$$

For equal active power and efficiency, the IM requires from the converter about 10%–15% less peak kVA rating.

When the cost of the converter per SRG cost is large, larger system costs with the SRG(M) are expected.

Note that the kW/peak kVA as defined in this example is not equivalent to *PF*, but it is a key design factor when the converter rating and costs are considered.

An equivalent *PF* for SRG(M) may be defined as follows [4]:

$$(PF)_{SRM} = \frac{\text{output shaft power}}{\text{input rms volt} * \text{ampere}}$$

For given power, this *PF* varies with speed, and for the very best designs in Reference 4, it is above 0.7 and up to 0.86.

However, it is the peak value of current, not its RMS value that determines the converter kVA rating.

From this viewpoint, AC drives seem slightly superior to SRG(M)s.

9.3 SRG(M) Modeling

It was proven through detailed finite element method (FEM) analysis that the interaction between phases in standard SRG is minimal. Consequently, the effects of various phases may be superposed:

$$V_{a,b,c,d} = R_s i_{a.b,c,d} + \frac{d\Psi_{a,b,c,d}(\theta_r, i_{a,b,c,d})}{dt} \tag{9.18}$$

A four-phase SRG(M) is considered here.

Only the family of flux/current/position curves for one phase is required, as periodicity with π/N_s exists. The $\Psi(\theta_r, i)$ curves may be obtained from experiments or through calculation via analytical methods or FEM. The torque per phase (Equations 9.1 and 9.2) becomes

$$T_{e,a,b,c,d} = \frac{\partial}{\partial \theta_r} \int_0^{i_{a,b,c,d}} \Psi_{a,b,c,d}(\theta_r, i_{a,b,c,d}) di_{a,b,c,d} \tag{9.19}$$

The total torque is

$$T_e = \sum_{a,b,c,d} T_{e,a,b,c,d} \tag{9.20}$$

The motion equations are

$$J 2\pi \frac{dn}{dt} = T_e - T_{load}; \quad \frac{d\theta_r}{dt} = 2\pi n \tag{9.21}$$

The phase i voltage Equation 9.18 may be written as follows:

$$V_i = R_s i_i + \frac{\partial \Psi_i}{\partial i}\frac{\partial i_i}{\partial t} + \frac{\partial \Psi_i}{\partial \theta_r}\frac{\partial \theta_r}{\partial t} \quad (9.22)$$

The transient inductance L_{ti} is defined as follows:

$$L_{ti} = \frac{\partial \Psi_i}{\partial i} = L_i(\theta_r, i_i) + i_i \frac{\partial L_i(\theta_r, i_i)}{\partial i_i}; \quad L_i = \frac{\Psi_i}{i_i} \quad (9.23)$$

The last term in Equation 9.22 represents a pseudo-emf E_i:

$$E_i = \frac{\partial \Psi_i}{\partial \theta_r} \cdot 2\pi n = K_E(\theta_r, i_i) \cdot 2\pi n \quad (9.24)$$

E_i is positive for motoring and negative for generating. V_i is considered positive and equal to V_{DC} when the DC source is connected through the active power switches to the SRG, it is zero when only one switch is turned off (soft commutation), and it is $-V_{DC}$ when both active power switches are turned off (hard commutation).

In a well-designed SRG(M), the $\Psi(\theta_r, i)$ family of curves shows notable nonlinearity (Figure 9.2). Consequently, both the transient inductance L_t and the emf coefficient K_E are dependent on rotor position and on current [5].

Typical emf coefficients K_E, calculated through FEM, are shown qualitatively in Figure 9.6. It should be noted that K_E, for constant current, is notably variable with rotor position; it is positive for motoring and negative for generating.

The voltage equation suggests an equivalent circuit with variable parameters (Figure 9.7a and b). The source voltage V_i is considered positive during phase energization and negative or zero during phase de-energization.

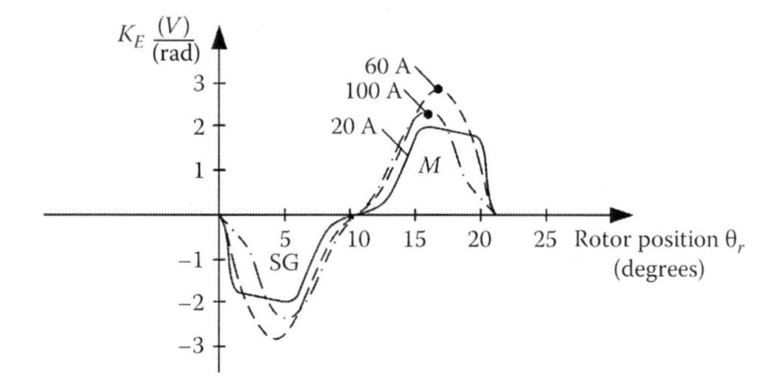

FIGURE 9.6 Electromagnetic force (emf) coefficient vs. rotor position for various currents.

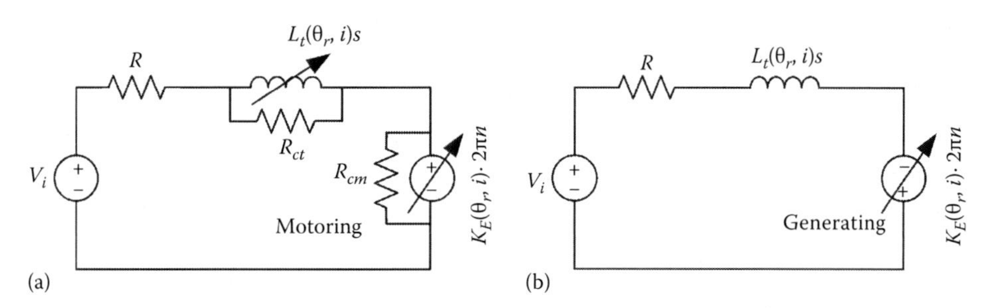

FIGURE 9.7 The equivalent circuit for (a) motoring and (b) generating.

It should be emphasized that the emf E is, in fact, a pseudo-emf (when the machine is magnetically saturated), as it contains a small part related to stored magnetic energy. Consequently, the instantaneous electromagnetic torque has to be calculated only from the coenergy (Equation 9.17), for a magnetically saturated machine [6].

Iron loss occurs in both the stator and the rotor, and this loss is due to current versus time variation and to motion at rectangular current in the phases [2]. Therefore, in the equivalent circuit, we should "hang" core resistances R_{cm} and R_{ct} in parallel to the transient inductance voltage and around the pseudo-emf (Figure 9.7). R_{cm} and R_{ct} reflect the core loss presence in the model.

9.4 Flux/Current/Position Curves

For refined design attempts and for digital system simulations of SRG for transient and control, the nonlinear flux/current/position family of curves has to be put in some analytical, easy-to-handle form. Even simpler expressions are suitable for control implementation.

Through FEM calculations, the family of such curves is obtained first, and then curve fitting is applied. The problem is that it is not enough to curve-fit $\Psi(\theta_r, i)$, but also to determine $i(\Psi, \theta_r)$ and, eventually, $\theta_r(\Psi, i)$. Polynomial or exponential functionals, fuzzy logic, artificial neural network (ANN), or other methods of curve-fitting were proposed for this function [2].

Examples of exponential approximations are as follows [7]:

$$\Psi(\theta_r, i) = a_1 \cdot (\theta_r) \cdot (1 - e^{-a_2 \cdot (\theta_r) \cdot i}) + a_3 \cdot (\theta_r) \cdot i \tag{9.25}$$

with

$$a_{1,2,3}^{(\theta_r)} = \sum_{K=0}^{\infty} A_{1,2,3}^{K} \cos(K \cdot N_r \cdot \theta_r) \tag{9.26}$$

or

$$\Psi_j = \Psi_{sat}(1 - e^{-i_j f_j(\theta_r)}); \quad i_j \geq 0 \tag{9.27}$$

$$f_j(\theta_r) = a_0 + \sum_{k=1}^{3} a_n \cdot \cos\left(k \cdot N_r \cdot \theta_r - (j-1)\frac{2\pi}{N_s}\right); \quad j = 1,2,3,4 \tag{9.28}$$

From Equation 9.27, the inverse function $i_j(\Psi_j, \theta_r)$ is as follows:

$$i_j = \frac{1}{f_j(\theta_r)} \ln\left(\frac{\Psi_s}{\Psi_s - \Psi_j}\right) \tag{9.29}$$

Ψ_s is the saturated value of the flux for the aligned position.

The coenergy (Equation 9.17) with Equation 9.27 becomes

$$W_{cj} = \int_0^i \Psi_j di_j = \int_0^i \Psi_{sat}(1 - e^{-i_j f_j(\theta_r)}) di_j = \Psi_{sat}\left[i_j - \frac{(1 - e^{-i_j f_j(\theta_r)})}{f_j(\theta_r)}\right] \tag{9.30}$$

Therefore, the instantaneous torque per phase is

$$T_{ej} = \frac{\partial W_{cj}}{\partial \theta_r} = -\Psi_{sat} \frac{\partial}{\partial \theta_r}\left[\frac{(1 - e^{-i_j f_j(\theta_r)})}{f_j(\theta_r)}\right] \tag{9.31}$$

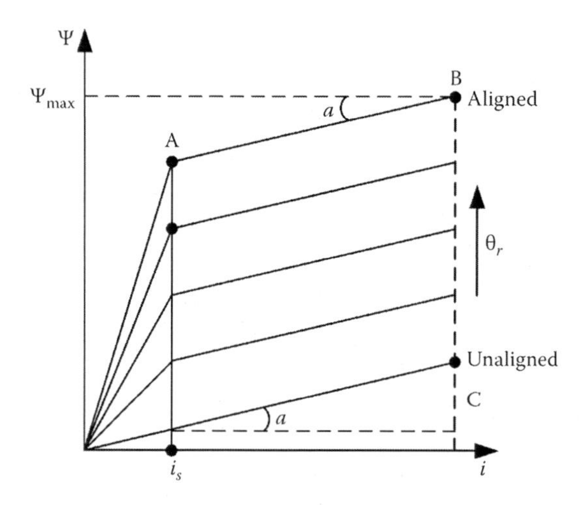

FIGURE 9.8 Piecewise linear approximations of the magnetization curves.

or

$$T_{ej} = \frac{-(\partial f_j/\partial \theta_r)}{f_j^2}(\Psi_{sat} i_j f_j + \Psi_j (1 - i_j f_j)) \quad (9.32)$$

For control design purpose, even a piecewise linear approximation of magnetization curves may be acceptable, as test results confirm that in practical designs, the magnetic saturation curve corner occurs at about the same current, independent of rotor position (Figure 9.8). This, in fact, implies that right after some stator/rotor pole overlapping, local saturation is achieved, and thus the flux varies almost linearly with rotor position.

Consequently,

$$\Psi_j = i_j\left(L_u + \frac{K_s(\theta_r - \theta_0)}{i_s \beta_s}\right) \quad \text{for } i \le i_s$$
$$\Psi_j = i_j L_u + \frac{K_s(\theta_r - \theta_0)}{\beta_s} \quad \text{for } i \ge i_s \quad (9.33)$$

The only variables to be found from the family of magnetization curves are i_s and K_s. The angle θ_0 corresponds to the rotor position where the rotor poles start overlapping the stator poles of the respective phase. While this simple approximation is tempting, because the continuity of flux at i_s is maintained, the continuity of the $\partial\Psi/di$ at i_s is not preserved. However, $\partial\Psi/d\theta_r$ is continuous at $i = i_s$. Knowing only the positions of A, B, and C on the graph in Figure 9.8 leads to unique values of K_s, i_s, and L_u. It suffices to calculate only the fully aligned and unaligned position magnetization curves for this approximation. A thorough analytical model, with FEM verifications, is developed in Reference 8.

9.5 Design Issues

The comprehensive design of SRG(M)s is complex [8]. A few basic issues are listed here:

- Motor and generator specifications
- Number of stator phases m and stator and rotor poles N_s, N_r
- Stator bore diameter D_{is} and stack length l_{stack} calculation
- Computation of stator and rotor pole and yoke geometry

- Number of turns per coil, conductor gauge, and connections
- Loss computation model
- Thermal model and temperature versus speed for specific duty-cycle operation modes
- Magnetization curve family and curve fitting for the design of the control system
- Peak and rated current, torque, and losses versus speed

Some of these subjects will be detailed in what follows.

9.5.1 Motor and Generator Specifications

The design specifications are tied to the application. For a generator-only application (wind generator, auxiliary power generator on aircraft, etc.), the motoring mode is excluded from start, while the DC output voltage power bus may be as follows:

- Independent, with passive and active loads
- With a battery backup

The DC voltage may be constant for stand-alone operation, but a variable when say, connected to a grid through an additional PWM inverter interface. For wind generation, a ±30% variation of speed around base speed is sufficient for most practical situations.

In addition, for the backup battery load case, the voltage varies from a minimum to a maximum value $V_{DC} = V_{DC}^{r}(1 \pm (0.2 - 0.25))$. The battery backup operation stabilizes the control requirements on the generator.

For automotive starter/generators, both motoring and generating are mandatory. The constant power speed range for motoring is generally larger than 3–4 to 1; but for generating (at lower power), much larger values are welcome, and a 10:1 ratio is not unusual (Figure 9.9).

The design DC voltage V_{DC} is very important when calculating the number of turns per winding and, consequently, the peak current for peak torque. A boost DC–DC converter might be added to raise the voltage to the battery maximum voltage and to reduce the peak machine and PWM converter current.

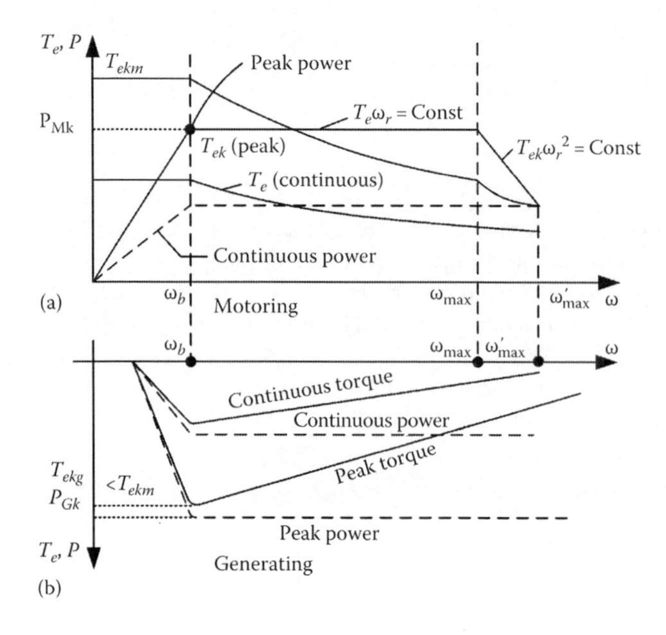

FIGURE 9.9 Typical torque and power vs. speed requirements for starter generators: (a) motoring and (b) generating.

9.5.2 Number of Phases, Stator and Rotor Poles: m, N_s, N_r

The fundamental frequency in each phase f_1 is as follows:

$$f_1 = n \times N_r; \quad n\text{—speed in rps} \tag{9.34}$$

The number of strokes per second f_s is

$$f_s = m \cdot f_1; \quad m\text{—number of phases} \tag{9.35}$$

The number of stator poles per phase is a multiple of two (pole pairs):

$$N_s = m \cdot 2K; \quad K\text{—integer} \tag{9.36}$$

The number of stator and rotor poles is generally related by

$$N_s - N_r = 2K \tag{9.37}$$

though other combinations are feasible.

Due to the increased complexity and costs in the PWM converter, only three- and four-phase SRG(M)s are considered practical for most applications above 1 kW power per unit.

Increasing the number of poles N_s, for given outer stator diameter and stack length, leads to a reduction in the stator and rotor yoke thickness, without leading to higher torque to the same extent.

The frequency f_1 increases and so does f_s, which determines the core losses that tend to increase.

On the other hand, the noise level tends to decrease notably as the radial force of the active phase is distributed around the rotor periphery more evenly.

It was established that, in general, the stator and rotor pole heights h_{ps}, h_{pr} per pole spans, b_{ps}, b_{pr}, should be as follows:

$$\frac{h_{ps}}{b_{ps}} \approx \frac{h_{pr}}{b_{pr}} \approx (0.7 - 0.8) \tag{9.38}$$

In addition, the pole span per air gap g should be

$$\frac{b_{ps}}{g} \approx \frac{b_{pr}}{g} > (17 - 20) \tag{9.39}$$

to yield enough high ratios between aligned and unaligned maximum inductance ($L_a^s / L_u \approx 3 - 8$) to guarantee acceptably high average torque (W_{mec}).

The air gap should be as small as mechanically feasible, that is, for acceptable vibration, deflection, and noise levels.

The maximum frequency f_1 is also related to the switching capacity of the PWM converter, as the commutation converter losses (Chapter 8) increase with frequency.

9.5.3 Stator Bore Diameter D_{is} and Stack Length

Though SRG(M)s operate at variable speed, the Epson's coefficient heritage may be used to calculate the stator bore diameter D_{is}, assuming first that λ = l stack/b_{ps} and the number of stator poles N_s are given. Alternatively, the shear rotor stress f_t may be imposed:

$$f_t = (1\text{–}10)\ \text{N/cm}^2$$
$$\lambda = \frac{l_{stack}}{b_{ps}} \approx 0.5 - 2.0; \quad b_{ps} \approx \frac{\pi D_{is}}{2N_s} \tag{9.40}$$

The design tangential peak shear rotor stress may increase with stator bore diameter. Once the peak torque T_{ek} is given through the specifications, the stator bore diameter D_{is} is as follows:

$$D_{is} = \sqrt[3]{\frac{4 \cdot N_s \cdot T_{ek}}{\pi^2 \cdot f_{tk} \cdot \lambda}} \tag{9.41}$$

Observing Equations 9.38 and 9.39, we still need to size the stator and rotor yokes to complete the stator geometry. In general, $h_{ys}\, h_{yr} > b_{ps}/2$, as half the stator pole flux, flows through the yokes. The peak torque is needed for motoring at low speeds, when the current is kept constant through chopping. Consequently, the energy cycle area is full. Using FEM or of an analytical model, the unaligned and aligned position flux curves are obtained. Φ_{pole} is the flux per stator pole for active phase, and W_cI_c is the corresponding coil magnetomotive force (mmf; Figure 9.10). The maximum flux per pole is related to the saturation flux density and the pole area, plus the leakage flux (by K_l):

$$\Phi_{p\max} = B_{sat} \cdot b_{ps} \cdot l_{stack} \cdot (1 + K_l) \tag{9.42}$$

With B_{sat} = 1.6–2 T, b_{ps}, l_{stack} known already, and $K_l \approx$ 0.1–0.2, the area W_{mec}^p in Figure 9.10 leads to the average peak torque:

$$T_{ek} = \frac{W_{mec}^p \cdot m \cdot N_r \cdot 2K}{2\pi} = \frac{W_{mec}^p \cdot N_s \cdot N_r}{2\pi} \tag{9.43}$$

with $2K$ equal to the number of stator poles per phase. As T_{ek} is fixed by specifications, the value of B_{sat} may be increased gradually until W_{mec} is high enough to satisfy Equation 9.43, for a certain $W_cI_{c\max}$.

The window area available for a coil A_{Wc} is as follows:

$$A_{Wc} \approx \frac{\left[(\pi/4)\left((D_{is} + 2h_{ps})^2 - D_{is}^2\right) - N_s \cdot b_{ps} \cdot h_{ps}\right]}{2N_s} \tag{9.44}$$

For a window filling K_{fill} = 0.5–0.6, typical for preformed coils, the maximum current density $j_{CO\max}$ is obtained:

$$j_{CO\max} = \frac{W_c I_{c\max}}{A_{Wc} K_{fill}} \tag{9.45}$$

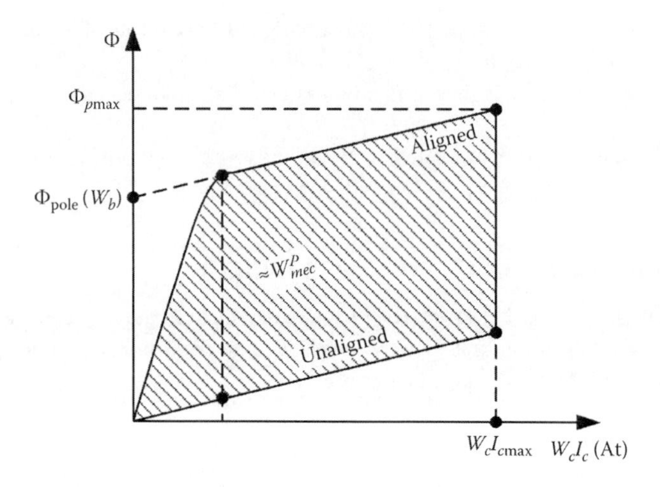

FIGURE 9.10 Extreme position magnetization curves.

Depending on the application, j_{COmax} may vary from 8–10 to 20–24 A/mm² for most strained automotive starter/generators with forced cooling. If j_{COmax} is out of this range, the design is restarted with a lower shear rotor stress f_{tk}.

9.5.4 Number of Turns per Coil W_c for Motoring

The 2K coils per phase may be connected many ways to form from 1 to 2K current paths in unity steps a = 1, 2, ... The number of turns in series per current path W_a is related to the total number of turns per phase by the following:

$$W_a = \frac{W_c \cdot N_s / m}{a} \tag{9.46}$$

The machine is designed so that the ideal emf E_{av}, for constant current and full angle conduction,

$$\theta_w = \theta_c - \theta_{on} \simeq \frac{2b_{ps}}{D_{is}} \tag{9.47}$$

at base speed n_b (rps), be a given fraction α_{PF} of the DC design voltage V_{DC} (Equation 9.11):

$$\frac{2\pi \cdot n_b}{\theta_w} \cdot (\Psi_{\max})_{path} = \frac{2\pi \cdot n_b}{\theta_w} \cdot \Phi_{p\max} \cdot W_a = \alpha_{PF} \cdot V_{DC} = E_{av} \tag{9.48}$$

From this expression, with $\Phi_{p\max}$ already calculated from Figure 9.10 and θ_w from Equation 9.47 and assigned values of n_b and α_{PF}, the number of turns W_a per current path is obtained. Subsequently, from Equation 9.46, the number of turns per coil W_c is calculated, and with $W_c I_{c\max}$ known from Figure 9.10 for peak torque (Equation 9.43), the maximum coil current I_{peak} is as follows:

$$I_{peak} = a \cdot I_{c\max} \tag{9.49}$$

The coefficient $\alpha_{PF} = E_{av}/V_{DC}$ at base speed is a measure of the machine power factor, as in AC machines. The α_{PF} best practical values gravitate around unity. When a large constant power speed range is required, $\alpha_{PF} < 1$, but then the number of turns W_a is smaller. Consequently, the peak current is larger, and inverter oversizing is required. In contrast, with $\alpha_{PF} > 1$, the constant power speed range decreases, but W_a increases.

It seems wise to start with $\alpha_{PF} = 1$ and then investigate the performance for small departures from this situation, hunting for the best overall performance for the constant power speed range.

Randomly modifying W_c would eventually lead to similar results, but the search may be too time-consuming. At this point, preliminary machine sizing is complete, and all parameters may be calculated. The average torque/speed envelopes for motoring and generating can be calculated up to the peak current and to the peak voltage V_{DC}. This way, the generator specifications may be checked after the machine is sized in terms of turns/coil W_c, for motoring.

In some applications, the generator mode is more demanding; thus, the W_c computation should be tied to generating. To do so, we first have to explore current waveforms for generator mode.

9.5.5 Current Waveforms for Generator Mode

In the generator mode, a phase is turned on at the angle θ_{on}^{g} around the maximum inductance rotor position, and then the controlled semiconductor rectifier (SCR) is turned off (commutated) at the angle θ_c, long before the phase inductance decreases to its minimum value (Figure 9.11a through c).

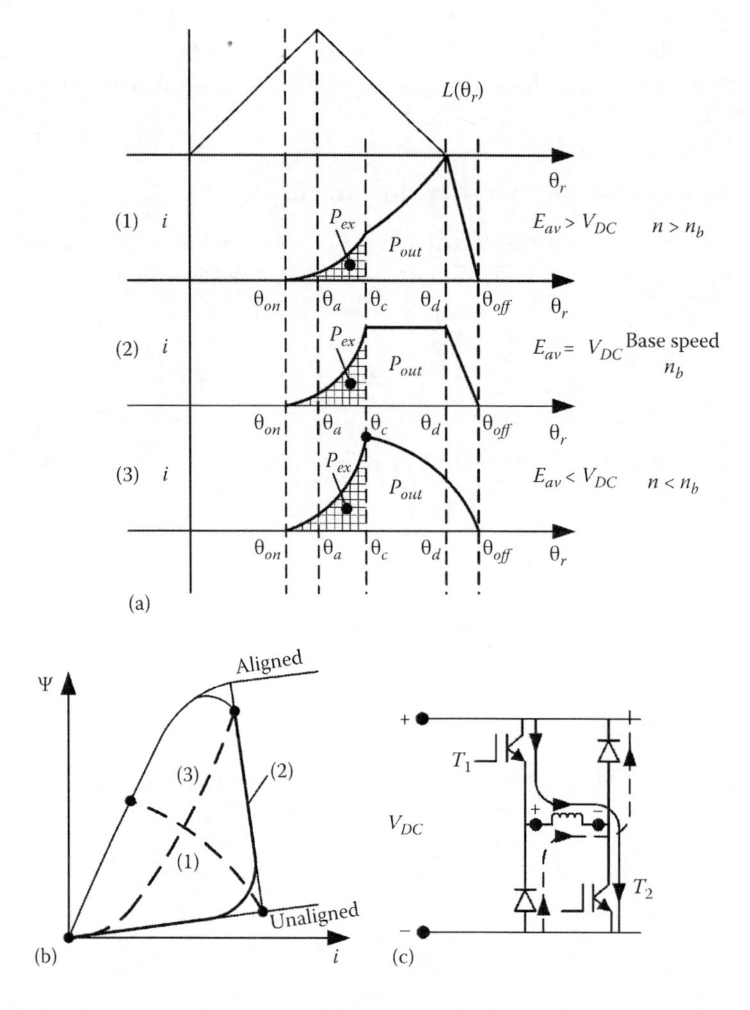

FIGURE 9.11 Current waveforms for (a) generator mode, (b) the corresponding energy cycles, and (c) the same peak current with pertinent converter.

The voltage equation for phase j (Equation 9.22) is as follows:

$$V_j = R_j I_j + L_{tj} \cdot \frac{di_j}{dt} + K_E \cdot 2\pi \cdot n; \quad K_E = i\frac{dL}{d\theta} < 0 \tag{9.50}$$

For simplicity, we may consider $L_t \approx L_u$(unaligned) = const, as it is the situation in very saturated conditions. The voltage $V_j = V_{DC}$ during phase energization (excitation) and $V_j = -V_{DC}$ during power delivery through the free-wheeling diodes (Figure 9.11a). Let us also neglect the stator resistance.

Figure 9.11a shows three cases that can be fully documented via Equation 9.50. In case (1), after T_1 and T_2 are turned off (phase energization is finished) at θ_c, the current still increases. As $V_j < 0$ and $E < 0$, the current increase is due to the fact that $|E| > |V_{DC}|$; that is, the emf is greater than the supply (magnetization) voltage. This case is typical for high speeds ($n > n_b$), when the torque is smaller (Figure 9.11b, cycle 1 area).

For case (2) in Figure 9.11a, it happens that at turn-off angle θ_c, $|E| = |V_{DC}|$ and thus, from Equation 9.50, with $R_s = 0$, $di/dt = 0$. Consequently, the current remains constant until the inductance reaches its minimum at θ_d. Energy cycle (2) in Figure 9.11b shows the large torque area.

In case (3) in Figure 9.11a, the maximum current is reached at θ_c, and after that, the current decreases steadily, because $|E| < |V_{DC}|$, which corresponds to low speeds. A smaller torque area is typical for case (3).

For constant DC voltage power delivery, the excitation energy W_{exc} per cycle (Figure 9.11a) is as follows:

$$W_{exc} = \frac{V_{DC}}{2\pi n} \int_{\theta_{on}}^{\theta_c} i \, d\theta \tag{9.51}$$

The delivered ideal energy/cycle, W_{out}, is

$$W_{out} = \frac{V_{DC}}{2\pi n} \int_{\theta_c}^{\theta_{off}} i d\theta \tag{9.52}$$

The excitation penalty ε is

$$\varepsilon = \frac{W_{exc}}{W_{out}} \tag{9.53}$$

Case (1) ($|E| > |V_{DC}|$) leads to a smaller excitation penalty than that for the other two cases, for the same net generated energy per cycle W_{out}. As we can see from Figure 9.11a and b, the excitation penalty has conflicting influences on the design. To have a small ε, we are tempted to design (control) the machine such that $|E| > |V_{DC}|$; however, the torque area (Figure 9.11b) is smaller, and the energy conversion ratio is not close to an optimum.

Case (2) ($E_{av} = V_{DC}$) produces the largest torque–energy cycle and, thus, seems more effective in energy conversion. Maintaining case (2) ($|E| = |V_{DC}|$) forces one to increase the DC voltage in proportion to speed. But, to deliver power at constant voltage, a voltage step-down DC–DC converter has to be added between the machine converter and the DC load (from V_{DC} to V_L). The peak value of current for case (1) occurs at the angle θ_d:

$$i_{peak} \approx \frac{\Psi_{peak}}{L_u} = \frac{1}{L_u}\frac{V_{DC}}{2\pi n}(\theta_{off} - \theta_d) \approx \frac{1}{L_u}\frac{V_{DC}}{2\pi n}(\theta_c - \theta_{on} + \theta_c - \theta_d) \tag{9.54}$$

An earlier turnoff (smaller θ_c) leads to a smaller peak current, even if $\theta_c - \theta_{on} = \text{const}$.

For case (2), the peak current occurs at θ_c:

$$i_{peak} = \frac{\Psi_{peak}}{L_{peak}} = \frac{1}{L} \cdot \frac{V_{DC}}{2\pi n} \cdot \frac{\theta_c - \theta_{on}}{\theta_d - \theta_c} \tag{9.55}$$

The phase inductance L_{peak} at $\theta_{peak} = \theta_c$ is considered to be proportional to $\theta_d - \theta_c$, because L varies almost linearly with rotor position. A constant dwell angle $\theta_c - \theta_{on}$, with early turnoff (smaller θ_c), leads to smaller peak current, and thus, lowers power delivery. Controlling the output power is thus possible by turn-on angle θ_{on} and turn-off angle θ_c.

When a voltage step-down DC–DC converter is not available or desirable, the generator mode should switch from case (3) below base speed, to case (2), and case (1) as speed increases. The energy conversion is not uniformly good, but the constant power speed range may be increased, especially if n_b for case (2) is lowered.

It may now be more evident that adopting $|E| \approx |V_{DC}|$ at the base speed n_b is also good for generator operation in terms of energy conversion. This is not so in terms of excitation penalty. In general, an excitation penalty $\varepsilon = 0.3$–0.4 is considered acceptable, though higher values may lead to higher energy conversion in the machine. The $|E| = |V_{DC}|$ condition, if maintained for the entire speed range (70%–140%, for wind generators), corresponds to the lowest generator peak current for given output power (energy per cycle) at maximum speed.

For wide constant speed range generating, it seems that choosing V_{DC} = cons = E at base speed is the solution. Designing the PWM converter at V_{DC} corresponding to base speed n_b pays off in terms of converter voltage rating.

9.6 PWM Converters for SRGs

As already alluded to in this chapter, each phase of SRG(M) is connected to a two-quadrant DC–DC converter (Figure 9.12).

The generator case is presented in Figure 9.12 for two main situations in terms of load—with and without backup battery. Only in the case with load backup battery may the SRG operate as a starter, supplied from the battery. In vehicular applications, this is the case for ICE starting and driving assistance. The generator mode appears, during exclusive ICE driving, for optimum battery recharge and load supply, and during regenerative vehicle braking. The low-voltage self-excitation battery is present and serves to initiate the raising of output voltage under no-load (self-excitation) only when the load backup battery is absent.

There are many other PWM converter configurations that use a smaller number of SCRs (m, or $m + 1$, instead of $2m$ pieces for the asymmetrical converter). A rather complete investigation of such converters is presented in Reference 2. The conclusion is that all have advantages and disadvantages. At least for SRGs, where the generator control, especially, may encounter instabilities in DC output voltage control, due to inadvertent self-excitation, the asymmetrical converter (Figure 9.12) is considered a very practical solution. This converter provides for the freedom of soft (one SCR off, T_1) or hard (both SCRs off) de-energization of phases, to reduce output voltage pulsations. In addition, to increase the generated power, an intermediary soft turning-off period (T_1 off), before the hard turning-off period (T_1, T_2 off) is feasible.

A DC–DC converter for voltage buck boost may be added between the load backup battery and the SRG side multiphase chopper (Figure 9.12). The battery voltage may be too small in some cases (42 V_{DC} in a mild hybrid vehicle), therefore it pays off to design the SRG and the machine-side converter at a higher and constant voltage, which should be at least equal to the battery maximum voltage $V_{b\max}$.

As dicussed in Chapter 8, for the PM-assisted reluctance synchronous machine (RSM) starter/generator PWM converter, it seems to pay off to include an additional DC–DC converter for voltage boost (in motoring) and step down (for generating), both in terms of silicon costs and system losses.

To stabilize the output voltage, it may be feasible, for stand-alone DC loads, to supply the excitation power (phase energization) from a separate (excitation) battery (Figure 9.13). As the excitation power may amount to 30% (or more) of the output power, the excitation battery has to be strong. To avoid adsorbing large currents from the excitation battery, the diode D_{oe} may be used [9]. The presence of the starting battery allows for operation during load faults and for clearing them out rather quickly (Figure 9.14) [9]. The battery is also used for starting the prime mover of the SRG when the latter works as a starter for rather rare starts (as for on board aircraft).

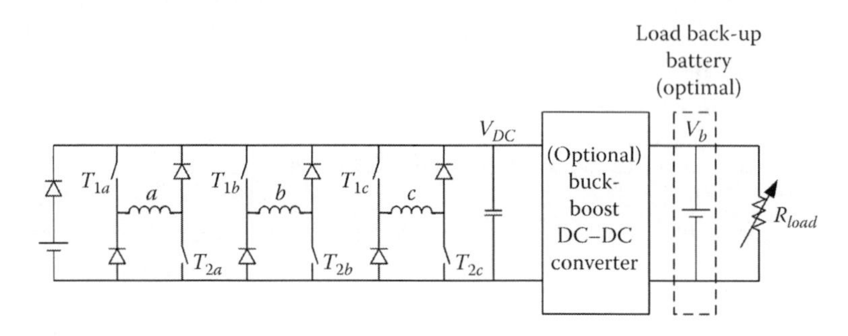

FIGURE 9.12 Three-phase switched reluctance generator (SRG) with asymmetrical converter, self-excitation, or load backup battery.

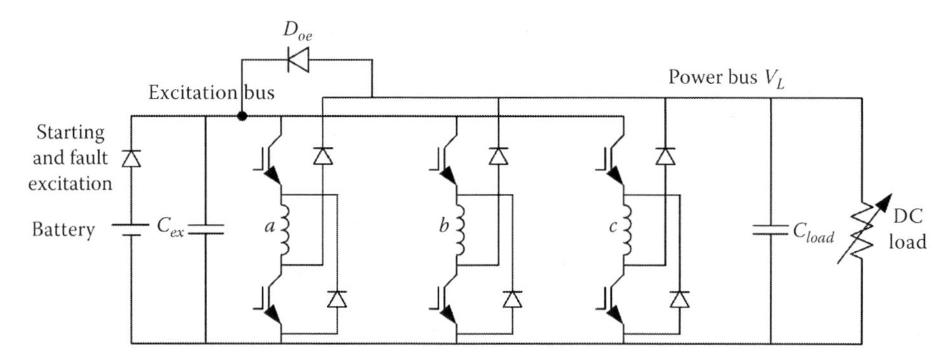

FIGURE 9.13 Switched reluctance generator (SRG) with separate excitation and power bus and fault clearing capability. (Adapted from Radun, A.V. et al., *IEEE Trans.*, IA-34(5), 1106, 1998.)

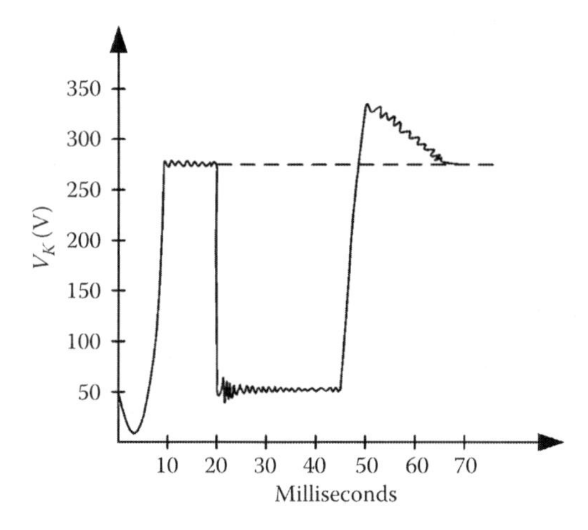

FIGURE 9.14 Voltage rise up, experiencing a fault (sharp load resistance decrease), clearing the fault, and voltage recovery.

Special means for the battery recharge are necessary with the scheme in Figure 9.13, but at a low power rating. When one phase goes off, the SRG may provide continuous operation, at low load power, though with higher output voltage ripple as only the *m*1, *m*2, and *m*3 phases are sound. This is a special feature of an SRG, with phases that are weakly coupled magnetically, even for heavy magnetic saturation conditions.

The converter for the case of SRG with AC output should contain one more stage: a DC–AC PWM converter (inverter) (Figure 9.15). For such applications with speed variation not more than ±25% around rated speed, it may be practical to design the SRG to be capable of delivering constant (rated) V_{DC}—DC link voltage from minimum speed upward.

The AC load may be independent or it could be a weak (or strong) power grid. The control of the PWM inverter should be adapted to each kind of AC load. The capacitor C_{DC} may be designed to provide all the reactive power that the AC load is expected to absorb. More information on SRGs at the power grid is provided in Reference 10.

It should be recognized that the machine-side converter is special for SRG in comparison with standard AC starter/alternators, but it contains about the same number of SCRs per phase (two for four-phase machines). However, it avoids shoot-through short circuits and is thus phase fault tolerant.

Finally, the danger of emf that is too high at largest speed, existent in PM generators, is not present with SRG.

We treated here three- and four-phase SRGs, but one- or two-phase SRGs may also be built for low powers or super high speeds [11].

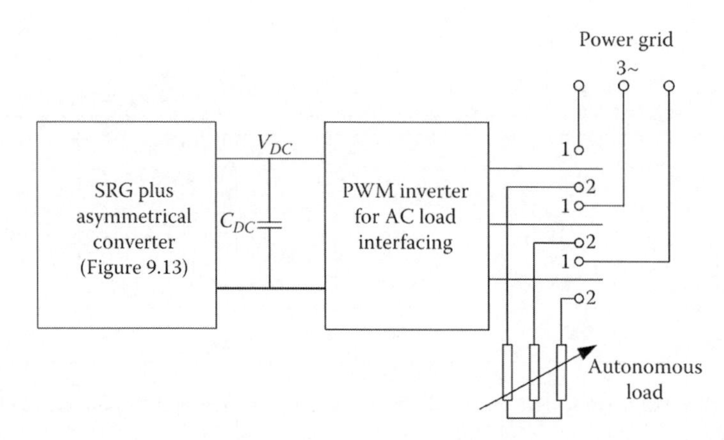

FIGURE 9.15 Switched reluctance generator (SRG) with additional alternating current (AC) load interface inverter.

9.7 Control of SRG(M)s

There is a very rich literature on SRM drives control, with pertinent updates in References 2, 4, and 12.

Though starter/generators experience longer generator than motor operation mode intervals, most attention has so far been placed on the motoring control.

We will distinguish two main high-grade control strategies for SRG(M)s:

- Feed-forward torque sharing control [2,4]
- Direct torque control [14–16]

Current profiling with efficient torque sharing in four-phase SRG(M)s is also considered here as a feed-forward—indirect—torque control, though executed under a current profile control.

The control of SRGs may be classified by the controlled variable in the following:

- Speed or voltage control for generation mode
- Torque control for starter/generators

Speed control for maximum wind power extraction is typical for wind generators, but a torque (power) interior closed loop may be used for faster response.

Torque control may be used for generating with load backup battery, as the reference torque may be calculated from the required (accepted) current by the battery state of charge. The control system may be implemented with or without a motion sensor (encoder or resolver).

For mainly generator operation, even with infrequent starting (motoring), the rather coarse (if any) position feedback may be used. During, drive-assisting operation mode on EHVs, fast nonhesitant robust and precise torque response is required from zero speed. In this latter case, either a precise encoder is used to measure the rotor position, or an advanced position estimator that works from zero speed and has initial rotor position estimation capability provided. As this kind of sensorless vector control was already demonstrated for PM-RSM starter/alternators (Chapter 8), such a standard behavior is expected from SRG(M), also.

In view of this, we will present, in what follows, representative methods with feed-forward and with direct torque control. And then, we will deal separately with observers for motion-sensorless control.

9.7.1 Feed-Forward Torque Control of SRG(M) with Position Feedback

In automotive applications, and in general for torque feed-forward control, the reference torque T_e^* is determined by the acceleration or braking pedal, corroborated with speed and battery state and load information. It is basically the average reference torque.

The magnetic saturation and the nonsinusoidal manner of phase inductance variation with rotor position make the relationship between torque and phase current highly nonlinear and speed dependent. Moreover, the phase current may be PWM controlled only below base speed, for motoring. To fully determine the torque produced by a phase, not only the current i_i should be referenced (for PWM current control below base speed), but also the turn-on, θ_{on}, and commutation (hard commutation), θ_c, angles have to be referenced (changed).

Above base speed, when only the single-pulse current operation mode is available, for motoring, only the $\theta_{on}^*, \theta_c^*$ angles have to be referenced for given T_e^*, battery voltage, and speed. The reference torque T_e^* versus speed envelope has to be known offline, to avoid control instabilities at high speeds for all battery voltages.

Feed-forward torque control is based on the off-line computation of $\theta_{on}^*, \theta_c^*$ for given T_e^*, speed n, and battery voltage, based on the $\Psi(i, \theta_r)$ curves and the nonlinear circuit model of SRG(M), as described in the previous section.

To reduce the complexity of the various look-up tables required for real-time control implementation, only one phase torque is considered in three-phase SRGs. Torque ripple minimization is obtained with current profiling below a certain speed, calculated for minimum battery voltage [15]. When, at low speeds, PWM current control is used, the latter is applied only to one phase at a time.

The relationship between $T_e^*, i_i^*, \theta_{on}^*, \theta_c^*$ for a large number of speed n and battery voltage V_{dc}^* levels, both for motoring and generating, are obtained offline from the machine nonlinear model via some analytical approximations.

In Reference 13, parabolic fitting curves are used:

$$
\begin{aligned}
\theta_{on}^*(T_e^*, n^*) &= \theta_z(n) + C_o(n) T_e^{m_o} \\
\theta_c^*(T_e^*, n^*) &= \theta_z(n) + C_c(n) T_e^{m_c} \\
i^* &= p_i(n) T_e + q_i(n)\sqrt{T_e}
\end{aligned}
\tag{9.56}
$$

The parameters θ_z, C_o, C_c, m_o, m_c, p_i, and q_i are dependent on speed n and also on battery voltage V_{DC}.

A linear dependence of p_i and q_i, on V_{DC}, may then be adopted [14]:

$$
\begin{aligned}
p_i &= k_p(n) + l_p(n) V_{DC} \\
q_i &= k_q(n) + l_q(n) V_{DC}
\end{aligned}
\tag{9.57}
$$

Above base speed, the current regulation is no longer imposed, and only the angles $\theta_{on}^*, \theta_c^*$ are imposed as dependent on speed.

It was proven [13] that the torque variation due to turn-on angle θ_{on} deviation is rather large, but it decreases when the torque increases.

Typical values of these off-line calculated control parameters for a 15 kW, 95 Nm, 12/8 (three-phase) SRG are shown in Figure 9.16 for motoring and in Figure 9.17 for generating, using the criterion of maximum torque for given current [13]. An extension of the torque/speed envelope is obtained when switching from this criterion to maximum torque/flux at high speeds [12]. The basic control scheme, shown in Figure 9.18, shows the torque to current i_i, angles θ_{on}^* and θ_c^*, and the current regulators. Typical torque

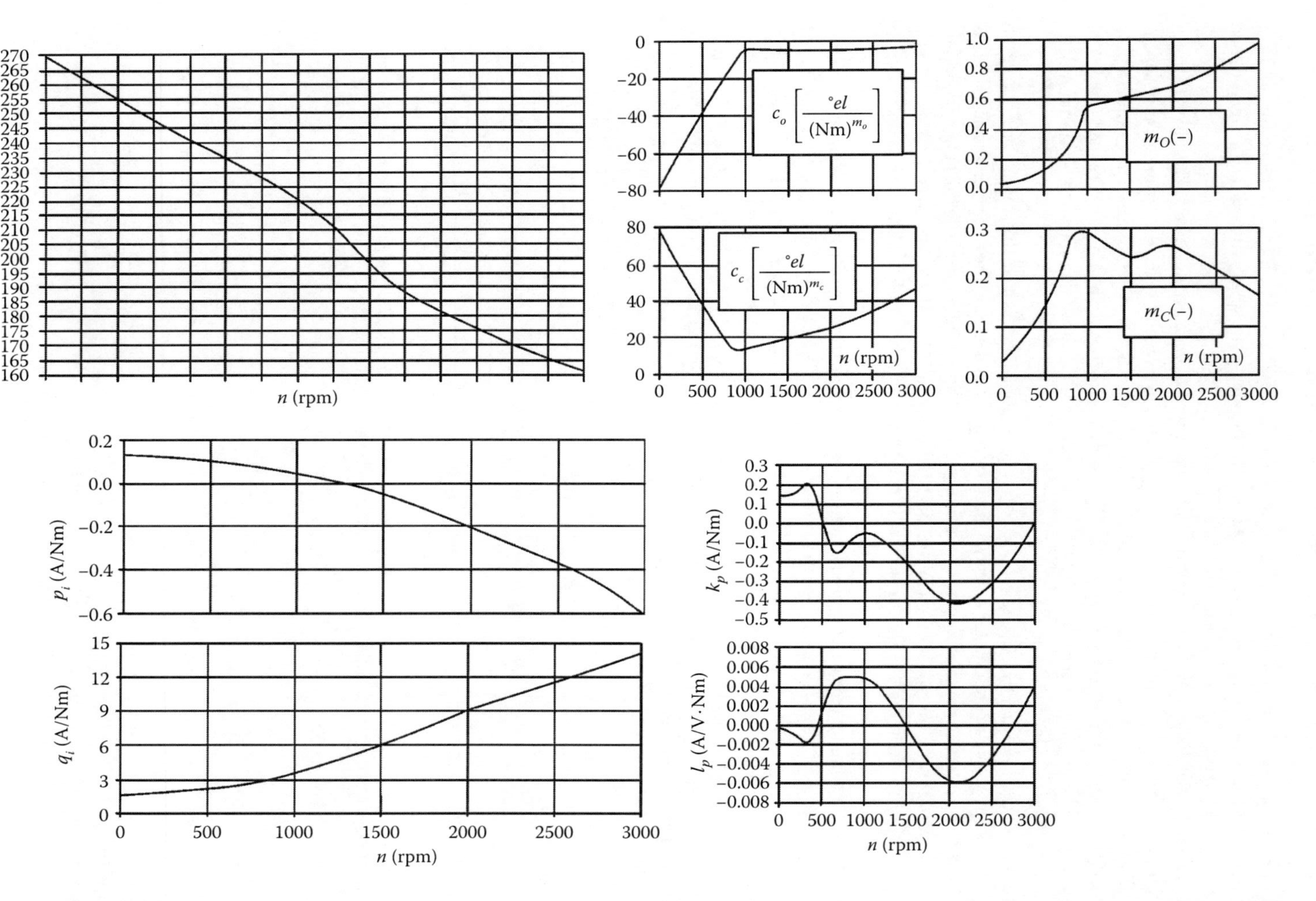

FIGURE 9.16 Offline calculated parameters in Equations 9.56 and 9.57 for vectoring. (Adapted from Bausch, H. et al., Torque control of battery-supplied switched reluctance drives for electrical vehicles, Record of ICEM-1998, Istanbul, Turkey, pp. 229–234, 1998.)

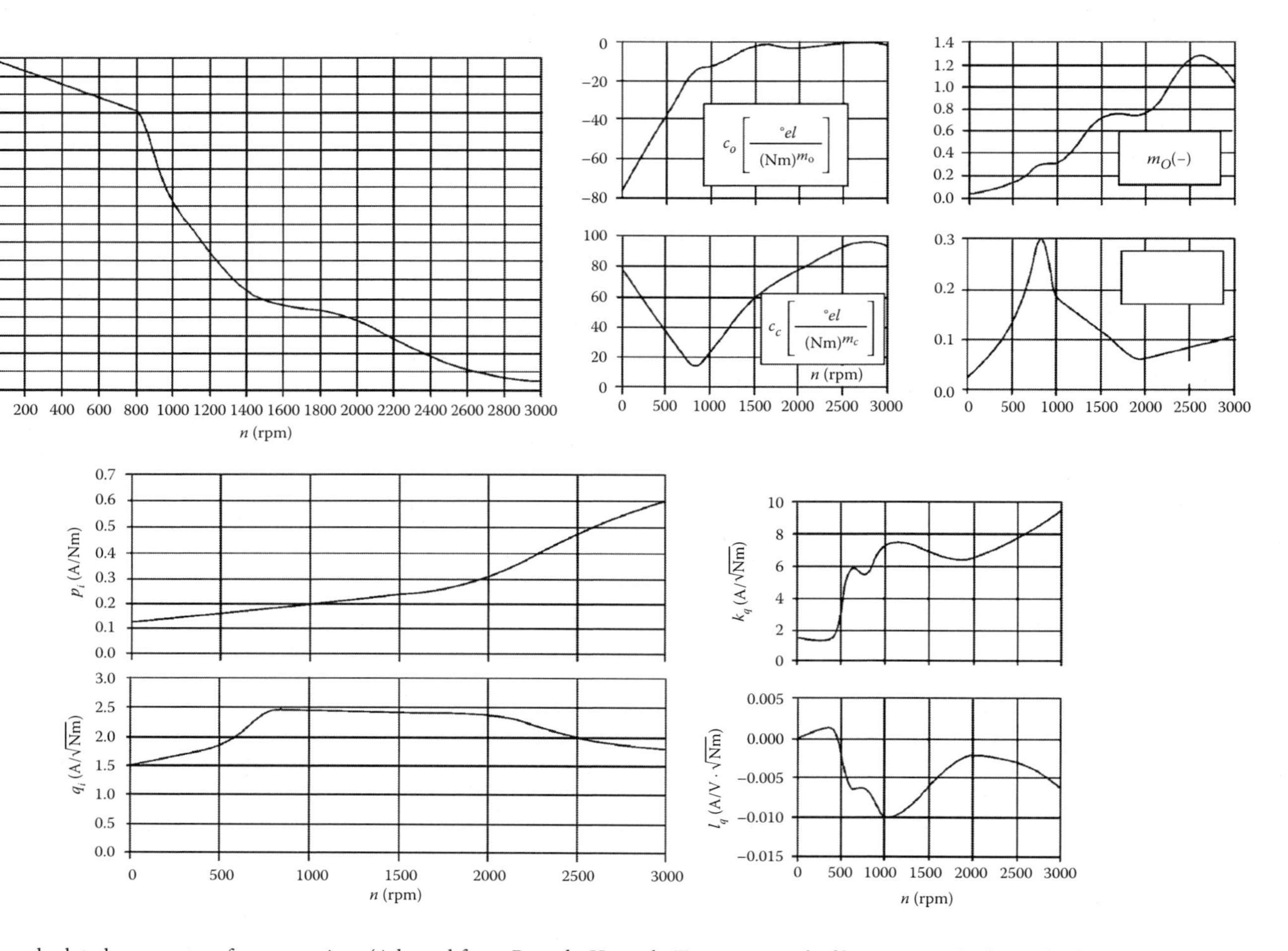

FIGURE 9.17 Offline calculated parameters for generating. (Adapted from Bausch, H. et al., Torque control of battery-supplied switched reluctance drives for electrical vehicles, Record of ICEM-1998, Istanbul, Turkey, pp. 229–234, 1998.)

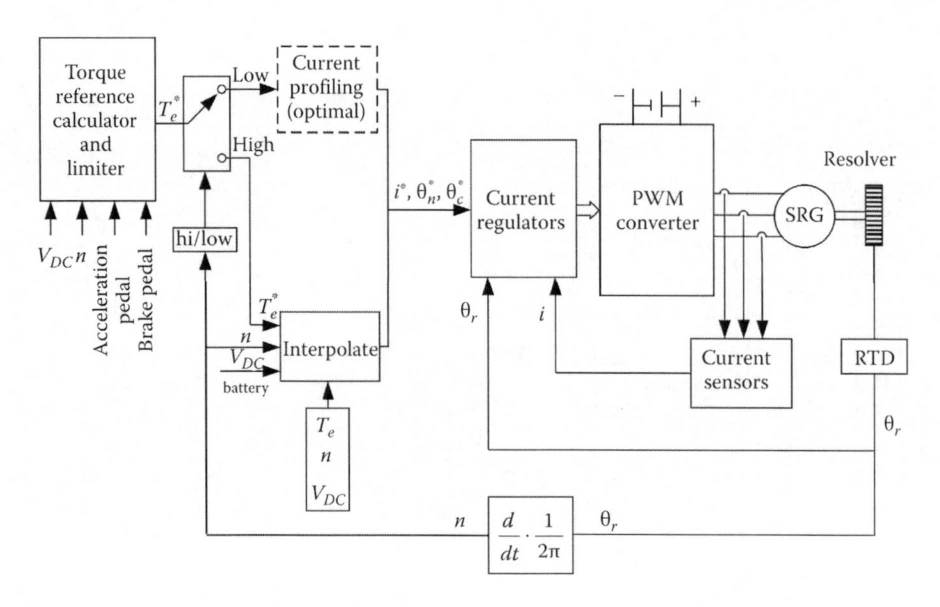

FIGURE 9.18 Feed-forward torque control of switched reluctance generator (SRG)(M). (Adapted from Bausch, H. et al., Torque control of battery-supplied switched reluctance drives for electrical vehicles, Record of ICEM-1998, Istanbul, Turkey, pp. 229–234, 1998.)

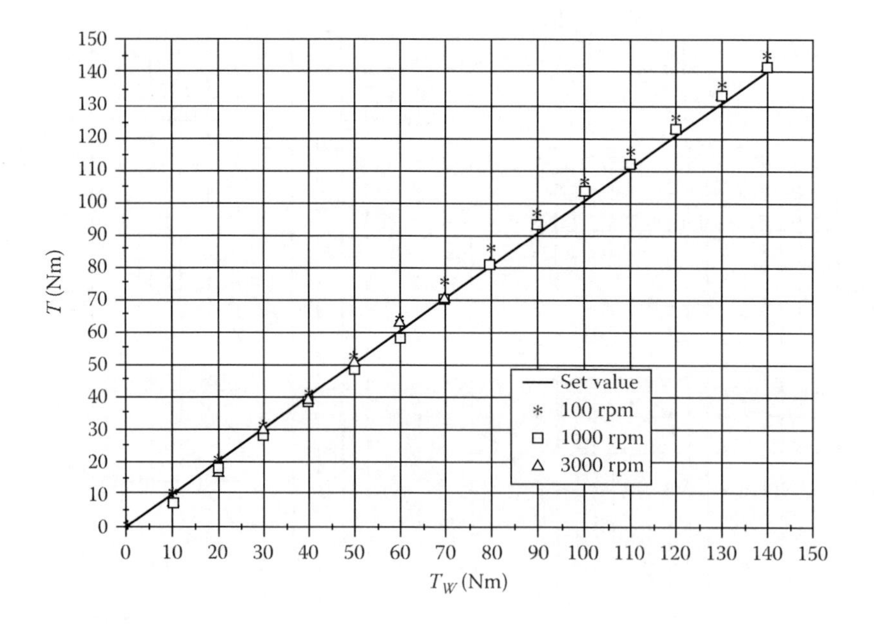

FIGURE 9.19 Torque-response precision.

response at three speeds is shown in Figure 9.19 with current and torque versus time for motoring and generating in Figures 9.20 and 9.21 [14]. Good-quality responses are shown in Figures 9.20 and 9.21 for single-pulse current control (high speed).

At low speeds, below 100 rpm in Reference 15, current profiling may be used to further reduce torque pulsation and noise. Typical such current profiles with simulated torque are shown in Figure 9.22 [15] for a 24/16 three-phase 350 Nm FRG(M) at 200 V_{DC} and at a peak current of 200 A.

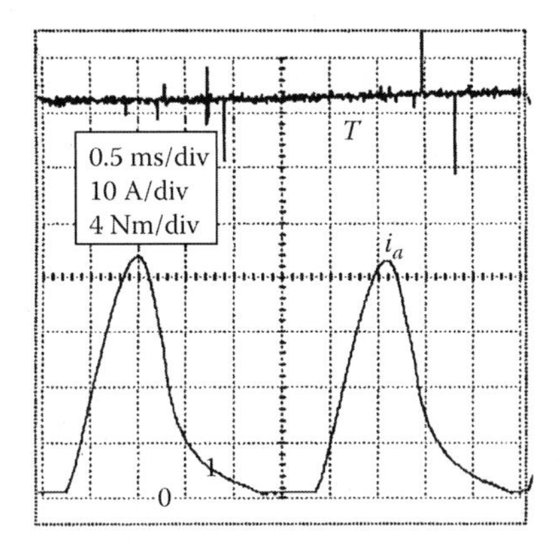

FIGURE 9.20 Current and torque for motoring at 1600 rpm and 32 Nm.

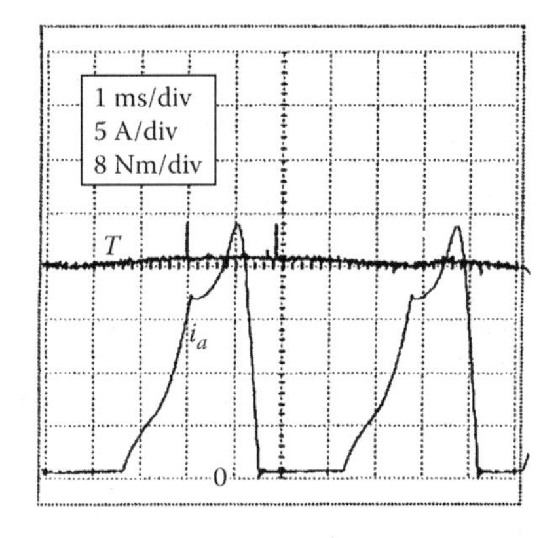

FIGURE 9.21 Current and torque for generating at 1600 rpm and 32 Nm.

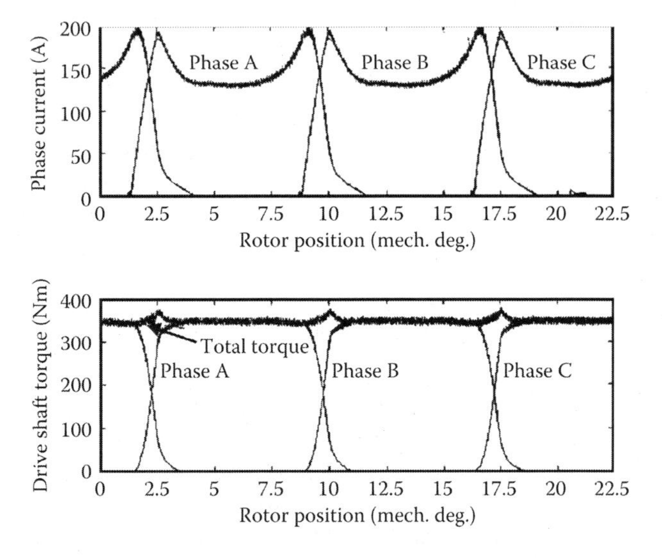

FIGURE 9.22 Simulated current profiles and torque.

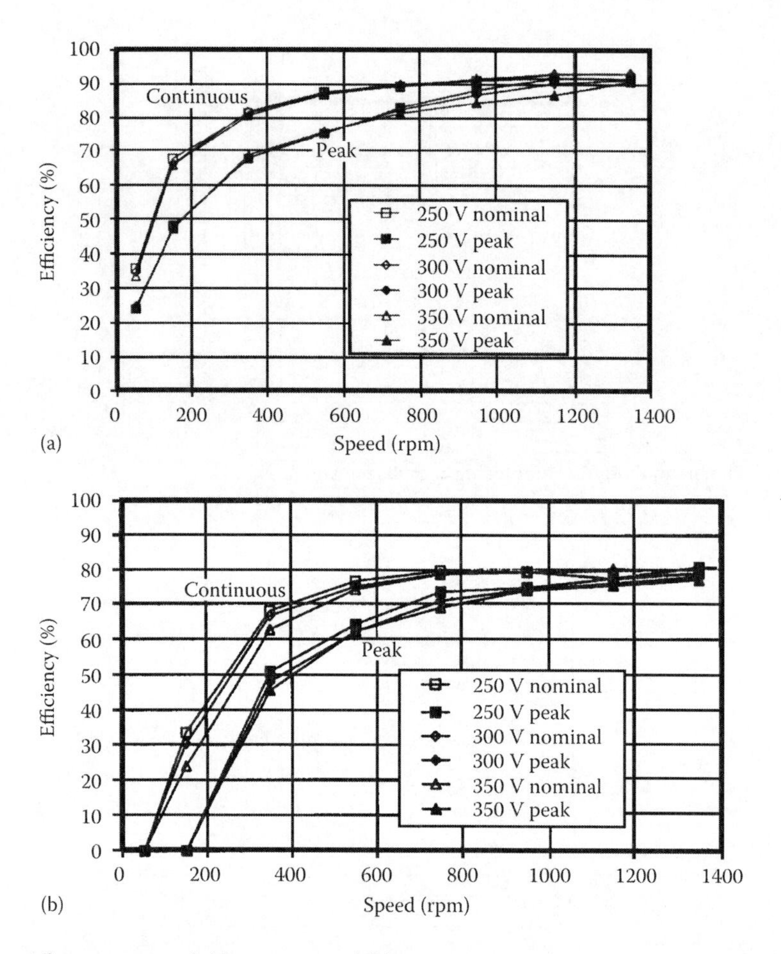

FIGURE 9.23 Efficiency vs. speed: (a) motoring and (b) generating.

While the offline computation effort of $T_e^*, i_i^*, \theta_{on}^*, \theta_c^*$ for given n and V_{DC} is done once, at implementation, it is very challenging in terms of memory for online control with 50 μs or so control decision cycles [15]. Rather good efficiency levels were demonstrated, however, by tests in Reference 15 (Figure 9.23a and b).

To further reduce the DC current ripple that is felt by the battery, a soft commutation stage ($V_{DC} = 0$, one SCR only off) per phase, prior to the hard commutation stage, up to a certain speed, may be implemented [12,15]. To avoid part of the tedious offline computation of $i_i^*, \theta_{on}^*, \theta_c^*$ for given T_e^*, n, V_{DC}, various online estimation methods, such as fuzzy logic and ANN, were proposed. A pertinent review of these control methods for motoring is given in Reference 12.

One step further in this direction is the concept of direct (close-loop) torque control of SRG(M).

9.8 Direct Torque Control of SRG(M)

Direct torque control [12,16,17] requires online torque estimation. Torque estimation, in turn, requires phase flux estimation. The reference torque is the average torque. For three-phase SRG, the average torque per energy cycle is determined basically for one active phase, but with four-phase machines, both active phases have to be considered. To save online computation effort, the flux and average torque of a single phase is estimated. However, this limits the average torque control response quickness to fast reference torque variations. Estimating the flux and torque of all phases would eventually bring superior results in torque response quickness and quality, but with a markedly larger hardware and software effort.

Average torque T_{avr} estimation per energy cycle thus seems to be a practical solution. The problem is that torque estimation has to be sound, even at zero speed, in order to provide adequate torque control at zero speed. The presence of a refined position sensor is an asset in this enterprise, as precision in $\theta_{on}^{*}, \theta_{c}^{*}$ control of less than 1° (mechanical) is required for 6/4 machines and less than 0.2° (mechanical) for 16/12 machines [17].

To calculate the average torque, the basic coenergy W_c formula of coenergy, via the mechanical energy per cycle W_{mec}, is used:

$$(W_{mec})_{cycle} = \int_{\theta_{on}}^{\theta_{off}} \frac{\partial W_c}{\partial \theta_r} d\theta_r = \int_{\theta_{on}}^{\theta_{off}} \int_{0}^{i} \frac{\partial \Psi(\theta_r, i)}{\partial \theta} di\, d\theta_r \tag{9.58}$$

The energy required for phase magnetization is eliminated from Equation 9.58, because it is "recovered" during phase demagnetization.

As expected, W_{mec} contains the core loss, therefore it is not exactly the shaft torque. The average machine torque T_{avr} is

$$T_{avr} = \frac{m \cdot N_r \cdot W_{mec}}{2\pi} \tag{9.59}$$

As the magnetization–demagnetization energies eliminate each other, W_{mec} may be calculated as follows:

$$W_{mec} = \oint_{cycle} \frac{d\Psi}{dt} i\, dt \tag{9.60}$$

The integral should be reset to zero when the phase current reaches zero, after each energy cycle.

But, the flux time derivative is straightforward:

$$\frac{d\Psi}{dt} = V_{ph} - R_s \cdot i \tag{9.61}$$

Simple as it may seem, but, when implemented, for low speeds, the phase voltage due to PWM is noisy, and temperature introduces notable errors. The situation may be ameliorated by some filtering of the phase voltage and by R_s adaptation.

The basic average torque observer T_{avr} is shown in Figure 9.24a and b.

More advanced W_{mec} estimators may be introduced to facilitate better performance at low speeds. They may make use of the machine approximate current model (flux/position/current), $\Psi(\theta_r, i)$, even

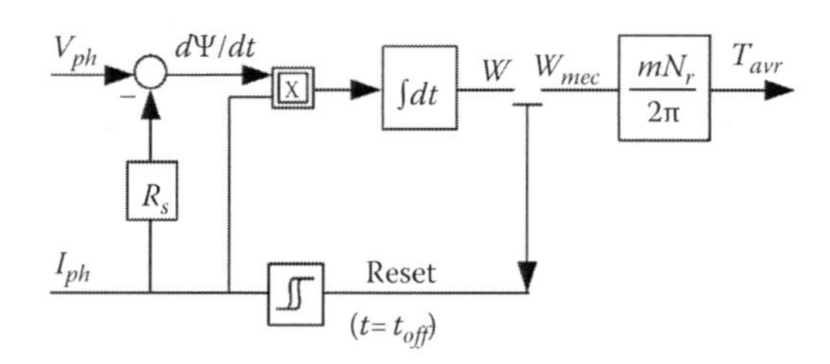

FIGURE 9.24 (a) Average torque estimator and (b) typical output at low speed (pulse-width modulator [PWM] current).

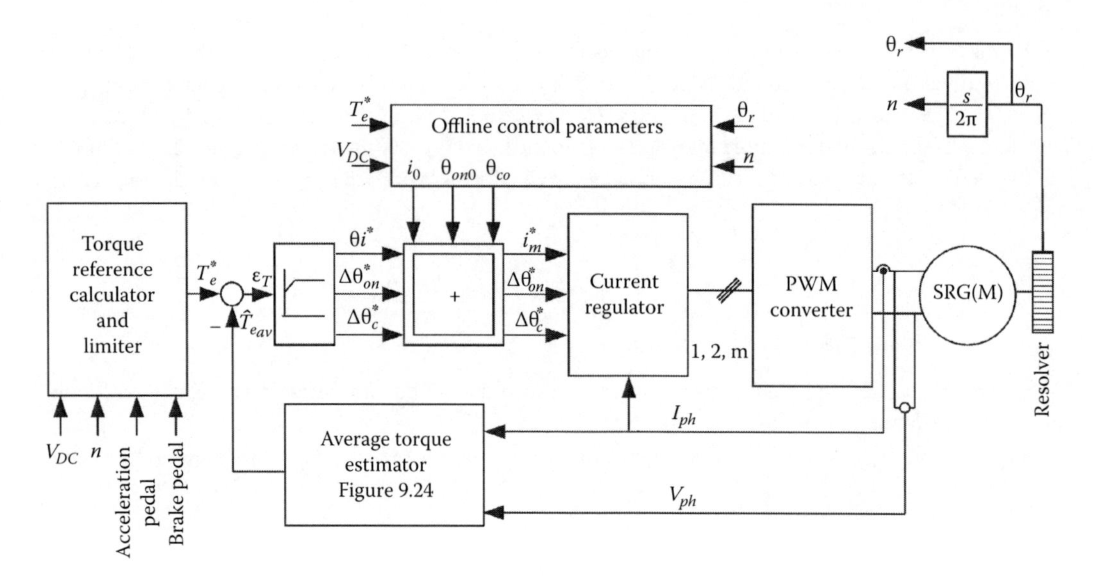

FIGURE 9.25 Direct average torque control of switched reluctance generator (SRG)(M).

through linear approximations (Equation 9.33). Analog implementation of the average torque estimator has proven efficient [17].

The torque error ε_T may be used to provide the stator current i^*, turn-on θ_{on}^* and turn-off θ_c^* increments through a PI regulator. These increments may then be added to the offline control parameters, as is done for feed-forward torque control (as discussed in the previous paragraph). A typical such direct torque control system is shown in Figure 9.25.

Good torque tracking control was claimed with this method in Reference 17, where, for simplicity, θ_{co} = constant.

The offline computation effort remains stern, while the direct average torque control adds more quickness and robustness to torque reference tracking, as many parameters vary in time. The battery voltage, V_{DC}, varies, for example, within 1 s, from its maximum to its minimum during peak motoring torque application for in-town driving of an electric vehicle (Figure 9.26) [17].

9.9 Rotor Position and Speed Observers for Motion-Sensorless Control

In the control schemes presented thus far, the rotor position and speed are provided by an encoder or a resolver with dedicated hardware for speed calculation.

As motion-sensorless control from zero speed, with signal injection for position estimation, was proven for PM-RSM starter/alternators (Chapter 8), the question is if the same can be done for SRG(M)s. There is a rich literature on motion-sensorless control of SRM, without and with signal injection (in the passive phase in general). A pertinent review may be found in Reference 18.

9.9.1 Signal Injection for Standstill Position Estimation

To identify the initial rotor position for a nonhesitant start, a voltage pulse is applied to one phase, and the current is acquired up to a certain value i_k. The corresponding flux Ψ_k in the winding is as follows:

$$\Psi_k \approx V_{DC} \cdot t_k \tag{9.62}$$

where t_k is the time for full voltage application.

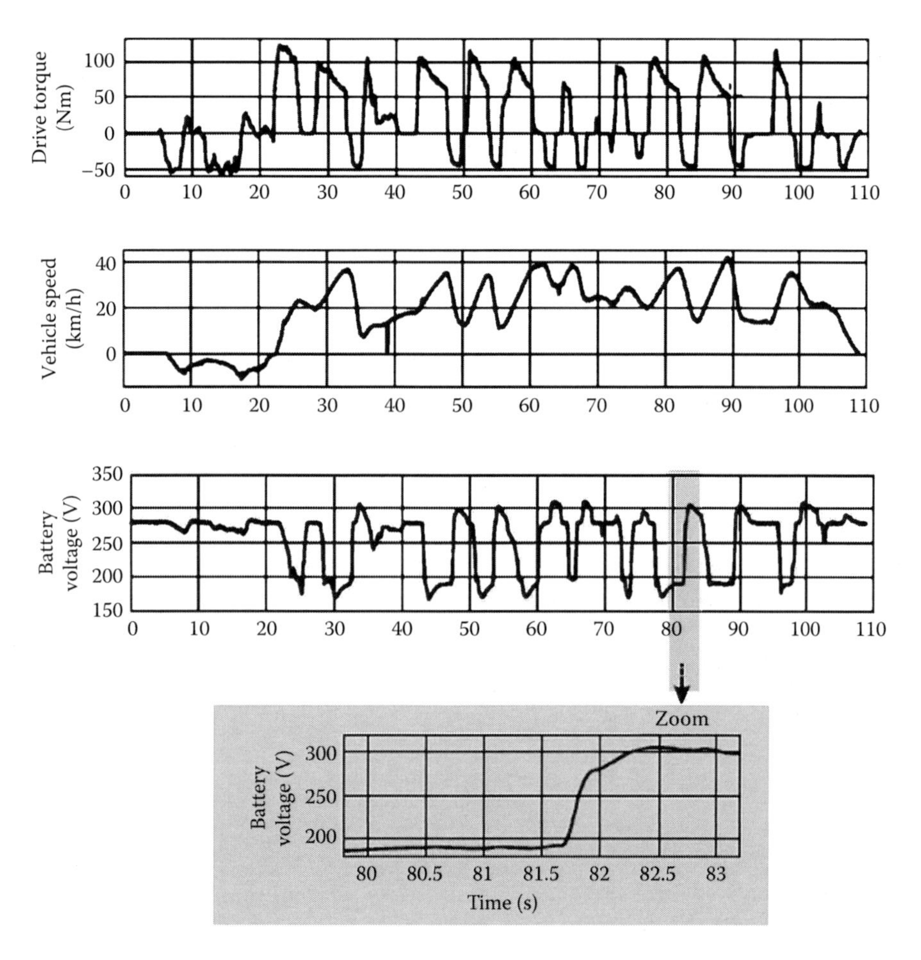

FIGURE 9.26 Torque, electrical vehicle (EV) speed, and battery voltage during standard town driving cycles.

For a given Ψ_k, i_k pair, from the already stored $\Psi(\theta_r, i)$, which is eventually curve fitted, the initial rotor position is estimated. While this method is correct in principle, it does not provide rotor position estimation at low speed and up to the speed when emf methods may be used. At low speeds, including standstill, choosing an idle phase, a small AC voltage V_{sens} signal may be injected. Neglecting the emf, the equation of idle phase is as follows:

$$V_{sens} = R_s i_j + L_j(\theta_r)\frac{di_j}{dt} \tag{9.63}$$

With a sinusoidal voltage injection of frequency ω_s:

$$V_{sens} = V_m \sin \omega_s t \tag{9.64}$$

The current in the idle phase is as follows:

$$I_{sens} = \frac{V_m}{\sqrt{R_s^2 + \omega_s^2 L_j^2(\theta_r)}} \sin(\omega_s t - \gamma) \tag{9.65}$$

$$\gamma = \tan^{-1}\frac{\omega_s L_j(\theta_r)}{R_s} \tag{9.66}$$

Both the amplitude and the phase of I_{sens} contain position information.

The measured inductance, obtained through a modulation technique, may be used again, in corroboration with an appropriate $\Psi(\theta_r, i)$ family of curves, to determine the rotor position. The sensing voltage signal has to be "moved" from one idle to the next idle phase, and special hardware is needed to produce the sinusoidal sensing voltage and so forth.

More advanced rotor position and speed observers may be used for low and high speeds without signal injection. A sliding mode such observer, for example, may be of the following form [14]:

$$\begin{aligned}\hat{\theta}_r &= \hat{\omega} + K_\theta \operatorname{sgn}(e_f)\\ \hat{\omega}_r &= K_\omega \operatorname{sgn}(e_f)\end{aligned} \tag{9.67}$$

$\hat{\theta}_r$ and $\hat{\omega}_r$ are the observed rotor position and speed

K_θ, K_ω are so-called innovation gains, dependent on position and, respectively, on speed

The error function of the observer e_f is as follows:

$$e_f = \sum_{j=1}^{m} \cos(\hat{\theta})(\hat{i}_j - i_j) \tag{9.68}$$

The estimated current is obtained after the phase flux is determined through integration:

$$\hat{\Psi}_j = \int (V_j - R_s i_j)dt \tag{9.69}$$

The current model based on known $\hat{\Psi}_j(\hat{\theta}_r, \hat{i}_j)$ retrieves, for already known $\hat{\theta}_r$, $\hat{\Psi}_j$, the estimated current $\hat{i}_j$. A combined voltage/current model may be used to estimate both $\hat{\Psi}_j$ and $\hat{i}_j$ at the same time, as done for AC machines (Chapters 7 and 8).

A motion-sensorless complete drive with interior torque close-loop control may be based on a parabolic relationship between flux Ψ and torque and a linear reduction of θ_{on} with speed and torque. The commutation angle θ_c varies from θ_c^i to θ_c^Ψ when the speed increases [14].

θ_c^i is the commutation (turn-off) angle for maximum torque per current; θ_c^Ψ is the commutation angle for maximum torque/flux. The latter criterion provides, for $\omega_r > \omega_b$, more available torque for given DC voltage, V_{DC}. Typical experimental results are shown in Figure 9.27a for responses at 95 rpm and in Figure 9.27b for those at 4890 rpm [14]. Due to integrator drift, sensorless operation at very low speeds (below 95 rpm in Reference 14) becomes difficult.

The combined voltage–current observer to estimate flux $\hat{\Psi}$ and current $\hat{i}$ may be used to reduce the speed for good position estimation, perhaps to a few (20) rpm [19,20]. Still, for safe standstill torque control at zero and a few rpm, a signal injection rotor position estimator is required. Reference 21 suggests a rotating signal injection position estimation that works from zero speed and is similar to its counterpart for AC drives.

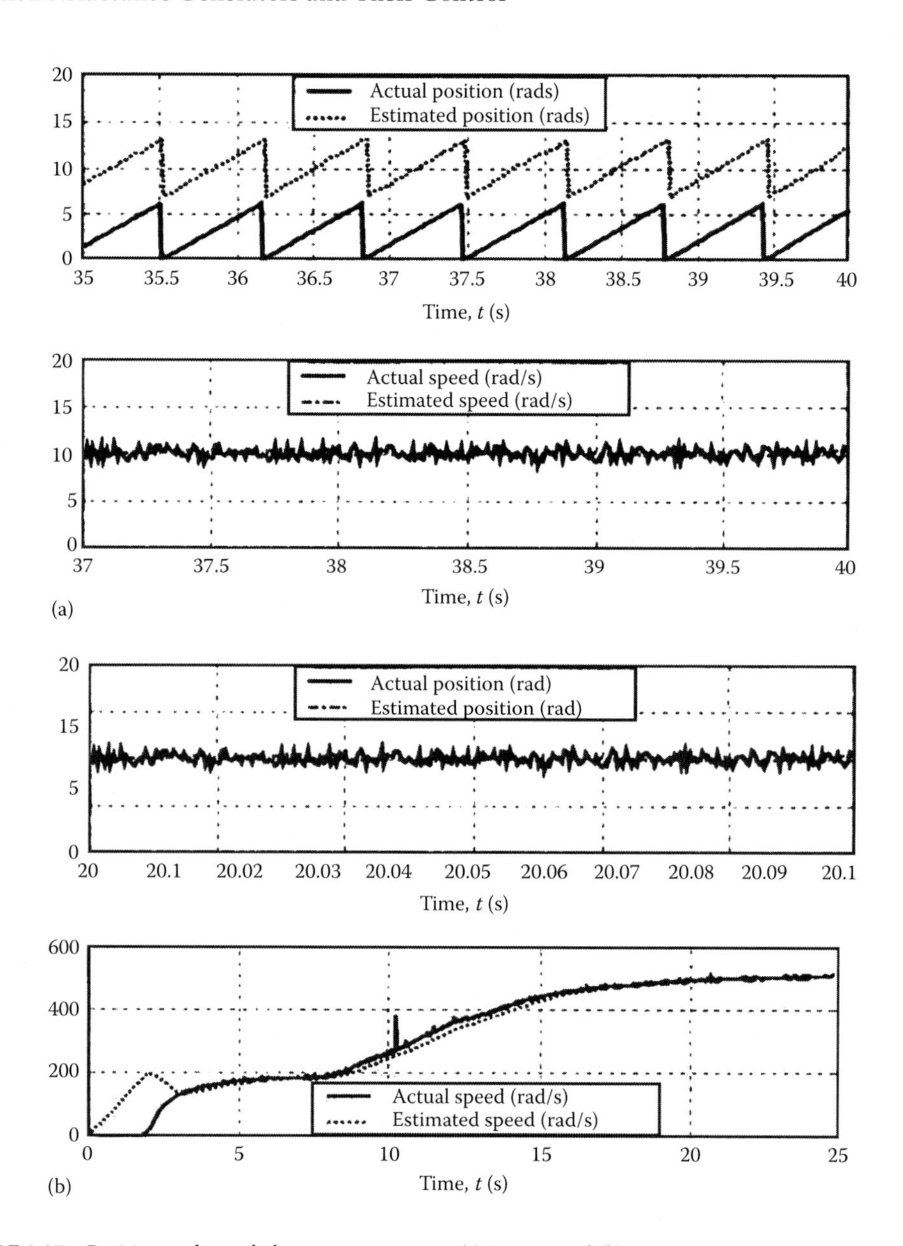

FIGURE 9.27 Position and speed observer responses at (a) 95 rpm and (b) 4890 rpm.

9.10 Output Voltage Control in SRG

Applications such as wind energy conversion or auxiliary power sources require constant DC voltage output for a limited speed range as mentioned before in this chapter.

A simplified control system for such applications relies on direct DC output voltage control with fixed or linearly decreasing with speed turn-on, θ_{on}, and commutation, θ_c, angles with rotor position feedback (Figure 9.28). An additional step-down chopper may be used to stabilize the DC output voltage control over a limited speed range.

The current regulator controls the level of machine energization through voltage PWM as the θ_{on} and θ_c angles are held constant. The self-excitation on no load is shown in Figure 9.29a to be slower than the

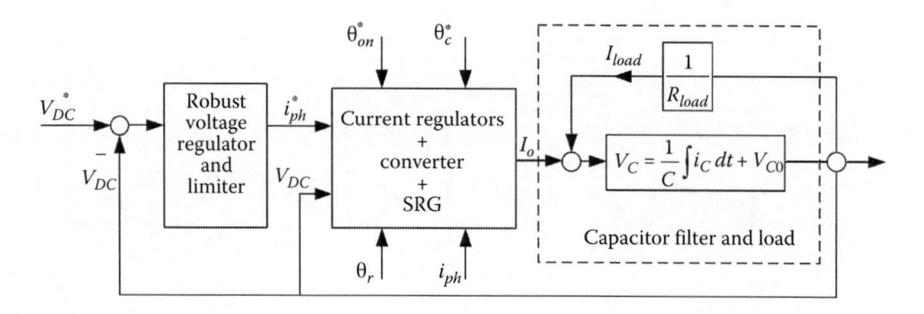

FIGURE 9.28 Direct current (DC) output voltage control of a switched reluctance generator (SRG).

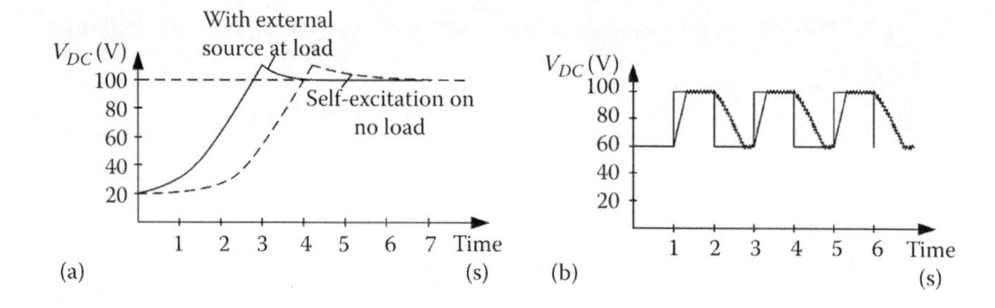

FIGURE 9.29 (a) The start-up and (b) voltage control.

start-up process on load with an external excitation source (a 12 V battery). Transient response to step reference DC output voltage shows rather fast response (Figure 9.29b) [22]. It is a stable voltage response with zero steady-state error. The low-voltage excitation source helps to self-excite the machine, especially when no remnant flux exists in the machine.

To produce more output, a limited soft turn-off stage—before hard turn-off at θ_c—may be introduced [23]. As a bonus, the noise level is also reduced.

9.11 Double Stator SRG with Segmented Rotor

SRG are known, in this radial air gap single stator configuration, with tooth-wound stator coils connected in multiple phases, to be inexpensive and, when optimally designed, show remarkable torque density (up to 45 Nm/L at 60 kW at 6000 rpm and above 92% efficiency in HEV propulsion systems with motor and generator attributes [24]).

But still, the ratio of the radial to tangential force density is much larger than in modern induction machines, DC excited or PM synchronous machines. This leads in presence of large torque pulsations at high speeds, to large noise and vibration.

In an effort to do away with this demerit, Reference 25 introduces a radial air gap dual stator SRM with a segmented–laminated rotor in between, with diametrical—span two-pole—like coils for the eight stator poles/six rotor poles, four-phase configuration (Figure 9.30) [25].

Here, more of the magnetic flux flows tangentially/laterally into the rotor's segments and thus the ratio between the radial and tangential (torque producing) force is reduced (Figure 9.31) [25].

It has to be mentioned that the "diametrical coils" (coil span = number of phases × stator pole slot pitch) in the stator windings of DS–SRM mean more copper weight and copper losses for more torque, so the efficiency will not increase notably. However, the torque density and the noise/vibration may be reduced substantially. Despite the manufacturing added complexity, DS–SRM seems worth pursuing for electric generators too, even in low speed direct drives but may be more in medium speed (up to 6000 rpm or so) applications.

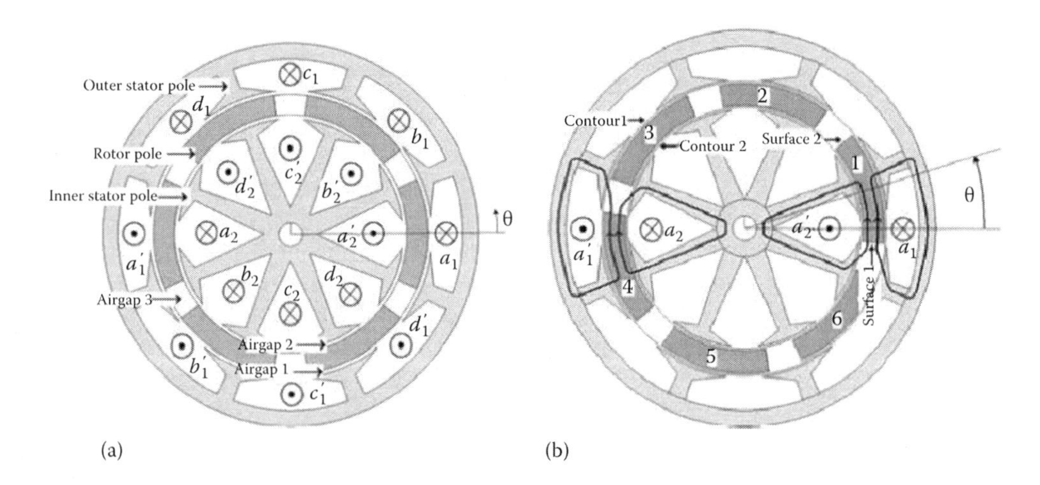

FIGURE 9.30 Dual stator SRG with four phases (a) and typical flux path (b).

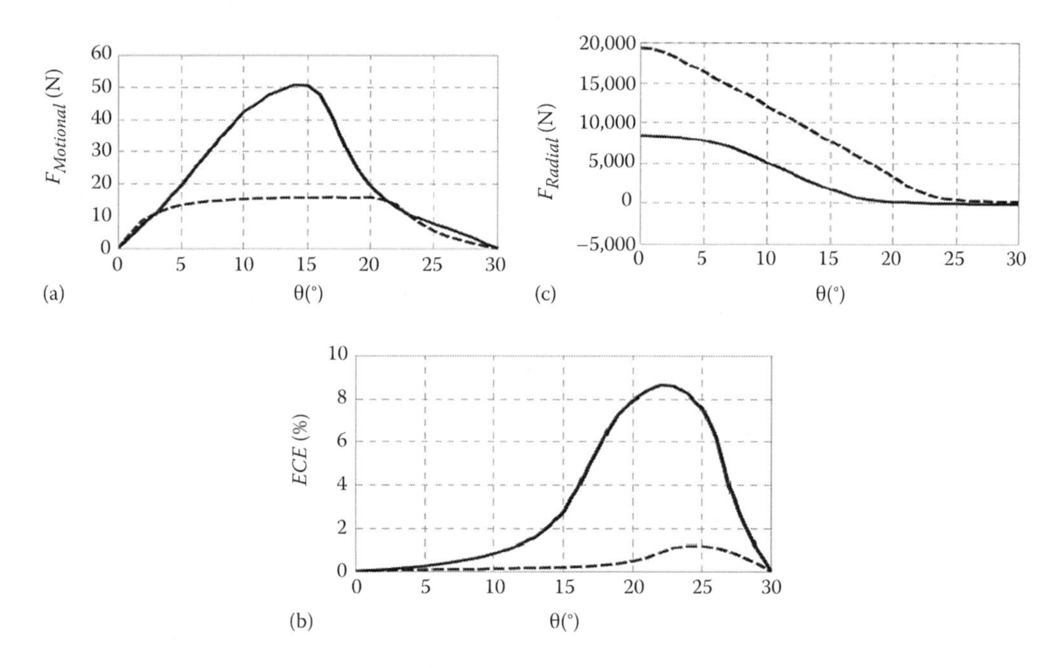

FIGURE 9.31 Torque producing (motional) force F_M vs. position for regular SRM and DS–SRM (a), radial forces F_R vs. position (b), and $F_M/(F_M + R_R)$ in%, for constant current in an 8/6 SRM (c).

An additional merit of DS–SRM could be lower commutation inductance, which in turn would be beneficial to faster current control and in reducing converter losses.

9.12 Summary

- Switched reluctance machines are of double-saliency type, with coils, around all stator poles, supplied in sequence with current controlled voltage pulses intact with rotor position. They have salient poles on the rotor and on the stator. Multiphase configurations are capable, with proper control, of rather smooth torque and self-starting capability from any initial position.

- Lacking windings or PMs on the rotor, SRG(M)s are rugged and suitable for hot environments (even up to 200°), and they are inexpensive.
- They work on the principle that the salient rotor poles are attracted to the energized phase (phases). As the rotor advances, one phase is turned off, while the next one is turned on at the adequate rotor position.
- SRMs are fully dependent on power electronics with rotor position-triggered control.
- Already, a handful of companies [3] commercialize SRM drives. As starter/generators, SRMs are vigorously proposed for automobiles, trucks, and aircraft applications.
- The machine configuration simplicity and ruggedness tend to be compensated for by the uniqueness, complexity, and added costs of the PWM converter system.
- In the absence of magnetic saturation, the inductance of each phase has a three-stage variation with rotor position: rising, descending, and constant. The three-stage period corresponds to $2\pi/N_r$ mechanical radians, where N_r is the number of rotor salient poles.
- In three-phase SRMs, only one phase has a nonconstant inductance versus rotor position for either motoring or generating at any rotor position. There are two such active phases at all rotor positions in four-phase machines.
- The number of stator poles is $N_s = m \cdot 2K$, where m is the number of phases and $2K$ is the number of poles per phase.
- N_s and N_r are related by, in general, $N_s = N_r + 2K$.
- Three or four phases are typical in industrial SRM (G)s with N_s/N_r = 6/4, 12/8, 24/16 for three phases and N_s/N_r = 8/6, 16/12, 32/24 for four phases. To reduce the electromagnetic noise, an even number of pole pairs K per phase is chosen.
- The flux interaction between phases, even in the presence of magnetic saturation, is small. Hence, fault tolerance is rightfully claimed for SRG(M)s.
- Each phase is motoring, while its inductance has a rising slope and is generating along the descending slope. Unfortunately, each phase has to be fully fluxed and defluxed every energy stroke (cycle); that is, N_r times per revolution.
- The current polarity is not relevant to torque sign, because there is no interaction between phases and no PMs. This property leads to special PWM converters with unipolar current control.
- The family of flux/position/curves per phase $\Psi(\theta_r, i)$ is the most important performance identifier for SRM (G)s. While for aligned position the magnetization curve has to be highly saturated, the unaligned rotor position magnetic saturation curve is linear.
- The area between these two extreme curves represents the maximum mechanical energy W_{mec} converted into an energy cycle. For magnetization, the energy area is W_{mag}. The energy conversion ratio is $ECR = W_{mec}/(W_{mec} + W_{mag})$, which is 0.5 short of magnetic saturation and 0.65–0.75 for heavy saturation.
- Indicative for SRG(M) performance, in connection with the PWM converter, is the kW/peak kVA, which is only 10%–15% lower than that for induction motor AC drives. However, 10%–15% extra cost in the converter can weigh a lot in the system costs.
- The more common kW/rms/kVA ratio, or equivalent power factor, is not as important as the ratio between peak instantaneous current and peak RMS current that tends to be larger for SRG(M)s.
- The circuit model is based on independent phase voltage equations. In addition, due to magnetic saturation, only a pseudo emf E may be defined, while the instantaneous torque has to be computed directly from $\partial W_c/\partial\theta_r$ (W_c is the coenergy per phase).
- The pseudo-emf E depends on rotor position and on current, but is positive for motoring and negative for generating (Figure 9.6).
- The pulsed-supply of phases leads to iron losses both in the stator and rotor and makes the core loss model rather involved.

- The transient inductance $L_t = \partial\Psi/\partial i$ is, under saturation conditions, close to its unaligned value. This explains the acceptably quick phase fluxing during generating. Magnetic saturation also limits the maximum flux in the machine to keep the voltage rating within reasonable limits.
- Exponential, polynomial, fuzzy logic, ANN, or even piece-wise linear approximations were proposed to curve-fit the $\Psi(\theta_r, i)$ family of magnetization curves that is so necessary, both for SRG performance simulation and for control design implementation.
- The SRG design issues depend on the speed range for constant power and on output voltage.
- SRG(M) specifications are mainly expressed in terms of torque/speed envelopes for motoring and generating, for given DC voltage source (output) value. The maximum value of the current i_{peak} is also specified.
- After sizing the SRG(M), an optimization design is pursued based on maximum torque/current below base speed n_b (rps) and maximum torque per total losses, to maximum torque/flux, at large speeds (above n_b).
- The base speed n_b corresponds to the case when the pseudo-emf E equals a given fraction $\alpha_{PF} \approx 1$ of the DC voltage, and the machine produces the peak torque under the conditions of maximum torque per ampere.
- When a wide constant power speed range is required, with $\alpha_{PF} = 1$, the minimum speed $n_{\min}$ may be chosen smaller than n_b, but at the price of a lower number of turns/coil and higher peak current, and, consequently, higher converter costs (converter oversizing as for the AC starter/alternators [Chapters 7 and 8]).
- The current is chopped below n_b, and single current pulse operation takes place above n_b for motoring.
- Single-pulse generation is feasible for $n > n_b$, but PWM modulation may be supplied for all speeds to lower the torque during generation.
- The flat top current generating corresponds roughly to $E = V_{DC}$ and may be maintained for a range of speeds only if V_{DC} is allowed to increase with speed. In this case, for constant output (load) voltage V_{load}, an additional step-down DC–DC converter is used to interface the load.
- $E = V_{DC}$ corresponds to very good energy conversion in the machine and converter.
- Phase energizing for generating takes place along descending slope of phase inductance and, thus, is slow, and it takes its toll of P_{exc} in the power. When the phase is turned off (commutated), the free-wheeling diodes become active and deliver power P_{out} to the load. For high speeds, $E > V_{DC}$, while for low speeds, $E \leq V_{DC}$, if single pulse generating is performed with turn-on θ_{on} and turn-off θ_c angles as controlled variables. The ratio $P_{exc}/P_{out} = \varepsilon$ is called the excitation penalty, which, for a well-designed machine, should lie in the interval of 0.3–0.5.
- The excitation penalty ε is not minimum when the energy conversion ratio in the machine is maximum ($E = V_{DC}$), but for $E > V_{DC}$.
- There are many unipolar current DC–DC multiphase converters suitable for SRG(M)s, but the asymmetrical one with two active power switches per phase is most used thus far, especially for generators, where both hard- and soft-switching are required to increase power output and reduce noise and vibration.
- In essence, when SRGs supply a passive DC load, self-excitation is difficult, and a special low-voltage battery with a series diode is provided for initiation. For safe and stable output control, it is possible to use a separate excitation bus (or battery) and a power bus that might flow power back to the excitation bus, to lower the power rating of the excitation battery (Figure 9.13).
- Control of SRG(M)s for starter/generator applications amounts to feed-forward or direct (close-loop) average torque with or without position sensors.
- Average torque estimation proves rather straightforward, as the mechanical energy per cycle W_{mec} equals the resultant electrical energy per cycle $W_{mec} = \int (d\Psi/dt) i\, dt$ at the moment when the current in the respective phase becomes zero. This is so because the energy for fluxing and defluxing

the phase cancel each other during a complete energy cycle. This is not so for continuous current control.

- The curve-fitted $\Psi(\theta_r, i)$ family is used offline to calculate the optimum θ_{on} and θ_c angles and current, for given reference average torque, speed, and battery voltage and state of charge.
- Thus far, no practical control system that fully avoids offline calculations was demonstrated for SRG(M)s in contrast to AC drives.
- Motion-sensorless control for starter/generators requires position estimation from zero speed. Therefore, a signal injection method is mandatory for zero and low speeds, while an emf method is to be used when the speed increases. In References 21 and 26, such an efficient method of saliency detection through quasirotating voltage vector injection and current response processing was proposed. It is a version to the one introduced earlier for AC drives.
- When only generator mode is required, simpler position estimators may be used, while direct DC output voltage control is performed.
- The control of SRG(M)s is more involved than for AC machines, but its capacity to perform over a wide constant power speed range, because phase defluxing is at hand, and machine ruggedness and low cost, make the former a tough competitor to the AC starter/generator systems.

References

1. T.J.E. Miller, *Switched Reluctance Motors and Their Control*, Oxford University Press, Oxford, U.K., 1993.
2. R. Krishnan, *Switched Reluctance Motor Drives*, CRC Press, Boca Raton, FL, 2001.
3. T.J.E. Miller, Optimal design of switched reluctance motors, *IEEE Trans.*, IE-49(1), 2002, 15–17.
4. M. Rahman, B. Fahimi, G. Suresh, A.V. Rajarathnam, and M. Eshani, Advantages of switched reluctance motor application to EV and HEV: Design and control issue, *IEEE Trans.*, IA-36(1), 2000, 111–121.
5. D.A. Torrey, Switched reluctance generators and their control, *IEEE Trans.*, EC-49(1), 2002, 3–14.
6. V.V. Athani and V.M. Walivadeker, Equivalent circuit for switched reluctance motor, *EMPS J.*, 22(4), 1994, 533–543.
7. M. Ilic-Spong, M. Marino, S. Peresada, and D. Taylor, Feedback linearizing control of switched reluctance motor, *IEEE Trans.*, AC-32(5), 1987, 371–379.
8. A.V. Radun, Design consideration of switched reluctance motor, In *Record of IEEE–IAS-1994 Annual Meeting*, 1994.
9. A.V. Radun, C.A. Ferreira, and E. Richter, Two-channel switched reluctance starter–generator results, *IEEE Trans.*, IA-34(5), 1998, 1106–1109.
10. D.E. Cameron and J.M. Lang, The control of high speed variable reluctance generators in electric power system, *IEEE Trans.*, IA-29(6), 1993, 1106–1109.
11. T. Sawata, Ph. Kjaer, C. Cossar, and T.J.E. Miller, Fault tolerant operation of single-phase switched reluctance generators, In *Record of IEEE–APEC-1997 Annual Meeting*, 1997.
12. I. Husain, Minimization of torque ripple SRM drives, *IEEE Trans.*, IE-49(1), 2002, 28–39.
13. H. Bausch, A. Grief, K. Kanelis, and A. Mickel, Torque control of battery-supplied switched reluctance drives for electrical vehicles, Record of ICEM-1998, Istanbul, Turkey, 1998, pp. 229–234.
14. M.S. Islam, M.M. Anwar, and I. Husain, A sensorless wide-speed range SRM drive with optimally designed critical rotor angles, In *Record of IEEE–IAS-2000 Annual Meeting*, 2000.
15. K.M. Rahman and S.E. Shultz, High-performance fully digital switched reluctance motor controller for vehicle propulsion, *IEEE Trans.*, IA-38(4), 2002, 1062–1071.
16. R.B. Inderka and R.W. De Doncker, Simple average torque estimation for control of switched reluctance machines, Record of EPE–PEMC-2000, Kosice, Slovakia, 2000, vol. 5, pp. 176–181.
17. R.B. Inderka, M. Menne, and R. De Donker, Control of switched reluctance drives for electric vehicle applications, *IEEE Trans.*, 49(1), 2002, 48–53.

18. M. Ehsani and B. Fahimi, Elimination of position sensor in switched reluctance motor drives: State of the art and future trends, *IEEE Trans.*, 49(1), 2002, 40–47.
19. P.P. Acarnley, C.D. French, and I.H. Al-Bahadly, Position estimation in switched reluctance drives, In *Record of EPE-1995 Conference*, Sevilla, Spain, 1995, pp. 3765–3770.
20. G. Lopez, P.C. Kjaer, and T.J.E. Miller, High grade position estimation for SRM using flux linkage current correction mode, In *Record of IEEE–IAS-1998 Annual Meeting*, 1998, pp. 731–738.
21. R.D. Lorenz and N.J. Nagel, Rotating vector methods for sensorless, smooth torque control of switched reluctance motor drive, In *Record of IEEE–IAS-1998 Annual Meeting*, 1998, vol. 1, pp. 723–730.
22. P. Chancharoensoon and M.F. Rahman, Control of a four-phase switched reluctance generator: Experimental investigations, Record of IEEE–IEMDC-2003, Madison, WI, 2003, vol. 1, pp. 842–848.
23. S. Dixon and B. Fahimi, Enhancement of output electric power in switched reluctance generators, Record of IEEE–IEMDC-2003, Madison, WI, 2003, vol. 2, pp. 849–856.
24. K. Kiyota and A. Chiba, Design of switched reluctance motor competitive to 60 kW IPMSM in third generation HEV, *IEEE Trans.*, IA-48(6), 2012, 2303–2309.
25. M. Abbasian, M. Moallem, and B. Fahimi, Double-stator switched reluctance machine (DS-RSM): Fundamentals and magnetic force analysis, *IEEE Trans.*, EC-25(3), 2010, 589–597.
26. B. Fahimi, A. Emadi, and R.B. Sepe Jr., Position sensorless control: Presenting a technology ready for switched reluctance machine drive applications, *IEEE–IA Magazine*, 10(1), 2004, 40–47.

10

Permanent Magnet Synchronous Generator Systems

10.1 Introduction

By permanent magnet synchronous generators (PMSGs), we mean here radial or axial air gap PM brushless generators with distributed ($q > 1$) or concentrated ($q \leq 1$) windings and rectangular or sinusoidal current control with surface PM or interior PM (IPM) rotors. Multiple pole transverse flux machines (TFMs) or flux switching (reversal) machines (FRMs) conceived for low speed and high torque density will be discussed in Chapter 11.

PMSG's output voltage amplitude and frequency are proportional to speed. In constant speed prime-mover applications, PMSGs might be capable of voltage self-regulation through proper design, that is, inset or interior PM pole rotors. Small speed variation (±10%–15%) may be acceptable for diode rectified loads with series capacitors and voltage self-regulation. However, most applications require operation at variable speed, and in this case, constant output voltage versus load, be it direct current (DC) or alternating current (AC), requires full static power conversion and closed-loop control.

Versatile mobile generator sets (gensets) use variable speed for fuel savings, and PMSGs with full power electronics control can provide high torque density, low losses, and multiple outputs (DC and AC at 50(60) or 400 Hz, single phase or three phase).

A high efficiency, high active power to peak kilovolt ampere (kVA) ratio allows for reasonable power converter costs that offset the additional costs of PMs in contrast to switched reluctance generators (SRGs) or induction generators (IGs) for the same speed.

For automotive applications, and when motoring is not necessary, PM generators may provide controlled DC output for a 10 to 1 speed range through a diode rectifier and a one insulated gate bipolar transistor (IGBT step-up DC–DC converter for powers above 2–3 kW. A series hybrid vehicle is a typical application here. Gas turbines run at super high speeds; 3.0 MW at 18 krpm to 150 kW at 80 krpm. Direct-driven super-high-speed PM generators, with their high efficiency and high power factor, seem to be the solution for such applications. With start-up facilities for bidirectional power flow, static converters allow for four-quadrant control at variable speed, with ±100% active and reactive power capabilities. Distributed power systems of the future should take advantage of this technology of high efficiency, reasonable cost, and high flexibility in energy conversion and in power quality.

Flywheel batteries with high kilowatts per kilogram (kW/kg), good kilowatt hours per kilogram (kWh/kg), and long life also use super-high-speed PMSGs with four-quadrant P and Q control. They are proposed for energy storage on vehicles and spacecraft and for power systems backup.

Diverse as they may seem, these applications are accommodated by only a few practical PMSGs classified as follows:

- With radial air gap (cylindrical rotor)
- With axial air gap (disk rotor)
- With distributed stator windings ($q > 1$)
- With concentrated windings ($q \leq 1$)
- With surface PM rotors
- With interior or inset PM rotors
- With rectangular current control
- With sinusoidal current control

In terms of loads, they are classified as follows:

- With passive AC load
- With DC load
- With controlled AC voltage and frequency at variable speed

The super-high-speed PM generators differ in rotor construction; they need a mechanical shell against centrifugal forces and a copper shield (damper) to reduce rotor losses. In addition, at high fundamental frequency (above 1 kHz), stator skin effect and control imply special solutions to reduce machine and static converter losses and overall costs.

As in most PMSGs, surface PM rotors are used; the latter will be given the most attention. An IPM rotor case will be covered in a single paragraph, when voltage self-regulation is acceptable due to almost constant speed operation. Basic configurations for stator and rotor will be introduced and characterized. A comprehensive analytical field model is introduced and checked through finite element method (FEM) field and torque production analysis. Loss models for generator steady-state circuit modeling are introduced for rectangular and sinusoidal current controls. Design issues and a methodology by example are treated in some detail.

PM generator control and its performance with direct AC loads, with rectified loads, and with constant voltage and frequency output at variable speed by four-quadrant AC–AC static power converter *P*–*Q* control systems are all treated in separate sections. Super-high-speed PM generator design and control are dealt with in some detail, with design issues as the focal point. Methods for testing the PMSGs to determine losses, efficiency, and parameters and a case study comprehensive PMSG control close the present chapter.

10.2 Practical Configurations and Their Characterization

PM brushless motor drives have become a rather mature technology with a sizeable market niche worldwide. Thorough updates of these technologies may be found in the literature [1–5].

For PMSGs, both surface PM and inset PM pole rotors are used. A typical cylindrical rotor configuration is shown in Figure 10.1. For IPM pole rotors, the magnetic reluctance along the direct (d) axis is larger than for the transverse (q) axis; thus, $L_d < L_q$—that is, inverse saliency, in contrast to electromagnetically excited pole rotors for standard synchronous machines.

The d axis falls along the PM field axis in the air gap. The rotor may be internal (Figure 10.1) or external (Figure 10.2) to the stator, in cylindrical rotor configurations. Interior rotors require a carbon fiber mechanical shield (retainer) against centrifugal forces for high-speed applications (above 50–80 m/s peripheral speed). In contrast, external rotors do not need such a retainer, but the yoke has to withstand high centrifugal forces in high-speed rotors.

For high-speed PMSGs with cylindrical rotors, characterized by a wide constant power speed range ($\omega_{max}/\omega_b > 4$), a hybrid surface PM pole and variable reluctance rotor may be used (Figure 10.3a through c). The surface PM pole rotors were proven capable of slightly more torque for given rotor external diameter

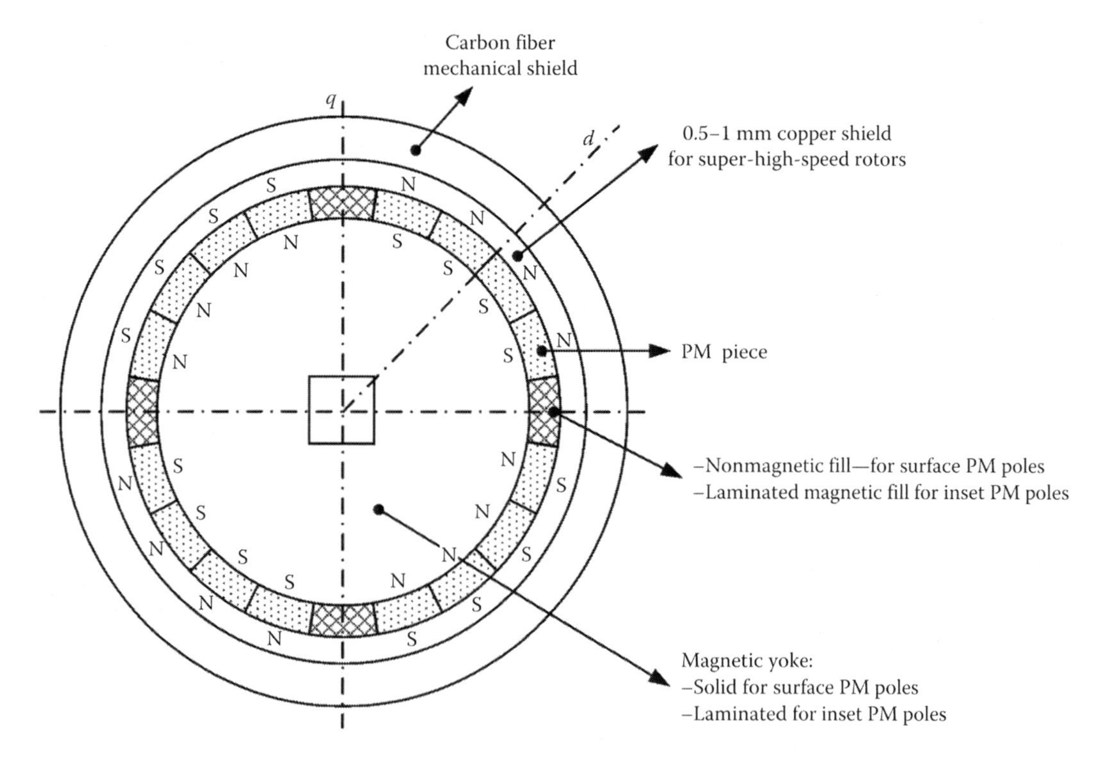

FIGURE 10.1 Four-pole surface permanent magnet (PM) and inset PM pole rotor configurations.

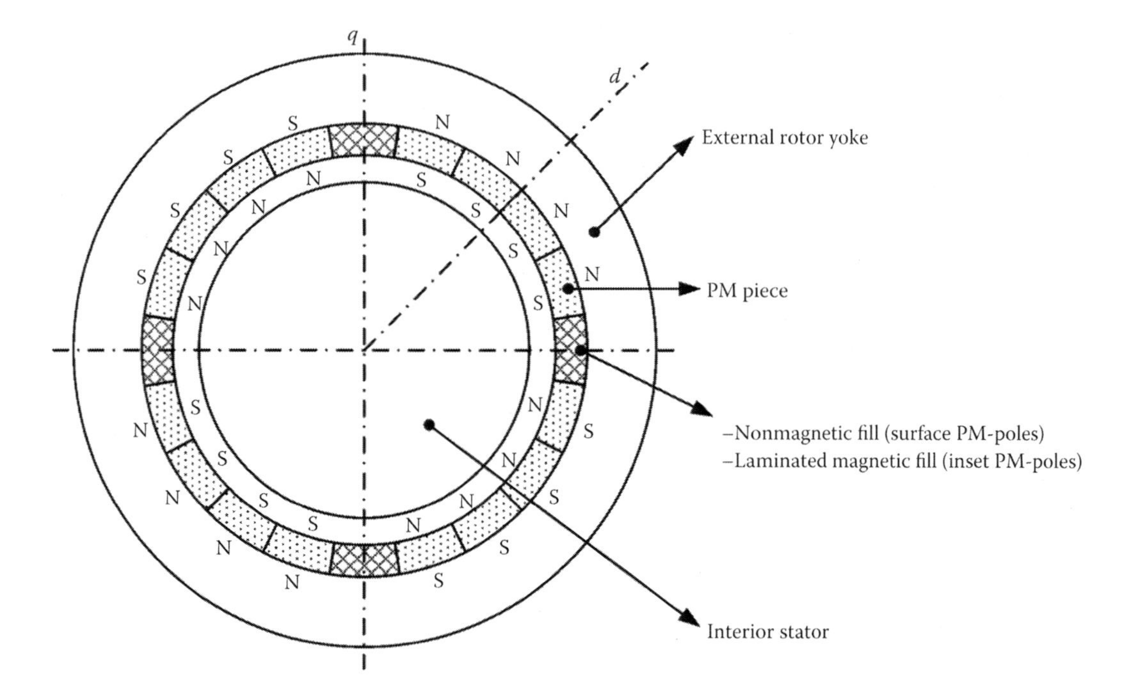

FIGURE 10.2 Four-pole cylindrical external rotor.

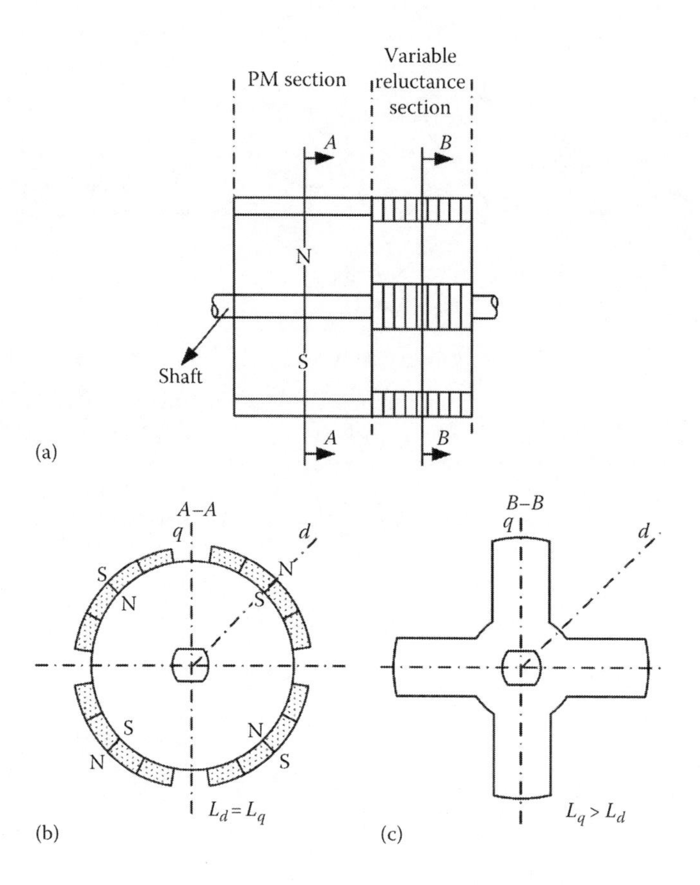

FIGURE 10.3 Four-pole PM rotor: (a) with surface PM pole, (b) A–A cross section and variable reluctance, and (c) B–B cross section.

and length and the same PM weight. For generators, they provide for small $L_d = L_q$ inductances in per unit (P.U.) terms, which should result in lower voltage regulation.

However, inset PM pole or hybrid rotors, due to inverse saliency ($L_d < L_q$), may produce zero voltage regulation at a certain resistive load that is fixed by design. In self-regulated PMSGs, driven at quasiconstant speed, such a design may prove practical.

In super-high-speed PMSGs with fundamental frequency above 1 kHz, the core loss and rotor loss are major concerns. To reduce rotor losses, the surface PM rotors are the natural choice. In such applications, copper shields surrounding the PM pole reduce the harmonics losses in the rotor PMs; moreover, solid back iron in the rotor is acceptable. This, in turn, leads to better rotor mechanical integrity.

Disk-shaped PM rotors were also proposed to increase the torque/volume in axial-length constraint designs. The typical rotor has the PMs embedded in a stainless iron disk; a mechanically resilient magnetic disk behind the PMs is also feasible, as it reduces the rotor harmonics losses (Figure 10.4a and b). The configurations in Figure 10.4 preserve the ideal zero axial force on the central part when the latter rotor is in the center position; thus, the bearings are spared large axial forces.

It is, in principle, possible to use only one of the two rotor parts in Figure 10.4b with a single stator in front of it, axially. But in this case, the axial (attraction) force between the rotor and the stator is very large, and the bearings have to handle it. Special measures to reduce vibration and noise are required in this situation.

In the interior rotor configuration, there is no rotor magnetic core in Figure 10.4a, and the axial rotor length and its inertia are small. In contrast, the two external stators will require a magnetic yoke (each of them). To reduce the stator's axial length, the number of poles must be increased. This renders the disk rotor PM generators, favorable for a large number of poles ($2p_1 > 6$) for fundamental frequency

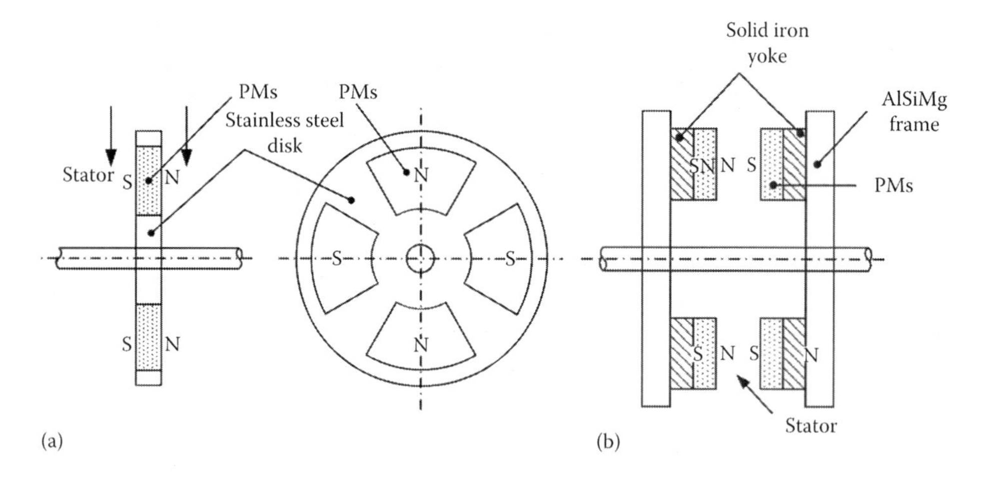

FIGURE 10.4 Four-pole disk-shape PM rotors: (a) interior rotor (with dual stator) and (b) double-sided rotor (with single and dual face stators).

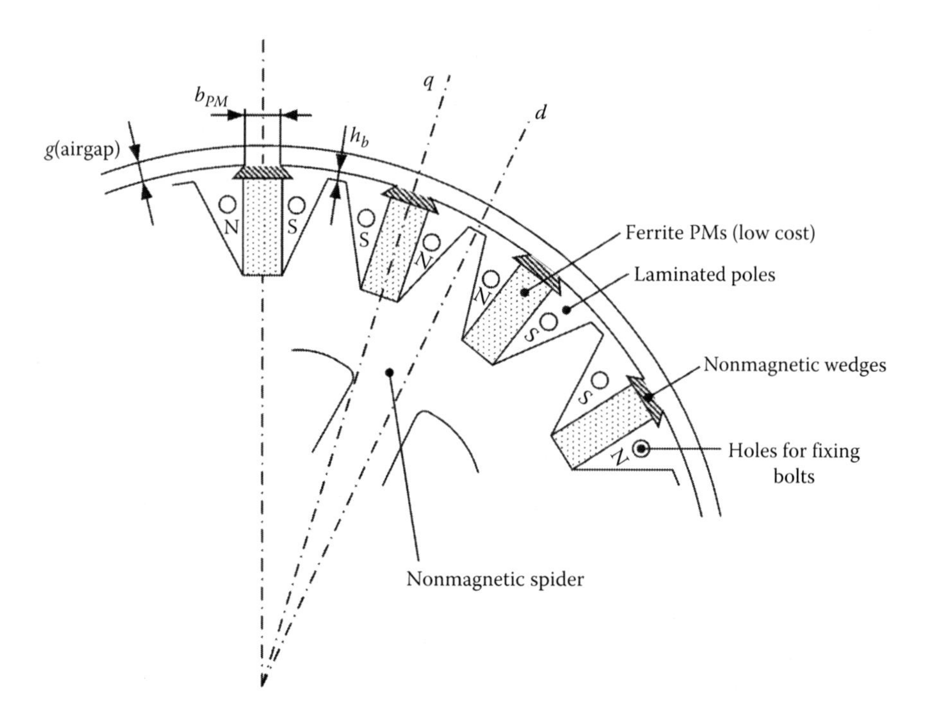

FIGURE 10.5 Ferrite PM rotor with flux concentration.

below 2 kHz. Multiple rotor and stator disk rotor configurations were proposed for low-speed, high-torque direct drives for elevators [5].

Though quite a few thorough comparisons between cylindrical and disk rotor PM synchronous machines were performed, no clear-cut "prescriptions" are available. However, it seems that depending on the number of poles ($2p_1 > 8, 10$) and if the stack length is short, the disk rotor PMSGs lead to higher torque/volume for equivalent losses and torque [6,7].

For large torque (diameter) applications, such as direct-driven wind generators, low-cost ferrite PMs were proposed to cut the costs [8]. Heavy PM flux concentration is required. A typical such rotor structure is shown in Figure 10.5.

Apparently, the configuration exhibits saliency ($L_d < L_q$) due to the presence of the PMs along the d-axis field path. In reality, the slim flux bridges (h_b) lead to their early saturation by the q-axis (torque) currents in the stator; thus, the saliency decreases with load to negligible values. A small machine inductance generally produces small voltage regulation.

To reduce the tangential fringing flux of the PMs in the air gap, $b_{PM} > (3–4)g$ (g is the mechanical gap between rotor and stator). For rotor diameters of 3–4 m, even at speeds below 30 rpm, typical for large wind turbines, a mechanical air gap of at least 5 mm is mechanically required with rigid frame designs. Therefore, the PM width b_{PM} should be at least 15–20 mm, and the pole pitch τ should not be less than 80–100 mm to provide smooth PM flux density distribution in the air gap.

The silicon iron laminated poles of the rotor lead to moderate rotor surface and pulsation harmonics core losses in the rotor.

Generally, stators are made of a laminated iron core with slots, but slotless cores may be adopted for high-speed (frequency) applications. The core may either be cylindrical or, again, disk-shaped (Figure 10.6a and b).

The radial and axial slots are stamped into radial and rolled silicon–iron lamination cores, respectively.

In super-high-speed applications, the core losses in the stator teeth may become so large that their elimination could prove beneficial, despite PM flux density reduction due to increased "magnetic" air gap. Slotless stators for cylindrical and disk-shaped rotor PMSGs are shown in Figure 10.7a and b.

The space left between Gramme-ring coils is required to fix the rolled lamination stator core of the disk rotor PMSG (Figure 10.7b) to the machine frame. Still, the stator structure is mechanically fragile, and care must be exercised in its mechanical design, to avoid inadmissible axial bending.

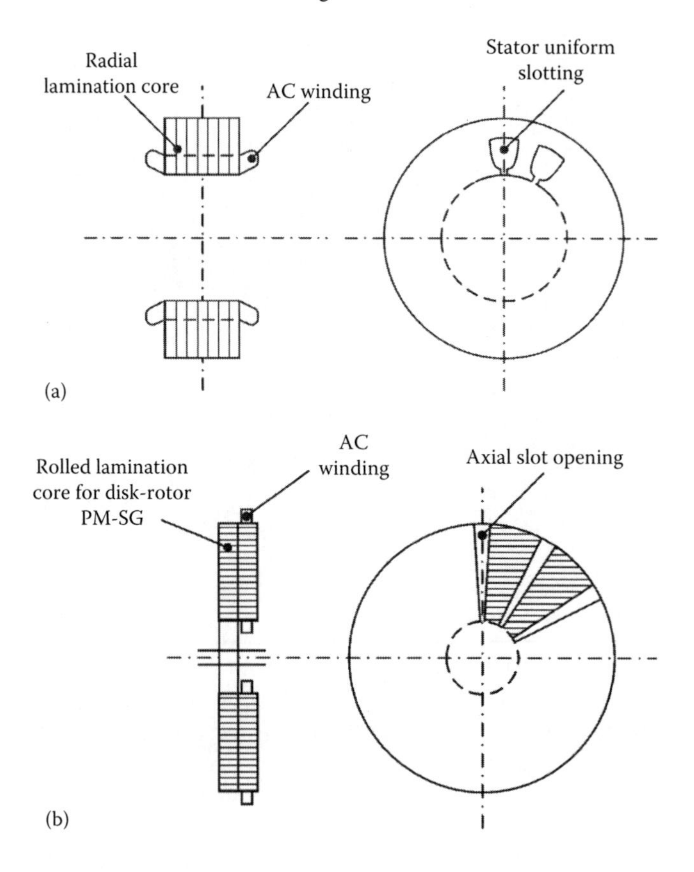

FIGURE 10.6 Slotted stator cores: (a) for cylindrical-rotor permanent magnet synchronous generator (PMSG) and (b) for disk-rotor PMSG.

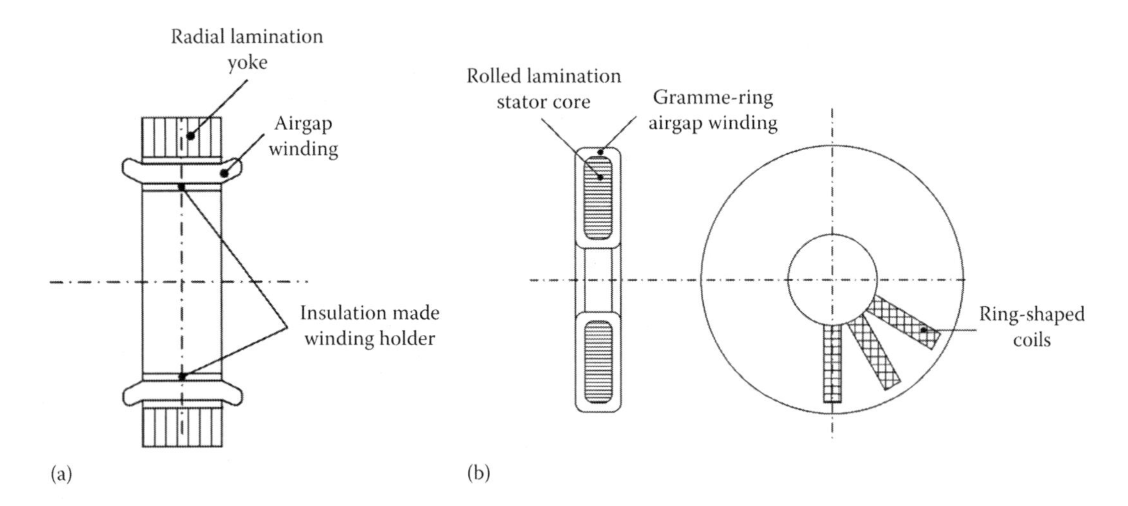

FIGURE 10.7 Air gap windings: (a) for cylindrical rotor and (b) for double-sided disk-shaped rotor.

The Gramme-ring winding is a special winding with short end connections, but a standard three-phase winding may be placed, as for the cylindrical rotor machine, in a dummy, insolation-made, slotted winding holder to reduce manufacturing time and costs.

As expected, being "washed" by the cooling air, the stator slotless windings may be better cooled; but due to the larger magnetic air gap, the winding loss and torque tend to be notably larger than that in slotted cylindrical stator configurations.

10.2.1 Distributed versus Concentrated Windings

Distributed single- or double-layer three-phase windings with $q = 2 \div 6$ are typical for PMSGs. Chording of coils in a ratio of about 5:6 is performed on double-layer windings to reduce the fifth and seventh stator magnetomotive force (mmf) space harmonics and their additional rotor surface and PM eddy current losses.

A large q (slot/pole/phase) leads to an increase in the order of the first slot harmonic, with amplitude that is thus reduced along with its additional rotor surface and PM eddy current losses. However, for large q, the end connections tend to be large, and thus the Joule losses tend to increase. Up to $q = 3$, a good compromise in terms of total losses may be secured.

Distributed windings are used to provide an almost sinusoidal electromagnetic field (emf) in the stator winding despite the nonsinusoidal (rather flat) PM flux density distribution in the air gap. Stator or rotor skewing may be added to further sinusoidalize the emf at the price of a 5%–7% reduction in emf root-mean-squared (RMS) value for one stator slot pitch skewing. Sinusoidal emf requires sinusoidal shape current to provide, ideally, rippleless interaction (PM to winding currents) torque.

The rather trapezoidal emf is obtained with three slots per pole ($q = 1$). If the PM air gap flux density distribution is flat in this case, trapezoidal distribution (ideally rectangular) current control is required to reduce torque pulsations between the commutation of phases. For rectified output PMSGs, the exploitation of the third harmonic in the emf can lead to increased power conversion; therefore, trapezoidal emf may be adopted in some designs.

In an effort to reduce stator manufacturing costs and the winding Joule losses, concentrated (non-overlapping) windings with $q < 1$ were adopted [9,10].

The number of stator slots N_s and the number of rotor PM poles $2p_1$ are close to each other. The closer to unity is the ratio $N_s/2p_1$, the better, as the number of periods of fundamental torque at zero current due to PMs and slot openings is equal to the lowest common multiplier (LCM) of N_s and $2p_1$. In addition, the equivalent winding factor for the winding is large if the LCM is large.

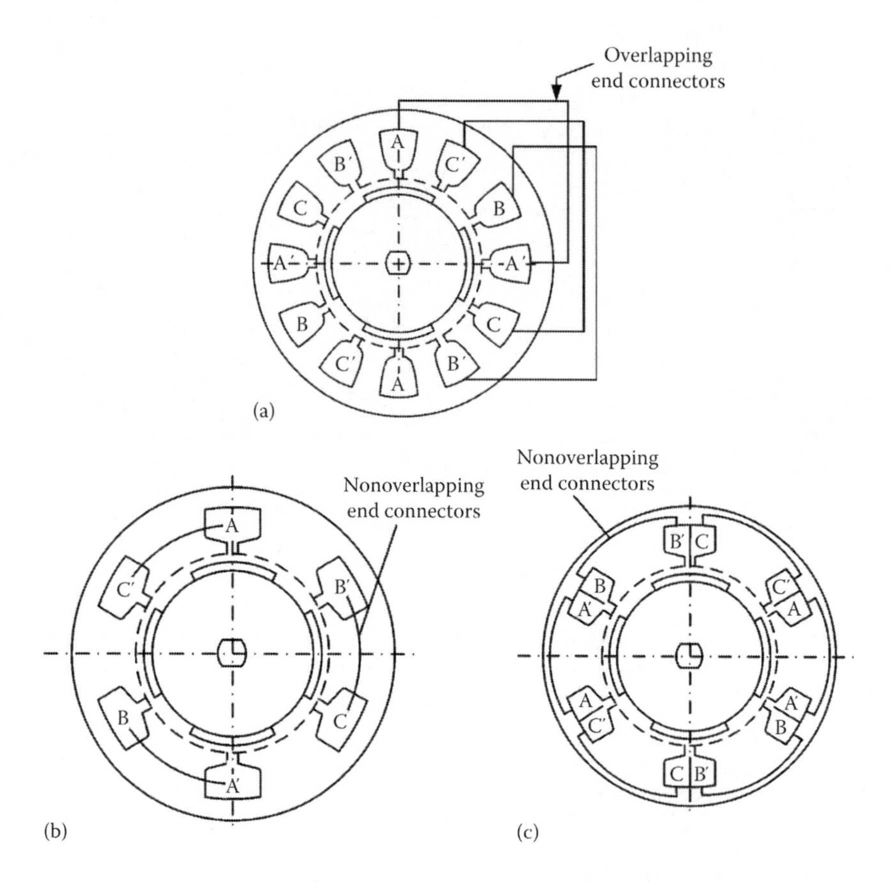

FIGURE 10.8 Four-pole winding layouts for permanent magnet synchronous generators (PMSGs): (a) distributed winding ($q = 1$), (b) single-layer concentrated winding, and (c) double-layer concentrated winding.

The nonoverlapping coil windings may be built in one layer or two layers (Figure 10.8a through c). The single-layer concentrated windings (Figure 10.8b) lead to only slightly higher winding factors (Table 10.1), but they imply longer end connections and, thus, longer stator frames in comparison to double-layer concentrated windings. The latter allow for a slightly less slot filling factor, as there are two coils in a slot and, inevitably, some airspace remains between the two coils [10].

The winding factor is defined as the ratio of the resulting emf E_{cp} per current path (or phase) divided by the product of the number of coils N_{cp} to their emfs E_c:

$$k_W = \frac{E_{cp}}{N_{cp} \cdot E_c} \tag{10.1}$$

Values of k_W 0.866 compare favorably with distributed windings. There are some side effects, such as the presence of both odd and even harmonics in the stator three-phase mmf. As such, harmonics count as leakage flux as they lead to an increase in the leakage inductance per current path (or phase). Some single-layer concentrated windings produce subharmonics that create additional core losses and further leakage inductance increases. This is not so, in general, for double-layer concentrated windings [10].

The cogging torque number of periods is given by the LCM of the N_s and $2p_1$. Some representative data are given in Table 10.2. As can be noted by comparing Tables 10.1 and 10.2, the maximum LCM does not always coincide with the maximum winding factor; thus, a compromise between the two is required, with the latter being more important. For $N_s/2p_1$ = 15/14, 8/9, 12/10, 12/14, 18/16, very good results are obtained.

TABLE 10.1 Winding Factors of Concentrated Windings

	2p—Poles															
N_s Slots	2		4		6		8		10		12		14		16	
	*	**	*	**	*	**	*	**	*	**	*	**	*	**	*	**
3	*	0.86	*	*	*	*	*	*	*	*	*	*	*	*	*	*
6	*	*	0.866	0.866	*	*	0.866	0.866	0.5	0.5	*	*	*	*	*	*
9	*	*	0.736	0.617	0.667	0.866	0.960	0.945	0.96	0.945	0.667	0.764	0.218	0.473	0.177	0.175
12	*	*	*	*	*	*	0.866	0.866	0.966	0.933	*	*	0.966	0.933	0.866	0.866
15	*	*	*	*	0.247	0.481	0.383	0.621	0.866	0.866	0.808	0.906	0.957	0.951	0.957	0.951
18	*	*	*	*	*	*	0.473	0.543	0.676	0.647	0.866	0.866	0.844	0.902	0.960	0.931
21	*	*	*	*	*	*	0.248	0.468	0.397	0.565	0.622	0.521	0.866	0.866	0.793	0.851
24	*	*	*	*	*	*	*	*	0.930	0.463	*	*	0.561	0.76	0.866	0.866

Source: Adapted from Magnussen, F. and Sadarangani, C., Winding factors and Joule losses of PM machines with concentrated windings, Record of IEEE–IEMDC-2003, Vol. 1, 2003, pp. 333–339.

Note: *, one layer; **, two layers.

TABLE 10.2 Lowest Common Multiplier (LCM) of N_s and $2p_1$

	$2p_1$							
N_s	2	4	6	8	10	12	14	16
3	6	3	6	3	6	3	6	3
6	*	6	*	6	*	6	*	6
9	*	9	*	9	*	9	*	9
12	*	12	*	12	*	12	*	12
15	*	15	*	15	*	15	*	15
18	*	18	*	18	*	18	*	18
21	*	21	*	21	*	21	*	21

A large LCM means that, in general, low cogging (zero current) torque is obtained without skewing. A few percent (5%–7%) output is saved this way.

As the number of poles $2p_1$ is increased, with $N_s \approx 2p_1$, both the winding factor and the LCM increase, securing good performance. This means that for some low-speed applications, concentrated windings are definitely favorable.

By adequate rotor geometry, the emf may be made quasi-sinusoidal, and thus vector and direct torque control are feasible. However, when the air gap has to increase due to mechanical reasons (higher diameter, etc.), the tangential fringing flux of the PMs tends to increase, reducing the emf. Poor use of PMs takes place and reduced output is the result. When the slots are thin (in small torque machines), there may not be enough mmf per slot to secure high torque density. Bearing these limitations in mind, concentrated windings may represent a practical solution to high-performance PMSGs in various applications.

The Gramme-ring windings in Figure 10.7 may be assimilated to a single-layer concentrated or distributed winding by making adequate connections of coils.

Note that thus far, we considered the stator cores as being made of single, stamped laminations. For outer stator diameter above 1 m, the stator core has to be made from a few sections, as is usually the case with SGs (Chapter 5). Modular stator cores are plagued with noncircularity eccentricities that create additional losses, torque pulsations, vibration, and noise.

10.3 Air Gap Field Distribution, emf, and Torque

The air gap field in PMSGs has two main sources: the PMs and the stator mmf.

The presence of slot openings and magnetic saturation, together with the rotor PM geometry, makes the computation of air gap field in PMSG a complex problem, solvable as it is only by two-dimensional (2D) or three-dimensional (3D) FEM.

However, at least for surface PM pole rotors, the influence of magnetic saturation may be neglected unless very high current loading is allowed for, in very high torque density designs. Once magnetic saturation is neglected, the superposition of PM and stator mmf fields in the air gap is acceptable.

Moreover, the effect of stator slot openings on air gap field distribution may be added by introducing a P.U. (relative) air gap performance function $\lambda(\theta)$, variable with angular position along the stator surface (Figure 10.9). Consequently, a 2D analytical field model may be used to calculate separately the PM and stator mmf produced air gap field distribution in the slotless machine, and then the slot-opening effect can be introduced by multiplication with $\lambda_{P.U.}(\theta)$:

$$B_{PM,mmf}(\theta) = B_{PM,mmf}^{slottless}(\theta - \theta_1) \cdot \lambda_{P.U.}(\theta) \tag{10.2}$$

where

θ_1 is the rotor displacement

θ and θ_1 are taken in electrical degrees or radians

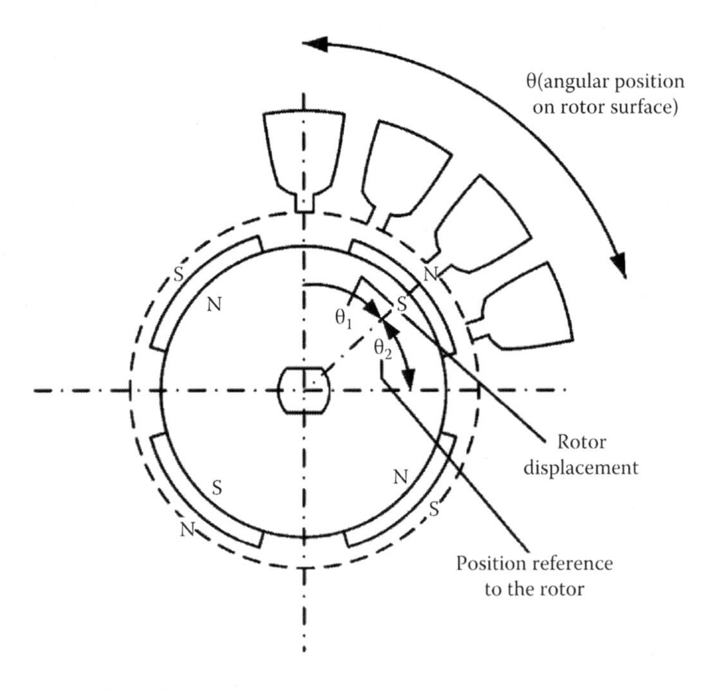

FIGURE 10.9 Position angles with rotor in motion.

Complete 2D analytical models could be built for the unsaturated PMSG with surface PM pole cylindrical rotors.

For the axial air gap (disk-shaped rotors), a similar 2D theory in the axial (z) and in polar coordinates (θ_1, θ) may be developed, while the variation of variables in the radial direction, r, is handled by an additional P.U. coefficient $\lambda_{P.U.}(r)$. A quasi-3D analytical model is thus obtained. A complete 2D analytical model for a cylindrical rotor is introduced in Reference 11, and it pursued further in Reference 12 for cogging torque and emf computation. For the disk-shaped rotor case, a quasi-3D complete analytical model is introduced in References 13 and 14 with emf and cogging torque thoroughly documented. In both cases, the PM equivalent mmf and the stator mmf distributions are decomposed in Fourier space harmonics, and superposition is applied afterward.

As these analytical models are fairly general, and could handle both distributed and concentrated stator windings, we will treat these here in some detail in view of their potential application to design optimization attempts.

The equivalent circuit for phase A in stator coordinates is shown in Figure 10.10. Core losses are neglected. E_{PM_a} represents the PM-induced voltage (emf) in phase a, which, evidently, is a bipolar motion-induced voltage that varies with the rotor position:

$$E_{PM_a} = \sum_{i=1}^{N_s/3} W_c \cdot B_{PMi} \cdot l_{stack} \cdot u \tag{10.3}$$

where

l_{stack} is the stack
u is the peripheral speed
B_{PMi} is the PM average flux density at slot i of phase a
W_c is the number of turns per coil
N_s is the stator slots

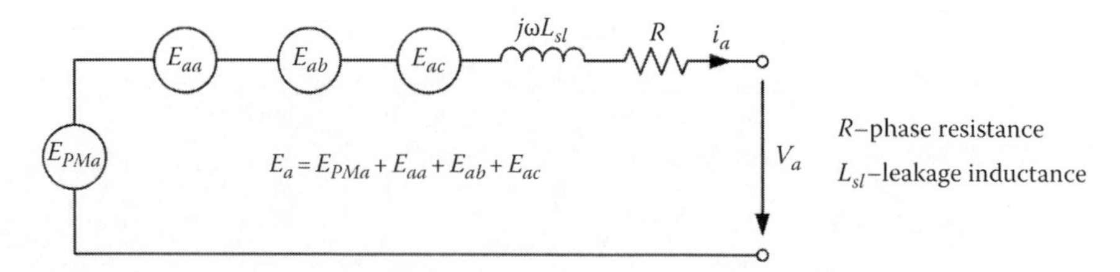

FIGURE 10.10 Steady-state phase equivalent circuit.

E_{aa}, E_{ab}, and E_{ac} are the self-induced and mutual-motion-induced voltages in phase a. They are produced by the stator currents:

$$
\begin{aligned}
E_{aa} &= \sum_{i=1}^{N_s/3}\left(\sum_{j=1}^{N_c} W_c \cdot B_{cij}^{a} \cdot l_{stack} \cdot u\right) \\
E_{ab} &= \sum_{i=1}^{N_s/3}\left(\sum_{j=1}^{N_c} W_c \cdot B_{bij}^{a} \cdot l_{stack} \cdot u\right) \\
E_{ac} &= \sum_{i=1}^{N_s/3}\left(\sum_{j=1}^{N_c} W_c \cdot B_{cij}^{a} \cdot l_{stack} \cdot u\right)
\end{aligned} \tag{10.4}
$$

where

B_{aij}^{a} is the average instantaneous magnetic flux density produced by coil i of phase a at slot j of the same phase a

B_{bij}^{a}, B_{cij}^{a} are the average instantaneous magnetic flux densities produced by coil i of phase $b(c)$ at slot j of phase a

That is, mutual stator emfs E_{aa}, E_{ab}, E_{ac} vary with rotor position if inset or interior (or variable reluctance) rotor poles exist.

When considering the following, the approximations in Figure 10.11a through d are operated:

- PM magnetization harmonics
- Current sheet distribution of a coil
- Stator slot openings

The rectangular PM magnetization (Figure 10.11) may be decomposed in Fourier series:

$$M(\theta_2) = \sum_{n=1,3,5}^{\infty} 2\frac{B_r}{\mu_0} \cdot \alpha_p \cdot \frac{\sin(n\pi\alpha_p/2)}{n\pi\alpha_p/2} \cos np_1\theta_2 \tag{10.5}$$

where

$\alpha_p = \tau_m/\tau$ is the magnet/pole pitch ratio
p_1 is the pole pairs
B_r is the remnant flux density of the PMs
$\theta_2 = \theta - \theta_1$

The stator current sheet produced by a coil (Figure 10.11b) may also be approximated in Fourier series:

$$I(\theta) = \frac{4W_c i}{\pi w_s} \sum_{n=1}^{\infty} \frac{1}{n} \cdot \sin\frac{nw_s}{2R_s} \cdot \cos\left(n \cdot \left(\frac{\theta - \tau_c}{2}\right)\right) \tag{10.6}$$

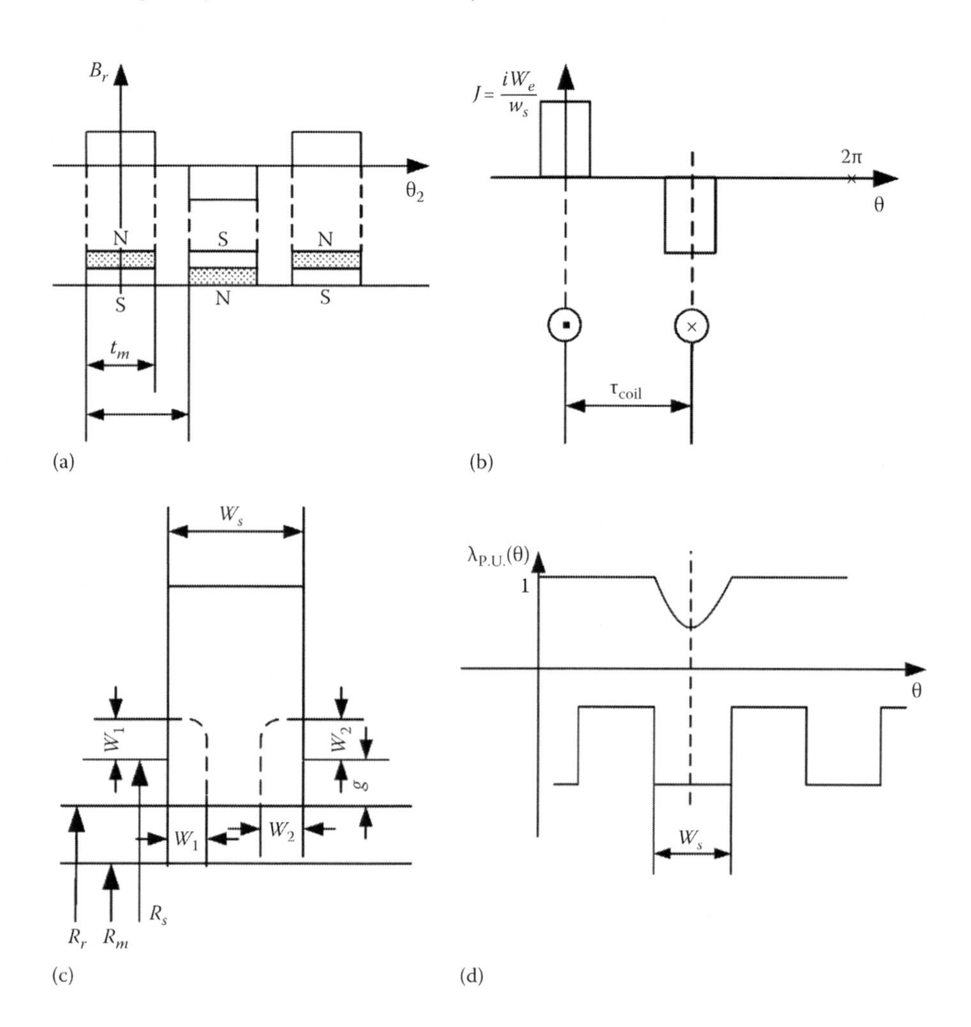

FIGURE 10.11 (a) PM magnetization, (b) current sheet, (c) slot flux line ideal distributions, and (d) $\lambda_{P.U.}(\theta)$.

where

i is the instantaneous current in the coil
w_s is the slot opening
R_s is the interior stator radius
τ_c is the coil span angle

The relative air gap permeance (Figure 10.11c) $\lambda_{P.U.}(\theta)$ may be determined by considering semicircular flux paths in the slots with radii equal to the shortest distance to tooth edges:

$$\lambda_{P.U.}(\theta) = \frac{g + (l_m/\mu_r)}{g + (l_m/\mu_r) + (\pi/2)w_1(\theta)} \tag{10.7}$$

Away from slot openings, $\lambda_{P.U.}$ is unity, and in front of slot openings, has a variable subunity value, dependent on w_1, which is the distance to the closest slot edge. For semiclosed slots, w_s refers to the slot opening of the air gap and not to the actual slot width, and thus the air gap permeance pulsations with θ are notably smaller, as expected. Conformal mapping may be used also to calculate $\lambda_{P.U.}(\theta)$.

Using Equation 10.2, the expressions of $B_{PM}^{slotless}(\theta-\theta_1)$ and $B_{mmf}^{slotless}(\theta-\theta_1)$ are obtained with Equations 10.5 and 10.6 and the 2D field model [11]:

$$B_{PM}^{slotless}(\theta-\theta_1)=\sum_{n=1,3,5}^{\infty}4\cdot\frac{B_r}{\mu_r}\cdot\frac{\sin(n\pi\alpha_p/2)}{\pi\alpha_p}\cdot\frac{p_1}{(np_1)^2-1}$$
$$\cdot\left[\frac{(np_1-1)+2(R_r/R_m)^{np_1+1}-(np_1+1)\cdot(R_r/R_m)^{2np_1}}{(\mu_r+1/\mu_r)\cdot(1-(R_r/R_s)^{2np_1})-(\mu_r-1/\mu_r)((R_m/R_s)^{2np_1}-(R_r/R_m)^{2np_1})}\right]$$
$$\cdot\left(\left(\frac{r}{R_s}\right)^{np_1-1}\cdot\left(\frac{R_m}{R_s}\right)^{np_1+1}+\left(\frac{R_m}{r}\right)^{np_1+1}\right)\cos np_1(\theta-\theta_1)\quad \text{for } p_1\neq 1 \tag{10.8}$$

where μ_r is the P.U. value of PM permeability: 1.03–1.08. For $np_1 = 1$, Equation 10.8 has a zero denominator and should be changed to the following [11]:

$$B_{PM}^{slotless}(\theta-\theta_1)=\sum_{n=1,3,5}^{\infty}2\cdot\frac{B_r}{\mu_r}\cdot\frac{\sin(n\pi\alpha_p/2)}{n\pi\alpha_p}\cdot\frac{p_1}{(np_1)^2-1}$$
$$\cdot\left[\frac{(R_m/R_s)^2-(R_r/R_s)^2+(R_r/R_s)^2\ln(R_m/R_r)^2}{((\mu_r+1)/\mu_r)\cdot(1-(R_r/R_s)^2)-((\mu_r-1)/\mu_r)((R_m/R_s)^2-(R_r/R_m)^2)}\right]$$
$$\cdot\left(1+\frac{R_s}{r}\right)\cos(\theta-\theta_1) \tag{10.9}$$

For the interior rotor,

$$R_m=R_s-g;\quad R_r=R_s-g-h_m \tag{10.10}$$

while for the exterior rotor,

$$R_m=R_s+g;\quad R_r=R_s+g+h_m \tag{10.11}$$

Consequently, Equations 10.8 and 10.9 have to be modified, as now $R_s < R_r < R_m$:

$$B_{PM}^{slotless}(\theta-\theta_1)=-\sum_{n=1,3,5}^{\infty}4\cdot\frac{B_r}{\mu_r}\cdot\frac{\sin(n\pi\alpha_p/2)}{\pi\alpha_p}\cdot\frac{p_1}{(np_1)^2-1}\cdot$$
$$\cdot\left[\frac{(np_1-1)(R_m/R_r)^{2np_1}+2(R_m/R_r)^{np_1-1}-(np_1+1)}{((\mu_r+1)/\mu_r)\cdot(1-(R_s/R_r)^{2np_1})-((\mu_r-1)/\mu_r)((R_s/R_m)^{2np_1}-(R_m/R_r)^{2np_1})}\right]$$
$$\cdot\left(\left(\frac{r}{R_m}\right)^{np_1-1}+\left(\frac{R_s}{R_m}\right)^{np_1-1}\cdot\left(\frac{R_s}{R_r}\right)^{np_1+1}\right)\cos np_1(\theta-\theta_1)\quad \text{for } np_1\neq 1 \tag{10.12}$$

The above expressions account for the rotor curvature and should be particularly instrumental for low rotor diameter (low torque) PMSGs.

The armature reaction flux density in the air gap, produced by a single stator coil with its sides placed in slots k and l, is as follows [12]:

$$B_{kl}(\theta) \cong \frac{2\mu_0 W_c i}{\pi} \cdot \sum_{n=1}^{\infty} \frac{1}{r} \cdot \frac{\sin(n(w_s/2R_s))}{(n(w_s/2R_s))} \cdot \sin\left(\frac{n\tau_c}{2}\right)\left(\frac{r}{R_s}\right)^n \cdot \frac{(1+(R_r/r)^{2n})}{1-(R_r/R_s)^{2n}} \cdot \cos\left(n\left(\frac{\theta-\tau_{coil}}{2}\right)\right) \quad (10.13)$$

For external rotors, a similar formula could be obtained with R_s/r instead of r/R_s, r/R_r instead of R_r/r and R_s/R_r instead of R_r/R_s.

Ultimately, both PM and mmf air gap flux density expressions in Equations 10.9 and 10.10 have to be multiplied by the P.U. air gap permeance function $\lambda_{P.U.}(\theta)$ of Equation 10.7 with the contributions of all coils for each phase added up (Equations 10.3 and 10.4).

Denoting by E_a, E_b, and E_c the total emf per phases a, b, and c, the electromagnetic torque T_e is, simply,

$$T_e = \frac{(E_a i_a + E_b i_b + E_c i_c)}{\omega_1} \cdot p_1; \quad \omega_1 = \Omega_r \cdot p_1 \approx \frac{U}{R_m} \cdot p_1 \quad (10.14)$$

The instantaneous electromagnetic torque includes the interaction torque between the PM flux and stator current and the so-called reluctance (no PM) torque.

The torque at zero current (cogging torque) is not, however, included in Equation 10.14. The cogging torque is, in fact, produced by the tangential forces on the slot walls:

$$T_{cog}(\theta) \cong \frac{\pi l_{stack} R_s}{2\mu_0 N_s} \cdot \sum_{m=1}^{N_s}\left[B_{PM}^2\left(\frac{2\pi m}{N_s}+\theta_1\right) \cdot (R_m + g_\alpha) ssg\right] \quad (10.15)$$

$g_\alpha = 0$ and $ssg = 0$ outside the slot opening; $g_\alpha = w_1 + g$ and $ssg = 1$ on the left side of the slot opening; and $g_\alpha = w_2 + g$, and $ssg = -1$ on the right side of the slot opening.

Typical results that compare PM, mmf air gap flux densities, cogging, and total torque obtained from the analytical model and from 2D FEM are shown in Figure 10.12a through d [12]. The 2D analytical model looks adequate to describe air gap flux density pulsations due to slot openings both for the PM and armature contributions. Also, it correctly portrays the cogging torque and the total torque variation with rotor position (or with time). Any shape of current waveform can be dealt with through its time harmonics.

The model, however, ignores magnetic saturation and cannot directly handle inset or interior PM rotor configurations, though a notably adapted model could do it.

As already alluded to, a similar quasi-3D analytical model was developed for disk-shaped (axial flux) PM rotor configurations [13,14].

The 2D analytical model for the axial air gap machine is developed in the axial circumferential coordinates (z, $\theta - \theta_1$), while the correction for the radial direction is handled [13] as follows (Figure 10.13a and b)

$$g(r) = \frac{1}{\pi}\left[\tan\left(\frac{(r-R_{ai})}{a}\right) - \tan^{-1}\left(\frac{(r-R_{ae})}{a}\right)\right] \quad (10.16)$$

with R_{ai} and R_{ae} the inner and outer PM radii, and

$$a = \frac{\alpha(R_{ae}-R_{ai})}{\tan(\beta\pi/2)} \quad (10.17)$$

with α and β as correction coefficients. By making the computations for 15–20 values of the radius and averaging them, while taking into account that the pole pitch τ and the PM span τ_m vary with radius r, similar results may be obtained.

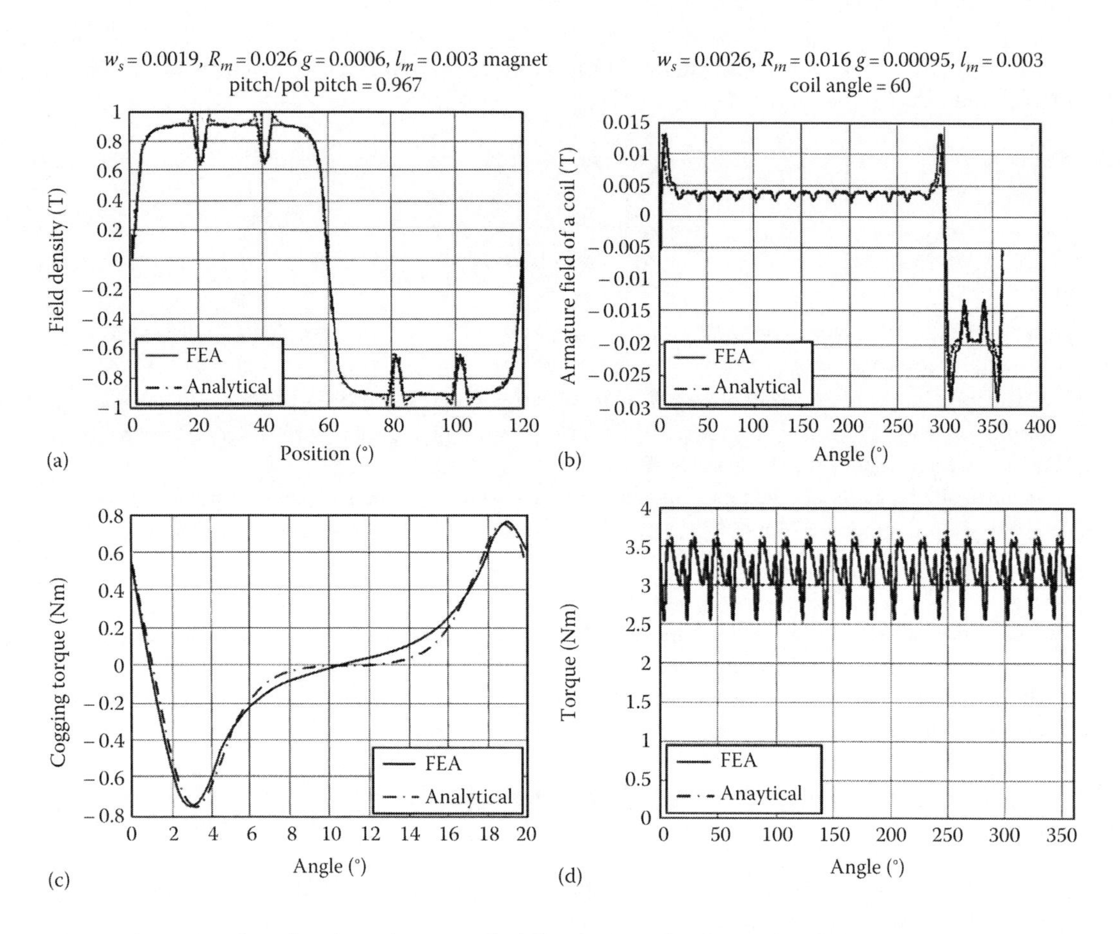

FIGURE 10.12 Analytical vs. finite element method (FEM) air gap flux density distributions and torques for sinusoidal current: (a) PM flux density, (b) armature flux density, (c) cogging torque, and (d) total torque.

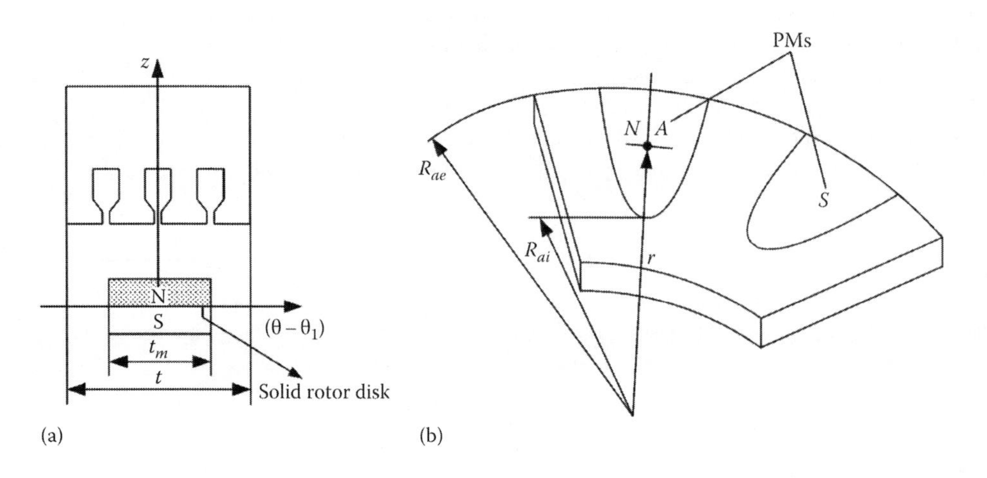

FIGURE 10.13 Disk-rotor permanent magnet synchronous generator (PMSG) model: (a) two-dimensional geometry $(z, \theta - \theta_1)$ and (b) radial (third) dimension r.

The emf and cogging torque waveforms obtained through such a model [14] are compared with 3D FEM results (Figure 10.14a and b). There is some 11% difference between analytical and 3D FEM results, probably due to the fringing flux through stator slot necks, which is neglected. Unfortunately, this fringing (leakage) PM flux may be calculated only through FEM. FEM-based corrections for fringing are possible in an iterative way or through applying a fudge factor derived from experiments. The same assertion should hold for the cylindrical rotor analytical model. The situation is delicate, especially with small slot openings.

Note that while a complete 2D (3D) analytical field model was developed for both cylindrical and disk-shaped rotors, its Fourier series nature makes it useful for design optimization (Figure 10.14a and b). It is still a bit too complicated for preliminary design, when a more simplified circuit model with analytically defined parameters is required. Finally, the 2D (3D) FEM may be used for model design optimization.

But before dealing with the circuit model, let us explore first core loss modeling, as they use FEM, also, in advanced representations.

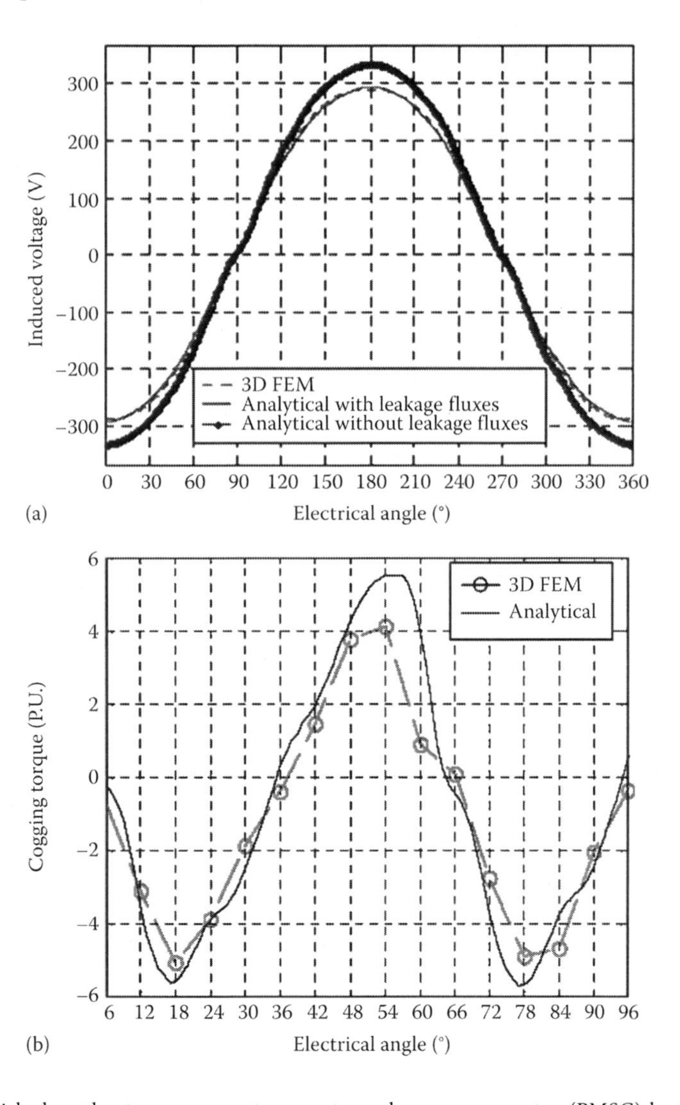

FIGURE 10.14 Disk-shaped rotor permanent magnet synchronous generator (PMSG) by three-dimensional analytical vs. three-dimensional finite element method (FEM) models: (a) electromagnetic field (emf) waveforms and (b) cogging torque.

10.4 Stator Core Loss Modeling

There are two ways to approach the complex problem of stator core losses in PMSGs. The first approach starts with a simplified flux distribution and a simplified core loss formula and eventually advances to using FEM for refinements.

In the second approach, FEM distribution of flux density evolution with time is considered in each core volume element, which gives stator current/time waveforms. Then, the hysteresis and eddy current losses are added up to obtain the total core losses by again using some simplified core loss formulas.

Recent research efforts showed that not only is the flux density not sinusoidal in time at a point in the stator core but also its vector direction changes in time at that point, more in the yoke and less in the teeth. This way, the traveling component of flux density time variation was proven to be important in computing the core loss. The data from AC magnetization (from Epstein probe) have to be corrected for eddy current losses due to rotation. The presence of slot openings and the distribution of windings in slots lead to time harmonics in the stator core flux density, even for distributed windings and sinusoidal currents at constant speed. Also, the on-load core losses tend to differ from no-load core losses; thus, the simplified core loss standard formulas [3,5] may be used only in very preliminary design stages. The presence of slot openings and the distribution of windings in slots lead to time harmonics in the stator core flux density, even for distributed windings and sinusoidal currents at constant speed. Finally, the importance of "rotation" on core loss tends to increase for concentrated windings in comparison with distributed windings.

All of these aspects tend to suggest that a realistic approach should start first with FEM core loss estimates, and then their curve fitting to even more simplified formula may be operated, for design optimization tasks.

10.4.1 FEM-Derived Core Loss Formulas

It has been known for a long time that the combined alternative and traveling fields produce more core losses than alternating fields of the same frequency and amplitude. The quantization of this phenomenon has not been, in general, taken into consideration.

FEM has brought in a possibility to account for rotational field losses in core losses of brushless machines (Figure 10.15a through e) [15]. In different points, such as a, b, c, d, e, the flux density has traveling field components of various values, depending on their locations in the stator teeth and yoke. Further on, at some points, the situation is different for given points for concentrated windings (Figure 10.16a and b) [15].

It is evident that the short/long axis ratio c of the elliptical variation of B_r/B_o for each harmonic is different for distributed and concentrated windings (Figures 10.15a through d and 10.16a and b).

For the distributed winding, the traveling field components ($c > 0.3$) are notable at the teeth bottom and in part of the yoke; while for concentrated winding, they also show up near the teeth top and throughout the yoke. Larger core losses are expected for concentrated windings. A general FEM-derived formula for total stator core losses is of the following form [15]:

$$P_{Fe\text{-}total} = \sum_{j=1}^{n}\left\{\sum_{i=1}^{m} g_i\left[\left(\varepsilon_h\left(\frac{jf}{100}\right)B_{mji}^{\alpha} + \varepsilon_{eddy}\left(\frac{jf}{100}\right)^2 B_{mji}^{\alpha}\right)\cdot(1+\gamma c_{ij})\right]\right\} \tag{10.18}$$

where

j is the order of the flux density time harmonic
i is the finite element number
g_i is the mass of i finite element
f is the frequency
γ is a correcting constant
c is the short-to-long axis ratio of field vector hodograph in element i for the harmonic j

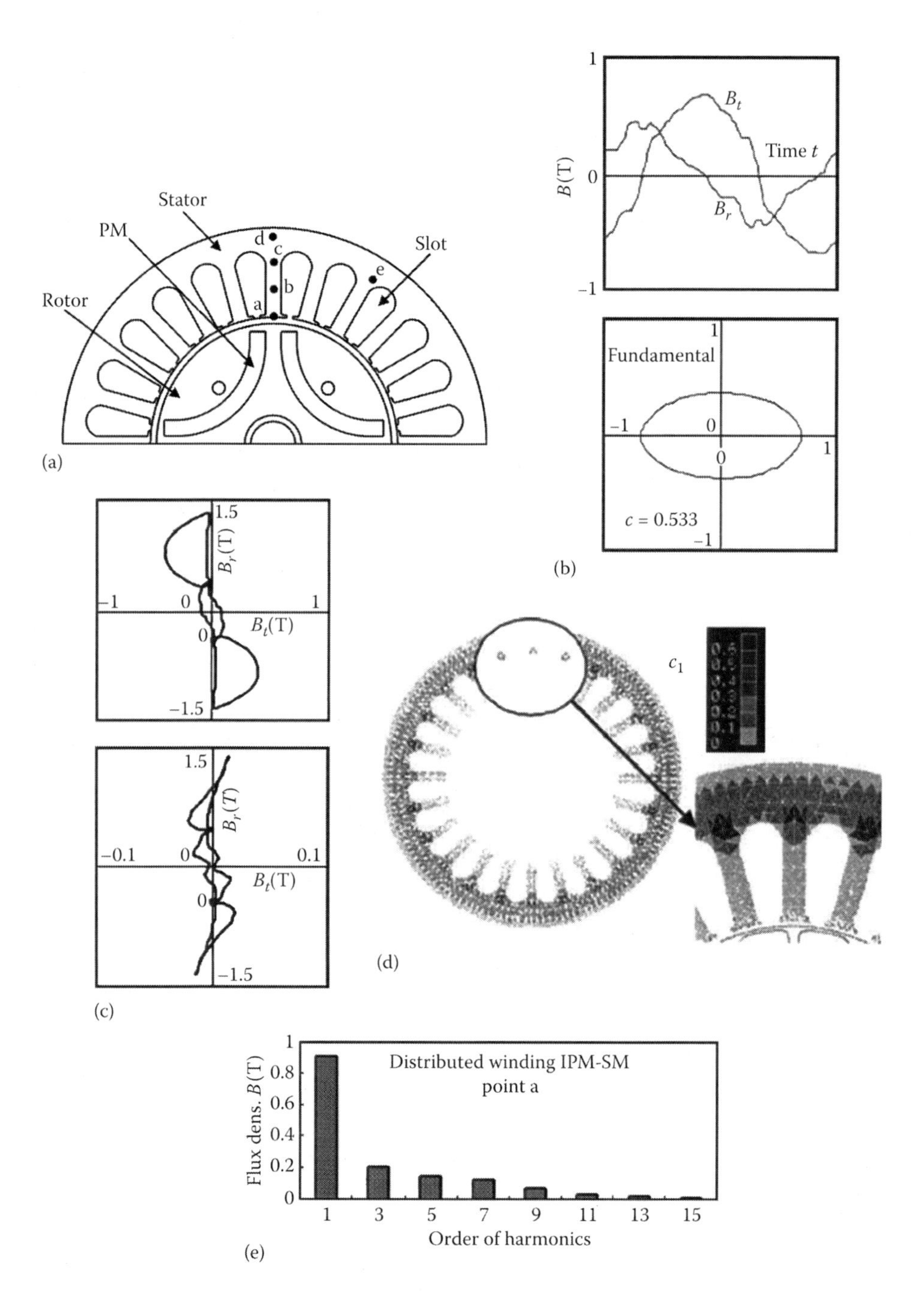

FIGURE 10.15 Distributed winding interior permanent magnet synchronous generator (IPM-SG) core flux density: (a) the points a, b, c, d, and e, (b) flux density vector hodograph in point a, (c) elliptical variation for each harmonic, (d) short-to-long axis *c* coefficient, and (e) harmonics spectrum of flux density.

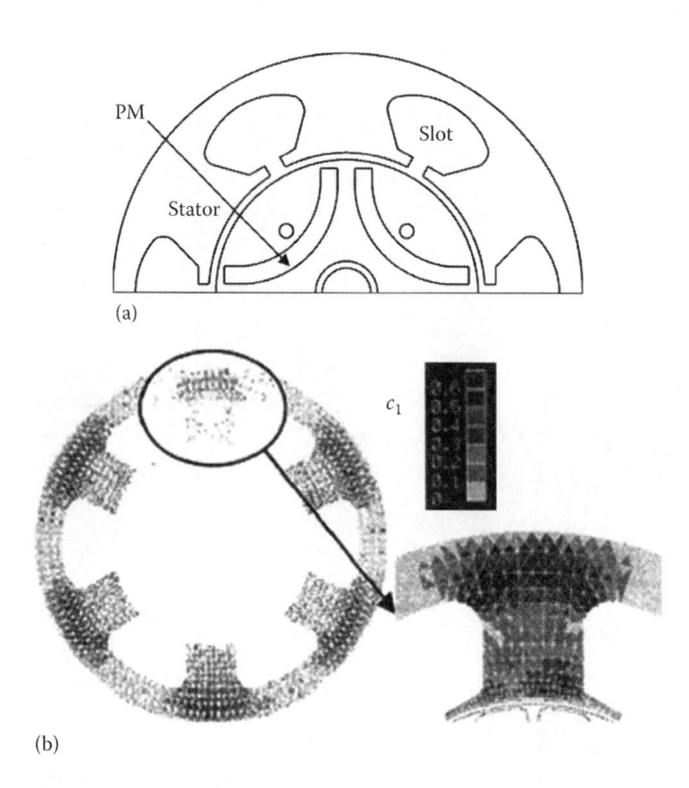

FIGURE 10.16 Concentrated winding interior permanent magnet synchronous generator (IPM-SG): (a) cross section and (b) distribution of short-to-long axis rate c.

In case stator current time harmonics are considered, the same formula is applied for each current harmonic frequency.

Comparison between experimental losses and calculated ones by conventional (analytical) models and Equation 10.18—for the distributed and concentrated windings—are shown in Figures 10.17 and 10.18 [15].

The increase in precision by considering traveling field components is evident.

In addition, the contribution of harmonics in core losses is shown in Figures 10.19 and 10.20 for sinusoidal current.

As expected, the third flux density harmonic is responsible for about 25% of core loss for the concentrated winding. Extending the use of concentrated windings at high speed should be done with care, as core losses tend to be larger than those for distributed windings.

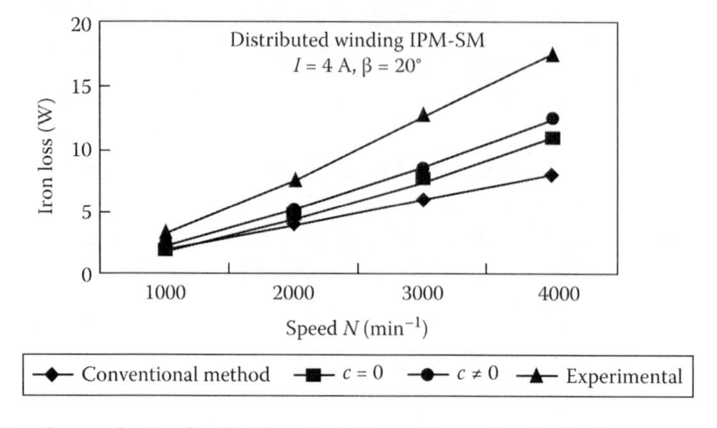

FIGURE 10.17 Core loss evaluation for distributed winding with speed under load.

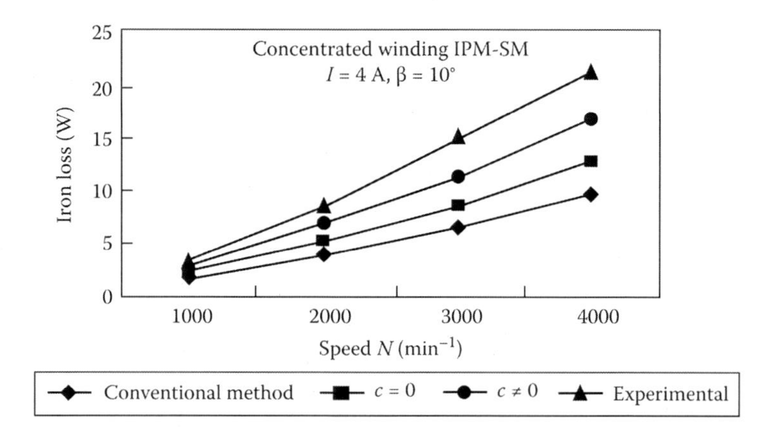

FIGURE 10.18 Core loss evaluation for concentrated windings with speed under load.

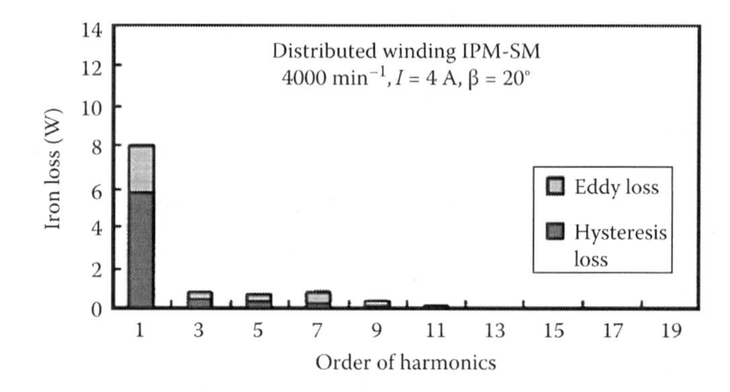

FIGURE 10.19 Flux density harmonics contributions to core losses (distributed winding).

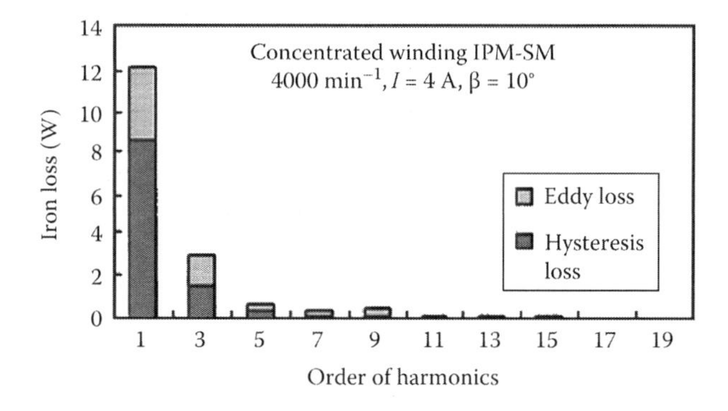

FIGURE 10.20 Core loss harmonics breakdown for concentrated winding.

Alternatively, by adding the core loss of radial and tangential flux density time variations in each finite element for N samples in a time period and overall finite mass elements and separately for eddy currents and hysteresis losses, another total core loss formula may be obtained [16]:

$$P_{Fe\text{-}total} = \sum_{i=1}^{m} \frac{g_i}{N} \sum_{k=1}^{N} k_e \left[\left(\frac{B_r^{k+1} - B_r^k}{\Delta t} \right)^2 + \left(\frac{B_\theta^{k+1} - B_\theta^k}{\Delta t} \right)^2 \right] + \sum_{i=1}^{m} \frac{g_i}{T} \cdot \frac{K_h}{2} \cdot \left[\sum_{j=1}^{N_{pr}^i} \left(B_{mr}^{ij}\right)^2 + \sum_{j=1}^{N_{p\theta}^i} \left(B_{m\theta}^{ij}\right)^2 \right] \tag{10.19}$$

where

Δt is the sampling time (there are N sampling times in one period T)
B_r^k, B_θ^k are the radial and tangential instantaneous flux density components in element I
$B_{mr}^{ij}, B_{m\theta}^{ij}$ are the amplitudes of radial and axial flux densities in element i for the hysteresis cycle j
k_e, k_h are experimental constants from the Epstein frame obtained from the slope and intercept variation of core losses/frequency versus frequency straight line

Equation 10.19 may also be applied for the computation of rotor core losses.

The FEM-derived formulas account for actual flux density evolution in all parts of the magnetic circuit, either for no-load or for on-load conditions. In addition, the case of rectangular or sinusoidal current waveforms may be treated with the same method.

Finally, IPM or surface PM rotors can be accommodated by simplified formulas.

10.4.2 Simplified Analytical Core Loss Formulas

Analytical core loss formulas for stator core loss in PMSGs were derived first for distributed windings and sinusoidal currents. It was implicit that the flux density varies sinusoidally in time in both stator teeth and yoke, and also, that their value is the same in all points of the core. Both these approximations are coarse, as discussed in the previous paragraph. The no-load core losses were evaluated first. A simple formula that even ignores the hysteresis losses is as follows:

$$P_{Fe}/\text{kg} \approx 2.8 \cdot k_e \cdot \omega^2 \cdot B_g^2 \tag{10.20}$$

$\hat{B}_g^2$ is the peak flux density in the air gap. As this formula cannot discriminate between lightly saturated or heavily saturated teeth and yoke designs, it is somewhat surprising that it is still mentioned in the literature today [5]. But as long as the core losses are notably smaller than copper losses, notable errors in the former do not produce design disasters. This is not so in optimized designs, where winding and nonwinding losses tend to be of about the same value, or in high-speed PMSGs.

For $q = 1$ distributed windings and rectangular current, the PM air gap flux density distribution produces different, linear, flux density variations in the stator teeth and in the yoke (Figure 10.21) at no load [3]. The total eddy current loss formula at no load for such cases is of the following form:

$$P_{eddy}/\text{kg} = \frac{4R}{\pi p_1} k_e \omega^2 \left(\frac{2(w_t + w_s)^2}{w_t^3} + \frac{\alpha R}{p \cdot hys^2} \right) B_g^2 \tag{10.21}$$

where

R is the rotor radius
w_t is the tooth width
w_s is the slot width
hys is the stator yoke radial thickness
p_1 is the pole pairs
ω is the electrical fundamental frequency

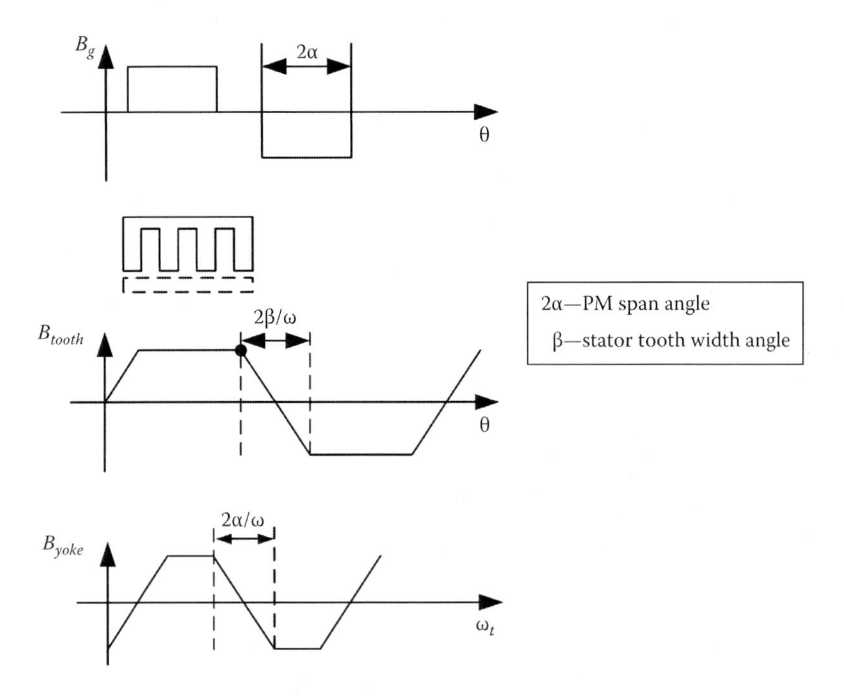

FIGURE 10.21 Ideal teeth and yoke flux densities variation with rotor position for rectangular air gap flux density.

A rough approximation of Equation 10.21 is as follows [16]:

$$P_{eddy} \approx 6 \cdot k_e \cdot \left(\omega^2 \cdot \hat{B}_g^2\right) \tag{10.22}$$

Equations 10.20 and 10.22 show that the rectangular air gap flux density in the air gap is characterized by notably larger core loss than a sinusoidal one. Consequently, when core loss becomes important, sinusoidal air gap flux density distribution should be targeted. In the above analytical expressions, the armature reaction field was not considered. In heavily loaded and, especially in IPM rotors, such an approximation is hardly practical.

Additionally, it is to be noted that eddy current core losses depend on the flux density waveform (or its variation velocity). The hysteresis core losses depend essentially only on the peak values of flux density. This is how simplified analytical formulas for eddy current and hysteresis core loss for rectangular flux density variation in time and IPM-SG on load were derived [17]:

$$P_{eddy} = P_{etooth} + P_{eyoke} \tag{10.23}$$

$$P_{eteeth}/\text{kg} \approx \frac{k_e}{T}\int_0^T \left(\frac{dB_t}{dt}\right)^2 dt = k_e \frac{\omega^2}{\pi}\left[\frac{2}{p_1 \cdot \beta}\left((B_{gm} + B_{dm} - \hat{B}_d \cos\alpha_t)^2 + \hat{B}_q^2 \sin^2\alpha_t\right.\right.$$

$$\left.\left. + \hat{B}_d^2\left(\frac{\alpha_t - \sin 2\alpha_t}{2}\right) + \hat{B}_q^2\left(\frac{\alpha_t + \sin 2\alpha_t}{2}\right)\right)\right] \tag{10.24}$$

$$P_{eyoke}/\text{kg} \approx \frac{k_e}{\pi}\left(\frac{R_s\omega}{p_1 \cdot hys}\right)^2\left[2p_1\alpha(B_{gm}+B_{dm})^2+\hat{B}_d^2\left(p_1\alpha+\frac{\sin 2p_1\alpha}{2}\right)\right.$$

$$\left.+\hat{B}_q^2\left(\frac{2p_1\alpha-\sin 2p_1\alpha+p_1\delta+\sin p_1\alpha}{2}\right)-4(B_{gm}+B_{dm})\hat{B}_d\sin p_1\alpha\right] \tag{10.25}$$

where

B_{gm} is the PM air gap flux density
B_{dm} is the peak air gap flux density at the rotor surface
$\hat{B}_d$ is the peak air gap flux density due to d-axis current
$\hat{B}_q$ is the peak air gap flux density due to q-axis current
α is the pole magnet mechanical angle
$\alpha_t = p_1(\alpha-\beta/2)$
β is the slot pitch mechanical angle
R_s is the stator interior radius

Finally, the hysteresis losses are expressed by the usual formula:

$$P_{ht,y}/\text{kg} = k_h \cdot \omega_1 \cdot \hat{B}_{t,y}^2 \tag{10.26}$$

Though expressions such as Equations 10.24 through 10.26 tend to account for the load (armature reaction) influence on core losses, they still cannot discriminate directly between designs with various teeth thickness (geometry) or local magnetic saturation effects. However, as shown in Figure 10.22 [17], the load has a strong influence on stator core losses. In terms of core losses, FEM verifications at the design stage are highly recommended for critical operation modes. Experiments should follow whenever possible.

Note that it was suggested that core losses be represented in the circuit model of PMSG by an equivalent resistance R_c of the following form:

$$R_c = \frac{1}{(1/R_{eddy})+(1/\omega_1 \cdot r_{hys})} = \frac{(2/3)E_{total}^2}{P_{Fe\,total}} \tag{10.27}$$

E_{total} is the total emf in the d–q model of PMSG. The term ω_1 accounts for the hysteresis loss linear variation with frequency. E_{total} is also proportional to frequency. Eddy current core losses are proportional to

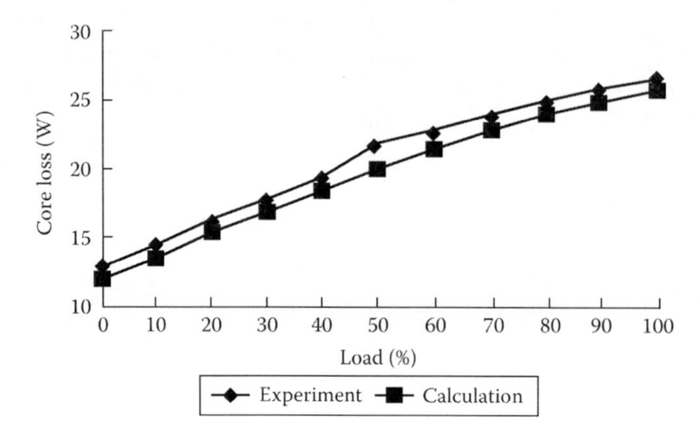

FIGURE 10.22 Core loss vs. load at 2500 rpm for a four-pole three-phase 600 W interior permanent magnet (IPM) synchronous machine.

frequency squared. Such formulas should also be used with care, as, in fact, R_{eddy} and r_{hys} vary in turn with load conditions, and so forth.

For transients, it seems better to neglect stator core losses; while for steady state, it is more practical to add them, after their FEM computation or measurement, to estimate efficiency or temperatures in the machine.

Note that rotor losses were mentioned only with respect to rotor core and in relation to FEM. In principle, eddy currents occur in the PMs due to armature field space harmonics. In high-speed machines, they are notable [26].

10.5 Circuit Model

10.5.1 Phase Coordinate Model

In essence, the circuit model of a PMSG starts with the phase voltage equations in stator coordinates:

$$\begin{aligned} i_a R_s - V_a &= -\frac{d\Psi_a}{dt} \\ i_b R_s - V_b &= -\frac{d\Psi_b}{dt} \\ i_c R_s - V_c &= -\frac{d\Psi_c}{dt} \end{aligned} \tag{10.28}$$

$$\begin{vmatrix} \Psi_a \\ \Psi_b \\ \Psi_c \end{vmatrix} = \begin{vmatrix} L_{sl} + L_{aa}(\theta_{er}) & L_{ab}(\theta_{er}) & L_{ca}(\theta_{er}) \\ L_{ab}(\theta_{er}) & L_{sl} + L_{bb}(\theta_{er}) & L_{bc}(\theta_{er}) \\ L_{ca}(\theta_{er}) & L_{bc}(\theta_{er}) & L_{sl} + L_{cc}(\theta_{er}) \end{vmatrix} \cdot \begin{vmatrix} i_a \\ i_b \\ i_c \end{vmatrix} + \begin{vmatrix} \Psi_{PMa}(\theta_{er}) \\ \Psi_{PMb}(\theta_{er}) \\ \Psi_{PMc}(\theta_{er}) \end{vmatrix} \tag{10.29}$$

where θ_{er} is the rotor PM axis angle to stator phase a axis/electrical angle.

The self-inductance and mutual inductance of the stator depend sinusoidally on θ_{er} only for IPM rotors, in distributed winding IPM rotor machines. For surface PM pole rotors, stator inductances are independent of θ_{er}. However, the presence of slot openings introduces an additional dependence of stator inductances on $N_s\theta_{er}$ ($N_s\theta_{er}$ is the number of stator slots) for IPM rotors.

For concentrated windings, the stator self-inductance and mutual inductance perform in a similar way, with respect to rotor pole configurations, but their values tend to be larger than for distributed windings and equivalent machine geometries. However, the end-turn leakage inductances are notably smaller for concentrated windings.

The PM flux linkages in the stator phases $\Psi_{PMa}(\theta_{er})$, $\Psi_{PMb}(\theta_{er})$, and $\Psi_{PMc}(\theta_{er})$ are either sinusoidal or they also contain space harmonics.

The trapezoidal distribution may be considered this way:

$$\Psi_{PMa}(\theta_{er}) = \Psi_{PM1}(\theta_{er}) + \Psi_{PM2}(2\theta_{er} - \gamma_2) + \Psi_{PM3}(3\theta_{er} - \gamma_3) + \cdots \tag{10.30}$$

The even harmonics occur only with PM pole shifting, adopted to reduce cogging torque. In addition, space subharmonics may occur. They have to be avoided for practical designs.

For surface PM pole rotors, the general expressions of E_{aa}, E_{ab}, E_{ac}, and Ψ_{PMa} developed earlier may be used to calculate L_{aah}, L_{abh}, L_{cah}, Ψ_{PMa}, ($i_b = i_c = 0$)

$$L_{aah} = \frac{E_{aa}}{\omega_1 i_a}; \quad L_{abh} = \frac{E_{ab}}{\omega_1 i_a}; \quad L_{cah} = \frac{E_{ac}}{\omega_1 i_a}; \quad \Psi_{PMa} = \frac{|E_{PMa}|}{\omega_1} \tag{10.31}$$

The analytical method in Section 10.2 may be applied to estimate all parameters in Equations 10.29 and 10.30 for distributed and concentrated windings for surface PM pole rotor configurations.

FEM field distributions may produce even better values for inductances and $\Psi_{PMa,b,c}(\theta_{er})$ expressions, and they are also recommended for IPM rotors.

The instantaneous interaction torque expression, in the absence of magnetic saturation, is as follows:

$$T_e = -\frac{\partial W_m}{\partial \theta_{er}} \tag{10.32}$$

$$W_m = \frac{1}{2}L_{aah}i_a^2 + \frac{1}{2}L_{bbh}i_b^2 + \frac{1}{2}L_{cch}i_c^2 + L_{abh}i_a i_b + L_{bch}i_b i_c + L_{cah}i_c i_a$$
$$+\Psi_{PMa}(\theta_{er})i_a + \Psi_{PMb}(\theta_{er})i_b + \Psi_{PMc}(\theta_{er})i_c \tag{10.33}$$

Though it is feasible to use the phase coordinate model, especially for complex digital simulations in the presence of static power converters, and, even for trapezoidal emf distributions, the *dq* model is preferred when simpler models for control design are required. The PM flux linkage harmonics are introduced to simulate its departure from a sinusoidal distribution.

The *dq* model with emf (PM emf) space harmonics operates as a general tool for design and control, for both sinusoidal and trapezoidal emf and distributed or concentrated windings, either for surface pole PM or for interior PM rotor pole configurations. For concentrated windings and IPM pole rotors, sinusoidal PM flux linkage distributions are targeted, and again, the *dq* model is feasible.

10.5.2 *dq* Model of PMSG

The *dq* model is based on the assumption that the stator self-inductance and mutual inductance are either constant or vary sinusoidally with the rotor position ($2\theta_{er}$). In general, the PM flux linkages $\Psi_{PMa,b,c}(\theta_{er})$ in the stator phases also vary sinusoidally, but with θ_{er}. Eventual harmonics in v may be treated in the *dq* model also, and they are expected to create time pulsations in the torque with sinusoidal currents at speed and frequency $\omega_1 = \omega_r$:

$$|L_{abc}(\theta_{er})| = \begin{vmatrix} L_{sl}+L_0+L_2\cos 2\theta_{er} & M_0+L_2\cos\left(2\theta_{er}+\frac{2\pi}{3}\right) & M_0+L_2\cos\left(2\theta_{er}-\frac{2\pi}{3}\right) \\ M_0+L_2\cos\left(2\theta_{er}+\frac{2\pi}{3}\right) & L_{sl}+L_0+L_2\cos\left(2\theta_{er}-\frac{2\pi}{3}\right) & M_0+L_2\cos 2\theta_{er} \\ M_0+L_2\cos\left(2\theta_{er}-\frac{2\pi}{3}\right) & M_0+L_2\cos 2\theta_{er} & L_{sl}+L_0+L_2\cos\left(2\theta_{er}+\frac{2\pi}{3}\right) \end{vmatrix}$$

$$M = -\frac{L_0}{2} \text{ for distributed windings} \tag{10.34}$$

For PM machines, $L_2 < 0$, as they have the PMs placed along axis *d* and exhibit "inverse saliency," in contrast to standard synchronous machine excitation:

$$\begin{vmatrix} \Psi_{PMa}(\theta_{er}) \\ \Psi_{PMb}(\theta_{er}) \\ \Psi_{PMc}(\theta_{er}) \end{vmatrix} = \begin{vmatrix} \Psi_{PM1}\cos\theta_{er}+\cdots \\ \Psi_{PM1}\cos\left(\theta_{er}-\frac{2\pi}{3}\right)+\cdots \\ \Psi_{PM1}\cos\left(\theta_{er}+\frac{2\pi}{3}\right)+\cdots \end{vmatrix} \tag{10.35}$$

The matrix form of the phase coordinates model is as follows:

$$|i_{a,b,c}||R_s|-|V_{a,b,c}|=-\frac{d|\Psi_{a,b,c}|}{dt} \tag{10.36}$$

$$\Psi_{a,b,c}=|L_{a,b,c}(\theta_{er})||i_{a,b,c}|+\Psi_{PMa,b,c}(\theta_{er}) \tag{10.37}$$

The Park transformation $P(\theta_{er})$ is used to derive the dq model:

$$|P(\theta_{er})|=\frac{2}{3}\begin{vmatrix}\cos(-\theta_{er}) & \cos\left(-\theta_{er}+\frac{2\pi}{3}\right) & \cos\left(-\theta_{er}-\frac{2\pi}{3}\right)\\ \sin(-\theta_{er}) & \sin\left(-\theta_{er}+\frac{2\pi}{3}\right) & \sin\left(-\theta_{er}-\frac{2\pi}{3}\right)\\ \frac{1}{2} & \frac{1}{2} & \frac{1}{2}\end{vmatrix} \tag{10.38}$$

The Park transformation from stator to rotor coordinates is, in Equation 10.38, valid for the trigonometric motion direction and axis q in front of axis d by 90° (electrical degrees) (Figure 10.23):

$$\begin{vmatrix}i_d\\ i_q\\ i_0\end{vmatrix}=|P(\theta_{er})|\begin{vmatrix}i_a\\ i_b\\ i_c\end{vmatrix} \tag{10.39}$$

The same transformation is valid for Ψ_{dq0}, V_{dq0}.

Finally, for sinusoidal $\Psi_{PMa,b,c}(\theta_{er})$ distributions,

$$\begin{aligned}i_dR_s-V_d&=-L_d\frac{di_d}{dt}+\omega_rL_qi_q\\ i_qR_s-V_q&=-L_q\frac{di_q}{dt}-\omega_r(L_di_d+\Psi_{PM1})\end{aligned} \tag{10.40}$$

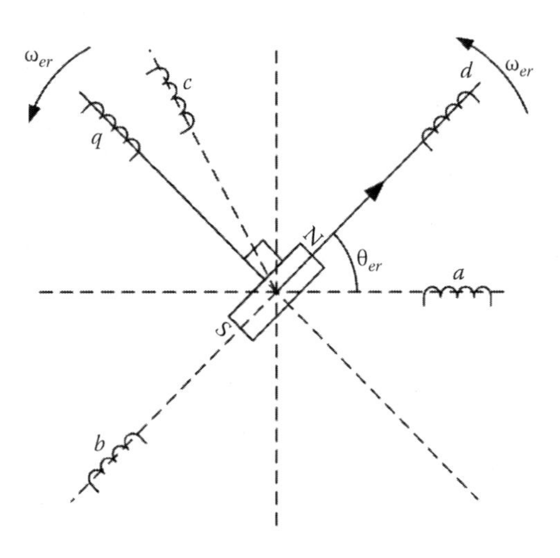

FIGURE 10.23 Three-phase to dq transformation.

with

$$\bar{\Psi}_s = \Psi_d + j\Psi_q \Psi_d = \Psi_{PM1} + L_d i_d \Psi_q = L_q i_q$$
$$\bar{V}_s = V_d + jV_q \bar{i}_s = i_d + ji_q \tag{10.41}$$

The so-called space-vector (or complex variable) model of a PMSG is obtained:

$$\bar{i}_s R_s - \bar{V}_s = -\frac{d\bar{\Psi}_s}{dt} - j\omega_r \bar{\Psi}_s \tag{10.42}$$

The torque is obtained from power balance in Equation 10.37:

$$T_e = p_1 \frac{P_e}{\omega_r} = \frac{3}{2} p_1 \mathrm{Re}\left(j\bar{\Psi}_s \bar{i}_s^*\right) = \frac{3}{2} p_1 (\Psi_d i_q - \Psi_q i_d) = \frac{3}{2} p_1 (\Psi_{PM1} + (L_d - L_q) i_d) i_q \tag{10.43}$$

In addition,

$$L_d = L_{ls} + \frac{3}{2}\left(L_0 - |L_2|\right) L_q = L_{ls} + \frac{3}{2}\left(L_0 + |L_2|\right) \tag{10.44}$$

The winding losses P_{copper} are as follows:

$$P_{copper} = \frac{3}{2} R_s \left(i_d^2 + i_q^2\right) \tag{10.45}$$

The expressions in Equation 10.40 lead to the dq circuit model in Figure 10.24, where the association of signs corresponds to motoring. For generating, i_q changes sign (is negative) and so does the torque. The corresponding general vector diagrams for motoring and generating are shown in Figure 10.25a and b.

Under steady state, $s = 0$ (rotor coordinates), and the circuit model and vector diagram further simplify.

For steady state and sinusoidal phase voltages,

$$V_{abc} = V_1 \sqrt{2} \cos\left(\omega_r t - (i-1)\frac{2\pi}{3}\right) \tag{10.46}$$

According to the Park transformation, where $\theta_{er} = \omega_r t + \theta_0$ (ω_r = constant) and Figure 10.25b,

$$V_d = V_1 \sqrt{2} \cos\theta_0; \quad \theta_0 = -\left(\frac{\pi}{2} + \delta_V\right)$$
$$V_q = -V_1 \sqrt{2} \sin\theta_0 \tag{10.47}$$

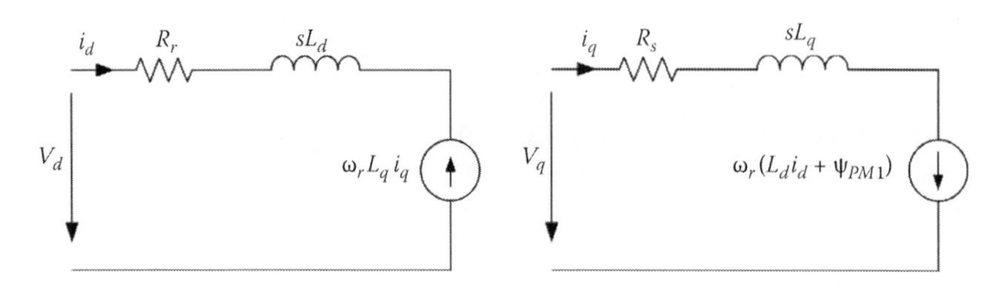

FIGURE 10.24 *dq* equivalent circuits of permanent magnet synchronous generator induction machine (PMSGIM).

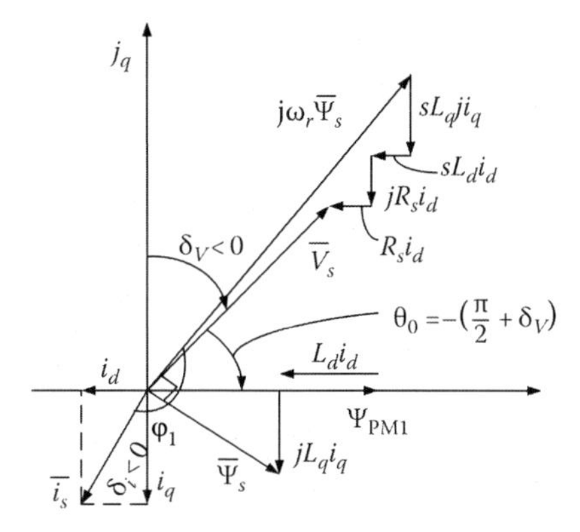

FIGURE 10.25 General vector diagrams of permanent magnet synchronous generator (PMSG).

$\delta_V > 0$ for motoring and $\delta_V < 0$ for generating, or $V_d < 0$, $V_q > 0$ for motoring, and $V_d > 0$, $V_q > 0$ for generating. Also, $i_d < 0$ and $i_q > 0$ for motoring, and $i_q < 0$ for generating. The power factor angle is $\varphi_1 < 90°$ for motoring, and $\varphi_1 > 90°$ for generating. To produce reactive power in generating, $\varphi_1 > 180°$.

Example 10.1

Consider a surface PM pole rotor PMSG with $R_s = 1\ \Omega$, $L_s = L_d = L_q = 0.05$ H, $\Psi_{PM1} = 0.5$ Wb, and $p_1 = 2$ pole pairs that have to deliver power into a three-phase resistance of $R_L = 10\ \Omega$/phase at the speed $n_1 = 3000$ rpm.

Calculate the phase voltage, current, and power delivered to the load resistance and also the voltage and current power angles δ_V, δ_i; justify the results. Change the load resistance to obtain more output power.

Solution

The symmetric three-phase load may be transformed into a dq load, maintaining $R_L = 10\ \Omega$.

Thus, the generator turned Equation 10.35, for steady state, is as follows:

$$i_d(R_s + R_L) = \omega_r L_s i_q \text{ (A)}$$
$$i_q(R_s + R_L) = -\omega_r(\Psi_{PM1} + L_d i_d) \text{ (A)}$$
$$\omega_r = 2\pi p_1 n_1 = 2\pi \cdot 2 \cdot \frac{3000}{60} = 200\pi$$

Equation (A) allows us to calculate i_d and i_q:

$$i_d(R_s + R_L) = \omega_r L_s i_q \text{ (A)}$$
$$i_q(R_s + R_L) = -\omega_r(\Psi_{PM1} + L_d i_d) \text{ (A)}$$
$$\omega_r = 2\pi p_1 n_1 = 2\pi \cdot 2 \cdot \frac{3000}{60} = 200\pi$$

As the dq voltage equation was written for motoring,

$$V_d = -R_L i_d = -10 \times (-8.9) = 89 \text{ V}$$
$$V_q = -R_L i_q = -10 \times (-3.12) = 31.2 \text{ V}$$

The phase voltage (RMS) V_1 is as follows:

$$V_1 = \frac{1}{\sqrt{2}}\sqrt{V_d^2 + V_q^2} = \frac{1}{\sqrt{2}}\sqrt{89^2 + 31.2^2} = 66.88 \text{ V}$$

In addition,

$$I_1 = \frac{1}{\sqrt{2}}\sqrt{I_d^2 + I_q^2} = \frac{1}{\sqrt{2}}\sqrt{8.9^2 + 3.12^2} = 6.688 \text{ A}$$

The active power, for resistive load P_{out}, is as follows:

$$P_{out} = \frac{3}{2}\cdot(V_d i_d + V_q i_q) = 3\cdot V_1 \cdot I_1 = 3\cdot 66.88\cdot 6.688 = 1342 \text{ W}$$

The no-load voltage per phase E_{phase} (RMS) is as follows:

$$E_{phase}\,(\text{RMS}) = \frac{\omega_r \Psi_{PM1}}{\sqrt{2}} = \frac{200\pi\cdot 0.5}{\sqrt{2}} = 222.695 \text{ V}$$

The internal voltage power angle δ_V (Figure 10.25b) is as follows:

$$\delta_V = -\tan^{-1}\frac{V_d}{V_q} = -\tan^{-1}\frac{(-8.9)}{-3.12} = -70.68°$$

The voltage power angle δ_V should be equal to the d–q current power angle δ_i, as the load power factor is unity (resistive load):

$$\delta_i = -\tan^{-1}\frac{i_d}{i_q} = -\tan^{-1}\frac{89}{31.2} = -70.68°$$

The "gigantic" voltage regulation (from 222.695 to 66.88 V) is due to the too large $L_d = L_q$. This implies a machine that is too small for the load resistance in consideration.

This fact is certified by $|i_d| > |i_q|$.

Besides resistive loads, diode rectifier loads are also characterized by almost unity power factor; but in this case, the diode commutation process further reduces the rectified voltage as the load increases.

The vector diagram for the case in point is shown in Figure 10.26a and b.

Changing the resistance load to a larger value $R_L = 30\ \Omega$, we obtain the following:

$$i_d = \frac{-100\pi}{((31^2/10\pi) + 10\pi)} = -5.06 \text{ A} \quad i_q = -5 \text{ A}$$

Again,

$$I_1 = \frac{1}{\sqrt{2}}\sqrt{I_d^2 + I_q^2} = \frac{1}{\sqrt{2}}\sqrt{(-5.06)^2 + (-5)^2} = 5.03 \text{ A}$$

The phase voltage (RMS) V_1 is

$$V_1 = R_L \cdot I_1 = 30\cdot 5.03 = 150.90 \text{ V}$$

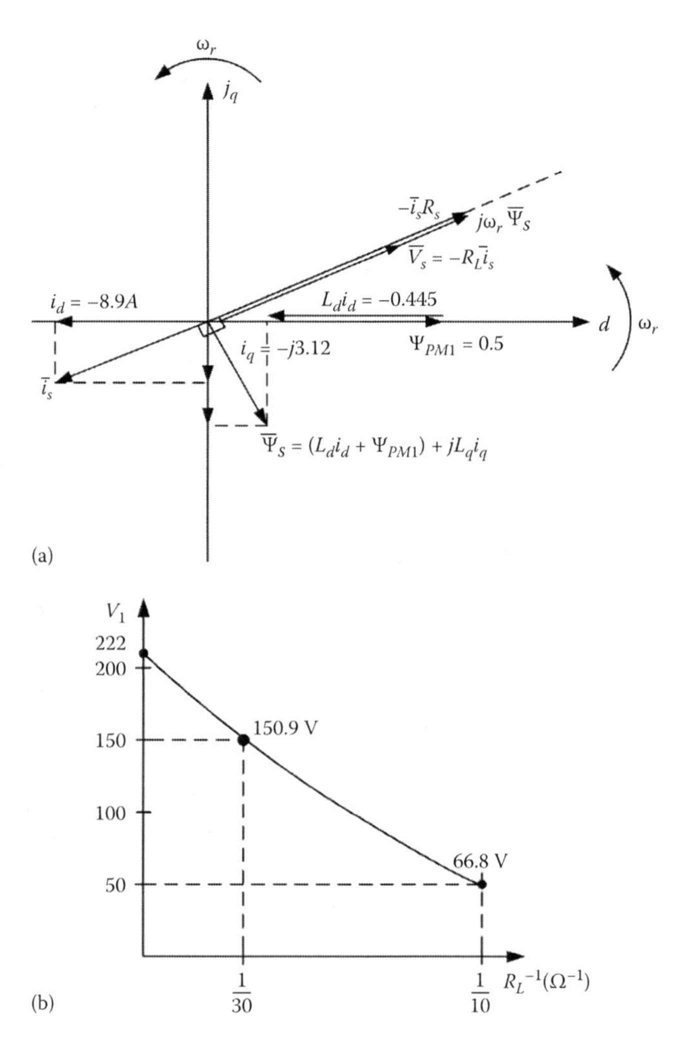

FIGURE 10.26 (a) Unity power factor generator vector diagram for the case in Example 10.1 and (b) voltage vs. R_L^{-1}.

The output power is as follows:

$$P_{out} = 3V_1 I_1 = 3 \cdot 150.90 \cdot 5.03 = 2277 \text{ W}$$

A large load resistance per phase led to more output power, as phase voltage ended up larger. The voltage power angle δ_V is now

$$\delta_i = \delta_V = -\tan^{-1}\frac{i_d}{i_q} = -\tan^{-1}\frac{(-5.06)}{(-5.00)} = -45.34°$$

This is about the maximum power that can be delivered by the machine to a resistive load, as $I_d \approx I_q (L_d = L_q)$.

Now the result is more reasonable, but the voltage regulation is still large. Moreover, as rated voltage power angles range around $\delta_V = 30°–35°$, the rated voltage regulation runs in the range of 30%, which is reasonable for a synchronous machine, even with surface PM pole rotor.

10.6 Circuit Model of PMSG with Shunt Capacitors and AC Load

As visible in Example 10.1, the voltage regulation is notable even for surface PM pole rotor PMSGs with resistive load. IPM-SGs were proposed to make use of inverse saliency ($L_q > L_d$c) to bring the terminal voltage back to the no-load value for a single (designed) load level. Capacitors in parallel or in series may be connected at PMSG terminals to compensate for increased machine required reactive power as the load level increases.

Parallel capacitors are considered here (Figure 10.27).

To model the PMSG plus shunt capacitors and load, using the *dq* model, the capacitors and the load effects have to be transformed into the *dq* model.

In phase coordinates, the capacitor and load equations are as follows:

$$\frac{dV_{a,b,c}}{dt} = \frac{(i_{a_s,b_s,c_s} - i_{a_L,b_L,c_L})}{C} \tag{10.48}$$

$$V_{a,b,c} = -(R_L + sL_L)i_{a_L,b_L,c_L} \tag{10.49}$$

The sign (–) written for motoring association of signs in Equation 10.48 reflects the fact that the PMSG *dq* model is transformed into *dq* rotor coordinates by using the inverse Park transform $(3/2)[P(\theta_{er})]^T$ (Equation 10.38) to obtain [18]:

$$\begin{aligned} sV_d &= (I_d - I_{dL})\frac{1}{C} + \omega_r V_q \\ sV_q &= (I_q - I_{qL})\frac{1}{C} - \omega_r V_d \end{aligned} \tag{10.50}$$

$$\begin{aligned} sI_{dL} &= -\frac{1}{L_L}V_d - \omega_r I_{qL} - \frac{R_L}{L_L}i_{qL} \\ sI_{qL} &= -\frac{1}{L_L}V_q + \omega_r I_{dL} - \frac{R_L}{L_L}i_{qL} \end{aligned} \tag{10.51}$$

Now we simply add the PMSG *dq* model equations derived in the previous paragraph:

$$\begin{aligned} s\Psi_d &= V_d - R_s i_d + \omega_r \Psi_q, \quad \Psi_q = L_d i_q \\ s\Psi_q &= V_q - R_s i_q - \omega_r \Psi_d, \quad \Psi_d = L_d i_d + \Psi_{PM1} \end{aligned} \tag{10.52}$$

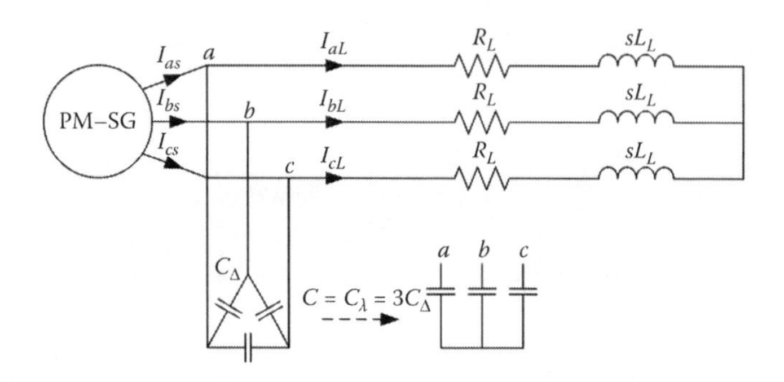

FIGURE 10.27 Permanent magnet synchronous generator (PMSG) with shunt capacitors and AC load.

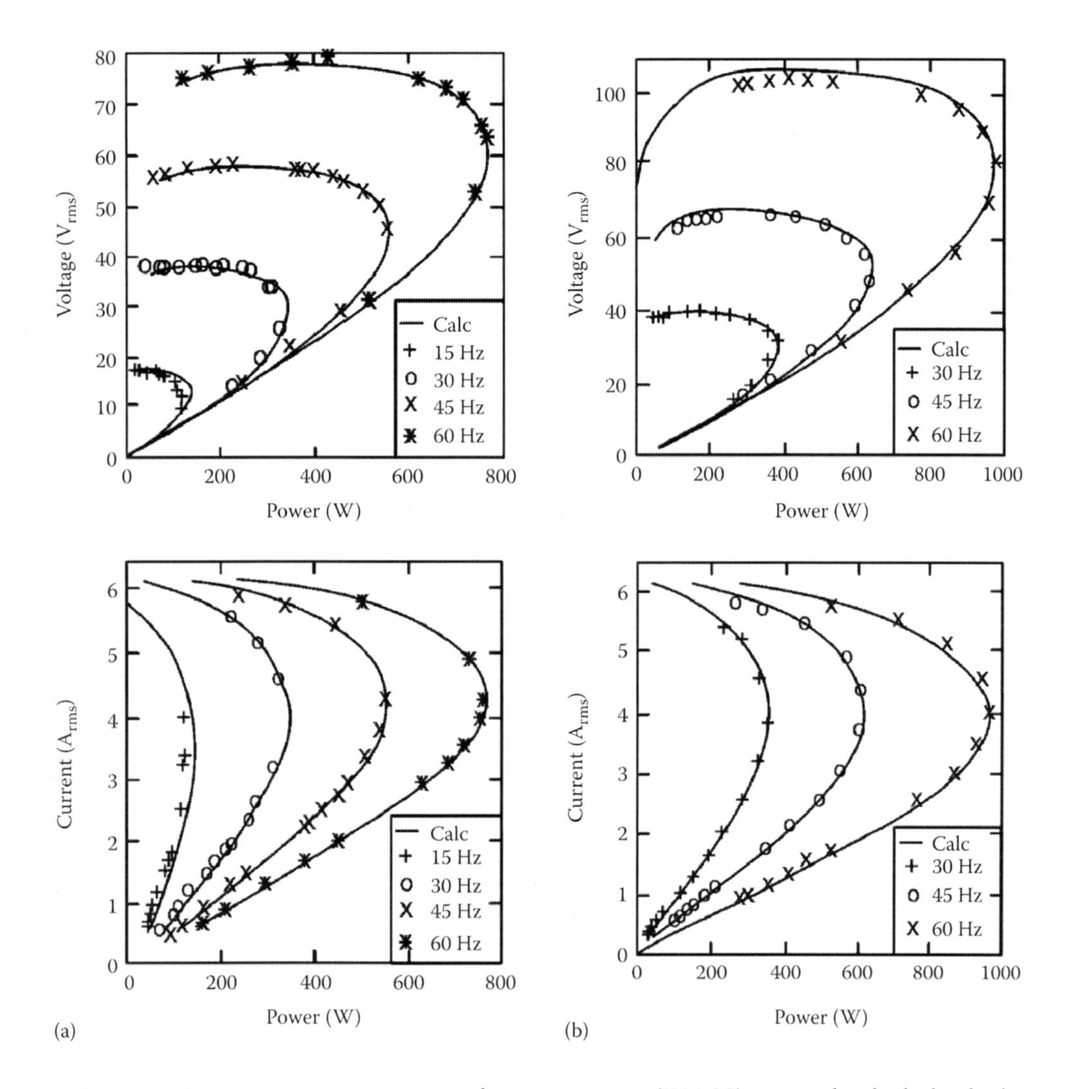

FIGURE 10.28 Interior permanent magnet synchronous generator (IPM-SG) measured and calculated voltage/power: (a) without shunt capacitors and (b) with shunt capacitors.

The currents i_d and i_q may be expressed as functions of Ψ_d and Ψ_q and are thus dummy variables. The actual variables are I_{dL}, I_{qL}, V_d, V_q, Ψ_d, and Ψ_q. For steady state, $s = 0$ (rotor coordinates). While for surface PM pole rotors, $L_d = L_q \approx$ const, for IPM rotors both $L_d \neq L_q$ and Ψ_{PM1} may vary with magnetic saturation.

In Reference 19, it is assumed that L_d, L_q, and Ψ_{PM1} depend on the stator current vector i_s, that is cross-coupling saturation is implicit and based on the assumption of two unique d–q axis curves. As both L_d and L_q decrease with i_s, the voltage regulation drops with increased loads.

A maximum power point on the voltage/power curve is obtained. The shunt capacitors at constant speed move this maximum to higher values for even smaller voltage regulation as in Figure 10.28a and b [19]. For constant $L_d = L_q$ and Ψ_{PM1} parameters, Equations 10.50 through 10.52 yield straightforward solutions for V_d, V_q, i_d, i_q, i_{dL}, and i_{qL} for given C, L_L, R_L, and ω_r.

When magnetic saturation is considered, an iterative solution is required to account for the curve-fitted $L_d(i_s)$, $L_q(i_s)$, and $\Psi_{PM1}(i_s)$ nonlinear functions [10].

10.7 Circuit Model of PMSG with Diode Rectifier Load

Diode rectifiers are used in many PMSG applications both for DC loads with or without battery backup and as a first stage in dual-stage AC–AC pulse-width modulated (PWM) converters with constant frequency and voltage output, for variable speed.

Let us consider here the case of diode rectifier filter and DC load (Figure 10.29).

As the *dq* circuit model of the PMSG with shunt capacitors was developed in a previous section, here we will add the diode rectifier filter DC load equations.

The input and output relationships for lossless diode rectifier based on the existence function model [10] are as follows:

$$\begin{aligned} &V_{DC} = V_a \cdot S_{aV} + V_b \cdot S_{bV} + V_c \cdot S_{cV} \\ &I_{aL} = I_{DC} \cdot S_a; \quad I_{bL} = I_{DC} \cdot S_b; \quad I_{cL} = I_{DC} \cdot S_c \end{aligned} \tag{10.53}$$

The diode rectifier voltage and current switching functions are

$$\begin{aligned} S_{a,b,c} &= \frac{2\sqrt{3}}{\pi}\cos\left(\omega_r t - (i-1)\frac{2\pi}{3}\right); \quad i = 1,2,3 \\ S_{a,b,c,V} &= \frac{2\sqrt{3}}{\pi}\cos\left(\frac{\mu}{2}\right)\cdot\cos\left(\omega_r t - (i-1)\frac{2\pi}{3}\right); \quad i = 1,2,3 \end{aligned} \tag{10.54}$$

For steady state [20],

$$\cos\mu = 1 - \frac{2\omega_r L_c I_{DC}}{V_{LL}\sqrt{2}} \quad V_{LL}\text{—line voltage (RMS) value} \tag{10.55}$$

The commutation angle μ increases with the rectified current I_{DC} and with increasing machine commutation inductance. In the absence of a damper cage on the rotor, $L_c \approx (L_d + L_q/2)$. So, from the view point of lowering the voltage drop along the diode rectifier due to machine inductances, it is beneficial to place a strong damper cage on the PM rotor. The maximum ideal value of μ should be less than $(\pi/3)(60°)$.

Approximately, the rectified DC voltage V_{DC} is related to machine line voltage [20], under steady state:

$$V_{DC} = V_{LL}\frac{3\sqrt{2}}{\pi} - \frac{3\omega_r L_c}{\pi} I \tag{10.56}$$

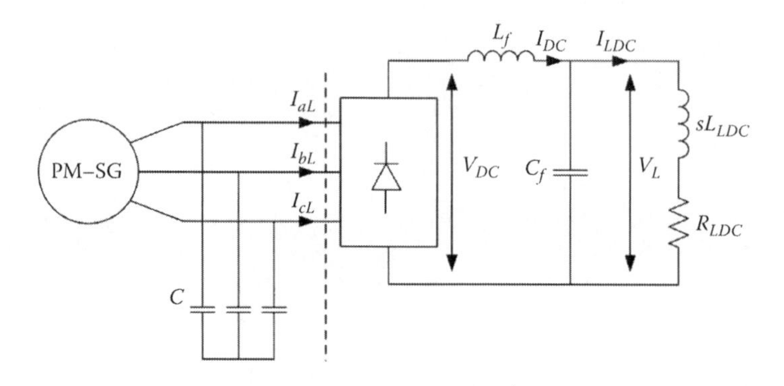

FIGURE 10.29 Permanent magnet synchronous generator (PMSG) with shunt capacitors, diode rectifier, filter, and load.

As the load was replaced by the diode rectifier, the shunt capacitor equations in *dq* coordinates become thus [19]:

$$\begin{aligned} sV_d &= \frac{1}{C} I_d + \omega_r \cdot V_q \\ sV_q &= \frac{1}{C}\left(I_q - \frac{2\sqrt{3}}{\pi} I_{DC} \right) - \omega_r \cdot V_d \\ V_{DC} &= V_q \frac{2\sqrt{3}}{\pi} \cos\left(\frac{\mu}{2}\right) \end{aligned} \tag{10.57}$$

The DC filter and load equations are added:

$$\begin{aligned} sI_{DC} &= \frac{1}{L_f}(V_{DC} - V_L) \\ sV_L &= \frac{1}{C_f}(I_{DC} - I_L) \\ sI_L &= \frac{1}{L_{L_{DC}}}(V_L - R_{L_{DC}} I_L) \end{aligned} \tag{10.58}$$

Under steady state, $s = 0$; thus, all time derivatives are zero. With the voltage drop in the diode rectifier neglected, the load resistance $R_{L_{DC}}$ may be seen as a star-connected phase AC resistance $R_L = (\pi^2/18)R_{L_{DC}}$ in parallel with the shunt capacitors at PMSG terminals as follows:

$$\begin{aligned} 3R_L \left(\frac{I_{ph}^{peak}}{\sqrt{2}} \right)^2 &= R_{L_{DC}} \cdot I_{DC}^2 \\ I_{ph}^{peak} &\approx \frac{2\sqrt{3}}{\pi} I_{DC} \end{aligned} \tag{10.59}$$

Due to diode commutation of machine commutation inductances (L_c), the DC load voltage decreases notably with load [19], especially if $L_c = (L_d + L_q/2)$ (no damper cage) when the speed (frequency) is large, as in automotive applications.

With a more complete representation of a diode rectifier, the generator actual output voltage and current waveforms may be obtained [19] (Figure 10.30a and b).

The presence of shunt capacitors may generate a kind of resonance phenomena with quasiperiodic oscillations with bounded dynamics in the generator line voltage and current, especially for light load. These oscillations are visible as amplitude variations in generator line voltage and phase current and in the DC load voltage (Figure 10.31a through c) [19]. Special active or passive measures have to be taken to eliminate these oscillations.

10.8 Utilization of Third Harmonic for PMSG with Diode Rectifiers

A trapezoidal AC voltage (emf) is often suggested for increased output for PMSG with rectifier DC loads. Two different rectifier topologies, as illustrated in Figure 10.32a and b, are of interest.

For a purely sinusoidal system with *n* phases, neglecting the voltage drop in the rectifier, the ideal relationship between the AC and DC variables is as follows [16]:

$$\begin{aligned} V_{DC} &= \sum_{i=1} V_{di} = \frac{2n}{\pi} V_1; \quad V_1\text{—peak phase voltage} \\ P_{DC} &= P_{ac} = V_{DC} I_{DC} = \frac{2n}{\pi} V_1 \cdot I_{ac_{rms}}; \quad n\text{—number of phases} \end{aligned} \tag{10.60}$$

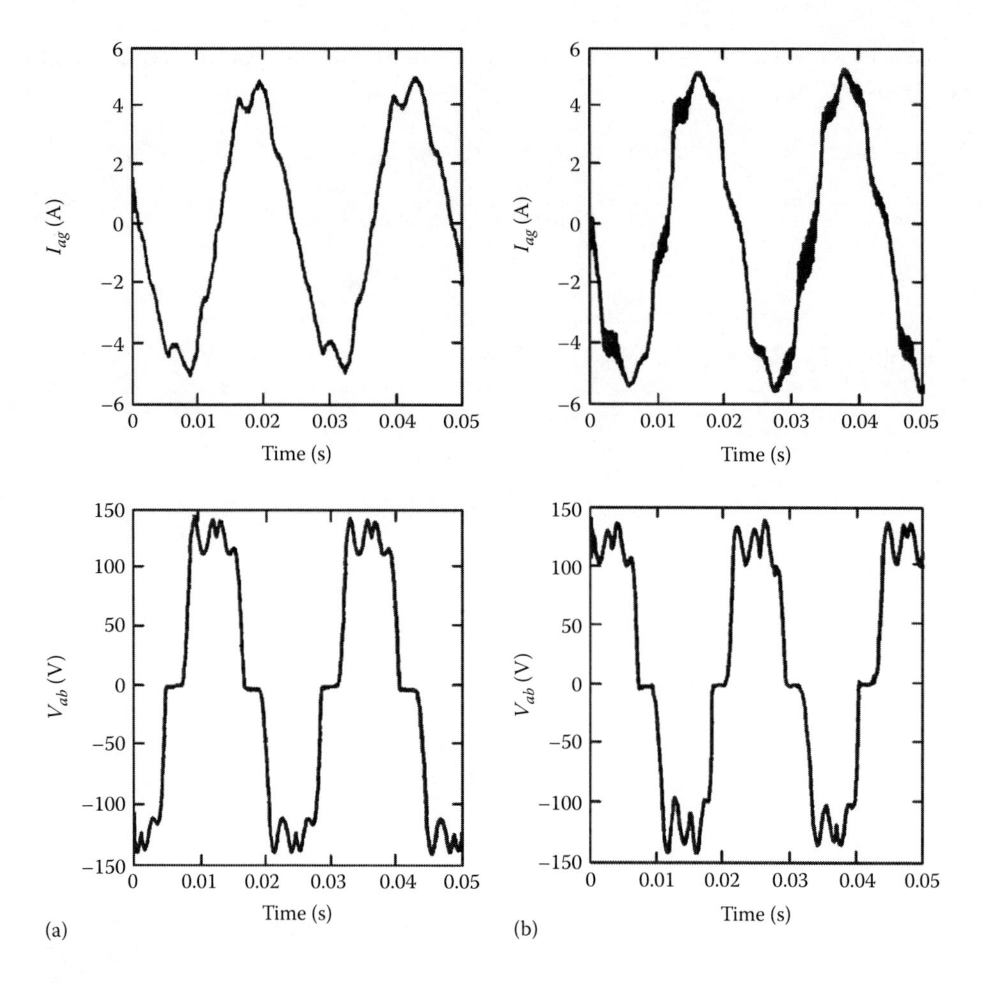

FIGURE 10.30 (a) Simulated and (b) measured generator phase current and line voltage with shunt capacitor, diode rectifier, and direct current (DC) load: $R_L = 22\ \Omega$, $n = 1350$ rpm.

for the series rectification, and

$$V_{DC} = \frac{2n}{\pi}\sin\frac{\pi}{n}V_1$$
$$P_{DC} = \frac{\sqrt{2}(n)^{3/2}}{\pi}\sin\frac{\pi}{n}I_{ac_{rms}}\cdot V_1 \tag{10.61}$$

for polyphase bridge rectification.

The situation changes notably when a third harmonic component $k_h V_1$ is present in the phase emfs of the PMSG.

For series rectification, simply,

$$V_{DC} = \frac{2n}{\pi}\left(1+\frac{K_h}{3}\right)V_1$$
$$P_{DC} = V_{DC}I_{DC} = \frac{2n}{\pi}\left(1+\frac{K_h}{3}\right)I_{ac_{rms}}V_1 \tag{10.62}$$

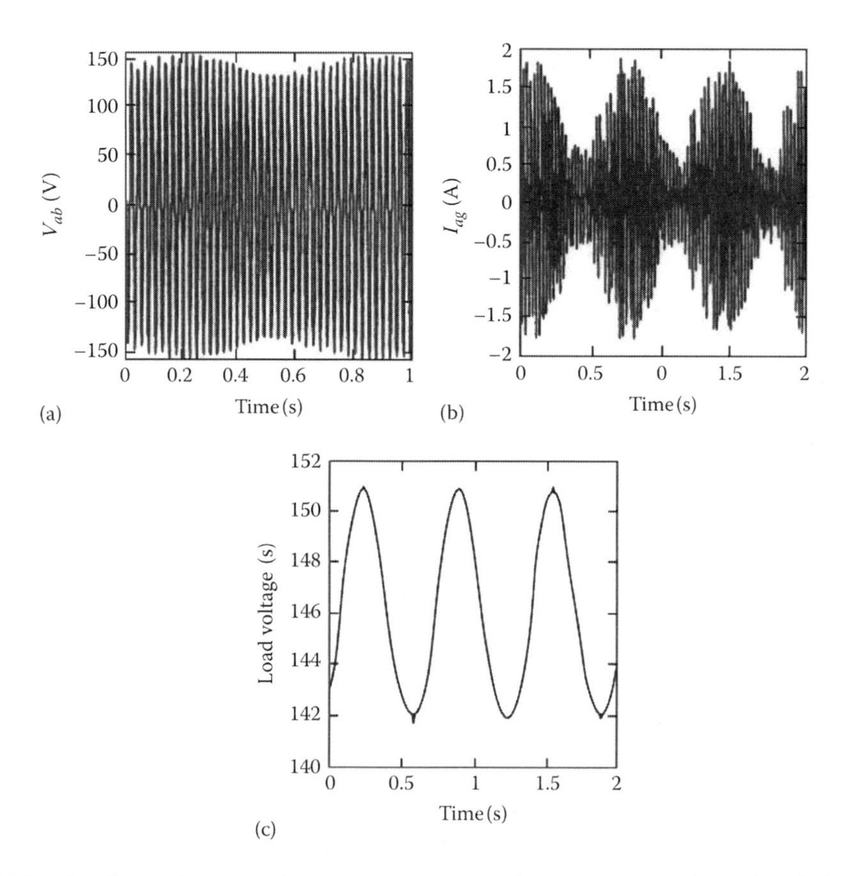

FIGURE 10.31 Oscillatory dynamics of permanent magnet synchronous generator (PMSG) with shunt capacitors and diode rectifier load: (a) generator line voltage, V_{ab}; (b) generator phase current, i_a; and (c) DC load voltage.

For the polyphase rectification, the third emf harmonic leads to multiple commutations; and thus, the DC output formulas are more involved [16] only for the voltage:

$$I_{DC} = \sqrt{\frac{n}{2}} I_{ac_{rms}} \tag{10.63}$$

There are basically five multiple commutation modes for polyphase rectification 0, 1, 2, 3, and 4 [16]:

$$\begin{aligned}
V_{DC0} &= \frac{2n}{\pi}\left[\sin\frac{\pi}{n} - \frac{k_h}{3}\sin\left(\frac{3\pi}{n}\right)\right]\cdot V_1 \\
V_{DC1} &= \frac{2n}{\pi}\left\{\sin\theta - \sin\left(\theta - \frac{2\pi}{n}\right) - \sin\frac{\pi}{n} - \frac{k_h}{3}\left[\sin\frac{3\pi}{n} + \sin 3\theta - \sin\left(3\theta - \frac{6\pi}{n}\right)\right]\right\}\cdot V_1 \\
V_{DC2} &= \frac{2n}{\pi}\left\{\sin\frac{2\pi}{n} - \sin\frac{\pi}{n} + \frac{k_h}{3}\left[\sin\frac{3\pi}{n} - \sin\frac{6\pi}{n}\right]\right\}\cdot V_1 \\
V_{DC3} &= \frac{2n}{\pi}\left\{(2\cos\theta - 1)\sin\frac{2\pi}{n} - \sin\frac{\pi}{n} + \frac{k_h}{3}\left[\sin\frac{3\pi}{n} - (2\cos 3\theta - 1)\sin\frac{6\pi}{n}\right]\right\}\cdot V_1 \\
V_{DC4} &= \frac{2n}{\pi}\left\{\sin\frac{3\pi}{n} - \sin\frac{2\pi}{n} + \frac{k_h}{3}\left[\sin\frac{6\pi}{n} - \sin\frac{9\pi}{n}\right]\right\}\cdot V_1
\end{aligned} \tag{10.64}$$

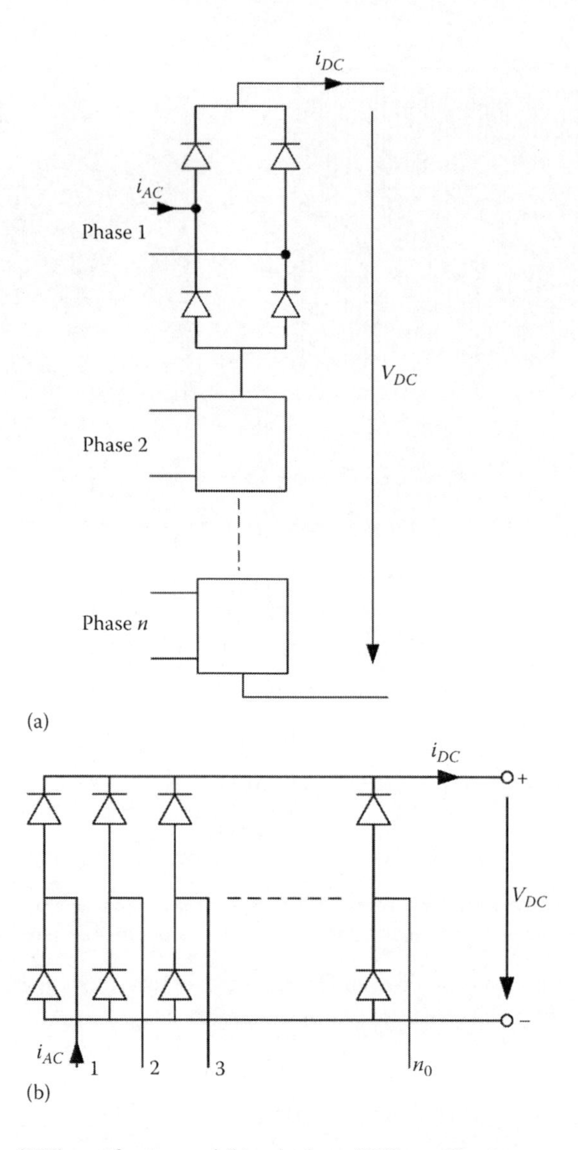

FIGURE 10.32 (a) Series (SRS) rectification and (b) polyphase (PRS) rectification.

with θ from

$$\sin\left(\theta - \frac{\pi}{n}\right) \cdot \sin\frac{\pi}{n} = k_h \sin\left(2\theta - \frac{3\pi}{n}\right) \cdot \sin\frac{3\pi}{n}; \quad \frac{\pi}{n} < \theta < \delta; \; n > 5 \tag{10.65}$$

The angle δ is the offset angle of peak phase voltage value with respect to π/2. It should be noted that a three-phase system does not show crossover (multiple commutations). A six-phase system will exhibit crossovers.

The power gain due to the presence of a third harmonic $k_h V_1$ in the PMSG phase emf is as follows [16]:

$$P_{gain}^{peak} = \frac{1 + k_h/3}{\cos\delta - k_h \cos 3\delta} \tag{10.66}$$

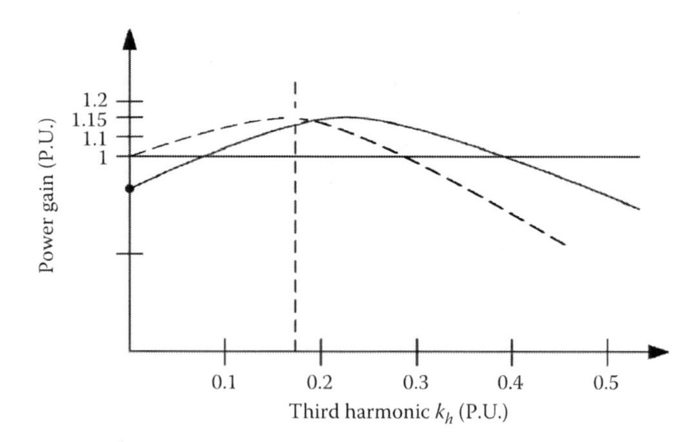

FIGURE 10.33 Power gain for three-phase rectification for constant peak voltage. The solid line represents series rectification and the broken line represents three-phase bridge rectification.

for constant peak voltage, and

$$P_{gain}^{rms} = \frac{1 + k_h/3}{\sqrt{1 + k_h^2}} \tag{10.67}$$

for constant RMS voltage.

For constant peak voltage, the majority of additional power is due to increases in the fundamental; while for constant RMS voltage, it is due to increased third harmonic.

The highest power gain is obtained with a three-phase system for constant peak voltage, and typical results are shown in Figure 10.33 [16]. In the three-phase rectification, a fourth diode leg is added for the null to allow for tapping the third harmonic power.

A maximum of about 15% power gain may be obtained with both rectification methods at distinct levels of third harmonic (k_h) (Figure 10.33).

As already alluded to, the third harmonic presence in the emf (that is, trapezoidal emf) leads to additional core losses. Still, the power gain generally offsets this loss increase, and the efficiency stays constant or even increases by 1%–2% for a 15% power gain.

Note that multiphase (e.g., nine-phase) PMSG windings, in groups of three, may be used to build a combination of series/polyphase rectification connections, in order to reduce DC output voltage pulsations (in amplitude and frequency) and thus allow for a smaller capacitor filter in the DC voltage link.

The presence of DC voltage pulsations and DC link capacitor filter were not considered in this section.

10.9 Autonomous PMSGs with Controlled Constant Speed and AC Load

Thermal-engine-driven autonomous PMSGs in the tens to thousands of kilowatts may be controlled for constant speed to deliver constant voltage and frequency for AC loads at moderate generator system costs. To reduce generator control system costs, fixed capacitors are connected at generator terminals. They are accompanied by thyristor-controlled three-phase inductances (Figure 10.34). For 400 Hz outputs, typical for aircraft, the size of inductors and capacitors is smaller than for 50(60) Hz outputs.

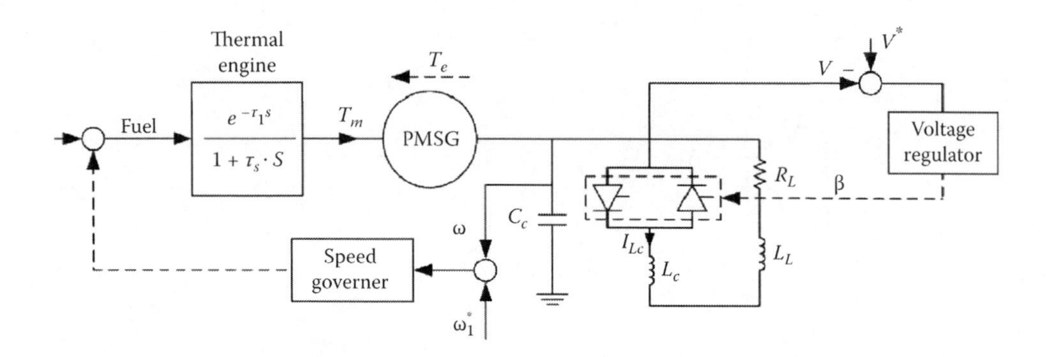

FIGURE 10.34 Permanent magnet synchronous generator (PMSG) with constant frequency (speed) and AC output voltage regulation.

In principle, a frequency regulator acts upon the fuel admission valve to keep the frequency (speed) constant. A voltage regulator controls the ignition angle β in the thyristor-controlled reactor and fixed capacitor volt-ampere reactive (VAR) compensator, to keep the voltage dynamically constant with load.

The inductor current i_{Lc} [21] is as follows:

$$i_{Lc} = \frac{V_g}{\omega L_c} \frac{(2\beta - \sin 2\beta)}{\pi} c \tag{10.68}$$

The equivalent (variable) inductance is a function of β:

$$L_{Lc} = \frac{\pi L_c}{2\beta - \sin 2\beta} \tag{10.69}$$

To investigate the steady state and transients of the system in Figure 10.34, the dq model is used. For steady state, the dq model preserves the phase capacitance C, inductance L_{Lc}, and impedance Z_L along dq axes, as the machine phase resistance and synchronous inductance are preserved.

The capacitor $\underline{I}_{Cc}$, inductor $\underline{I}_{Lc}$, generator $\underline{I}_s$, and load $\underline{I}_L$ currents are visible in Figure 10.35a. For any given value of load (L_L, R_L), capacitance C_c, inductance L_{Lc} (β), frequency, and machine parameters, the equivalent circuits or the vector diagram produce the four above-mentioned currents (Figure 10.35b). Then, the voltages V_d and V_q are obtained, and the respective active and reactive powers are calculated [21]:

$$\begin{aligned} I_d &= I_L \sin(\phi_L + \delta_V) - (I_{Cc} - I_{Lc})\cos\delta_V \\ I_q &= I_L \cos(\phi_L + \delta_V) + (I_{Cc} - I_{Lc})\sin\delta_V \end{aligned} \tag{10.70}$$

$$\begin{aligned} V_s &= \sqrt{V_d^2 + V_q^2} = \frac{E_0}{\cos\delta_V \left[1 + X_d(K_{xL} - X_{Ce}) + R_s K_{rL} + K_{\delta V}\right]} \\ \tan\delta_V &= \frac{X_q K_{rL} - R_s(X_{Ce} - K_{xL})}{1 + X_q(K_{xL} - X_{Ce}) + R_s K_{rL}} \end{aligned} \tag{10.71}$$

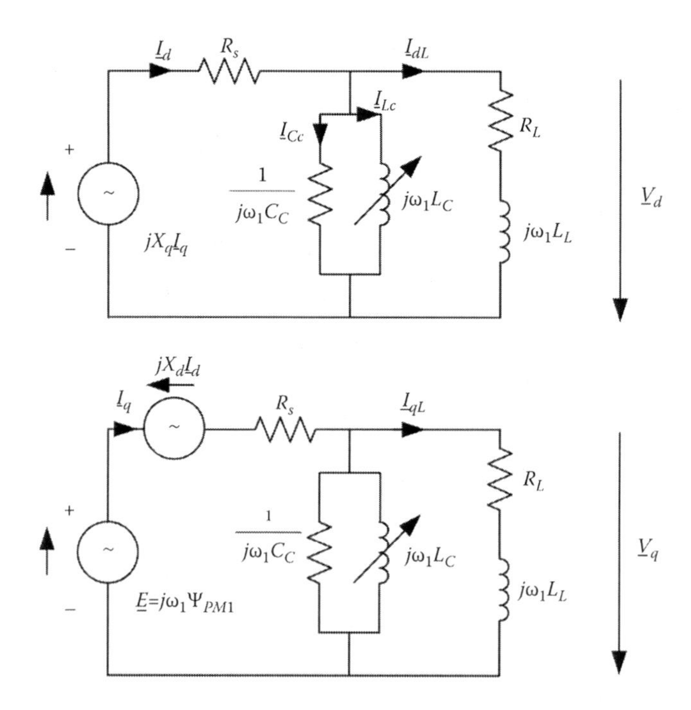

FIGURE 10.35 Steady-state equivalent circuits in phasor variables decomposed in dq components: (a) and (b).

with

$$K_{r_L} = \frac{R_L}{Z_L^2}; \quad K_{x_L} = \frac{X_L}{Z_L^2}; \quad Z_L = \sqrt{R_L^2 + X_L^2}$$

$$X_{Ce} = \frac{1}{X_c} - \frac{1}{X_{Lc}}; \quad K_{\delta_V} = [X_d K_{r_L} - R_s(K_{x_L} - X_{Ce})]\tan\delta_V$$

Once the terminal V_s and the voltage power angle δ_V are calculated, all the other variables (currents, powers) are straightforwardly determined as follows:

$$I_L = \sqrt{I_{dL}^2 + I_{qL}^2}; \quad I_{Lc} = \frac{V_s}{\omega_1 L_{Lc}}; \quad I_{Cc} = V_s \cdot \omega \cdot C_c$$
$$I_s = \sqrt{I_d^2 + I_q^2} = \frac{V_s}{Z_L} \tag{10.72}$$

Given the range of load impedance Z_L and its power factor angle φ_L variation range, the required capacitor and inductance that provide for constant terminal voltage for constant speed may be determined. When defining the variable inductance value L_C, the angle β is from Equation 10.69, and thus, the voltage drop compensation conditions may be investigated as a function of the ignition angle β.

For all load power factor angles (except for $\cos\varphi_L = 1$), there are two values of ignition angle—one for leading power factor and one for lagging power factor.

While the design of C_C, L_C for steady state is the first step, the investigation of transients and control aspects is crucial.

The dq model, in the rotor coordinates, is used for the scope. For the sake of completeness, the system equations are given below—first for the load and the VAR compensator:

$$\begin{aligned}
sL_d &= \left(\frac{V_d - R_L i_{dL}}{L_L}\right) + \omega_r i_{Lq} \\
sL_q &= \left(\frac{V_q - R_L i_{qL}}{L_L}\right) - \omega_r i_{Ld} \\
sV_d &= \frac{i_{Cd}}{C_c} + \omega_r V_q \\
sV_q &= \frac{i_{Cq}}{C_c} - \omega_r V_d \\
si_{d_{Lc}} &= \frac{V_d}{L_c} + \omega_r i_{q_{Lc}} \\
si_{q_{Lc}} &= \frac{V_q}{L_c} - \omega_r i_{d_{Lc}}
\end{aligned} \tag{10.73}$$

The dq model of PMSG is described by Equation 10.52, with i_d, i_q as dummy variables. The motion equation has to be added:

$$\frac{J}{p_1}\frac{d\omega_r}{dt} = T_m - T_e; \quad T_e = \frac{3}{2}p_1(\Psi_d i_q - \Psi_q i_d) \tag{10.74}$$

The above equations are first linearized and so is the thermal engine (say diesel engine) equation:

$$\tau_s \cdot s \cdot \Delta T_m + \Delta T_m = \Delta F(t - \tau_1) \tag{10.75}$$

where

τ_1 is the engine dead time (sampling time, approximately)

τ_s is the electrohydraulic actuator time constant

While independent speed (frequency) and voltage regulators may be adopted, more advanced state feedback, or output feedback, controls were introduced for the scope, with good results [21]. Typical transients with optimal state feedback control are shown in Figure 10.36, based on the results from Reference 21.

A notable short-lived reduction in the terminal voltage is noticed. However, this event is followed by a fast recovery. The fuel rate increases steadily as the load increases. The speed deviation transients are small, while the thyristor ignition angle varies markedly to accommodate the new, larger, load current at about the same load power factor.

Given the numerous nonlinearities in the system, nonlinear (such as variable structure) control may be applied to further reduce the voltage surges during transients.

10.10 Grid-Connected Variable-Speed PMSG System

Variable-speed use is good for extracting more prime-mover power (as in wind turbines) or for providing optimum efficiency for the prime mover by increasing its speed with power. Variable speed also allows for a more flexible generator system. For wind turbines, a battery may be added to store the extra wind energy that is not momentarily needed for the existing loads (or local power grid).

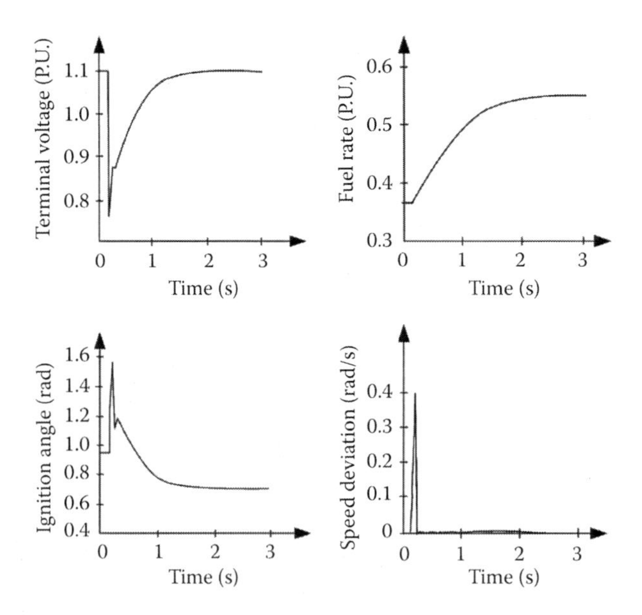

FIGURE 10.36 State feedback controller response when the load impedance varies suddenly from $(1.8 + j1.35)$ P.U. to $(1.2 + j0.9)$ P.U. (From Rahman, M.A. et al., *IEEE Trans.*, EC-11(2), 324, 1996.)

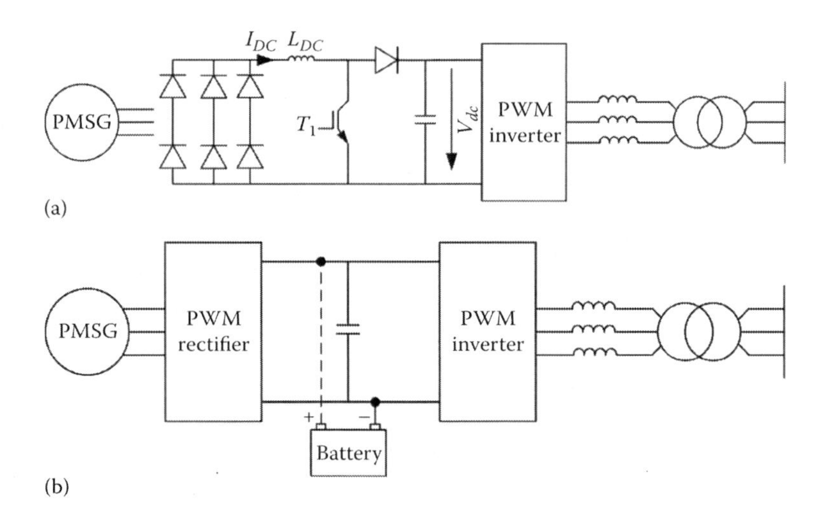

FIGURE 10.37 Variable-speed permanent magnet synchronous generator (PMSG) system with constant output voltage and frequency: (a) with diode rectifier and boost direct current (DC)–DC converter and (b) with pulse-width modulator (PWM) rectifier.

There are two main ways of handling the necessity of stable (constant) DC link voltage at variable speed as illustrated in Figure 10.37a and b.

For the case of the diode rectifier, the generator is designed with terminal voltage at maximum speed to allow for a small voltage boost in the DC–DC converter for all situations.

Full swing boost is required at minimum speed.

10.10.1 Diode Rectifier and Boost DC–DC Converter Case

In the absence of a rotor cage, the dq (space-vector) model of PMSG (after Equation 10.52), for the generator mode association of signs, is as follows:

$$\bar{V}_s = -\bar{i}_s R_s - j\omega_r \bar{\Psi}_s$$
$$\bar{\Psi}_s = \Psi_d + j\Psi_q;\quad \bar{i}_s = i_d + ji_q;\quad \bar{V}_s = V_d + jV_q \tag{10.76}$$

$$\bar{\Psi}_s = \Psi_{PM} + L_d i_d + jL_q i_q \tag{10.77}$$

As for steady state ($s = 0$), the diode rectifier imposes almost unity power factor at the fundamental frequency:

$$Q_1 = 0 = \frac{3}{2}(V_d i_q - V_q i_d)c \tag{10.78}$$

Consequently, from Equations 10.76 through 10.78, the following condition is obtained:

$$L_q i_q^2 + L_d i_d^2 - \Psi_{PM} i_d = 0 \tag{10.79}$$

Here, $i_d > 0$.

The electromagnetic torque T_e is as follows:

$$T_e = \frac{3}{2} p_1 (\Psi_{PM} - (L_d - L_q) i_d) i_q;\quad L_d < L_q;\quad i_d > 0 \tag{10.80}$$

with $L_d = L_q = L_s$, Equation 10.79, when used in Equation 10.80, yields the following:

$$\frac{\partial T_e}{\partial i_d} = \frac{\partial}{\partial i_d}\left(\frac{3}{2} p_1 \Psi_{PM} \sqrt{\frac{i_d \Psi_{PM} - L_s i_d^2}{L_s}}\right) = 0 \tag{10.81}$$

with the following solution:

$$i_{dk} = \frac{\Psi_{PM}}{2L_s} \tag{10.82}$$

Therefore, the maximum torque T_{ek} is

$$(T_{ek})_{\cos\varphi_1 = 1} = \frac{3}{2} p_1 \frac{\Psi_{PM}^2}{2L_s};\quad L_s = L_m + L_{sl} \tag{10.83}$$

The larger the stator leakage inductance L_{sl}, the lower the maximum torque at unity power factor. In addition, for maximum torque, from Equations 10.79 and 10.82,

$$i_{qk} = i_{dk} = \frac{\Psi_{PM}}{2L_s} \tag{10.84}$$

But, the condition in Equation 10.84 corresponds, in general, to maximum torque/current, which is to be expected, as the power factor was taken as unity, for the diode rectifier case. With base torque T_b defined as follows,

$$T_b = \frac{P_b}{\omega_{rb}} \cdot p_1 \tag{10.85}$$

the maximum torque is

$$\frac{(T_{ek})_{\cos\varphi_1 = 1}}{T_b} = \frac{1}{2X_s(\text{P.U.})} \tag{10.86}$$

As expected, the lower the X_s (in P.U.), the larger the peak torque. Reducing the stator leakage inductance becomes a very important design task.

The control of power flow to the power grid depends on both, the DC link voltage level (controlled by the PWM inverter on the power grid side) and on DC–DC boost converter output DC voltage.

For a wind turbine, the maximum wind power extraction occurs for a given $\omega_r(T_m^*)$ reference curve up to maximum power value, when stall regulation occurs. The DC current may then be regulated to prescribe the modulation index of the IGBT in the DC–DC boost converter (Figure 10.38).

To limit the DC–DC converter chopping frequency, no external means are required, as the large commutation inductance (L_s) of the generator is connected in series with L_{DC} during on states of T_1 [22,23].

As already mentioned in this section, an active machine end (PWM) rectifier may replace the diode rectifier plus the boost DC–DC converter. In this case, the voltage at the machine terminals, at lower speeds, may be lower than the constant-kept capacitor voltage in the DC link by the action of the grid-side PWM converter. In essence, the capacitor in the DC link provides, in a controlled manner, the required reactive power injection into the machine in addition to the commutation reactive power in the rectifier.

The PMSG may produce more torque than for the case of the diode rectifier but does this also for higher stator currents, as required by the apparent power of the generator.

The back-to-back PWM converter system with DC capacitor link may provide both variable speed operation and reactive power controlled delivery at constant AC output voltage and frequency.

The machine-side converter is torque (speed) controlled along the q axis and is constant flux controlled along the d axis field-oriented control (FOC) or in direct torque and flux control (DTFC) schemes.

On the other hand, the load-side converter is controlled differently for stand-alone operation.

In power grid operation, the grid-side PWM converter keeps the DC link voltage constant and delivers a certain amount of reactive power, as desired. With an adequate design for the DC link capacitor, up to ±100 reactive and active power control is feasible with PMSGs at a lower cost of the capacitor in comparison with the induction generator with cage rotor.

For stand-alone loads, an L–C power filter is mandatory between the load-side PWM inverter and the loads. Now the output voltage frequency is set at a constant value, while the AC output voltage

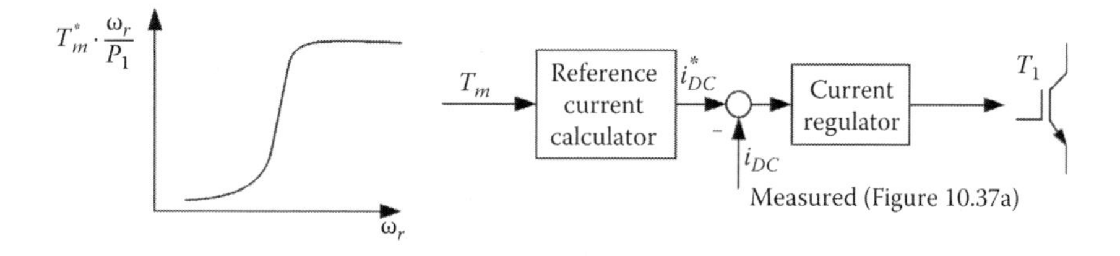

FIGURE 10.38 Boost direct current (DC)–DC converter control for a wind-turbine permanent magnet synchronous generator (PMSG).

amplitude, after the output filter, is controlled through the DC link capacitor voltage closed-loop control via, again, the load-side converter.

The PMSG, once in rotation, with the machine-side converter idle, charges the DC link capacitor through its diodes to offer conditions for self-excitation of the system on no load.

A backup battery and a solar panel may work together with the PMSG systems with wind or hydroturbines to compensate for the limited availability, dependent on time of day, of wind (hydro) energy.

10.11 PM Genset with Multiple Outputs

Medium-power (tens to hundreds of kilowatts) gensets were traditionally designed to be rugged and cost effective, though heavier, for special applications. In essence, their speed, when driven by a diesel engine, is in the 1800–3600 rpm range; while for gas turbines, it may be in the tens of thousands of revolutions per minute.

Gensets may be used as auxiliary power sources on ground (for aircraft land service), as standby power supply units, or on trucks, vessel, or aircraft as medium-power auxiliary power sources. The multiple output possibility stems from the capability of the PWM inverter on the load side to be controlled for various voltages, frequencies, and numbers of phase outputs (Figure 10.39) [24].

The diode rectifier, plus the boost DC–DC converter, is used to produce constant DC link voltage over a wide speed range (up to about 2/1 range).

The bidirectional buck-boost DC–DC converter is used to supplement the 28 V_{DC} loads and recharge the 28 V_{DC} backup battery, and in addition, to provide a DC voltage boost whenever necessary during load or speed transients.

The standard PWM 6 IGBT voltage source inverter is controlled on a menu basis, to produce AC output at 50 or 60 Hz, or 400 Hz (for example), and with single-phase, dual-phase, or three-phase attributes, according to needs.

Variable speed is used to reduce fuel consumption (Figure 10.40) [24] or to reduce thermal engine pollution, by reducing speed when required power is reduced.

As expected, there is minimum fuel consumption for almost constant power at a certain speed (2900 rpm in Figure 10.40).

The machine-side diode rectifier is standard for a three-phase PMSG, but the boost DC–DC converter may take various configurations. A representative one is shown in Figure 10.41 [24].

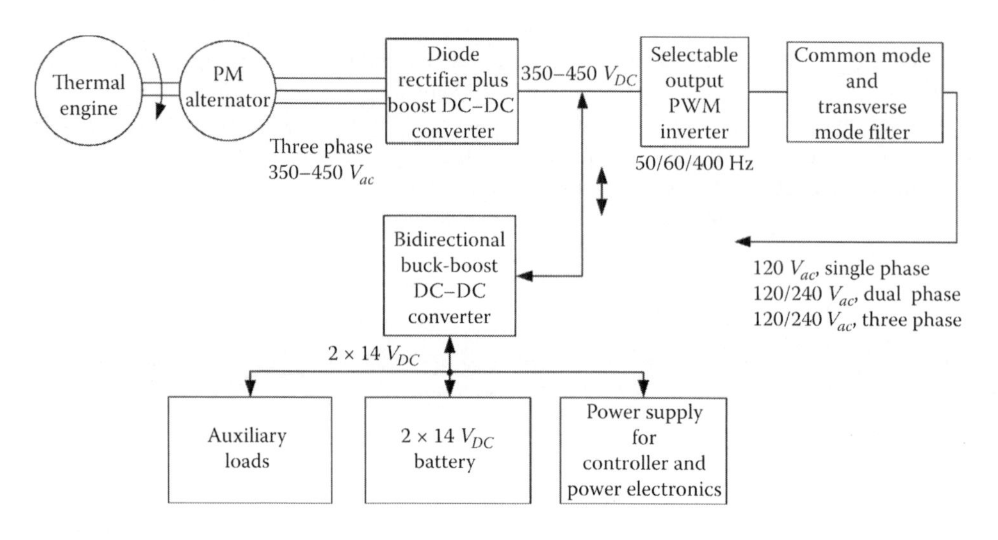

FIGURE 10.39 A PM genset with multiple outputs.

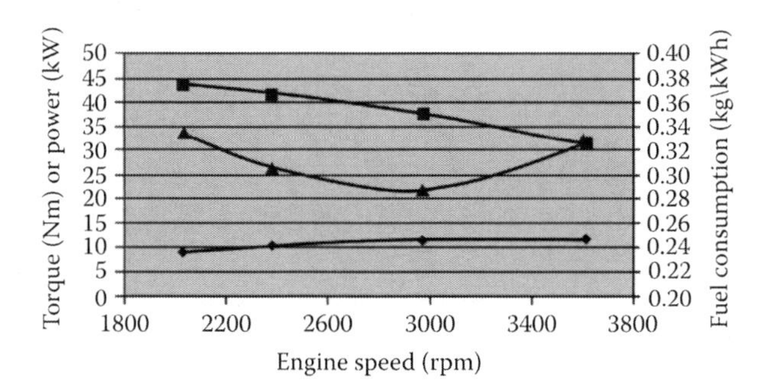

FIGURE 10.40 Peak torque (▲), power (■), and fuel consumption (♦) of a diesel engine permanent magnet (PM) genset.

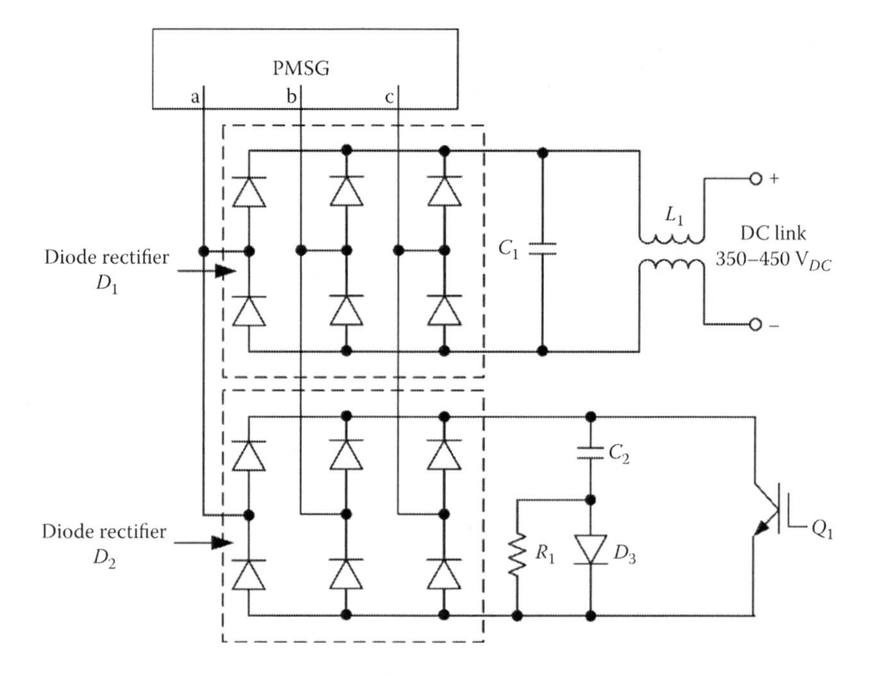

FIGURE 10.41 Diode rectifier plus boost direct current (DC)–DC converter.

Given the rather large value of generator commutation (subtransient, almost synchronous) inductance, the latter is used for voltage boost. The accumulation of energy in the machine inductances is performed through the additional diode rectifier D_2, which is short-circuited by the IGBT, Q_1.

Subsequently, Q_1 is disengaged, and the higher voltage obtained at the output of the main diode rectifier D_1 charges the capacitor C_1 in the DC link to boost the DC link voltage. The snubber C_2–R_1 limits the *di*/*dt* spikes at Q_1 when the latter is disengaged. The alternator and load neutrals are connected to the center of two capacitors C_3 in the DC link (Figure 10.42). The neutral connection allows for unbalanced load currents to flow through the PMSG in parallel with the capacitors C_3, and thus eliminates the necessity that the latter have low fundamental frequency impedance.

The inductance L_1 (Figure 10.41) avoids the charging of capacitors C_3 to the half-wave rectified line to neutral voltage. Poor power factor and high crest currents in the generator are thus avoided.

The three-phase differential (L_2, C_4) and common-mode (L_3, C_5) filters are also shown in Figure 10.42 [24]. They are meant to achieve near sine wave output (load) voltage (<3% total harmonic distortion [THD]). Line-to-neutral inverter output voltage pre- and post-filter output voltages shown in Figure 10.43 are fully indicative of a filter's beneficial effect [24].

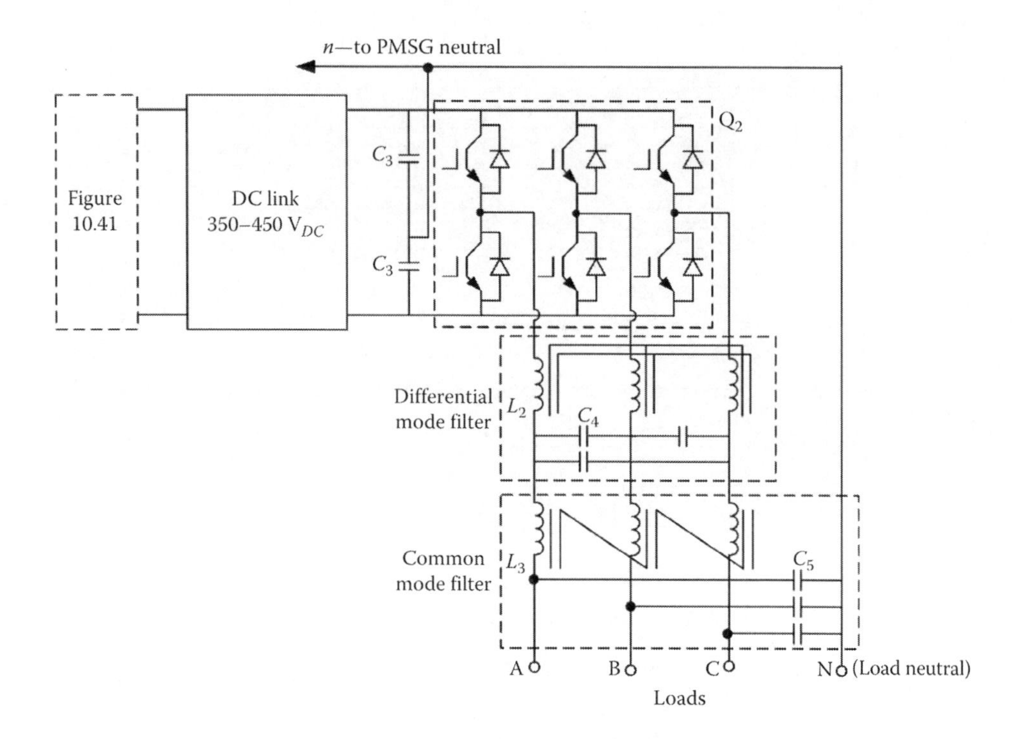

FIGURE 10.42 The permanent magnet synchronous generator (PMSG) inverter and filters.

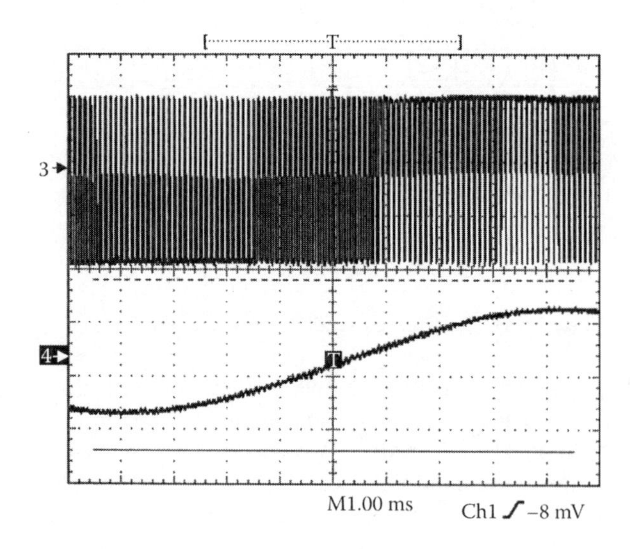

FIGURE 10.43 Pre- and post-filter line-to-neutral load voltage (V_{AN}).

The bidirectional DC–DC converter converts power from (to) 350 V_{DC} to (from) 28 V_{DC} to supply auxiliary loads and to recharge the engine-starting battery that supplies, in turn, the engine starter (a DC PM motor).

There are many competitive bidirectional buck-boost DC–DC converter configurations for such a high voltage ratio, but they all make use of a high-voltage-frequency transformer.

Such a typical configuration making use of low power switch count is shown in Figure 10.44 [24].

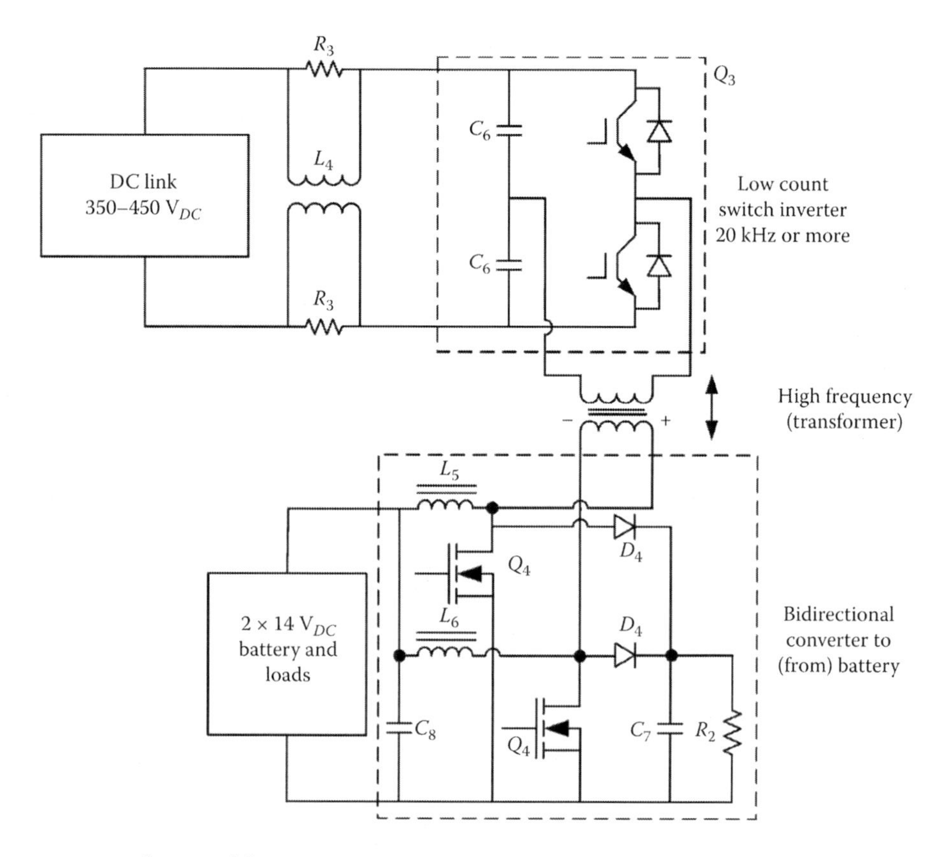

FIGURE 10.44 Bidirectional direct current (DC)–DC power converter.

For high-to-low-voltage operation modes, the DC link high voltage is first passed through the damper R_3–L_4 to prevent oscillations between capacitors C_6 and C_3 (of the PWM three-phase converter). The capacitors C_6 make up for the missing half of the low count switch inverter/rectifier and provide a one half reduction in voltage, up front. The transformer T_1 may provide a one-half to one-third voltage reduction ratio. A second reduction of one-half of the voltage is performed by a current doubler through the diodes D_4 and filter L_5, L_6. Thus, the low count switch converter Q_3 is operated at a maximum duty ratio of less than 0.85 [24].

When the DC link voltage is below 350 V_{DC} (340 V_{DC} and less), the low-to-high-voltage conversion is engaged.

When both Q_4 metal-oxide semiconductor field-effect transistors (MOSFETs) are turned on, the secondary of the transformer T_1 is short-circuited, and the inductances L_5, L_6 are energized. When only one of them is on, the input voltage plus the energy stored in one inductor is applied to the secondary of the transformer. This in turn steps up the voltage that is then rectified through the diodes in the low switch count converter Q_3. When both Q_4 are off, diodes D_4 discharge the energy stored in L_5 and L_6 to capacitor C_7, which is discharged by the resistance R_2.

Even under complete loss of power from the generator, the bidirectional boost DC–DC converter should be capable of holding the DC link high voltage for a few seconds. In general, the voltage boost operation mode is needed to complement the generator power short-lived current peaks and thus maintain the load voltage.

A fast load power increase is, for example, quickly accommodated before the thermal engine speed is increased to sustain it by additional, controlled fuel injection.

10.12 Super-High-Speed PM Generators: Design Issues

By super-high speeds, we mean here speeds above 15,000 rpm. Power decreases with speed to limit the rotor peripheral speed. It is up to 100–170 kW at 70–80,000 rpm and goes up to 1–5 megawatt (MW) at 12,000–18,000 rpm. Small- to medium-power super-high-speed microturbine-driven PM generators are envisaged for distributed power generation or on board series hybrid vehicles, and for other usage. Drastic reductions in size and weight make such generator systems very attractive. The main design features are as follows:

- Electromagnetic design (sizing)
- Mechanical design (verification)
- Thermal design

The starting point is related to specifications. They depend on application, and here is an example:

- Rated electric power: $P_{en} = 80$ kW
- Rated speed: $n_n = 80{,}000$ rpm
- Minimum speed at rated power: $n_{min} = 60{,}000$ rpm
- Rated power factor: $\cos\varphi_n = 0.8$
- Number of poles: $2p_1 = 4$
- Rated phase voltage (RMS): 220 V

In the electromagnetic design (sizing), due to the severe rotor mechanical constraints, the latter should be considered first.

10.12.1 Rotor Sizing

Two main rotor configurations are to be considered—the cylindrical rotor and disk-shaped rotor (Figure 10.45a and b). In general, the disk-shaped PM rotors perform better with a larger number of poles $2p_1 > 6, 8$. Multiple disk-shaped rotors are used to increase power for a given speed. As $2p_1 = 6, 8, 12, 14, 16$, the stator fundamental frequency tends to be larger, so the maximum speed with $2p_1 = 16$ would be 15,000 rpm for $f_1 = 2.0$ kHz, and 30,000 rpm for $2p_1 = 8$. The maximum peripheral speed of the rotor, say 300 m/s, is another limiting factor for the maximum allowable power and speed. The cylindrical rotors are used, in general, with $2p_1 = 2, 4$. Better power/volume performance is claimed with multiple disk rotor PMSGs.

As mentioned at the beginning of this chapter, the copper thin shields are crucial in reducing the rotor eddy currents produced by the space and time stator field harmonics in the PMs and in the rotor back iron. Solid back iron is acceptable with copper shields, which leads to more rugged rotors. Typical material properties for the rotor are as follows:

- *Retaining ring*: Carbon fiber + epoxy resin: 2100 MPA
- *Shaft*: Nonmagnetic iron: 1300 MPA
- *Rotor mass*: Magnetic iron: 1800 MPA
- *Interpolar spacer*: Nonmagnetic iron: 1300 MPA

To size the cylindrical rotor PMSG, the following main variables have to be determined:

- Inner stator diameter D_{is} and stack length
- Rotor geometry details and mechanical validations
- Stator sizing and thermal verification

A similar path is to be traveled for the disk rotor (axial air gap) PMSG, but with adapted (new) formulas.

The analytical preliminary design has to be followed by FEM comprehensive verification and design optimization.

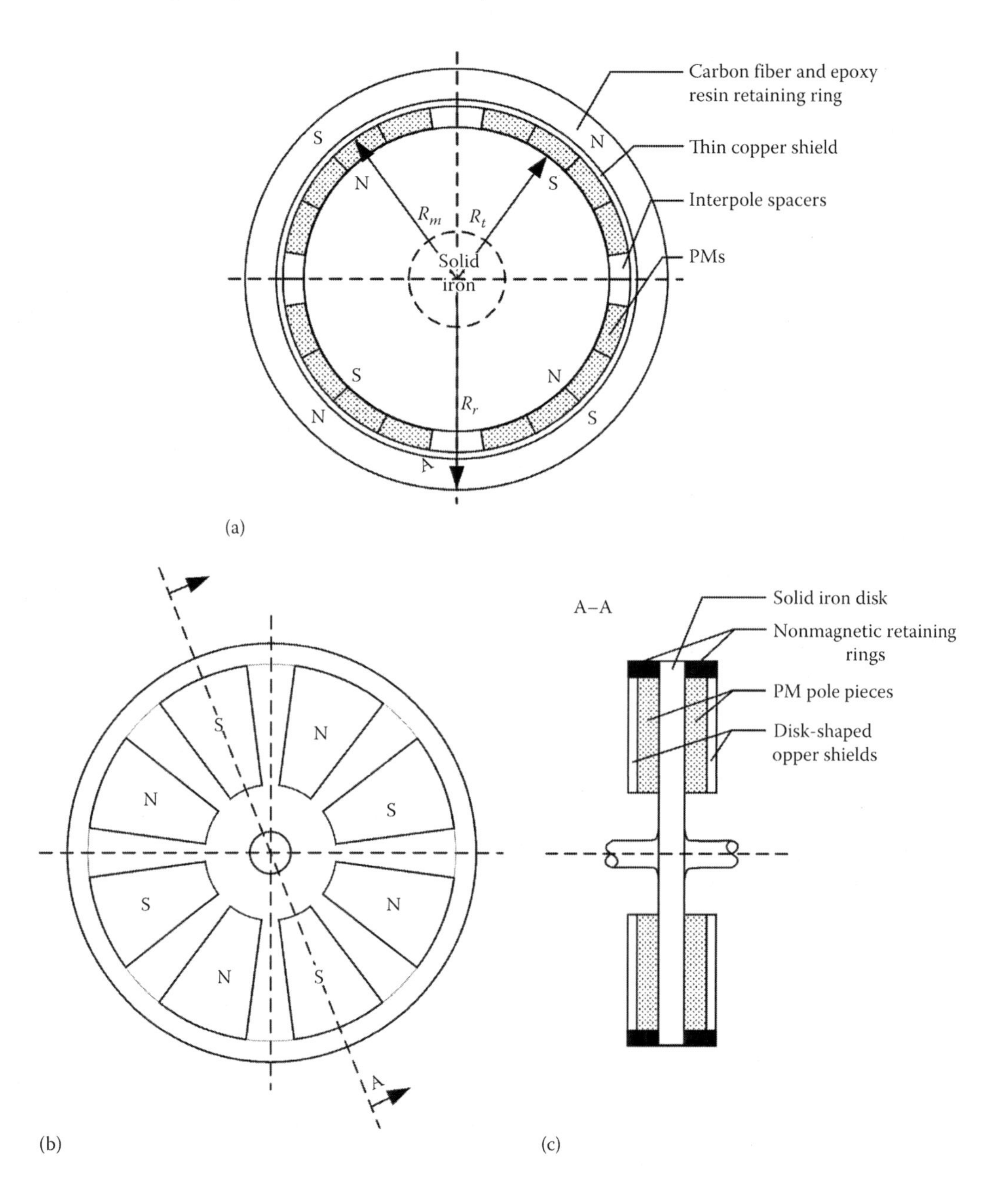

FIGURE 10.45 Super-high-speed PM rotors: (a), (b) cylindrical and (c) disk-shaped.

The specific tangential force f_{tk} in the range of 0.5–3 N/cm² is used here to obtain the preliminary value for stator interior diameter D_{is}, for given λ:

$$f_{tn} = T_{en} \cdot \frac{2}{D_{is}} \cdot \frac{1}{\pi \cdot D_{is} \cdot \lambda \cdot D_{is}}; \quad \lambda = \frac{l_{stack}}{D_{is}} \tag{10.87}$$

$$T_{en} \approx \frac{P_{en}}{\eta_n \cdot 2\pi n_n} c \tag{10.88}$$

where η_n is the rated (forecasted) efficiency. For super-high-speed PMSGs, $\lambda \approx$ 1–3.0. The rated efficiency is assigned a value of 0.9–0.95 at this early stage of design. In addition, f_{tn} is higher when the rated torque increases.

The mechanical air gap g should be larger than usual to reduce the space harmonics of the stator current magnetic flux at the rotor copper shield outer surface caused by the stator mmf and slot opening. This way, the losses in the copper shield will be reduced.

In general, the stator slot opening W_{es} should be the same as the radial distance h_{tg} between the stator and the PMs:

$$h_{tg} = g + b_{rr} + b_{copper} > W_{es} \tag{10.89}$$

where

$b_{rr} = 1.0\text{–}3$ mm retaining ring thickness
$b_{copper} = 0.5\text{–}1.5$ mm
$g = 1\text{–}2$ mm

For the same reason, whenever possible, the stator slot pitch τ_s should be of the size of the total (magnetic air gap) h_{Mg}:

$$h_{Mg} = h_{tg} + h_{PM} \approx \tau_s \tag{10.90}$$

with h_{PM} equal to the PM thickness.

In general, the PM air gap flux density B_{gPM} in the air gap is quasi rectangular:

$$B_{gPM} \cong \frac{B_r}{(1+k_f)} \cdot \frac{h_{PM}}{h_{Mg}} \tag{10.91}$$

The fringing factor k_f accounts for the PM flux lines that close into the air gap or through stator slot necks, instead of embracing the stator slots. As h_{tg} gets larger, k_f increases. The value of $k_f = 0.3\text{–}0.7$ are not uncommon, but they have to be checked through FEM calculations.

In general, $B_{gPM} = 0.3\text{–}0.45$ T in super-high-speed PMSGs, as the fundamental frequency is in the 600 Hz–2.5 kHz range, and thus, core losses tend to be large.

The PM flux per half a pole pitch $\Phi_p/2$ is as follows:

$$\frac{\Phi_p}{2} = \Phi_{yr} = B_{gPM} \cdot \frac{\tau_{PM}}{2} \cdot l_{stack} \cdot k_{fill} \tag{10.92}$$

k_{fill} is the lamination filling factor. Even with thin (0.1 mm or so) laminations, the insulation coating should keep $k_{fill} > 0.85\text{–}0.9$. τ_{PM} is the PM span length on one rotor pole.

Half the PM flux/pole passes through the rotor back iron, with the radial dimension h_{yr} that needs to be as follows:

$$h_{yr} \geq \frac{\Phi_{yr}}{l_{stack} \cdot B_{yc0}} \tag{10.93}$$

B_{yc0} is the rotor yoke design flux density. Generally, $B_{yc0} \cong 0.8\text{–}1.0$ T (here at no load) to leave room for armature reaction field on load but still maintain B_{yc} on load below 1.2–1.3 T.

The shaft diameter D_{shaft} is, basically,

$$D_{shaft} = D_{is} - 2h_{PMg} - 2h_{yr} \tag{10.94}$$

This value has to be checked for mechanical strength.

The rotor mechanical stress verification starts with the computation of spin-off stress σ_{max}, which contains the PM plus copper shield and its own weight contributions:

$$\sigma_{max} = \frac{F_E}{A} = \frac{\omega_r^2(\gamma_{PM} V_M R_m) + \gamma_{copper} V_{copper}(R_m + b_{copper})}{b_{rr} l_{stack}} + \gamma_{ring} \cdot R_t^2 \cdot \omega_r^2 < \frac{\sigma_{max}}{k_{safe}} \tag{10.95}$$

The safety coefficient $k_{safe} = 2–3$; R_m and R_t are visible in Figure 10.45a.

The retaining ring radial elongation due to the applied stress should also be limited:

$$\Lambda = \frac{\sigma_{max}}{E} \tag{10.96}$$

E equals the modulus of elasticity of the ring.

Critical flexural speeds of the rotor also have to be checked, as gravity bends the shaft, and centrifugal and magnetic uncompensated radial forces produce mechanical radial oscillations. There may be more than one critical rotor speed below the rated (superhigh) speed. With one critical speed ω_k, the latter should be much less than the rated speed ω_n: $\omega_k < \omega_n/4$, in general. Driving through critical speeds should be quick to avoid large vibration and noise.

10.12.2 Stator Sizing

The stator sizing means, essentially, the slotting and winding design once the stator bore diameter and stack length are already calculated.

Typically, for super-high-speed PMSGs, distributed double-layer windings with chorded coils are used, as the number of poles is limited.

In addition, when the fundamental frequency goes up to 2.2–2.5 kHz, the skin effect winding losses require special treatment, through design, in order to maintain the skin effect losses within 30%–35% of fundamental copper losses. Sinusoidal current control seems to be the preferred choice, and when a diode rectifier is connected at machine terminals, above 500 Hz fundamental frequency, this is the actual situation anyway, due to commutation overlapping.

Rectangular current control with trapezoidal emf may also be practical, as fundamental frequency and power go up, and thus, the limited switching frequency in the PWM converter is better suited for such six-pulse/cycle operation.

Let us suppose here sinusoidal current control with sinusoidal emf. The phase emf fundamental E_1 per current path is as follows:

$$E_1 = \pi\sqrt{2}\Phi_p W_a K_{W_1} f_1 \tag{10.97}$$

where

W_a is the number of turns per current path (in series)
K_{W_1} is the winding factor for the fundamental
f_1 is the stator frequency

With q_1 slots/pole/phase, W_a (for two coils per slot) is as follows:

$$W_a = \frac{2n_c p_1 q_1}{a} \tag{10.98}$$

where

n_c is the number of turns per coil
a is the current path count

Denoting by V_1 the rated phase voltage (RMS value), the design ratio (E_1/V_1) depends on the type of machine end converter, overload, type of load, and speed range. Generally, at rated speed, $E_1/V_1 \approx 1.1–1.3$.

Once the winding type is chosen (two layers, chorded coils, and q_1 as given), the winding factor for the fundamental, K_{W_1}, is as follows:

$$K_{W_1} = \frac{\sin \pi/6}{q_1 \sin \pi/6q_1} \cdot \sin\left(\frac{Y}{\tau} \cdot \frac{\pi}{2}\right) \quad (10.99)$$

Consequently, from Equation 10.97, the number of turns per path W_a (with a given) can be calculated.

The rated current I_{1n} comes directly from the specifications:

$$I_{1n} = \frac{P_{en}}{(\cos\varphi_n) 3V_{1phn}} \quad (10.100)$$

The number of Ampere-turns per slot F_{1slot} is

$$F_{1slot} = 2n_c \frac{I_{1n}}{a} \quad (10.101)$$

Adopting a design current density j_{con} = 3–8 A/mm² for air cooling and higher for liquid cooling, the conductor-filled area of the slot A_{su} is as follows:

$$A_{su} = \frac{F_{1slot}}{k_{fills} \cdot j_{con}} \quad (10.102)$$

$k_{fills} \approx 0.3–0.45$ is the slot filling factor. Smaller values of k_{fills} have to be considered for direct cooling of conductors in slot by pipes (or hollow conductors).

The tooth width W_t is imposed by the tooth flux density B_t:

$$B_t \approx \frac{B_{gPM} \cdot \tau_s}{W_t} \leq 1.0 - 1.2\ \text{T}; \quad \tau_s = \frac{\pi D_{is}}{6p_1 q} \quad (10.103)$$

The stator yoke height h_{ys} is about equal to the rotor yoke height h_{yr}:

$$h_{ys} \approx \frac{\Phi_p}{2l_{stack} B_{ys}}, \quad B_{ys} < (0.9 - 1.2)\ \text{T} \quad (10.104)$$

At $\tau_s/W_1 \approx 1.8 - 2.2$ and knowing A_{su} leads to complete sizing of the slot. Approximately, the slot useful (conductor-filled) height h_{su} is obtained from the following:

$$A_{su} \approx \frac{h_{su}}{2}\left(W_s + \frac{\pi(D_{is} + 2h_{su} + 2h_{ss})}{6p_1 q_1} - (\tau_s - W_s)\right) \quad (10.105)$$

$$D_{os} = D_{is} + 2(h_{su} + h_{ss}) + 2h_{ys} \quad (10.106)$$

Note that if it is required to increase the machine inductance, the winding may be placed only in the upper part of the slot (Figure 10.46). In this latter case, Equation 10.95 has to be modified to produce the correct value of the conductor-filled slot height $h'_{su} < h_{su}$.

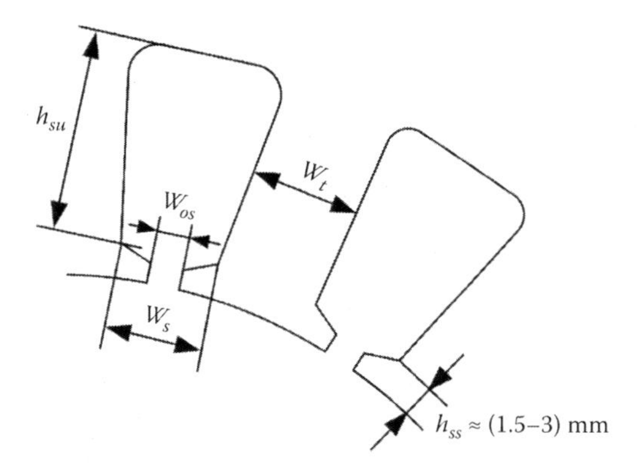

FIGURE 10.46 Stator slotting.

As the machine has PM poles on the rotor, it is characterized by a single synchronous inductance L_m:

$$L_m = \frac{6\mu_0 (W_a K_{W_1})^2 \cdot \tau \cdot L(1 + K_{fringe})}{\pi^2 p_1 g_{Mt}} \tag{10.107}$$

The leakage inductance L_{sl} [25] has to be added to complete the synchronous inductance expression $L_s = L_m + L_{sl}$. K_{fringe} accounts for the flux fringing in the air gap in the radial plane and in the axial plane. The smaller the τ/g_{Mt} and λ ratios, the larger the k_{fringe}. FEM flux distribution calculations are required to calculate k_{fringe} with precision.

10.12.3 Losses

The loss components in PMSG are as follows:

- Fundamental stator core losses
- Fundamental stator copper losses
- Rotor electric losses
- Mechanical losses
- Additional stator copper losses
- Additional stator core losses

Due to the large fundamental frequency and the contents in time harmonics of stator currents, the additional stator core and copper losses are notable. In addition, the rotor electric losses have to be reduced to avoid PM overheating due to eddy currents in the PMs caused by stator space field harmonics and time current harmonics. The thin copper shield is a solution to the problem that results not only in PM thermal protection but also in lower total rotor electric losses [26]. As a bonus, the rotor yoke may be made of solid iron without inflicting notable additional yoke losses through eddy currents.

A complete machine model to account for the rotor electric losses and additional stator core losses may be analytically developed.

The fundamental stator core and copper losses are straightforward [25], but the additional winding losses caused by the skin effect are of special interest. When the fundamental frequency rises above 500–600 Hz, full transposition of elementary paralleled conductors or the Litz wire are the ideal solutions to reducing the skin effect copper losses to negligible values. However, such a solution is costly.

It may as well serve the purpose, up to 2.2 kHz fundamental frequency and up to 200 kW, to parallel elementary conductors and use only 180° transposition (twisting turns only at one end-connection side).

A 20%–30% increase in fundamental copper losses due to skin effect ("self-inflicted" and through circulating eddy currents) may be obtained this way with a two-layer winding with chorded coils.

For powers above 200 kW and fundamental frequency above 0.8–1.0 kHz, refined transposition, as done in large SGs, may be necessary to secure less than 33% of skin-effect copper losses in excess to fundamental copper losses.

Mechanical losses are also notable at high speeds, and they require special treatment for an in-depth investigation. They depend on the rotor diameter, length, peripheral speed, mechanical air gap, and bearing type (mechanical, with air or active magnetic).

A rough approximation for p_{mec} is as follows:

$$p_{mec} = C_0 \cdot D_r \cdot l_{rotor}(\pi \cdot D_r \cdot n)^2; \quad D_r = D_{is} - 2g \tag{10.108}$$

The constant $C_0 \approx 5$ W s^2/m^4 for air-cooled generators. D_r is the rotor diameter, and n is the rotor speed (rps). For additional information on super-high-speed PM synchronous machines, see the literature [27–33].

Note that the PM generator/motor for flywheel energy storage is a typical case of a super-high-speed such machine [34].

10.13 Super-High-Speed PM Generators: Power Electronics Control Issues

The power electronics control of super-high-speed PM generators depends on the following:

- Operation modes: generating only or self-starting (motoring)
- Constant power operation speed range at constant output voltage and frequency
- Power level and fundamental rated frequency

Let us discriminate between machines below 2–3 MW where standard six-leg PWM low-voltage source IGBT converters (up to 690 V line voltage [RMS]) are already produced, and powers up to 5–6 MW at 15,000–18,000 rpm, envisaged for the near future where, most probably, medium voltage (2–5 kV) multilevel inverters will be required. In general, in the MW power range, generator-only operation is more likely.

Then, with the fundamental frequency in the 0.6–1.2 kHz range, above 2 MW, in a medium-voltage PMSG design (3.0–6 kV line voltage RMS), a general control system configuration comprises the following (Figure 10.47):

- A front machine end diode rectifier without or with a boost converter
- A multilevel voltage-source inverter for constant frequency and voltage output

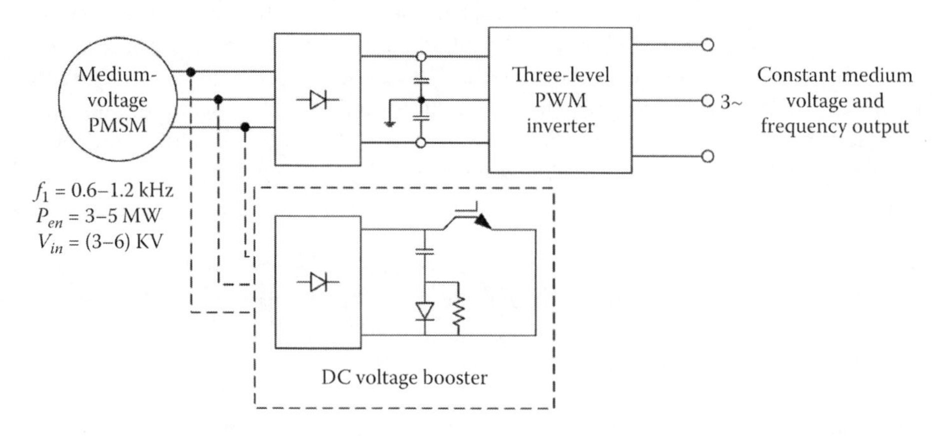

FIGURE 10.47 A 3–5 MW medium-voltage super-high-speed permanent magnet synchronous generator (PMSG) ($f_1 \approx 0.6$–1.2 kHz) with direct current (DC) voltage booster and three-level pulse-width modulated (PWM) inverter.

The DC voltage booster is mandatory if the speed range for constant power is larger than 110%–120%. In Figure 10.47, the DC voltage booster makes use of the generator inductances for energy storage through a fast recovery additional diode rectifier short-circuited in a controlled manner through an oversized IGBT power switch at a frequency of at least 8–10 kHz, as the fundamental frequency may reach 1.2 kHz. This way, the voltage in the DC link is kept constant over the entire speed range of interest and for all load (and overload) conditions.

Three-level PWM inverters represent a strong emerging technology with already notable industrial success. There are quite a few basic categories of multilevel PWM inverters [35]:

- Cascaded multicell inverters
- Flying capacitor inverters
- Diode-clamp (neutral point clamp) inverters

All of these categories [35] have merits and demerits, but the diode-clamp multilevel PWM converters seem to be the first-choice solution for the case in point because of more industrial experience with MW drives at 2.3, 3.3, 4.16, and 6 kV.

A typical three-level diode-clamp inverter leg is shown in Figure 10.48 with only AC terminal *a* visible. Because the inverter is now used as a power source, careful output filtering of the output is required. Such output filters are different, in general, for power grid and stand-alone operation.

Vector control or direct active and reactive power control may be applied here as done for two-level standard PWM converters [36–39].

For powers below 1.5 MW, low-voltage (690 V line RMS voltage) PMSGs of super-high speeds may be built.

A two-level (standard) PWM voltage source IGBT converter may be used for this case. For four-quadrant operation (self-starting), the use of two back-to-back PWM voltage source converters is the first choice (Figure 10.49).

The control of the cascaded AC–AC converter is similar to the case of the cage-rotor induction generator. In essence, for power grid operation, the grid-side converter may be vector controlled to keep the DC voltage rather constant (along axis q) and to produce a certain value of reactive power (along axis d).

The control of the machine-side converter for this situation may be implemented as vector control or as direct torque control. For a wind turbine, a torque/speed reference produces, for each value of speed, a certain optimum reference torque. The machine stator flux reference may be set constant with speed, and finally, direct torque and flux control may be performed (Figure 10.50).

The speed of the prime mover should be close-loop controlled through the speed governor, having the optimum speed for the required power known offline (Figure 10.50a).

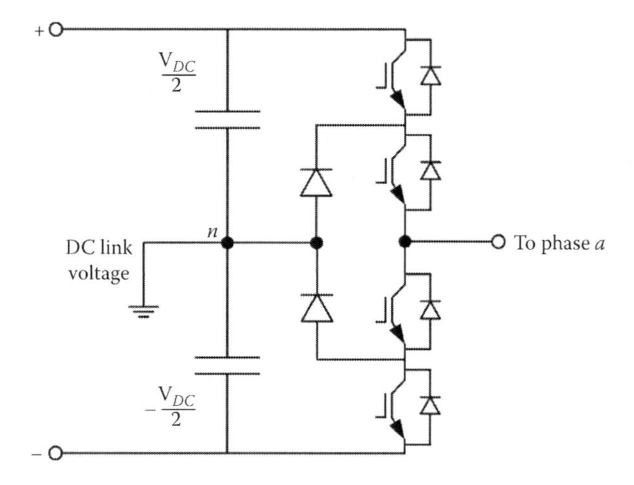

FIGURE 10.48 Three-level diode-clamped pulse-width modulated (PWM) inverter.

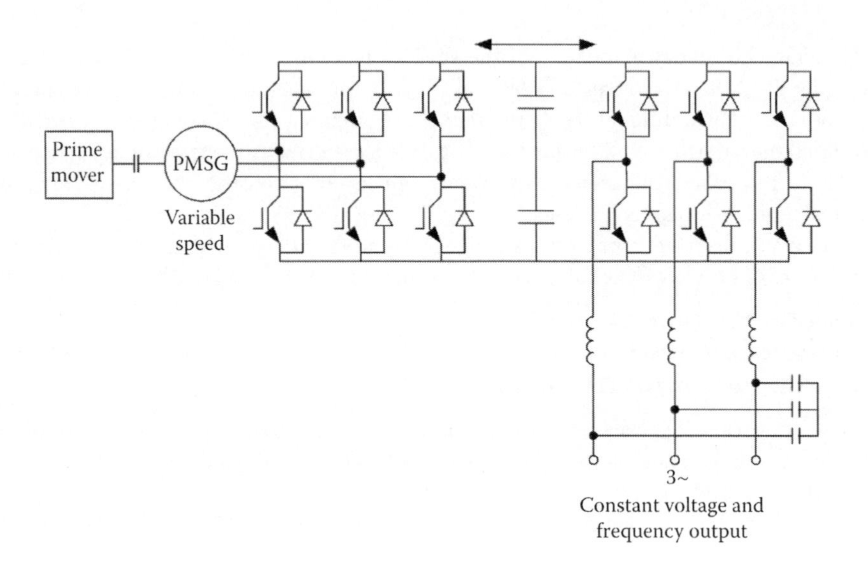

FIGURE 10.49 Four-quadrant cascaded alternating current (AC)–AC converter.

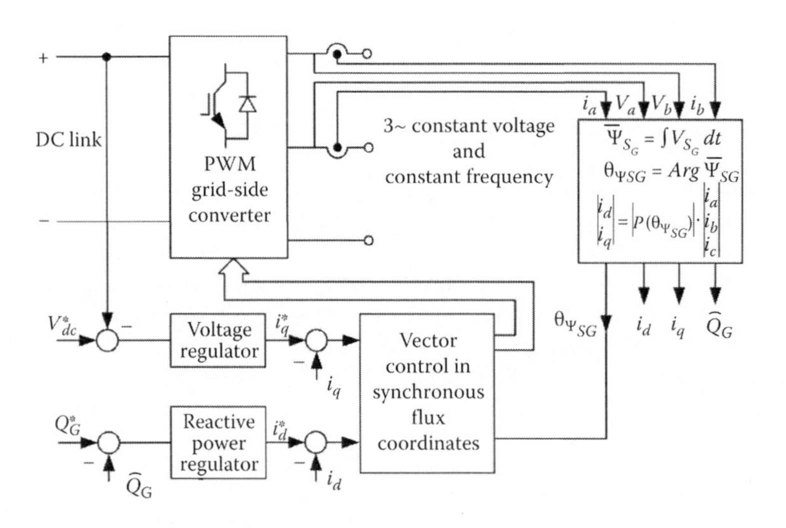

FIGURE 10.50 Vector control of grid-side converter.

As the fundamental frequency goes up to 1–2.5 kHz, the question arises, again, as to whether trapezoidal current control [39] in the generator is not more practical in reducing the commutation losses in the machine-side converter, especially if a DC–DC converter is added to control the DC link voltage for variable speed when the PM generator works alone or is connected to a weak local power system.

10.14 Design of a 42 V_{DC} Battery-Controlled-Output PMSG System

In the previous section, dedicated to super-high-speed PMSGs, a few design (sizing) issues were treated in some detail. In general, PMSG sizing has to be performed with full consideration of converter costs and losses as also calculated for the critical duty cycles of motoring and generating versus speed.

For example, the design of a PM reluctance synchronous generator/motor for a mild hybrid vehicle with a 42 V_{DC} battery and a 6:1 motor/generator constant power speed range (from 1000 to 6000 rpm) [40,41] is quite different from the design of an automotive PMSG with a 6:1 constant power speed range in generating only.

The increase in electric power demand on board of pure internal combustion engine (ICE) vehicles today seems to establish the PMSG with single IGBT control and 42 V_{DC} battery for powers above 1.5 kW as a potential replacement for the claw-pole-rotor generator. For 50 million cars per year, the production potential seems staggering.

Fast power control at higher efficiency is the main merit of the automotive PMSG.

In view of the above, in what follows, we will treat in some detail the design of the 42 V_{DC} battery controlled output hypothetical automotive PMSG.

The design process generally includes the following:

- Design initial data
- Topologies of interest
- Electromagnetic design
- Thermal verification
- Design output data

10.14.1 Design Initial Data

For the case in point, only the rated power P_{en} and the battery voltage V_{DC} are defined from the start.

A 10:1 speed range is typical, but it may be required to deliver only 50% of P_{en} at minimum speed $n_{\min}$ and 100% P_{en} above $2n_{\min}$ and up to $n_{\max}$ [40]. A more stringent requirement corresponds to 100% P_{en} delivery at $n_{\min}$ [40]. It is well understood that in this latter case, the machine has to deliver twice as much torque in comparison to that in the first case.

As the torque "dictates" the machine size, together with losses, and as volume is severely constrained, the 50% P_{en} at $n_{\min}$ requirement is tempting and corresponds to existing practice with claw-pole-rotor alternators on board vehicles.

10.14.2 Minimum Speed: $n_{\min}$

The second important initial data include the speed range.

The ICE idle speed is around 800 rpm. As the larger the PMSG speeds, the lower its volume; the use of a belt transmission to raise the PMSG speed seems to be a practical solution. But, what should be the belt transmission ratio k_{belt}?

In Reference 40, to determine $n_{\min}$ of PMSG, the total losses in the generator are made functions of speed n and are then minimized to find the optimum $n_{\min}$.

In essence, the PMSG losses are divided into the following:

- p_{trans}: by vehicle deceleration (mechanical)
- p_{rota}: by generator deceleration (mechanical)
- p_d: in the rectifier
- p_{iron}: iron losses
- p_{co}: copper losses
- p_{fan}: fan losses

The p_{trans} and p_{rota} are to be calculated as corresponding to the standard town driving cycle.

In general, the torque of the generator is proportional to machine volume (r^3), and thus, at minimum generator speed $n_{\min}$,

$$P \sim r^3 n_{\min} \tag{10.109}$$

Consequently, from Equation 10.109,

$$r \approx n_{\min}^{-1/3} \tag{10.110}$$

Therefore, the generator mass m is

$$m \sim n_{\min}^{-1} \tag{10.111}$$

The translational energy of the generator is reduced when the vehicle decelerates from U_ν to $U_{\nu-1}$:

$$p_{tran} = \sum_{\nu} \frac{m}{2}\left(U_\nu^2 - U_{\nu-1}^2\right) \tag{10.112}$$

In addition, when the generator decelerates from n_ν to $n_{\nu-1}$, the losses related to this rotation deceleration are as follows:

$$p_{rota} = 2\pi^2 J \sum_{\nu}\left(n_\nu^2 - n_{\nu-1}^2\right) \tag{10.113}$$

with J equal to the rotor inertia.

The whole town driving cycle has to be considered to find average values for p_{trans} and p_{rota}:

$$p_{trans} \sim m \sim n_{\min}^{-1} \tag{10.114}$$

$$p_{rota} \sim J \sim r^5 \sim n_{\min}^{-5/3} \tag{10.115}$$

The iron losses in the generator have two components: hysteresis and eddy current losses. As hysteresis losses are proportional to speed (frequency), and the mass is proportional to $n_{\min}^{-1}$, it follows that the former are independent of speed. The eddy current losses, however, are proportional to mass and speed squared, and thus,

$$p_{iron} \approx m n_{\min}^2 \sim n_{\min} \tag{10.116}$$

The copper losses depend on the skin effect if the frequency goes higher than 500–600 Hz, but they may be assumed to be proportional to the radius of the rotor:

$$p_{co} \sim r \sim n_{\min}^{-1/3} \tag{10.117}$$

Finally, the fan losses are considered to be proportional to r^4:

$$p_{fan} \sim r^4 \sim n_{\min}^{4/3} \tag{10.118}$$

If the coefficients in the above formulas are known, the total losses in generator $\sum p$ may be minimized with respect to $n_{\min}$:

$$\sum p = C_{tran} n_{\min}^{-1} + C_{rota} n_{\min}^{1/3} + C_{iron} n_{\min} + C_{co} n_{\min}^{-1/3} + C_{fan} n_{\min}^{4/3} \tag{10.119}$$

As the information to define the constants in Equation 10.119 is not very reliable, it is safe to choose a few values of $n_{\min}$ and compare the final design results. However, it seems that $n_{\min}$ = 1800–3000 rpm is a realistic range. The belt transmission ratio thus varies as k_{belt} = (1800/800)–(3000/800) = 2.25–3.75. With the increase in material performance and power electronics switching frequency, the tendency should be toward $n_{\min}$ = 3000 rpm with $n_{\max}$ = 30,000 rpm.

10.14.3 Number of Poles: $2p_1$

Increasing the number of poles of the PMSG tends to reduce the stator coil end turns and the stator yoke height (and weight) but leads to higher fundamental frequency f_1:

$$f_1 = p_1 \cdot n \tag{10.120}$$

The higher frequency generally leads to an increase in core loss despite a reduction of yoke height. Therefore, the copper losses tend to decrease, but the core losses tend to increase with frequency for given speed.

The fast recovery diodes used in the diode rectifier, most likely to be used in the PMSG system, together with a single IGBT chopper, allow for a maximum fundamental frequency in the range of $f_{1\max}$ = 1.5–2.5 kHz.

For $n_{\max}$ = 30,000 rpm and $2p_1$ = 10 poles, the frequency $f_{1\max} = 5 \cdot 30 \cdot 10^3/60 = 2.5$ kHz.

As, currently, super-high-speed AC drives with fundamental frequency in this range have been built up to 100 kW and more, there is enough credibility to pursue this way of thinking for car alternators.

Concerning the stator core laminations used for the scope, it was found that even 0.36 mm thick 3.1% silicon laminations with special thermal treatment may cause only 30 W/kg core losses at 1 T and 800 Hz. Reducing the silicon lamination thickness to around 0.1–0.12 mm, while maintaining a 0.8–0.85 filling factor with a super thin insulation coating, can keep the specific iron losses to about 30 W/kg, at 1 T and 2.5 kHz fundamental frequency. Finally, it is worth considering fabricating the stator core from soft magnetic composites (SMCs), with permeability that has reached μ_{rel} = 500 P.U. [42] at 1 T. The hysteresis losses in such materials are independent of frequency, while the eddy current losses increase linearly with frequency. The SMC losses become smaller than those for silicon laminations above 500–700 Hz.

Though the relative permeability is still notably smaller than that for silicon steel laminations, the use of thick surface PM poles on the rotor of the PMSG leads to acceptable air gap flux density reduction as long as the flux path length in iron is not too long. Consequently, a large number of poles favors the application of SMC.

The possibility of compressing the stator core with windings in slots through electromagnetic force [43] in case of SMC, may lead to further increases in slot filling factor and to geometry reduction.

The above rationale leads to the conclusion that the number of poles $2p_1$ may be as large as 10 for PMSGs dedicated to automotive applications with fundamental maximum frequencies up to 2.5 kHz.

10.14.4 Rotor Configuration

The rotor configuration may be of cylindrical shape (with radial air gap) or of disk shape (with axial air gap). There are pros and cons for both geometries [40,41]. However, it is considered here that the cylindrical rotor is more rugged, and the stator is easier to fabricate. Further, the surface PM rotor configuration was proven to produce higher torque per PM volume. In addition, it is the configuration that allows for a solid rotor yoke (back iron), provided a thin copper screen is placed over the PMs. In Section 10.12.1, this configuration was introduced (Figure 10.45).

The surface PM poles on the rotor are to be made of parallelepipedic PMs, tangentially and axially, in order to reduce the costs of the PMs. The copper thin screen over the PMs is meant to reduce the PM and back iron eddy current losses due to stator mmf and slot opening space harmonics and due to stator current time harmonics at the cost of some, but reduced losses in the copper screen, which, at high frequencies, behaves like a predominantly reactive circuit.

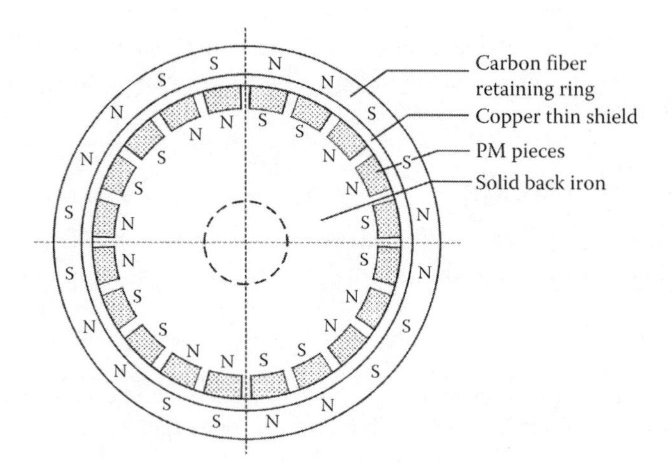

FIGURE 10.51 Ten-pole PM rotor.

The retaining carbon fiber epoxy resin ring is used to protect the PMs and the copper shield from centrifugal forces, as the rotor peripheral speed reaches 150–200 m/s for the case in point. The solid back iron of the rotor provides for a rugged rotor structure (Figure 10.51).

10.14.5 Stator Winding Type

The stator winding may be distributed, in general, for $2p_1 = 4, 6$ and $q = 2$ slots/pole/phase, when chorded coils with $Y/\tau = 5/6$ are used. To cut the costs of winding and copper end-turn losses, nonoverlapping coils may be used.

For $2p_1 = 8$, the number of stator slots N_s should be 9 or even 12, while for 10 poles, $N_s = 12$. In these cases, the equivalent winding factor (as shown in Table 10.1) is $K_{W_1} = 0.945$. The cogging torque is also low, as the smallest common multiplier (SCM) of N_s and $2p_1$ is rather high: 72 and 60, respectively.

Two coils are inserted in a slot, and thus the end-turn axial extension and, consequently, the frame axial length are further reduced (Figure 10.52a and b).

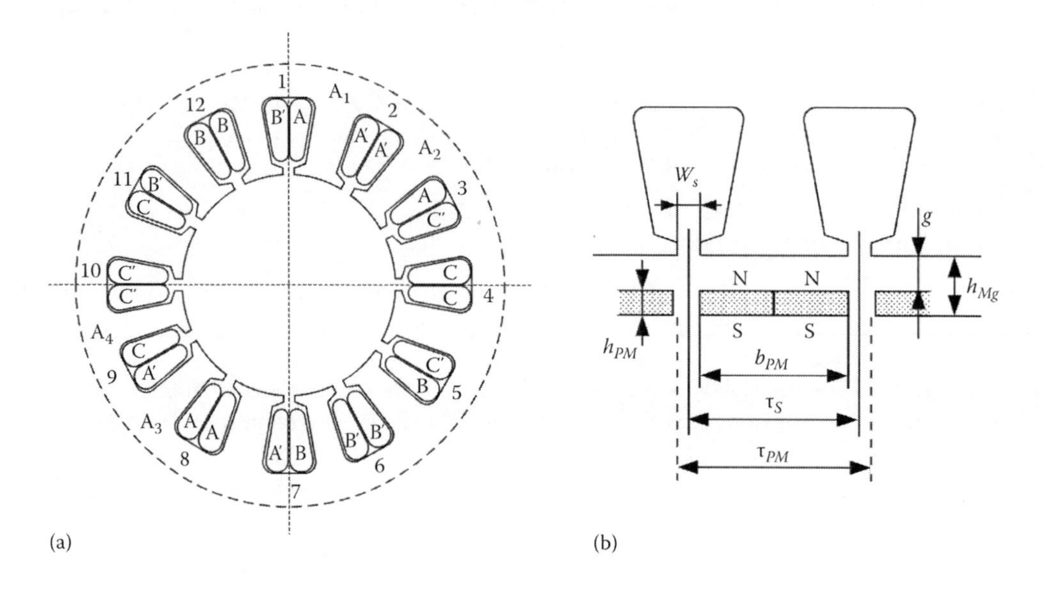

FIGURE 10.52 (a) Ten-pole twelve-slot two-layer winding and (b) PM slot geometry.

For the 12/10 ($N_s/2p_1$) combination (Figure 10.52),

$$\frac{\tau_s}{\tau_{PM}} = \frac{10}{12} \tag{10.121}$$

To further reduce cogging torque, in general, the PM span b_{PM} is taken around

$$b_{PM} \approx \tau_s - W_{os} \tag{10.122}$$

As the design of a PMSG for $2p1 = 4$ and distributed winding is presented in detail in Reference 42, we will concentrate more on the case of nonoverlapping windings.

It was noted that a sinusoidal emf may also be produced with this winding by carefully calibrating W_{os} to magnetic air gap h_{Mg} and PM span b_{PM}. In general, the slot opening W_{os} is smaller than the total magnetic air gap h_{Mg}, in order to reduce the slot opening reinforcement of stator mmf harmonics flux density as felt on the copper shield upper surface. This way, the copper shield losses are reduced.

In addition, a smaller slot opening means a relatively larger slot leakage inductance, which influences the maximum torque that can be produced by the PMSG, due to an enlarged synchronous inductance L_s.

10.14.6 Winding Tapping

For 5:1 constant power speed range, winding tapping may not be ruled out as a means to reduce the emf at maximum speed and thus allows for a lower voltage rating of the single IGBT chopper that will control the output DC current received by the battery and load.

It is a 5:1 speed range from $2n_{min}$ to n_{max}, while at n_{min}, only half the power is delivered. The simplest way to achieve the winding tapping is by dividing it into two halves. The two halves work in series up to about $n_{max}/2$. After that, only one-half remains active. To achieve this goal, two diode rectifier bridges and an electromagnetic power switch are required (Figure 10.53).

The diode rectifier D_{low} works below $n_{max}/2$ with power switching (PS) closed, while D_{high} works at high speeds (above $n_{max}/2$) with PS open.

For a 12/10 combination winding, one of the two neighboring coils pertaining to each phase is grouped in one of the two winding halves A_1–A_2 and A_3–A_4 in Figure 10.53. In this way, a uniform mmf reaction field is secured even when only half of the winding is active above $n_{max}/2$.

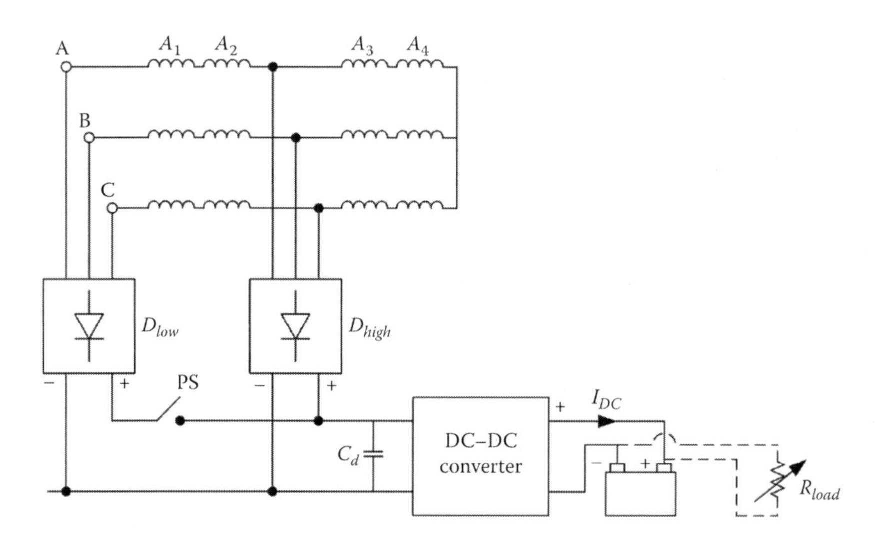

FIGURE 10.53 Winding tapping with two diode rectifiers and an electromagnetic power switch (PS).

10.14.7 PMSG Current Waveform

Depending on the machine inductance L_s, DC link capacitor C_d, and the fundamental frequency (speed), the generator phase current may be continuous (almost sinusoidal) or discontinuous (trapezoidal). The DC–DC converter controls the current in the generator, according to load needs.

In Reference 40, it is inferred that the generator current is discontinuous, though only test waveforms at 3000 rpm are exhibited to prove it. Moreover, the distributed windings in Reference 40 are slotless, and thus, the machine inductance L_s is smaller than usual.

The winding tapping increases the phase current above $N_{max}/2$ when it otherwise tends to be smaller, because the emf increases with speed.

10.14.8 Diode Rectifier Imposes Almost Unity Power Factor

Considering that the emfs per phase are sinusoidal and $L_s = L_d = L_q$, the d–q model (Equations 10.76 and 10.77) applies, and for almost zero Q_1 (reactive power), Equation 10.79 becomes

$$L_s\left(i_d^2 + i_q^2\right) - \Psi_{PM} i_d = 0 \tag{10.123}$$

The maximum torque conditions (Equations 10.81 and 10.82) apply to give the following:

$$\begin{aligned} i_{dk} &= \frac{\Psi_{PM}}{2L_s} = i_{qk} = \frac{i_{sk}}{\sqrt{2}} \\ T_{ek} &= \frac{3}{2} p_1 \frac{\Psi_{PM}^2}{2L_s} \end{aligned} \tag{10.124}$$

10.14.9 Peak Torque-Based Sizing

At n_{min}, corresponding to the ICE idle speed (n_{min}/k_{belt}) with half the rated power delivered, and with full power at $2n_{min}$, it seems that the design for torque has to be performed for n_{min} and $P_{en}/2$:

$$T_{ek} = \frac{(P_{en}/2)}{\eta_{n_{min}} 2\pi n_{min}} \tag{10.125}$$

The maximum torque condition (Equation 10.124) may be used for sizing the machine. An assigned value of efficiency at n_{min} (0.7 or so) and T_{ek} has to be put forward only to be corrected later in design.

For sinusoidal current, the voltage equation (Equation 10.76) yields the vector diagram in Figure 10.54:

$$\begin{aligned} &\overline{I}_s R_s + \overline{V}_s = -j2\pi p n_{min} \overline{\Psi}_s = -j\omega_{rmin} \overline{\Psi}_s \\ &\overline{\Psi}_s = \Psi_{PM} + L_s i_{dk} + jL_s i_{qk}; \quad i_{dk} = i_{qk} = \frac{\Psi_{PM}}{2L_s} \end{aligned} \tag{10.126}$$

As known, V_s is equal to $V_{1ph}\sqrt{2}$, where V_{1ph} is the phase RMS voltage.

10.14.10 Generator-to-DC Voltage Relationships

The relationship between V_s and the diode rectifier voltage V_{DC} is as follows:

$$V_{DC} \approx \frac{3\sqrt{6}}{\pi} E_{1ph} - \Delta V_{diode} - \frac{3}{\pi} L_c \omega_{1min} I_{DC} \tag{10.127}$$

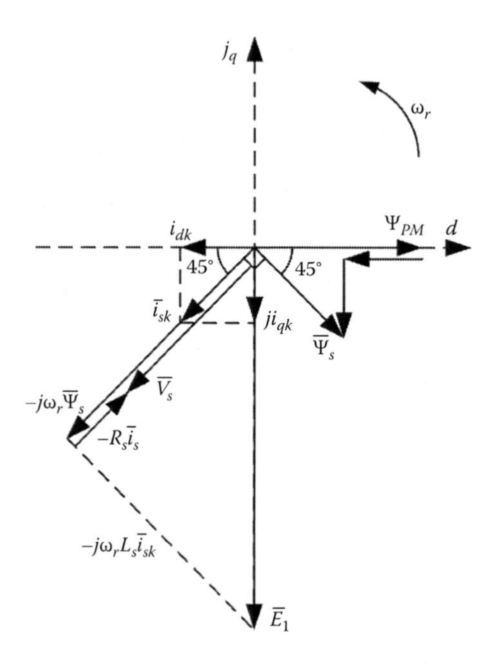

FIGURE 10.54 Vector diagram at maximum torque and unity power factor.

ΔV_{diode} is the voltage drop along the diodes in series between two PMSG phases. L_c is the commutation inductance of PMSG. L_c corresponds to the subtransient inductance $L_s'' \approx L_s$. However, it should be maintained that the voltage drop along L_c in Equation 10.127 is valid with discontinuous current and E_{1ph}:

$$E_{1ph}\sqrt{2} = \omega_r \Psi_{PM} \tag{10.128}$$

For sinusoidal current, the commutation voltage drop in the diode rectifier is "replaced" by the voltage drop along the machine reactance $\omega_1 L_s$, and then,

$$V_{DC} \approx \frac{3\sqrt{6}}{\pi} V_{1ph} - \Delta V_{diode} \tag{10.129}$$

$$\begin{aligned} V_{1ph} &= -R_s \frac{I_{sk}}{\sqrt{2}} + \sqrt{\left(\frac{\omega_{r\min} \cdot \Psi_{PM}}{\sqrt{2}}\right)^2 - \frac{\omega_{r\min}^2 L_{sk}^2 I_{sk}^2}{2}} \\ I_{sk} &= i_{dk}\sqrt{2} = \frac{\Psi_{PM}}{2L_s}\sqrt{2} \end{aligned} \tag{10.130}$$

Equations 10.127 through 10.130, together with the power balance, are crucial for the design at $\omega_{r_{\min}}$ and T_{ek}:

$$\frac{P_{en}}{2} = V_{DC} I_{DC} = 3V_{1ph} \cdot \frac{I_{sk}}{\sqrt{2}} - \Delta V_{diode} I_{DC} \tag{10.131}$$

10.14.11 Ψ_{PM}, L_s, and R_s

The stator PM flux Ψ_{PM} is

$$\Psi_{PM} = \phi_{P\max} \times W_A K_{W_1} \tag{10.132}$$

The maximum PM flux per phase pole (tooth) is as follows:

$$\phi_{P\max} = B_{gPM}^{a} \cdot \tau_s \cdot l_{stack} \cdot K_{Fc} \tag{10.133}$$

B_{gPM}^{a} is the average air gap PM flux density below the stator slot pitch (for the nonoverlapping coil winding) when the PM pole has its axis along that stator tooth (pole) axis. In addition, l_{stack} is the magnetic length of the stator stack; W_a is the number of turns in series per phase (or per current path). For the 12/10 ($N_s/2p_1$) combination, $W_1 = 4n_c c$ if all the coils per phase are connected in series.

Again, the equivalent winding factor for the 12/10 ($N_s/2p_1$) combination is 0.945 according to Table 10.1.

The cyclic machine inductance contains the self-inductance and a contribution of the mutual inductance between adjacent phases.

Because for the 12/10 ($N_s/2p_1$) combination half of the flux is closed through adjacent phases, the cyclic inductance L_s becomes as follows:

$$L_s \approx L_{sl} + \frac{3}{2} L_{ss} \tag{10.134}$$

L_{ss} is the air gap self-inductance per phase:

$$L_{ss} = \frac{\mu_0 \left(W_a K_{W_1}\right)^2 \tau_s l_{stack}}{\left(N_s/3\right) h_{Mg} K_c (1+K_s)} (1 + K_{fringe}) \tag{10.135}$$

where

h_{Mg} is the total magnetic air gap
K_c is the Carter coefficient
K_s is the magnetic core saturation coefficient

The leakage inductance L_{sl} is calculated with a standard expression and essentially contains slot leakage and end-turn leakage:

$$L_{sl} = c_{sl} \cdot W_a^2 \tag{10.136}$$

L_{ss} tends to be larger than for the nonoverlapping windings with equivalent data by as much as 20%–30% or more, mainly due to additional differential leakage inductance additions. In view of the large total magnetic air gap, the magnetic saturation coefficient k_s is likely to be smaller than 0.25.

The Carter coefficient accounts for slot openings and is likely to be close to unity, as $W_{os} < h_{Mg}$ ($1 < K_c < 1.1$).

As the slot openings are rather small and the total magnetic air gap is large, L_{sl} is a good part (up to, maybe, 66%, in some cases) of L_{ss}.

The stator phase resistance R_s is as follows:

$$R_s = \frac{N_s}{3} \rho_{c0} \cdot \frac{l_{coil}}{A_{copper}} (1 + K_{skin}); \quad l_{coil} \approx 2 l_{stack} + \tau_s \left(1 + \frac{\pi}{2}\right) \tag{10.137}$$

The cross section of the conductor from which the stator turns are made is as follows:

$$A_{copper} = \frac{I_{sk}/\sqrt{2}}{aj_{co_k}} \tag{10.138}$$

where j_{co_k} is the peak current density adopted for peak torque ($P_{en}/2$ at $n_{\min}$).

Given the large maximum frequency, elementary conductors in parallel should be used to reduce skin effect. Moreover, at least 180° transportation (twisting of wires only in one coil end-connection zone) is required to reduce the circulating currents due to proximity skin subeffect in paralleled conductors. In general, k_{skin} has to be brought to below 0.33, as done for large SGs that face a similar problem at low frequency due to very large conductor cross section.

The design approach in the previous paragraph corroborated with the methodology in this paragraph should provide a solid basis for a practical design. We mention here only that peak torque T_{ek} expressions (Equations 10.124 and 10.125) are the basis for stator bore diameter and voltage and power expressions (Equations 10.128 and 10.129) and are to be used to find the number of turns per phase (coil).

We alluded to the design methodology in Reference 41 and stipulated above a few new design insights for the envisaged application. We now stop here, leaving the interested reader the freedom to further our pursuit the way he or she feels fit.

10.15 Methods for Testing PMSGs

PMSGs represent a rather novel technology and thus special standards for their testing are not yet available. They may be, however, assimilated with SGs with standards that are available (Institute of Electrical and Electronics Engineers [IEEE] standard 115/1995). Even for SGs, the standards refer to constant speed operation.

On the other hand, methods for testing PM synchronous motors were proposed for over two decades [44,45].

The methods of testing PMSGs can be classified by the operation modes:

- Standstill modes
- No-load and short-circuit tests
- Load tests

or by the scope:

- For loss and efficiency assessment
- For parameters estimation

Standstill tests are used only for parameter estimation, while the others may be used either for parameter or for loss (efficiency) assessment.

The types of tests required for parameter (L_d, L_q) estimation depend on the following:

- The presence (or absence) of damper (copper shield) on the rotor
- The surface PM or IPM rotor pole configuration, which helps to determine if magnetic saturation is or is not important

We will proceed with standstill tests and then continue with no-load and short-circuit tests to end with load tests and emphasize which parameters or losses can be identified there.

10.15.1 Standstill Tests

The DC current decay and frequency response in axial d and q standstill tests were standardized and are described in the IEEE standard 115/1995, dedicated to general SG.

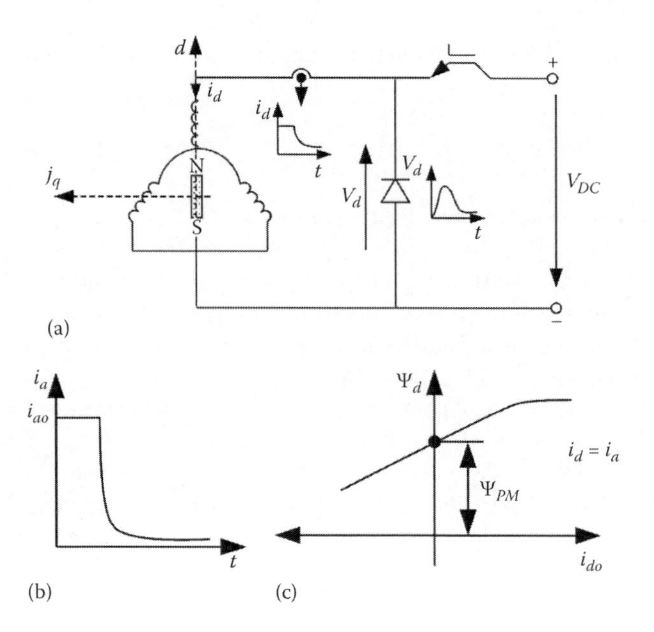

FIGURE 10.55 Standstill tests in axis d: (a) the scheme, (b) the current decay, and (c) d-axis magnetization curve.

The DC current decay tests may produce the magnetization curves along axes d and q by integrating the voltage drop along stator resistance (Figures 10.55a through c and 10.56a through c):

$$L_d(i_d)i_{d0} = \int_0^\infty R_s i_d dt + \frac{2}{3}\int_0^\infty V_d dt; \quad i_d(t) = i_a(t) \tag{10.139}$$

$$\Psi_d(i_{d0}) = \Psi_{PM} + L_d(i_{d0})i_{d0} \tag{10.140}$$

The rotor is placed in axis d by itself when the stator is DC current fed with the connection of phases as in Figure 10.55a. The rotor is then stalled into the d-axis position.

The tests in axis q are made without moving the rotor from the initial position by changing the connection of phases, as in Figure 10.56:

$$\Psi_q(i_{q0}) = L_q(i_{q0})i_{q0} = \frac{\sqrt{3}}{2}\left[\int_0^\infty R_s i_b dt + \frac{1}{2}\int_0^\infty V_d dt\right] \tag{10.141}$$

The experiment has to be done for a few initial values of DC current—positive and negative in axis d (along PMs) and only positive in axis q.

Note the following:

- The diode voltage drop has to be considered in the integral, as it may not be negligible, and thus, may introduce errors above 10%–15%. The integration process has been ended at 2%–3% of initial current value, to reduce error.
- The stator resistance R_s may vary from one test to another; and thus, it should be updated before each current decay test by measuring V_{DC} and $i_a(i_b)$ and calculating $R_s = 2V_{DC}/3I_{a0}$; $R_s = V_{DC}/2I_{b0}$.
- The magnetic hysteresis may also influence the result. After each test, a few positive and negative (±) decreasing current pulses may be applied to cancel the hysteresis effects.

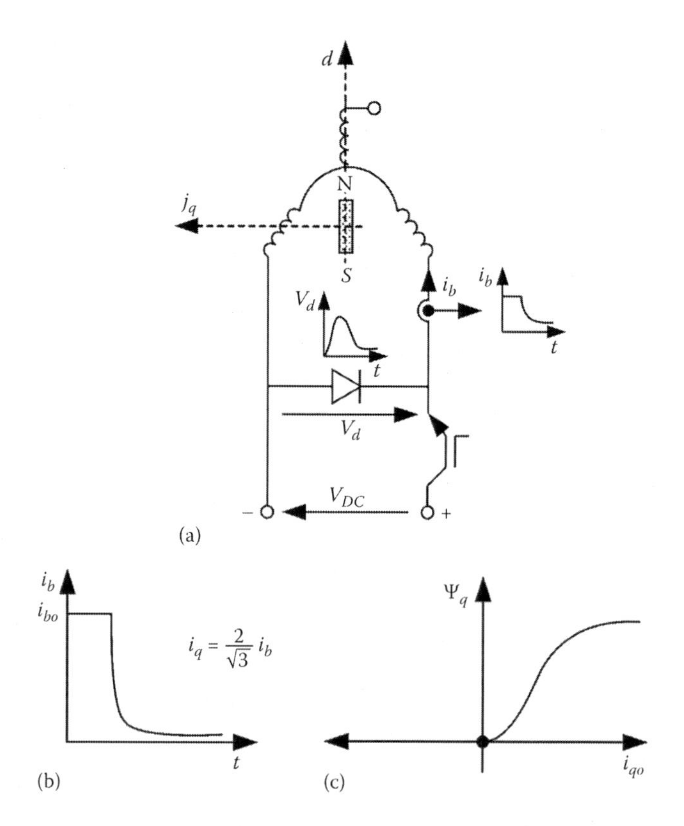

FIGURE 10.56 Standstill tests in axis q: (a) the scheme, (b) the current decay, and (c) q-axis magnetization curve.

- The variable voltage (current) source with fast turnoff may be a standard voltage-source PWM inverter controlled with a variable modulation index to produce only the conduction of phases in Figures 10.55a and 10.56a.
- In a surface PM pole rotor PMSG, $L_d = L_q$ and the d and q standstill DC current decay tests may be used to verify that this is the situation or to uncover an IPM rotor.
- Typical DC current decay test results, obtained according to the above methodology, are shown in Figure 10.57a and b for a 140 N·m peak torque, for a high-saliency IPM rotor SG.

In axis d, the flux Ψ_d changes sign only, apparently demagnetizing the PMs. But this is not the case, as the leakage inductance L_{sl} is about equal to the magnetization axis inductance L_{dm} ($L_d = L_{dm} + L_{sl}$). Consequently, most negative flux in axis d is leakage flux.

FEM calculations with current components in both axes d and q led to unique magnetization curves in axes d and q as functions of total stator current i_s (Figure 10.58).

The cross-magnetization, visible mostly in IPM machines, is still mild and may be described by the unique magnetization curves concept [46]:

$$L_d(i_s) = \frac{\left(\Psi_d^* - \Psi_{PM}\right)}{i_s}$$

$$i_s <> 0 \quad \text{for } \Psi_d^* <> \Psi_{PM};\ \Psi_d = \Psi_{PM} + L_d(i_s)\cdot i_d \qquad (10.142)$$

$$L_q(i_s) = \frac{\Psi_q^*}{i_s};\quad \Psi_q = L_q(i_s) i_q$$

Equation 10.142 allows for the computation of dq inductances after the unique magnetization curves are identified from FEM or from load tests, as explained later in this section.

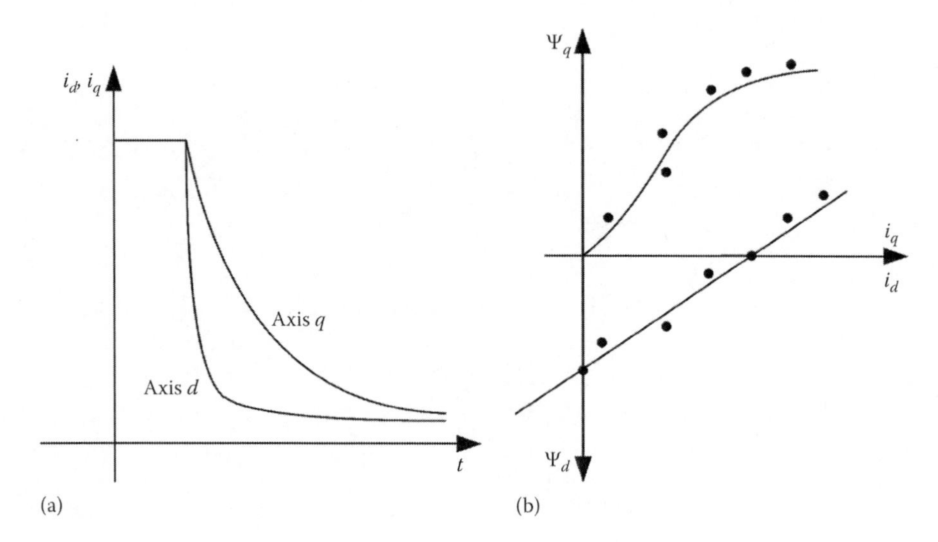

FIGURE 10.57 (a) Test results in direct current (DC) decay tests in axis d, axis q and (b) for a 140 N·m peak torque high-saliency interior permanent magnet (IPM) machine.

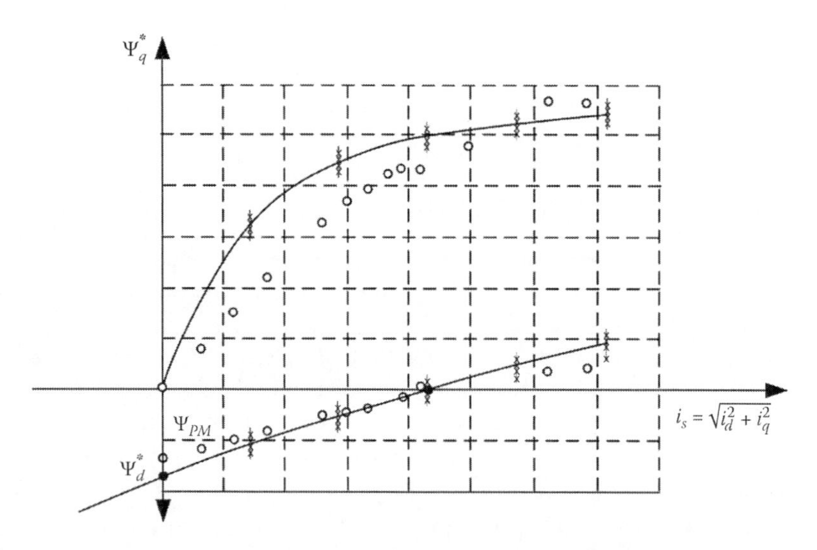

FIGURE 10.58 Finite element method (FEM)-extracted and experimental unique magnetization curves in axes d and q vs. total stator current i_s for an interior permanent magnet synchronous generator (IPM-SG). (Adapted from Boldea, I. et al., *IEEE Trans.*, IA-40(2), 492, 2004.)

In general, L_d (along the axis of PMs) may be considered constant, even in IPM machines, while L_q is a function of i_s (total stator current, when $i_d < i_q$) as obtained from standstill DC current decay tests.

For surface PM pole rotor machines, $L_d = L_q$, and both are constant or mildly variable as a function of i_s, rather than of i_d or i_q.

Let us note that, in reality, the above-described magnetization curve in axis d may not be drawn unless the PM flux linkage Ψ_{PM} is determined from a different test. At standstill, only if torque may be measured, the Ψ_{PM} may be determined from the following:

$$T_e = \frac{3}{2} p_1 (\Psi_d(i_s) i_q - \Psi_q(i_s) i_d) \tag{10.143}$$

If the rotor position θ_{er} and stator DC current i_a, with the phase connection as in Figure 10.55a, are known,

$$i_d = i_s \cos\theta_{er}; \quad i_q = i_s \sin\theta_{er} \tag{10.144}$$

From the two unique magnetization curves, for given i_d, i_q, and measured torque T_e, $\Psi_d(i_s)$ is obtained with a few values of i_s and θ_{er}.

To measure the torque at standstill, either a torquemeter is used or a PM DC motor is loading the PMSG to maintain zero speed. The DC motor current i_{DCm} is proportional to the torque, and calibration of i_{dcm} to torque for the PM DC motor is straightforward: $T_{e_{dcm}} = k_T \cdot i_{dcm}$.

The frequency response tests at a single frequency may not be enough to determine the transient inductances L_d', L_q' of the PMSG. The PMs represent a weak damper winding, while the copper rotor shield is also a mild one. We have to operate with frequencies in the range of those caused by the space mmf slot-opening harmonics and the current time harmonics corresponding to the entire speed range of the machine. For super-high-speed PMSGs, such frequencies are in the range of kilohertz and tens of kilohertz. For more details on frequency response tests on super-high-speed PMSGs, see Reference 26.

10.15.2 No-Load Generator Tests

No-load generator tests need a small power prime mover. Today, variable-speed induction motor drives, with torque and speed estimation, are available off the shelf and may be used to drive the PMSG on no load at various speeds (Figure 10.59).

The no-load generator tests at various speeds may be used to identify the mechanical plus core losses at those speeds:

$$p_{mec} + p_{iron} \approx \hat{T}_e^c \cdot \frac{\hat{\omega}_{rmin}}{p_1} \tag{10.145}$$

with p_1 equal to the number of pole pairs in the induction machine (IM) driver.

$\hat{T}_e^c$ is corrected for mechanical loss torque of the IM. To do so, the IM drive is first driven uncoupled from the PMSG, and the torque is again estimated. The errors of this estimation are not small, however. Alternatively, the IM drive losses may be segregated in advance.

On no-load, with IM electrical speed measured, the PMSG electric speed ω_r is as follows:

$$\omega_r = \omega_{rIM} \frac{(p_1)_{PMSG}}{(p_1)_{IM}} \tag{10.146}$$

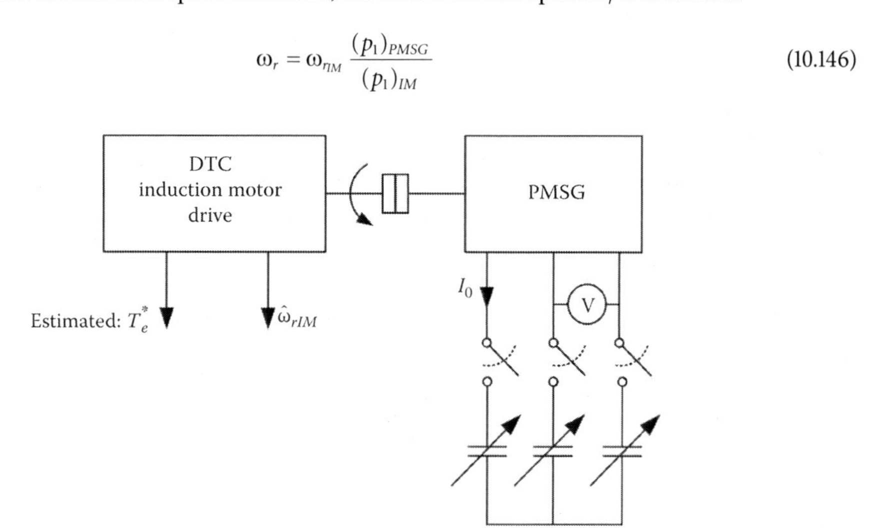

FIGURE 10.59 Generator no-load test arrangement.

The PM flux linkage Ψ_{PM} is

$$\Psi_{PM} = \frac{V_{0line(RMS)}\sqrt{2}}{\sqrt{3}\cdot\omega_r} \quad (10.147)$$

It is feasible to connect variable capacitors (Figure 10.59) to modify the magnetization state, mainly along axis d:

$$\omega_r\Psi_d = \omega_r(\Psi_{PM} + L_d I_d) = \frac{I_d}{\omega_r C}; \quad I_d = I_{os}\sqrt{2} \quad (10.148)$$

Consequently, the $\Psi_d(i_d)$ curves may be obtained with variable capacitors only.

The core losses also change due to i_d armature mmf presence, and this eventually is sensed by using, again, Equation 10.145.

The separation of p_{mec} from p_{iron} may not be done through no-load generator tests unless the capacitor is variable. But many times, this separation is not necessary.

10.15.3 Short-Circuit Generator Tests

The same arrangement as those for no-load generator tests may be used for short-circuit steady-state tests, to be operated at a few frequencies. Above a certain frequency,

$$I_{sc3} = \frac{I_d}{\sqrt{2}} \approx \frac{\omega_r\Psi_{PM}}{\sqrt{2}\omega_r L_{du}} = \frac{V_{0l(RMS)}}{\sqrt{3}\omega_r L_{du}} \quad (10.149)$$

L_{du} is the unsaturated value of d-axis inductance.

The core losses are small in this test, but besides mechanical losses and stator-winding losses, stray-load losses in the rotor may be measured in this test if p_{mec} is already somehow known:

$$\sum p = p_{mec} + \frac{3}{2}R_s(1 + k_{skin})I_d^2 + p_{stray}(\omega_r, I_s) \quad (10.150)$$

The rotor losses p_{stray} are produced by the space harmonics of the stator mmf augmented by the stator slot openings and by the time harmonics of the stator currents (mmf). It may be argued that we lumped into p_{stray} the stator surface losses due to air gap PM flux density harmonics caused by stator slot opening. This is true, but this is the core of the stray load losses definition.

Considering that the mechanical losses are known, and also the stator resistance at low frequency is known, the skin effect coefficient k_{skin} (in Equation 10.150) may be obtained through a test without the rotor in the air gap. If this test is performed at frequency f_1 up to maximum frequency with a PWM voltage source inverter, only the winding losses are important, and thus, with $(R_s)_{DC}$ known, current and power measured, $k_{skin}(f_1)$ may be determined.

10.15.4 Stator Leakage Inductance and Skin Effect

The test with the rotor absent in the air gap (Figure 10.60a) may also be used to separate the reactive power:

$$Q_3 = \frac{3\omega_1 L_{sl3} I_{s0}^2}{2} \quad (10.151)$$

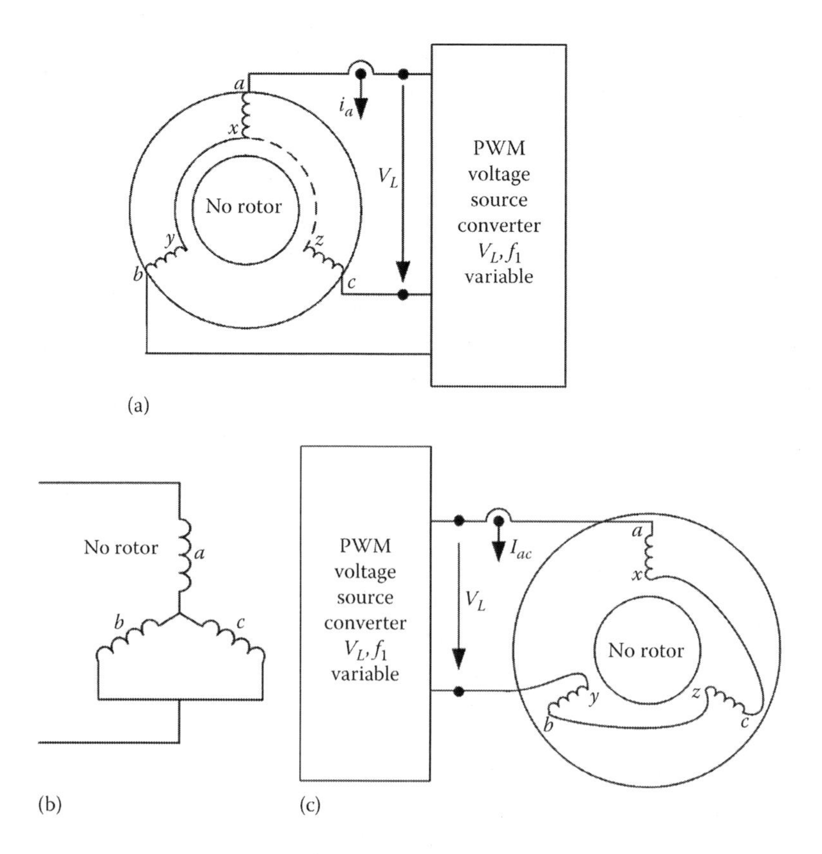

FIGURE 10.60 AC tests without rotor in place: (a) three-phase, (b) a–bc connection, and (c) phases in series.

The machine inductance is only marginally larger than the stator leakage inductance. One more test in the same situation, with all phases in series (Figure 10.60b), provides for the homopolar inductance L_{so}:

$$Q_{10} = \frac{3\omega_1 L_{so} I_{s0}^2}{2} \tag{10.152}$$

L_{so} is smaller than L_{sl}. Consequently, from the two tests, the stator leakage inductance L_{sl} may be safely calculated as an average of the two:

$$L_{sl} = \frac{(L_{sl3} + L_{slo})}{2} \tag{10.153}$$

If the null point of windings is not available, it is possible to connect phase a in series with phase b and c in parallel, as in Figure 10.60c, and repeat the tests. Again, the inductance measured is around the leakage inductance. The reactive power is as follows:

$$Q_1 \approx \frac{3}{2}\frac{\omega_1 L_{sl_{abc}} I_{so}^2}{2} \tag{10.154}$$

Note that I_s is the current space-phasor amplitude $I_s = I_1 \cdot \sqrt{2}$, where I_1 is the stator-phase RMS current. If the stator current contains time harmonics, the leakage inductance has to be calculated by segregating and using only the current and reactive power fundamentals.

10.15.5 Motor No-Load Test

The motor no-load test at variable voltage is standardly used to segregate mechanical losses from iron losses in motors. It may also be used for the PMSG; but if a notable stator mmf reaction occurs, the iron losses on load are notably larger than on no load.

Running the PMSG as a motor, even at no load (mechanical no load), implies using either a PWM converter with position-triggered control or driving it somehow near synchronism and with self-synchronization to the power grid.

The total losses (active power) in this test are as follows:

$$\sum p = p_{mec}(\omega_r) + p_{iron}\left(V_s^2\right) + \frac{3}{2}R_s I_{som}^2 \quad (10.155)$$

During the test for variable voltage V_s and the same speed (frequency, ω_r), p_{mec} remains constant, but p_{iron} varies with V_s^2 (if eddy current core losses are notably larger than hysteresis losses).

When representing Equation 10.155 on a graph (Figure 10.61), its intercept with the vertical axis represents the mechanical losses for the given speed ω_r.

The no-load motoring test may also be used to approximately obtain the d-axis (PM) magnetization curve as follows:

$$V_s \approx \omega_r \Psi_d(i_d); \quad i_d \approx i_{som} \quad (10.156)$$

Next, the machine current i_d changes sign. In general, at rated voltage $V_{sn} < V_0 = \omega_{rn}\Psi_{PM}$, the reduction of V_{sn} during the test keeps the machine overexcited. Consequently, the armature reaction reduces the total flux in the machine (i_d stands for demagnetizing).

Note that generating for the same voltage and speed would mean higher total emf and, thus, heavier magnetic saturation. Operating the motor on no load at a higher than rated generator voltage by 3%–4% might produce more realistic results.

10.15.6 Generator Load Tests

The load tests should generally be performed for the same conditions as in the real operation mode. That is, if a diode or an active front-end rectifier is used in the application, it has to be there during the tests. However, it is also feasible, for novel configurations and proof-of-principle prototypes, to use a simplified test arrangement (Figure 10.62a and b).

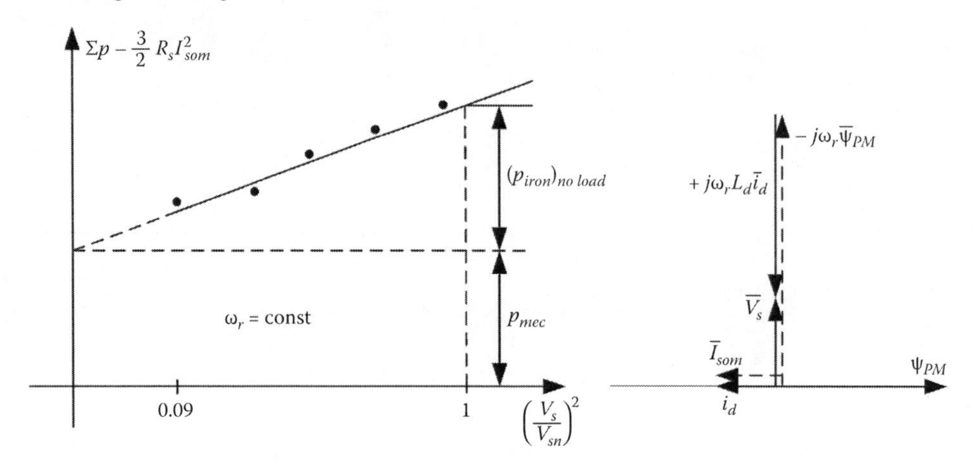

FIGURE 10.61 Segregation of mechanical losses in the no-load motoring test.

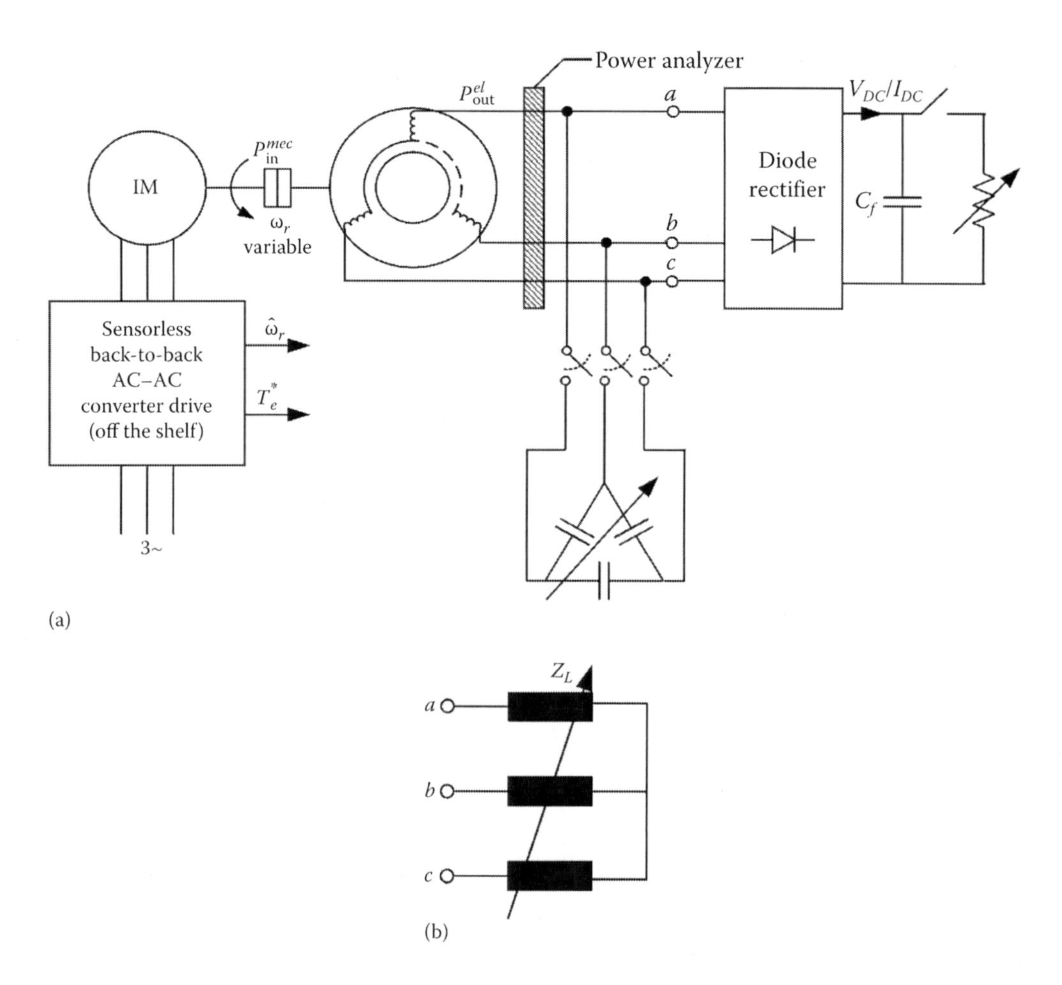

FIGURE 10.62 Load testing arrangements: (a) with diode rectifier load and (b) with AC load.

The variable capacitor bank is used to simulate PMSG voltage boosting to keep, for example, the load voltage constant at constant speed, when the load increases.

The on-load tests are used to verify temperatures and to determine the efficiency as the input P_{in}^{mec} and output P_{out}^{el} are measured:

$$\eta_{PMSG} = \frac{P_{out}^{el}}{P_{in}^{mec}} \tag{10.157}$$

High-precision power measurements are required to render the real value of efficiency.

The DC voltage versus DC current $V_{DC}(I_{DC})$, for given speed and various capacitors or for various speeds and loads, may be obtained with diode rectifier load. For the AC load Z_L, with given R_L/L_L ratio, the generator voltage V_1 versus phase current I_1 can also be obtained. $V_1(P_{out}^{el})$ can be obtained as in stand-alone tests (some results were shown in Sections 10.6 and 10.7). The capacitor C_f on the DC side is replaced by a battery if this is the case in the application.

The load testing, basically with resistive inductive AC or DC load, may also be used to detect at least one (q axis) of the dq magnetization curves if the other (d axis) is known from no-load tests.

Let us consider the steady-state equations in the dq model:

$$\begin{aligned} i_d R_s + V_d &= +\omega_r \Psi_q; \quad \Psi_q = L_q i_q \\ i_q R_s + V_q &= -\omega_r \Psi_d; \quad \Psi_d = \Psi_{PM} + L_d i_d \end{aligned} \tag{10.158}$$

From Equation 10.158 and Figure 10.63,

$$V_d = -V_s \cos\delta_V < 0; \quad V_q = -V_s \sin\delta_V < 0; \quad 0 < \delta_V < 90° \tag{10.159}$$

$$\begin{aligned} I_d &= -I_s \sin(\delta_V + \varphi_1) < 0; \quad I_s = I_1\sqrt{2} \\ I_q &= -I_s \cos(\delta_V + \varphi_1) < 0; \quad V_s = V_1\sqrt{2} \end{aligned} \tag{10.160}$$

$$\begin{aligned} \Psi_d &= \frac{V_s \sin\delta_V + I_s R_s \cos(\delta_V + \varphi_1)}{\omega_r} > 0 \\ \Psi_q &= \frac{-V_s \cos\delta_V - I_s R_s \sin(\delta_V + \varphi_1)}{\omega_r} < 0 \end{aligned} \tag{10.161}$$

By measuring V_1, I_1 and cos φ_1 (from P_1, V_1, I_1), Equation 10.161 still contains three unknowns Ψ_d, Ψ_q, and δ_V.

The zero power generator plus capacitor test produces the expression of Ψ_d:

$$\Psi_d = \Psi_{PM} + L_d i_d; \quad \frac{\sqrt{2}V_0}{\omega_{r0}} = \Psi_{PM} \tag{10.162}$$

with L_d being basically constant.

The unsaturated inductance L_d may also be provided from the short-circuit test:

$$L_d = \frac{\omega_r \Psi_{PM}}{\omega_r I_{sc3}\sqrt{2}} = \frac{E}{\omega_r I_{sc3}} \tag{10.163}$$

E is the emf (PM-induced voltage) or the no-load voltage for given speed (frequency).

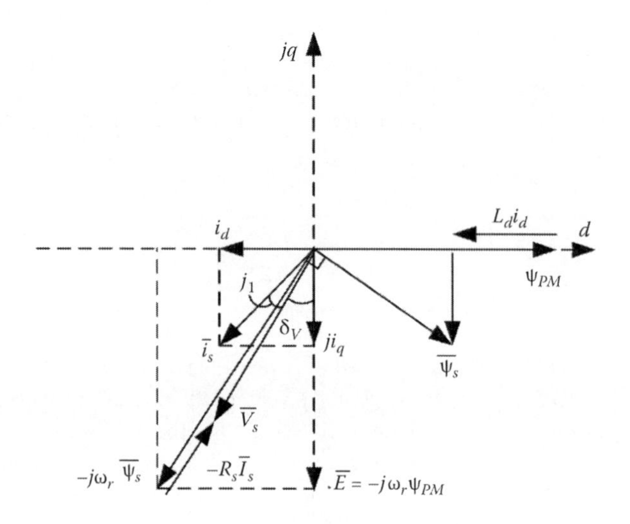

FIGURE 10.63 Vector diagram with permanent magnet synchronous generator (PMSG) delivering active and reactive power.

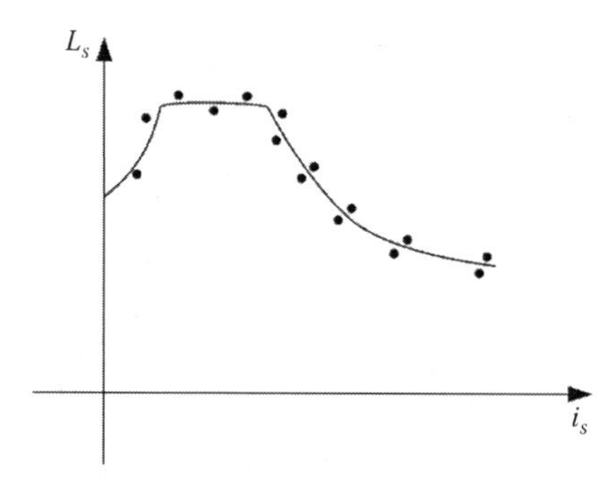

FIGURE 10.64 Unique inductance L_s vs. total current i_s for various (i_d, i_q) combinations.

With Equation 10.162 in Equation 10.161, and L_d and Ψ_{PM} known,

$$\omega_r \Psi_{PM} + I_s(\omega_r L_d \sin(\delta_V + \varphi_1) - R_s \cos(\delta_V + \varphi_1)) = V_s \sin\delta_V \tag{10.164}$$

It is now obvious that at least iteratively: the power voltage angle δ_V may be calculated from Equation 10.164. In Reference 47, Equation 10.164 is manipulated toward a second-order algebraic equation in $x = \cos\delta_V$ ($\sin\delta_V = \sqrt{1-x^2}$), whereby with the condition $\delta_V + \varphi_1 < 90°$, the solution is chosen. Once δ_V is known, from Equation 10.161, the magnetization curve along axis q, $\Psi_q(i_s)$, for various i_d, i_q combinations, may be obtained.

In a surface PM pole rotor, $L_d = L_q = L_s$, even if the machine gets saturated. Then,

$$\begin{aligned} \Psi_d &= \Psi_{PM} + L_s(i_d, i_q) i_d \\ \Psi_q &= L_s(i_d, i_q) i_q \end{aligned} \tag{10.165}$$

with Ψ_{PM} found from the no-load test, and after introducing Equation 10.165 in Equation 10.161, we obtain two equations with two unknowns—$L_s(i_d, i_q)$ and δ_V.

A simple iterative solution of Equation 10.161, now under the form as follows,

$$I_s[\omega_r L_s \sin(\delta_V + \varphi_1) - R_s \cos(\delta_V + \varphi_1)] = V_s \sin\delta_V - \omega_r \Psi_{PM} \tag{10.166}$$

$$I_s\left[\omega_r L_s \cos(\delta_V + \varphi_1) - R_s \sin(\delta_V + \varphi_1)\right] = V_s \cos\delta_V \tag{10.167}$$

produces both variables $L_s(i_d, i_q)$ and δ_V. Then, it may be checked if the unique magnetization curve concept really applies for the case, by representing $L_s(I_s)$ (Equation 10.142) with $i_s = \sqrt{i_d^2 + i_q^2}$, as in Figure 10.64.

If the experimental values of L_s calculated for various (i_d, i_q) do not fall around a unique $L_s(i_s)$ function, the family of curves $L_s(i_d, i_q)$ is to be used. The unique $L_s(i_s)$ function, if applicable, however, simplifies the digital simulations for transients and control design. Virtual loading of PMSG may also be used by the existing of respective method of IMs, with PWM inverter control.

10.16 Grid to Stand-Alone Transition Motion-Sensorless Dual-Inverter Control of PMSG with Asymmetrical Grid Voltage Sags and Harmonics Filtering: A Case Study

In this paragraph, a rather comprehensive sensorless control system for the PMSG with a back voltage source PWM converter is described in some detail [31] (Figure 10.65).

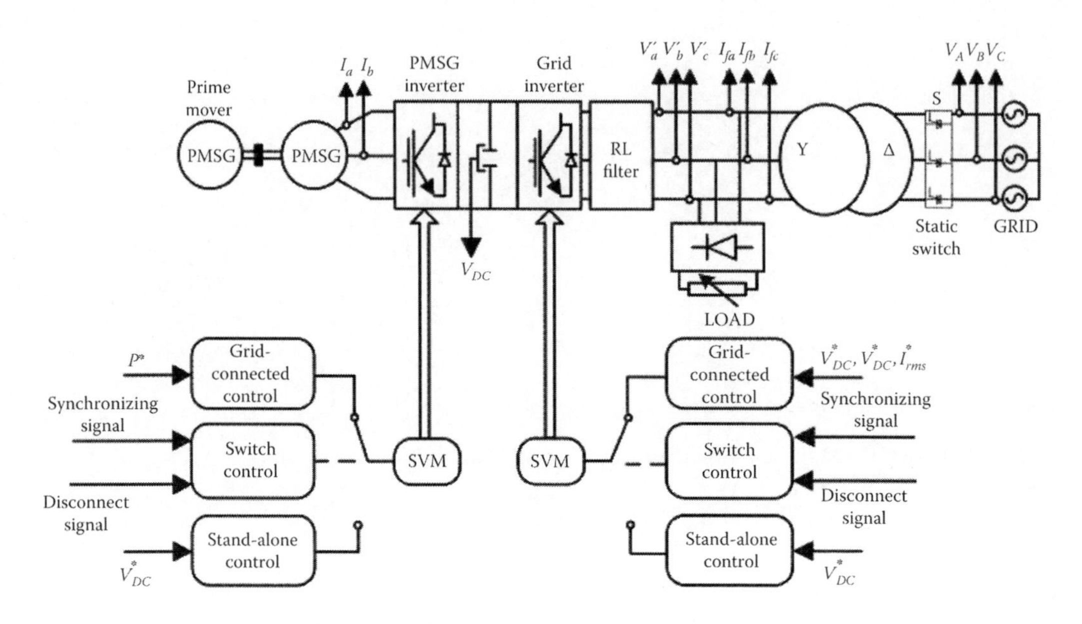

FIGURE 10.65 Back-to-back PWM converter PMSM for transfer from grid to stand-alone and back operation modes.

10.16.1 Voltage Sags Ride-Through Capability

A major drawback when using grid-connected voltage source converters (VSCs) is their sensitiveness to grid disturbances. This is especially true for variable speed wind turbines, which are often located in rural areas and connected to the grid by long overhead lines, easily subject to faults. Short-duration grid disturbances often result in forced stoppage of the turbine, thus in production losses. In the wide range of power quality disturbances, the interest focuses on voltage dips, which can severely affect the performance of the VSC. Most faults result in dips characterized not only by a positive-sequence voltage component, but also by negative- and zero-sequence voltage components. Low pass, band stop and notch filters can detect positive and negative sequence in synchronous frame, but their response is too slow. When utility frequency is not constant, the usage of phase-locked-loop (PLL) closed-loop adaptive methods, are indeed the most representative especially implemented in synchronous reference frame. In case the utility voltage is distorted with high-order harmonics, the synchronous reference filter, SRF-PLL, can still operate satisfactorily if its bandwidth is reduced in order to reject and cancel out the effect of these harmonics on the output. But under voltage unbalance, however, the bandwidth reduction is not an acceptable solution, since the overall dynamic performance of the PLL system would become unacceptably deficient. This drawback can be overcome by using a PLL based on the decoupled double synchronous reference frame (DSRF-PLL) [48] or other additional improvements in order to provide a clean synchronization signal [49].

In References [50,51] a virtual flux-based solution using second-order generalized integrator (SOGI) configured as a quadrature signal generator (QSG) for voltage sensor-less power control of VSC is proposed with good results.

The block diagram of the grid-side inverter control system is shown in Figure 10.66.

Its main components are: PI DC-link voltage controller, PI current controllers, band-pass filter (BPF) based on D-module filter to extract the line voltage positive-sequence V^+, and PLL-based observer to estimate the positive-sequence angle θ_e. The major objective is the independent control of the active and reactive power by using current control loops for $i_f(i_d, i_q)$, implemented in the synchronous reference frame with the d-axis aligned to the θ_e angle, for robust control including the case of

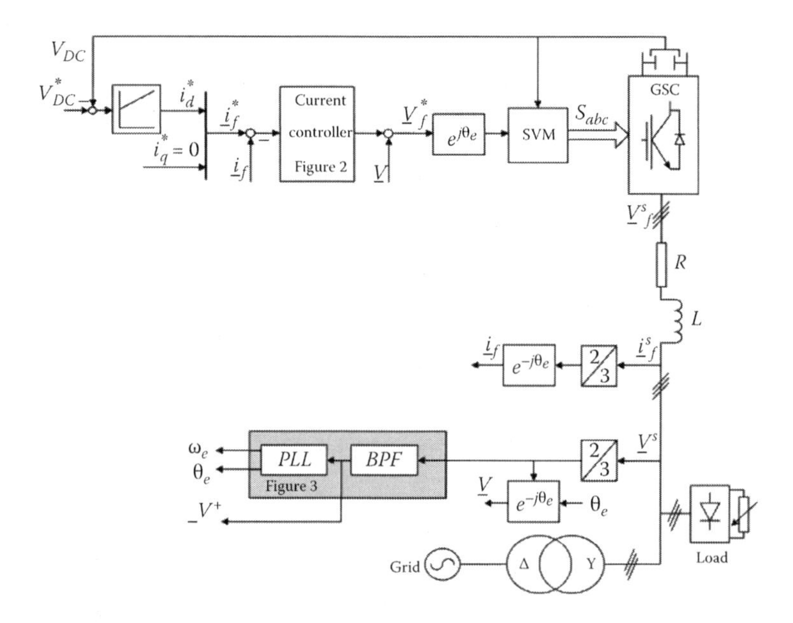

FIGURE 10.66 Grid-side inverter control system.

asymmetric grid voltages. The PI DC-link voltage (V_{DC}) controller gives the i_d^* reference, meanwhile, $i_d^* = 0$ to achieve unity power factor.

$$i_d^* = k_{p\_V_{DC}} \left(\frac{1 + k_{i\_V_{DC}}}{s} \right) \left(V_{DC}^* - V_{DC} \right) \tag{10.168}$$

The PI controller gains $k_{p\_V_{DC}} = 0.02$ A/V and $k_{i\_V_{DC}} = 5\ \mathrm{s}^{-1}$ are tuned for slow-dynamic response to avoid interference between the DC-link voltage loop and the current loops. The DC-link voltage reference is $V_{DC}^* = 600$ V.

The $i_f(i_d, i_q)$ current loop control (Figure 10.67) provides pole-zero cancellation for the plant. It employs PI controllers having complex-coefficients with cross-coupling decoupling, and with feedforward disturbance compensation, that is, the line voltage $V(V_d, V_q)$, to avoid load influence:

$$\underline{V}_f^* = \left(K_p + \frac{(K_i + j\omega_e K_p)}{s} \right) \left(\underline{i}_f^* - \underline{i}_f \right) + \underline{V}, \quad \frac{K_p}{K_i} = \frac{L}{R} \tag{10.169}$$

where K_p and K_i are the PI gains and $\omega_e = 2\pi 50$ rad/s.

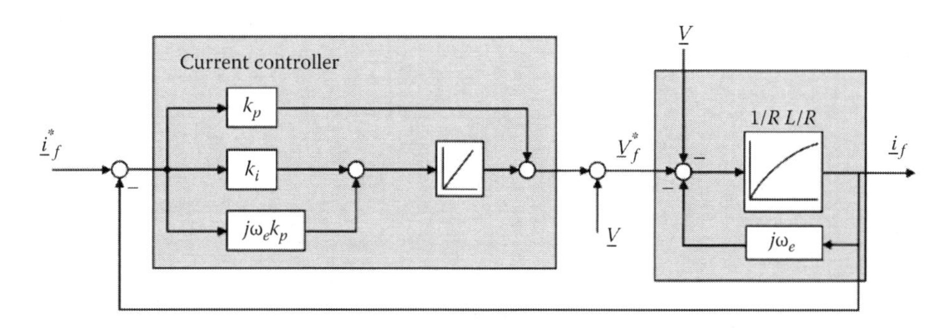

FIGURE 10.67 Current control loop in synchronous reference frame.

10.16.1.1 Line Voltage Positive Sequence with *D*-Module Filter

The angle θ_e used for the current control in the synchronous reference frame, including the case of asymmetric grid voltages, is obtained as shown in Figure 10.68a. In stationary reference frame, a BPF with *D*-module filter extracts from the line voltage vector V^s the positive-sequence V^{+s} and a PLL-based observer estimates its angle θ_e. Vector (two-input/output) *D*-module filters, shown in Figure 10.68, are a class of filters that can separate the positive- and negative-sequence frequencies of signals, which is crucial to solve the problem here. The *D*-module filter design is based on the following property: if the parameters of a single input-output filter *F*(*s*) are designed for a LPF with an appropriate bandwidth, then the associated vector (two input/output) filter in *D*-module will be a BPF with the central frequency ω_c and the same bandwidth. In this application, a second-order scalar filter with the following transfer function is considered:

$$F(s) = \frac{b_2 s^2 + b_1 s + b_0}{s^2 + a_1 s + a_0} \tag{10.170}$$

The *D*-module filter is very robust to distortion of sinusoidal input signals, that is, it produces almost pure sinusoidal output even for distorted input [52]. As in distributed power systems (with nonlinear loads) the voltages are somewhat distorted. The *D*-module filter is a proper way to build a line-voltage positive-sequence estimator.

The BPF with *D*-module filter $F(D(s, \omega_c))$, shown in Figure 10.68a, employs the inverse $D^{-1}(s, \omega_c)$ module function from Equation 10.171 with the realization shown in Figure 10.68b. Specifically, *F*(*s*) is selected as a second-order LPF with $b_2 = b_1 = 0$ and $b_0 = a_0$. For a chosen bandwidth $\omega_0 = 20$ rad/s ($\pm$1.5 Hz frequency deviation) and for a damping factor $\xi = 0.7$, the corresponding BPF has the parameters: $a_0 = \omega_0^2$, $a_1 = 2\xi\omega_0$, with $\omega_e = 2\pi 50$ rad/s.

$$D(s,\omega_c) = \begin{bmatrix} s & -\omega_c \\ \omega_c & s \end{bmatrix} = sI + \omega_c J \tag{10.171}$$

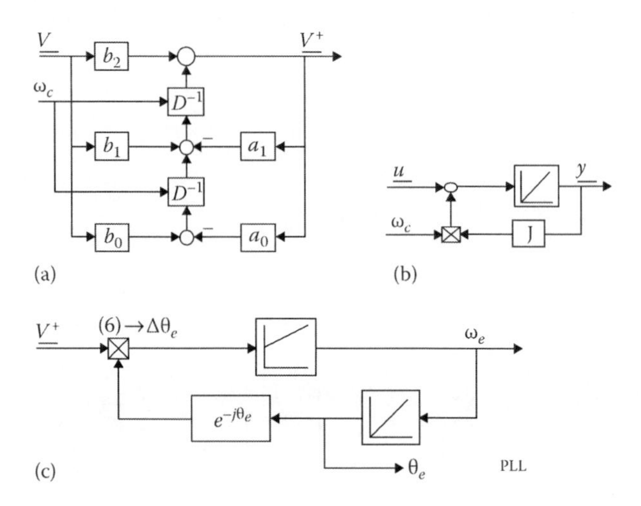

FIGURE 10.68 (a) Second-order vector BPF $F(D(s, \omega_c))$ with filter in *D*-module, (b) Realization of inverse $D^{-1}(s, \omega_c)$ module, and (c) PLL observer structure. (Adapted from Borowy, B.S. and Salameh, Z.M., *IEEE Trans.*, EC-12(1), 73, 1997.)

10.16.1.2 Line Voltage Angle Estimation

A PLL-based observer (Figure 10.68c) estimates the phase θ_e and the frequency ω_e of the positive-sequence line-voltage vector V^{+s}, in order to maintain the line voltages synchronism even in the case of asymmetric grid voltages. In essence, the angle estimation error $\Delta\theta_e = \theta_{line} - \theta_e$ is obtained from the imaginary component $\mathrm{Im}(\underline{V}^{+s}, \underline{V}_1)$, where V_1 is a unity vector.

$$\underline{V}^{+s} = V^{+}e^{j\theta_{line}} = V_{\alpha}^{+} + jV_{\beta}^{+}, \quad V_1 = e^{-j\theta_e} \tag{10.172}$$

$$\Delta\theta_e \cong \sin\Delta\theta = \frac{\mathrm{Im}\left(\underline{V}^{+s} \cdot \underline{V}_1\right)}{V^{+}} = \frac{\left(V_{\beta}^{+}\cos\theta - V_{\alpha}^{+}\sin\theta\right)}{V^{+}} \tag{10.173}$$

The PLL compensator for driving the error $\Delta\theta_e$ to zero is

$$\omega_e = k_p\left(\frac{1+k_i}{s}\right)\frac{V_{\beta}^{+}\cos\theta_e - V_{\alpha}^{+}\sin\theta_e}{V^{+}}, \quad \theta_e = \int \omega_e dt \tag{10.174}$$

The PI compensator gains k_p, k_i of the PLL-based observer are tuned based on the generalized Kessler symmetrical-optimum method [53]. From Figure 10.68c and Equation 10.173, considering the sampling delay h, the equivalent plant transfer function associated to the PI compensator and the PI gains k_p, k_i are

$$H(s) = \frac{V^{+}}{s(hs+1)}, \quad k_p = \frac{1}{mhV^{+}}, \quad k_i = \frac{1}{m^2h}, \quad \xi = \frac{m-1}{2} \tag{10.175}$$

where m is a tuning parameter depending on the damping factor ξ (phase reserve). For a small angle-estimation overshoot, $\xi = 2$ is selected and with $V^{+} = 310$ V and $h = 110$ μs, the PI gains are tuned as $k_p = 6$ and $k_i = 300\ \mathrm{s}^{-1}$.

10.16.2 Stand-Alone PMSG Control: Harmonic and Negative- Sequence Voltage Compensation under Nonlinear Load

In the stand-alone mode one of the main challenges is to reduce harmonic pollution caused by nonlinear loads, which may cause additional losses and heating in the electrical equipment, failures of the sensitive equipments, resonances and interference with electronic equipment, premature aging.

A multiloop controller with voltage differential feedback, is proposed in Reference [54] for stand-alone VSCs.

The voltage control is divided in two distinct parts: fundamental voltage control and harmonic voltage control (Figure 10.69). In the first case, the d-axis, respectively, q-axis voltage controllers are standard PI controllers, and they receive as inputs the reference voltage, respectively, $V_q^{*} = 0$. The DC-link voltage and current limiter controllers will decrease the reference voltage only when the DC voltage is above a designated level or the load current amplitude exceeds a certain level. The filter acts as a harmonic voltage source, injecting harmonics into the load voltages with the same amplitude and opposite phase to the load voltages. The control structure for the grid side inverter is shown in Figure 10.69b. Since coordinate rotation provides a frequency shift by $-\omega_{sa}$, in fundamental frame, harmonics orders $k = 6n \pm 1$ become $k = 6n$. The harmonic voltage controller, is realized as superposition of individuals controllers, each design for a specific pair of harmonics ($k = 6, 12, 17, 19, 23$) of positive and negative sequence. Due to intense computational effort needed to implement the proposed harmonic compensation algorithm and the sample frequency set during the experiments at 9 kHz, the implementation of the PR controller tuned on 18th harmonic did not produce the expected results. Therefore, in order to compensate above twelfth-order, PR resonant controllers, implemented in stationary reference frame, have been used for each of the

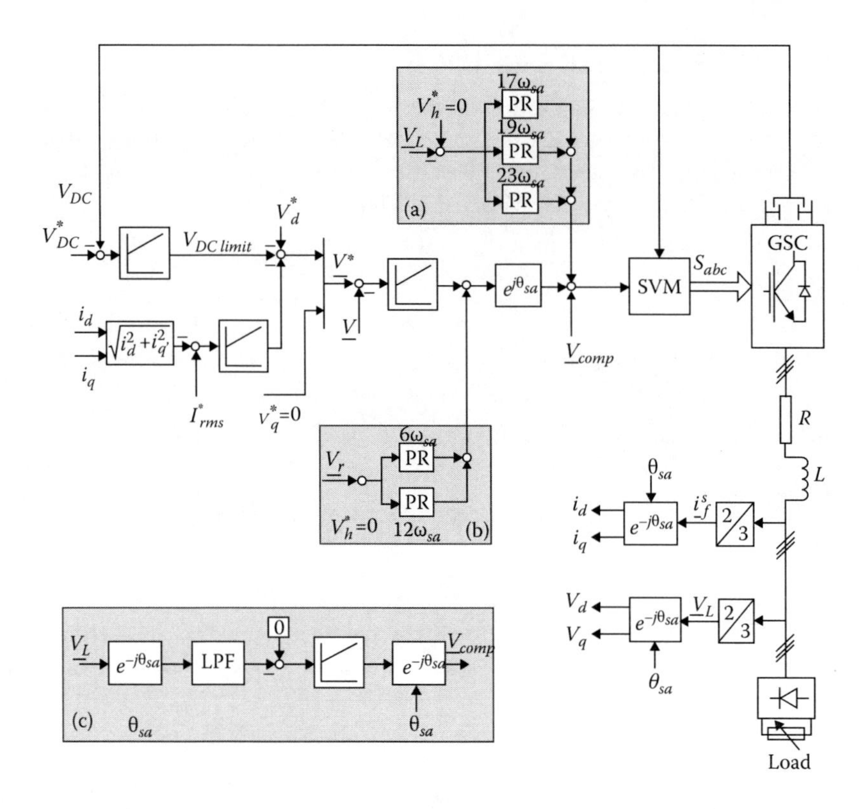

FIGURE 10.69 Stand-alone control with harmonic and inversed component voltage compensation: (a) general scheme, (b) harmonic voltage compensator, and (c) negative-sequence voltage compensator.

odd 17th, 19th, 23th harmonics. The harmonic compensator is of proportional resonant type (PR), defined for harmonics. The harmonic compensator is of proportional resonant type (PR), defined as [55] follows:

$$H_{PR} = 2\frac{K_{pk}s^2 + K_{ik}s}{s^2 + (k\omega_e)^2}, \quad k = 6n \tag{10.176}$$

When an unbalanced AC load is connected (or an output terminal of the load-side inverter is open), in order to keep the voltage symmetry necessary for remaining AC load, the negative sequence voltage is compensated. We can write the magnitude of the fundamental voltage as a sum of its harmonics:

$$V_L = V_{50} \cdot e^{j\theta_{sa}} + V_{-50} \cdot e^{-j\theta_{sa}} + V_{250} \cdot e^{j \cdot 5 \cdot \theta_{sa}} + \cdots \tag{10.177}$$

The voltages are additionally rotated with inversed angle, therefore Equation 10.177 becomes:

$$V_L = V_{50} \cdot e^{j\theta_{sa}} \cdot e^{j\theta_{sa}} + V_{-50} \cdot e^{-j\theta_{sa}} \cdot e^{j\theta_{sa}} + V_{250} \cdot e^{-j \cdot 5 \cdot \theta_{sa}} \cdot e^{j\theta_{sa}} + \cdots \tag{10.178}$$

10.16.3 Seamless Switching Transfer from Stand-Alone to Grid (and Back)

Among these technical issues, the grid side inverter capability to work both in grid-connected but also in stand-alone mode as well as the protection in the case of the fault event are considered to be the most challenging in terms of control strategy and limited fault current. In the case of sensitive and mission-critical industrial loads, maintaining a continuous, uninterrupted AC power is of utmost importance. Seamless transfer methods from grid-connected to stand-alone and vice versa for critical loads with a zero-load current stage, are described in Reference [56]. The proportional–integral, trapezoidal, sinusoidal, and staircase frequency variation techniques are analyzed in Reference [57] to find the best approach for minimizing the total harmonic distortion. Other method proposed in Reference [58] is based on introducing a disturbance at the inverter output and observing the behavior of the voltage at the point of common coupling (PCC), which depends on the impedance connected to the PCC in an islanding situation.

10.16.3.1 Transition from Stand-Alone to Grid-Connected Mode

Initially, the grid converter is running in stand-alone mode. When grid is recovered (grid voltages suddenly occur), a PLL observer will generate the grid frequency and phase. If the amplitude of the voltages and frequency for the recovered grid are within the limits defined by standards (Figure 10.70a), a synchronizing signal will be automatically generated, in order to match the supply-side inverter output phase of voltages, still in stand-alone operation, with the phase of grid voltages. The error $\Delta\theta = \theta - \theta_{sa}$ can be approximated, considering small arguments, as follows:

$$\Delta\theta = \theta - \theta_{sa} \cong \sin(\theta - \theta_{sa}) \tag{10.179}$$

A PI controller (Figure 10.70b) is employed to compensate the error between those two. In order to achieve a smooth synchronization, slow dynamic is imposed to controller. When the error is between −0.01 and +0.01 rad, the voltages are considered to be synchronized ($\theta_{comp} = 0$) and the control of the supply-side inverter will be automatically switched from stand-alone to grid-connected control mode. The online estimation of phase angle of stand-alone voltages (Figure 10.70c) will remain *on* even after the system is running in grid connected mode, in order to avoid load current interruption during future switching from power grid to stand-alone.

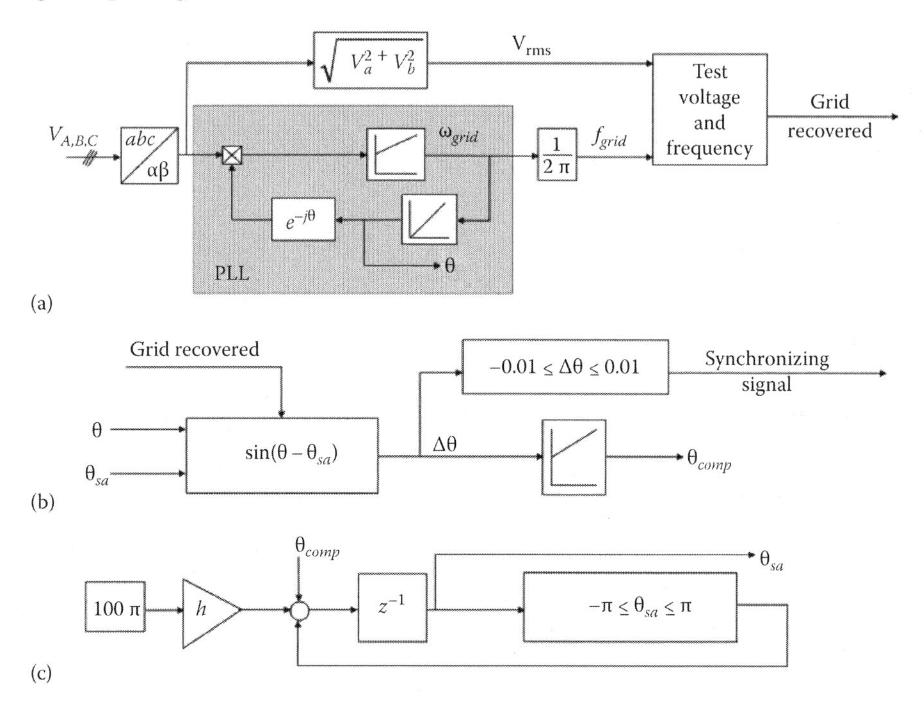

FIGURE 10.70 Transfer algorithm from stand-alone to control grid connected mode: (a) check if the grid parameters (voltage and frequency) are valid, (b) if the grid is recovered start the synchronization, and (c) stand-alone phase angle calculation.

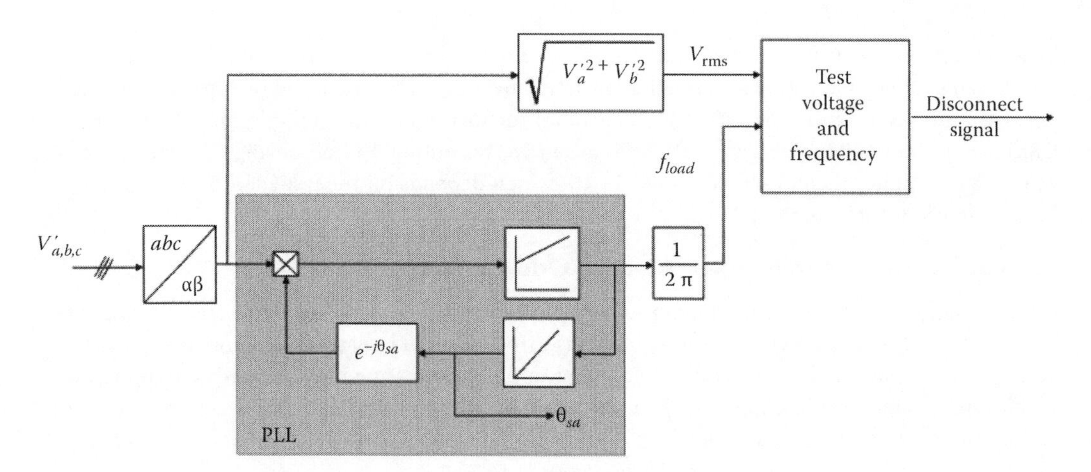

FIGURE 10.71 Transfer algorithm from grid connected to stand-alone control.

10.16.3.2 Transition from Grid-Connected to Stand-Alone Mode

Assume the grid inverter is operating in current controlled mode and that the future stand-alone load is connected to the grid in parallel. Therefore the stand-alone load remains connected all the time at the supply-side converter output terminals. A fault event is simulated by opening the static switch S (Figure 10.65). In this case, because we still have current control of supply-side converter (corresponding still to grid connection, Figure 10.66), there is an increase in load voltage level, and when this exceeds a certain threshold or the frequency is out of range of 49.5–50.5 Hz, the system switches automatically to stand-alone mode (Figure 10.71).

Because the initial value of the phase angle for stand-alone voltages coincides with the last value of the grid phase angle the transition is smooth, without current interruption or major current spikes.

10.16.4 PMSG Motion-Sensorless Control System

10.16.4.1 PMSG Modeling

The space-vector model of PMSG, equipped with interior permanent magnets, in rotor reference frame (no superscript), is follows:

$$-\underline{V}_s = \underline{i}_s R_s + \frac{d\underline{\Psi}_s}{dt} + j\omega_r \underline{\Psi}_s \tag{10.180}$$

$$\underline{\Psi}_s = \Psi_d + j\Psi_q = \lambda_{PM} + L_d i_d + jL_q i_q \tag{10.181}$$

where

$V_s = V_d + jV_q$, $i_s = i_d + ji_q$, and $\underline{\Psi}_s$ are the stator voltage, current and flux vectors, respectively
R_s is the stator phase resistance
L_d and L_q are the dq inductances
λ_{PM} is the PM flux
ω_r is the electrical rotor speed

The interior PM rotor case was considered not only because it is more general but also because inset PM rotors that have rather small magnetic saliency ($1 > L_d/L_q > 0.7$) allows for an easier placing of PMs on the rotor especially if the PM poles are made from a few PM cubicles, in part to reduce the PM eddy current losses due to stator mmf space and time harmonics and in part, to produce segmental skewing and thus reduce cogging torque.

10.16.4.2 Active Power and Current Control

The proposed PMSG motion-sensorless control system is depicted in Figure 10.72a. The PMSG inverter is active power controlled along the q-axis when the system works in grid connected mode and voltage controlled when it is in stand-alone mode. The current reference i_q^* for grid connected mode is obtained from the power control loop, which employs a PI controller that has as input the error between the active power reference P^* and the active power delivered in DC-link: $V_{DC}i_{DC}$.

$$i_q^* = k_{p\_pow}\left(1+\frac{k_{i\_pow}}{s}\right)(P^* - V_{DC}i_{DC}) \tag{10.182}$$

For stand-alone mode the reference q-axis current is obtained from the DC-link control loop, which employs a PI controller that has as input the error between the reference and measured DC-link voltage:

$$i_q^* = k_{p\_V_{DC}}\left(1+\frac{k_{i\_V_{DC}}}{s}\right)\left(V_{DC}^* - V_{DC}\right) \tag{10.183}$$

The PI gains k_{p_pow} = 0.002 A/W, k_{p_VDC} = 0.002 V/W and $k_{i\_pow} = k_{i\_VDC}$ = 25 s^{-1} are tuned for slower-dynamic response to avoid interference between the active power loop and the generator current loops.

While it is clear why regulating i_q it means regulating torque and power, it is not so clear how to reference i_d^* in presence of some magnetic saliency (Figure 10.73).

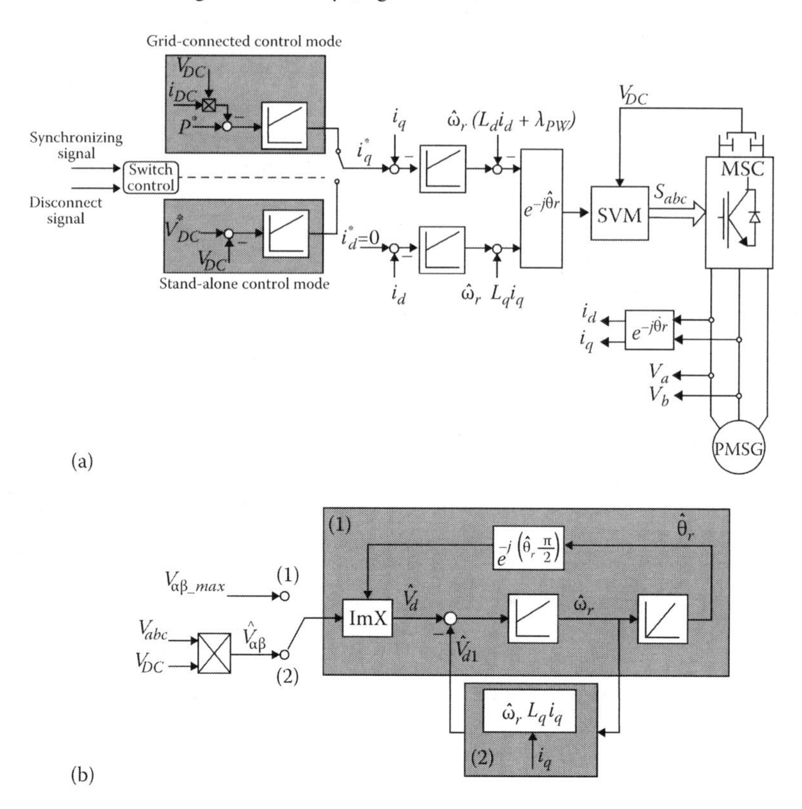

FIGURE 10.72 Motion-sensorless control of PMSG: (a) active power and current controllers and (b) rotor speed and position observer. (From Fatu, M. et al., Motion sensorless bidirectional PWM converter PMSG control with sensorless switching from power grid to stand-alone and back, in: *Proceedings of the IEEE Power Electronics Specialists Conference (PESC)*, 2007, pp. 1239–1244.)

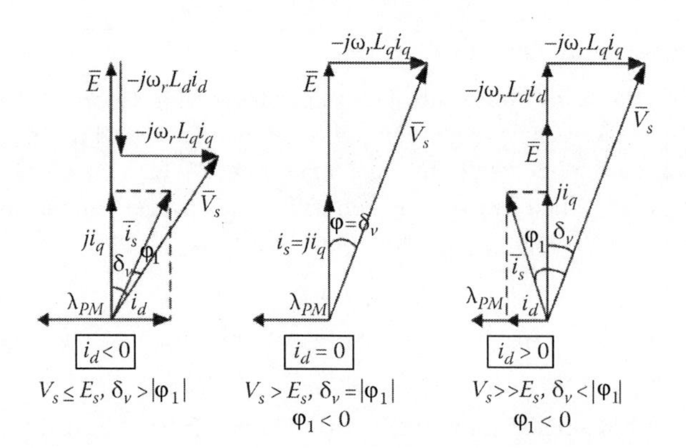

FIGURE 10.73 Vector diagram for steady state (small saliency ratio, strong magnets).

There are three main possibilities:

- $i_d^* < 0$ with positive (generating) reluctance torque ($L_d < L_q$)
- $i_d^* = 0$ with zero reluctance torque
- $i_d^* > 0$ with negative (motoring) reluctance torque

In terms of max torque/current (or per winding losses) $i_d^* < 0$ is to be preferred if $L_d/L_q < 1$ (generating reluctance torque). However, there is one more issue: the PWM rectifier operates only as a voltage booster, to maintain controllability at all considered generator speeds. Therefore, in general, $E = \omega_r \lambda_{PM} < V_s$. Now checking the vector diagram for steady state in Figure 10.73, for $i_d < 0$, $i_d > 0$, we noted that $i_d^* > 0$ allows for the largest V_s (or DC-link) voltage, that is more stable operation at variable speed (δ_v is larger), but implies negative (motoring) reluctance torque and then larger winding losses per given torque (power and speed). Also, we noted that $i_d^* < 0$ may lead to $V_s < E_s$ situations when the PWM rectifier on machine-side approaches diode rectifier operation and thus loses power (torque) control. Both $i_d^* < 0$ and $i_d^* > 0$ controls require some rather complex strategy to plan, in relation to PMSG speed, power and parameters.

On the other hand, $i_d^* = 0$ provides for $V_s > E_s$ operation, as required by PWM rectifier control, it is rather simple and the "lost" reluctance torque is not so important because the saliency ratio is not far from unity. If $L_d/L_q < 1/3$ the situation changes and $i_d^* < 0$ control is more practical because E is small in P.U. ($E/V_s < 0.3$ P.U). In general due to limited speed range control, strong magnets ($E/V_s > 0.7$ P.U at base speed) are used when low saliency in the rotor is used to reduce rotor cost.

Note: If a diode rectifier with a boost DC–DC converter is used instead of the PWM rectifier, then the fundamental power factor would be almost unity, $i_d^* < 0$ control is implicit. This rational leads us to set i_d^* to zero for the case in point. The current controllers with emf decoupling are implemented in rotor reference frame, and produce the reference voltage vector $\underline{V}_s^* = V_d^* + jV_q^*$:

$$V_d^* = k_{p\_crt}\left(\frac{1+k_{i\_crt}}{s}\right)\left(i_d^* - i_d\right) + \hat{\omega}_r i_q L_q \tag{10.184}$$

$$V_q^* = k_{p\_crt}\left(\frac{1+k_{i\_crt}}{s}\right)\left(i_q^* - i_q\right) - \hat{\omega}_r (i_d L_d + \lambda_{PM}) \tag{10.185}$$

where the PI gains are $k_{p\_crt} = 25$ V/A and $k_{i\_crt} = 75$ s^{-1}. The integral k_{i_crt} gain is tuned for pole-zero cancellation of the PMSG electrical time constant: $k_{i\_crt} \sim R_s/L_{dq}$.

10.16.4.3 Rotor Position and Speed Observer

The proposed observer for rotor position $\hat{\theta}_r$ and speed $\hat{\omega}_r$ estimation, depicted in Figure 10.72b, has two operation regimes [53].

1. *The first operation regime* is for PMSG at startup and low speed, no loading ($i_s = 0$). From Equations 10.180 and 10.181, $\hat{\theta}_r$ is obtained from the stator-voltage vector angle θ_v:

$$\underline{V}_s = -j\omega_r\lambda_{PM}, \quad V_d = 0, \quad \hat{\theta}_r = \theta_v + \frac{\pi}{2} \tag{10.186}$$

A PLL-based observer extracts the rotor position and speed estimations from the measured stator voltage vector $\underline{V}^0_{\alpha\beta}$. The PLL error $\Delta\theta$ is given by the imaginary component of the vector product in Equation 10.187, where V_s is the stator voltage vector amplitude, and finally, $\Delta\theta$ is given by Equation 10.188. The PLL goal is to lead $\Delta\theta$ to zero, that is equivalent to lead $\hat{V}_d$ to zero (Equation 10.187).

$$\Delta\theta \cdot V_s \approx \operatorname{Im}\left[\underline{V}^0_{\alpha\beta} e^{-j(\hat{\theta}_r - \pi/2)}\right], \quad \Delta\theta = \theta_r - \hat{\theta}_r, \tag{10.187}$$

$$\Delta\theta \cdot V_s \approx V_\alpha \cos\hat{\theta}_r + V_\beta \sin\hat{\theta}_r = \hat{V}_d \tag{10.188}$$

The PI compensator gains of the PLL-based observer from Figure 10.72b are tuned using the procedure (Equation 10.175) for $\xi = 3$, as $k_{p\_est} = 0.6$ and $k_{i\_est} = 200\ \text{s}^{-1}$.

Note that the initial rotor position is not needed, and the estimation is robust to PMSG parameter variations, that is, the rotor position error is independent of them. This estimation is used only to prepare the initial rotor speed/position used in the next regime.

2. *The second operation regime* is for PMSG with load, imposing the condition that $i_d^* = 0$. From Equations 10.175 through 10.181, the d-axis stator voltage estimated component $\hat{V}_{d1}$ is:

$$\hat{V}_{d1} = \hat{\omega}_r L_q i_q \tag{10.189}$$

Now, a correction is used in PLL, based on the two expressions of V_d, that is, $\hat{V}_d$ (Equation 10.188) and $\hat{V}_{d1}$ (Equation 10.189) which implements an internal minor control loop shown at the bottom side of Figure 10.72b. This is equivalent to a model reference adaptive system (MRAS), with the reference $\hat{V}_d$ (Equation 10.188), and the adaptive model $\hat{V}_{d1}$ (Equation 10.189). The MRAS goal is to lead the V_d error to zero by using a PI controller:

$$\hat{\omega}_r = k_{p\_est}\left(\frac{1 + k_{i\_est}}{s}\right)\left(\hat{V}_d - \hat{V}_{d1}\right) \tag{10.190}$$

During the loading regime, the stator vector voltage $V_{\alpha\beta}$ can be reconstructed from the duty cycles d_a, d_b, d_c of the PMSG voltage source-inverter and the measured DC-link voltage V_{DC}:

$$\underline{V}_{\alpha\beta} = \frac{V_{DC}(2d_a - d_b - d_c)}{3} + \frac{jV_{DC}(d_b - d_c)}{\sqrt{3}} \tag{10.191}$$

The novel observer is robust to PMSG parameters variation, depending only on L_q (Equation 10.189). This observer is really suitable for PMSG sensorless control in wind turbines applications.

10.16.5 Test Platform and Experimental Results

The test-platform setup consists in a 12 N·m PMSG connected to the grid by two back-to-back Danfoss VLT 5005 voltage source inverters, and mechanically driven by a PMSM prime mover with Simovert Masterdrive (Figure 10.74). The grid is replaced here by a programmable three-phase AC power source (California Instruments 5005), capable of creating voltage faults. Since the AC power source cannot receive power, a resistive nonlinear load has been connected to line.

10.16.5.1 Voltage Sags Ride-Through

The AC power source is programmed to perform single-phase voltage sag from 230 to 120 V_{rms}, from 5.2 to 5.7 s in the primary side of the transformer. The voltages in the secondary side of the transformer are shown in Figure 10.75. The faulty phase and its current are presented in Figure 10.76. The estimated and measured rotor speeds are shown in Figure 10.77a, and the error between those two in Figure 10.77b. Note that during transient, the speed observer shows fairly good response. This is a clear sign that the

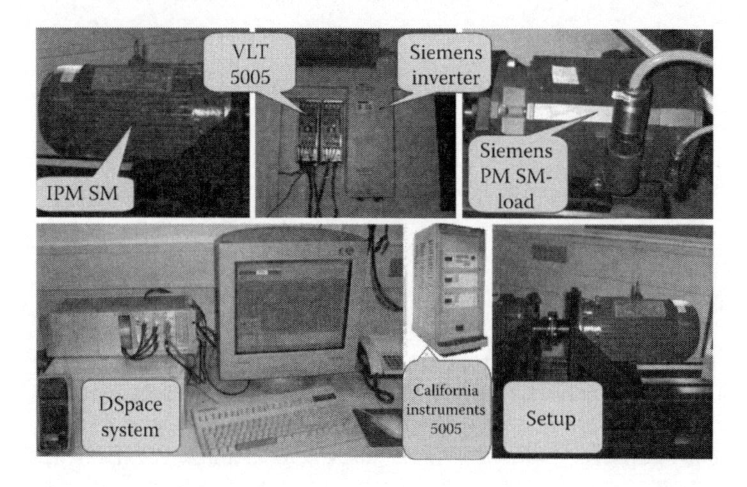

FIGURE 10.74 Test platform.

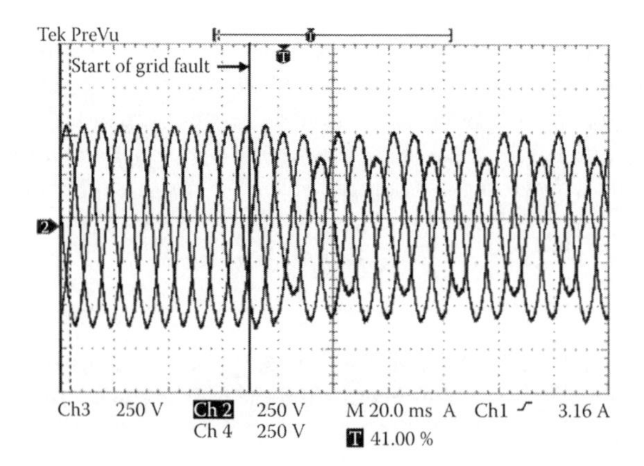

FIGURE 10.75 Grid voltages after transformer with single-phase voltage sag (250 V/div). (Adapted from Fatu, M. et al., Voltage sags ride-through of within sensorless controlled dual-converter PMSG for wind turbines, Record of IEEE–IAS 2007, 2007.)

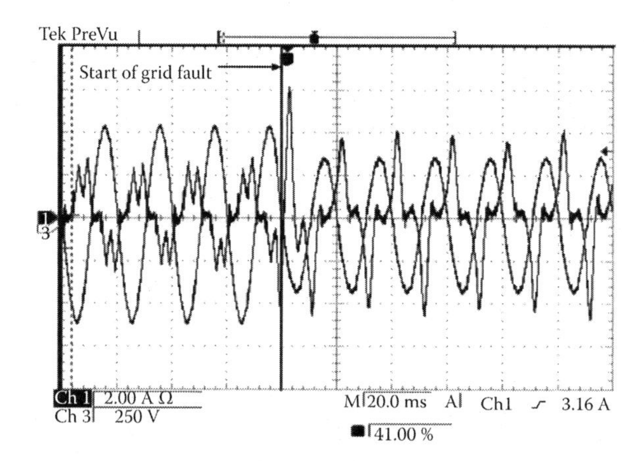

FIGURE 10.76 Oscilloscope grid-side phase voltage and current for single-phase voltage sag (250 V/div, 5 A/div). (Adapted from Fatu, M. et al., Voltage sags ride through of within sensorless controlled dual-converter PMSG for wind turbines, Record of IEEE–IAS 2007, 2007.)

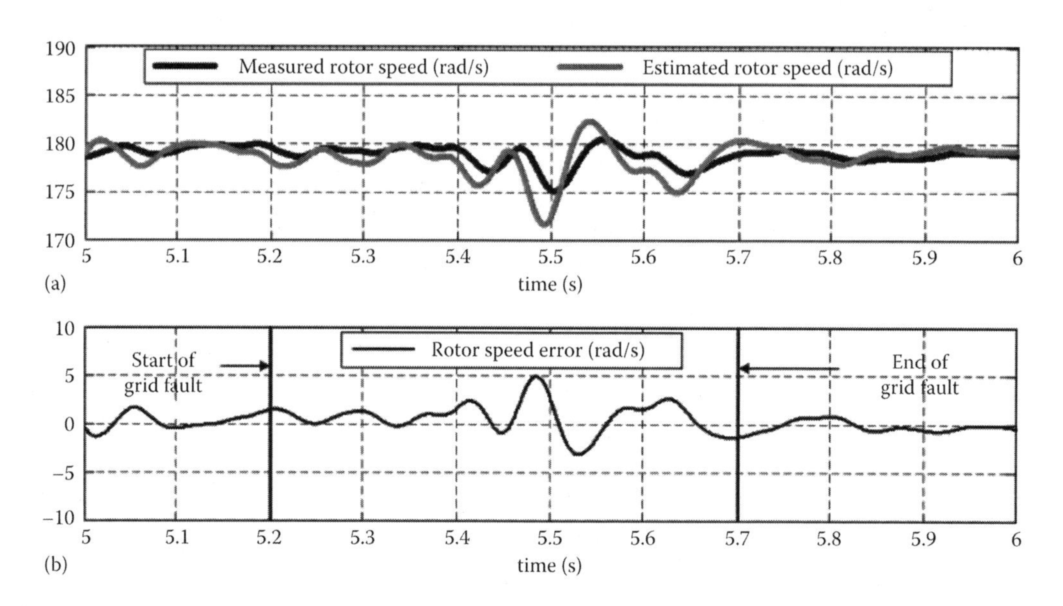

FIGURE 10.77 (a) Estimated and measured rotor speed and (b) rotor speed error (rad/s) for single-phase voltage sag. (Adapted from Fatu, M. et al., Voltage sags ride through of within sensorless controlled dual-converter PMSG for wind turbines, Record of IEEE–IAS 2007, 2007.)

position and speed observer are robust to large DC-voltage variations and to power variation in the generator. The PMSG phase *A* current is presented in Figure 10.78a, and the reference and measured *q*-axis current in Figure 10.78b. There are no significant transients during the fault. Another type of grid fault is the three-phase voltage sag situation. In this case, all three phases register the same magnitude of voltage drops. The voltage waveforms and one phase current are illustrated in Figure 10.79. The magnitude of all three voltages drops from 230 to 110 V_{rms} from 6.5 to 7 s, but no phase jump is registered, and the system recovers rapidly in one cycle. The DC-link voltage shown in Figure 10.80 is kept almost constant at 600 V during the fault.

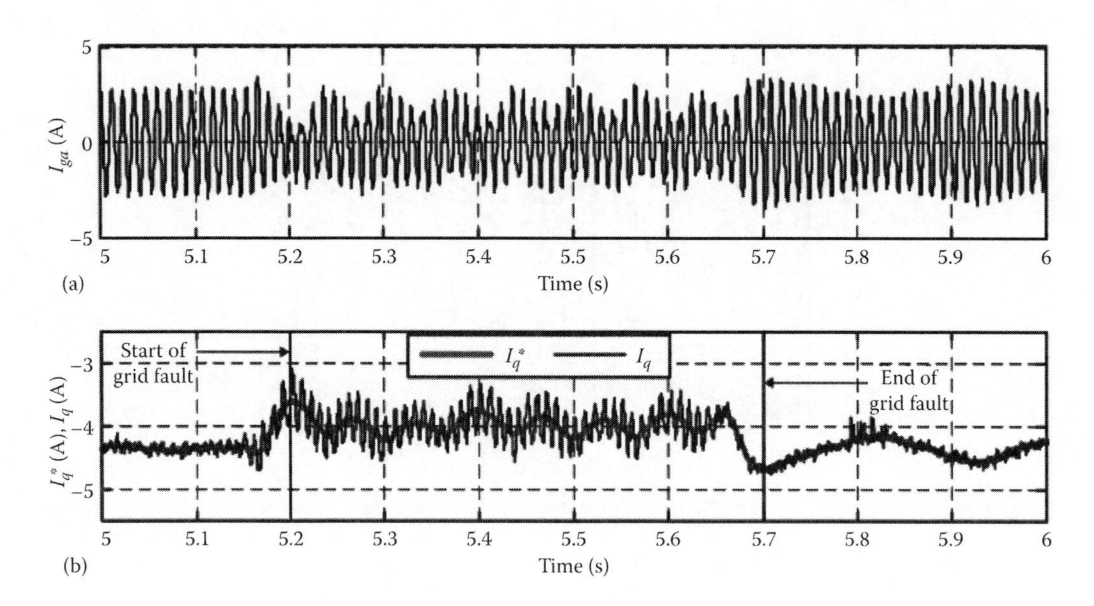

FIGURE 10.78 (a) PMSG phase current and (b) PMSG q-axis reference and measured current for single-phase voltage sag. (Adapted from Fatu, M. et al., Voltage sags ride through of within sensorless controlled dual-converter PMSG for wind turbines, Record of IEEE–IAS 2007, 2007.)

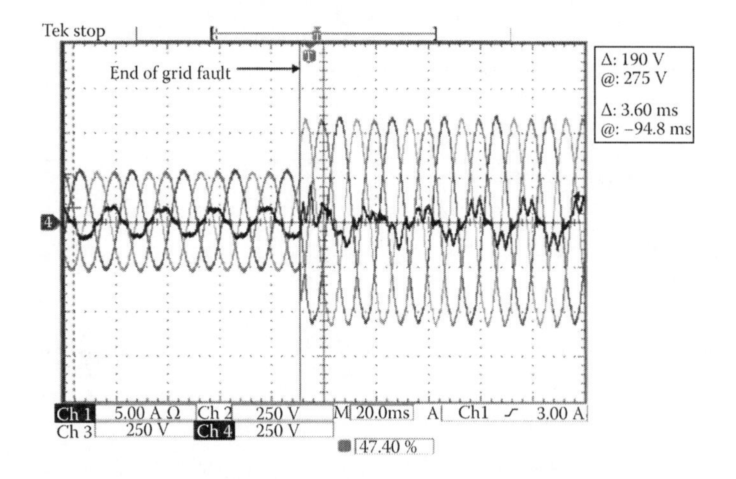

FIGURE 10.79 Grid voltages after transformer and phase current (250 V/div, 3 A/div) during three-phase voltage sags. (Adapted from Fatu, M. et al., Voltage sags ride through of within sensorless controlled dual-converter PMSG for wind turbines, Record of IEEE–IAS 2007, 2007.)

10.16.5.2 Harmonic and Negative-Sequence Voltage Compensation under Nonlinear Load

For stand-alone the phase voltage and its current waveforms, together with the voltage harmonic spectrum, without harmonic compensation, are presented in Figure 10.81.

A notable reduction of load voltage harmonics content is evident. The THD factor for the uncompensated harmonic voltages is 5.5% and with harmonic compensation was reduced to 1.5% (Figure 10.82).

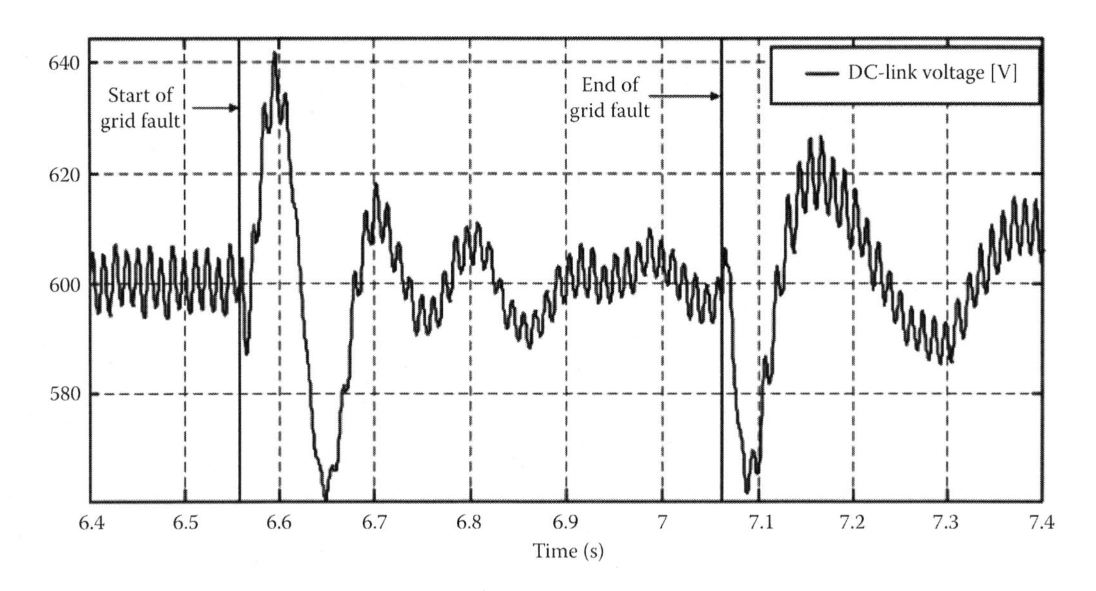

FIGURE 10.80 DC-link voltage during three-phase voltage sags. (Adapted from Fatu, M. et al., Voltage sags ride through of within sensorless controlled dual-converter PMSG for wind turbines, Record of IEEE-IAS 2007, 2007.)

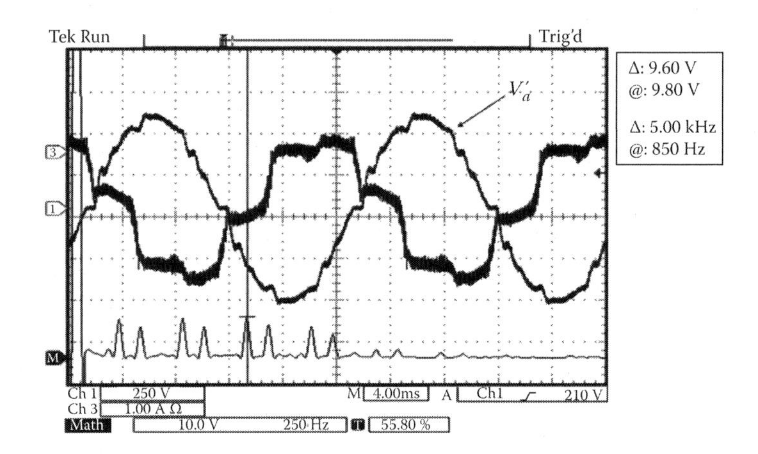

FIGURE 10.81 Nonlinear load current, voltage wave and voltage harmonic spectrum for a rectified load without harmonics voltage compensation. (From Fatu, M. et al., Novel motion sensorless control of stand-alone PMSG: Harmonics and negative sequence voltage compensation under nonlinear load, Record of EPE-2007, Aalborg, Denmark, 2007.)

10.16.5.3 Transition from Stand-Alone to Grid-Connected Mode

At time t = 10.8 s the grid is connected through switch S and subsequently rendered as recovered. A synchronization signal (Figure 10.83a) is automatically initiated in order to match the phases between load voltages and grid voltages. When the error $\Delta\theta$ is between $0.01 \leq \Delta\theta \leq 0.01$ (Figure 10.83b), the system switches from stand-alone control to grid connected control at t = 11.3 s (Figure 10.83a). It can be noted that there are no spikes in load voltage during this process or in the transition moment (Figure 10.83c), but there are some reasonable transients in the load current (Figure 10.83d). The PMSG currents are shown in Figure 10.83e without any significant transients when then system is switching from one control strategy to the other.

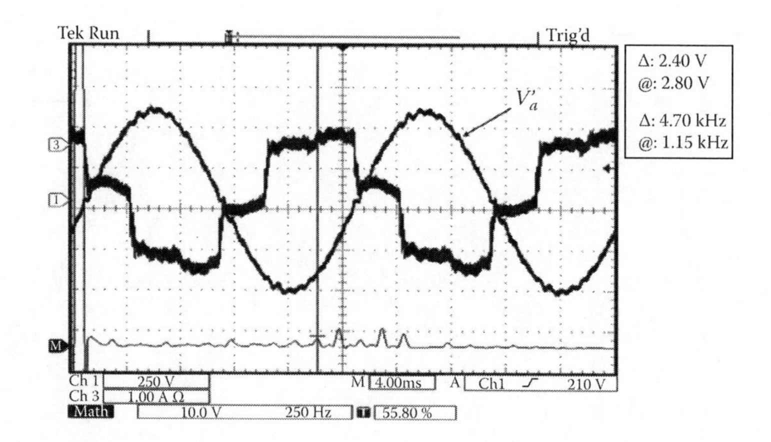

FIGURE 10.82 Nonlinear load current, voltage wave, and voltage harmonic spectrum for a rectified load with synchronous voltage compensation. (From Fatu, M. et al., Novel motion sensorless control of stand-alone PMSG: Harmonics and negative sequence voltage compensation under nonlinear load, Record of EPE-2007, Aalborg, Denmark, 2007.)

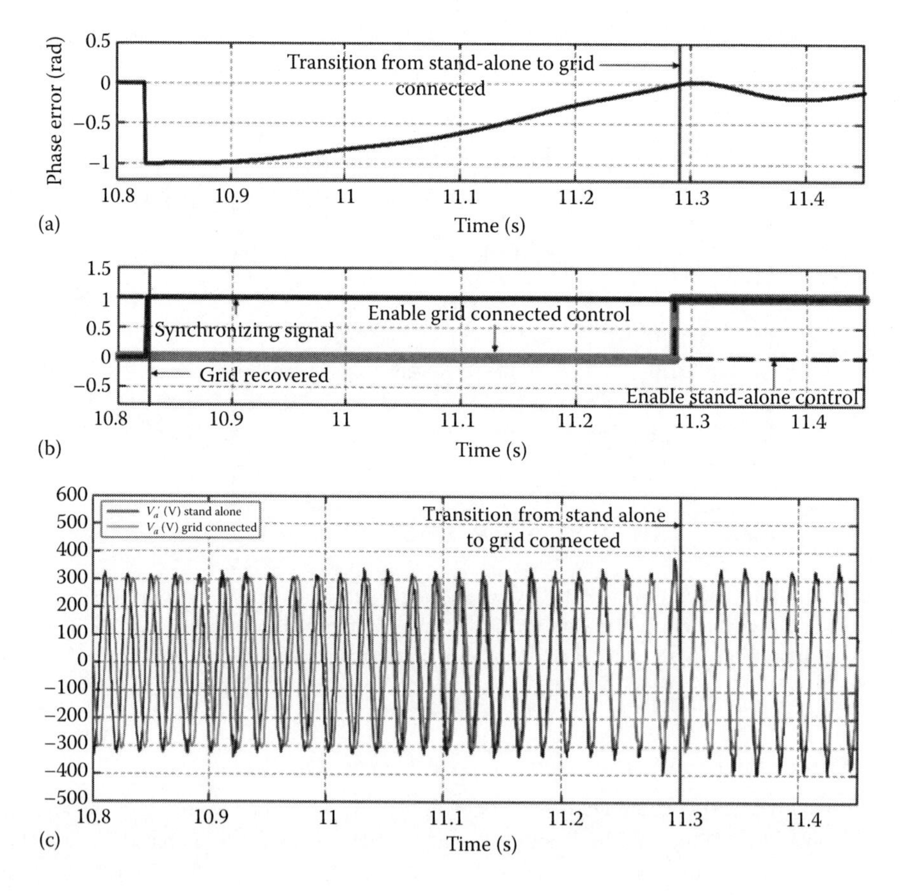

FIGURE 10.83 Measured stand-alone grid connected transition: (a) phase error; (b) enable grid control/stand-alone signals, synchronizing signal; and (c) phase A load voltage (V_a'), phase A grid voltage (V_a). (Adapted from Fatu, M. et al., Motion sensorless bidirectional PWM converter PMSG control with sensorless switching from power grid to stand-alone and back, in: *Proceedings of the IEEE Power Electronics Specialists Conference* [*PESC*], 2007, pp. 1239–1244.)

(*Continued*)

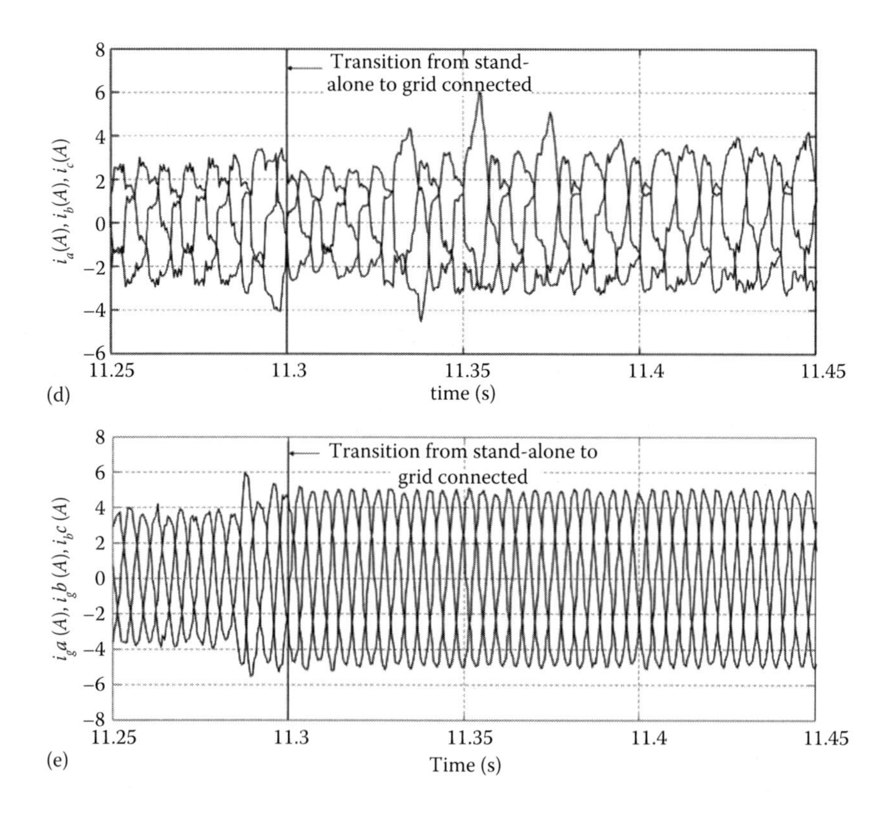

FIGURE 10.83 (*Continued*) Measured stand-alone grid connected transition: (d) load currents and (e) PMSG currents. (Adapted from Fatu, M. et al., Motion sensorless bidirectional PWM converter PMSG control with sensorless switching from power grid to stand-alone and back, in: *Proceedings of the IEEE Power Electronics Specialists Conference [PESC]*, 2007, pp. 1239–1244.)

10.16.5.4 Transition from Grid-Connected to Stand-Alone Mode

A fault event is simulated by opening the static switch S (Figure 10.65) at time $t = 15.8$ s. In this case, because we still have current control of supply-side converter (corresponding still to grid connection, Figure 10.65), there is an increase in load voltage level (Figure 10.84c), and when this exceeds a certain threshold or the frequency is out of range (Figure 10.84b), the system switches automatically to stand-alone mode. Figure 10.84d and e shows the load currents and PMSG currents, which both do not show any inrush spikes during entire transition period, indicating a smooth transition. It may be noted that there are no important transients or generator current interruption during transition from grid connected to stand-alone mode. The estimated and actual rotor speed as well as the rotor speed error for steady state at 180 rad/s is depicted in Figure 10.84f and g showing good tracking performance of position/speed observer used for motion-sensorless control of PMSG.

10.16.6 Conclusion

This case study aimed at documenting through analysis, design and experimental tests, the PMSG connected at weak power grids during riding through asymmetric faults, power quality, seamless insularization, and reconnection to the grid. The presentation focuses on the following key attributes of the proposed control system:

- Operate in stand-alone and grid connection with smooth transition between the two operation modes
- Handle unbalanced and nonlinear loads in both operation modes

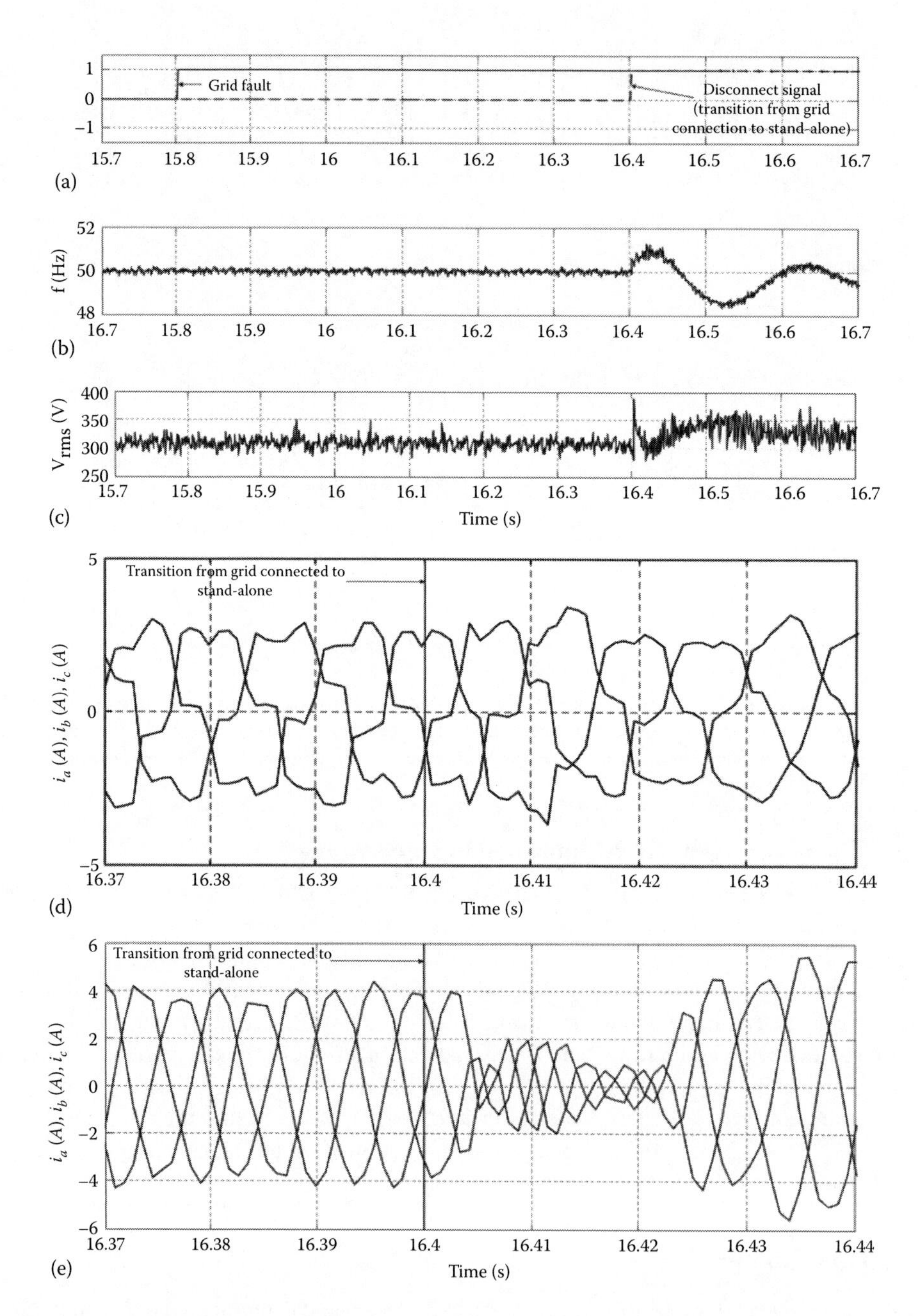

FIGURE 10.84 Measured transition from grid-connected stand-alone control mode: (a) disconnect signal, (b) f_{grid}, (c) V_{rms}, (d) grid currents, and (e) PMSG currents. *(Continued)*

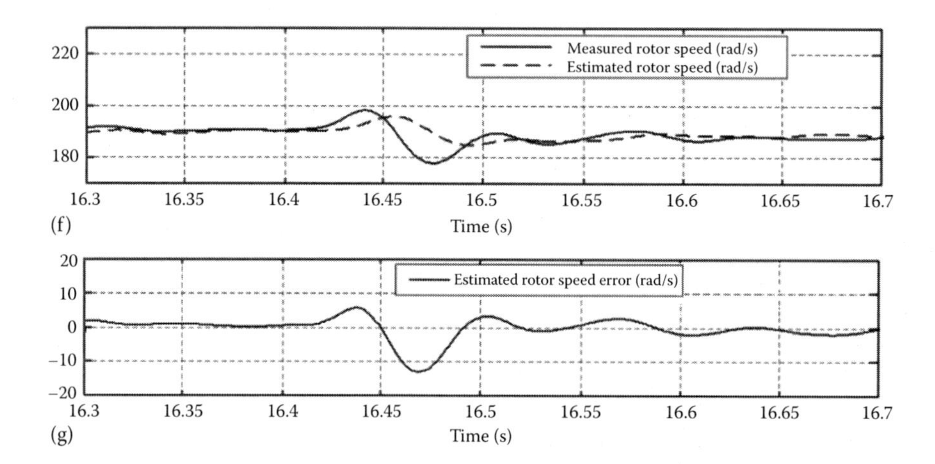

FIGURE 10.84 (*Continued*) Measured transition from grid-connected stand-alone control mode: (f) estimated and measured rotor speed and (g) rotor speed estimation error. (Adapted from Fatu, M. et al., Motion sensorless bidirectional PWM converter PMSG control with sensorless switching from power grid to stand-alone and back, in: *Proceedings of the IEEE Power Electronics Specialists Conference* [PESC], 2007, pp. 1239–1244.)

- Compensate harmonics through the grid-side converter
- Smooth and fast transition through asymmetric voltage transients
- Band-pass filter based on vector D-module filter to extract the grid voltage positive-sequence PLL based observer for rotor position and speed estimation in PMSG motion-sensorless control, without using emf integration and initial rotor position, depending only on L_q, with good estimation during voltage sags, transition from grid connected to stand alone and vice versa

10.17 Note on Medium-Power Vehicular Electric Generator Systems

By medium-power vehicular electric generator systems, we mean here those systems with the powers in the hundreds of kilowatts to tens of megawatts used on aircraft and vessels. In terms of operation principles, they may be synchronous with excitation or PMs or induction or switched reluctance types.

All these types were treated in notable detail in previous chapters. This is why here we will dwell only on advanced configurations most suitable for aircraft, trains, and vessels.

The electric power on board of aircraft is expected to reach soon 500 kW per engine in a more electric aircraft.

Diesel engine long-haul or commuter trains with electric propulsion require better generators.

Electric propulsion is also typical in low dead weight vessel, and up to 20–30 MVA SGs are used as power sources to supply the electric propulsion motors through their power electronics control.

Up to 3 MVA generators with 690 V (line voltage, RMS value) output may be adopted.

Above these power levels, medium-output voltage generators up to 6 kV (line voltage, RMS value) are applicable.

Aircraft generators operate at speeds up to 15–60 krpm and powers up to 250–500 kW/unit. For diesel engine electric trains and vessel, the speeds are generally in the range of 750–3000 rpm, and powers per unit from a few hundred kilowatts to MVA and tens of MVA.

Besides, starter generators on engine shafts at powers up to 250–500 kW/unit, emergency generators in the range of tens of kilowatts are also needed on aircraft. Three or four 250 kW generators have to be connected in parallel, or islanded, as required, for optimum use of energy and best reliability and power quality.

SGs with brushless excitation control for constant 400 Hz and constant voltage output are the standard on aircraft. Once full power electronics on board is accepted, PM or switched reluctance generators with the DC power bus at 270 V_{DC}, for example [46,59,60], might come quickly.

SGs with excitation regulation for voltage control and speed regulation for frequency control are also standard on diesel engine electric trains and vessels.

However, at least up to 3–5 MVA multiple-pole PM generators with frequency up to 500 Hz, together with boost DC–DC converters to secure a constant DC voltage bus, may be adopted for trains and small vessel applications.

It should now be noted that the above schemes were already discussed in the chapters on SG performance, design, and control, in this chapter and in Chapter 9 on switched reluctance generators.

This is why here we do not pursue further the medium-power vehicular generator systems.

10.18 Summary

- By PMSG, we understand here a radial and axial air gap PM brushless generator with distributed or concentrated stator winding and surface PM or interior PM pole rotors.
- The PMSG applications span from wind turbine and microhydroturbine direct-driven electric generators, to automotive starter generators, or generators only for automotive applications, mobile or standby gensets with multiple output, and super-high-speed gas turbine generator systems for distributed power systems or for stand-alone applications.
- The market for PMSGs is gaining momentum worldwide, and the potential for fast growth seems very good.
- Rotors for PMSGs are built in cylindrical or disk shapes with surface PM or interior poles. Hybrid surface PM plus variable reluctance rotor sections may be combined for wide constant power speed range applications.
- For super-high speeds (above 15,000 rpm), a thin copper shield over the PMs and a carbon fiber resin retaining ring may be needed to reduce rotor electrical losses and provide for mechanical rigidity.
- Besides distributed windings, for powers up to 100 kW at 80,000 rpm, nonoverlapping (tooth-embrace) coil windings may be applied, as they show high fundamental winding factors, lower copper losses, but slightly higher core losses as the number of poles is larger.
- The maximum fundamental frequency in super-high-speed PMSGs may go up to 2.5 kHz for up to 150 kW and up to 1 kHz for powers in the 1–5 MW range.
- For surface PM pole rotor PMSGs, in the absence of magnetic saturation, a comprehensive analytical model was built to explore steady state and transient performance. The space harmonics of the stator mmf air gap field with consideration of slot openings, through a permeance P.U. function, are included; thus, a realistic resultant air gap flux density distribution is built. A similar model was described for the axial air gap rotor PMSG, where an additional radial flux distribution change P.U. factor was introduced to make the model more realistic.
- Traditionally, core losses were calculated by analytic formulas under no-load conditions. It was recently shown that additional substantial core losses occur under load conditions. Analytical formulas for this case were also proposed, but were only based on FEM-extracted time evolution of flux density in the finite element added contributions. Good correlation with test results was shown by such a combined method. The rotational character of the magnetic field in various machine core regions is very important to consider.
- For the investigation of transients, the phase coordinate model is first used. Then, the *dq* model is shown to be a practical method for the study of transients and control performance and design.
- The *dq* model may be used when the emf is sinusoidal, and the phase inductances are either independent of or vary sinusoidally with rotor position. For the PMSG with shunt capacitors and AC

load or diode rectifier load, the *dq* model is instrumental in investigating the PMSG steady state and transient performance. A potential oscillatory dynamics effect has to be eliminated in such schemes.
- When a PWM voltage source inverter is added to the diode rectifier, the PMSG system may deliver constant frequency and voltage power to a local power grid or to separate loads, at variable speed.
- It was proven that if a third harmonic in the stator emf of PMSG is produced, the latter may increase the output by as much as 15% with a three-phase diode rectifier for both series and bridge connections, with a loss of efficiency of about 1% only. The addition of the fourth diode leg to the three-phase bridge diode rectifier seems to pay off. A larger number of phases do not seem to produce better effects due to repetitive commutations.
- A variable inductance in parallel with a capacitor at machine terminals may produce very good dynamic performance in PMSG with constant speed controlled diesel engine prime movers and constant voltage output with load. A thyristor variac in series with the fixed inductances is controlled to regulate the load voltage at constant value. State feedback control systems perform well for such applications.
- When PMSGs have a diode rectifier at their terminals, the equivalent fundamental power factor angle in the machine φ_1 is forced to a small value: from zero at low frequencies to 10°–15° at 1 kHz or so. In these conditions, the maximum torque, for $L_d = L_q = L_s$ (surface PM rotor), is as follows:

$$T_e = \frac{3}{2} p_1 \frac{\Psi_{PM}^2}{2L_s} \quad \text{or} \quad t_e\,(\text{P.U.}) = \frac{1}{2X_s\,(\text{P.U.})}$$

  This limitation may serve as a key design relationship, when sizing a PMSG of lower synchronous reactance X_s (P.U.) is desirable.
- Gensets for mobile or standby power based on PMSGs may be built with multiple outputs, provided full power electronics control is provided. Variable speed is used to reduce fuel consumption. In such conditions, a DC voltage booster is required, besides the PWM voltage source inverter on the load side. As a 28 V_{DC} battery is used for the starter of the prime mover, the latter is used through a three-stage voltage booster to supplement the DC link voltage when it is too low momentarily (for a few seconds to minutes) until the speed of the engine recovers [24].
- The 50(60), 400 Hz three- and single-phase output selection is a special asset of the new gensets, in addition to their reduced mass and volume in comparison with the standard ones that make use of voltage-controlled SGs at constant output frequency (and speed).
- PMSGs were recently proposed for automobiles at 42 V_{DC} to replace the existing claw-pole alternators, as more power is needed on board and at a higher efficiency (less fuel), with faster availability under load variations.
- PMSGs are being introduced for super-high-speed gas turbines for powers up to a few MW to reduce volume and cost, while providing fast active and reactive power control (at variable speed) on local power grids or separate loads.
- Mechanical constraints expressed in terms of maximum peripheral speed or rotor diameter versus speed have to be observed for super-high-speed PMSGs.
- In super-high-speed PMSGs (rotor peripheral speed above 150–200 m/s), the fundamental frequency may go up to 2.5 kHz for 100–200 kW and 1 kHz in the MW power/unity range. Higher frequency leads to a lower volume machine with lower copper losses but larger iron losses. It also causes large commutation losses in the PWM converter used for control.
- A super-high-speed rotor may run through one to three critical speeds before reaching the rated speed [45]. These critical speeds have to be driven through quickly, to avoid excessive noise and vibration.
- Super-high-speed PMSGs in the MW power range and 12,000–15,000 rpm require, in addition to the diode rectifier, with or without a DC link voltage booster, a PWM medium-voltage multilevel converter. Such converters are now available up to tens of MW and may also be used for the scope.

- Back-to-back two-level voltage source (six-legged) PWM converters [61] may be used for powers up to 1 MW or so to provide full four-quadrant operation with motoring, for starting the turbine.
- The electromagnetic design of a PMSG depends essentially on the operating modes, power, and output voltage range. Due to such variations, it seems practical to use a given rotor shear stress range (0.5–5) N/cm^2 to size the stator internal diameter D_{is} and the stack length for the maximum torque required in the design. From high-speed voltage constraints, the number of turns per phase is calculated after looking for the minimization of peak current for maximum torque at minimum speed. Winding tapping may be used for wide constant power speed range, to reduce the peak current for given output voltage; that is, to reduce the PWM converter costs.
- PMSGs are a novel technology and thus, special standards for their testing are not yet available. Many of the standard tests (IEEE standard 115/1995) for regular synchronous machines may be applied to PMSGs as well. Standstill, no-load, and short-circuit generator, no-load motor, and load tests may all provide information on PMSG energy conversion performance (losses, efficiency, power at given voltage and speed) and parameter (Ψ_{PM}, L_d, L_q, R_s) estimation.
- Only mild cross-coupling saturation effects were found in PMSGs, and thus, they may be described by two unique magnetization curves along axes d and q: $\Psi_d^*(i_s)$, $\Psi_q^*(i_s)$ with the following inductances [62]:

$$L_d(i_d) = \frac{\left(\Psi_d^*(i_s) - \Psi_{PM}\right)}{i_s}; \quad L_q(i_s) = \frac{\Psi_q^*(i_s)}{i_s}$$

and

$$\Psi_d = L_d(i_s)i_d + \Psi_{PM}; \quad \Psi_q = L_q(i_s)i_q$$

where i_s is the total stator current $i_s = \sqrt{i_d^2 + i_q^2}$.
- The PMSG systems—with full power electronics digital control—may be considered a novel technology with exceptional growth potential in the near future for renewable energy conversion distributed power systems and for vehicular technologies [61,63].

References

1. T.J. Miller, *Brushless PM and Reluctance Motor Drives*, Clarendon Press, Oxford, U.K., 1989.
2. S.A. Nasar, I. Boldea, and L.E. Unnewehr, *PM, Reluctance and Selfsynchronous Motors*, CRC Press, Boca Raton, FL, 1993.
3. T.J.E. Miller and J. Hendershot, *Design of PM Brushless Motors*, Clarendon Press, Oxford, U.K., 1994.
4. D.C. Hanselman, *Brushless PM Motor Design*, Marcel Dekker, New York, 1997.
5. J.F. Gieras and M. Wing, *Permanent Magnet Motor Technology*, Marcel Dekker, New York, 2002.
6. K. Sitapati and R. Krishnan, Performance comparisons between radial and axial field PM brushless machines, *IEEE Trans.*, IA-37(5), 2001, 1219–1226.
7. A. Covagnino, M. Lazzari, F. Profumo, and A. Tenconi, A comparison between axial flux and radial flux PM synchronous motors, *IEEE Trans.*, IA-38(6), 2002, 1517–1524.
8. E. Spooner and A. Williamson, Modular PM wind-turbines generators, In *Record of IEEE–IAS-1996 Annual Meeting*, 1996, Vol. 1, pp. 497–502.
9. J. Cros and P. Viarouge, Synthesis of high performance PM motors with concentrated windings, *IEEE Trans.*, EC-17(2), 2002, 248–253.
10. F. Magnussen and C. Sadarangani, Winding factors and Joule losses of PM machines with concentrated windings, Record of IEEE–IEMDC-2003, 2003, Vol. 1, pp. 333–339.

11. Z.Q. Zhu, D. Howe, E. Bolte, and B. Ackerman, Instantaneous magnetic field distribution brushless PM d.c. motors, Parts I, II, III, IV, *IEEE Trans.*, MAG-29(1), 1993, 124–158.
12. A.B. Proca, A. Keyhani, A. El-Antably, W. Lu, and M. Dai, Analytical model for permanent magnet motors with surface mounted magnets, *IEEE Trans.*, EC-18(3), 2003, 386–391.
13. J. Azzouzi, G. Barakat, and B. Dakyo, Quasi 3D analytical modeling of magnetic field of axial flux PMSM, Record of IEEE–IEMDC-2003, 2003, Vol. 3, pp. 1941–1946.
14. A. Parviainen, M. Niemala, and J. Pyrhonen, Modeling of axial flux PM machines, Record of IEEE–IEMDC-2003, 2003, Vol. 3, pp. 1955–1961.
15. L. Ma, M. Sanada, S. Morimoto, and Y. Takeda, Prediction of iron loss in rotating machines with rotational losses included, *IEEE Trans.*, MAG-39(4), 2003, 2036–2041.
16. A. Munoz-Garcia and D.W. Novotny, Characterization of third harmonic-induced voltages in PM generators, *Record of IEEE–IAS-1996 Annual Meeting*, 1996, Vol. 1, pp. 525–532.
17. D.-H. Che, H.-K. Jung, and D.-J. Sim, Multiobjective optimal design of interior PM synchronous motors considering improved core loss formulae, *IEEE Trans.*, EC-14(4), 1999, 1347–1352.
18. V. Honsinger, Performance of poliphase PM machines, *IEEE Trans.*, PAS-99(6), 1980, 1510–1517.
19. O. Ojo and J. Cox, Investigation into the performance characteristic of an IPM generator including saturation effects, Record of IEEE–IAS-1996, 1996, Vol. 1, pp. 533–540.
20. I. Boldea and S.A. Nasar, *Electric Drives*, CRC Press, Boca Raton, FL, 1998.
21. M.A. Rahman, A.M. Osheiba, and T.S. Radwan, Modelling and controller design of an insulated diesel engine permanent magnet synchronous generator, *IEEE Trans.*, EC-11(2), 1996, 324–330.
22. D.C. Aliprantis, S.A. Papathanassiou, M.P. Papadopoulos, and A.G. Kladar, Modeling and control of variable-speed wind turbine equipped with PM synchronous generator, Record of ICEM-2000, Espoo, Finland, 2000, Vol. 1, pp. 558–562.
23. B.S. Borowy and Z.M. Salameh, Dynamic response of a stand-alone wind energy conversion system with battery energy storage to a wind gust, *IEEE Trans.*, EC-12(1), 1997, 73–78.
24. L.M. Tolbert, W.A. Peterson, T.J. Theiss, and M.B. Scudiere, Gen-sets, *IEEE–IA Magazine*, 9(2), 2003, 48–54.
25. I. Boldea and S.A. Nasar, *Induction Machine Handbook*, CRC Press, Boca Raton, FL, 2001, Chapters 7, 9, 11.
26. H. Polinder, On the losses in a high-speed permanent-magnet generator with rectifier with special attention to the effect of a damper cylinder. PhD thesis, University of Delft, Delft, the Netherlands, 1998.
27. A. Diop and U. Schröder, Experimental losses determination of high speed synchronous machines 70 kW–45,000 rpm with magnetic bearings, Record of ICEM-1992, Manchester, U.K., 1992, Vol. 3, pp. 2106–2110.
28. A. Castagnini, M. Garavaghia, F. Moriconi, and G. Secondo, Development of a high speed and power synchronous PM motor, Record of ICEM-1992, Manchester, U.K., 1992
29. B.P. James and D.A.T. Al Zahawi, A high speed alternator for small scale gas turbines CHP unit, In *Record of International Conference on Electrical Machines and Drives*, IEE Publication No. 412, 1995.
30. A. Castagnini and I. Leone, Test results of a very high speed PM brushless motor, Record of ICEM-2002.
31. Z.J.J. Offringa, R.W.P. Kerkenaar, and J.L.F. Van der Veen, A high speed 1400 kW PM generator with rectifier, Record of ICEM-2000, Espoo, Finland, 2000, Vol. 1, pp. 308–313.
32. V. Callea, S. Elia, M. Pasquali, M.B. Sabene, and E. Santini, Ultra high-speed electric generator for gas turbine series utility, In *Record of EVS Congress*, Berlin, Germany, 2001.
33. W. Vetter, A. Collotti, and K. Reichert, A new motor/generator for flywheel applications, Record of ICEM-1998, Istanbul, Turkey, 1998, Vol. 1, pp. 348–352.
34. R. Hebner, J. Beno, and A. Walls, Flyweel batteries come around again, *IEEE Spectrum*, 4, April 2002, 46–51.

35. J. Rodrigues, Multilevel inverters II—Guest Editorial and special section, *IEEE Trans.*, IE-49(4 and 5), 2002, 722–888, 946–1100.
36. G. Podder, A. Joseph, and A.K. Unnikrishnan, Sensorless variable-speed controller for existing fixed-speed wind power generator with unity power-factor operation, *IEEE Trans.*, IE-50(5), 2003, 1007–1015.
37. E.A.A. Coelho, P.C. Cortizo, and P.F.D. Garcia, Small-signal stability for parallel-connected inverters in stand-alone A.C. supply systems, *IEEE Trans.*, IA-38(2), 2002, 533–542.
38. B.H. Bae, S.-K. Sul, J.-H. Kwon, and J.-S. Byeon, Implementation of sensorless vector control for super-high-speed PM–SM of turbo-compressor, *IEEE Trans.*, IA-39(3), 2003, 811–818.
39. K.-H. Kein and M.-J. Youn, DSP-based high-speed sensorless control for a brushless D.C. motor using a D.C. link voltage control, *EPCS J.*, 30(9), 2002, 889–906.
40. H.J. Gutt and J. Müller, New aspects for developing and optimizing modern motor car generators, *Record of IEEE–IAS-1994 Annual Meeting*, 1994, Vol. 1, pp. 3–7.
41. M. Comanescu, A. Keyhani, and M. Dai, Design and analysis of a 42 V permanent magnet generator for automotive applications, *IEEE Trans.*, EC-18(1), 2003, 107–112.
42. D. Gay, Composite iron powder for A.C. electromagnetic applications: History and use, In *Record of SMMA Fall Conference*, Chicago, IL, October 4–6, 2000.
43. E.A. Knoth, Motors for 21st century, In *Record of SMMA-2000, Fall Conference*, Chicago, IL, October 4–6, 2000.
44. T.J.E. Miller, Methods for testing permanent magnet polyphase A.C. motors, Record of IEEE–IAS-1981, 1981, Vol. 1, pp. 494–499.
45. J.D. Ede, Z.Q. Zhu, and D. Howe, Rotor resonances of high speed PM brushless machines, *IEEE Trans.*, IA-38(6), 2002, 1542–1548.
46. J.A. Weimer, The role of electric machines and drives in the more electric aircraft, Record of IEEE–IEMDC, 2003, Vol. 1, pp. 11–15.
47. H.P. Nee, L. Lefeure, P. Thelin, and J. Soulard, Determination of D and Q reactances of PM synchronous motors without measurements of the rotor position, *IEEE Trans.*, IA-36(5), 2000, 1330–1335.
48. P. Rodríguez, J. Pou, J. Bergas, I. Candela, R. Burgos, and D. Boroyevich, Double synchronous reference frame PLL for power converters, In *Proceedings of the IEEE Power Electronics Specialists Conference (PESC'05)*, 2005, pp. 1415–1421.
49. N.R.N. Ama, F.O. Martiz, L. Matacas Jr., and F. Kassab Jr., Phase-locked loop based on selective harmonics compensation for utility applications, *IEEE Trans. Power Electron.*, 28(1), January 2013, 144–153.
50. J.A. Suul, A. Luna, P. Rodriguez, and T. Undeland, Voltage-sensor-less synchronization to unbalanced grids by frequency-adaptive virtual flux estimation, *IEEE Trans. Power Electron.*, 59(7), July 2012, 2910–2923.
51. J.A. Suul, A. Luna, P. Rodriguez, and T. Undeland, Virtual-flux-based voltage-sensor-less power control for unbalanced grid conditions, *IEEE Trans. Power Electron.*, 27(9), September 2012, 4071–4087.
52. S. Shinnaka, A new characteristics-variable two-input/output filter in D-module—Designs, realizations, and equivalences, *IEEE Trans. Ind. Appl.*, 38(5), September/October 2002, 1290–1296.
53. M. Fatu, C. Lascu, R. Teodorescu, I. Boldea, and G.D. Andreescu, Voltage sags ride through of within sensorless controlled dual-converter PMSG for wind turbines, Record of IEEE-IAS, 2007.
54. Q. Lei, F. Zheng, and S. Yang, Multiloop control method for high-performance microgrid inverter through load voltages and current decoupling with only voltage feedback, *IEEE Trans. Power Electron.*, 26(3), March 2011, 953–960.
55. C. Lascu, L. Asiminoaei, I. Boldea, and F. Blaabjerg, High performance current controller for selective harmonic compensation in active power filters, In *Proceedings of the OPTIM 2006*, Brasov, Romania, 2006 (IEEEXplore).

56. Z. Yao, Z. Wang, L. Xiao, and Y. Yang, A novel control strategy for grid-interactive inverter in grid-connected and stand-alone modes, In *Twenty-First Annual IEEE Applied Power Electronics Conference and Exposition (APEC'06)*, 2006.
57. Md. Nayeem Arafat, S. Palle, Y. Sozer, and I. Husain, Transition control strategy between stand-alone and grid-connected operations of voltage source inverters, *IEEE Trans. Ind. Appl.*, 48(5), September/October 2012, 1516–1525.
58. D. Velasco, C. Trujillo, G. Garcera, and E. Figueres, An active anti-islanding method based on phase-PLL perturbation, *IEEE Trans. Power Electron.*, 26(4), April 2011, 1056–1066.
59. P.H. Mellor, S.G. Burrow, T. Sawata, and M. Holme, A wide speed range permanent magnet generator for future embedded aircraft generation systems, Record of IEEE–IEMDC-2003, 2003, Vol. 3, pp. 1308–1313.
60. C.A. Ferreira, S.R. Jones, W.S. Heglund, and W.D. Jones, Detailed design of a 30 kW switched reluctance starter-generator system for gas turbine application, *IEEE Trans.*, IA-31(3), 1995, 553–561.
61. A.S. Meris, N.A. Voues, and G.B. Giannakopoulos, A variable speed wind energy conversion scheme for connecting to weak A.C. systems, *IEEE Trans.*, EC-14(1), 1999, 122–127.
62. I. Boldea, L. Tutelea, and C.I. Pitic, PM-assisted reluctance synchronous motor/generator (PM-RSM), for mild hybrid vehicles: Electromagnetic design, *IEEE Trans.*, IA-40(2), 2004, 492–498.
63. M. Fatu, F. Blaabjerg, and I. Boldea, Grid to standalone transition motion-sensorless dual-inverter control of PMSG with asymmetrical grid voltage sages and harmonics filtering, *IEEE Trans.*, PE-29(7), 2014, 3463–3472.
64. M. Fatu, L. Tutelea, R. Teodorescu, F. Blaajberg, and I. Boldea, Motion sensorless bidirectional PWM converter PMSG control with sensorless switching from power grid to stand-alone and back, In *Proceedings of the IEEE Power Electronics Specialists Conference (PESC)*, 2007, pp. 1239–1244.
65. M. Fatu, L. Tutelea, R. Teodorescu, and I. Boldea, Novel motion sensorless control of stand-alone PMSG: Harmonics and negative sequence voltage compensation under nonlinear load, Record of EPE-2007, Aalborg, Denmark, 2007.

11

Transverse Flux and Flux Reversal Permanent Magnet Generator Systems

11.1 Introduction

There are certain applications, such as direct-driven wind generators, that have very low speeds (15–50 rpm), and then others, as micro hydro generators, with speeds in the range of up to 500 rpm and powers up to a few megawatts (MWs) for which permanent magnet (PM) generators are strong candidates, provided the size and the costs are reasonable.

Even for higher speed applications, but for lower power, today's power electronics allow for acceptable current waveforms up to 1–2.5 kHz fundamental frequency f_{1n}.

Increasing the number of PM poles 2_{p1} in the PM generator to fulfill the standard condition,

$$f_{1n} = p_1 \cdot n_n; \quad n_n\text{—speed (rps)} \tag{11.1}$$

thus becomes necessary.

Then, the question arises as to whether the PM synchronous generators (SGs) with distributed windings are the only solution for such applications, when the lowest pole pitch for which such windings can be built is about $\tau_{min} \approx 30$ mm, for three slots/pole. And, even at $\tau = 30–60$ mm, is the slot aspect ratio large enough to allow for a high enough electrical loading to provide for high torque density? The apparent answer to this latter question is negative.

The nonoverlapping coil-winding concept (detailed in Chapter 10) is the first candidate that comes to mind for pole pitches $\tau > 20$ mm for large torque machines (hundreds of Newton meters [Nm]), but when the pole pitch $\tau = 10$ mm or so, they are again limited in electrical loading per pole, though there is about one slot per pole ($N_s \approx 2_{p1}$). In an effort to increase the torque density, the concept of a multipole span coil winding can be used, which becomes practical, especially when the number of PM poles $2_{p1} > 10–12$.

Two main breeds of PM machines were proposed for high numbers of pole applications: transverse flux machines (TFMs) and flux switching (or reversal) machines (FRMs).

The TFMs are basically single-phase configurations with single circumferential coil per phase in the stator, embraced by U-shaped cores that create a variable reluctance structure with 2_{p1} poles. A 2_{p1} pole surface or an interior pole PM rotor is added (Figure 11.1a and b). Two or three such configurations placed along the shaft direction would make a two-phase or three-phase machine [1].

There are many other ways to embodying the TFM, but the principle is the same. For example, the concept can be extended by placing the PM pole structure in the stator, around the circumferential stator coil, while the rotor is a passive variable reluctance structure with axial or radial air gap (Figure 11.2a and b) [2].

The PM flux concentration is performed in the stator, in Figure 11.2 configurations, but the PM flux paths in the rotor run both axially and radially; thus, the rotor has to be made of a composite magnetic material (magnetic powder).

The TFMs with rotor or stator PM poles are characterized by the fact that the PM fluxes of all North Poles add up at one time in the circumferential coil, and then, after the rotor travels one PM pole span

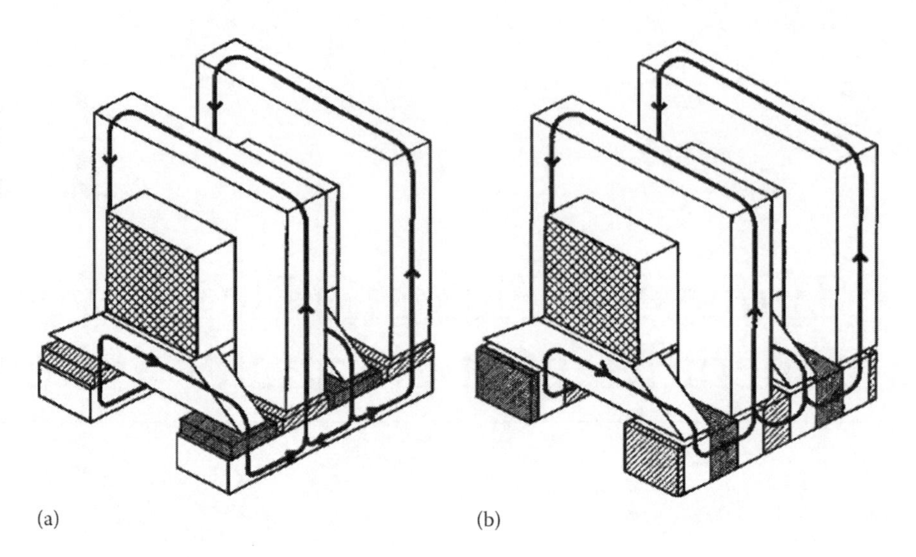

FIGURE 11.1 The single-sided transverse flux PM machine: (a) with surface PM pole rotor and (b) with rotor PM flux concentration (interior PM poles).

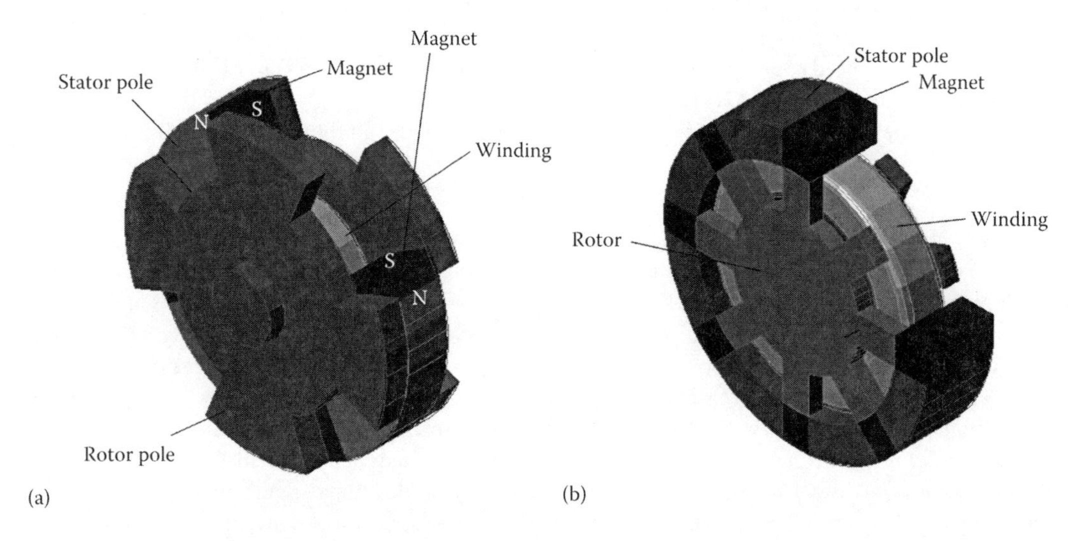

FIGURE 11.2 Transverse flux machine (TFM) with stator PMs: (a) with axial air gap and (b) with radial air gap.

angle, all South Poles add up their flux in the coil. Thus, the PM flux linkage in the coil reverses polarity $2p_1$ times per rotor revolution and produces an electromagnetic force (emf) E_s:

$$E_s = -W_1 \cdot p_1 \cdot \frac{\partial \phi_{PM}^t}{\partial \theta_r} \cdot \frac{d\theta_r}{dt}; \quad \frac{\partial \phi_{PM}^t}{\partial \theta_r} \approx B_g p_1 l_{stack} \cdot \tau \cdot \sin p_1 \theta_r \tag{11.2}$$

where

- l_{stack} is the axial length of the U-shaped core leg
- θ_r is the mechanical angle
- B_g is the PM air gap average flux density
- τ is the pole pitch
- ϕ_{PM}^t is the total flux per one turn coil
- p_1 pole pairs

From Equation 11.2,

$$E_s \approx -W_1 \cdot p_1^2 \cdot B_g \tau \cdot l_{stack} \cdot 2 \cdot \pi \cdot n = -W_1 \cdot B_g \cdot \tau \cdot p_1 \cdot l_{stack} \cdot \omega_1 \cdot \sin p\theta_r \tag{11.3}$$

Equation 11.2 serves to prove that for the same coil and machine diameter ($2_{p_{1\tau}}$ = constant), the number of turns, and stator core stack length, if the number p_1 of U cores is increased, the emf is increased, for a given speed. This effect may be called torque magnification [2], as torque T_e per phase (coil) is as follows:

$$T_{e_{phase}} = \frac{E_s(\theta_r) \cdot I_1(\theta_r)}{2 \cdot \pi \cdot n}; \quad \omega_1 = 2 \cdot \pi \cdot n \cdot p_1 \tag{11.4}$$

The structure of the magnetic circuit of the TFM is complex, as the PM flux paths are three-dimensional either in the stator or in the rotor or in both.

Soft composite materials may be used for the scope, as their core losses are smaller than those in silicon laminations for frequencies above 600 Hz, but their relative magnetic permeability is around 500 μ_0 at 1.0 T. This reduces the magnetic anisotropy effect, which is so important in TFMs. This is why, thus far, the external rotor TFM with the U- and I-shaped stator cores located in an aluminum hub is considered the most manufacturable (Figure 11.3a and b) [3]. Still, the additional eddy current losses in the aluminum hub carrier are notable.

Though double-sided TFMs with rotor PM flux concentration were proposed (Figure 11.4a and b) to increase the torque per PM volume ratio, they prove to be difficult to manufacture.

While increasing the torque/volume and decreasing the losses per torque are key design factors, the power factor of the machine is essential, as it defines the kilovolt ampere of the converter associated with the TFM for motoring and generating.

The ideal power factor angle φ_1 of the TFM (or of any PM synchronous machine [SM]) may be defined, in general, as follows:

$$\varphi_{1n} = \tan^{-1} \frac{\omega_1 \cdot L_s \cdot I_{sn}}{E_s(\omega_1)} \tag{11.5}$$

where

L_s is the synchronous inductance of the machine

I_s is the phase current

Equation 11.5 is valid for a nonsalient pole machine behavior (surface PM pole machines).

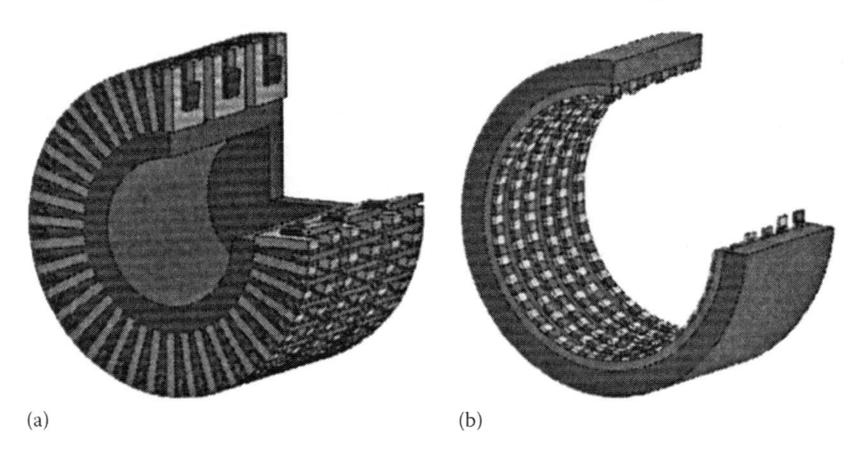

(a) (b)

FIGURE 11.3 The three-phase external PM-rotor transverse flux machine (TFM): (a) internal stator and (b) external rotor.

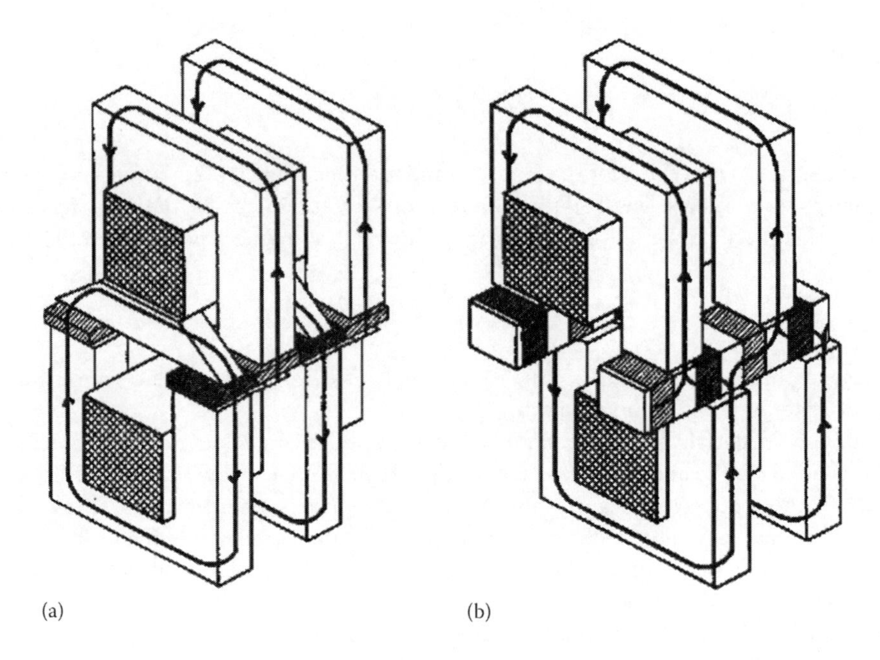

FIGURE 11.4 Double-sided transverse flux machine (TFM) (a) without and (b) with rotor PM flux concentration.

For the interior PM pole TFM with flux concentration and a diode rectifier, the machine is forced to operate at about unity power factor. In such conditions, what is the importance of φ_{1n} as defined in Equation 11.5? It is to show the power factor angle of the machine for peak torque (pure I_q current control for the nonsalient pole machine). For generality, the presence of a front-end pulse-width modulated (PWM) converter is necessary to extract the maximum power and experience the φ_{1n} power factor angle conditions. The power factor at rated current $\cos\varphi_{1n}$ is a crucial performance index.

In any case, the larger the machine inductance voltage drop per emf, the lower the power factor goodness of the machine. From this viewpoint, the PM flux concentration, though it leads to larger torque/volume, also means a higher inductance, inevitably. The claw-pole stator cores were proposed to improve the torque/volume, but the result is still modest [3], due to low power factor. Consequently, only at the same torque density as in the surface PM pole machine, the TFM with PM flux concentration can eventually produce the same power factor at better efficiency and with a better PM usage. While this rationale in power factor is valid for all PM generators, the problem of manufacturability remains heavy with TFMs.

In order to produce a more manufacturable machine, the three-phase flux reversal PM machine (FRM) was introduced [4]. FRM stems from the single-phase flux-switch generator [5] and is basically a doubly salient stator PM machine. The three-phase FRM uses a standard silicon laminated core with $6k$ large semiclosed slots that hold $6k$ nonoverlapping coils for the three phases.

Within each stator coil large pole, there are $2n_p$ PM poles of alternate polarity. Each such PM pole spans τ_{PM}. The large slot opening in the stator spans 2/3 τ_{PM}. The rotor has a passive salient-pole laminated core with N_r poles. To produce a symmetric, basically synchronous, machine,

$$2\cdot N_r\cdot\tau_{PM}=\left(2\cdot n_p\cdot 6\cdot k+6\cdot k\cdot\frac{2}{3}\right)\cdot\tau_{PM}\tag{11.6}$$

where

$k = 1, 2, \ldots$

$n_p = 1, 2, \ldots$

A typical three-phase FRM is shown in Figure 11.5a and b.

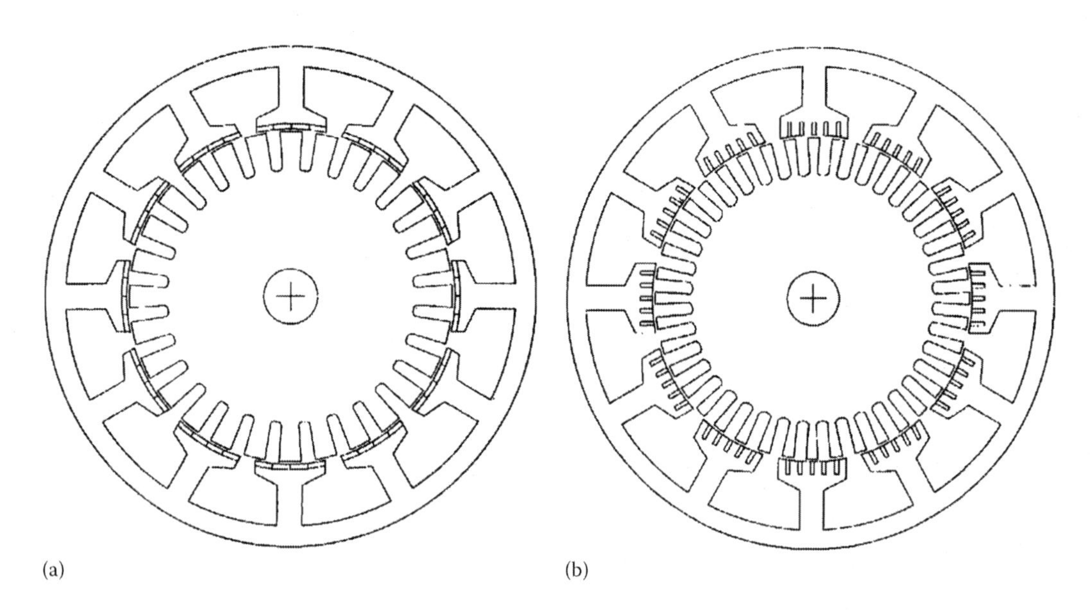

FIGURE 11.5 Three-phase flux reversal machine (FRM) with (a) stator surface PMs ($N_s = 12$; $n_p = 2$; $N_r = 28$) and (b) inset PMs ($N_s = 12$; $n_p = 3$; $N_r = 40$).

As for the TFM, the PM flux linkage in the phase coils changes polarity (reverses sign) when the rotor moves along a PM pole span angle. The structure is fully manufacturable, as it "borrows" the magnetic circuit of a switched reluctance machine. However, the problem is, as for the TFM, that the PM flux fringing reduces the ideal PM flux linkage in the coils to around 30%–60%, in general. The smaller the pole pitch τ_{PM}, the larger the fringing and, thus, the smaller the output. There seems to be an optimum thickness of the PM h_{PM} to pole pitch τ_{PM} ratio, for minimum fringing.

On the other hand, the machine inductance L_s tends to be reasonably small, as end connections are reasonable (nonoverlapping coils) and the surface PM poles secure a notably large magnetic air gap.

Still, in Reference 4, for a 700 Nm peak torque at 7 N/cm^2 force density, the power factor would be around 0.3. In order to increase the torque density for reasonable power factor, PM flux concentration may be performed in the stator (Figure 11.6) or in the rotor (Figure 11.7). Flux switching and flux reversal concepts are in fact similar (dual polarity PM flux in the AC coils).

The FRM with stator PM flux concentration is highly manufacturable, but as the pole pitch τ_{PM} gets smaller, because the coil slot width w_s is less than τ_{PM}, the power factor tends to be smaller. For τ_{PM} = 10 mm and $w_s = \tau_{PM}$ and a $4w_s$ height, with j_{peak} = 10 A/mm^2 and slot filling factor k_{fill} = 0.4, the slot magnetomotive force (mmf) $W_c I_{peak}$ is as follows:

$$W_c \cdot I_{peak} = w_s(\text{mm}) \cdot 4 \cdot w_s(\text{mm}) \cdot k_{fill} \cdot j_{peak} = 10 \cdot 4 \cdot 10 \cdot 0.4 \cdot 10 = 1600 \text{ A turns/slot}$$

Larger slot mmfs could be provided for the TFM and FRM with stator surface PM poles. However, the configuration in Figure 11.7 allows for the highest PM flux concentration, which may compensate for the lower $W_c I_{peak}$ and allow for lower cost PMs, because the radial PM height is generally larger than 3–4 τ_{PM}. As with any PM flux concentration scheme, the machine inductance remains large. But, for not so large a number of poles, the machine's easy manufacturability may pay off. On the other hand, the FRM with rotor PM flux concentration configuration (Figure 11.7) provides for large torque density, because, additionally, the allowable peak coil mmf may be notably larger than that for the configurations with stator PM flux concentration (Figure 11.6).

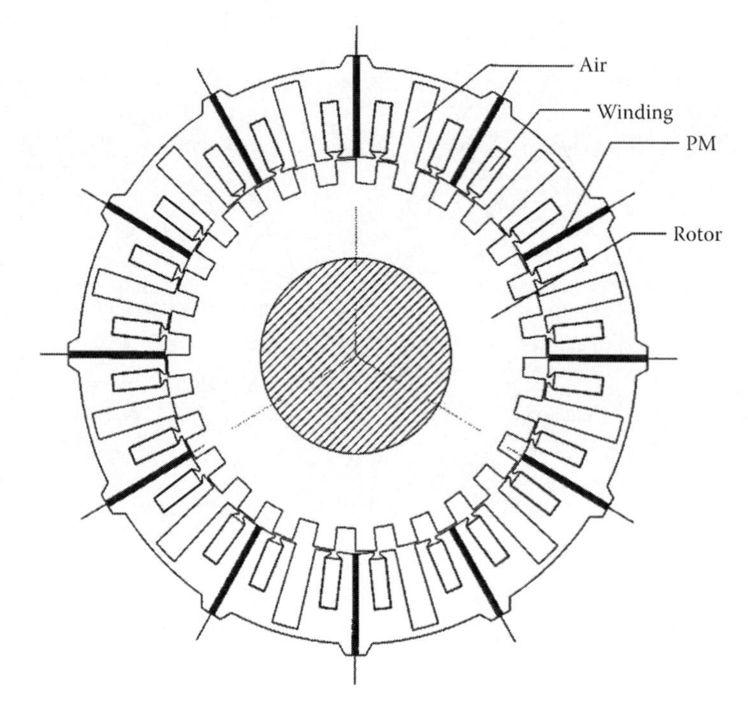

FIGURE 11.6 Three-phase flux switching (reversal) machine (FS(R)M) with stator PM flux concentration.

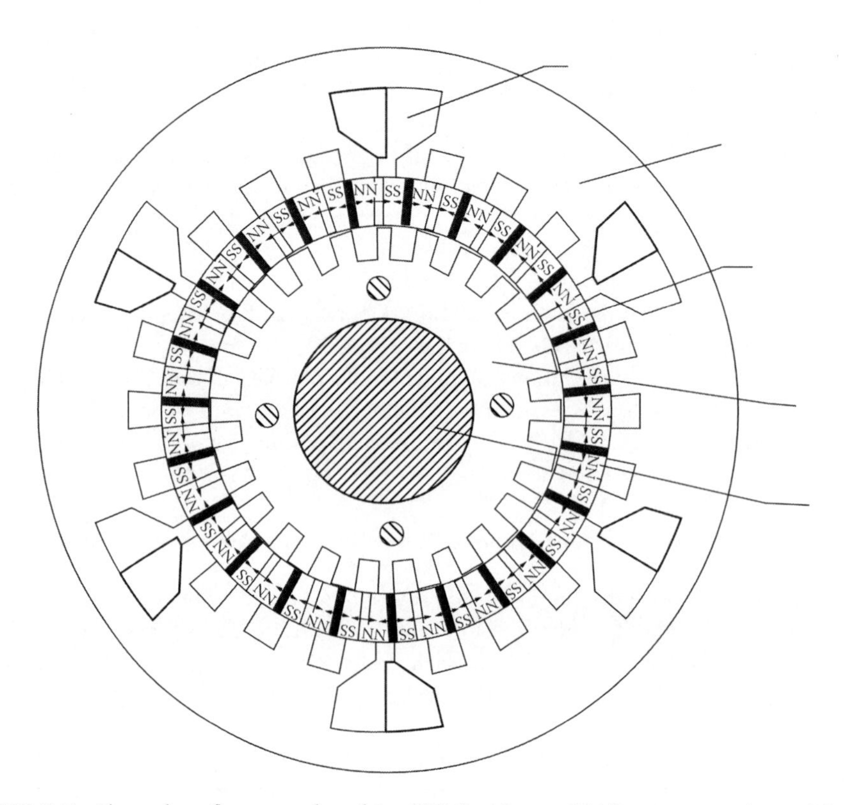

FIGURE 11.7 Three-phase flux reversal machine (FRM) with rotor PM flux concentration and dual stator.

The rotor mechanical rigidity appears, however, to be lower, and the dual stator makes manufacturability a bit more difficult. Still, conventional stamped laminations can be used for both stator and rotor cores.

To provide more generality to the analysis that follows, we will consider only the three-phase TFMs and FRMs, which may work both as generators and motors with standard PWM converters and position-triggered control.

11.2 Three-Phase Transverse Flux Machine: Magnetic Circuit Design

The configuration for one phase may be reduced to the one in Figure 11.1a, but inverted, to have an external rotor. The iron behind the PMs on the rotor may be, in principle, solid iron, which is a great advantage when building the rotor.

In addition, the stator U-shaped and I-shaped cores (Figure 11.4a) may be made of silicon laminations. The aluminum carriers that hold tight the stator I- and U-cores are the main new frame elements that have to be fabricated by precision casting.

Apparently, the circumferential coil has to be wound turn by turn on a machine tool after the U-cores have been implanted in the aluminum carrier. Then, the I-cores are placed one by one in their locations on top of the coil (Figure 11.8).

It was shown that, in order to reduce the cogging torque, the stator U- and I-cores of the three phases have to be shifted by 120 electrical degrees with respect to each other [6]. In such a case, the three phases are magnetically independent, though the PMs are axially aligned on the rotor for all three phases.

As can be seen in Figure 11.3a, the PM flux paths are basically three-dimensional in the rotor but only bidimensional in the stator. However, as the flux paths are basically radial in the PMs and circumferential-radial in the rotor back iron, the rotor back iron may be made of mild solid steel. The bidimensional flux paths in the stator allow for the use of transformer laminations in the U- and I-cores.

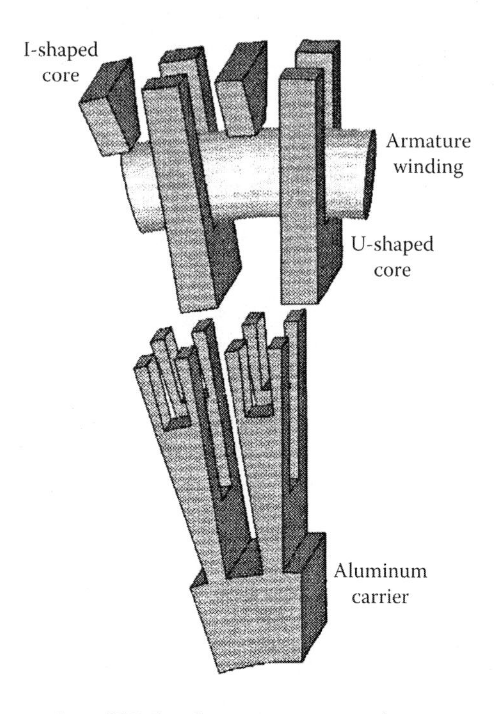

FIGURE 11.8 Transverse flux machine (TFM)—aluminum carrier with interior stator cores and coil.

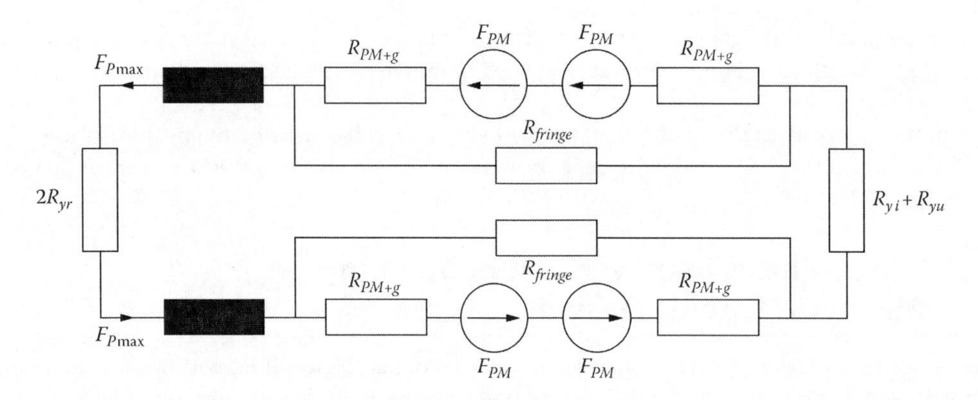

FIGURE 11.9 The magnetic equivalent circuit for PMs in the position for maximum flux in the stator coil.

There is, however, substantial PM flux fringing, which crosses the air gap and closes the path between the stator U- and I-cores in the circumferential direction. To reduce it, both U- and I-cores should expand circumferentially less than a PM pole pitch τ_{PM}: b_u, $b_i < \tau_{PM}$. In addition, to reduce axial PM flux fringing, the axial distance between the Ucore legs l_{slot} should be equal to or larger than the magnetic air gap ($l_{slot} > g + h_{PM}$). But, l_{slot} is, in fact, equal to the width of the open slot where the stator coil is placed.

The U-core yoke h_{yu} and the I-core h_{yi} heights should be about equal to each other and around the value of 2/3 l_{stack} in order to secure uniform and mild magnetic saturation in the stator iron cores. In addition, the rotor yoke radial thickness h_{yr} should be around half the PM pole pitch, as half of the PM flux goes through the rotor magnetic yoke (Figure 11.9).

We will now approach performance through an analytical method, accounting approximately for magnetic saturation in the stator and iron cores.

Each PM equivalent mmf $F_{PM} = H_c h_{PM}$ (H_c is the coercive field of the PM material) "is responsible" for one magnetic air gap $h_{gM} = g + h_{PM}$. This way, the equivalent magnetic circuit on no load (zero current) is as in Figure 11.9:

$$R_{PM+g} = \frac{h_{PM} + g}{\mu_0 \cdot b_{PM} \cdot l_{stack} \cdot (b_{PM} + b_u)/2} \tag{11.7}$$

where

R_{PM+g} is the PM and air gap magnetic reluctance
b_{PM} is the PM width ($b_{PM}/\tau_{PM} = 0.66$–1.0)
b_u is the U-core width
R_{fringe} is the magnetic reluctance of the PM fringing flux between the stator U- and I-cores through the air gap, corresponding to one leg of the U- and I-cores

To a first approximation,

$$R_{fringe} \approx \left(\frac{h_{PM}}{\mu_0 \cdot (b_{PM}/2)} + \frac{\tau_{PM} - b_{yu}}{\mu_0 \cdot (h_{yi}/2)} \right) \cdot \frac{1}{l_{stack}}; \quad b_{yu} = b_{yi} \tag{11.8}$$

Straight-line magnetic flux paths are considered between U- and I-cores up to the height of the I-core (Figure 11.8). The axial fringing may be added as a reluctance in parallel to R_{fringe}. The stator U- and I-core reluctances R_{yu} and R_{yi} are as follows:

$$R_{yu} = \frac{2 \cdot (h_{slot} + h_{yi})}{\mu_{cu} \cdot b_u \cdot l_{stack}} + \frac{l_{stack} + l_{slot}}{\mu_{yu} \cdot h_{yu} \cdot b_u} \tag{11.9}$$

$$R_{yi} \approx \frac{l_{slot}}{\mu_{yi} \cdot h_{yi} \cdot b_i} + \frac{4 \cdot l_{stack}}{\mu_{yi} \cdot h_{yi} \cdot b_i}; \quad b_i \approx b_u \tag{11.10}$$

Additionally,

$$R_{yr} = \frac{\tau_{PM}}{\mu_{yr} \cdot b_{yr} \cdot l_{stack}} \tag{11.11}$$

where μ_{cu}, μ_{yu}, μ_{yi}, and μ_{yr} are the magnetic permeabilities dependent on magnetic saturation. As the PM equivalent mmf F_c is given, once all the PM geometry and material properties are known, an iterative procedure is required to solve the magnetic circuit in Figure 11.9. To start with, initial values are given to the four permeabilities in the iron parts: $\mu_{cu}(0)$, $\mu_{yu}(0)$, $\mu_{yi}(0)$, and $\mu_{yr}(0)$. With these values, the flux in the stator and rotor core parts and ϕ_{PMax} are computed.

But, the average flux densities in various core parts are straightforward, once ϕ_{PMax} is known:

$$\begin{aligned} \phi_{PMax} &= b_{yu} \cdot l_{stack} \cdot B_{cu} = b_{yu} \cdot h_{yu} \cdot B_{yu} = h_{yi} \cdot b_i \cdot B_{yi} \\ \frac{\phi_{PMax}}{2} &= B_{yr} \cdot h_{yr} \cdot l_{stack} \end{aligned} \tag{11.12}$$

Once the average values of flux densities B_{cu}, B_{yu}, B_{yi}, and B_{yr} are computed, and from the magnetization curve of the core materials, new values of permeabilities $\mu_{cu}(1)$, $\mu_{yu}(1)$, $\mu_{yi}(1)$, and $\mu_{yr}(1)$ are calculated. The computation process is reinitiated with renewed permeabilities:

$$\mu(1) = \mu_i(0) + c(\mu_i(1) - \mu_i(0)); \quad c = 0.2 - 0.3 \tag{11.13}$$

such as to speed up convergence.

The computation is ended when the largest relative permeability error between two successive computation cycles is smaller than a given value (say, 0.01). This way, the maximum value of the PM flux per pole (U-core), ϕ_{PMax}, in the coil, is obtained.

The PM flux per pole varies from $+\phi_{PMax}$ to $-\phi_{PMax}$ when the rotor moves along a PM pole angle ($2\pi/2_{P1}$ mechanical radians). What is difficult to find out analytically, even with a complicated magnetic circuit model with rotor position permeances, is the variation of Φ_{PM} with rotor position from $+\phi_{PMax}$ to $-\phi_{PMax}$ and back.

It was shown by three-dimensional (3D) finite element method (FEM) that the PM flux per pole varies trapezoidally with rotor position (Figure 11.10).

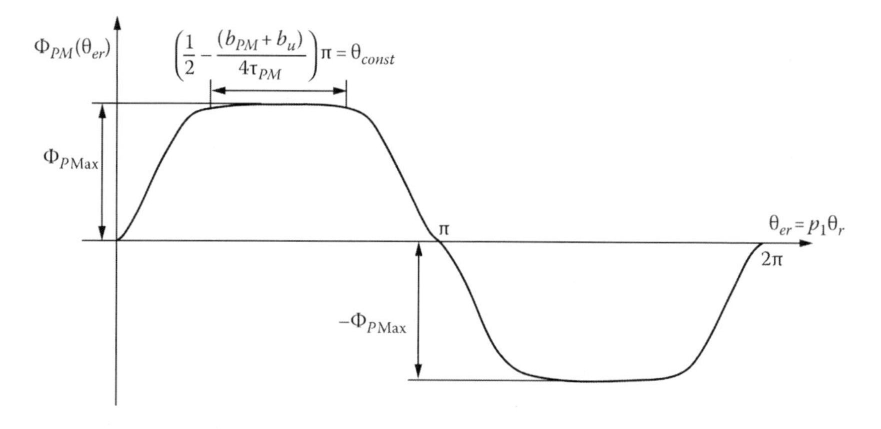

FIGURE 11.10 PM flux per stator pole vs. rotor position.

We may consider, approximately, that the electrical angle θ_{econst}, along which the PM flux stays constant (Figure 11.10) is rather small and is as follows [3]:

$$\theta_{econst} \approx \left(\frac{1}{2} - \frac{(b_{PM} + b_{u})}{4 \cdot \tau_{PM}} \right) \cdot \pi \,(\text{rad}) \tag{11.14}$$

Magnetic saturation alters not only ϕ_{PMax} but also the value of θ_{econst}, tending to flatten the trapezoidal waveform and bringing it closer to a rectangular waveform of lower height. But, local magnetic saturation plays an even more important role when the machine is under load.

In general, the computation of the force density (in N/cm²) on the rotor surface is done via the Maxwell stress tensor through FEM. The magnetic saturation presence is evident when the force is calculated for various rotor positions for constant phase current (Figure 11.11).

The PM flux per pole and the emf, obtained through 3D FEM, looks as shown in Figure 11.12.

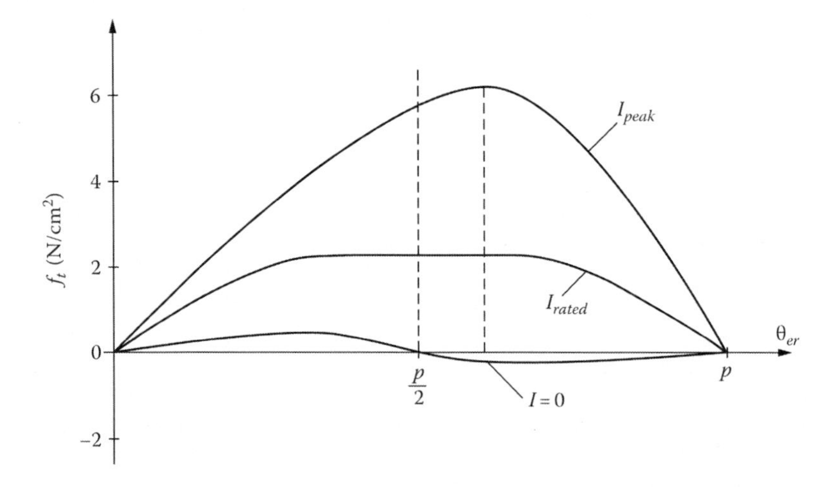

FIGURE 11.11 Typical force density (N/cm²) vs. rotor position and various constant current values.

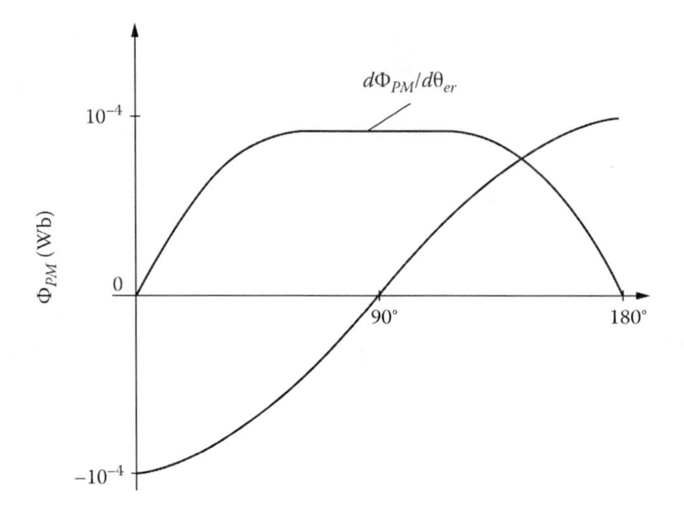

FIGURE 11.12 Typical finite element method (FEM)-extracted PM flux and its derivative vs. rotor position.

The instantaneous emf per phase E is as follows (Equations 11.3 and 11.4):

$$E = W_1 \cdot p \cdot \frac{d\phi_{PM}(\theta_{er})}{d\theta_{er}} \cdot 2 \cdot \pi \cdot n \cdot p \tag{11.15}$$

$$\theta_{er} = \omega_r \cdot t \tag{11.16}$$

Therefore, the emf per phase has a waveform in time, which emulates the waveform of $d\Phi_{PM}(\theta_{er})/d\theta_{er}$. This derivative maximum decreases when the PM pole pitch decreases due to the increase in fringing flux, when the number of poles increases. Consequently, there should be an optimum number of PM poles for a given rotor diameter that produces maximum emf for given W_1, stack length, and mechanical gap g.

If the phase emf is considered through its fundamental, the average torque per phase, when the emf and current are phase shifted by angle γ_1, is as follows:

$$(T_e)_{aph} = \frac{E_s \cdot I_1 \cdot \cos\gamma_1}{2 \cdot \pi \cdot n} = \frac{W_1 \cdot p^2 \cdot \phi_{PMax} \cdot I_1 \cdot \cos\gamma_1}{\sqrt{2}} \tag{11.17}$$

where I_1 is the root mean squared (RMS) phase current, and

$$\phi_{PM}(\theta_r) = \phi_{PMax} \cdot \cos(p \cdot \theta_r) \tag{11.18}$$

11.2.1 Phase Inductance L_s

The phase inductance L_s is composed from the leakage inductance L_{sl} and the main path inductance L_m. As the magnetic circuits of the three phases are separated, there is no coupling inductance between phases. L_s refers to one phase only. The air gap inductance L_m is, approximately,

$$L_m = \mu_0 \cdot \frac{W_1^2}{R_{mag}} \approx \mu_0 \cdot \frac{W_1^2 \cdot p \cdot (b_u + b_{PM}/2) \cdot l_{stack} \cdot (1 + k_f)}{4 \cdot (g + h_{PM}) \cdot (1 + k_s)} \tag{11.19}$$

where k_s is an equivalent magnetic saturation coefficient that considers the contribution of the iron parts to the total mmf along a flux path. k_f is a fringing coefficient that takes care of the fringing flux in the air gap: $k_f < 0.2$ in general.

In a similar manner, we may treat the maximum flux per pole:

$$\phi_{PMax} = B_{gi} \cdot \frac{b_{PM} + b_u}{2} \cdot l_{stack} \cdot \frac{1}{(1 + k_s)(1 + k_{fringe})} \tag{11.20}$$

k_{fringe} is the PM fringing flux coefficient (very different from k_f) that can go as high as two ($k_{fringe} = 0.7$–2), while k_s is the magnetic saturation coefficient. In general, k_s has to be calculated when current is present in stator phase. The leakage inductance may be calculated approximately from slot leakage flux, extended between U-cores, because I-cores "create" a moderate "slot leakage effect":

$$L_{sl} \approx \mu_0 \cdot W_1^2 \cdot \frac{h_s}{3 \cdot l_{slot}} \cdot p \cdot b_u \cdot (1 + k_i) \tag{11.21}$$

k_i accounts for the leakage inductance between the U-cores. In general, $k_i < 0.2$–0.3.

As the total iron area, seen along the air gap, by the stator coil, is independent of the number of poles, both L_m and L_{sl} are independent of the number of poles 2_{p1}. At the same time, the emf E and the torque $(T_e)_{aphase}$ per phase are proportional to the number of poles if the increasing fringing (k_{fringe}) with the number of poles is not considered.

11.2.2 Phase Resistance and Slot Area

The phase resistance R_s is straightforward:

$$R_s \approx \frac{\pi \cdot (D_{is} + 2 \cdot h_{yi} + h_s) \cdot W_1^2}{\sigma_{copper} \cdot (W_1 \cdot I_1 / j_{con})} \tag{11.22}$$

The ampere-turns per phase (RMS value), $W_1 I_1$, may be calculated from Equation 11.17 once the torque per phase requirement is given.

The rated (continuous) current density j_{con} depends on the machine duty cycle, type of cooling, and design optimization criterion (maximum efficiency or minimum machine cost, etc.). In general, j_{con} = 4–12 A/mm². The window area in the slot A_{slot} is as follows:

$$A_{slot} = h_s \cdot l_{slot} = \frac{W_1 \cdot I_1}{j_{con} \cdot k_{fill}} \tag{11.23}$$

The total slot filling factor k_{fill} is, in general, $k_{fill} = 0.4–0.6$. The larger values correspond to the preformed coils eventually made of rectangular cross-sectional conductors.

Example 11.1

Consider sizing a three-phase TFM generator with surface PM interior rotor and single-sided stator (with U- and I-cores) that has the following specifications: T_{en} = 200 kNm and n_n = 30 rpm.

Solution

While part of the design formulas are included in the previous paragraphs, some new ones are introduced here. They are mainly related to the stator bore diameter D_{is} with given l_{stack}/D_{is} ratio ($\lambda = l_{stack}/D_{is} = 0.05–0.1$).

We use the tangential force density f_t = 2–8 N/cm² to determine the interior stator diameter D_{is}:

$$D_{is} = \sqrt[3]{\frac{T_e}{3 \cdot \pi \cdot \lambda \cdot f_t}}; \quad T_e = \frac{D_{is}}{2} \cdot f_t \cdot \pi \cdot D_{is} \cdot 3 \cdot 2 \cdot l_{stack} \tag{11.1.1}$$

In our case, with $\lambda = 0.05$, f_t = 2.66 N/cm²:

$$D_{is} = \sqrt[3]{\frac{200{,}000}{3 \cdot \pi \cdot 0.05 \cdot 2.66 \cdot 10^4}} = 2.525 \text{ m} \tag{11.1.2}$$

The stack length is as follows:

$$l_{stack} = \lambda D_{is} = 0.05 \cdot 2.515 = 0.1257 \text{ m} \tag{11.1.3}$$

We now have to choose the air gap g, first.

For such a large diameter, an air gap of 1.5–5 mm is required for mechanical rigidity reasons. Let us consider $g = 4 \cdot 10^{-3}$ m. The ideal air gap flux density B_{gi} produced by the PM magnets in the air gap is as follows:

$$B_{gi} = B_r \cdot \frac{h_{PM}}{h_{PM} + g} \tag{11.1.4}$$

with B_r = 1.3 T, $\mu_{rem} = 1.04\ \mu_{rem}$ at 100°C (very good NeFeB magnets: B_{r0} = 1.37 T at 20°C).

We need to choose a large PM height h_{PM}, because the actual air gap flux density will be notably reduced by the fringing ($k_{fringe} \approx 2$). So, $h_{PM} = 3 \cdot g = 3 \cdot 4 \cdot 10^{-3} = 12 \cdot 10^{-3} = 1.2 \cdot 10^{-2}$ m. Consequently, from Equation 11.1.4,

$$B_{gi} = \frac{1.3 \cdot 12}{12 + 4} = 0.975\text{T} \tag{11.1.5}$$

Now the maximum flux per pole may be calculated if we adopt the number of poles 2_{p_1}.

In general, the PM pole pitch τ_{PM} should be higher than the total (magnetic) air gap: $g + h_{PM} = 1.6 \cdot 10^{-2}$ m. The number of poles 2_{p_1} is, thus,

$$2 \cdot p = \frac{\pi \cdot D_{is}}{\tau_{PM}} = \frac{\pi \cdot 2.515}{0.02} = 394.8 \tag{11.1.6}$$

Let us consider $2p = 400$ poles and $\tau_{PM} = 1.974 \cdot 10^{-2}$ m. The fundamental frequency at 30 rpm is $f_n = p_1 \cdot n_n = 200 \cdot 30/60 = 100$ Hz, which is a reasonable value in terms of core losses. Now, the maximum PM flux per pole from Equation 11.20, with $k_{fringe} = 2.33$; $k_s = 0.1$; $b_{PM}/\tau_{PM} = 0.85$; $b_u/\tau_{PM} = 0.8$ is as follows:

$$\begin{aligned} \phi_{\text{PMax}} &= B_{gi} \cdot \frac{b_{PM} + b_u}{2} \cdot l_{stack} \cdot \frac{1}{(1 + k_{fringe}) \cdot (1 + k_s)} \\ &= 0.975 \cdot 1.628 \cdot 10^{-2} \cdot \frac{0.2515}{2} \cdot \frac{1}{(1 + 2.33) \cdot (1 + 0.1)} = 0.5986 \cdot 10^{-3}\text{ Wb} \end{aligned} \tag{11.1.7}$$

We now turn directly to the torque expression to find the ampere turns per phase $W_1 I_1$ (RMS value) from Equation 11.17, with $\gamma_1 = 0$ (emf and current in phase):

$$W_1 \cdot I_1 = \frac{T_{en} \cdot \sqrt{2}}{3 \cdot p_1^2 \cdot \phi_{\text{PMax}}} = \frac{200 \cdot 10^3 \cdot \sqrt{2}}{3 \cdot 200^2 \cdot 0.5986 \cdot 10^{-3}} = 3885\,\text{Aturns/coil} \tag{11.1.8}$$

This is a reasonable value.

The slot area required for the coil for $j = 6$ A/mm^2 and $k_{fill} = 0.6$, from Equation 11.23, is as follows:

$$A_{slot} = \frac{W_1 \cdot I_1}{j_{con} \cdot k_{fill}} = \frac{3885}{6 \cdot 0.6} = 1079.16\,\text{mm}^2 \tag{11.1.9}$$

Taking

$$l_{slot} = 2(g + h_{PM}) = 32\text{ mm} \tag{11.1.10}$$

the slot height h_s is

$$h_s = \frac{A_{slot}}{l_{slot}} = \frac{1079}{32} = 33.72\,\text{mm} \tag{11.1.11}$$

The fact that $h_s/l_{slot} \approx 1$ will lead to a smaller leakage inductance, which is favorable for a good power factor.

Consider $h_{yi} = h_{yu} = 2/3 \cdot l_{stack} = 0.083$ m. Then, from Equation 11.22, the stator phase resistance R_s is

$$R_s = \frac{2.3 \cdot 10^{-8} \cdot \pi \cdot (2.515 + 2 \cdot 0.083 + 0.03372)}{(3885/6 \cdot 10^6)} \cdot W_1^2 = 0.3026 \cdot 10^{-3} \cdot W_1^2 \quad (11.1.12)$$

The total winding losses for the machine, P_{con}, are as follows:

$$P_{con} = 3 \cdot R_s \cdot I_1^2 = 3 \cdot 0.3206 \cdot 10^{-3} \cdot 3885^2 = 13.70\,\text{kW} \quad (11.1.13)$$

The electromagnetic power, P_{elm}, is

$$P_{elm} = T_e \cdot 2 \cdot \pi \cdot n_n = 200 \cdot 10^3 \cdot 2 \cdot \pi \cdot \frac{30}{60} = 628\,\text{kW} \quad (11.1.14)$$

The winding losses are about 2% of input power.

Now we need to calculate the synchronous inductance $L_s = L_m + L_{sl}$. From Equation 11.20,

$$L_m = \frac{\mu_0 \cdot W_1^2 \cdot p \cdot (b_u + b_{PM}/2) \cdot l_{stack} \cdot (1+k_f)}{4 \cdot (g + h_{PM}) \cdot (1+k_s)}$$

$$= \frac{1.256 \cdot 10^{-6} \cdot 200 \cdot 1.62 \cdot 10^{-2} \cdot (0.2515/2) \cdot (1+0.1)}{4 \cdot 1.6 \cdot 10^{-2} \cdot (1+0.1)} \cdot W_1^2 = 0.8 \cdot 10^{-5} \cdot W_1^2 [H] \quad (11.1.15)$$

In addition, from Equation 11.21, L_{se}, the leakage inductance is as follows:

$$L_{sl} = \mu_0 \cdot W_1^2 \cdot \frac{h_s}{3 \cdot l_{slot}} \cdot (1+k_i) \cdot p_1 \cdot b_u$$

$$= 1.256 \cdot 10^{-6} \cdot \frac{0.03372}{3 \cdot 0.032} \cdot (1+0.3) \cdot 200 \cdot 0.016 \cdot W_1^2 = 1.834 \cdot 10^{-6} \cdot W_1^2 \quad (11.1.16)$$

Therefore,

$$L_{sl} = L_m + L_{se} = (0.8 \cdot 10^{-5} + 1.834 \cdot 10^{-6}) \cdot W_1^2 = 0.9834 \cdot 10^{-5} \cdot W_1^2 \quad (11.1.17)$$

The emf E (RMS value) is as follows:

$$E = \frac{T_{en} \cdot 2 \cdot \pi \cdot n}{3 \cdot I_1 \cdot \cos\gamma} = \frac{628 \cdot 10^3}{3 \cdot I_1} = \frac{209.3 \cdot 10^3}{I_1} \quad (11.1.18)$$

The ideal power factor angle φ_1 (for $\gamma = 0$) is

$$\varphi_{1n} = \tan^{-1} \frac{\omega_1 \cdot L_s \cdot I_1}{E} = \tan^{-1} \left(\frac{2 \cdot \pi \cdot 100 \cdot 0.9834 \cdot 10^{-5}}{208.33 \cdot 10^3} \cdot W_1^2 \cdot I_1^2 \right) = 24.1° \quad (11.1.19)$$

$$\cos\varphi_{1n} = 0.912$$

The very good value of the power factor angle for E and I in phase is a clear indication of machine volume reduction reserve (the specific tangential force is only 2.66 N/cm²). As at 30 rpm the machine is more expensive than the converter, a smaller volume (cost) and efficiency with a power factor down to 0.7 may be worth considering.

The data of the preliminary design are given in Table 11.1.

TABLE 11.1 Example Design Summary

Electromagnetic torque	$T_e = 200$ kNm
Speed	$n_n = 30$ rpm
Force density	$f_t = 2.66$ N/cm^2
Current density	$j_{con} = 9$ A/mm^2
Stator interior diameter	$D_{is} = 2.515$ m
Stator stack U-core leg	$l_{stack} = 0.1257$ m
Mechanical gap	$g = 4 \cdot 10^{-3}$ m
Permanent magnet radial thickness	$h_{PM} = 12 \cdot 10^{-3}$ m
Number of permanent magnet poles	$2 \cdot p = 400$
Fundamental frequency	$f_n = 100$ Hz
Permanent magnet pole pitch	$\tau_{PM} = 1.974 \cdot 10^{-3}$ m
Axial room between U-core legs	$l_{slot} = 32 \cdot 10^{-3}$ m
U-core width	$b_u/\tau_{PM} = 0.8$
Permanent magnet width per pole	$b_{PM}/\tau_{PM} = 0.85$
U-core yoke height	$h_{yu} = 2/3 \cdot l_{stack}$
I-core yoke height	$h_{yi} = h_{yu}$
Slot useful height	$h_s = 33.72 \cdot 10^{-3}$ m
Total U-core interior height	$h_s + h_{yi}$
External stator diameter	$D_{os} = 2.621$ m
Total axial length per phase	$l_{slot} + 2 \cdot l_{stack} = 0.032 + 0.2515$ m $= 0.2835$ m
Total axial length for three phases	$l_{axial} = 3 \cdot (l_{slot} + 2 \cdot l_{stack}) = 0.85$ m
Reduction of permanent magnet flux due to fringing	$1/(1 + k_{fringe}) = 1/3$
Stator resistance per phase	$R_s = 0.454 \cdot 10^{-3} \cdot W_1^2$
Stator inductance per phase	$L_s = 0.9834 \cdot 10^{-5} \cdot W_1^2$
emf at $f_1 = 100$ Hz	$E = 53.61 \cdot W_1$ (rms)
Stator copper losses	$P_{con} = 20.55$ kW
Electromagnetic power	$T_e \cdot 2 \cdot \pi \cdot n_n = 628$ kW
Power factor angle (with E and I in phase)	$\varphi_{1n} \approx 24°$; $\cos \varphi_n = 0.912$

11.3 TFM: The *dq* Model and Steady State

Let us consider here that TFM is provided with a surface PM rotor, and thus the synchronous inductance L_s is independent of rotor position. We adopt this configuration, as the power factor may be higher due to smaller inductance L_s, even though at slightly smaller torque density than for the PM flux concentration configurations. Though the emf waveform is not quite sinusoidal, we consider it here as sinusoidal.

The *dq* model is thus straightforward, if the core losses are neglected for now:

$$\bar{I}_s \cdot R_s + \bar{V}_s = -\frac{d\bar{\Psi}_s}{dt} - j \cdot \omega_r \cdot \bar{\Psi}_s$$

$$\bar{\Psi}_s = \psi_d + j \cdot \psi_q; \quad \psi_d = \psi_{PM} + L_s \cdot I_d; \quad \psi_q = L_s \cdot I_q \tag{11.24}$$

$$\bar{I}_s = I_d + j \cdot I_q; \quad \bar{V}_s = V_d + j \cdot V_q; \quad \psi_{PM} = \phi_{PMax} \cdot W_1 \cdot p_1$$

where

W_1 is the turns per coil (phase)

p_1 is the pole pairs:

$$T_e = \frac{3}{2} \cdot p_1 \cdot (\psi_d \cdot I_q - \psi_q \cdot I_d) = \frac{3}{2} \cdot p_1 \cdot \psi_{PM} \cdot I_q$$

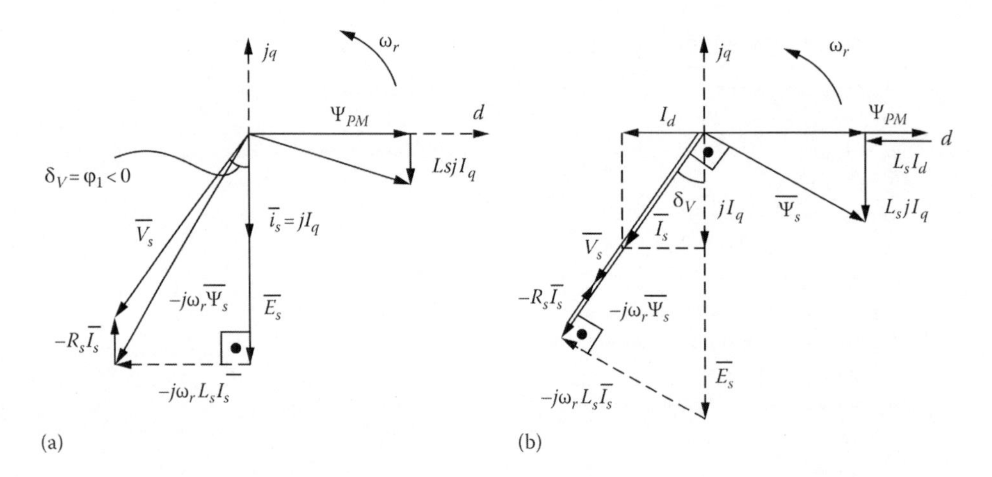

FIGURE 11.13 The transverse flux machine (TFM) vector diagram for steady state: (a) pure I_q control and (b) with power factor control (with diode rectifier).

For steady state $d/dt = 0$, and the vector diagram for generating is as shown in Figure 11.13a and b for pure I_q control ($I_d = 0$) and for unity power factor (with diode rectifier). Pure I_q control for L_s = constant corresponds to maximum torque per current, but the machine requires reactive power for magnetization. A controlled front rectifier is required.

For the diode rectifier, the current always contains an I_d component (Figure 11.13b), and thus the torque is produced with more losses. However, the reactive power required is zero, and the diode rectifier is less expensive. If constant direct current (DC) link voltage operation at variable speed is required, either an active front or PWM converter is provided on the machine side or a diode rectifier plus a DC–DC boost converter is required. The core losses might be considered, mainly for steady state, as follows:

$$p_{iron} = \frac{3}{2} \cdot \frac{\omega_r \cdot \psi_s^2}{R_{iron}} \tag{11.25}$$

In Equation 11.25, the core losses are considered to be produced by the resultant flux in the machine, in the presence of stator current. The core losses p_{iron} might be determined from tests that segregate the iron losses, or they may be calculated from analytical or FEM models, as shown in Chapter 10.

The efficiency may be defined as follows:

$$\eta = \frac{P_{2el}}{P_{2el} + p_{mec} + p_{iron} + p_{copper} + p_{rotor} + p_{Al}} \tag{11.26}$$

The rotor losses p_{rotor} refer to the eddy current losses in the PMs, which may be neglected only for $f_1 \leq$ 100 Hz, in general. The stray load losses are included in p_{iron} and p_{copper}, but the eddy current losses in the aluminum carriers of the U- and I-cores are individualized as p_{Al} and may be calculated only through 3D FEM eddy current models [7].

When the fundamental frequency goes up, above 100 Hz, p_{iron} and p_{rotor} become important and have to be dealt with great care (see Chapter 10 for more details).

Example 11.2

Consider the TFM design in Example 11.1 and calculate the number of turns per phase for a line voltage $V_{Ln} = 584$ V (rms)—star connection—at rated speed for E, I in phase, and, for the same number of turns W_1 and unity power factor, calculate the performance for the same current as before.

Solution

We simply use the vector diagram (Figure 11.13a), where ψ_{PM} is

$$\psi_{PM} = \frac{E(\text{rms})\cdot\sqrt{2}}{\omega_r} = \frac{53.61\cdot W_1\cdot\sqrt{2}}{2\cdot\pi\cdot 100} = 0.1203\cdot W_1$$
$$V_{sn} = \frac{V_{Ln}(\text{rms})\cdot\sqrt{2}}{\sqrt{3}} = \frac{584\cdot\sqrt{2}}{\sqrt{3}} = 476\,\text{V} \tag{11.2.1}$$

From Figure 11.13a,

$$E_s = E(\text{rms})\cdot\sqrt{2} = V_s\cdot\cos\varphi_1 + R_s\cdot I_s \tag{11.2.2}$$

or

$$75.59\cdot W_1 = 476\cdot 0.933 + 0.454\cdot 10^{-3}\cdot W_1^2\cdot I_1\cdot\sqrt{2} \tag{11.2.3}$$

With $W_1 I_1 = 3885$ A turns/coil, we obtain $W_1 \approx 6$ turns/coil. As the core and mechanical losses have not been considered, the efficiency will reflect only the presence of the already calculated copper losses:

$$\eta = \frac{P_{1elm} - p_{con}}{P_{1elm}} = \frac{P_2}{P_{1elm}} = \frac{628 - 13.70}{628} = \frac{614.3\,\text{kW}}{628\,\text{kW}} = 0.978 \tag{11.2.4}$$

Alternatively,

$$\eta_{en} = \frac{\sqrt{3}\cdot V_{Ln}\cdot I_1\cdot\cos\varphi_1}{P_{1elm}} \tag{11.2.5}$$

with

$$I_1 = \frac{W_1\cdot I_1}{W_1} = \frac{3885}{6} = 647.5\,\text{A} \tag{11.2.6}$$

$$\eta_{en} = \frac{\sqrt{3}\cdot 584\cdot 647\cdot 0.933}{628\cdot 10^3} = \frac{609}{628} = 0.971 \tag{11.2.7}$$

The difference in efficiency from these two formulas is only due to calculation errors, as the model is the same.

Now, for the second case, we have to note that if we consider the same phase current (RMS) I_1 and voltage V_{Ln}, the vector diagram in Figure 11.13b provides the following:

$$\left(V_s + R_s\cdot I_1\right)^2 + \omega_1^2\cdot L_s^2\cdot I_1^2 = E_s^2 \tag{11.2.8}$$

with $W_1 = 6$ turns, $I_1 = 647.5$ A, we can calculate, from Equation 11.2.8, the terminal voltage:

$$V_s' = \sqrt{E_s^2 - \omega_1^2 \cdot L_s^2 \cdot I_s^2} - R_s \cdot I_s$$
$$= \sqrt{(75.59 \cdot 6)^2 - (2 \cdot \pi \cdot 100 \cdot 1.5044 \cdot 10^{-5})^2 \cdot 6^4 \cdot 647.5^2}$$
$$-0.3026 \cdot 10^{-3} \cdot W_1^2 \cdot I_1 \cdot \sqrt{2} = 396.48 - 9.945 = 386.63 \text{ V} \tag{11.2.9}$$

$$V_L' = \frac{V_s' \cdot \sqrt{3}}{\sqrt{2}} = 386.63 \cdot \sqrt{\frac{3}{2}} = 473.40 \text{ V} < 584 \text{ V} \tag{11.2.10}$$

The line terminal voltage for the same current decreased notably for unity power factor.

Now, the output power is only

$$\left(P_2'\right)_{\cos\varphi_1 = 1} = \sqrt{3} \cdot 473.40 \cdot 647.5 = 530.29 \text{ kW} \tag{11.2.11}$$

Therefore, for the same number of turns $W_1 = 6$ and the same conductor cross section, current, and speed, with a diode rectifier, the generator will produce 20% less power. The additional reactive power provided through the active front PWM converter, when E and I are in phase, allows for delivering power in better conditions (higher voltage and power).

Note on the Control of TFM

Once a three-phase machine is considered, the control may be implemented with sinusoidal current (for almost sinusoidal emf) or with trapezoidal current (for trapezoidal emf), with third harmonic tapping, if a diode rectifier is used. The control system, with or without motion sensors, is practically the same as that for standard PMSGs, as detailed in Chapter 10. This is why we do not elaborate here on TFM control.

The methods used to reduce the cogging torque are similar to those used for standard PMSGs.

It is to be seen if the TFM will make it to the markets soon, at least as a low-speed high-torque directly driven generator.

11.4 Three-Phase FR-PM Generator: Magnetic and Electric Circuit Design

As mentioned in the section on TFMs, the large PM flux fringing and, consequently, the low power factor are the main obstacles in the way for producing very good performance, even at very low speeds. The FRMs face similar problems, but they are easier to manufacture [8–11].

We have already introduced the PM flux concentration schemes for FRM with stator (Figure 11.6) and rotor PMs (Figure 11.7). We will, however, treat in detail here only the surface stator PM configuration, as it is similar to the TFM investigated in the previous paragraph and is characterized by a smaller inductance (in P.U.), though PM flux fringing is large. We will first present a conceptual design and complete FEM investigations of a 200 Nm, 128 rpm (60 Hz) FRM. Then, through a numerical example, an FRM preliminary design methodology is presented for 200 kNm, 30 rpm, 100 Hz system, as in Examples 11.1 and 11.2. We will refer to very low speeds, as we feel that for speeds above 500–600 rpm, PMSGs with distributed or fractional (with tooth-wound coils) windings are already well established (Figure 11.14).

The low-speed operation at a frequency of 50–60 Hz (to secure moderate core loss but still make good use of iron) implies a large number of "poles" or "pole pairs" or "periods" of the flux

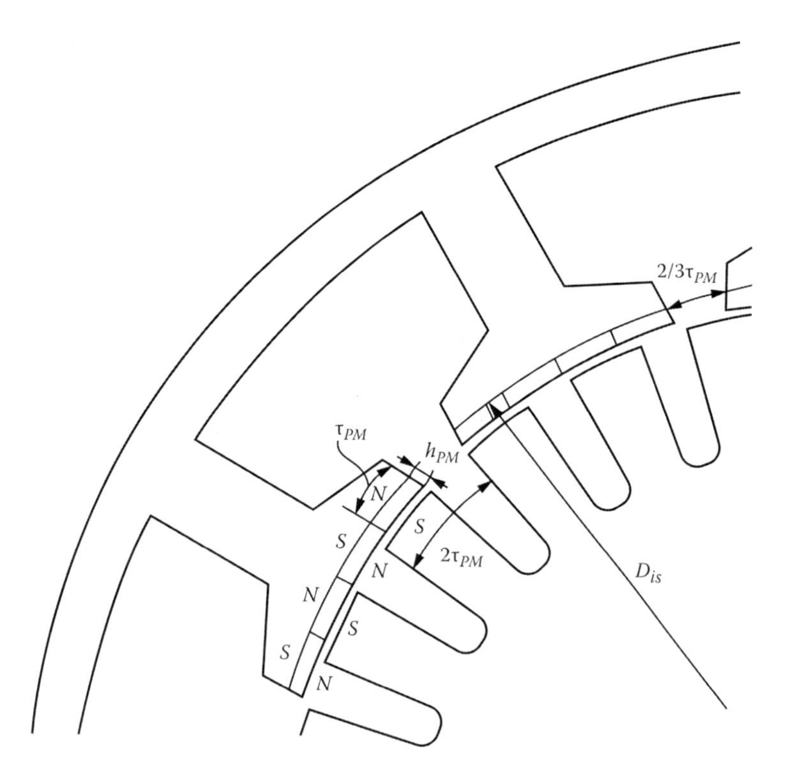

FIGURE 11.14 Surface PM FRM geometry details ($N_s = 12$, $n_p = 2$, $N_r = 28$).

reversal along the machine periphery. In the FRM with N_r rotor salient poles, the speed n and frequency f_1 are related by the following:

$$f_1 = n \cdot N_r \tag{11.27}$$

The two-pole pitch angle corresponds to two PM poles of alternate polarities on the stator, that is,

$$2 \cdot \tau_{PM} = 2 \cdot \tau_{rot} = \frac{\pi \cdot D_r}{N_r} \tag{11.28}$$

where D_r is the rotor diameter. A stator pole may accommodate $2 \cdot n_p$ PMs. The electrical angle of the space between two neighboring stator pole coils should be 120 electrical degrees or $120°/N_r$ geometrical degrees or $2 \cdot \tau_{PM}/3$. Thus,

$$N_s \cdot \left(2 \cdot n_p \cdot \tau_{PM} + \frac{2 \cdot \tau_{PM}}{3}\right) = \pi \cdot (D_r + 2 \cdot g) \approx 2 \cdot \tau_{PM} \cdot N_r \tag{11.29}$$

where

N_s is the number of stator poles
g is the mechanical air gap

Eliminating τ_{PM} from Equation 11.29 yields the following:

$$N_s \cdot \left(n_p + \frac{1}{3}\right) = N_r \tag{11.30}$$

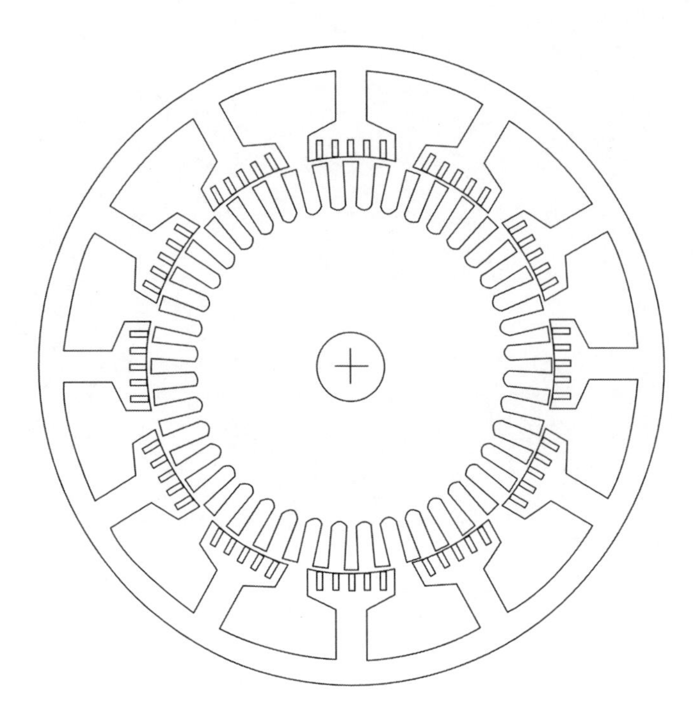

FIGURE 11.15 Low-speed flux reversal machine (FRM) with inset PMs on stator.

The typical configuration for $N_s = 12$ is shown in Figures 11.4 and 11.15. The PMs on the stator pole shoes are very close to the air gap. We may say it has pole PMs.

The FRM for low-speed drives has the following distinct features:

- It uses conventional stamped laminations on both the stator and the rotor.
- It has no PMs or windings on the rotor.
- It has PMs on the stator, where their temperature can be easily monitored and controlled.
- The stator has concentrated coils that are easy to manufacture.
- The lowest pole pitch τ_{PM} is to be larger than (PM + air gap) thickness [8–11] to limit flux fringing in PM utilization, as is the case with the transverse PM flux machine that has the PMs placed on the rotor.
- The higher the rotor diameter (or torque), the larger the maximum number of pole pairs $(2 \cdot \tau_{PM})$, and thus, the lower the speed at 50 (60) Hz.
- The pole PM FRM has lower inductances than the inset PM FRM (Figure 11.15), as expected. Also, the cogging torque of the latter is intrinsically smaller.
- For the inset PM FRM, the air gap should be as small as mechanically feasible, and the PM thickness $h_{PM} > 6g$. Also, the stator and rotor teeth should be equal, while the rotor interpole will span the rest of the double-pitch τ_{PM}:

$$\tau_{PM} = b_{t1} + h_{PM}; \quad 2 \cdot \tau_{PM} = b_{t2} + b_{s2}; \quad b_{t2} = b_{t1} \tag{11.31}$$

- Moreover, $b_{t1}/h_{PM} > 2/3$ provides enough saliency. Thus, the inset PM FRM with an air gap of 0.2 mm, $h_{PM} = 1.5$ mm, would allow for $b_{t1\min} = 5$ mm or a pole pitch $\tau_{PM} = 5 + 1.5 = 6.5$ mm, which is 65% of the one considered for the pole PM configuration. Further reductions of the air gap and magnet thickness may lead to even smaller pole pitches. Thus, at 60 Hz, the speed is accordingly reduced to a value depending on the rotor diameter.

- The inset PM FRM has the PMs parallel to the stator magnet flux lines and is much more difficult to demagnetize. As a bonus, the flux in the magnets varies less (especially under load). Thus, the eddy currents induced with PMs are notably smaller than those for the pole PM configuration, that has PMs that directly experience the stator current additional field.
- The much lower total (magnetic) air gap of the inset PM FRM leads to much larger inductance, which is limited by heavy saturation of the core for high currents (overload). Therefore, the stator current limits are mainly governed by stator temperature and magnetic oversaturation in the inset PM configuration, and stator temperature and PM demagnetization limit the currents for the pole PM FRM. Anyway, high tangential force density leads to lower power factor and thus to higher kVA ratings in the inverter.

11.4.1 Preliminary Geometry for 200 Nm at 128 rpm via Conceptual Design

Consider the pole PM configuration only, with $N_r = 28$ rotor poles, $N_s = 12$ stator poles, with $2n_p = 4$ PMs per stator pole. The designated primary frequency is $f_1 = 60$ Hz. It corresponds to the rated speed $n_n = (f_1/N_r)\cdot 60 = 128.5$ rpm. We start the preliminary design from the specific tangential force f_t to calculate the torque T_e:

$$T_e = f_t \cdot \pi \cdot D_r \cdot l_{stack} \cdot \frac{D_r}{2}; \quad f_{tn} = 1.5 - 4\ \text{N/cm}^2 \tag{11.32}$$

where
D_r is the rotor diameter
l_{stack} is the stack length

A stack-length-to-rotor-diameter D_r ratio λ is defined as follows:

$$\lambda = \frac{l_{stack}}{D_r} = 0.2 - 1.5 \tag{11.33}$$

The lower values of λ are justified for a large number of stator poles. With $N_s = 12$, $\lambda = 1.055$, $T_e = 200$ Nm, and $f_{tn} = 2$ N/cm² (continuous duty), from Equation 11.33, we obtain the following:

$$D_r = \sqrt[3]{\frac{2\cdot T_{en}}{\pi\cdot f_{tn}\cdot\lambda}} = \sqrt[3]{\frac{2\cdot 200}{\pi\cdot 2.0\cdot 10^4\cdot 1.055}} = 0.182\ \text{m} \tag{11.34}$$

We may choose $D_r = 0.180$ m to obtain $l_{stack} = 1.055 \cdot 0.18 = 0.19$ m. The PM flux per stator pole ϕ_{PM} is as follows:

$$\phi_{PM} = \frac{l_{stack}\cdot B_{PMi}\cdot \tau_{PM}\cdot n_{pp}}{1 + k_{fringe}} \tag{11.35}$$

B_{PMi} is the ideal flux density in the air gap as given by

$$B_{PMi} = B_r \cdot \frac{h_{PM}}{h_{PM} + g} \tag{11.36}$$

where
$h_{PM} = 2.5$ mm
$g = 0.5$ mm

At 75°C, $B_r = 1.21$ T, $H_c = 851 \cdot 10^3$ A/m for neodymium–iron–boron (NdFeB) PMs. k_{fringe} is a flux fringing coefficient to be determined by FEM. We choose here an initial safe value $k_{fringe} = 2.33$. The flux is to vary almost sinusoidally with the rotor position θ_r:

$$\phi_{PM}(\theta_r) = \phi_{PM} \cdot \sin(N_r \cdot \theta_r) \tag{11.37}$$

Therefore, the PM flux per stator pole derivative with θ_r is

$$\frac{d\phi_{PM}(\theta_r)}{d\theta_r} = \frac{l_{stack} \cdot B_{PMi} \cdot \tau_{PM} \cdot n_{pp} \cdot N_r \cdot \cos(N_r \cdot \theta_r)}{1 + k_{fringe}} \tag{11.38}$$

There are $N_s/3$ stator poles per phase and, with all coils per phase in series, according to Equation 11.30, the emf amplitude per phase E_m is

$$E_m = \frac{N_s}{3} \cdot n_c \cdot \frac{2 \cdot \pi \cdot n \cdot l_{stack} \cdot \pi \cdot D_r \cdot n_{pp} \cdot B_{PMi}}{1 + k_{fringe}} \tag{11.39}$$

where n_c is the number of turns per coil. For a given sinusoidal current I (RMS), in phase with the emf (I_q current control), the torque T_e is, again,

$$T_e = \frac{3}{2} \cdot E_m \cdot I \cdot \frac{\sqrt{2}}{2 \cdot \pi \cdot n}. \tag{11.40}$$

In our case, we find $n_c I_n \sqrt{2}$:

$$n_c \cdot I_n \cdot \sqrt{2} = n_c \cdot \frac{2}{3} \cdot T_e \cdot \frac{2 \cdot \pi \cdot n}{E_m} = \frac{2}{3} \cdot 200 \cdot \frac{1}{4 \cdot 0.19 \cdot \pi} \cdot \frac{1}{0.18 \cdot 0.3 \cdot 1.21}$$

$$\approx 0.855 \cdot 10^3 \text{ A turns/coil.} \tag{11.41}$$

For E_m, we used Equation 11.39.

Choosing a rated current density $j_{cn} = 3.5$ A/mm² and a slot fill factor $k_{fill} = 0.4$, and noting that there are two coils per slot, the slot area A_{slot} is as follows:

$$A_{slot} = \frac{2 \cdot n_c \cdot I}{j_{cn} \cdot k_{fill}} = \frac{\sqrt{2} \cdot 0.855 \cdot 10^3}{3.5 \cdot 10^6 \cdot 0.4} = 0.863 \cdot 10^{-3} \text{m}^2 \tag{11.42}$$

The slot detailed geometry may now be calculated if the stator pole (tooth) width is first found. As the current loading in such low-speed machines tends to be large, and the coupling between the PM and the coils is not so strong, an analytical dimensioning of its magnetic core is hardly practical, although apparently standard methods could be used. The rather low rated current density is a safety precaution for the torque capability (above 200 Nm) exploration by FEM.

11.4.2 FEM Analysis of Pole-PM FRM at No Load

The geometry with PM height $h_{PM} = 2.5$ mm and air gap $g = 0.5$ mm was, in fact, obtained after many FEM calculations with thinner and thicker PMs and eventually larger air gaps. The maximum PM flux, calculated in the middle of the coil, was found for $h_{PM} = 2.5$ mm and $g = 0.5$ mm (PM pitch is equal to 10 mm). The PM flux per stator pole with straight rotor poles is given in Figure 11.16. It should be noted that the waveforms are symmetric and almost sinusoidal [4].

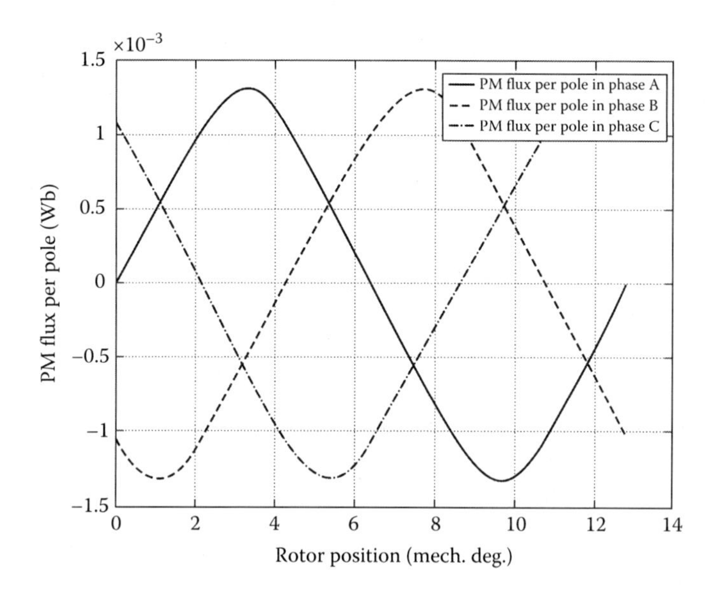

FIGURE 11.16 PM flux per pole vs. rotor position.

However, the cogging torque, although not large (less than 10% rated torque), for general purposes, may be further reduced. Skewing the rotor by various angles leads to cogging torque reduction (Figure 11.17a). For 1.8° mechanical skewing, the best result is obtained (Figure 11.17b).

A 7% reduction in the maximum PM flux per phase is encountered with 1.80 mechanical degrees of skewing (Figure 11.17c). We consider this to be acceptable, as the cogging torque was reduced below 1.5%,

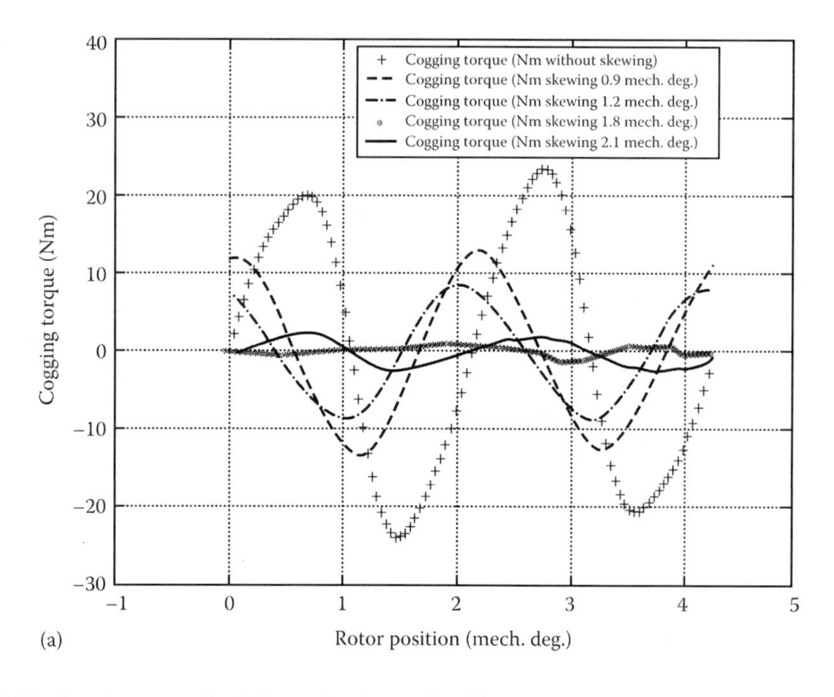

FIGURE 11.17 Cogging torque for different skewing angles: (a) cogging torque vs. position. *(Continued)*

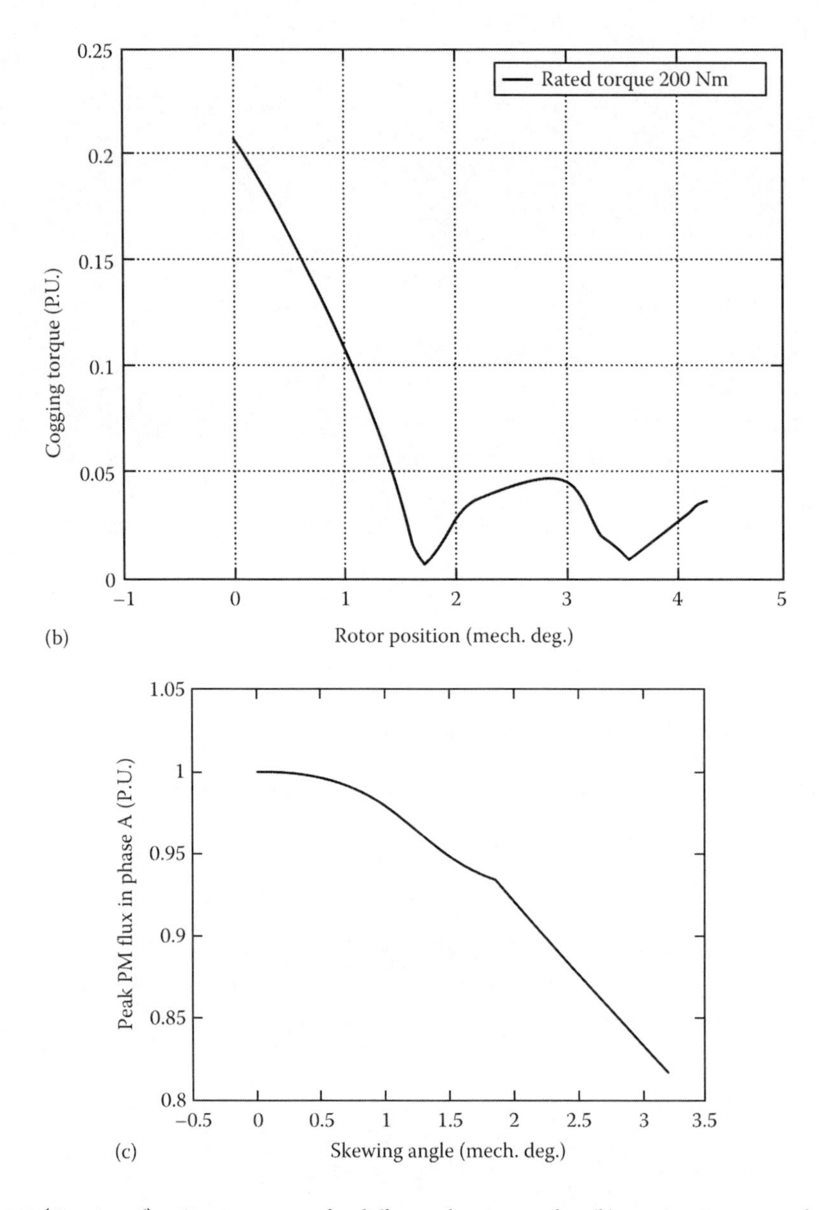

FIGURE 11.17 (*Continued*) Cogging torque for different skewing angles: (b) cogging torque vs. skewing angle, and (c) peak PM flux in a phase coil vs. skewing angle.

which qualifies FRM for high performance. The skewing has the additional effect of shifting the phase PM flux (as expected). This effect is important when designing the control system.

After filtering the phase PM fluxes obtained from FEM, their derivative with rotor position was performed. As demonstrated in Figure 11.18a and b, there are serious grounds to use vector (sinusoidal) control, as the emfs are proportional to $d\Phi/d\theta$ and are almost sinusoidal.

11.4.3 FEM Analysis at Steady State on Load

We may simulate steady-state load conditions by FEM if we let sinusoidal currents flow in the three phases. Therefore, in fact, by gearing the currents to rotor position such that the current in phase A is

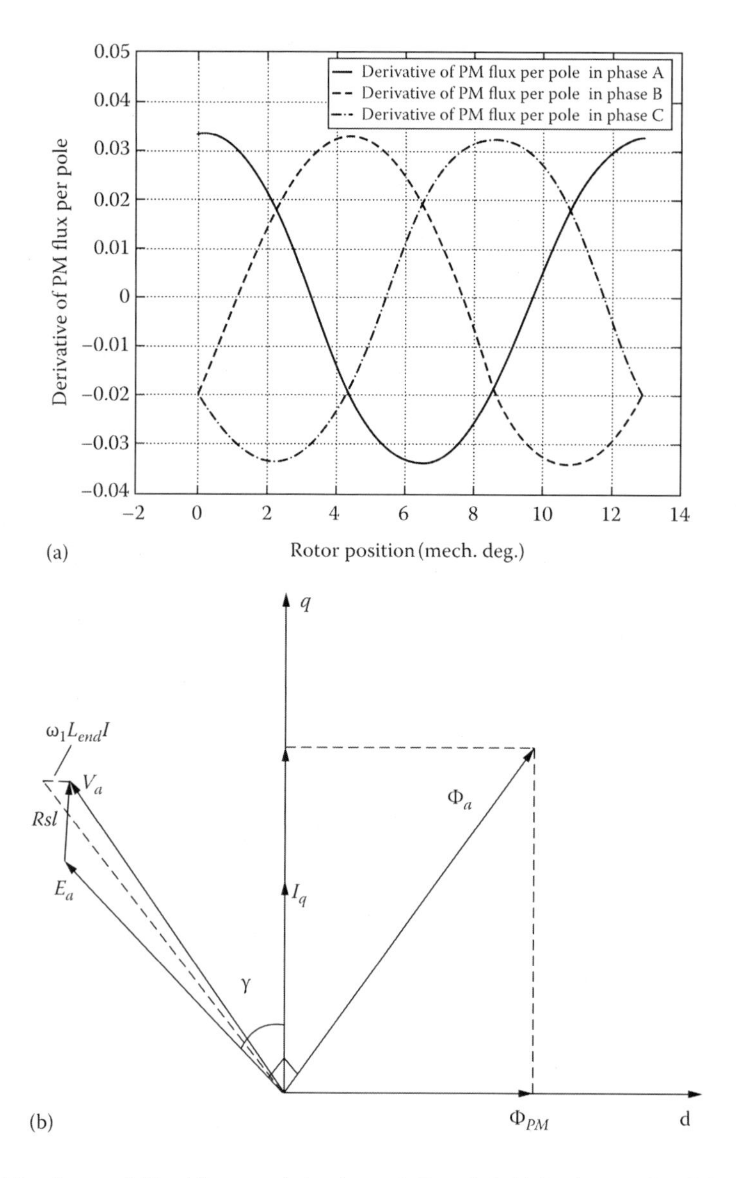

FIGURE 11.18 Waveforms of $d\Phi_{PM}/d\theta_r$ per pole in phases A, B, and C: (a) fundamental and third harmonic and (b) vector diagram with pure I_q control.

maximum when the PM flux in phase A is zero, we produce pure q-axis control. Figure 11.19a shows the phase currents versus position. They are "in phase" with the flux derivatives (Figure 11.18a).

For various rotor positions, the flux per pole of phases A, B, and C with all currents present with instantaneous values, which are position dependent, FEM flux distribution, was calculated. The flux per pole derivatives with θ_r for the three phases are determined and shown in Figure 11.19b. It should be noted that, as expected, the on-load total flux derivatives are not in phase with the currents, as they, in fact, represent the total derivatives, and also because current varies. That is to say, the total E(A, B, C) under steady state is proportional to $d\Phi_{A,B,C}/d\theta_r$:

$$E^t_{A,B,C}(\theta_r) = \frac{-d\phi_{A,B,C}(\theta_r)}{d\theta_r} \cdot \frac{N_s}{3} \cdot n_c \cdot 2 \cdot \pi \cdot n \text{ V (rms)} \tag{11.43}$$

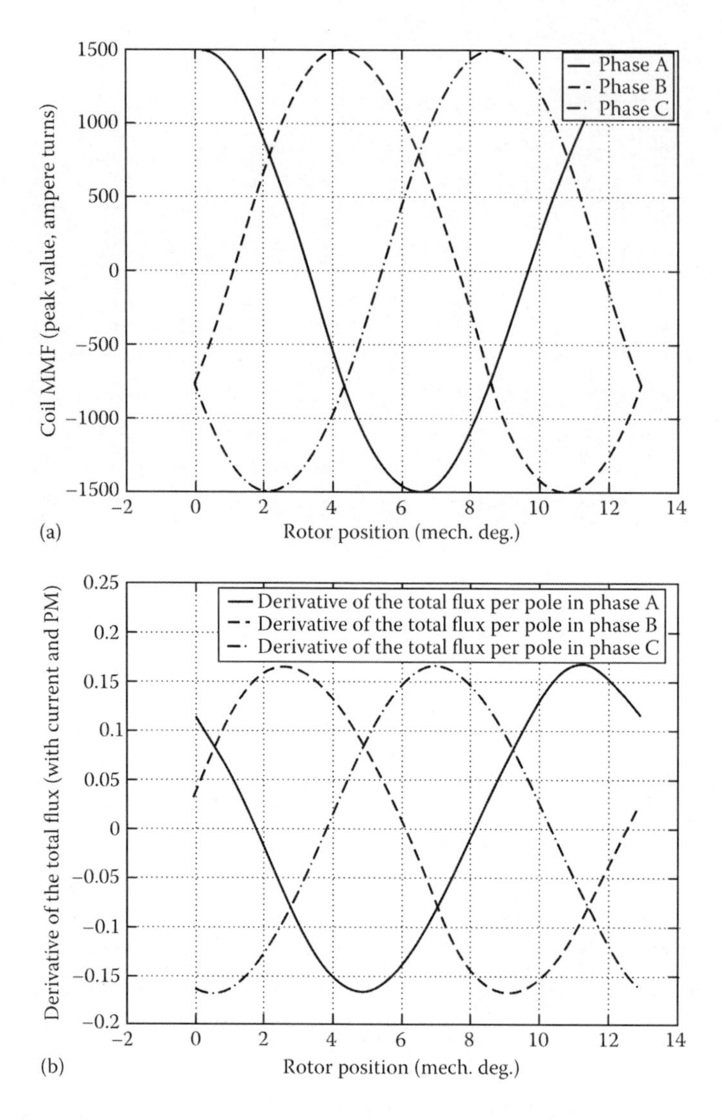

FIGURE 11.19 (a) Phase currents and (b) total flux derivatives with respect to rotor position θ_r.

Taking the maximum value of $E_A^t(\theta_r)$ and dividing it by $\sqrt{2}$, we may add the voltage phasor RI_A (Figure 11.17c) and obtain the phase voltage V_A:

$$V_A \approx \sqrt{(E_A^t(\theta_r)_{\max} + R_s \cdot I \cdot \cos\gamma)^2 + R_s \cdot I^2 \cdot \sin^2\gamma} \tag{11.44}$$

The only component missing is the end-connection leakage inductance component L_{end}. This inductance produces an additional voltage drop $\omega_1 \cdot L_{end} \cdot I$ ($\omega_1 = 2 \cdot \pi \cdot n \cdot N_r = 2 \cdot \pi \cdot f_1$, Figure 11.17b). This way, the steady-state characteristics may be calculated for various current levels in the machine. The flux per pole in phase A for a coil peak mmf of 1500, 2500, 3500, and 5000 A turns demonstrates the effect of saturation on performance. This is more evident in instantaneous torque $T_e(t)$:

$$T_e(t) = \frac{N_s}{3} \cdot n_c \cdot \left[\frac{d\phi_A^t(\theta_r)}{d\theta_r} \cdot i_A(\theta_r) + \frac{d\phi_B^t(\theta_r)}{d\theta_r} \cdot i_B(\theta_r) + \frac{d\phi_C^t(\theta_r)}{d\theta_r} \cdot i_C(\theta_r) \right] \tag{11.45}$$

The results are shown in Figure 11.20a through d. A few remarks are in order.

- Tangential force densities up to 7.108 N/cm^2 are practical, but at a low power factor (less than 0.3 in our case).
- The current-related torque pulsations increase with current due to saturation and the presence of small reluctance torque (Figure 11.20a).
- The torque/pole ampere turns (Figure 11.20c) decrease with current, especially for mmfs above 3500 A turns/pole.
- Even for 5000 A turns/pole, the PMs are not demagnetized (Figure 11.21).
- The current-related torque pulsations are 3% of the average torque at 3500 A turns/pole.
- If the cogging torque peak value is considered, the total torque pulsation/amplitude is 3.01% at 3500 A turns. This is not a large value and may be further reduced by introducing a small harmonic in the current reference waveform to compensate for it.

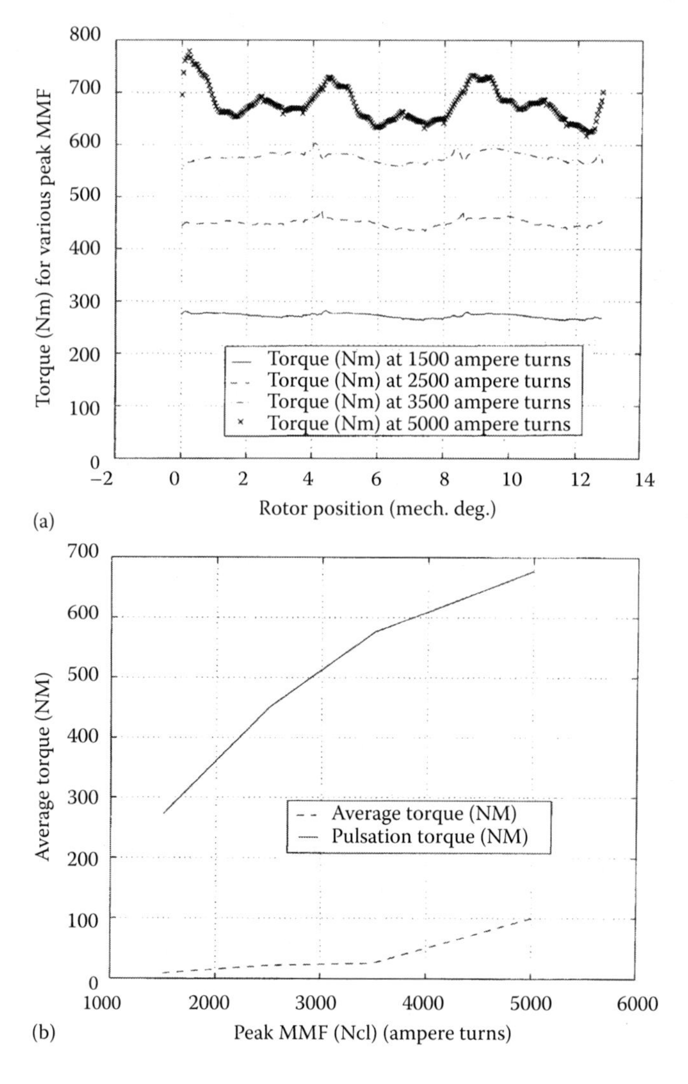

FIGURE 11.20 Curves representing current-related torque vs. position for various coil peak magnetomotive force (mmf) (sinusoidal currents in all phases): (a) and (b) average and pulsation torque vs. mmf. *(Continued)*

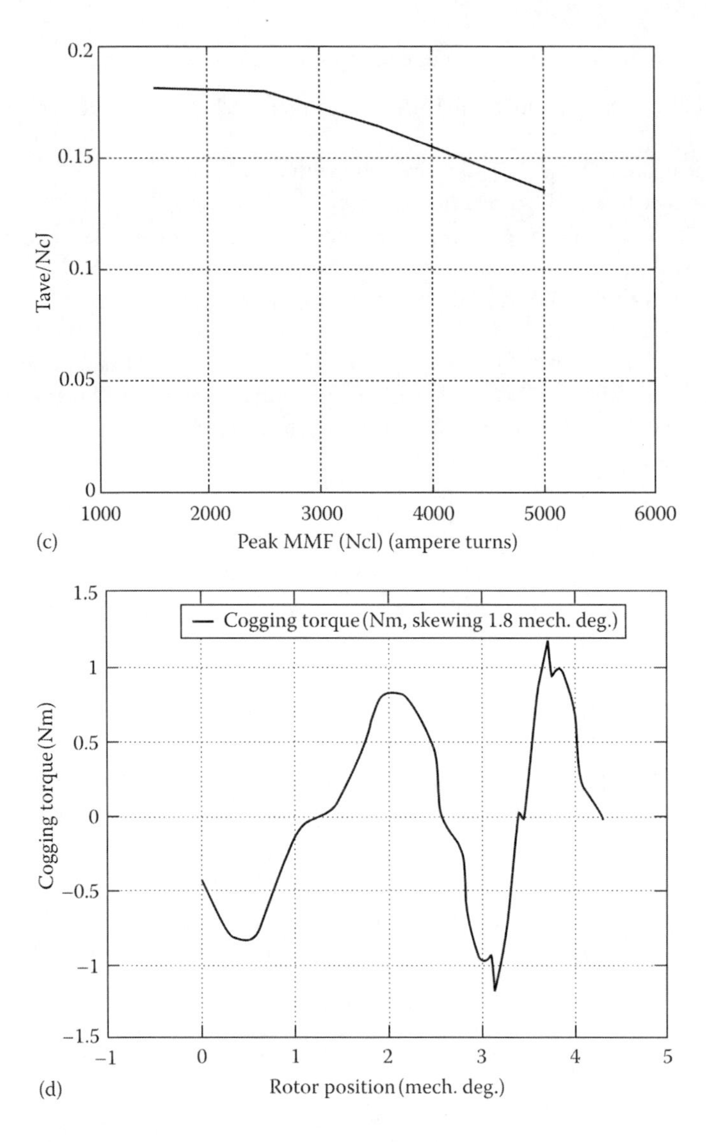

FIGURE 11.20 (*Continued*) Curves representing current-related torque vs. position for various coil peak magnetomotive force (mmf) (sinusoidal currents in all phases): (c) average torque/pole mmf, and (d) cogging torque vs. position.

As the PM height h_{PM} = 2.5 mm, and the air gap g = 0.5 mm, the question arises if, for 5000 A turns, the PMs do not get demagnetized. Through FEM, the flux density distributions in the PM (half North Pole and half South Pole) lower and upper borders were calculated (Figure 11.21). The PMs do not get demagnetized mainly because of the large fringing effects in a large magnetic gap structure. In other words, the not so good coupling between the PMs and the stator coils allows for a high current loading.

While the torque (tangential force) density is important, so is the ratio between stator losses and torque (W/Nm). For our case, and rated current (j_{con} = 3.5 A/mm²), the coil loss is as follows:

$$P_{coil} = 3 \cdot R_s \cdot I_n^2 = 3 \cdot \rho_{co} \cdot \frac{l_{coil} \cdot j_{con}}{n_c \cdot I_n} \cdot \frac{N_s}{3} \cdot n_c^2 \cdot I_n^2 = 3 \cdot k_R \cdot n_c^2 \cdot I_n^2$$

$$= 3 \cdot 2.1 \cdot 10^{-8} \cdot 0.46 \cdot 3.5 \cdot 10^6 \cdot 4 \cdot \frac{855}{\sqrt{2}} = 244.95\,\text{W} \tag{11.46}$$

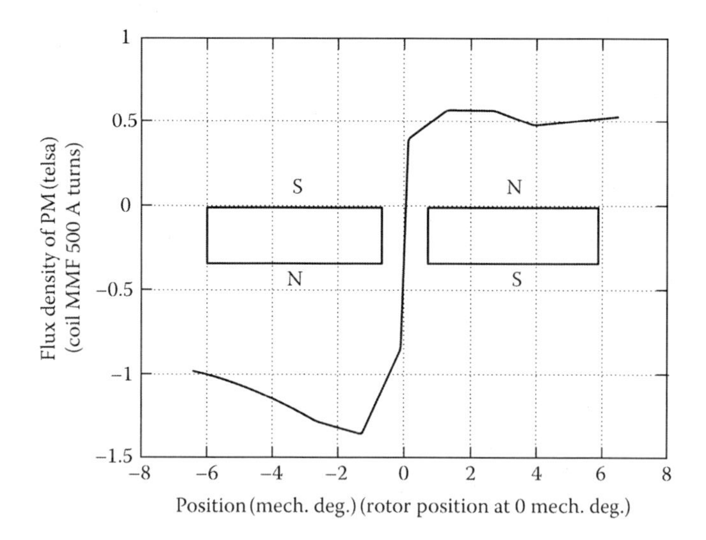

FIGURE 11.21 Flux density on the PM upper and lower border vs. position for 5000 A turns/pole and $\theta_r = 0$.

and

$$l_{coil} \approx 2 \cdot (l_{stack} + 2 \cdot n_p \cdot \tau_{PM}) = 2 \cdot (0.19 + 2 \cdot 2 \cdot 0.1) = 0.46 \text{ m} \tag{11.47}$$

The rated torque T_{en} is considered to be 200 Nm, and, thus, the ratio $P_{coil}/T_{en} = 244.95/200 = 1.225$ W/Nm. This is a very good result, but the force density is moderate:

$$f_{tn} = \frac{2 \cdot T_{en}}{\pi \cdot D_r^2 \cdot l_{stack}} = \frac{2 \cdot 200}{\pi \cdot 0.179^2 \cdot 0.189^2} = 2.1 \cdot 10^4 \text{ N/m}^2 = 2.1 \text{ N/cm}^2 \tag{11.48}$$

This situation corresponds to a pole peak value of mmf of 885 A turns. For 5000 A turns/pole, the torque is 677 Nm (7.1085 N/cm^2), so the torque increases by 3.385 times, but the losses increase $(5000/855)^2$ = 34.2 times. In addition, the current density for peak torque is 16.95 A/mm^2. Such a current density may be sustained for continuous operation only with liquid stator cooling.

For the peak torque, the efficiency with core and PM eddy current losses neglected, as they are relatively small, is as follows:

$$\eta_k \approx \frac{T_{ek} \cdot \Omega_n}{T_{ek} \cdot \Omega + p_{coil}} = \frac{677 \cdot 2 \cdot \pi \cdot (128/60)}{677 \cdot 2 \cdot \pi \cdot (128/60) + 244.95 \cdot 23.47} = 0.52 \tag{11.49}$$

For rated conditions, however,

$$\eta_n \approx \frac{T_{en} \cdot \Omega_n}{T_{en} \cdot \Omega + p_{coil}} = \frac{200 \cdot 2 \cdot \pi \cdot (128/60)}{200 \cdot 2 \cdot \pi \cdot (128/60) + 244.95} = 0.91 \tag{11.50}$$

The ideal power factor angle φ_1 is (for $n_c I\sqrt{2} = 855$ A turns, from Equation 11.6),

$$\varphi_1 = \tan^{-1} \frac{\omega_1 \cdot L_s \cdot I \cdot \sqrt{2}}{E_{peak}} = \tan^{-1} 2.2 = 65°$$

Or cos $\varphi_1 = 0.41$. Considering the low speed of 128 rpm, the efficiencies are reasonable. Note that the 0.52 efficiency at very heavy current overload implies a large voltage drop along stator resistance, to be considered when calculating the number of turns/coil (n_c) for a given inverter voltage value. The power factor is already low, as in general, the magnetic air gap is small, and $L_s I_s$ is still large.

11.4.4 FEM Computation of Inductances

As already mentioned, the inductance needed for the circuit model is the transient inductance:

$$L_{AAt} = \left(\frac{\Delta\phi_A}{\Delta i_A}\right); \quad L_{ABt} = \left(\frac{\Delta\phi_B}{\Delta i_A}\right); \quad \theta_r, i_B, i_C = \text{const.} \tag{11.51}$$

Therefore, only the current amplitude in phase A was modified by a relatively small value, while the values of currents in the other phases remain unchanged. When the rotor position is varied, the instantaneous values of the currents in all phases vary accordingly. The self-inductance and mutual transient inductance, L_{AAt} and L_{ABt}, respectively, are shown in Figure 11.22a and b for two-phase current amplitudes (corresponding to 2500 and 2600 peak mmf/pole). Two remarks are appropriate [4]:

1. The variation of transient inductance with θ_r may be neglected.
2. The mutual transient inductance is almost 10 times smaller than the self-inductance.

Repeating the FEM computation for a low value of current in the machine, the average self-inductance and mutual inductance are shown to vary notably with current due to magnetic saturation (Figure 11.23). To check the necessity of considering all currents present when calculating the transient inductances, the computation has also been done with current in phase A only. Figure 11.23 shows clearly that a correct estimation of transient inductance requires all phases to be energized. The same is true when calculating the torque from emfs and currents.

For the circuit model, we may use either the phase coordinate or the dq model. Let us first define the phase coordinate model:

$$i_{A,B,C} \cdot R_s - V_{A,B,C} = E_{A,B,C}(\theta_s, i_s) - L_t(i_s) \cdot \frac{di_{A,B,C}}{dt} \tag{11.52}$$

$$E_{A,B,C} = -\left(\frac{d\phi_{A,B,C}(\theta_r, i_s)}{d\theta_r}\right)_{i_s} \cdot \frac{N_s}{3} \cdot n_c \cdot 2 \cdot \pi \cdot n \tag{11.53}$$

11.4.5 Inductances and the Circuit Model of FRM

First, the synchronous inductance is

$$L_t(i_s) = L_{end} + [L_{AAt}(i_s) - L_{ABt}(i_s)] \tag{11.54}$$

$$i_s = \left|\frac{2}{3} \cdot [i_A(t) + i_B(t) \cdot e^{j\cdot(2\cdot\pi/3)} + i_C(t) \cdot e^{-j\cdot(2\cdot\pi/3)}] \cdot e^{-j\cdot N_r \cdot \theta_r}\right| \tag{11.55}$$

We use here the current space vector absolute value to account for saturation with the A, B, C model. In addition, analytical approximations in total emf and inductance dependence on θ_r and i_s are required:

$$\frac{d\phi(i_s, \theta_r)}{d\theta_r} = (a_0 + a_1 \cdot n_c \cdot I_s - a_2 \cdot n_c^2 \cdot I_s^2) \cdot [\cos(N_r \cdot \theta_{rA,B,C}) + a_3 \cdot \cos(3 \cdot N_r \cdot \theta_{rA,B,C})] \tag{11.56}$$

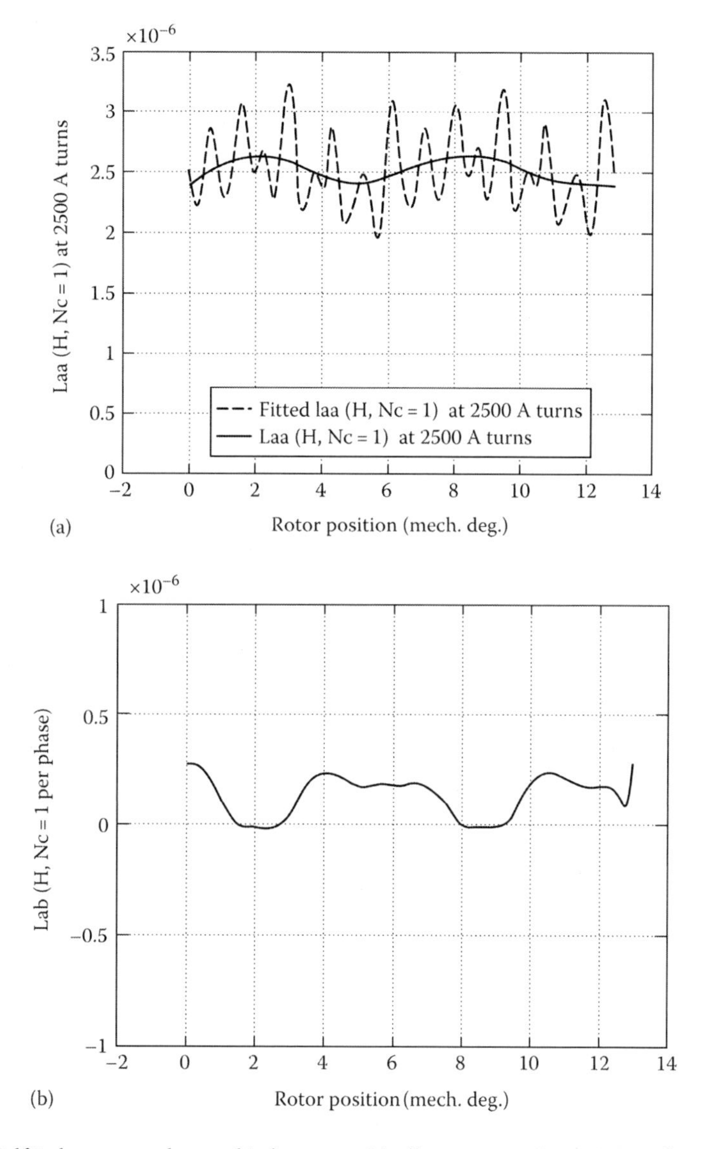

FIGURE 11.22 Self-inductance and mutual inductance with all currents in phases A, B, and C considered: (a) self-inductance and (b) mutual inductance.

where

$$\theta_{rA,B,C} = \theta_r + \gamma_0(n_c \cdot I_s) - (k-1)\cdot\frac{2\cdot\pi}{3}, \quad k = 1,2,3$$

and

$$\begin{aligned}\gamma_0(n_c \cdot I_s) &= \gamma_{\theta 0} + c_1 \cdot n_c \cdot I_s + c_2 \cdot n_c^2 \cdot I_s^2 \\ L_t(i_s) &\approx [L_{end} + b_1 - b_2 \cdot n_c^2 \cdot I_s^2]\cdot n_c^2\end{aligned} \quad (11.57)$$

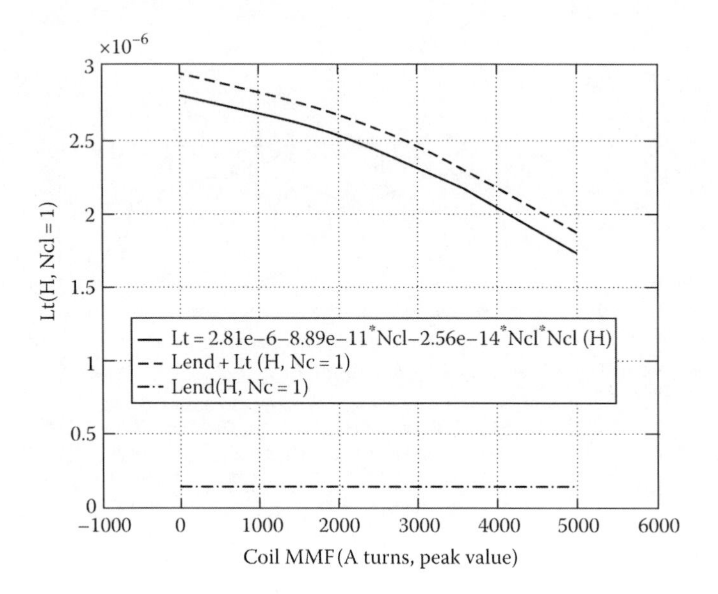

FIGURE 11.23 Average transient inductance dependence on current.

The torque equation T_e is as follows:

$$T_e = (E_A \cdot i_A + E_B \cdot i_B + E_C \cdot i_C) \cdot \frac{1}{2 \cdot \pi \cdot n} \quad (11.58)$$

The motion equations complete the model:

$$J \cdot 2 \cdot \pi \cdot \frac{dn}{dt} = T_e - T_{load}; \quad \frac{d\theta_r}{dt} = n \quad (11.59)$$

11.4.6 *dq* Model of FRM

The study of transients may be performed in A, B, C coordinates, but for control purposes, the *d*–*q* model is required. To simplify the *dq* model, we will neglect the third harmonic in the emfs, as it is below 5%. When implementing the control for very low torque pulsation, the current i_q may be supplied an additional component to cancel the already small torque pulsations. The *dq* model, in stator coordinates, is simply,

$$\bar{i}_s \cdot R_s - \bar{V}_s = -\frac{d\bar{\psi}_s}{dt} - j \cdot \omega_r \cdot \bar{\psi}_s; \quad \frac{d\bar{\psi}_s}{dt} = L_t \cdot \frac{d\bar{i}_s}{dt} \quad (11.60)$$

$$\bar{\psi}_s = \psi_d + j \cdot \psi_q; \quad \bar{i}_s = i_d + j \cdot i_q; \quad \bar{V}_s = V_d + j \cdot V_q \quad (11.61)$$

$$i_s = \sqrt{i_d^2 + i_q^2} \quad (11.62)$$

$$P(\theta_r) = \frac{2}{3} \begin{bmatrix} \cos(-\theta_r) & \cos\left(-\theta_r + \frac{2 \cdot \pi}{3}\right) & \cos\left(-\theta_r - \frac{2 \cdot \pi}{3}\right) \\ \sin(-\theta_r) & \sin\left(-\theta_r + \frac{2 \cdot \pi}{3}\right) & \sin\left(-\theta_r - \frac{2 \cdot \pi}{3}\right) \end{bmatrix}$$

$$\cdot \begin{bmatrix} \omega_r \cdot \psi_d(i_s) \\ \omega_r \cdot \psi_q(i_s) \end{bmatrix} = P(\theta_r) \cdot \begin{bmatrix} E_A(\theta_r, i_s) \\ E_B(\theta_r, i_s) \\ E_C(\theta_r, i_s) \end{bmatrix} \quad (11.63)$$

$$\begin{aligned}\psi_d(i_s) &= L_s(i_s)\cdot i_d + \psi_{PM}\\ \psi_q(i_s) &= L_s(i_s)\cdot i_q\end{aligned} \tag{11.64}$$

$$T_e = \frac{2}{3}\cdot N_r\cdot(\psi_d\cdot i_q - \psi_q\cdot i_d) \tag{11.65}$$

$$J\cdot 2\cdot\pi\cdot\frac{dn}{dt} = T_e - T_L;\quad \frac{d\theta_r}{dt} = 2\cdot\pi\cdot n\cdot p \tag{11.66}$$

Note that L_s is the average normal steady-state inductance. L_s and the transient inductance L_t both depend on i_s due to saturation (Figure 11.23):

$$\bar{V}_s = \frac{2}{3}\cdot\left(V_A(t) + V_B(t)\cdot e^{j\cdot(2\cdot\pi/3)} + V_C(t)\cdot e^{-j\cdot(2\cdot\pi/3)}\right) \tag{11.67}$$

As expected, with sinusoidal $E(\theta_r)$, the dq model will not exhibit θ_r other than in the Park transformation. The magnetic saturation is considered according to Equations 11.54 through 11.57.

Example 11.3

Consider the preliminary design of a FRM as generator for a torque T_{en} = 200 kNm at a speed n_n = 30 rpm for a frequency $f_n \approx 100$ Hz.

Solution

Using Equation 11.34, with f_{tn} = 2.66 N/cm² with λ = 0.3, we obtain directly the interior diameter, D_{ir}:

$$D_{ir} = \sqrt[3]{\frac{2\cdot T_{en}}{\pi\cdot f_{tn}\cdot\lambda}} = \sqrt[3]{\frac{2\cdot 200\cdot 10^3}{\pi\cdot 2.66\cdot 10^4\cdot 0.3}} = 2.5158\text{ m}$$

The stack length $l_{stack} = \lambda\cdot D_{ir}$ = 0.3·2.5158 = 0.7545 m. This is the same as in Example 11.1 for the three-phase TFM.

For 100 Hz, the number of rotor salient poles N_r is as follows (Equation 11.27):

$$N_{ri} = \frac{f_{ni}}{n_n} \approx \frac{100}{(30/60)} = 200$$

We are choosing the same PM material, NdFeB, with B_r = 1.3 T and μ_{rem} = 1.05 at 100°C.

The number of stator large poles (coils) N_s and the number of PM poles n_{pp} per stator large pole are as follows:

$$N_s\cdot\left(2\cdot n_{pp} + \frac{2}{3}\right) = 2\cdot N_r$$

and thus, with $N_s = 6\cdot k$ and n_{pp} = 4, we obtain N_s = 48 for N_r = 208. Consequently, the PM pole width τ_{PM} is as follows:

$$\tau_{PM} = \frac{\pi\cdot D_{ir}}{2\cdot N_r} = \frac{\pi\cdot 2.515}{2\cdot 208} = 0.01898\,\text{m}$$

and the frequency $f_n = n_n\cdot N_r$ = 30/60·208 = 104 Hz.

As the configuration is more rugged (it resembles the SRM topology), we may adopt a slightly smaller air gap of $g = 3$ mm.

We then assign the PM radial height h_{PM} a value, which is about three times the air gap $h_{PM} = 3 \cdot g = 9$ mm, to provide for high enough ideal maximum PM flux density in the air gap:

$$B_{gPMaxi} = B_r \cdot \frac{h_{PM}}{h_{PM} + g} = 1.3 \cdot \frac{9}{9+3} = 0.975\text{ T}$$

The actual maximum PM flux density in the air gap B_{gPMax} is as follows:

$$B_{gPMax} = B_{gPMaxi} \cdot \frac{1}{(1 + k_{fringe}) \cdot (1 + k_s)} \tag{11.3.1}$$

with the fringing coefficient $k_{fringe} = 2.33$ and the magnetic core contribution $k_s = 0.1$, B_{gPMax} becomes

$$B_{gPMax} = 0.975 \cdot \frac{1}{(1 + 2.33) \cdot (1 + 0.1)} = 0.266\text{ T}$$

Note that this is about the same as for TFM in Example 11.1.

This seems a small value; but at this small pole pitch, it may be justified, given the large gap imposed by the large rotor diameter.

The peak value of the phase emf, with N_s coils in series, and each having n_c turns, is as follows (Equation 11.39):

$$E_m = \frac{N_s}{3} \cdot n_c \cdot 2 \cdot \pi \cdot n \cdot l_{stack} \cdot \frac{\pi \cdot D_r}{2} \cdot n_{pp} \cdot B_{gPMax}$$

$$= \frac{48}{3} \cdot 2 \cdot \pi \cdot \frac{30}{60} \cdot 0.7545 \cdot \pi \cdot \frac{2.515}{2} \cdot 4 \cdot 0.266 \cdot n_c = 159.2 \cdot n_c$$

The RMS ampere turns per coil $n_c I_n$ for rated torque are as follows (Equation 11.41):

$$n_c \cdot I_n = \frac{2}{3} \cdot \frac{n_c}{\sqrt{2}} \cdot T_{en} \cdot \frac{2 \cdot \pi \cdot n}{E_{...}} = \frac{2 \cdot 200 \cdot 10^3 \cdot 2 \cdot \pi \cdot 1/2}{3 \cdot \sqrt{2} \cdot 159.2} = 1865\text{ A turns(rms)/coil} \tag{11.3.2}$$

The area required in the slot to accommodate two coils A_{slot} is

$$A_{slot} = \frac{2 \cdot n_c \cdot I_n}{j_{con} \cdot k_{fill}} = \frac{2 \cdot 1865}{9 \cdot 10^6 \cdot 0.4} = 1.036 \cdot 10^{-3}\text{m}^2 = 1036\text{ mm}^2$$

where $j_{con} = 9$ A/mm^2 and $k_{fill} = 0.4$.

The slot bottom width b_1 is, approximately,

$$b_1 = \frac{2 \cdot \pi \cdot R_1}{N_s} - b_{ps} = \frac{2 \cdot \pi \cdot 1284.5}{48} - 54 = 144\text{ mm}$$

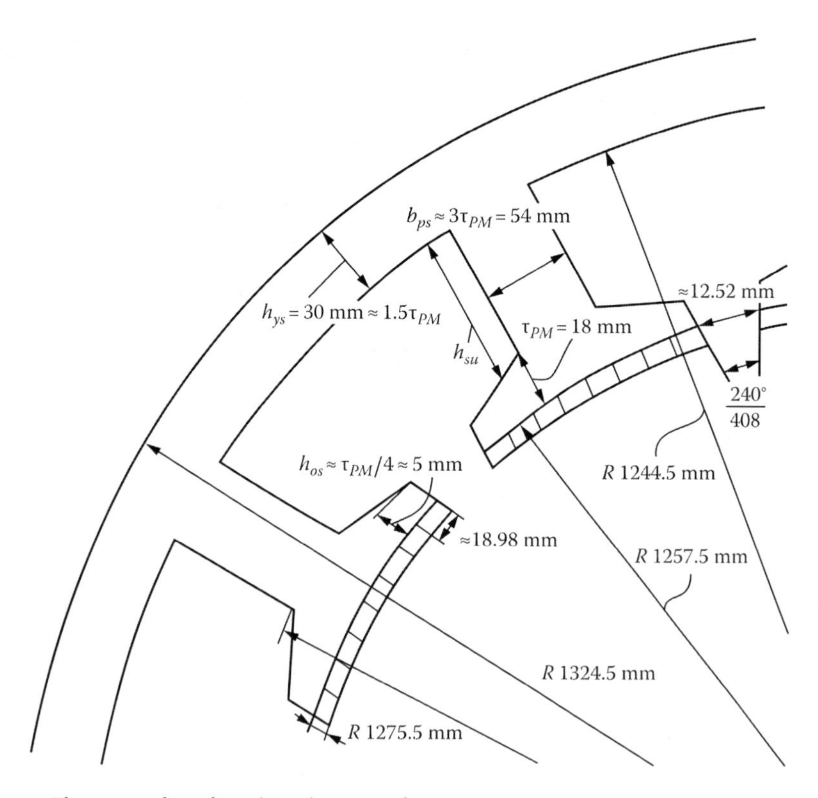

FIGURE 11.24 Flux reversal machine (FRM) stator pole geometry.

The pole body width b_{ps} is approximated having in mind the low useful PM flux density in the air gap. The same is valid for the stator yoke height h_{ys}, which is chosen as 30 mm for mechanical rather than magnetization constraints (Figure 11.24).

The rotor poles (Figure 11.14) should be tall enough to create enough saliency for the rather large magnetic air gap ($h_{PM} + g = 12$ mm).

This only shows that even a slot height of $h_{su} = 10$ mm would provide enough room for the two coils that require 1036 mm^2 in all (Figure 11.25).

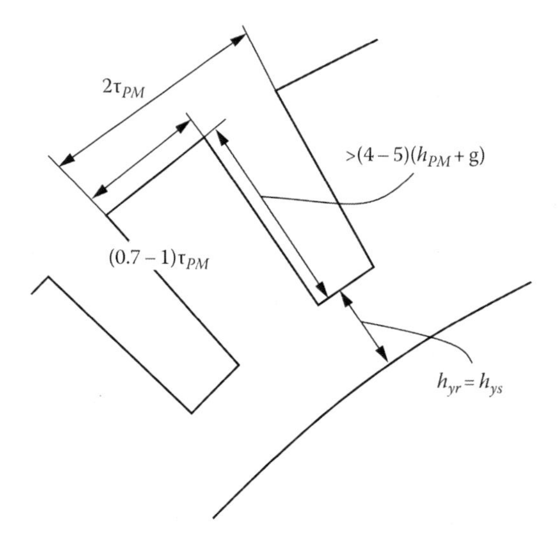

FIGURE 11.25 Rotor salient-pole geometry.

The machine resistance per phase R_s is

$$R_s = \frac{N_s}{3} \cdot \rho_{co} \cdot \frac{l_{coil} \cdot n_c^2}{(n_c \cdot I / j_{con})} = \frac{48}{3} \cdot \frac{2.3 \cdot 10^{-8} \cdot 1.77 \cdot n_c^2}{(1865/9 \cdot 10^6)} = 3.14 \cdot 10^{-3} \cdot n_c^2$$

where the coil average length l_{coil} is

$$l_{coil} = 2 \cdot l_{stack} + 2 \cdot b_{ps} + \pi \cdot \frac{2 \cdot n_{pp} \cdot \tau_{PM} - b_{ps}}{2}$$

$$= 2 \cdot 0.7545 + 2 \cdot 0.054 + \frac{\pi}{2} \cdot (2 \cdot 4 \cdot 0.019 - 0.054) = 1.77 \text{ m}$$

The copper losses for rated current (torque) are as follows:

$$p_{con} = 3 \cdot R_s \cdot I_n^2 = 3 \cdot 3.14 \cdot 10^{-3} \cdot (n_c \cdot I_n)^2 = 9.42 \cdot 10^{-3} \cdot 1865^2 = 32.76 \text{ kW}$$

Note that the simplicity of the manufacturing process in FRM is paid for dearly in copper losses, which increase from 13.70 kW in the TFM to 32.76 kW in the FRM.

As there are no aluminum carriers to hold the stator core, their eddy current losses are also absent in FRM, but still the copper losses of FRM are larger in a machine with the same size and torque.

The machine inductance L_s is again made up of two components: the air gap inductance L_m and the leakage inductance L_{sl}. The leakage inductance is moderate, as the slot aspect ratio h_{su}/b_1 is small; and thus, we concentrate on the air gap inductance:

$$L_m \approx \frac{N_s}{3} \cdot \mu_0 \cdot n_c^2 \cdot \frac{\tau_{PM} \cdot n_{pp}}{g + h_{PM}} \cdot \frac{1 + k_f}{1 + k_s} \cdot l_{stack} \tag{11.3.3}$$

k_f is a fringing effect coefficient $k_f < 0.2$, which accounts for the large air gap above the rotor interpole contribution to the coil self-flux linkage:

$$L_m = \frac{48}{3} \cdot 1.256 \cdot 10^{-6} \cdot \frac{0.01898 \cdot 4}{(3+9) \cdot 10^{-3}} \cdot \frac{1 + 0.2}{1 + 0.1} \cdot 0.7545 \cdot n_c^2 = 0.1046 \cdot 10^{-3} \cdot n_c^2$$

Considering that the leakage inductance represents 10% of L_m, $L_s = (1 + 0.1) \cdot 0.115 \cdot 10^3 \cdot n_c^2$. The power factor angle for pure I_q control is, again (Equation 11.6),

$$\varphi_1 = \tan^{-1}\left(\frac{\omega_1 \cdot L_s \cdot I \cdot \sqrt{2}}{E_m}\right) = \tan^{-1}\left(\frac{2 \cdot \pi \cdot 104 \cdot 0.115 \cdot 10^{-3} \cdot n_c^2 \cdot I \cdot \sqrt{2}}{159.6 \cdot n_c}\right) = \tan^{-1} 1.2375 = 51°$$

$$\cos \varphi_1 = 0.6285$$

Remember that the power factor for the same output conditions was 0.912 for the TFM. We also have to point out that the PM flux fringing ratio of 0.33 (i.e., 33% of ideal PM flux reaches the coils) is a bit too severe for the FRM, because the rotor poles are one pole pitch apart tangentially, while for the TFM, the axial distance between neighboring U- and I-shaped cores is two to three times smaller. Finally, the PM and air gap height were reduced by 25% for FRM. In-depth FEM studies are required to document the best solution of the two. However, with four air gaps per coil, TFM is expected to show smaller inductance.

Example 11.4: Flux Switch Machine with Stator PM-Flux Concentration

Let us consider again a directly driven wind generator system operating again at T_{en} = 200 kNm and n = 30 rpm, $f_n \approx$ 100 Hz. The design of the magnetic circuit, losses, and power factor angle for pure I_q control are required.

Solution

The machine geometry (Figure 11.6) is represented with two poles only (Figure 11.26).

Using the PM flux concentration, we adopt a larger force density f_{tn} than in previous design examples to offset the effect of the inductance increase due to the fringing effect: f_{tn} = 6 N/cm².

To yield the same diameter as before, $\lambda = l_{stack}/D_{is}$ is decreased accordingly to λ = 0.133:

$$D_{is} = \sqrt[3]{\frac{2 \cdot T_{en}}{\pi \cdot f_{tn} \cdot \lambda}} = \sqrt[3]{\frac{2 \cdot 200 \cdot 10^3}{\pi \cdot 6 \cdot 10^4 \cdot 0.133}} = 2.515 \text{ m}$$

The stack length is, however,

$$l_{stack} = \lambda \cdot D_{is} = 0.133 \cdot 2.515 = 0.3345 \text{ m}$$

The number of rotor salient poles for around 100 Hz should be around N_r = 200, calculated as follows (Figure 11.26):

$$2 \cdot N_r \cdot \tau = 6 \cdot k \cdot \left(4 \cdot \tau + \frac{2 \cdot \tau}{3}\right) = 28 \cdot k \cdot \tau \tag{11.4.1}$$

with N_r = 186; k = 14; $f_1 = N_r \cdot n$ = 186·30/60 = 93 Hz.

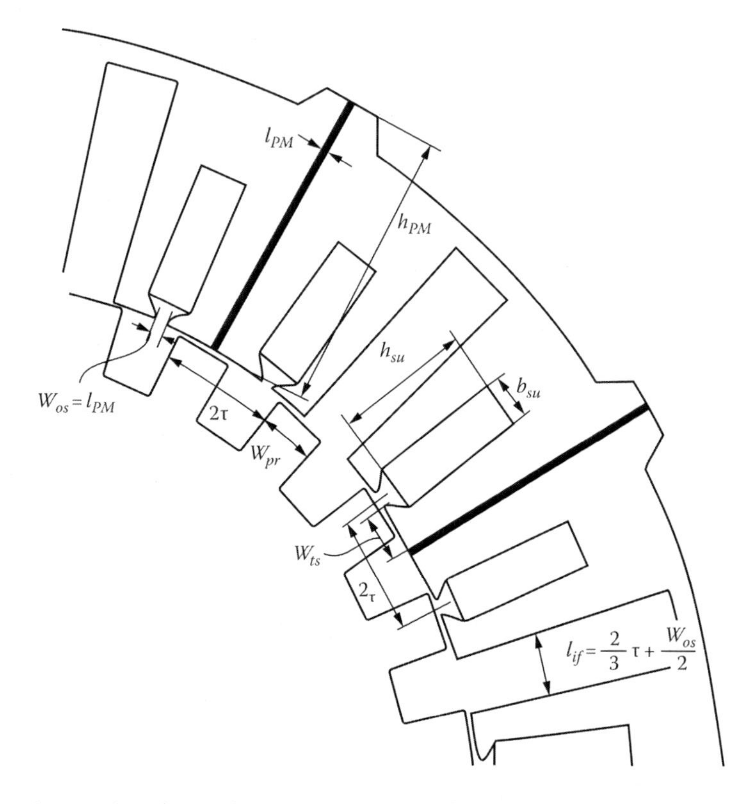

FIGURE 11.26 Flux switching (reversal) generator (FS(R)G) with stator PM flux concentration.

The air gap $g = 2.0$ mm, and the pole pitch τ is

$$\tau = \frac{\pi \cdot D_{is}}{2 \cdot N_r} = \frac{\pi \cdot 2.515}{2 \cdot 186} = 0.02014 \text{ m}$$

The slot opening, W_{os}, equal to the PM thickness, l_{PM}, is considered to be 8 mm. Therefore, the stator tooth $W_{ts} = \tau - W_{os} = 20 - 8 = 12$ mm. The rotor salient pole W_{pr} may be as wide as the pole pitch or smaller:

$$W_{pr} = (0.8 \div 1.0) \cdot \tau \tag{11.4.2}$$

The ideal PM flux density in the air gap B_{gPMax} is now as follows:

$$B_{gPMaxi} = \frac{B_r}{(W_{ts} / h_{PM} + (\mu_{rec} / \mu_0) \cdot (2 \cdot g / \tau_{PM})} \tag{11.4.3}$$

This expression is derived from

$$\begin{aligned} &B_m \cdot h_{PM} = B_g \cdot W_{ts} \\ &H_m \cdot l_{PM} + \frac{B_q}{\mu_0} \cdot 2 \cdot g = 0 \\ &B_m = B_r + \mu_{rec} \cdot H_m \end{aligned} \tag{11.4.4}$$

with $B_r = 1.3$ T and $\mu_{rec} = 1.05 \cdot \mu_0$.

The fringing effect will reduce the ideal PM air gap flux density to a smaller value, as in previous cases:

$$B_{gPMax} = \frac{B_{gPMaxi}}{(1 + k_{fringe}) \cdot (1 + k_s)} \tag{11.4.5}$$

The fringing coefficient depends again on the air gap g/W_{ts} ratio, l_{PM}/g ratio, and on the degree of saturation. With $k_{fringe} = 1.5$, $k_s = 0.2$, we set the actual air gap flux density B_{gPMax} at a reasonable value, say $B_{gPMax} = 0.7$ T.

In this case, from Equation 11.4.5,

$$B_{gPMaxi} = 0.7 \cdot (1 + 1.5) \cdot (1 + 0.2) = 2.1 \text{ T}$$

This is a high value, but we should remember that it is only a theoretical one. From Equation 11.4.3, we may size the PM height h_{PM}:

$$2.1 = \frac{1.3}{(12 / h_{PM}) + (1.05 \cdot \mu_0 / \mu_0) \cdot (2 \cdot 20 / 8)}; \quad h_{PM} = 120 \text{ mm}$$

The PM maximum flux per coil turn, Φ_{PMaxc}, is as follows:

$$\phi_{PMaxc} = B_{gPMax} \cdot W_{ts} \cdot l_{stack} = 0.7 \cdot 0.012 \cdot 0.3345 = 2.81 \cdot 10^{-3} \text{ Wb} \tag{11.4.6}$$

The maximum flux linkage $\psi_{PMphase}$ is

$$\psi_{PMphase} = 2 \cdot k \cdot \phi_{PMaxc} \cdot n_c = 2 \cdot 14 \cdot 2.81 \cdot 10^{-3} \cdot n_c = 0.07867 \cdot n_c \tag{11.4.7}$$

The emf per phase (peak value) E_m is

$$E_m = 2 \cdot \pi \cdot f_1 \cdot \psi_{PMphase} = 2 \cdot \pi \cdot 93 \cdot 0.07867 \cdot n_c = 45.949 \cdot n_c \quad (11.4.8)$$

The ampere turns per slot $n_c I_n$ may be calculated from Equation 11.3.2:

$$n_c \cdot I_n = \frac{2 \cdot n_c}{3 \cdot \sqrt{2}} \cdot T_{en} \cdot \frac{2 \cdot \pi \cdot n}{E_m} = \frac{2 \cdot n_c}{3 \cdot \sqrt{2}} \cdot \frac{200 \cdot 10^3 \cdot 2 \cdot \pi \cdot 30/60}{45.949 \cdot n_c} = 6.4589 \cdot 10^3 \text{ A turns/coil}$$

The slot width is about equal to the pole pitch (Figure 11.26). $W_{su} = (1\text{–}1.2) \cdot \tau_{PM} \approx 1.1 \cdot 20 = 22$ mm. This way, the stator tooth average width is around half the pole pitch τ_{PM}.

As the total slot height $h_{su} \approx h_{PM} - 2/3 \cdot W_{ts} - 0.005 = 103 \cdot 10^{-3} - 2/3 \cdot 12 \cdot 10^{-3} - 5 \cdot 10^{-3} = 90 \cdot 10^{-3}$ m, the current density required to host the coil j_{con} is as follows:

$$j_{con} = \frac{n_c \cdot I_n}{h_{su} \cdot W_{su} \cdot k_{fill}} = \frac{6458.9}{90 \cdot 22 \cdot 0.5} = 6.524 \text{ A/mm}^2$$

The stator coil resistance is

$$R_{sc} = \rho_{co} \cdot \frac{l_{coil}}{(n_c \cdot I_n / j_{con})} \cdot n_c^2 = \frac{2.3 \cdot 10^{-8} \cdot 0.818}{(6458.9/6.524 \cdot 10^6)} \cdot n_c^2 = 1.9 \cdot 10^{-5} \cdot n_c^2$$

$$l_{coil} = 2 \cdot l_{stack} + 4 \cdot \tau_{PM} + \pi \cdot W_{su} = 2 \cdot 0.3345 + 4 \cdot 0.020 + \pi \cdot 0.022 = 0.818 \text{ m} \quad (11.4.9)$$

For the entire phase,

$$R_s = 2 \cdot k \cdot R_{sc} = 2 \cdot 14 \cdot 1.9 \cdot 10^{-5} \cdot n_c^2 = 0.532 \cdot 10^{-3} \cdot n_c^2$$

The stator copper losses are then

$$P_{con} = 3 \cdot R_s \cdot I_n^2 = 3 \cdot 0.532 \cdot 10^{-3} \cdot 6458.9^2 = 66.58 \text{ kW}$$

A few remarks are in order:

- The FRM with stator PM flux concentration cannot appropriately take full advantage of the principle of PM flux magnification.
- The machine size was reduced (the stack length was 0.7545 m for the surface PM stator FRM). This reduction in size is paid for by larger copper losses (66.58 kW for an input power of 628 kW). There is notably more copper in this machine.
- It is possible to redo the design for a smaller force density; but for larger stack length, the same interior stator diameter D_{is} is needed to notably reduce the copper losses. Still, the inductance seems to be higher as each coil "entertains" two air gaps.
- The FRM with stator PM flux concentration seems to be restricted to a smaller number of poles per rotor and small torque machines, where the fabrication costs may be reduced due to its relatively more rugged topology.

11.4.7 Notes on Flux Reversal Generator Control

With its easy rotor skewing, an FRM can produce a rather sinusoidal emf and is thus eligible for field orientation or direct power (torque) control as any PM SG (see Chapter 10 for details and results).

Example 11.5: FRG with Rotor PM Flux Concentration

Let us now consider the FRG with rotor PM flux concentration (Figure 11.27) illustrated here with only a few poles (Figure 11.27). For the same data as in Example 11.3, prepare an adequate design.

Solution

The two stator cores and the rotor core are all made up of standard stamped laminations. Only the external rotor is provided with windings to save room in the rotor (Figure 11.7), and thus make it more rugged mechanically. Each stator large pole now contains $2 \cdot n_p + 1$ poles and $2 \cdot n_p$ interpoles. It is clearly visible that for maximum flux per large pole, all the PMs in the rotor are active. This is a better PM utilization in addition to PM flux concentration.

First, let us keep $f_{tn} = 6$ N/cm² and $\lambda = 0.133$. Then, D_{is} (stator interior diameter) stays the same as in Example 11.3; $D_{is} = 2.515$ m.

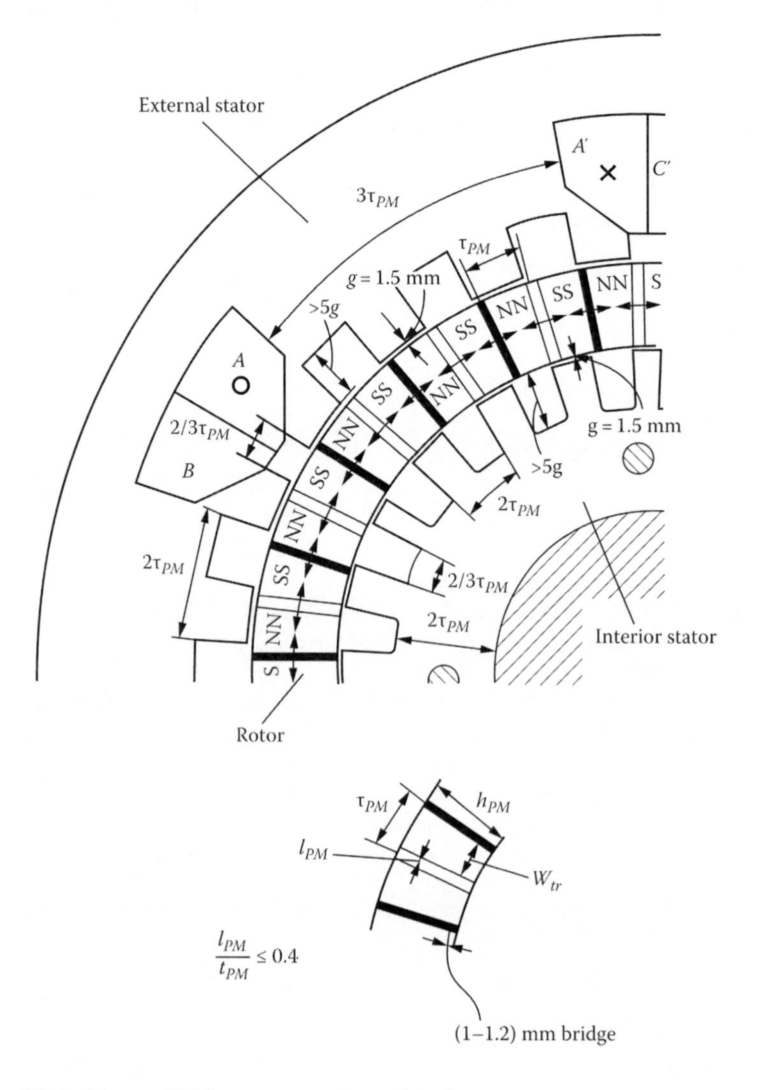

FIGURE 11.27 FRM with rotor PM flux concentration and dual stator.

The number of stator poles $6 \cdot k$ is now as follows:

$$6 \cdot k \cdot \left(2 \cdot n_{pp} + 1 + 2 \cdot n_{pp} + \frac{2}{3}\right) = 2 \cdot N_r \approx 400 \tag{11.5.1}$$

with $n_{pp} = 2$, $k = 7$, and $2 \cdot N_r = 406$ poles.

Therefore, the PM pole pitch $\tau_{PM} = \pi \cdot D_{is}/2 \cdot N_r = \pi \cdot 2.515/406 = 0.01945$ m.

The ideal PM flux density in the air gap B_{gPMaxi} is as follows:

$$B_{gPMaxi} = \frac{B_r}{(W_{tr}/2 \cdot h_{PM}) + (\mu_{rec}/\mu_0) \cdot (2 \cdot g/l_{PM})}; \quad B_r = 1.3\,\text{T}; \quad \mu_{rec} = 1.05 \cdot \mu_0 \tag{11.5.2}$$

In contrast to Equation 11.4.3, the factor $2 \cdot h_{PM}$ instead of h_{PM} is used in Equation 11.5.2, because two PM magnets cooperate in the rotor tooth W_{tr}.

With the actual air gap PM flux density $B_{gPMax} = 0.9$ T, the ideal air gap PM flux density B_{gPMaxi} is as follows:

$$B_{gPMaxi} = B_{gPMax} \cdot (1 + k_{fringe}) \cdot (1 + k_s) = 0.9 \cdot (1 + 1.5) \cdot (1 + 0.2) = 2.7\,\text{T} \tag{11.5.3}$$

With this value in Equation 11.4.9, the ratio h_{PM}/W_{tr} is obtained (the PM thickness $l_{PM} = 0.4 \cdot \tau_{PM} = 0.4 \cdot 20 = 8$ mm):

$$2.7 = \frac{1.3}{(W_{tr}/(2 \cdot h_{PM})) + ((1.05 \cdot \mu_0)/\mu_0) \cdot (3/8)} \tag{11.5.4}$$

It follows that $2 \cdot h_{PM}/W_{tr} = 11.4$.

Consequently, the PM radial height $h_{PM} = 11.4 \cdot W_{tr}/2 = 5.7 \cdot 12 = 68.40$ mm.

This is a notable reduction in PM weight for a better effect in comparison with Example 13.3. The maximum PM flux linkage per phase is obtained by adapting Equations 11.4.6 and 11.4.7 for the case in point:

$$\begin{aligned} \psi_{PMphase} &= 2 \cdot k \cdot W_{ts} \cdot B_{gPMax} \cdot l_{stack} \cdot (2 \cdot n_{pp} + 1) \cdot n_c \\ &= 2 \cdot 7 \cdot 0.012 \cdot 0.9 \cdot 0.3345 \cdot (2 \cdot 2 + 1) \cdot n_c = 0.2528 \cdot n_c \end{aligned} \tag{11.5.5}$$

Consequently, the peak value of emf per phase E_m is now calculated from Equation 11.4.7:

$$E_m = 2 \cdot \pi \cdot f_1 \cdot \phi_{PMphase} = 2 \cdot \pi \cdot \frac{30}{60} \cdot 203 \cdot 0.2528 \cdot n_c = 161.19 \cdot n_c \tag{11.5.6}$$

By now, we have all the data necessary to calculate the rated mmf per coil $n_c I_n$ (Equation 11.3.2):

$$n_c \cdot I_n = \frac{2}{3 \cdot \sqrt{2}} \cdot n_c \cdot T_{en} \cdot \frac{2 \cdot \pi \cdot n}{E_m} = \frac{2}{3 \cdot \sqrt{2}} \cdot n_c \cdot \frac{200 \cdot 10^3 \cdot 2 \cdot \pi \cdot 30/60}{161.19 \cdot n_c} = 1842.1\,\text{A turns}$$

A current density $j_{con} = 6$ A/mm² may be adopted, as the slot useful area A_{slot} is

$$A_{slot} = \frac{2 \cdot n_c \cdot I_n}{j_{con} \cdot k_{fill}} = \frac{2 \cdot 1842.1}{6 \cdot 10^6 \cdot 0.4} = 1.535 \cdot 10^{-3}\,\text{m}^2 = 1535\,\text{mm}^2$$

There is plenty of room to locate such a slot with low height, and thus with low slot leakage inductance contribution.

The phase resistance R_s (Equation 11.4.9) is as follows:

$$R_s = 2 \cdot k \cdot \rho_{co} \cdot \frac{l_{coil}}{((n_c \cdot I_n)/j_{con})} \cdot n_c^2 = \frac{2 \cdot 7 \cdot 2.3 \cdot 10^{-8} \cdot 1.1045}{(1842.1/6 \cdot 10^6)} \cdot n_c^2 = 1.1684 \cdot 10^{-3} \cdot n_c^2$$

$$l_{coil} \approx 2 \cdot l_{stack} + 2 \cdot \frac{\pi \cdot D_{is}}{6 \cdot k} = 2 \cdot 0.3345 + 2 \cdot \pi \cdot \frac{2.515}{6 \cdot 7} = 1.045 \text{ m}$$

The copper losses p_{con} are

$$p_{con} = 3 \cdot R_s \cdot I_n^2 = 3 \cdot 1.1684 \cdot 10^{-3} \cdot 1842.1^2 = 11.7925 \text{ kW}$$

This displays less copper loss than for the TFM (13.7 kW), while the machine axial total length is about half in the TFM for the same diameter. There is still one more problem: the power factor. For maximum I_q current control, the PM flux is zero in that phase; thus, the air gap inductance is rather large, as the PMs "do not stay in the way" of coil mmf flux.

Consequently,

$$L_m \approx \mu_0 \cdot 2 \cdot k \cdot n_c^2 \cdot (2 \cdot n_{pp} + 1) \cdot \frac{W_{tr} \cdot l_{stack}}{2 \cdot g \cdot (1 + k_s)}$$

$$= 1.256 \cdot 10^{-6} \cdot 2 \cdot 7 \cdot n_c^2 \cdot (2 \cdot 2 + 1) \cdot \frac{0.012 \cdot 3345}{2 \cdot 1.5 \cdot 10^{-3} \cdot (1 + 0.2)} = 0.98 \cdot 10^{-4} \cdot n_c^2 \tag{11.5.7}$$

$$L_s = (1 + 0.1) \cdot L_m = 1.078 \cdot 10^{-4} \cdot n_c^2$$

Finally, the power factor angle φ_1 is as follows (with $n_c I_n = 1842.1$ A turns):

$$\varphi_1 = \tan^{-1}\left(\frac{\omega_1 \cdot L_s \cdot I_n \cdot \sqrt{2}}{E_m}\right) = \tan^{-1}\left(2 \cdot \pi \cdot \frac{203}{2} \cdot \frac{1.078 \cdot 10^{-4} \cdot n_c^2 \cdot I_n \cdot \sqrt{2}}{161.2 \cdot n_c}\right) = \tan^{-1} 1.107 = 48°$$

$$\cos\varphi_1 = 0.67$$

Let us try to reduce copper losses further by reducing the number of stator coils with $k = 2$ (four coils per phase). We obtain $n_p = 8$ and $2 \cdot N_r' = 6 \cdot k\ (4 \cdot n_p + 1 + 2/3) = 404$. The PM pole pitch τ_{PM} remains about the same at 0.01947 m.

Repeating the design, the PM flux per phase (Equation 11.5.5) is

$$\psi_{PMphase} = 0.2528 \cdot n_c \cdot \frac{(2 \cdot n_{pp}' + 1) \cdot k'}{(2 \cdot n_{pp} + 1) \cdot k} = 0.2528 \cdot n_c \cdot \frac{(2 \cdot 8 + 1) \cdot 2}{(2 \cdot 2 + 1) \cdot 7} = 0.2455 \cdot n_c$$

and the emf (peak value) E_m' (Equation 11.5.6) is as follows:

$$E_m' = 2 \cdot \pi \cdot N_r \cdot n \cdot \phi_{PMphase} = 2 \cdot \pi \cdot \frac{202}{2} \cdot 0.2455 \cdot n_c = 155.76 \cdot n_c$$

Therefore, the coil mmf $(n_c I_n)'$ is

$$(n_c \cdot I_n)' = 1842.1 \cdot \frac{161.19}{155.76} = 1906.26 \text{ A turns (rms)}$$

with the same $j_{con} = 6$ A/mm^2, the stator resistance R_s' becomes

The copper losses p'_{con} are now

$$p'_{con} = 3 \cdot R'_s \cdot (n_c \cdot I_n)'^2 = 3 \cdot 0.575 \cdot 10^{-3} \cdot 1906.26^2 = 6.266\,\text{kW (about 1\%)}$$

Note that the copper losses were reduced, by half, at the price of thicker stator and iron yokes. Unfortunately, the inductance stays about the same, therefore the power factor stays about the same, around 0.67.

The following are some final remarks:

- In comparison with the FRM with stator surface PMs, the machine axial length was reduced by half, at the same interior and outer diameter, smaller copper losses, but same power factor.
- The PM weight is only slightly larger than in the TFM with rotor PM flux concentration.
- In comparison with the TFM, the machine length is half at about the same interior and outer stator diameter, the copper losses are only 30% less, but the power factor is 0.67 in comparison with the 0.912 of TFM.
- Reducing the number of stator coils could be another way to further improve FRM performance.
- The FRM is considered more manufacturable than the TFM.
- It could be argued that the double-sided TFM with flux concentration can produce even better results. This is true, but in a less manufacturable topology.
- Though in the numerical examples of this chapter the fringing flux coefficients were chosen conservatively low (0.3–0.4), full FEM studies are required to validate the performance claims to precision, on a case-by-case basis [12]. This way, by increasing the air gap through optimal design, even the power factor could be enhanced above 0.8, for a slightly larger machine volume.

11.5 High Power Factor Vernier PM Generators

A Vernier PM (VPM) machine has high torque density due to the so-called magnetic gear effect [13]. In addition, it has low pulsing torque due to its more sinusoidal emf waveform compared to that of a regular PM machine, therefore it is suitable to low-speed, direct-drive applications.

The objective of this paragraph is to investigate a new VPM topology that has advantages of high torque density and low torque ripple, while it overcomes low power factor of the state of art VPM machine. The novel topology-dual-stator, spoke-array PM rotor, VPM (DSSAVPM) machine is shown in Figure 11.28. The dual stator in this embodiment has 4-pole (2 ps = 4) diametrical AC 3 phase windings, placed in Ns = 12 slots ($q = 1$ slot/pole/phase) and a rotor in between with 2pr = 22 spoke-array PM poles (Ns + ps = pr).

11.5.1 Power Factor of VPM Machine

11.5.1.1 Power Factor

The surface-mounted VPM machine is often driven by $Id = 0$. If the resistance is neglected, the phasor diagram can be simplified as shown in Figure 11.29 and then power factor can be given by

$$PF = \frac{1}{\sqrt{1 + (L_s I / \psi_m)}} \tag{11.68}$$

where

Ψ_m is the magnet flux linkage
I is the RMS phase current
L_s is the synchronous inductance

Therefore, power factor is determined by $L_s I/\Psi_m$.

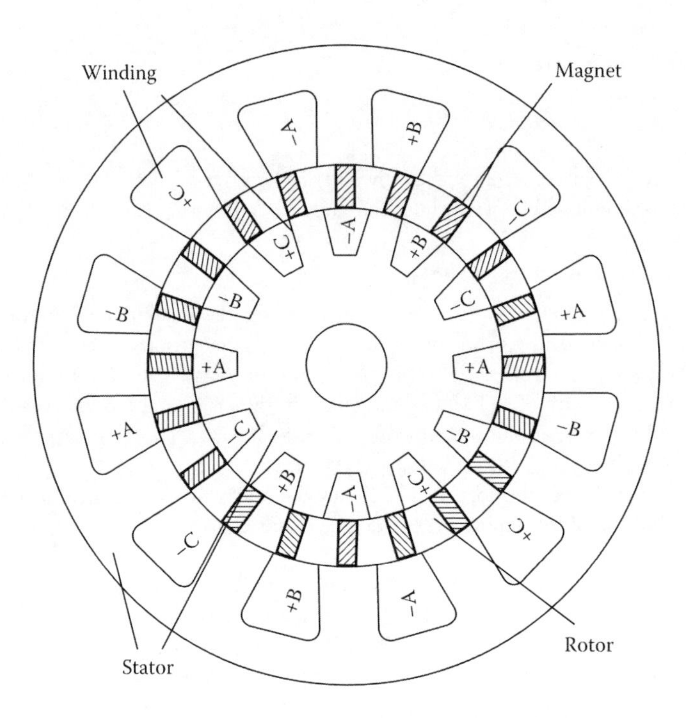

FIGURE 11.28 Dual-stator, spoke-array, Vernier permanent magnet (VPM machine). (Adapted after Li, D. et al., High power factor vernier permanent magnet machines, Record of IEEE-ECCE, Denver, CO, 2013.)

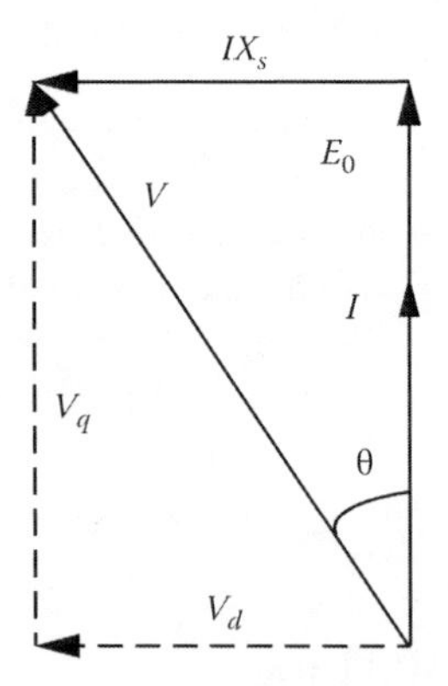

FIGURE 11.29 Phasor diagram when $Id = 0$.

Due to the special stator slot and rotor pole combination of the VPM machine, the armature field pole pitch is much larger than rotor pole pitch. Only half of magnets contribute to the flux per armature field pole pitch during one armature field pole as shown in Figure 11.30, and the other half magnets mainly produce flux leakage. This is typical for surface PM rotors. All these reasons reduce the fundamental flux density Bg, and the power factor of VPM machines.

In order to quantitatively investigate power factor of VPM machines, two FEA models, one 44-rotor pole, 4-armature pole VPM machine and one regular 4-pole PM machine, have been built, and the VPM machine's size data is listed in Table 11.2 [14]. The two machines have same stator structure, winding configuration and magnet thickness (Figure 11.30).

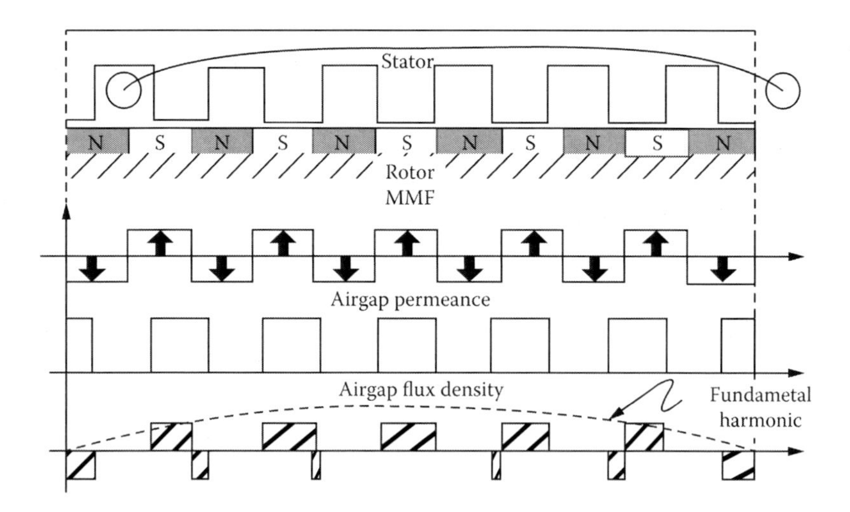

FIGURE 11.30 VPM machines stator teeth effect on the flux density distribution.

TABLE 11.2 Main Specifications of Single-Side VPM Machine

	Single-Side VPM Machine
Rotor pole pairs	22
Armature pole pairs	2
Number of turns in series per phase	88
Slot number	36
Axial lengh (mm)	50
Outside radius of stator (mm)	310
Inside radius of stator (mm)	244.5
Airgap length (mm)	1.5
Magnet radius (mm)	243
Outside radius of rotor (mm)	229
Inside radius of rotor (mm)	200
Ratio of slot opening to slot pitch	0.6
PM material (100°C)	N40UH
Steel material	35TW250

Source: Li, D. et al., High power factor vernier permanent magnet machines, Record of IEEE-ECCE, Denver, CO, 2013.

11.5.2 DSSA: VPM for Higher Power Factor

The dual shifted stator radial air gap DSSA–VPM (similar in a way to PM–rotor flux concentration FRM!) uses all magnets all the time holding the prospects of higher cos ϕ_1 (Figure 11.32).

The FEA results are summarized in Table 11.3, where the power factor of DSSA–VPM is 0.91.

An axial air gap version may be even more practical.

The design parameters and size data of prototype are listed in Table 11.4.

The DSSA–VPM machine has higher torque density than a single-stator surface PM machine, and smooth torque waveform. Therefore, it is suitable in the direct-drive applications. However, the inner stator has to use a cantilever structure due to the sandwich structure of dual stators and rotors.

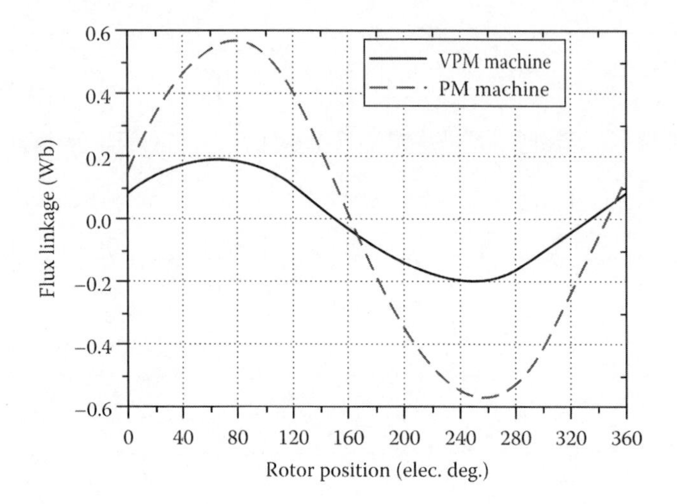

FIGURE 11.31 Flux linkage of the PM machine and VPM machine. (Adapted after Li, D. et al., High power factor vernier permanent magnet machines, Record of IEEE-ECCE, Denver, CO, 2013.)

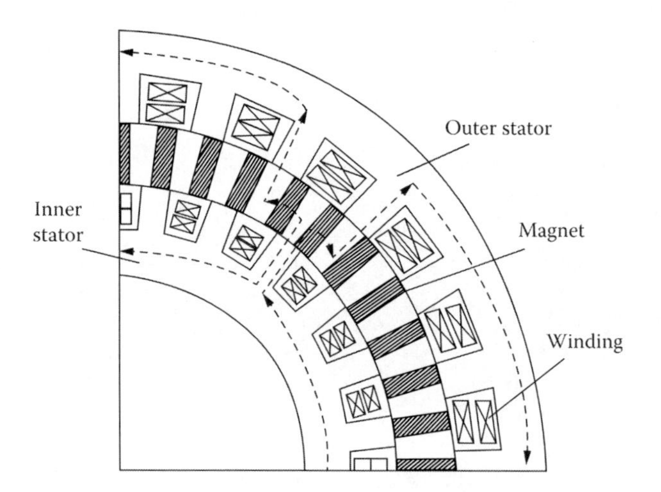

FIGURE 11.32 Flux line of DSSA VPM machine.

TABLE 11.3 Performance of DSSA–VPM Machine

Outside linear loading (A/cm)	120
Inside linear loading (A/cm)	100
Torque (Nm)	1000
Power factor	0.91
Magnet volume (cm^3)	801
Torque per magnet volume (Nm/cm^3)	1.73
D axis inductance (mH)	5.2
Q axis inductance (mH)	5.6

Source: Li, D. et al., High power factor vernier permanent magnet machines, Record of IEEE-ECCE, Denver, CO, 2013.

TABLE 11.4 Size Data DSSA–VPM

	Value
Torque (Nm)	1000
Speed (rpm)	30
Efficiency	90
Power factor	0.85
Number of slots	24
Number of rotor poles	44
Outsider stator outer diameter (mm)	620
Outside stator inner diameter (mm)	489
Outer airgap length (mm)	1.5
Rotor inner diameter (mm)	410
Inside stator outer diameter (mm)	407
Inside stator inner diameter (mm)	270
PM thickness (mm)	14
Outside stator teeth width (mm)	29.5
Inside stator teeth width (mm)	17.8

11.6 Summary

- This chapter investigates two special PM brushless generators with large numbers of poles and nonoverlapping stator coils.
- One is called the TFM, and it uses circular single coils per phase. The PMs are on the rotor or on the stator, with or without PM flux concentration. The flux paths are generally three dimensional.
- The other one is called the FRM, and it uses the switched reluctance machine standard laminated core. The PMs may be placed on top of stator poles in a large even-numbered $2 \cdot n_p$ with alternate polarity.
- In both, TFM and FRM, the PM flux linkage in the stator coils reverses polarity, but generally, for the same diameter and pole pitch (number of poles per periphery), the TFM coil embraces more alternate PM poles and is destined for better torque magnification.
- As the PM pole pitch decreases with a large number of poles, the PM fringing flux increases to the point that the latter becomes overwhelming. In essence, $\tau_{PM} > g + h_{PM}$, where g is the mechanical gap, and h_{PM} is the PM radial thickness to secure a less than 66% reduction of flux in coil due to fringing. Reductions of only 35%–40% are reported in Reference 2 for a small machine.
- The large fringing translates into poor use of PM, core material and high current loading for good torque density, higher copper losses, and lower power factor.
- The torque in Newton meters per watts of copper losses and the power factor $\cos \varphi_1$ are defined as performance indexes, which are independent of speed (frequency), to characterize TFM and FRM. Also, the total cost has to be considered, as these machines use less copper but more iron and PM materials than standard machines.
- The chapter develops preliminary electromagnetic models for TFM and FRM and then uses the same specifications: a 200 kNm, 30 rpm, 100 Hz generator in four different designs—a TFM with surface PMs and three FRM designs (one with surface PM stator, one with stator PM flux concentration (called also a flux switching machine), and one with rotor PM flux concentration).

- All four topologies could be designed for the specifications, but the TFM, for the same volume, was slightly better than FRM with the surface PM stator in power factor. However, a reduction in volume by one half, for about the same copper losses and lower power factor, was obtained with the PM rotor flux concentration FRM. It may be argued that the TFM could also be built with PM flux concentration. This is true, but the already low manufacturability of TFM is further reduced this way.
- It is too early to discriminate between TFM and FRM, as one seems slightly better in torque/copper losses for given volume, but the other is notably more manufacturable, for a very high number of poles.
- In terms of control, the FRM is easily capable (through skewing) of producing sinusoidal emfs and is thus directly eligible for standard field orientation or direct power (torque) control (see Chapter 10 for more details on this issue).
- New PM generator/motor configurations such as DSSA–VPM that depart from the standard PM synchronous generators/motors (Chapter 10) are still being proposed in the search for better very low speed directly driven generator systems with full power electronics control [14].
- High-speed PMSGs up to a few megawatts per unit, with multiphase stator windings are also being considered to match gas turbines for cogeneration [15].
- Optimal design of large PMSGs (for directly driven wind or hydro applications) is a very hot subject too [16].

References

1. H. Weh and H. May, Achievable force densities, Record of ICEM-1986, München, Germany, 1986, vol. 3, pp. 1101–1111.
2. J. Luo, S. Huang, S. Chen, and T.A. Lipo, Design and experiments of a novel axial circumferential current permanent magnet machine (AFCCM) with radial airgap, *Record of IEEE–IAS, Annual Meeting*, Chicago, IL, 2001.
3. G. Henneberger, and I.A. Viorel, *Variable Reluctance Electrical Machines*, Shaker-Verlag, Aachen, Germany, 2001, Chapter 6.
4. I. Boldea, J. Zhang, and S.A. Nasar, Theoretical characterization of flux reversal machine in low speed drives—The pole-PM configuration, *IEEE Trans.*, IA-38(6), 2002, 1549–1557.
5. S.E. Rauch and L.J. Johnson, Design principles of flux switch alternator, *AIEE Trans.*, 74(part III), 1955, 1261–1268.
6. A. Njeh, A. Masmoudi, and A. El Antably, 3D FEA based investigation of cogging torque of claw pole transverse flux PM machine, Record of IEEE–IEMDC, 2003, vol. I, pp. 319–324.
7. M. Bork, R. Blissenbach, and G. Henneberger, Identification of the loss distribution in a transverse flux machine, Record of ICEM, Istanbul, Turkey, 1998, vol. 3, pp. 1826–1831.
8. I. Boldea, E. Serban, and R. Babau, Flux-reversal stator-PM single phase generator with controlled DC output, Record of OPTIM, Poiana Brasov, Romania, 1996, vol. 4, pp. 1124–1134.
9. R. Deodhar, S. Andersson, I. Boldea, and T.J.E. Miller, The flux reversal machine: A new brushless doubly-salient PM machine, *IEEE Trans.*, IA-33(4), 925–934.
10. I. Boldea, C.X. Wang, and S.A. Nasar, Design of a three-phase flux reversal machine, *Electr. Mach. Power Syst.*, 27, 1999, 849–863.
11. C.X. Wang, S.A. Nasar, and I. Boldea, Three-phase flux reversal machine, *Proc. IEEE*, EPA-146(2), 1999, 139–146.
12. I. Boldea, L. Tutelea, and M. Topor, Theoretical characterization of three phase flux reversal machine with rotor-PM flux Concentration, In *13th International Conference on Optimization of Electrical and Electronic Equipment* (*OPTIM*), Brasov, Romania, 2012, pp. 472–476.
13. A. Toba and T.A. Lipo, Novel dual-excitation PM Vernier PM machine, Record of IEEE–IAS, Phoenix, AZ, 1999, vol. 4, pp. 2539–2544.

14. D. Li, R. Qu, and T. Lipo, High power factor vernier permanent magnet machines, Record of IEEE-ECCE, Denver, CO, 2013.
15. F. Luise, A. Tessarolo, F. Agnolet, S. Pieri, M. Scalabrin, and P. Raffin, A high performance 640 kW, 10000 rpm Halbach-array PM slotless motor with active magnetic bearings. Part I: Preliminary and detailed design, Record of IEEE–SPEEDAM, 2014.
16. J.A. Tapia, J. Pyrhonen, J. Puranen, P. Lindh, and S. Nyman, Optimal design of large permanent magnet synchronous generators, *IEEE Trans.*, MAG-49(1), 2013, 642–650.

12

Linear Motion Alternators

12.1 Introduction

This chapter mainly deals with linear oscillatory motion electric generators. The linear excursion is in general within a few centimeters. Free-piston Stirling engines (SEs), linear internal combustion single-piston engines, and direct wave energy engines (with meter range excursions) are proposed as prime movers.

Thus far, the Stirling engine [1] has been used for spacecraft and for residential electric energy generation (Figure 12.1) [1]. The linear internal combustion engine (ICE) was recently proposed for series hybrid electric vehicles.

While rotary motion electric generators are multiphase machines, in general, the linear motion alternators (LMAs) tend to be single-phase machines, because the linear oscillatory motion imposes a change of phase sequence for the change in the direction of motion. A three-phase LMA may be built from three single-phase LMAs.

Though at first LMAs with electromagnetic excitation were proposed as single-phase synchronous generators (SGs), more recently, permanent magnet (PM) excitation took over, and most competitive LMAs now rely on PMs.

A brief classification of LMAs into three categories seems in order:

1. With coil mover (and stator PMs)
2. With PM mover
3. With iron mover (and stator PMs)

One configuration in each category is treated separately in terms of principle and performance equations. The dynamics and control will be treated once for all configurations. A special kind of linear motion generator with progressive motion, to provide on-board energy on maglev (magnetically levitated) vehicles with active guideway, is also briefly discussed.

12.2 LMA Principle of Operation

PM-LMAs with oscillatory motion are, generally, single-phase machines with harmonic motion:

$$x = x_m \sin \omega_r t \tag{12.1}$$

The electromagnetic force (emf) is, in general,

$$e(t) = -\frac{\partial \Psi_{PM}}{\partial x}\frac{dx}{dt} \tag{12.2}$$

where Ψ_{PM} is the PM flux linkage in the phase coils.

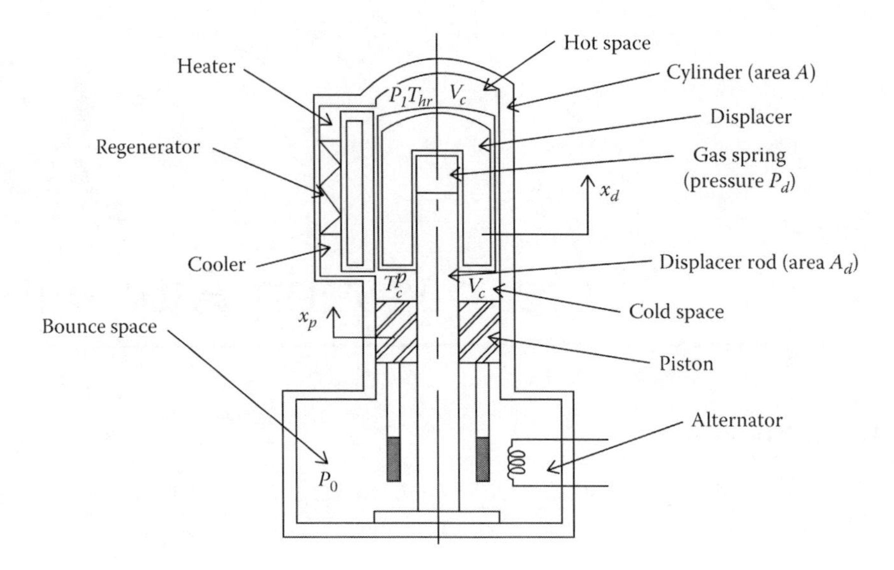

FIGURE 12.1 Stirling engine—linear alternator.

With Equation 12.1 in Equation 12.2,

$$e(t) = -\frac{\partial \Psi_{PM}}{\partial x} x_m \omega_r \sin \omega_r t \tag{12.3}$$

To obtain a sinusoidal emf waveform, Equation 12.3 yields the following:

$$\frac{\partial \Psi_{PM}}{\partial x} = C_e \tag{12.4}$$

This means that the PM flux linkage in the phase coils has to vary linearly with mover position. Flux reversal is most adequate for the scope (Figure 12.2a and b).

The excursion length is $2x_m$, but x varies sinusoidally between x_m and $-x_m$.

The ideal harmonic linear motion and PM flux linkage linear variation with mover position are met only approximately in practice.

In essence, the $\Psi_{PM}(x)$ flattens toward excursion ends (Figure 12.2b), which leads to the presence of third, fifth, and seventh harmonics in $e(t)$. In addition, mainly due to magnetic saturation variation with instantaneous current and mover position, even harmonics (second and fourth) may occur in $e(t)$.

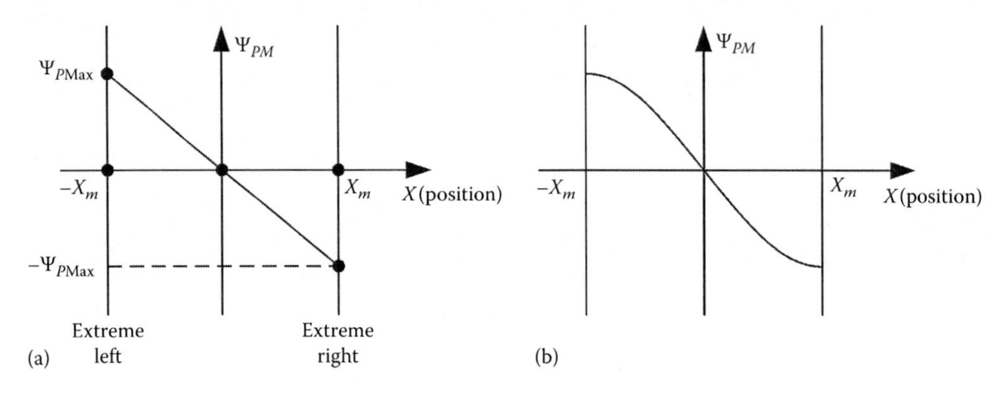

FIGURE 12.2 PM flux linkage vs. mover position: (a) ideal and (b) real.

Through finite element method (FEM), the emf harmonics content for harmonic motion may be fully elucidated. The electromagnetic force $F_e(t)$ is as follows:

$$F_e(t) = \frac{e(t) \cdot i(t)}{dx/dt} = -\frac{\partial \Psi_{PM}}{\partial x} i(t) \tag{12.5}$$

Ideally, with $d\Psi_{PM}/dx = C_e$, the trust varies as the current does.

The highest interaction electromagnetic force per given current occurs (Equation 12.5) when the $e(t)$ and $i(t)$ are in phase with each other.

For $d\Psi_{PM}/dx$ = constant, from Equation 12.4, it follows that $e(t)$ is in phase with the linear speed. Therefore, for the highest trust/current, the current also has to be in phase with the linear speed.

This rationale is valid if the phase inductance L_s is independent of mover position, that is, if the reluctance thrust F_r is zero.

In the presence of PMs, a cogging force, F_{cog}, occurs for zero current. This force, if existent, should be zero for the mover in the middle position and maximum at excursion ends, in order to behave like a "magnetic spring" and, thus, be useful in the energy conversion ($F_{cog} \approx -K_{cog} \cdot x$) (Figure 12.3).

For the ideal LMA, with L_s = const, $e(t) = E_m \cos\omega_r t$, $E_m = C_e \cdot x_m \cdot \omega_r$ harmonic motion, $x = x_m \sin\omega_r t$ (provided by the regulated prime mover), and perfectly linear cogging force characteristic, under steady state, the voltage equation is, in complex variables,

$$\underline{V_1} = -(R_s + j\omega_r L_s)\underline{I_1} + \underline{E_1} \tag{12.6}$$

with

$$E_1 = E_m e^{jrt} \quad \underline{V_1} = V_1\sqrt{2}e^{j(\omega_r t - \delta_V)} \tag{12.7}$$

The phase voltage of the power grid $\underline{V_1}$ is phase shifted by the voltage power angle δ_V with respect to the emf $\underline{E_1}$.

The phasor diagram of Equations 12.6 and 12.7 is shown in Figure 12.4 for the general case when $\underline{I_1}$ is not in phase with $\underline{E_1}$. The operation is similar to that for an SG, although a single-phase one.

The delivered power $\underline{S}$ is as follows:

$$\underline{S} = \left(\underline{V_1} \cdot \underline{I_1}^*\right) = P_1 + jQ_1 \tag{12.8}$$

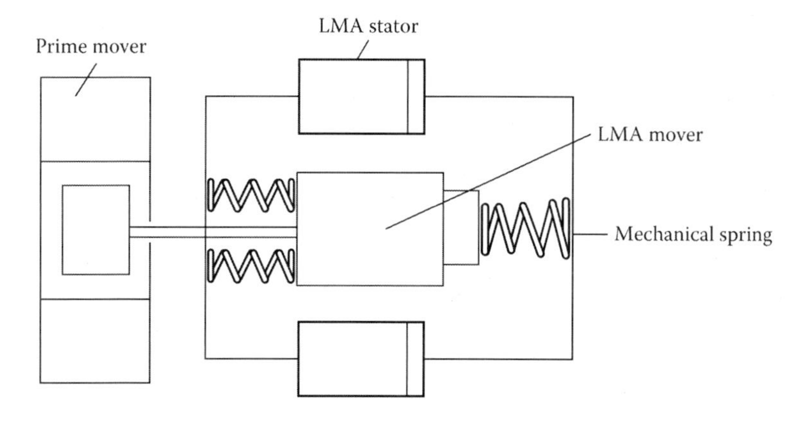

FIGURE 12.3 The prime mover and linear motion alternator (LMA) system.

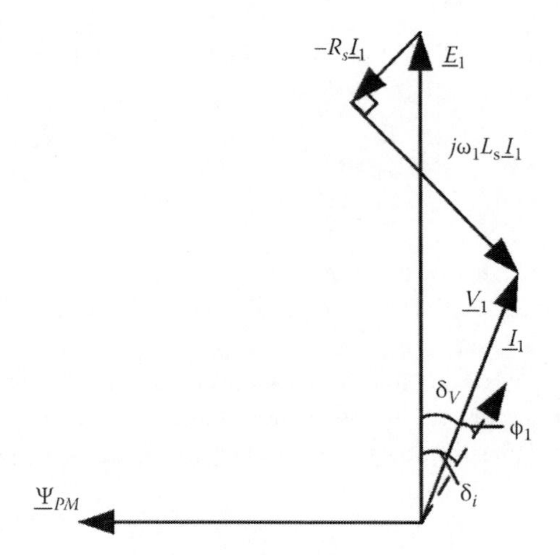

FIGURE 12.4 Phasor diagram of linear motion alternator (LMA).

The delivered active average power P_1 is

$$P_1 = \text{Real}\left(\underline{V}_1 \cdot \underline{I}_1^*\right) = \frac{E_m}{\sqrt{2}} I_1 \cos\delta_1 - R_s I_1^2 \quad (12.9)$$

To deliver power to a power grid of voltage V_1 (root-mean-squared [RMS] value), the RMS value of emf $E_m/\sqrt{2}$ should be considerably larger than the former:

$$E_m > V_1\sqrt{2} \quad (12.10)$$

due to the inductance L_s and resistance R_s voltage drop. A series capacitor may be used to compensate for these voltage drops, at least partially. The power pulsates with double-source frequency $2\omega_r$, as for any single-phase source.

We considered from the start that the frequency of the power grid voltage $\underline{V}_1$ is the same ω_r, as that of the emf, given by the harmonic motion (Equations 12.1 through 12.3).

The synchronization process would be similar to the case of rotary synchronous machines, but we have to regulate the motion amplitude and frequency so as to fulfill the condition of equal frequency and $\underline{E}_1 = \underline{V}_1$. Subsequently, to load the generator, the motion amplitude has to be increased in order to increase the delivered power.

The frequency remains constant, as the power grid is considered much stronger than the LMA. Alternatively, the stand-alone operation of LMA is not constrained in frequency, but the output is increased by increasing the motion amplitude, below the maximum limit x_{max}.

12.2.1 Motion Equation

The equation of motion is, essentially, as follows:

$$m_t \cdot \frac{d^2x}{dt^2} = F_{mec} - F_e + F_{cog} + F_{spring} \quad (12.11)$$

The mechanical openings force F_{spring} is

$$F_{spring} = -K \cdot x \quad (12.12)$$

With m_t equal to the total moving mass:

$$F_{cog} = -C_{cog} \cdot x \tag{12.13}$$

$$F_e(t) = C_e \cdot i(t) \tag{12.14}$$

For steady-state harmonic motion, the prime mover should ideally cover only the electromagnetic force F_e:

$$F_e(t) = F_{mec}(t) \tag{12.15}$$

Then, to fulfill Equation 12.11, we also need to know that

$$m_t \frac{d^2x}{dt^2} + (K + C_{cog})x = 0 \tag{12.16}$$

as $x = x_m \sin \omega_r t$, it follows that

$$\left(-m_t x_m \omega_r^2 + (K + C_{cog})x_m\right)\sin \omega_r t = 0 \tag{12.17}$$

and finally,

$$\omega_r = \sqrt{\frac{(K + C_{cog})}{m_t}} \tag{12.18}$$

Equation 12.18 spells out the mechanical resonance condition. Therefore, the electrical frequency (equal to the mechanical one) should be equal to the spring proper frequency.

In this case, the prime mover has to provide only the useful electromagnetic power, while the mechanical springs do the conversion of kinetic energy at excursion ends, securing the best efficiency conditions.

If linear, the cogging spring-type force helps the mechanical springs. In reality, the cogging force drops notably at excursion ends, leading to the nonlinear spring characteristic that limits the maximum stable motion amplitude to $(0.80–0.95)x_m$; x_m is the ideal maximum motion amplitude (where the flux linkage Ψ_{PM} is maximum).

Now the basic principles are elucidated, and we may proceed to the first category of LMAs—those with coil movers and stator PMs.

12.3 PM-LMA with Coil Mover

The PM-LMA with coil mover stems from the microphone (and loud speaker) principle (Figure 12.5). The ring-shaped PM is placed within a cylindrical space in the air gap. The air gap is partially filled with the coil mover. The coil mover is made of ring-shaped turns placed in an electrical insulation keeper that is mechanically rugged enough to withstand large forces. As the coil moves back and forth axially, the PM field induces an emf in the coil that is proportional to the linear speed dx/dt, the flux density B_{gPM}, the average length of the turn, and the number of active turns n_c' per PM length l_{PM}:

$$e(t) = -\frac{dx}{dt} \cdot n_c' \cdot l_{av} \cdot B_{PM_{av}} \tag{12.19}$$

The total number of turns per coil n_c corresponds to the total coil length, which surpasses the PM length l_{PM} by the stroke length $l_{stroke} = 2x_{\max}$:

$$n_c = n_c' \frac{(l_{stroke} + l_{PM})}{l_{PM}} \tag{12.20}$$

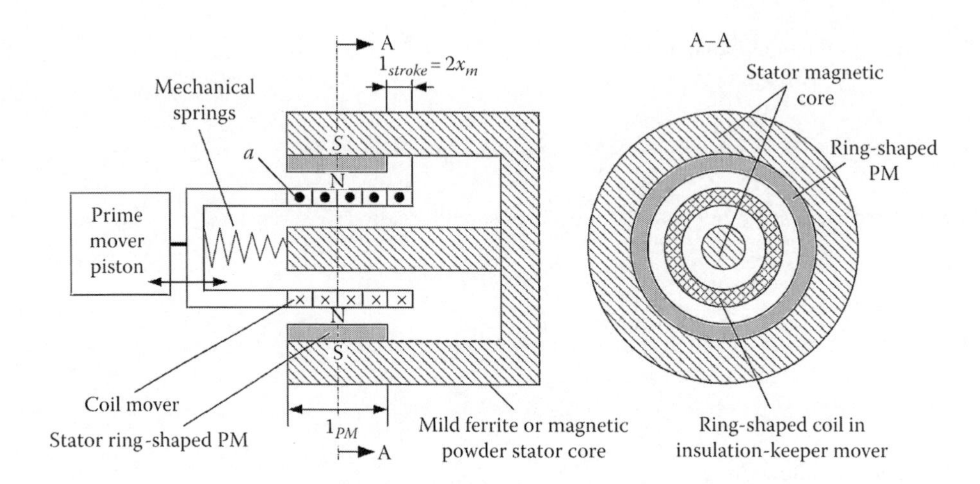

FIGURE 12.5 Homopolar linear motion alternator (LMA) with coil mover.

This means that only a part of the coil is active in terms of emf, while the whole coil intervenes with its resistance and inductance. Alternatively, the PMs may be longer than the coil by the stroke length.

As the motion is considered harmonic, the emf $e(t)$ is sinusoidal, unless magnetic saturation on load or PM flux fringing influences B_{PMav}, making it slightly dependent on the mover position.

It should also be noted that the homopolar character of the PM magnetic field leads to a large magnetic core of the stator, while the placement of mechanical springs is not an easy task either.

The mechanical rigidity of the coil mover with flexible electrical terminals is not high. On the good side, the mover weight tends to be low, and the copper use and heat transfer surface are large, allowing for large current densities and, thus, lower mover size, at the price of larger copper losses.

The average linear speed in LMA is rather low. For example, for speed at $f_1 = 60$ Hz and $x_m = 10$ mm, the maximum speed $U_{\max} = x_m\omega_r = 0.01 \cdot 2\pi \cdot 60 = 3.76$ m/s. The average speed $U_{av} = 4x_m f_1 = 4 \cdot 0.01 \cdot 2\pi \cdot 60$ m/s. For given electromagnetic power, say $P_{elm} = 1200$ W, it would mean an average electromagnetic force $F_{e_{av}} = P_{elm}/U_{av} = 1200/2.4 = 500$ N.

The total magnetic air gap h_M, which includes the two mechanical gaps g, plus the coil radial height h_{coil}, and the PM height h_{PM} is as follows:

$$h_M = h_{PM} + h_{coil} + 2g < (2 - 2.2)h_{PM} \tag{12.21}$$

The condition in Equation 12.21 preserves large enough PM flux density levels in the air gap with a good use of PMs.

For larger powers (forces), heteropolar LMA, with multiple coils connected in counter-series or in antiparallel, may be conceived as multipolar structures (Figure 12.6). As visible in Figure 12.6, the mover now contains a magnetic core that makes it more rugged, but it adds weight. The rather small pole pitch of the PM placing makes, however, the thickness of the mover core rather small, in general, $l_{PM} \approx (1\text{–}2)\, l_{stroke}$. As the actual PM flux density is below $B_r/2$, and only half of the PM pole flux flows through the back core, the mover core thickness is, in general, $h_{cr} \le l_{PM}/4$, even for $B_r = 1.2$ T. The larger the l_{PM}/l_{stroke}, the better the copper utilization, but this comes at the price of additional iron in the back cores. An optimum situation in terms of efficiency, another in terms of costs, and yet another in terms of mover weight is felt to exist when such a coil mover LMA is designed.

Note that the increase in mover weight poses severe limitations on the frequency of stable oscillatory motion by the prime-mover control.

As the homopolar LMA is treated in detail elsewhere [1], we will concentrate here on the multipolar (heteropolar) LMA with coil plus iron mover.

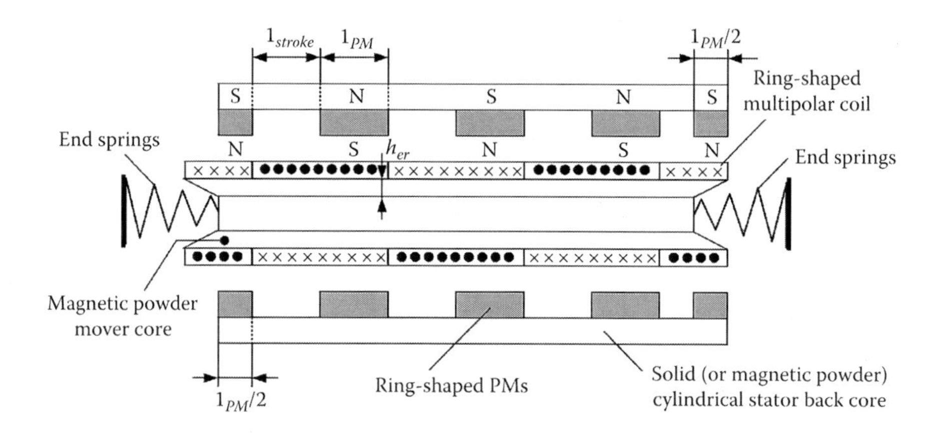

FIGURE 12.6 Heteropolar (multipolar) permanent magnet linear motion alternator (PM-LMA) with coil mover in extreme left position.

12.4 Multipole LMA with Coil Plus Iron Mover

The tubular configuration in Figure 12.5 that constitutes the multipole LMA with coil plus iron mover is suitable for high-force applications, because as the force increases, both the mover external diameter and the number of poles $2p_1$ may be increased. This is not the case with the homopolar configuration (Figure 12.3), where only the mover diameter may be increased when more force is needed. In addition, the force at zero current (cogging force F_{cog}) and the reluctance force (F_r) are both zero. The PM flux distribution and force production may be calculated with precision by two-dimensional (2D) FEM.

The stator core may be made of solid mild iron, while the mover core below the coils has to be fabricated from magnetic powders or from ring-shaped thin laminations with a filling factor of 0.95 or more to allow for large enough PM flux density in the air gap.

Magnetic saturation may be considered a second-order effect, as the coils are placed in air to reduce the mover weight and also to reduce machine inductance.

An approximate analytical model is always useful for a preliminary investigation. Let us pursue it here.

The detailed pole (section) geometry is as shown in Figure 12.7.

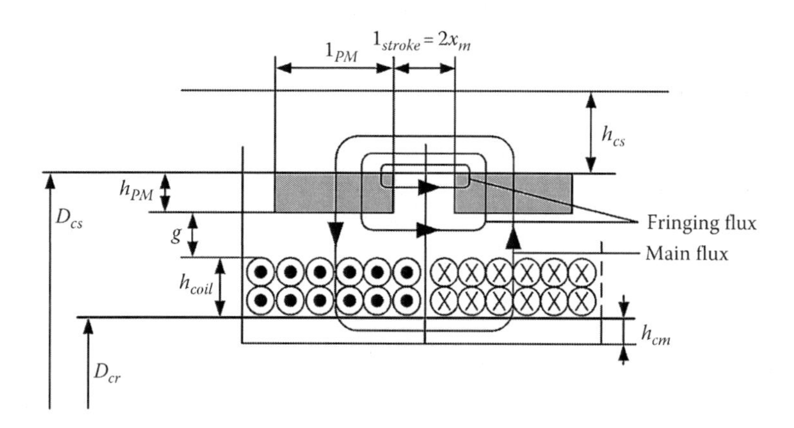

FIGURE 12.7 Geometrical details of multipole permanent magnet linear motion alternator (PM-LMA) with coil plus iron mover.

The air gap PM flux density in the air gap B_{gPM} is

$$B_{gPM} \approx \frac{B_r \cdot h_{PM} \cdot \mu_{rec}}{h_{PM} + (g + h_{coil}) \cdot \mu_{rec}} \cdot \frac{1}{(1 + k_{fringe}) \cdot (1 + K_s)}, \tag{12.22}$$

where k_{fringe} is the fringing factor that accounts for the PM flux leaking through the air gap and between neighboring PMs, axially.

The fringing coefficient k_{fringe} depends on $(1 + (g + h_{coil})/h_{PM})$ and on l_{stroke}/h_{PM}, besides the magnetic saturation in the mover and in the stator back cores.

Though approximate analytical expressions for k_{fringe} may be derived, FEM should be used to obtain reliable information in this matter.

In general, however, for a good design, $k_{fringe} < 0.3–0.5K_s$ takes care of magnetic saturation and is generally less than 0.05–0.15 in a well-designed machine.

The emf in the $2p_1$ coils in series, E, is as follows:

$$E(t) = B_{gPM} \cdot U(t) \cdot \pi \cdot D_{avc} \cdot 2p_1 \cdot n_c \cdot \left(\frac{l_{PM}}{l_{PM} + l_{stroke}} \right) \tag{12.23}$$

In Equation 12.23, only the part of the coils under the direct influence of PMs is considered to produce force, while the total number of turns/coil is n_c; D_{av} is the average coil-turn diameter.

The electromagnetic force F_e is as follows:

$$F_e(t) = B_{gPM} \cdot n_c \cdot \left(\frac{l_{PM}}{l_{PM} + l_{stroke}} \right) \cdot \pi \cdot D_{avc} \cdot 2p \cdot i(t) \tag{12.24}$$

The machine inductance L_s and resistance R_s are

$$L_s \approx \frac{1}{8} \cdot 2p \cdot \mu_0 \cdot n_c^2 \cdot \pi \cdot D_{avc} \frac{l_{PM} + l_{stroke}}{(h_{PM} + g + h_{coil})}$$
$$R_s = \rho_{co} \cdot \pi \cdot D_{avc} \cdot \frac{n_c^2 \cdot 2p}{(I_n n_c') / j_{con}} \tag{12.25}$$

where

I_n is the RMS value of rated current

j_{con} is the rated current density in the design

For forced air cooling, j_{con} = 6–10 A/mm²; otherwise, it is 4–6 A/mm². The factor 1/8 comes from the linear variation of current-produced field with position. The core losses occur mainly in the mover core; but, at least for electrical frequencies $f_1 \leq 50(60)$ Hz, they are likely to be only a fraction (10%–15% at most) of copper losses.

The machine behaves like a single-phase nonsalient pole PM alternator; and thus under steady state, the voltage equation in complex variables is as known (Equations 12.6 and 12.7):

$$\underline{V}_1 = -(R_s + j\omega_1 L_s)\underline{I}_1 + \underline{E}_1 \tag{12.26}$$

The general phasor diagram in Figure 12.4 remains valid.

As the stator and mover cores handle half the pole flux, the core depths h_{cs} and h_{cr} are as follows:

$$h_{cs} \approx \frac{B_{gPM}(1+K_{load})\cdot l_{PM}/2}{B_{cs}};$$
$$\frac{\pi\cdot\left(D_{cr}^2-\left(D_{cr}-2h_{cm}\right)^2\right)}{4}\cdot B_{cr} = \frac{B_{gPM}\cdot\pi\cdot(D_{cr}-2h_{PM})}{2}\cdot(1+k_{load})\cdot l_{PM} \quad (12.27)$$

The influence of circularity was considered in the mover, because when the mover diameter gets smaller, the latter has a notable influence.

The coefficient $K_{load} = 0.3–0.5$ accounts for the increase in flux density due to coil current for the case when the E equals in amplitude $I\omega_1 L_s$. This corresponds approximately to a power factor of $\cos\varphi_1 = 0.707$ (Figure 12.4), which is considered a good compromise between force density and energy conversion performance:

$$\tan\varphi_1 = \frac{I_1\omega_1 L_s}{E} \approx 1 \quad (12.28)$$

Example 12.1

Consider a multipolar PM-LMA with coil iron mover and the following specifications:

- Electrical output: $P_n = 22.5$ kW at $V_{1n} = 220\ V_{ac}$ (RMS)
- Efficiency: $\eta > 0.92$
- Load power factor: unity
- Frequency: $f_1 = 60$ Hz
- Stroke length (imposed by the prime mover): $l_{stroke} = 30$ mm
- Harmonic motion $x = \frac{l_{stroke}}{2}\sin 2\pi\cdot f_1\cdot t$

The task is to design the machine and calculate its performance.

Solution

We shall remember that a good design requires mechanical springs attached to the mover, sized at resonance, $2\pi f_1 = \sqrt{K/m_t}$, where K is the spring coefficient, and m_t is the total mover mass.

There are two ways of handling the unity power factor load: with or without a capacitor (in series with the machine phase).

In order to provide the largest stroke length for lower copper losses, not only are resonant conditions required, but also, emf $\underline{E}_1$ and current $\underline{I}_1$ should be in phase.

In this case, however, the power factor is mandatory lagging. Consequently, a compensating capacitor is necessary:

$$\underline{V}_1 = -\left[R_s + j\left(\omega_1 L_s - \frac{1}{\omega_1 C_s}\right)\right]\underline{I}_1 + \underline{E}_1 \quad (12.1.1)$$

with

$$\omega_1 L_s = \frac{1}{\omega_1 C_s} \quad (12.1.2)$$

$$V_1 = E_1 - R_s I_1 \quad (12.1.3)$$

The compensating capacitor C_s makes the alternator behave like a direct current (DC) generator in the sense that only the resistive voltage drop counts. The short circuit has to be avoided.

Once these fundamental design aspects are elucidated, we may proceed to dimensioning.

- The electromagnetic force F_{en} is

$$F_{en} = \frac{P_n}{\eta_n \cdot U_{ave}}; \quad U_{ave} = 2 \cdot l_{stroke} \cdot f_1 \tag{12.1.4}$$

or,

$$U_{ave} = 2 \cdot 30 \cdot 10^{-3} \cdot 60 = 3.6 \text{ m/s}$$

$$F_{en} = \frac{22.5 \cdot 10^3}{0.92 \cdot 3.6} = 6793.5 \text{ N}$$

- The mover core external diameter D_{cr} may be obtained based on a given specific force f_{tn}, for a given number of poles $2p_1$:

$$\pi \cdot D_{cr} = \frac{F_{en}}{f_{tn} \cdot 2p_1 \cdot l_{PM}} \tag{12.1.5}$$

The PM pole length l_{PM} is not yet known, and a value should be adopted in relation to the stroke length l_{stroke}. Let us consider $l_{PM}/l_{stroke} = 2$, and the number of poles $2p = 4$, with $f_{tn} = 4$ N/cm^2.

- The PM air gap flux density B_{gPM} (Equation 12.22; for neodymium–iron–boron [NdFeB] PMs with $B_r = 1.2$ T and $\mu_{re} = 1.07$) has to first be assigned an approximate value to be adjusted later, as is shown in what follows.
- Let us consider $K_{fringe} = 0.2, K_s = 0.05$ and $h_{PM\mu rec}/(h_{PM\mu rec} + g + h_{coil}) = 1/2$.

$$B_{gPM} \approx 1.2 \cdot \frac{1}{2} \cdot \frac{1}{(1+0.2)(1+0.05)} = 0.4762 \text{ T}.$$

From the force equation (Equation 12.24), the ampere-turns per coil $n_c I_{av}$ is obtained ($l_{PM} = 2l_{stack} = 2 \cdot 30 = 60$ mm):

$$F_{eav} = B_{gPM} \frac{l_{PM}}{l_{PM} + l_{stack}} \cdot \pi \cdot D_{avc} \cdot 2p \cdot n_c \cdot I_{av} \tag{12.1.6}$$

For sinusoidal current (corresponding to harmonic motion),

$$I_{av} = I_n \sqrt{2} \cdot \frac{2}{\pi}, \quad I_n\text{—RMS value} \tag{12.1.7}$$

- The average turn diameter is not known exactly; but if we consider D_{cr} instead of D_{avc}, the computation renders conservative results, as obtained from Equation 12.1.6:

$$6793.5 = 0.4762 \cdot \frac{1}{1+(1/2)} \cdot \pi \cdot 0.2313 \cdot n_c \cdot \frac{2\sqrt{2}}{\pi} I_n$$

$n_c I_n = 8201$ A turns/coil. (There are $2p_1 = 4$ coils.)

- These coil ampere-turns spread over a length $l_{PM} + l_{stroke} = 60 + 30 = 90$ mm. The filling factor K_{fill} for the coil, in such a premade coil, could be up to 0.6. Consequently, the height of the coil h_{coil} is

$$h_{coil} = \frac{n_c I_n}{j_{con} K_{fill} l_{PM}} = \frac{8201}{9 \cdot 10^6 \cdot 0.6 \cdot 0.06} = 38 \text{ mm} \tag{12.1.8}$$

- The total air gap $g = 2 \cdot 1.5$ mm = 3 mm. With $\mu_{rec} = 1.07$ from the adopted ratio 1/2,

$$\frac{h_{PM}}{h_{PM} + (g + h_{coil})/\mu_{rec}} = \frac{1}{2} \tag{12.1.9}$$

we obtain $h_{PM} \approx 44$ mm.

Both the coil height h_{coil} and the PM height h_{PM} seem to be reasonable for the force and power involved.

- To size the capacitor C_s, the machine inductance is calculated from Equation 12.25:

$$L_s \approx \frac{1}{8} \cdot 2p_1 \cdot \mu_0 \cdot n_c^2 \cdot \pi \cdot D_{avc} \cdot \frac{(l_{PM} + l_{stroke})}{(h_{PM} + g + h_{coil})(1 + K_s)}$$

$$= \frac{1}{2} 4 \cdot 1.256 \cdot 10^{-6} \cdot \pi \cdot \frac{0.2313 + 2 \cdot 0.0380.06 + 0.03}{0.044 + 0.003 + 0.038} n_c^2 = 0.55 \cdot 10^{-6} \cdot n_c^2 H$$

The stator resistance R_s (Equation 12.25) is as follows:

$$R_s = \rho_{CO} \frac{\pi \cdot D_{avc} \cdot n_c^2 \cdot 2p_1}{I \cdot n_c / j_{con}}$$

$$= \frac{2.3 \cdot 10^{-8} \cdot \pi \cdot 0.2693 \cdot 4 \cdot n_c^2}{(8201/9 \cdot 10^6)} = 8.5374 \cdot 10^{-5} \cdot n_c^2 \tag{12.1.10}$$

The electrical time constant T_e is

$$T_e = \frac{L_s}{R_s} = \frac{0.55 \cdot 10^{-6} \cdot n_c^2}{1.5 \cdot 5.69 \cdot 10^{-5} \cdot n_c^2} = 0.00644 \text{ s}$$

- The copper losses P_{con} are, simply,

$$P_{con} = R_s I_n^2 = 1.5 \cdot 5.69 \cdot 10^{-5} \cdot 8201^2 = 5.74 \text{ kW}$$

As can be seen, P_{con} are much larger than 10% of output; thus, the efficiency target is too large.

The current density j_{con} may be decreased to increase efficiency; but in this case, the size of the coil height (h_{coil}) and, consequently, the PM height have to be increased also. This is feasible for the case in point.

- There are many ways of continuing this design toward optimization, based on minimum material costs, or on $\eta \cdot \cos\varphi$, etc., separately or combined.
- Here, we verify the power factor angle φ_1 first (Equation 12.28), with E_1 from

$$E_1 = \frac{F_{en}(U_{1\max}/\sqrt{2})}{I_n} = C_E \cdot n_c \tag{12.1.11}$$

$$C_e = \frac{F_{en} \cdot U_{\max}}{\sqrt{2} \cdot n_c \cdot I_n} = \frac{6793.5 \cdot 0.03 \cdot \pi \cdot 60}{\sqrt{2} \cdot 8201} = 3.32$$

so,

$$\tan\varphi_1 = \frac{I\omega_1 L_s}{E_1} = \frac{I \cdot \omega_1 \cdot 0.55 \cdot 10^{-6} \cdot n_c^2}{3.32 \cdot n_c} = 0.512$$

and

$$\varphi_1 = 27.11°, \quad \cos\varphi_1 = 0.89$$

This is an almost practical design value.

We may compensate the whole reactive power of L_s through a capacitor C_s:

$$C_s = \frac{1}{\omega_1^2 \cdot L_s} = \frac{10^6}{(2\pi 60)^2 \cdot 0.55 \cdot n_c^2} = \frac{12.806}{n_c^2}\ \mathrm{F} \tag{12.1.12}$$

The reactive power in the capacitor Q_{Cs} is

$$Q_{Cs} = \frac{I_n^2}{\omega_1 C_s} = \frac{(n_c I_n)^2 2.25}{2\pi \cdot 60 \cdot 12.806} = 13.93\ \mathrm{kVAR} \tag{12.1.13}$$

If the machine works in the presence of series capacitor C_s, then the voltage equation (Equation 12.1.3) applies:

$$V_n = 220 = -R_s I_n + E_1 = -5.69 \cdot 1.5 \cdot 10^{-5} \cdot n_c \cdot (n_c I_n) + 3.32 n_c \tag{12.1.14}$$

The number of turns $n_c \approx 77$ turns/coil.

The wire diameter d_{co} is

$$d_{co} = \sqrt{\frac{4 \cdot I_n \cdot n_c}{\pi \cdot j_{con} \cdot n_c}} = \sqrt{\frac{4 \cdot 8201}{\pi \cdot 9 \cdot 10^6 \cdot 77}} = 3.929 \cdot 10^{-3}\ \mathrm{m} \tag{12.1.15}$$

This is way too large a value; thus, quite a few conductors in parallel are required, as the current $I_n = n_c I_n / n_c = 8201/77 = 106.50$ A.

Alternatively, a copper sheet may be used, as the conductor area required A_{co} is as follows:

$$A_{co} = \frac{I_n}{j_{con}} = \frac{106.50}{9 \cdot 10^6} = 11.83\ \mathrm{mm}^2 \tag{12.1.16}$$

Building the coils 1.183 mm thick may make 10 mm wide rectangular conductors feasible.

- As the power factor is large enough, it is possible to give up the capacitor. In the absence of a capacitor, the full voltage equation (Equation 12.1.1), with current and emf in phase and $C_s = \infty$, is to be used to calculate the number of turns per coil. A value larger than 77 will be found. At the same $n_c I_n$ and voltage V_1, less power will be delivered, though.

Note on Performance Characteristics

Once the machine parameters are known, the voltage and force equations may be applied to calculate steady-state performance, as in any synchronous machine.

A few remarks on the coil plus iron mover PM-LMA are in order:

- As all LMAs, the coil plus iron mover LMA, with coils in an insulation keeper, is a low-speed machine for which it is advisable to work at mechanical resonance frequency; that is, electrical frequency is equal to mechanical resonance frequency.
- The coil layer provides an extra air gap; thus, the inductance of the machine is reasonably low. Consequently, the power factor is good for a specific force $f_t = 4$ N/cm^2 in the 22.5 kW, 3.6 (average) m/s at 60 Hz, design example.
- The reluctance and cogging forces are theoretically zero.
- The main reliability disadvantage is the presence of flexible electrical terminals to extract the electric power from the mover. As the speed is small, even copper ring brushes may be used instead.
- The copper losses are large, as each coil has to cover not only a PM pole span but also the full stroke length. It is feasible to make the PMs longer than the coils at the price of additional PM costs.
- To make the machine more rugged, we may let the PM part move and the coils be at standstill, while putting the PM part inside the coil. But then, a PM-mover-type configuration is obtained.

12.5 PM-Mover LMAs

A few PM-mover LMAs are shown in Figures 12.8 through 12.11. The configuration in Figure 12.8 contains, basically, a single ring-shaped stator coil surrounded by a U-laminated (or magnetic powder) core and a two-pole cylindrical mover with surface PMs enclosed in an insulation retainer.

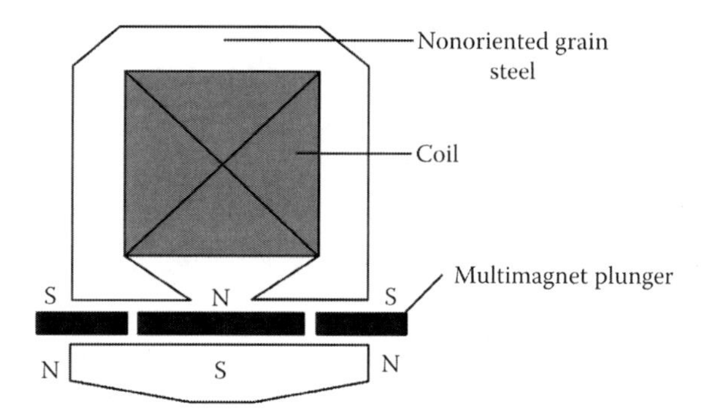

FIGURE 12.8 Linear motion alternator (LMA) with radial air gap and tubular PM mover.

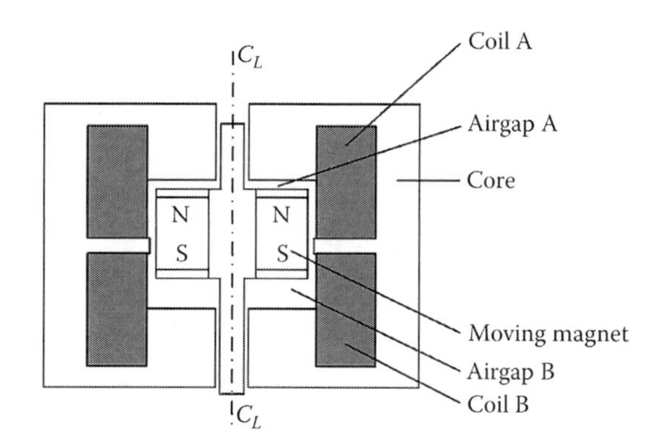

FIGURE 12.9 LMA with axial air gap and PM mover.

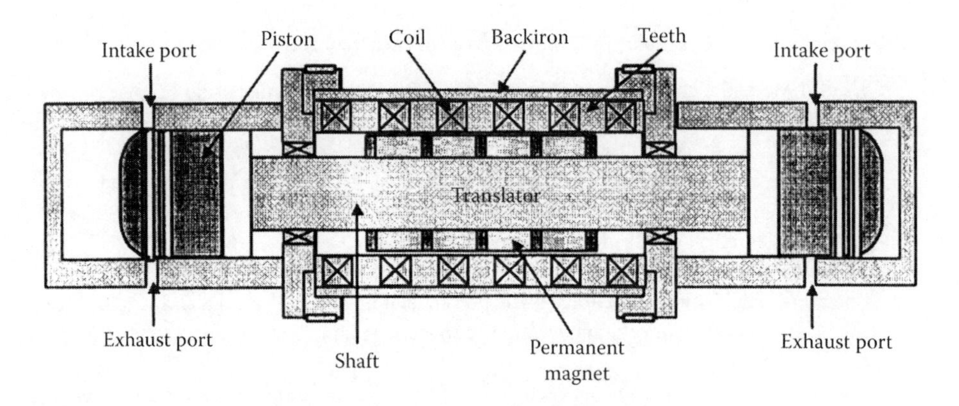

FIGURE 12.10 LMA with PM multipolar mover and air core stator winding.

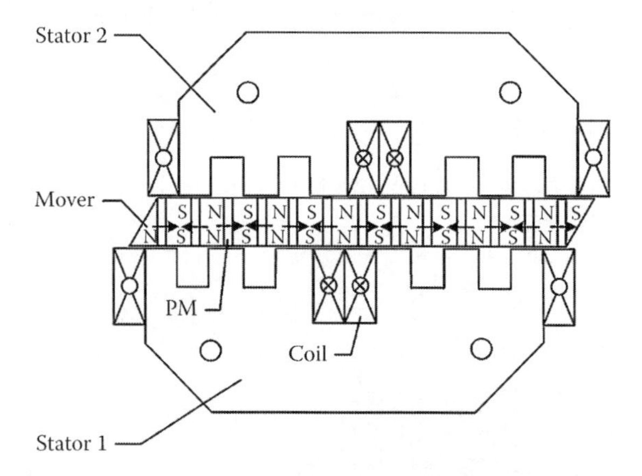

FIGURE 12.11 Flux reversal linear motion alternator (FR-LMA) with flat PM flux concentration.

The interior stator may also be made of U-shaped laminations or of magnetic powder. The magnetic powder makes the machine more manufacturable, but it essentially reduces the force, for the same PM weight and geometry, by about 30% (for $\mu_{re} = 500$ in magnetic powder).

As the mover advances from the extreme left position by half the PM pole pitch $\tau/2$, the PM flux linkage in the coil switches polarity. Therefore, we may call the configuration a flux reversal machine [2], though it was proposed earlier than the flux reversal machines (see Reference 1).

The fragility of the mover and the leakage flux of the PM half-poles are, together with the manufacturing difficulties, the main liabilities of this otherwise moderate force density, good efficiency LMA.

The axial air gap PM-mover tubular LMA in Figure 12.9 is good for small stroke length (up to 10 mm maximum) but retains the manufacturability liabilities of the configuration in Figure 12.8. Additionally, it has a smaller usage of coils with flux linkage that switches from maximum to minimum without changing polarity.

The tubular configuration in Figure 12.10 is a typical multipole PM mover air multicoil stator, with the number of coils larger than the number of PM poles by one, in general.

This type of LMA behaves basically as the one in the previous paragraph in all ways, but the placement of the air coils on the stator removes the necessity of flexible electrical terminals.

These configurations do not take advantage of PM flux concentration; but instead, they show a reasonably small machine inductance.

In an attempt to increase the force density at the cost of larger inductance, the flat-mover PM flux concentration concept was matched with a multiple teeth (Figure 12.11) variable reluctance stator. There are two or four concentrated coils (Figure 12.11) in this machine.

The concept of flux reversal is used [3] but with a PM mover. The two stators are shifted by a PM pole pitch (or one stroke length). As the mover moves from the extreme left to the extreme right position, the PM flux linkage in the coils reverses polarity with the contribution of all PMs all the time.

Alternatively, the mover PM flux could be closed in a plane transverse to the motion direction when the concept of rotary transverse flux PM machine is applied (Figure 12.12a and b), with or without PM flux concentration [4,5].

The transverse flux linear motion alternator (TF-LMA) configuration with PM flux concentration (Figure 12.12a) works as does the flux reversal linear motion alternator FR-LMA (Figure 12.11) but requires a special frame to hold the U-shaped stator cores and the stator coils. In the FR-LMA, the stator cores, premade of axial laminations and assembled in one rugged stack (which does not need spacers, etc.), make it essentially more manufacturable. At the price of additional stator core weights, though.

In contrast, the surface PM mover TF-LMA (Figure 12.12b) makes poor use of PMs. It has a smaller inductance that leads to a reasonable power factor, but at a lower force density than for the configuration with PM flux concentration.

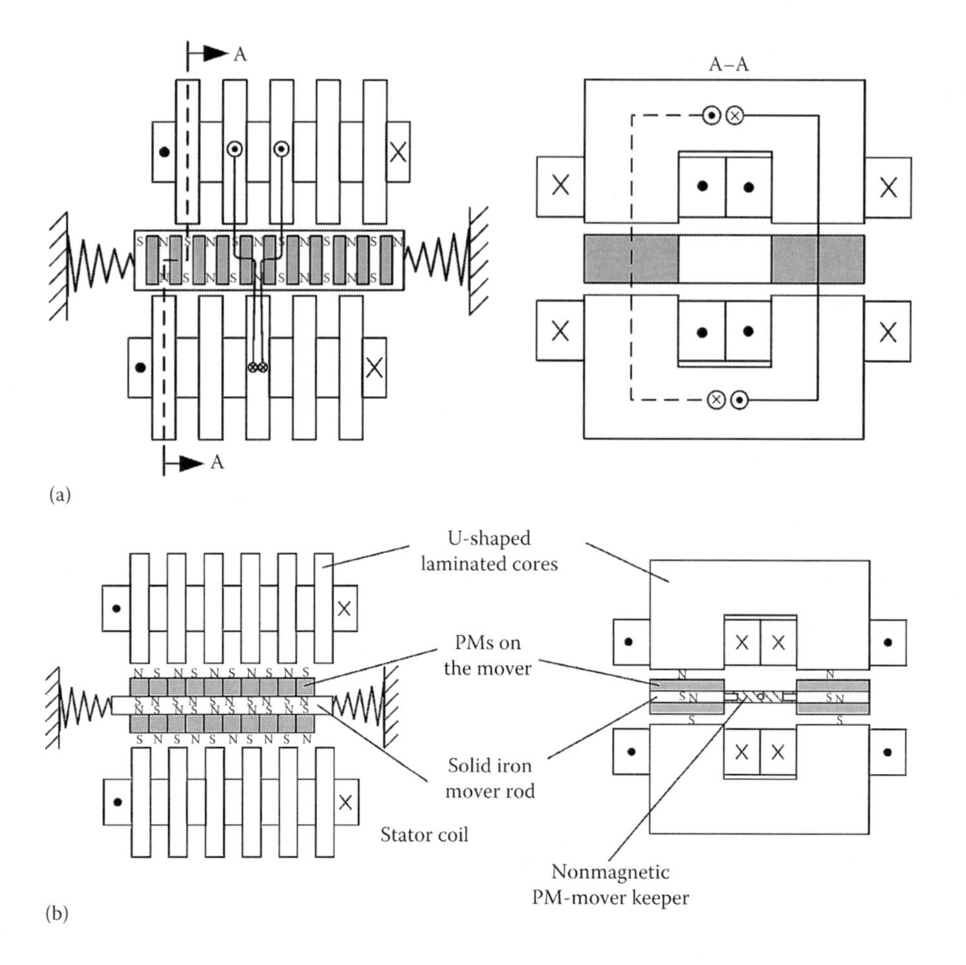

FIGURE 12.12 Generic PM mover TF-LMAs: (a) with PM flux concentration and (b) without PM flux concentration.

Double-sided flat configurations were chosen for both TF-LMAs and FR-LMAs to secure ideal zero mover to stator total normal (attraction) force.

In the following paragraphs, we will deal with the theory and design of the tubular PM mover single-coil LMA (Figure 12.8) and the FR-LMA with flux concentration (Figure 12.11).

The TF-LMA with PM flux concentration works similarly to the FR-LMA with flux concentration and, thus, is not pursued further in this chapter.

12.6 Tubular Homopolar PM Mover Single-Coil LMA

The basic configuration for this type of LMA (Figure 12.8) has a long fragile mover with external PM half-poles that produce, through their time-variable leakage flux, eddy current losses in the machine frame. Additional frame is needed to reduce the eddy currents by placing the conducting frame parts away from the PM leakage fields. Alternatively, those fields may be magnetically contained without causing additional forces or losses.

This is why a single PM pole rotor may be preferred, at the price of halving the air gap PM flux density (Figure 12.13). A large diameter short-length aspect ratio should be aimed for, as the active length is just two stroke lengths ($2l_{stroke}$), to keep the mover mechanically rugged-rigid enough. In addition, a large outer stator bore diameter allows for an interior stator coil for better volume utilization.

The homopolar PM mover exhibits a dynamically unstable central position. The mechanical springs are responsible for holding the mover in the central position when the machine is turned off.

As the total length of iron in front of a PM (radially) is the same, irrespective of mover position, the cogging force is small and does not have to be considered for preliminary designs.

Also, the machine inductance is independent of mover position: L_s = constant.

The influence of magnetic saturation is small due to the large magnetic air gap.

Applying Ampere's and Gauss's laws, we may find the average air gap PM flux density in the air gap above the PM in the extreme right position (Figure 12.14):

$$B_{gPM} \approx B_r \cdot \frac{h_{PM}}{h_{PM} + \mu_{rec}(4g + h_{PM})} \cdot \frac{1}{(1 + K_{fringe})(1 + K_s)} \tag{12.29}$$

Thick magnets are required ($h_{PM} \gg 4g$) to provide a reasonable average PM flux density.

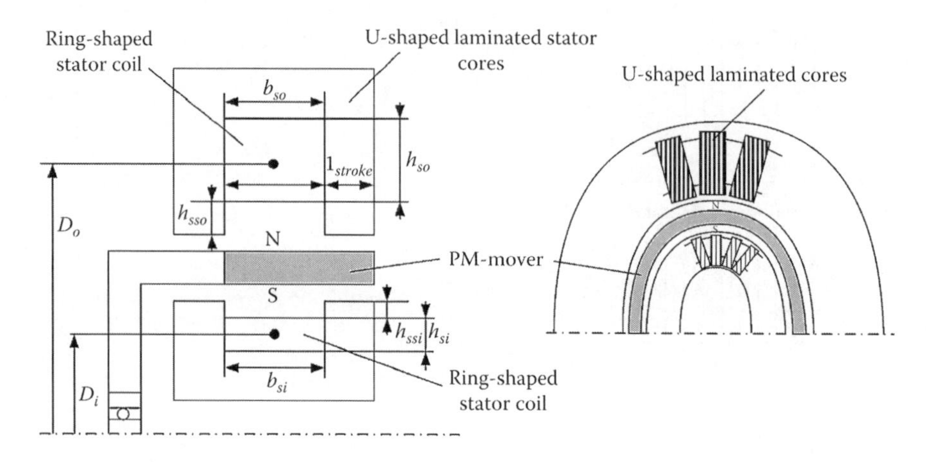

FIGURE 12.13 Unipolar PM rotor LMA (extreme right position).

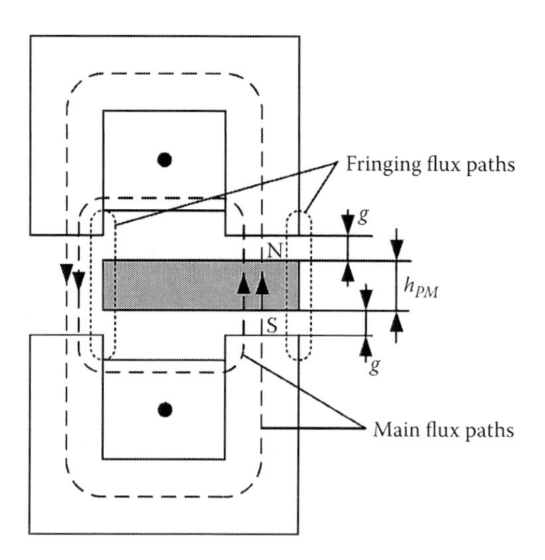

FIGURE 12.14 PM flux paths.

Adopting harmonic motion and a linear variation of PM flux linkage in the coils from $\Psi_{P\max}$ to $-\Psi_{P\max}$, the emf varies sinusoidally in time:

$$E(t) = -2\pi f_1 B_{gPM} l_{stack} \pi D_{mav}(N_o + N_i)\cos(2\pi f_1 t). \tag{12.30}$$

where
D_{mav} is the mover average diameter
N_o and N_i are the outer and inner coil number of turns

The machine inductance has two components, L_l and L_m, corresponding to slot leakage flux and main flux path:

$$L_l = \mu_0 N_i^2\left(\frac{h_{si}}{3b_{si}} + \frac{h_{ssi}}{b_{si}}\right)\pi D_i + \mu_0 N_o^2\left(\frac{h_{so}}{3b_{so}} + \frac{h_{sso}}{b_{so}}\right)\pi D_o \tag{12.31}$$

$$L_m \approx \frac{\mu_0 \pi D_{mav}(N_o + N_i)^2 l_{stack}}{(1+K_s)(4g + h_{PM}(1+\mu_{rec}))} \tag{12.32}$$

where D_i and D_o are inner and outer coil average diameters; all the other dimensions are as shown in Figure 12.13.

For harmonic motion,

$$x = \frac{l_{stack}}{2}\sin(2\pi f_1 t) \tag{12.33}$$

the instantaneous speed $u(t)$ is

$$u(t) = \frac{dx}{dt} = \frac{l_{stack}}{2} 2\pi f_1 \cos(2\pi f_1 t) \tag{12.34}$$

For linear flux/position dependence, $E(t)$ is in counterphase to the linear speed $u(t)$, which explains the negative sign (−) in Equation 12.30.

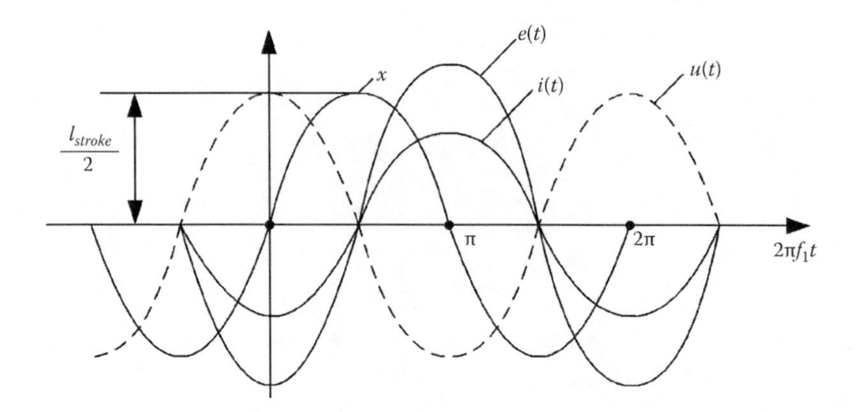

FIGURE 12.15 Position (x), speed (u), electromagnetic field (emf) (e), and current $i(t)$ for harmonic motion ideal condition (linear PM flux/position dependence).

For sinusoidal current, instantaneous force $F_e(t)$ is as follows:

$$F_e(t) = \frac{E(t) \cdot i(t)}{u(t)} = -\frac{B_{gPM} \cdot l_{stack} \cdot \pi \cdot D_{mav}(N_o + N_i)}{l_{stack}/2} \cdot I\sqrt{2}\cos(2\pi f_1 t) \tag{12.35}$$

Again, the largest force/ampere occurs for the current in phase with emf. The emf, current, speed, and position variations in time for $\gamma_1 = 0$ and harmonic motion are shown in Figure 12.15.

In reality, toward the end of excursion, the flux rise with position slows, and basically, a third harmonic occurs in the emf in addition to the fundamental.

Because of magnetic saturation, even harmonics may occur in the emf, with the force being asymmetric with respect to the middle mover position. Through FEM, such aspects could be treated with reasonable accuracy.

Note that the voltage equation and steady-state performance for various values of stroke length for autonomous or constant voltage power grid loads are as in the previous paragraph dedicated to prime coil mover PM-LMA. A numerical design example follows.

Example 12.2

Consider a homopolar PM mover LMA with outer and inner coils and the following data:

- Mean diameter of inner coil $D_i = 0.3$ m
- Stroke length $l_{stroke} = 30$ mm
- Mechanical air gap $g = 1$ mm
- PM radial thickness $h_{PM} = 10$ mm
- Outer and inner slot dimensions as follows: $b_{so} = b_{si} = 30$ mm, $h_{so} = 45$ mm, $h_{si} = 15$ mm, $h_{sso} = h_{ssi} = 3$ mm, $B_r = 1.2$ T, $\mu_{rec} = 1.05$

Calculate the following:

- The rated current for $j_{con} = 6$ A/mm^2 and a slot filling factor $k_{fill} = 0.5$
- The emf RMS value at $f_1 = 60$ Hz and full stroke length
- The machine inductance and resistance L_s, R_s
- For rated current and emf in phase with the current, determine the terminal voltage and active and reactive rated powers P_n, Q_n at 60 Hz

Solution

- The useful window area for the coils $A_t = (A_o + A_i)$ is as follows:

$$A_t = h_{so}b_{so} + h_{si}b_{si} = 30(45+15) = 1650\ \text{mm}^2$$

With a filling factor $K_{fill} = 0.5$, the total RMS magnetomotive force (mmf) $F_o + F_i$ is

$$F_o + F_i = (N_o + N_i)I_n = A_t K_{fill} j_{con} = 1650 \cdot 0.5 \cdot 6 = 4950\ \text{A turns}$$

Finally,

$$I_n = \frac{(N_o + N_i)I_n}{N_o + N_i} = \frac{4950}{40+15} = 90\ \text{A}$$

- The mean mover diameter D_{mav} is

$$D_{mav} = D_i + h_{si} + 2h_{ssi} + 2g + h_{PM} = 300 + 15 + 2 \cdot 3 + 2 \cdot 1 + 10 = 333\ \text{mm}$$

Simply, the mean diameter of outer stator coils D_o is

$$D_o = D_{mav} + h_{so} + 2h_{sso} + 2g + h_{PM} = 333 + 2 \cdot 1 + 10 + 2 \cdot 3 + 40 = 391\ \text{mm}$$

To calculate the emf, we also need the air gap PM average flux density B_{gPM} (Equation 12.29):

$$B_{gPM} = B_r \frac{h_{PM}}{h_{PM} + \mu_{rec}(4g + h_{PM})} \cdot \frac{1}{(1+K_{fringe})(1+K_s)}$$

$$= 1.2 \cdot \frac{10}{10 + 1.05(4 \cdot 1 + 10)} \cdot \frac{1}{(1+0.4)(1+0.05)} = 0.3305\ \text{T}$$

The fringing factor $K_{fringe} = 0.4$ and the magnetic saturation factor $K_s = 0.05$ are conservative values in this equation.

It should be noted that the flux density is rather small, despite the use of good PMs (B_r = 1.2 T). Fringing plays an important role, together with the homopolar magnet configuration, in this notable reduction in performance.

The emf is now as follows (Equation 12.30):

$$E(\text{rms value}) = \frac{2\pi f_1}{\sqrt{2}} B_{gPM} \cdot l_{stroke} \cdot \pi \cdot D_{mav} \cdot (N_o + N_i)$$

$$= \frac{2\pi \cdot 60}{\sqrt{2}} \cdot 0.3305 \cdot 0.03 \cdot \pi \cdot 0.333 \cdot (40+15) = 152.376\ \text{V}$$

- The machine inductance components L_m, L_l are straightforward (Equations 12.31 and 12.32):

$$L_m = \frac{\mu_0 \cdot \pi \cdot D_{mav} \cdot (N_0 + N_i)^2 \cdot l_{stroke}}{(1+K_s) \cdot (4g + h_{PM}(1+\mu_{rec}))} = \frac{1.256 \cdot 10^{-6} \cdot \pi \cdot 0.333 \cdot (40+15)^2 \cdot 0.03}{(1+0.05) \cdot (4 \cdot 1 + 10(1+1.05))10^{-3}} = 4.633 \times 10^{-3}\ \text{H}$$

$$L_l = \mu_0 N_i^2 \left(\frac{h_{si}}{3b_{si}} + \frac{h_{ssi}}{b_{si}} \right) \pi D_i + \mu_0 N_o^2 \left(\frac{h_{so}}{3b_{so}} + \frac{h_{sso}}{b_{so}} \right) \pi D_o$$

$$= 1.25610-6 \left[40^2 \left(\frac{40}{3 \cdot 30} \cdot \frac{3}{30} \right) \cdot 0.333 + 15^2 \left(\frac{15}{3 \cdot 30} \cdot \frac{3}{30} \right) \cdot 0.391 \right] = 1.229 \cdot 10^{-3} \text{ H}$$

$$L_s = L_m + L_l = (4.633 + 1.229) 10^{-3} = 5.8626 \cdot 10^{-3} \text{ H}$$

The total resistance R_s is

$$R_s = \rho_{co} \frac{(\pi D_i N_i + \pi D_o N_o) I_n}{I_n} j_{con} = 2.3 \cdot 10^{-8} \pi \cdot (0.333 \cdot 15 + 0.381 \cdot 40) \cdot 6 \cdot 10^6 = 0.09742 \, \Omega$$

The rated copper losses are $P_{con} = R_s I_n^2 = 0.09742 \cdot 90^2 = 780$ W

- To calculate the terminal voltage, the phasor diagram is required (Figure 12.16):

$$\underline{V}_1 = -\underline{I}_1 (R_s + j\omega_1 L_s) + \underline{E}_1$$

From Figure 12.16,

$$-\varphi_1 = -\delta_v = \tan^{-1} \left(\frac{\omega_1 L_s I_1}{E_1 - R_s I_1} \right) = \tan^{-1} \left(\frac{2\pi \cdot 60 \cdot 5 \cdot 8620 \cdot 10^{-3} \cdot 90}{152.376 - 1.09742 \cdot 90} \right) = \tan^{-1}(1.384) = 54.15°$$

$$\cos\varphi_1 = 0.5856$$

This is a moderately low power factor. The machine is absorbing considerable reactive power.

The terminal voltage V_1 is, simply,

$$V_1 = \frac{(E_1 - R_s I_n)}{\cos\varphi_1} = \frac{152.376 - 0.09742 \cdot 90}{0.5856} \approx 245.2 \text{ V}$$

The delivered active power P_n is

$$P_n = V_1 I_n \cos\varphi_1 = 245.2 \cdot 90 \cdot 0.5856 = 12.924 \text{ kW}$$

The copper losses are about 6% of rated output power.

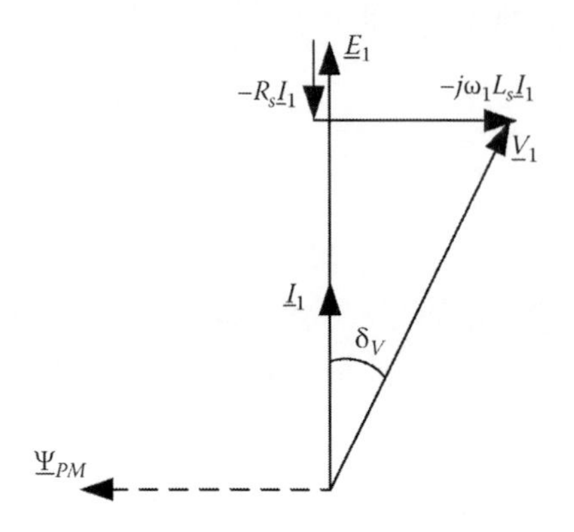

FIGURE 12.16 Phasor diagram for $\underline{E}_1$, $\underline{I}_1$ in phase.

The absorbed reactive power Q_n is

$$Q_n = V_1 I_n \sin\varphi_1 = 245.2 \cdot 90 \cdot \sin(-54.15°) = -17.88 \text{ kVAR}$$

A 17.88 kVAR series capacitor has to be added to compensate for this reactive power and thus make the LMA work with resistive voltage regulation.

12.7 Flux Reversal LMA with Mover PM Flux Concentration

The flat double-sided configuration in Figure 12.11 represents the flux reversal LMA with mover PM flux concentration (FR-LMA–FC). It is reproduced here in its cross-sectional form to assist in performance computation (Figure 12.17).

It should be noted first that the two identical stators are shifted by the maximum stroke length: l_{stroke}. The maximum stroke length is equal to PM pitch τ_{PM} and to half the small flat pitch τ_s in the stator poles: $\tau_{PM} = \tau_s/2 = l_{stroke}$.

To make the best use of copper in the coils, the coil turn geometry should be close to a quadrangle (which is closest to a circle). The PM flux linkage in the coils switches polarity from the extreme left position to the extreme right position by l_{stroke}.

All PMs are active all the time, but again, flux fringing is a severe limiting factor in performance.

The mechanical air gap should be small, to reduce fringing, but not too small, as the machine inductances increase too much. In principle, the machine inductance varies with mover position.

PM flux concentration is obtained as the h_{PM}/l_{stroke} ratio becomes greater than unity.

To overcome the natural increase in inductance, which is maximum in axis q (when the PM flux linkage in the coils is zero; Figure 12.15), the flux concentration has to be substantial, as it also has to cover for large fringing.

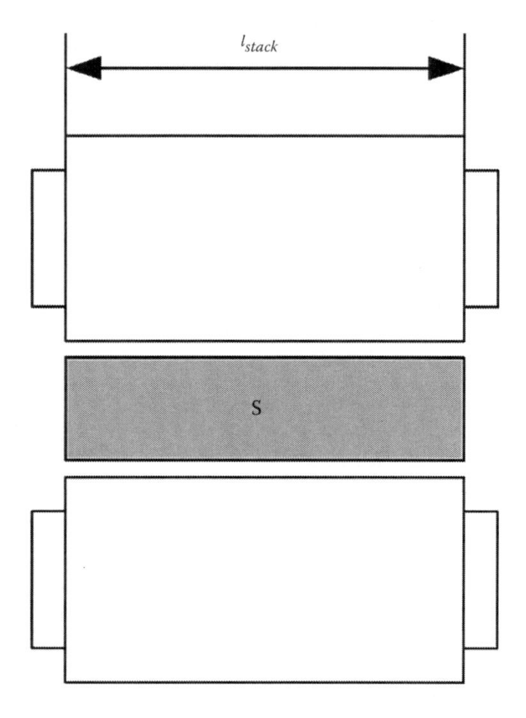

FIGURE 12.17 The FR-LMA with mover PM flux concentration (see also Figure 12.14).

Example 12.3

Let us consider an FR-LMA–FC with the following specifications:

- $P_n = 1$ kW
- $\eta_n \approx 0.93$
- $U_n = 120$ V
- $f_1 = 60$ Hz
- Stroke length: $l_{stroke} = 12$ mm

The task is to do preliminary sizing of such a machine.

Solution

First, we have to calculate the average air gap PM flux density under stator tooth for the mover position of maximum flux linkage in the stator coils:

$$B_{gPM}\tau_{PM} = \frac{2B_m l_{PM}}{(1+K_{fringe})} 2g\frac{B_{gPM}}{\mu_0} + H_m h_{PM} = 0 B_m = B_r + H_m \mu_{rec} \tag{12.3.1}$$

From Equation 12.3.1, after B_m and H_m are eliminated,

$$B_{gPM} = \frac{Br \cdot 2l_{PM}/\tau_{PM}}{(1+K_{fringe})\left(1+(\mu_{rec}/\mu_0)\cdot(4g \cdot l_{PM}/\tau_{PM} \cdot h_{PM})\right)} \tag{12.3.2}$$

with $2l_{PM}/\tau_{PM} = 5$, $K_{fringe} = 0.5$, $g = 10^{-3}$ m, $h_{PM} = 4 \times 10^{-3}$:

$$B_{gPM} = \frac{1.2 \cdot 5}{(1+0.5)(1+(1.05 \cdot 4 \cdot 10^{-3} \cdot 2.5)/(4 \cdot 10^{-3}))} = 1.103\ \text{T}$$

The maximum flux in a coil Φ_{max} is as follows:

$$\Phi_{max} = 2.5 \cdot B_{gPM} \cdot l_{stack} \cdot \tau_{PM} = 2.5 \cdot 1.103 \cdot l_{stack} \cdot 0.012 = 0.033 \cdot l_{stack}$$

The average force F_e is

$$F_e = \frac{P_n}{\eta_n \cdot U_{av}} = \frac{2000}{0.93 \cdot 2 \cdot 0.012 \cdot 60} = 746.7\ \text{N} \tag{12.3.3}$$

In addition, the total force has the following expression:

$$F_e = 4 \cdot \frac{2}{\tau_{PM}} \cdot \Phi_{PMax} \cdot \frac{1}{\sqrt{2}} \cdot (n_c I_n) \tag{12.3.4}$$

The stack length l_{stack} has to be determined based on the force density concept ($f_{tn} = 4$ N/cm^2 in our case):

$$\begin{gathered} F_e = f_{tn} \cdot 4 l_{stack} \cdot 5\tau_{PM} \\ l_{stack} = \frac{746.7}{4 \cdot 4 \cdot 10^4 \cdot 5 \cdot 0.012} = 0.078\ \text{m} \end{gathered} \tag{12.3.5}$$

Fortunately, l_{stack} is close to $5\ \tau_{PM} = 60$ mm.

From Equation 12.3.4, the ampere-turns per coil $n_c I_n$ can be calculated:

$$n_c I_n = \frac{F_e \cdot \tau_{PM} \cdot \sqrt{2}}{2 \cdot 4 \cdot 0.033 \cdot l_{stack}} = \frac{746.7 \cdot 0.012 \cdot \sqrt{2}}{0.264 \cdot 0.078} = 613.54 \text{ A-turns/coil} \quad (12.3.6)$$

For a large slot filling factor, $K_{fill} = 0.55$ (premade coils), the window area A_w is as follows:

$$A_w = \frac{2n_c \cdot I_n}{j_{con} \cdot K_{fill}} = \frac{2 \cdot 613.54}{6 \cdot 10^6 \cdot 0.55} = 371.8 \text{ mm}^2 \quad (12.3.7)$$

With the window width $W_s = 2\tau_{PM} = 24$ mm, the large slot height $h_s = A_w/W_s = 371.8/24 = 15.49$ mm. This small value is good for reducing the slot leakage inductance.

The emf is computable from the following:

$$E = \frac{F_e \cdot U_{av}}{I_n} = K_e \cdot n_c \quad (12.3.8)$$

or

$$K_e = \frac{746.7 \cdot 1.44}{613.54} = 1.752$$

Further on, the machine inductance L_s is obtained from its two components L_m and L_l:

$$L_m \approx 4\mu_0 \cdot n_c^2 \cdot \frac{3\tau_{PM} l_{stack}}{g} \cdot (1 + K_{fringe}) \cdot \frac{(\tau_{PM} - h_{PM})}{\tau_{PM}}$$
$$= 4 \cdot 1.256 \cdot 10^{-6} \cdot \frac{3 \cdot 0.012 \cdot 0.078}{1 \cdot 10^{-3}} (1 + 0.5) \cdot n_c^2 \cdot \frac{2}{3} = 1.3966 \cdot 10^{-5} n_c^2 \quad (12.3.9)$$

$$L_l \approx 4\mu_0 n_c^2 \left[l_{stack} \frac{h_s}{3W_s} + l_{end} \cdot 0.33 \right] \quad (12.3.10)$$

The turn end connection, length, l_{end}, is as follows:

$$l_{end} \approx l_{stack} + 10\tau_{PM} + \frac{3W_s}{2} = 0.078 + 10 \cdot 0.012 + \frac{3 \cdot 0.024}{2} = 0.234 \text{ m}$$

The factor 0.33 in Equation 12.3.10 represents the usual approximation used for single coils in induction machinery:

$$L_l = 4 \cdot 1.256 \cdot 10^{-6} \left[0.078 \cdot \frac{15.49}{3 \cdot 24} + 0.27 \cdot 0.33 \right] n_c^2 = 0.532 \cdot 10^{-6} n_c^2$$

The total resistance R_s with all coils in series R_s is as follows:

$$R_s = 4\rho_{co} \frac{l_{coil} \cdot n_c^2}{I_n \cdot n_c / j_{con}} = \frac{4 \cdot 2.3 \cdot 10^{-8} \cdot 0.324 \cdot n_c^2 \cdot 6 \cdot 10^6}{613.54} = 2.915 \cdot 10^{-4} n_c^2$$

The coil length l_{coil} is

$$l_{coil} \approx 2l_{stack} + 10\tau_{PM} + 4 \cdot \frac{W_s}{2} = 2 \cdot 0.078 + 10 \cdot 0.012 + 4 \cdot \frac{0.024}{2} \approx 0.324 \text{ m}$$

Let us consider that a series capacitor fully compensates the inductance L_s and, thus,

$$V_1 = -R_s \cdot I_n + E_1$$

$$120 = -2.915 \cdot 10^{-4} \cdot n_c \cdot (n_c \cdot I_n) + 1.752 \cdot n_c$$

$$\text{With } n_c \cdot I_n = 613.54 \text{ A turns}$$

$$n_c = 76 \text{ turns/coil}$$

Finally, $R_s = 1.6837\ \Omega$.

The rated current $I_n = n_c \cdot I_n / n_c = 613.54/76 = 8.0723$ A.

The machine maximum inductance, on axis q, is

$$L_s = L_m + L_l = (1.3966 \cdot 10^{-5} + 0.532 \cdot 10^{-6}) \cdot 76^2 = 0.08263 \text{ H}$$

The copper losses P_{con} are as follows:

$$P_{con} = R_s \cdot I_n^2 = 1.6537 \cdot 8.0723^2 = 109.7 \text{ W}$$

It is now visible that with this design, the efficiency target of 0.93 may not be reached, as the copper losses are already 10% of the output of about 1 kW: $P_n = I_n \cdot V_n = 8.073 \cdot 120 \approx 968$ W. Reducing the current density will result in increased efficiency.

It is important now to calculate the power factor angle φ_1 in the machine for rated output:

$$\varphi_1 \approx \tan^{-1}\frac{(I_n\omega_1 L_s)}{E_1} = \tan^{-1}\frac{(8.0723 \cdot 2\pi \cdot 60 \cdot 0.08203)}{1.752 \cdot 76} = \tan^{-1}(1.88) = 62°; \quad \cos\varphi_1 \approx 0.47$$

Note that the power factor is rather poor, as expected. Therefore, we will need a sizable capacitor to compensate for it:

$$Q_c = Q_{Ls} \approx \omega_1 \cdot L_s \cdot I_n^2 = 2\pi \cdot 60 \cdot 0.08263 \cdot 8.0723^2 = 2.029 \text{ kVAR(!)}$$

However, the machine size for 1 kW is acceptable with a total weight of less than 8 kg. What made it possible was heavy PM flux concentration with full use of PMs all the time.

FEM Analysis

For a similar prototype but with a 100 mm stack length, a detailed FEM analysis was performed [3].

The flux paths for the extreme right and middle positions are shown in Figure 12.17.

It is apparent that the flux is not quite zero in the middle position, as ideally it should be (Figure 12.18a and b).

The static forces versus position for constant DC ampere-turns per coils are shown in Figure 12.19.

It is clear that the force decreases toward the excursion ends and is not fully symmetrical with respect to the middle position. Then, when the current is sinusoidal and in antiphase with the speed (in phase with the emf), the force versus position looks as shown in Figure 12.20.

The presence of reluctance force due to mild magnetic saliency is evident in Figure 12.19, and it also appears in Figure 12.20. Figure 12.21 illustrates inductance versus position (for one turn per coil) and rated current.

The PM air gap flux density distribution for the extreme left position of the mover is apparent in Figure 12.22. It is clearly visible that the average design $B_{gPM} \approx 1.1$ T under the stator teeth is actually obtained. Therefore, the analytical predictions were realistic ($K_{fringe} = 0.5$).

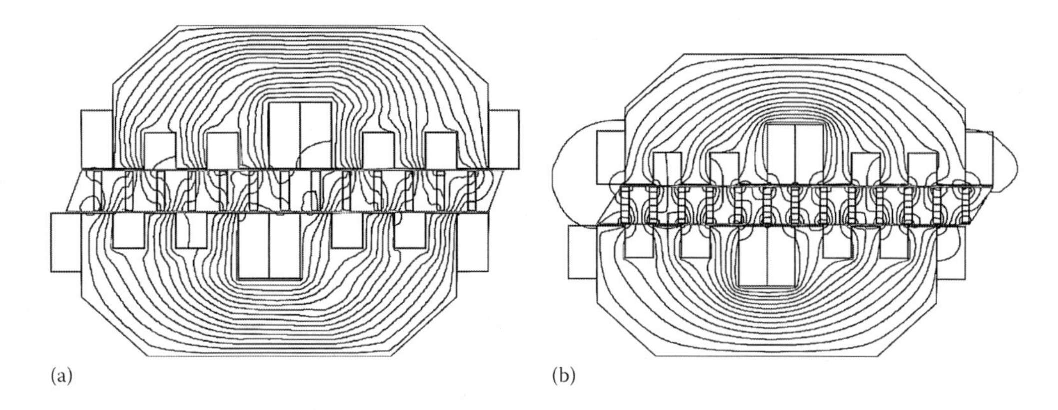

FIGURE 12.18 Flux paths (at zero current) for (a) extreme right and (b) middle position.

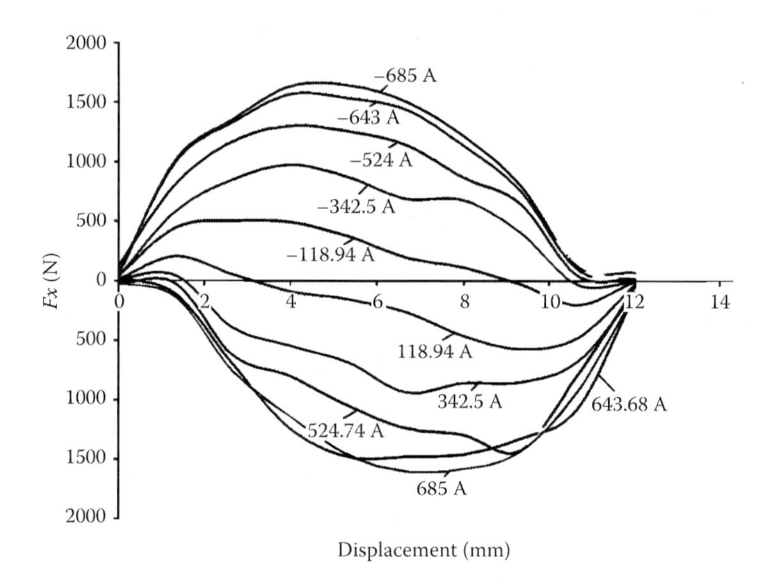

FIGURE 12.19 Static force vs. position for constant coil magnetomotive force (mmf) (RMS).

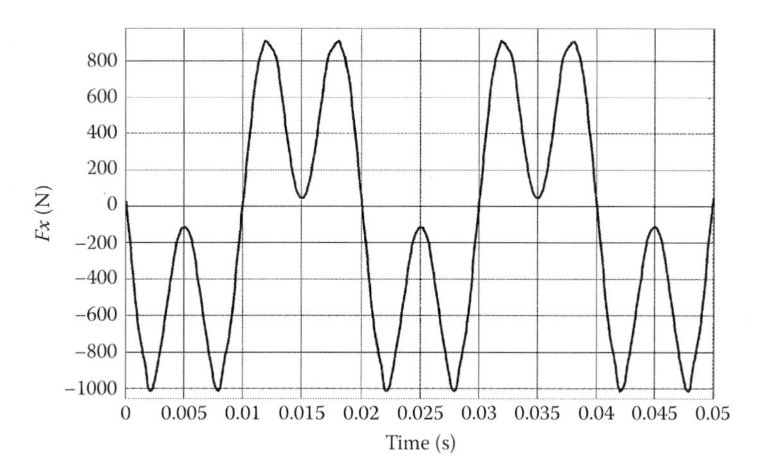

FIGURE 12.20 Thrust vs. position for sinusoidal current.

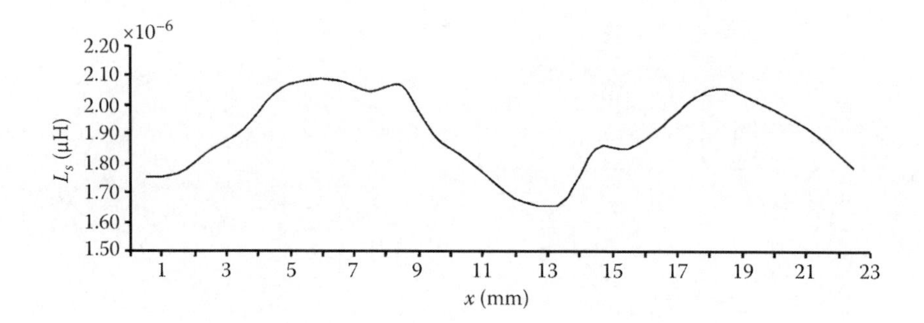

FIGURE 12.21 Inductance vs. position (for one turn per coil) and rated current.

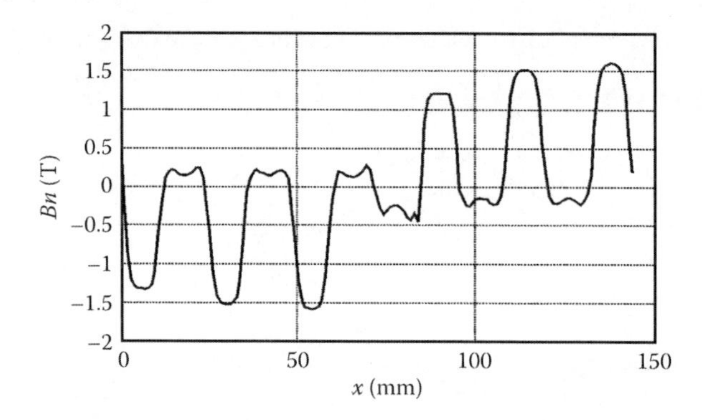

FIGURE 12.22 PM air gap flux density distribution at extreme left position of the mover.

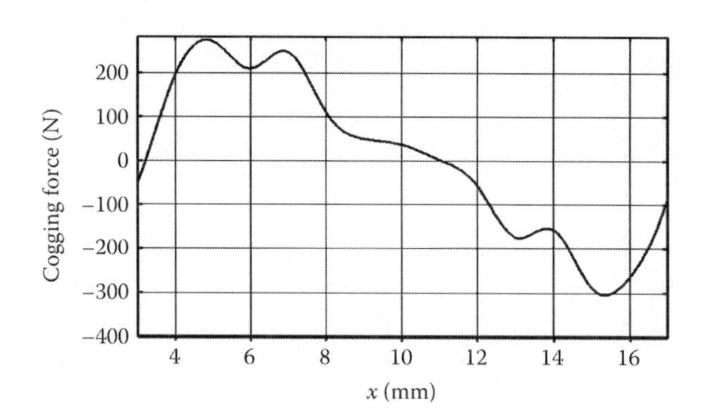

FIGURE 12.23 The cogging force.

Finally, the peak cogging force in Figure 12.23 of about 270 N is about 35% of the average electromagnetic force of 746 N. It is not zero exactly in the middle position, but it is monotonous (as in a spring) over almost 11 mm out of the 12 mm maximum stroke length. With strong springs to cover for this cogging force nonlinearity toward stroke ends, the operation is expected to be stable over the maximum stroke length of τ_{PM} = 12 mm. The cogging force saves some of the otherwise larger required mechanical spring material.

It may be stated that these analytical and FEM investigations corroborate well enough to be a solid base for refined designs, optimization, transients, and control.

12.8 PM-LMAs with Iron Mover

A few PM-LMAs with iron mover were proposed (Figure 12.24a through f). They may be classified into flux switch (Figure 12.24a through c) and flux reversal (Figure 12.24d) configurations. Flux switch refers here to homopolar change.

While configurations in Figure 12.24a through d are tubular, the ones in Figure 12.24e and f are flat and double sided.

It should be noted that the tubular configuration in Figure 12.24d [2] and the flat ones in Figure 12.24e and f [5] are based on the same, flux reversal, principle.

As both have about the same merits and demerits, we will treat here in some detail the tubular version, as it seems to be more manufacturable, and it also offers more flexibility in design (the number of radial poles may be $2p_1 = 4, 6, 8$). The lamination design does not depend on the stroke length because the flux paths are essentially in the radial plane, which is transverse to the motion direction. Another stator PM flux concentration configuration is introduced in Figure 12.24f.

12.9 Flux Reversal PM-LMA Tubular Configuration

The tubular configuration of the flux reversal PM-LMA (Figure 12.24d) takes advantage of the basic topology of the switched reluctance machine.

It conventionally uses stamped radial laminations with $2p_1$ radial poles. Along the axial direction, $2n$ PMs are attached to the stator poles with alternate polarity and a pitch of τ_{PM}.

The mover is also made of stamped laminations, but axially, only each other pole is filled with laminations with the interpole made of an insulating low-density material, to reduce mover weight.

The number of radial poles $2p_1$, the number of axial poles $2n$, the stator bare diameter D_{is}, and the PM thickness h_{PM} are the main variables to consider in a design, in addition to the PM pole pitch τ_{PM}, equal to the stroke length ($l_{stroke} = \tau_{PM}$). Then other variables, such as the stator pole angle $\alpha_p = (0.5\text{–}0.6)\pi/p_1$, the coil height h_{coil}, and the stator back core radial height h_{coil} come into play (Figure 12.25).

The blessing of circularity is no small an advantage in securing an ideally zero radial force for zero eccentricity and in facilitating easy framing.

The PM flux in the stator $2p_1$ coils reverses polarity (when the mover moves by l_{stroke} from the extreme left to the extreme right position) from $+\Phi_{PM}$ to $-\Phi_{PM}$ per one radial × one axial pole lengths.

On the other hand, the PM flux of axial PM pole varies from $\Phi_{P\max}$ to $\Phi_{P\min}$, and thus,

$$\Phi_{PM} = \Phi_{PMax} - \Phi_{PMin} \tag{12.36}$$

How to maximize Φ_{PM} for given D_{is}, $2p_1$, and l_{stroke} is a complex problem. However, it was realized [1] that when the mover laminated pole length is $l_m = \tau_{PM}/2$ and $(g + h_{PM})/\tau_{PM} = 0.5$, the maximum force density f_t (N/cm^2) is obtained. For minimum copper losses per N of force, however, $l_m/\tau_{PM} = 1$ (pole/interpole).

Though much of PM flux lines flow in the radial plane, due to alternate polarity along the axial direction, some of them flow in the axial plane as fringing flux. This effect has to be calculated in a three-dimensional (3D) FEM or quasi-2D FEM [6].

An analytical model is developed first, and then test results on a prototype are shown.

12.9.1 The Analytical Model

First, the maximum and minimum flux per axial pole Φ_{PMax} may be written as follows:

$$\Phi_{PMax} = \frac{G_{\max} I_{PM}}{1+K_s}; \quad I_{PM} = H_c \cdot h_{PM}$$
$$\Phi_{PMin} = \frac{G_{\min} I_{PM}}{1+K_s} \tag{12.37}$$

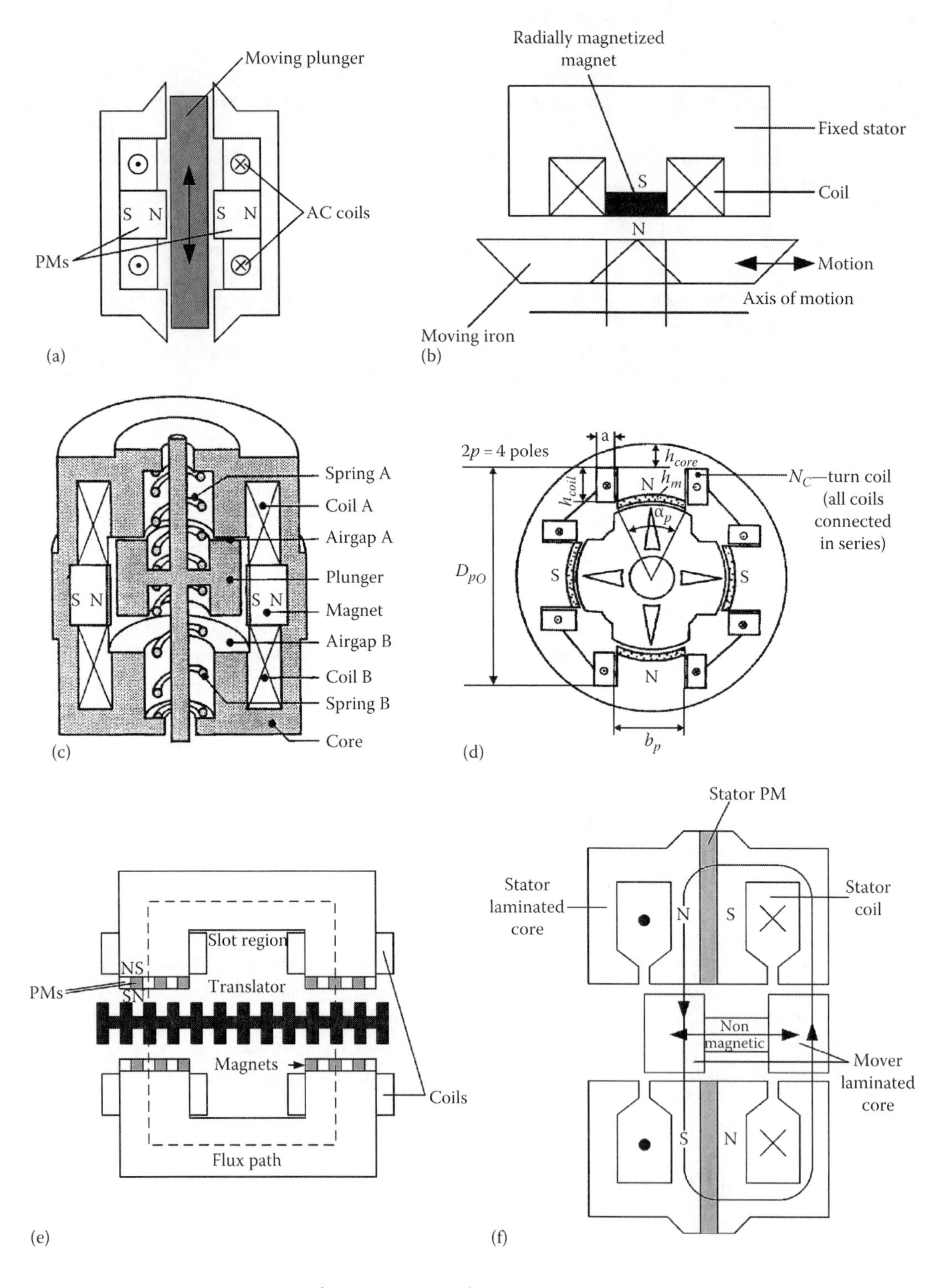

FIGURE 12.24 Permanent magnet linear motion accelerators (PM-LMAs) with iron movers: (a) with radial air gap, (b) with pulsed flux, (c) with axial air gap, (d) tubular with flux reversal, (e) flat with flux reversal, and (f) flat with PM flux stator concentration.

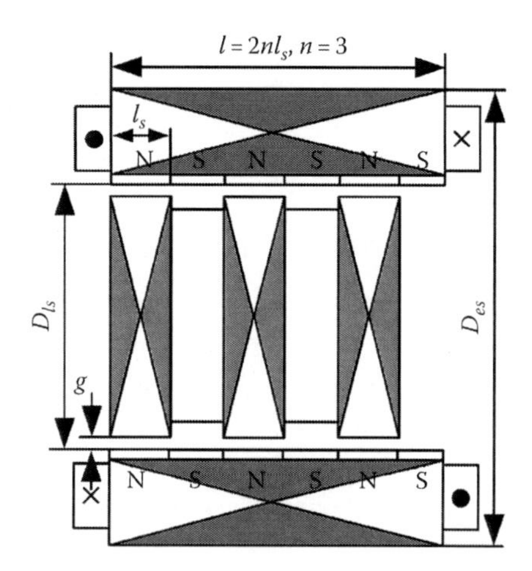

FIGURE 12.25 Main geometry variables in a $2p_1 = 4$ radial and $2n = 6$ axial poles.

The magnetic permeances G_{max} and G_{min} are as follows [2]:

$$
\begin{aligned}
G_{\max} &= \mu_0\pi\left(D_{is} - 2g + \frac{(g + h_{PM})}{2}\right)\cdot\frac{l_{stroke}}{g + h_{PM}}\cdot\frac{\alpha_p}{360} \\
G_{\min} &= G_{\min 1} + G_{\min 2} \\
G_{\min 1} &\approx 3.3\mu_0\frac{1}{2}(D_{is} - g + h_{PM})\cdot\frac{\alpha_p}{360} \\
G_{\min 2} &\approx 4\mu_0\left(\frac{1}{2}D_{is} - \sqrt{\frac{l_{stroke}(g + h_{PM})}{2}}\right)\cdot\left(\ln\left(\frac{l_{stroke}}{2(g + h_{PM})}\right)\right)\cdot\frac{\alpha_p}{360}
\end{aligned}
\tag{12.38}
$$

If $l_{stroke} \leq 2(g + h_{PM})$, then $G_{\min 2} = 0$. $G_{\min}$ accounts approximately for flux axial fringing.

The emf, $e(t)$, is now, simply,

$$
e(t) = \frac{d\Phi_{PM}}{dx}\cdot N\cdot 2p_1 n\cdot\frac{dx}{dt} \approx \frac{2\Phi_{PM}}{l_{stroke}}\cdot 2pn\cdot N\cdot u(t) \tag{12.39}
$$

The machine inductance, independent of mover position, L_s is

$$
\begin{aligned}
L_s &= L_m + L_l \\
L_m &= C_m\cdot N^2;\quad C_m = 4p_1 n(G_{\max} + G_{\min})
\end{aligned}
\tag{12.40}
$$

The leakage inductance L_l is

$$
\begin{aligned}
L_l &= C_l\cdot N^2 \\
C_l &= 4p_1\mu_0(\lambda_s l_{cs} + \lambda_{end} l_{end})
\end{aligned}
\tag{12.41}
$$

where

$$l_{cs} = 2n \cdot l_{stroke}$$

$$l_{end} = b_p + \frac{\pi a}{2}$$

b_p is the stator pole width in the radial plane
a is the coil width in radial plane
l_{cs} is the coil in slot length
l_{end} is the coil end connection length

The slot and end connection geometrical permeance coefficients λ_s and λ_{end} are as follows:

$$\lambda_s \approx \frac{h_{coil}}{3a}; \quad \lambda_{end} \approx \frac{\lambda_s}{2}; \quad h_{coil}\text{—coil height} \tag{12.42}$$

Finally, the phase resistance with all coils in series, R_s, is

$$R_s = 2p_1 \cdot \rho_{co} \cdot \frac{l_{coil} \cdot N^2}{I_n \cdot N / j_{con}}; \quad l_{coil} = \left(2n \cdot l_{stroke} + b_p + \frac{\pi a}{2}\right) \times 2 \tag{12.43}$$

where

I_n is the rated RMS current
j_{con} is the design current density
N is the turns per coil

The instantaneous force F_e is written, with Equation 12.39, as follows:

$$F_e = \frac{e(t) \cdot i(t)}{u(t)} = \frac{2\Phi_{PM}}{l_{stroke}} \cdot 2p_1 n \cdot N \cdot i(t) = C_F \cdot i(t) \tag{12.44}$$

This model is based on the assumption that the PM flux in the stator coil (per axial–radial PM pole) Φ_{PM} varies linearly with mover position. In reality, at least a third harmonic may be detected in the emf once the excursion length comes close to the maximum (ideal) stroke length. Consequently, the force F_e will not be strictly proportional to current (C_F will decrease with the mover on departure from the middle position).

Proof of these phenomena is shown, through tests, on a 125 W, 120 V, 60 Hz linear alternator with l_s = 10 mm [2], in Figures 12.26 and 12.27.

For the prototype, the measured cogging F_{cog} (at zero current), electromagnetic force F_e (current only), and total force F_t are shown in Figure 12.28. It is to be noted that for constant current, the electromagnetic

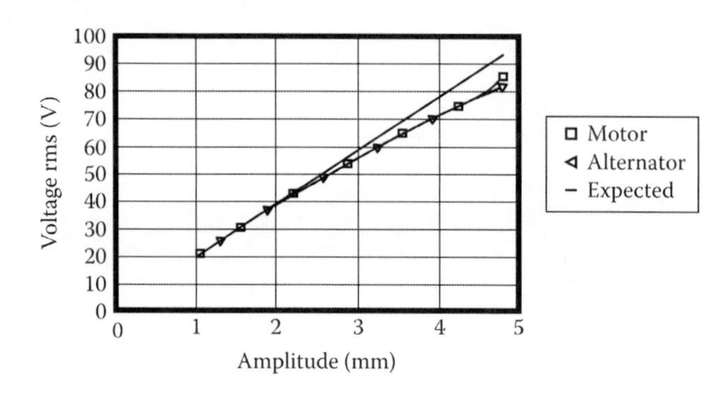

FIGURE 12.26 Electromagnetic force (emf) vs. mover position for l_{stroke} = 10 mm (test results).

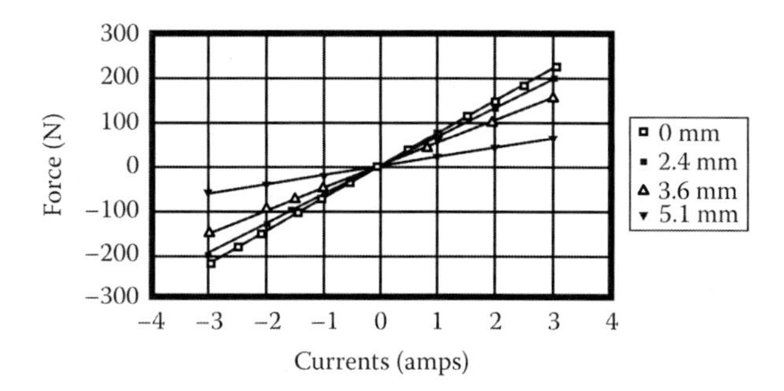

FIGURE 12.27 Current-only force vs. current for various motion amplitude $x_m = l_{stroke}/2$ (test results).

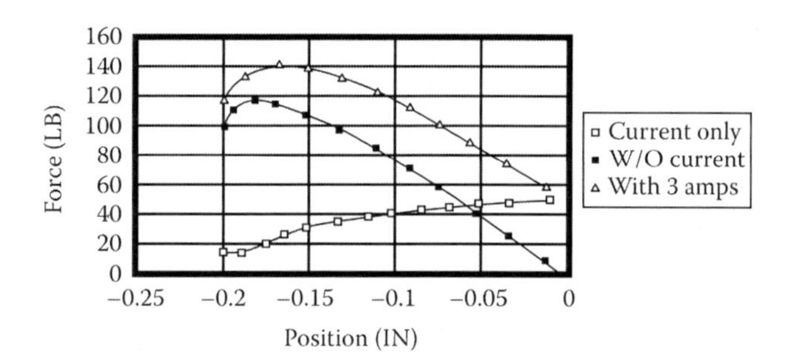

FIGURE 12.28 Force components vs. mover position.

force F_e decreases notably toward the excursion end. The cogging force is large, but it increases steadily up to more than 90% of full stroke length (2 × 5 = 10 mm).

Note again that the cogging force behaves like an additional mechanical spring. The actual springs should only complement the cogging force to secure stable oscillations up to ideal full stroke length. It should also be borne in mind that toward the stroke end, the current in the coil decreases to zero; and thus, the corresponding decrease of the force constant C_F does not produce severe damage in performance (in average force).

A capacitor is added to compensate (partially or totally) for the machine inductance voltage drop.

For a resistive passive load, the measured terminal voltage versus current is shown in Figure 12.29 for three values of the capacitor, all below full compensation conditions.

Electrical efficiencies above 85% were measured in back-to-back motor/generator loading experiments of the 125 W prototype.

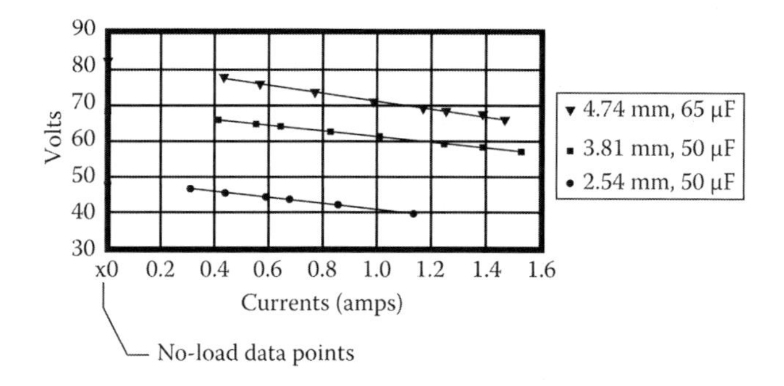

FIGURE 12.29 Alternator voltage vs. current (resistive load) for various constant motion amplitudes and capacitor pairs (test results).

Note that a design example for such a 1 kW alternator is given in Reference 1 and, thus, is not repeated here.
We investigated a number of practical configurations for PM-LMAs and conclude the following:

- All operate as single-phase synchronous alternators.
- All should operate so that electrical and mechanical resonance and frequency are equal: $f_e = \sqrt{K_t/m_t}/2\pi$.
- For all, in general, mechanical springs, made of copper–beryllium flexures [3], are used to secure stable oscillations up to full stroke length.
- For all, to decrease the electrical power delivered on a given load resistance, the stroke length is generally reduced by adequate control in the prime mover, while f_e = constant, and so is the terminal voltage.

In what follows, we will dwell a while on the control and stability of PM-LMAs.

The phase coordinate state-space model of PM-LMA is straightforward, and thus it is not developed here though it is needed for transients and control design.

12.10 Control of PM-LMAs

The control of the PM-LMAs depends on the prime mover and load characteristics.

Typical prime movers for PM-LMAs are as follows:

- Stirling engines: free-piston engines [16]
- Spark-ignited linear internal combustion engine (ICE) [7]
- Very low-speed reciprocating wave machines (0.5–2.0 m/s average speed) [5]

The control of stroke should be instrumented through the prime mover, while the adaptation of alternator voltage to the load—independent or at power grid—is to be performed by power electronics.

The motion frequency may be maintained constant and equal to power grid frequency, securing operation at the mechanical resonance frequency. If the frequency of motion varies within some limits, then the complexity of power electronics has to increase, if the load demands constant frequency and voltage output.

12.10.1 Electrical Control

Thus far, we encountered only single-phase linear alternators.

The control solution refers to the following:

- Constant motion frequency applications
- Variable motion frequency applications

They handle the following:

- Stand-alone nondemanding loads
- Strong power grids

For constant motion frequency, the control is decoupled (Figure 12.30):

- Motion amplitude control via prime-mover governor to deliver more or less mechanical power
- Voltage adaptation if operation at the power grid or at a certain voltage is required

The power error is the input of a power regulator followed by the stroke length, l^*_{stroke}, regulator. Finally, the fuel injection rate is modified to produce the required motion amplitude, l^*_{stroke}, according to electric power requirements.

As the frequency is constant, and connection to the grid is required, the motion phase may be modified slowly to provide the synchronization to grid conditions. The voltage adapter may be an autotransformer with variable ratio via a small servomotor. Alternatively, a soft-starter may be used to reduce the transients during connection to the grid. An additional resistance may be added before

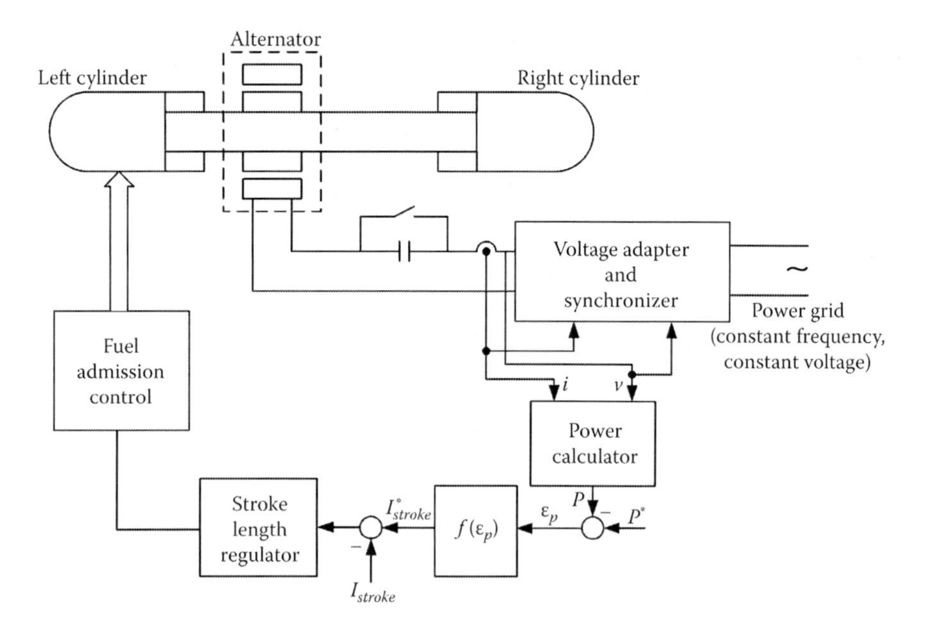

FIGURE 12.30 Control at power grid.

synchronization, in series with the alternator, and then short-circuited a few seconds later, as is done with wind induction generators.

When the motion frequency is allowed to vary, a front-end rectifier plus inverter may be applied (Figure 12.31). In this case, the series capacitor is no longer required, as the DC link capacitor does the job. The DC link voltage may now be larger than the LMA emf.

A bidirectional converter makes the interface between the variable voltage and frequency of the PM-LMA and the load requirements. The control of the load-side converter is different for independent load in contrast to power grid operation. The subject was presented for three-phase PM generators in Chapter 10. It should

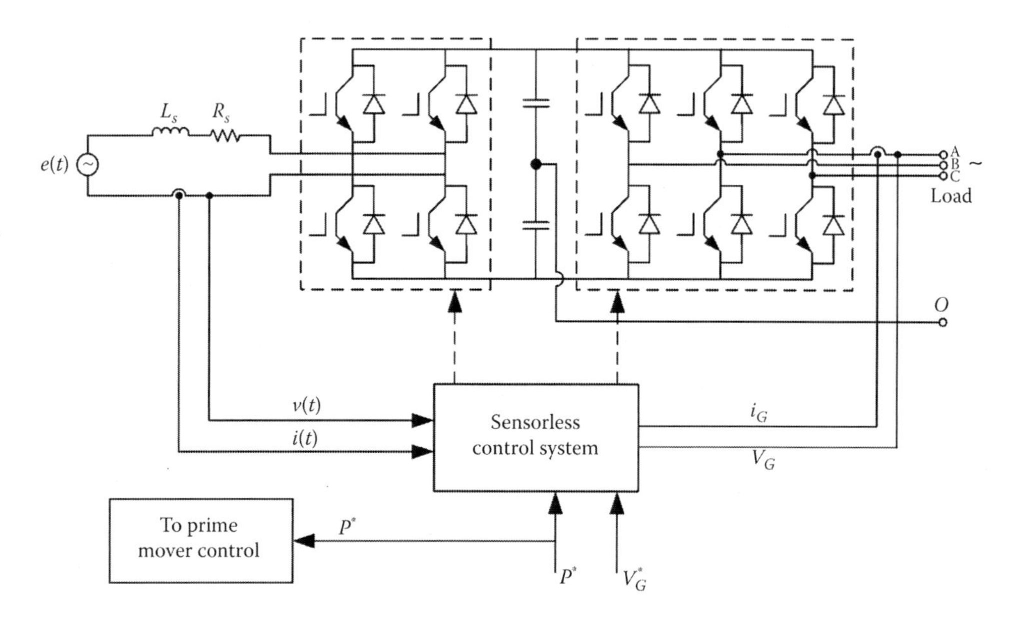

FIGURE 12.31 Permanent magnet linear motion alternator (PM-LMA) with bidirectional power converter.

be similar for the single-phase alternators. The bidirectional converter also allows for motoring and, thus, could help in starting or stabilizing the prime mover's motion when the operation takes place at the power grid or with a backup battery. In this case, the smooth connection to the power grid is implicit.

Note that for ocean-wave linear translator alternators, three-phase configurations were proposed [5]. When the linear motion changes directions, the phase sequence changes, but the active front rectifier can implicitly handle this situation. A three-phase bidirectional converter is required in this case. The control of such a system is similar to the case of rotary PM generators (Chapter 10).

12.10.2 Spark-Ignited Gasoline Linear Engine Model

The spark-ignited gasoline linear engine model was developed in Reference 8 (Figure 12.30).

The basic force balance equation of the system is as follows:

$$P_L(x)A_B - P_R(x) - F(x) = m_t x \tag{12.45}$$

where

$P_L(x)$ is the instantaneous pressure in the left cylinder
$P_R(x)$ is the instantaneous pressure in the right cylinder
A_B is the bore area
$F(x)$ is electromagnetic, friction, and mechanical spring (if any) force
m_t is the total translator mass

Under ideal (frictionless) no-load operation, no heat is required to sustain the motion; the forces from compressing and expanding gas in the engine will maintain motion by themselves.

The natural frequency of the engine is thus met.

In the absence of mechanical springs and of cogging force of the alternator, the force balance at no load becomes [9] as follows:

$$A_B P_1 \left(\frac{2r}{r+1}\right)^n \left[\left(1+\frac{x}{x_m}\right)^{-n} - \left(1-\frac{x}{x_m}\right)^{-n}\right] = m_t \ddot{x} \tag{12.46}$$

where

P_1 is the intake pressure
x_m is the mid-position point (half-stroke length: $x_m = l_{stroke}/2$)
r is the compression ratio

Equation 12.46 produces almost harmonic motion [9].

As expected, the real prime mover departs from this situation, and the motion frequency is slightly reduced under load. The presence of mechanical springs (and of cogging force) tends, however, to bring back the frequency to resonance conditions.

12.10.3 Note on Stirling Engine LMA Stability

The stability of an LMA with Stirling engine prime mover in operation at constant voltage is treated in Reference 1. In the absence of mechanical springs and of cogging force, but with full capacitor compensation, the motion frequency is varied, and the steady-state stability condition is checked:

$$W_d < W_e \tag{12.47}$$

where

W_d is the mechanical (input) energy per cycle of ω (frequency)
W_e is the electrical (output) energy per cycle of ω

A complete study of stability with closed-loop stroke control, mechanical springs, cogging force, and with power electronics interface to independent load or at power grid is still to come; but before that, see References [10,11].

12.11 Progressive-Motion LMAs for Maglevs with Active Guideway

Thus far, we discussed only LMAs with linear oscillatory motion. The LMAs may also be used with progressive linear motion. A typical case is the magnetically levitated trains (maglevs) with active guideway. By active guideway, we mean that the maglev propulsion is provided by linear synchronous motors with superconducting or conventional DC excitation on board [12–14]. Auxiliary power on board, in the absence of electric power mechanical collectors, has to be obtained through electromagnetic induction (contactless) and used in corroboration with battery storage.

For the maglev with electromagnetic controlled excitation, the German Transrapid 06 and 08 (Figure 12.32a through c), the three-phase cable winding, which is located along the guideway (face down), in an open-slot laminated core, is fed section after section in synchronism with the vehicle (excitation rotor poles) such that the emfs are in phase with the phase currents.

The suspension is produced by the control of the excitation current on board, from zero speed. The propulsion is controlled from on-ground power stations, provided with variable frequency bidirectional alternating current (AC)–AC power electronics converters.

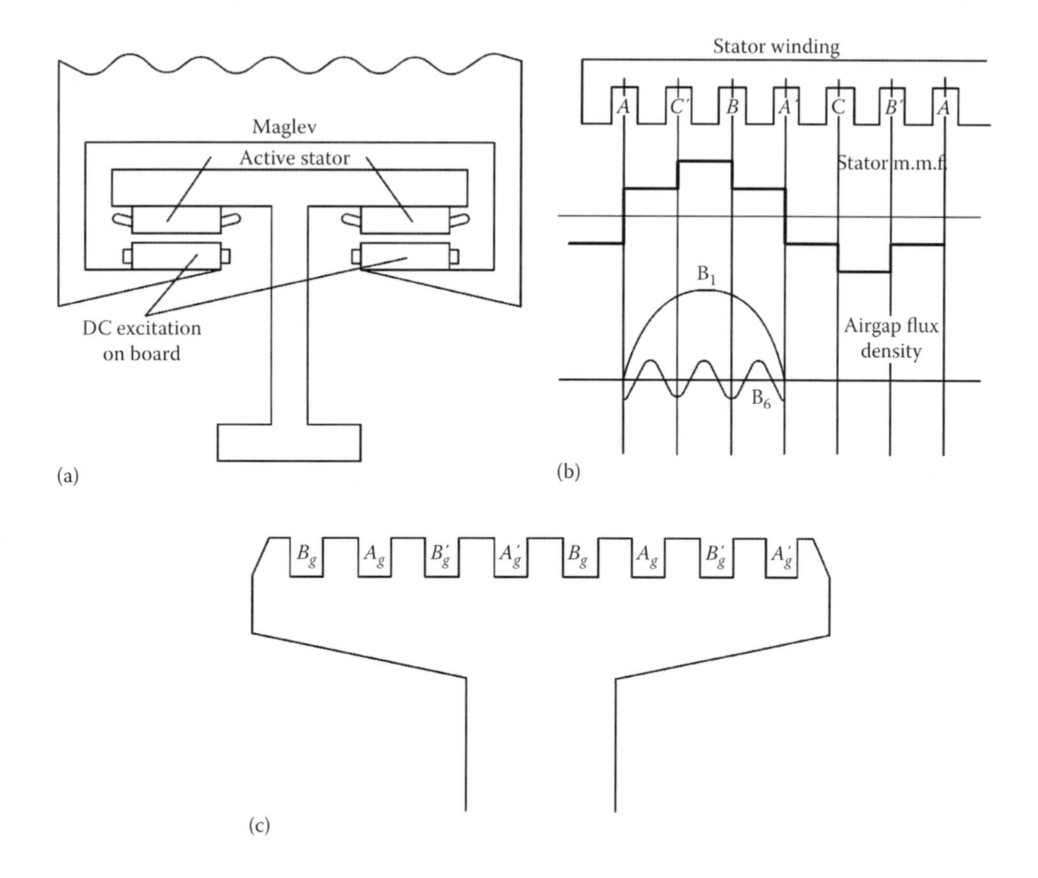

FIGURE 12.32 The Shanghai Transrapid Maglev Line: (a) the structure, (b) open-slot stator armature air gap flux density (B6), and (c) linear generator winding.

The armature (stator) mmf, in the presence of open slots, produces an air gap flux density that pulsates visibly with the stator slot pitch periodicity.

With three slots per pole (Figure 12.32b), the pole pitch of the flux density pulsation due to slot opening is $\tau_g = \tau/6$.

If the emf produced by this flux density harmonics is to be extracted, we need to plant, on the salient poles of the inductors on board the maglev, a two- or three-phase winding with the pole pitch τ_g. For a two-phase winding, we need to plant at least six smaller slots on the inductor pole.

For a two-phase winding, three coils per phase are thus obtainable, as the inductor pole span is about $2\tau/3$ (Figure 12.32c).

A diode rectifier with the two phases in series, a filter and a DC–DC converter, may represent the solution to interface the linear generator to the battery and to the load in parallel with it.

It was shown that such a linear generator can produce sufficient energy on board for excitation control and for auxiliary loads, above the half-rated speed of the vehicle. Under this speed, the battery takes over. When the vehicle is in a stop station, the batteries are charged from an on-ground energy system.

In essence, the linear generator investigated here is a synchronous linear biphase machine with independently variable excitation current. It is the task of the diode rectifier plus the DC–DC converter, eventually, of the boost-buck type, to extract most of the available energy in the slot-opening air gap flux density harmonic, produced by the stator current.

For the superconducting Japanese maglev (MLX01) [14], a similar solution may be feasible, as the configuration is somewhat similar (Figure 12.33 [14]).

The track is provided laterally along a U-shaped concrete track, with a three-phase cable propulsion winding with three "slots" per pole, and pole pitch τ, and figure-eight-shaped levitation short-circuited coils with a span of $\tau/2$ to decouple the two windings.

On board, there is a row of superconducting coils with pole pitch τ.

The superconducting coils on board produce levitation and guidance through the currents induced by motion in the figure-eight-shaped stator short-circuited (levitation) coils. At the same time, they interact with the three stator phase windings for propulsion.

Now it is known that the levitation guidance through induced currents is based on repulsive forces that provide statically stable operation and, within some conditions, some negative damping of levitation and guidance motions.

To further improve negative damping, an actively controlled coil system on board was used [12]. But using a linear generator for the purpose is better, because both energy production on board and damping are provided by the same hardware.

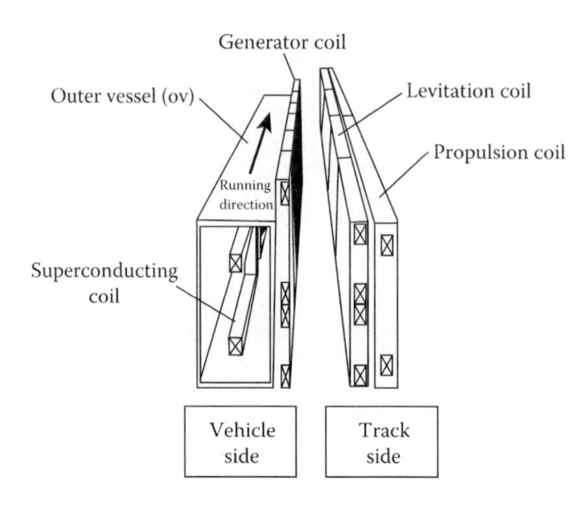

FIGURE 12.33 Superconducting maglev with linear superconducting generator.

This is how figure-eight-shaped coils are placed on board to form a three-phase winding. They have to sense a harmonic field of levitation coil currents. The linear generator uses superconducting coils to induce, through vehicle motion, currents in the levitation guidance ground coils. In turn, a harmonic field of the stator levitation coil currents induces emf in the linear generator winding.

A power electronics converter is needed as an interface between the linear generator and the battery (Figure 12.34).

The maximum output of the linear generator to the fixed voltage battery and load, increases with speed (Figure 12.35) [14].

Details of the control of such a system are given in Reference 14, with reference to the superconducting maglev MLX01.

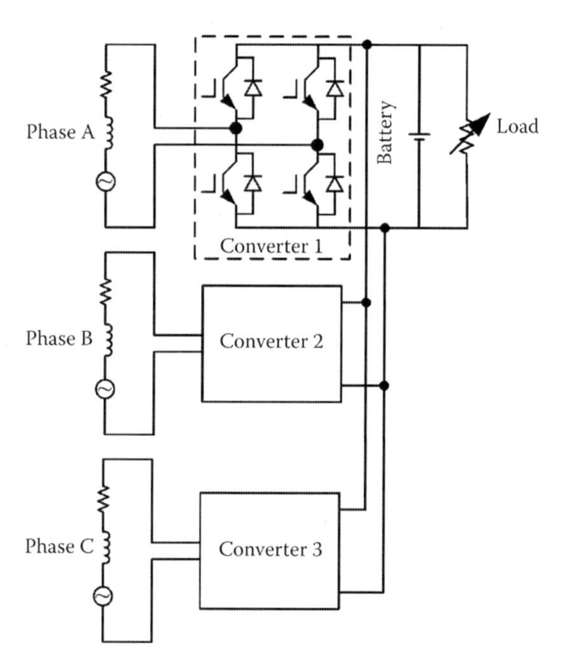

FIGURE 12.34 Three single-phase converters for the superconducting linear generator for maglev.

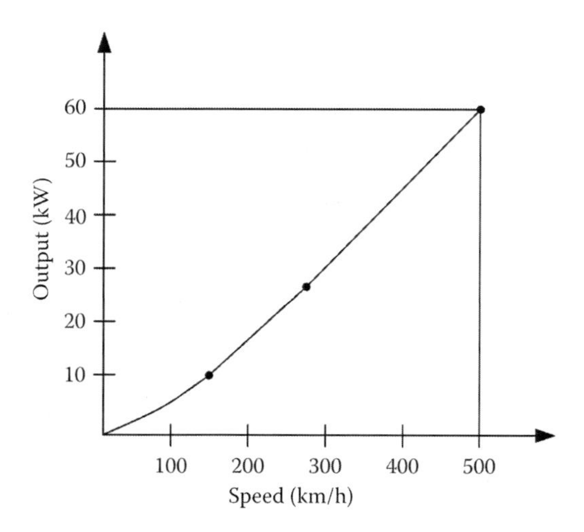

FIGURE 12.35 Output vs. speed of superconducting linear generator on maglevs.

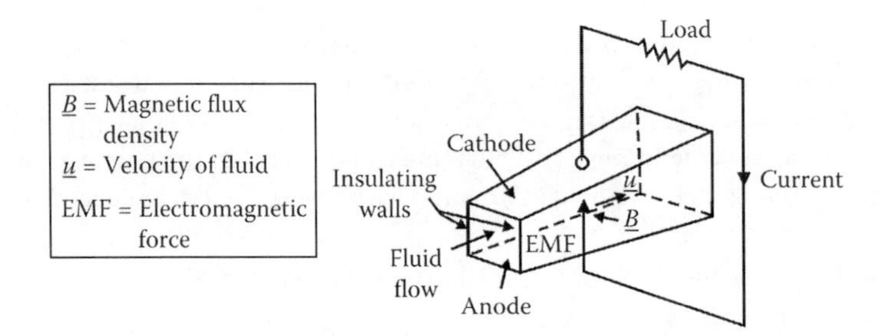

FIGURE 12.36 Plasma direct current (DC) linear magnetohydrodynamic (MHD) generator.

12.11.1 Note on Magnetohydrodynamic Linear Generators

The MHD linear generators were proposed decades ago for direct conversion of ionized plasma at 3000 K heat to linear plasma motion at 1000 m/s or so, and then, to electrical energy in a plasma DC linear superconducting generator (Figure 12.36) [12].

The magnetic field $\underline{B}$ has to be large and, thus, is produced via superconducting magnets. Perpendicular to $\underline{B}$ and to the direction of plasma speed $\underline{u}$, there are two electrodes that collect the emf E_{DC}:

$$E_{DC} = uBL \tag{12.48}$$

Despite the seeded plasma low conductivity ($\sigma_{plasma} \approx$ 40–50 Siemens), the large speed $u \approx$ 800–1000 m/s and the large flux density B = 4 T provide for acceptable electrical performance. For adiabatic thermal conditions, efficiency is above 55%.

It is still claimed that such a linear MHD generator could notably improve the total efficiency in thermal power plants. However, since the recent developments of dual-cycle gas turbines with total efficiency above 60%, the future of linear MHD generators for power systems may seem less promising.

For a detailed study of them, however, you may start with Reference 12, Chapter 5.

12.12 Summary

- LMAs directly convert linear oscillatory or progressive motion mechanical energy into electrical energy.
- Prime movers for LMAs thus far are free-piston Stirling engines, single- or dual-piston gas linear engines, or wave energy engines for oscillatory linear motion.
- Maglev (magnetically levitated) vehicles are typical applications for on-board linear progressive motion alternators that work with battery backup to provide auxiliary power. In addition, plasma MHD linear motion energy of seeded hot plasma to directly convert thermal energy to mechanical energy to electrical energy is feasible.
- The oscillatory LMAs are typically single-phase parametric machines, especially if the stroke length is generally below 100 mm.
- It was soon realized that equality between mechanical resonance and electrical frequency is key to high efficiency. Mechanical calibrated springs, as copper–beryllium flexures, are used to store and release mover kinetic energy at stroke ends.

  As the resonance frequency $f_1 = \sqrt{K/m_t}/2\pi$ decreases with mover mass, the reduction of the latter is a problem at f_1 = 60 Hz for powers in the kilowatts and tens of kilowatts range for stroke length l_{stroke} well below 100 mm. Consequently, reduction of machine and mover size is paramount. This is how permanent magnets come into play, with the added advantage of higher force per watt of losses.

- Therefore, basically, PM-LMAs are single-phase SGs. Though trapezoidal motion was tried [15], as trapezoidal motion implies small power pulsations, the harmonic (sinusoidal) speed profile is still in favor, mainly due to practical reasons.
- For sinusoidal motion, the PM flux linkage variation in the armature coils has to be linear to secure sinusoidal emf waveform. All efforts to achieve this goal should be made.
- There are numerous LMA configurations of practical interest, and they may be classified with respect to mover type and geometrical shape as follows:
  - With coil mover
  - With PM mover
  - With iron mover
  - Tubular
  - Flat, double sided
- There are many subdivisions of these main types, such as with air coils or in-slot coils or with PM flux paths within the motion plane or transverse to it, or without or with PM flux concentration.

  The aim for high force/volume is in contradiction to high force/watt of losses and to low-power factor angle $\varphi_1 = \tan^{-1}(\omega_1 L_s I_n / E)$. The latter is decisive in the power output level for given terminal voltage or in voltage regulation. Full compensation of machine inductance by a series capacitor is not uncommon to maximize the output for given geometry at the highest force/watt of losses.
- LMA configurations with PM flux concentration lead to large force/volume and force/watt, but a larger capacitor for reactive power compensation is required. We trade here PMs for capacitors when optimization is performed.
- Air coils, on stator or on mover, have the lowest electrical time constants T_e ($T_e < 10$ ms for powers in the 1–50 kW range) at good power factor and moderate force density (4 N/cm^2 at 22.5 kW).
- PM-mover and iron-mover LMAs are shown to produce the same force density for higher force/watt of losses but at moderate power factors.
- Realistic analytical models for a few configurations were applied for numerical examples of interest, and two of them were validated through FEM analysis. Efficiency above 85% at 100 W output was demonstrated with weight less than 15 kg/kW.
- The control of LMA connected to the power grid was demonstrated to be possible only via stroke control of the Stirling engine prime mover, with a full compensation capacitor [2].
- A more elaborated control is obtained through a bidirectional power electronics converter for interfacing with the load, as much as for the three-phase rotary PM generators (Chapter 6).
- Though very large force density (12 N/cm^2) [5] was recently claimed, as for rotary counterparts (Chapter 11), with some configurations, the corresponding power factor was about 0.2. This means very large capacitor energy storage (for reactive power compensation) that is not easy to justify.
- Low-cost and advanced power electronics control was introduced by LMAs [15].
- For progressive linear motion, maglev linear generators were proposed in the tens of kilowatts power range. They exploit the air gap flux density space harmonics of stator coils, be them the three-phase winding or the levitation short-circuited coils. With proper battery backup, they were shown to be able to cover the energy needs on board active guideway maglev vehicles.
- Seeded plasma linear progressive motion MHD DC brush generators were also proposed to improve the overall efficiency in thermal power plants above 50% without low energy heat delivery considered (10% over standard turbines). As the dual-cycle gas turbines, introduced recently, pushed the overall efficiency (thermal plus electrical) above 60%, the MHD generators may not aggressively enter power systems anytime soon. See Reference 12 for a detailed introduction to linear MHD generators.
- For more energy conversion via linear alternators, see References [16,17].

References

1. I. Boldea and S.A. Nasar, *Linear Electric Actuators and Generators*, Cambridge University Press, New York, 1997, Chapter 8.
2. I. Boldea, S.A. Nasar, B. Penswick, B. Ross, and R. Olan, New linear reciprocating machine with stationary permanent magnets, Record of IEEE–IAS 1996, Vol. 2, 1996, pp. 825–829.
3. I. Boldea, M. Topor, and J. Lee, Linear flux reversal PM oscillo-machine with effective flux concentration, Record of OPTIM-2004, Poiana Brasov, Romania, 2004.
4. D.H. Kang, D.H. Koo, I. Vadan, and Q. Cemuca, Influence of mechanical springs in the transverse flux linear oscillating motor operation, Record of Electromotion-2003, Marrakesh, Morocco, Vol. 1, 2003, pp. 410–415.
5. M.A. Mueller, N.J. Baker, P.R.M. Brooking, and J. Xiang, Low speed linear electrical generators for reversible energy applications, Record of LDIA–2003, Birmingham, U.K., 2003, pp. 29–32.
6. I. Boldea, C. Wang, B. Yang, and S.A. Nasar, Analysis and design of flux reversal linear machine, Record of IEEE–IAS 1998, Vol. 1, 1998, pp. 136–142.
7. W. Cawthorne, P. Famouri, and N. Clark, Integrated design of linear alternator/engine system for HEV auxiliary unit, Record of IEEE–IEMDC, 2001.
8. A. Bosic, J. Lindbäck, W.M. Arshad, P. Thelin, and E. Nordlund, Application of a free-piston generator in a series hybrid vehicle, Record of LDIA-2003, Birmigham, U.K., 2003, pp. 541–544.
9. N. Clark, S. Nandkumar, and P. Famouri, Fundamental analysis of a linear two-cylinder internal combustion engine, In *Record of SAE International Full Fuels and Lubricants Meeting and Exposition*, San Francisco, CA, 1998.
10. I. Boldea, *Linear Electric Magnetic, Drives and MAGLEVs Handbook*, CRC Press, Taylor & Francis Group, New York, 2013.
11. P. Zeng, C. Tong, J. Bai, B. Yu, Y. Sui, and W. Shi, Electromagnetic design and control strategy of an axially magnetized PM linear alternator for free piston Stirling engines, *IEEE Trans.*, IA-48(6), 2012, 2230–2239.
12. I. Boldea and S.A. Nasar, *Linear Motion Electromagnetic Systems*, John Wiley & Sons, New York, 1985.
13. J. Gieras, *Linear Synchronous Motors*, CRC Press, Boca Raton, FL, 1998.
14. H. Hasegawa, T. Murai, and T. Yamamoto, Study of a PWM converter for linear generator controlling zero-phase current, Record of LDIA-2003, Birmigham, U.K., 2003, pp. 227–230.
15. Y. Liu, M. Leksell, W.M. Arshad, and P. Thelin, Influence of speed and current profiles upon converter dimensioning and electric machine performance in a free-piston generator, Record of LDIA-2003, Birmigham, U.K., pp. 553–556.
16. R. Vermaak, M.J. Kamper, Experimental evaluation and predictive control of an air cored linear generator for direct-drive wave energy converters, *IEEE Trans.*, IA-48(6), 2012, 1817–1826.
17. L. Cappelli, F. Marinetti, G. Mattiazzo, E. Giorcelli, G. Bracco, S. Carbone, and C. Attaianese, Linear tubular permanent-magnet generators for the inertial sea wave energy converter, *IEEE Trans.*, IA-50(3), 2014, 1817–1828.